VISUALIZING
EARTH SCIENCE

VISUALIZING
EARTH SCIENCE

Zeeya Merali, PhD

Brian J. Skinner, PhD
Yale University

with contributions by
Alan Strahler, PhD
Boston University

In collaboration with
THE NATIONAL GEOGRAPHIC SOCIETY

CREDITS

VP AND PUBLISHER Jay O'Callaghan
MANAGING DIRECTOR Helen McInnis
EXECUTIVE EDITOR Ryan Flahive
DIRECTOR OF DEVELOPMENT Barbara Heaney
MANAGER, PRODUCT DEVELOPMENT Nancy Perry
DEVELOPMENT EDITOR Carolyn Smith
PROJECT EDITOR Joan Kalkut
ASSISTANT EDITOR Courtney Nelson
EDITORIAL ASSISTANTS Erin Grattan, Sean Boda
EXECUTIVE MARKETING MANAGER Jeffrey Rucker
MARKETING MANAGER Danielle Torio
PRODUCTION MANAGER Micheline Frederick
MEDIA EDITORS Lynn Pearlman, Bridget O'Lavin
CREATIVE DIRECTOR Harry Nolan
COVER DESIGNER Harry Nolan
INTERIOR DESIGN Vertigo Design
SENIOR PHOTO EDITOR Elle Wagner
PHOTO RESEARCHER, NATIONAL GEOGRAPHIC Stacy Gold
SENIOR ILLUSTRATION EDITOR Sandra Rigby
PRODUCTION SERVICES Camelot Editorial Services, LLC

Cover credits: Main photo: Stockbyte/SuperStock, Inc.; Thumbnails (from left to right): NASA/JPL/CALTECH/OLIVER KRAUSE/NG Image Collection; Creatas/MediaBakery; Photodisc/SuperStock, Inc.; John Eastcott and Yva Momatiuk/NG Image Collection; Photographers Choice RF/SuperStock, Inc.

This book was set in New Baskerville by Preparé, Inc., and printed and bound by Quebecor World. The cover was printed by Phoenix Color.

Copyright © 2009 John Wiley & Sons, Inc. All rights reserved. No part of this publication may be reproduced, stored in a retrieval system or transmitted in any form or by any means, electronic, mechanical, photocopying, recording, scanning or otherwise, except as permitted under Sections 107 or 108 of the 1976 United States Copyright Act, without either the prior written permission of the Publisher, or authorization through payment of the appropriate per-copy fee to the Copyright Clearance Center, Inc. 222 Rosewood Drive, Danvers, MA 01923, web site www.copyright.com. Requests to the Publisher for permission should be addressed to the Permissions Department, John Wiley & Sons, Inc., 111 River Street, Hoboken, NJ 07030-5774, (201) 748-6011, fax (201) 748-6008, Web site http://www.wiley.com/go/permissions.

To order books or for customer service, please call 1-800-CALL WILEY (225-5945).

ISBN-13: 978-0471-74705-5
BRV ISBN: 978-0470-41847-5

Printed in the United States of America
10 9 8 7 6 5 4

PREFACE

The goal of *Visualizing Earth Science* is *to bring the story of the way Earth works to the majority of students,* not just to those few who choose a career in the sciences. The groundbreaking Visualizing books by John Wiley and Sons in collaboration with the National Geographic Society and its vast image resource, provide an opportunity to tell Earth's story in a fresh new way, while maintaining the rigor and currency needed in science. Students will learn that features we see and experience in the environment around us result from the interactions of many separate phenomena and processes which extend from Earth's core to the outer reaches of the solar system.

This book is intended to serve as an introductory text primarily for undergraduate students who are not majoring in a scientific discipline. The accessible format of *Visualizing Earth Science,* which has been constructed under the assumption that students have little prior knowledge of Earth science, allows students to easily make the transition from an introduction to a topic to its more complex aspects. With its highly visual presentation, which mirrors the very nature of Earth itself, this book is appropriate for use in one-semester courses in Earth science of the kind offered at many institutions of higher learning.

ORGANIZATION

Visualizing Earth Science is structured around two related systems, the Earth system and the solar system. Earth is a system of interacting parts—geosphere, hydrosphere, atmosphere, and biosphere—and as modern science has made it possible to study the interactions within and between the parts in real time, the new field of Earth system science has emerged. But Earth is just one body among many in the still larger solar system, and from that larger system Earth receives the energy that determines weather and climate and powers the biosphere.

Chapter 1 introduces Earth system science, the solar system, the scientific method of investigation, and the reliance of the human population on natural resources, some of which are either locally or globally limited. Chapters 2, 3, and 4 deal with the materials of which the solid Earth is composed—minerals, rocks, and soils. Chapter 5 addresses the hydrosphere, and Chapter 6 regions of climatic extremes—glaciers, ice sheets, and deserts.

The continual restructuring of Earth as a result of forces driven by the internal heat energy is covered in Chapters 7 though 9. Chapter 7 discusses plate tectonics and the ever-changing landscape; Chapter 8 deals with

earthquakes and Earth's internal structure; and Chapter 9 discusses magma, volcanism, and other igneous manifestations.

Earth science is not just Earth system science. Earth science is also the past history of Earth, and Chapters 10 and 11 address this. Chapter 10 covers the determination of relative time and numerical time, and Chapter 11 presents a brief history of life on Earth, from the most ancient evidence to the present.

We walk on the solid Earth but we live in a fluid, the atmosphere, and are surrounded by oceans of seawater. Chapters 12 through 15 are devoted to the fluid regimes. Chapters 12 and 13 are devoted to the oceans and to the shoreline regime, where the ocean meets the land. Chapters 14 and 15 introduce the atmosphere, its structure, circulation, and weather systems.

The final chapters of the book discuss climate and the solar system. Climates past and present, discussed in Chapter 16, attract much attention today as we face the problem of possible climate changes induced by human activities. Space exploration continues to provide a flood of data and new understandings about the solar system, including hypotheses of its origin, its structure today, lessons for Earth from other planets, and the recent discovery that planetary systems exist around other nearby stars. These and other topics are discussed in Chapter 17.

Visualizing Earth Science is, as mentioned earlier, intended as a textbook for an introductory college-level course in Earth science. We do not expect that most of the students who read the book will go on to become Earth scientists, but we hope that all will come to have a better understanding of, and appreciation for, their home planet, its history, and its place in the solar system. For those students who do wish to take further courses in the field—and we hope there are many—we have provided a solid, sufficient, and challenging background to do so with confidence.

NATIONAL GEOGRAPHIC SOCIETY

Visualizing Earth Science offers an array of remarkable photographs, maps, media, and film from the National Geographic Society collections. Students using the book benefit from the long history and rich, fascinating resources of National Geographic.

Fact-Checking: The National Geographic Society has also performed an invaluable service in fact-checking *Visualizing Earth Science*. They have verified every fact in the book with two outside sources, to ensure that the text is accurate and up-to-date.

MEDIA AND SUPPLEMENTS

Visualizing Earth Science is accompanied by a rich array of media and supplements that incorporate the visuals from the textbook extensively to form a pedagogically cohesive package. For example, a Process Diagram from the book appears in the Instructor's Manual with suggestions on using it as a PowerPoint in the classroom; it may be the subject of a short video or an online animation; and it may also appear with questions in the Test Bank, as part of the chapter review, homework assignment, assessment questions, and other online features.

INSTRUCTOR SUPPLEMENTS

 Wiley*PLUS* offers a powerful online tool that provides instructors and students with an integrated suite of teaching and learning resources in one easy-to-use Web site. These resources include:

VIDEOS

A collection of videos, a number from the award-winning National Geographic Film Collection, have been selected by Robert Altamura of Florida Community College at Jacksonville to accompany and enrich the text. Each chapter includes at least one video clip, available online as digitized streaming video that illustrates and expands on a concept or topic to aid student understanding. Accompanying each of the videos are contextualized commentary and questions that can further develop student understanding. The videos are available in **Wiley*PLUS***.

POWERPOINT PRESENTATIONS AND IMAGE GALLERY

A complete set of highly visual PowerPoint presentations by Burair Kothari of Indiana University is available online to enhance classroom presentations. Tailored to the text's topical coverage and learning objectives, these presentations are designed to convey key text concepts, illustrated by embedded text art.

Image Gallery All photographs, figures, maps, and other visuals from the text are online and can be used as you wish in the classroom. These online electronic files allow you to easily incorporate them into your PowerPoint presentations as you choose, or to create your own overhead transparencies and handouts.

TEST BANK (AVAILABLE IN WILEY*PLUS* AND ELECTRONIC FORMAT)

The visuals from the textbook are also included in the Test Bank by Arthur Lee, Roane State Community College. The Test Bank contains approximately 1700 test items, at least 25 percent of which incorporate visuals from the book. The test items include multiple-choice and essay questions that test a variety of comprehension levels. The Test Bank is available in two formats: online in MS Word files and as a computerized test bank on a multiplatform CD-ROM. The easy-to-use test-generation program fully supports graphics, printed tests, student answer sheets, and answer keys. The software's advanced features allow you to create an exam to your exact specifications.

INSTRUCTOR'S MANUAL (AVAILABLE IN ELECTRONIC FORMAT)

The Instructor's Manual begins with a special introduction on *Using Visuals in the Classroom*, prepared by Matthew Leavitt of Arizona State University, in which he provides guidelines and suggestions on how to use the visuals in teaching the course. For each chapter, materials by Paul Cutlip of St. Petersburg College include suggestions and directions for using Web-based learning modules in the classroom and for homework assignments, as well as creative ideas for in-class activities.

WEB-BASED LEARNING MODULES

A robust suite of multimedia learning resources have been designed for *Visualizing Earth Science*, focusing on and using the visuals from the book. Available in **Wiley*PLUS***, the content is organized into *tutorial animations*. These animations visually support the learning of a difficult concept, process, or theory, many of them built around a specific feature such as a Process Diagram, Visualizing feature, or key visual in the chapter. The animations go beyond the content and visuals presented in the book, providing additional visual examples and descriptive narration.

ILLUSTRATED

A number of pedagogical features using visuals have been developed specifically for *Visualizing Earth Science*. Presenting the highly varied and often technical concepts woven throughout Earth science raises challenges for reader and instructor alike. This **Illustrated Book Tour** provides a guide to the diverse features that contribute to *Visualizing Earth Science's* pedagogical plan.

CHAPTER INTRODUCTIONS illustrate certain concepts in the chapter with concise stories about some of the world's most remarkable places and events of unusual interest. These narratives are featured alongside striking photographs. The chapter openers also include illustrated **CHAPTER OUTLINES** that use thumbnails of illustrations from the chapter to refer visually to the content.

GLOBAL LOCATOR MAPS, prepared specifically for this book by the National Geographic Society cartographers, accompany some photos. These locator maps help students visualize where the area depicted in the photo is situated on Earth.

VISUALIZING features are specially designed, multipart visual spreads that focus on a key concept or topic in the chapter, exploring it in detail or in broader context using a combination of photos and figures.

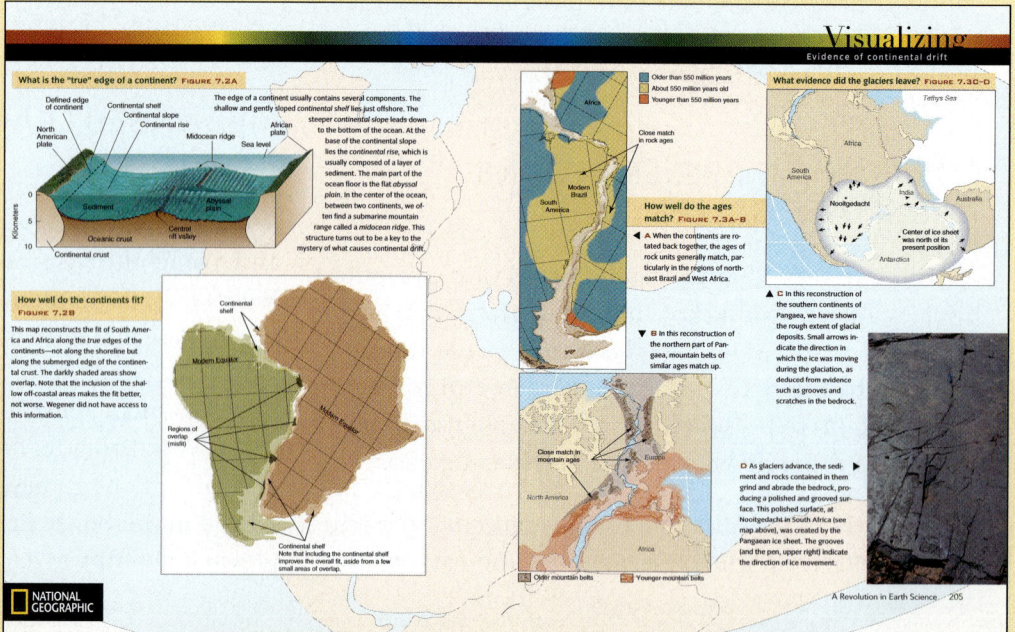

BOOK TOUR

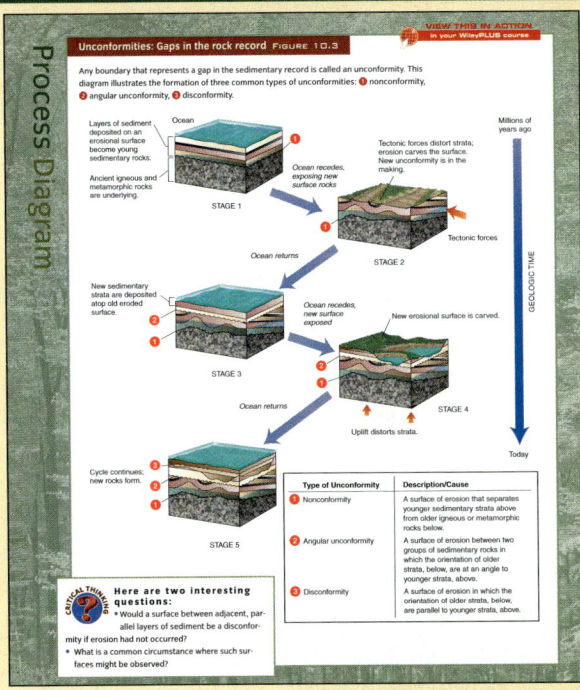

WHAT AN EARTH SCIENTIST SEES are features that highlight a concept or phenomenon, using photos and figures that would stand out to a professional in the field, and helping students to develop observational skills.

PROCESS DIAGRAMS present a series of figures or a combination of figures and photos that describe and depict a complex process, helping students to observe, follow, and understand the process.

HERE'S AN INTERESTING QUESTION asks students challenging questions related to the topics discussed in What an Earth Scientist Sees and in Process Diagrams.

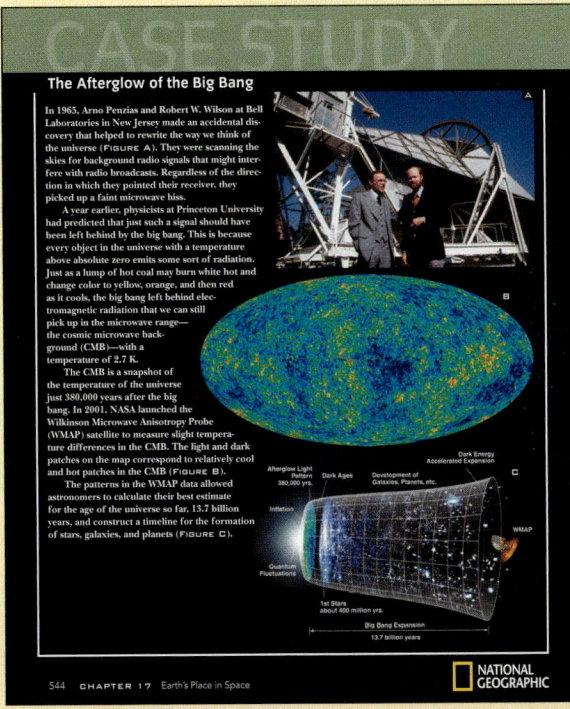

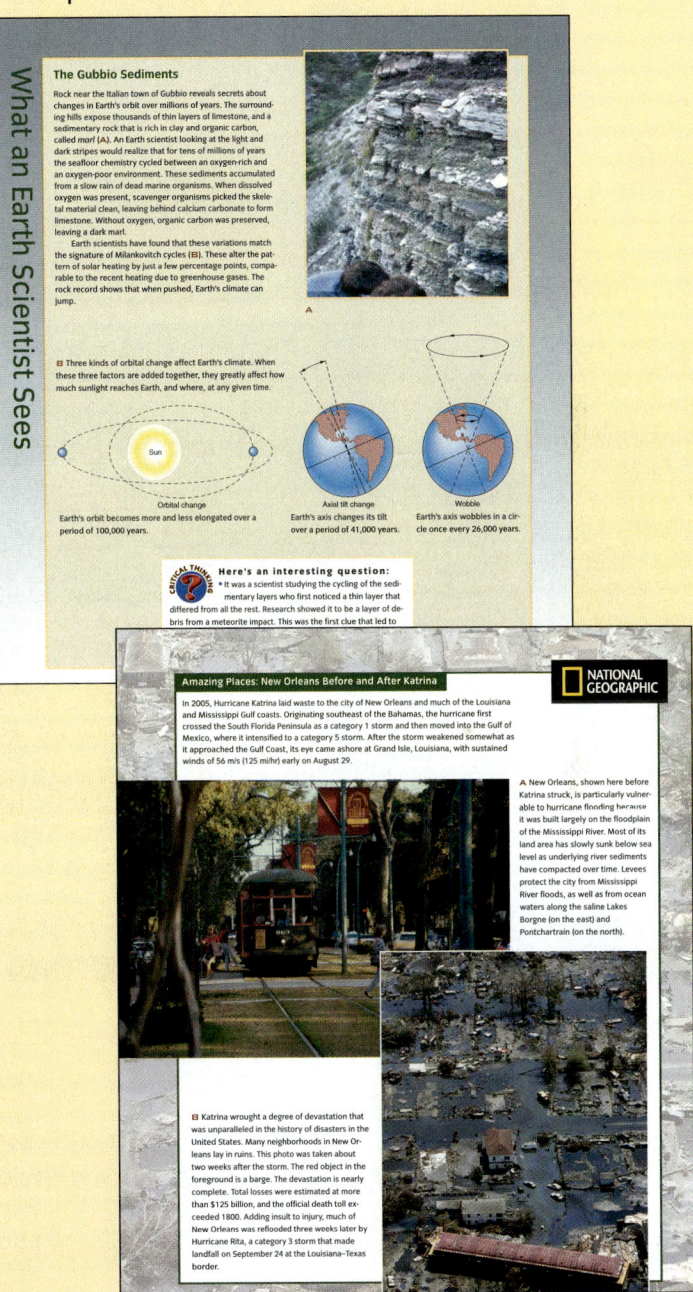

AMAZING PLACES is a feature that takes students to a unique location in the world that provides a vivid illustration of a theme in the chapter. Students could easily visit most of the Amazing Places someday and so continue their Earth science education after they finish this book.

CASE STUDIES are illustrated features that offer a wide variety of in-depth examinations that address important issues in Earth science.

OTHER PEDAGOGICAL FEATURES

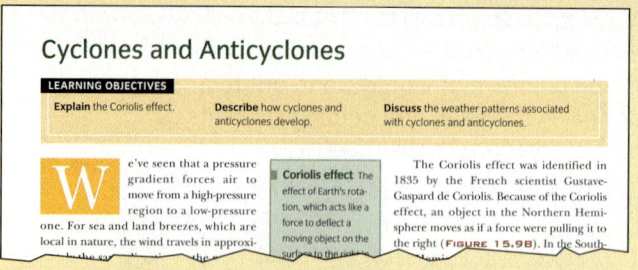

LEARNING OBJECTIVES at the beginning of each section indicate in behavioral terms that the student must be able to demonstrate mastery of the material in the chapter.

CONCEPT CHECK questions at the end of each section give students the opportunity to test their comprehension of the learning objectives.

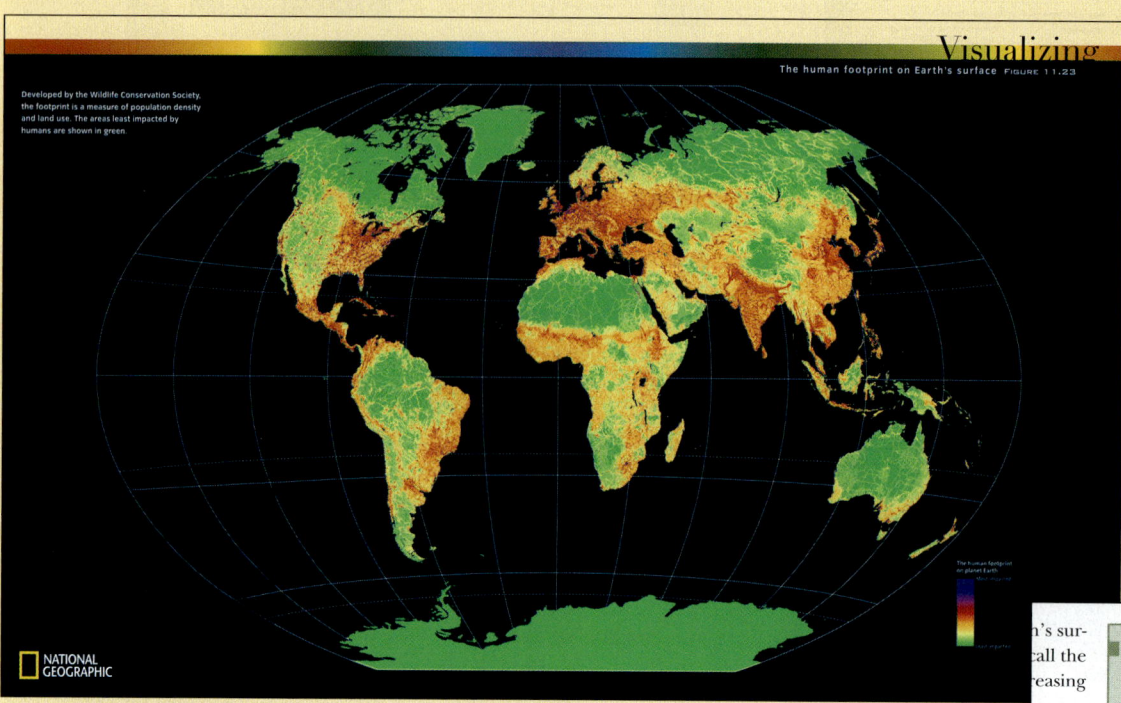

NATIONAL GEOGRAPHIC SOCIETY MAPS are featured throughout the book.

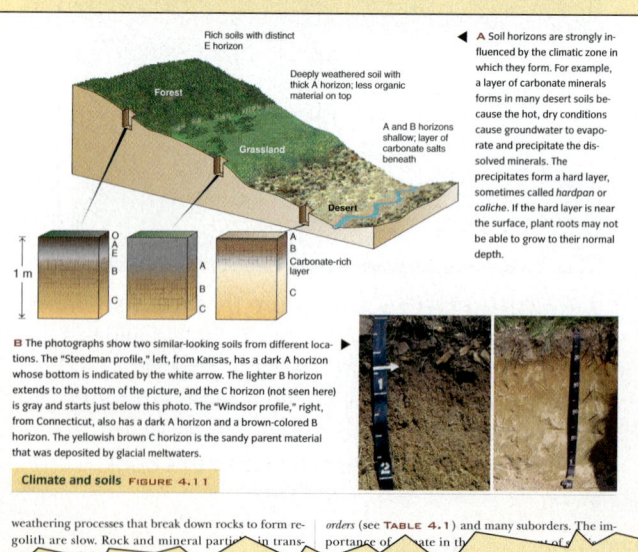

ILLUSTRATIONS AND PHOTOS support concepts covered in the text, elaborate on relevant issues, and add visual detail. Many of the photos originate from National Geographic's rich sources.

MARGIN GLOSSARY TERMS (in green boldface) introduce each chapter's most important terms. The second most important terms appear in **black boldface** and are defined in the text.

TABLES AND GRAPHS, with data sources cited at the end of the text, summarize and organize important information.

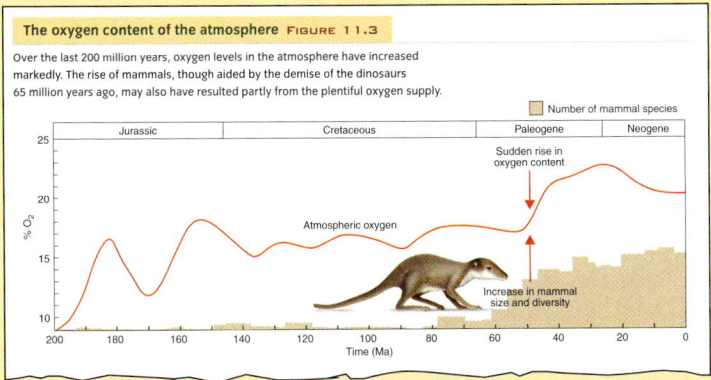

SELF-TESTS at the end of each chapter provide a series of multiple-choice questions, many of them incorporating visuals from the chapter, that review the major concepts.

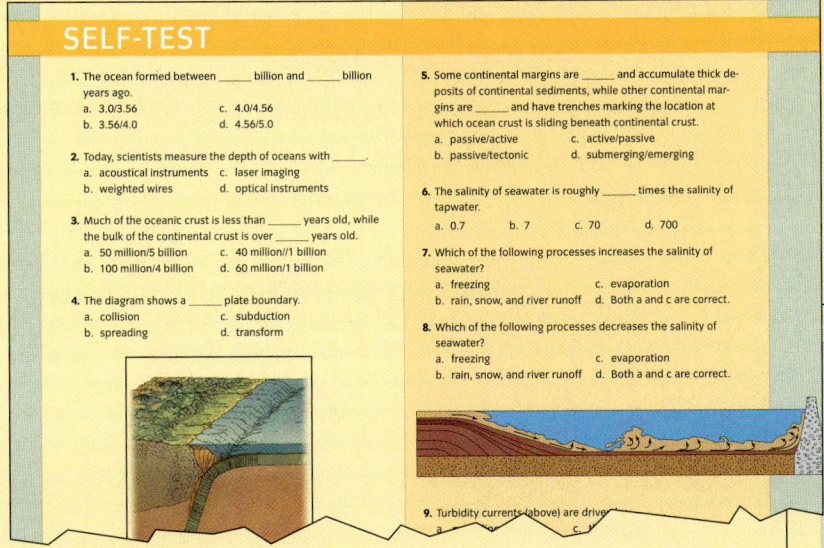

CRITICAL AND CREATIVE THINKING QUESTIONS encourage critical thinking and highlight each chapter's important concepts and applications.

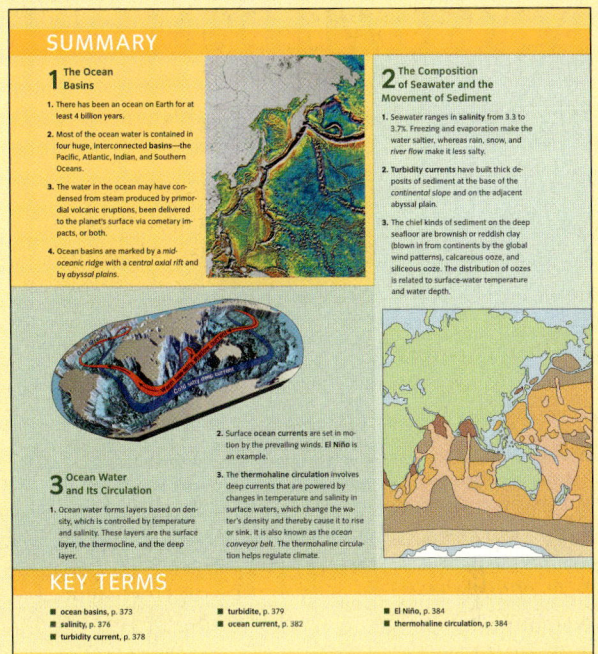

The **SUMMARY** revisits each learning objective and redefines each margin glossary term, featured in boldface here and included in a list of **KEY TERMS**. Students are thus able to study vocabulary words in the context of related concepts. Each portion of the Summary is illustrated with a relevant photo from its respective chapter section.

WHAT IS HAPPENING IN THIS PICTURE? are end-of-chapter features that present students with a photograph that is relevant to chapter topics but illustrates a situation students are not likely to have encountered previously. The photograph is paired with questions designed to stimulate creative thinking.

xi

ACKNOWLEDGMENTS

PROFESSIONAL FEEDBACK

Throughout the process of writing and developing this text and the visual pedagogy, we benefited from the comments and constructive criticism provided by the instructors and colleagues listed below. We offer our sincere appreciation to these individuals for their helpful reviews:

Robert Altamura
Florida Community College at Jacksonville

Steve Bennett
Western Illinois University

Callan Bentley
Northern Virginia Community College

Natalie Bursztyn
Bakersfield College

Marianne Caldwell
Hillsborough Community College

James Carew
College of Charleston

Claire Coyne
Santa Ana College

Constantin Cranganu
Brooklyn College of the City University of New York

Vincent Devlahovich
California State University, Northridge

Gerald Grams
Clark Atlanta University

Clay Harris
Middle Tennessee State University

George Hazelton
Chowan College

Joann Hochstein
Central Florida Community College

Paul Horton
Indian River Community College

Asaad Istephan
Madonna University

Linda Jones
Southwest Minnesota State University

Burair Kothari
Wilbur Wright College

Kristine Larsen
Central Connecticut State University

Jay Lennartsen
University of North Carolina, Greensboro

Michael Lewis
University of North Carolina, Greensboro

Zhaohui Li
University of Wisconsin, Parkside

Tim Long
Georgia Institute of Technology, Atlanta

Donald Lovejoy
Palm Beach Atlantic University

Steven Maier
Northwest Oklahoma State University

Ravi Nandigan
University of Texas, Brownsville

James F. Nugent
Salve Regina University

Guillermo Rocha
Brooklyn College of the City University of New York

Laura Sanders
Northeastern Illinois University

Steven Schimmrich
SUNY Ulster County Community College

John Tacinelli
Rochester Community and Technical College

Dave Thomas
Washtenaw Community College

HusanThompson
Loyola Marymount University

Craig Van Boskirk
Florida Community College at Jacksonville

Kim Van Scoy
University of the Ozarks

Jaehyung Yu
Texas A&M University, Kingsville

SPECIAL THANKS

We are extremely grateful to the many members of the editorial and production staff at John Wiley & Sons who guided us through the challenging steps of developing this book. Their tireless enthusiasm, professional assistance, and endless patience smoothed the path as we found our way. We thank in particular Ryan Flahive, who expertly launched and directed our process; Nancy Perry, Manager, Product Development; Helen McInnis, Managing Director, Wiley Visualizing, who oversaw the concept of the book; Carolyn Smith, Developmental Editor, for her careful editing of our book; Micheline Frederick, Production Manager, who stepped in whenever we needed expert advice; Jay O'Callaghan, Vice President and Publisher, who oversaw the entire project; and Jeffrey Rucker, Executive Marketing Manager for

Wiley Visualizing, and Danielle Torio, Marketing Manager for Geosciences, who adeptly represent the Visualizing imprint. We appreciate the expertise of Elle Wagner, Senior Photo Editor, in managing and researching our photo program and of Sandra Rigby, Senior Illustration Editor, in managing the illustration program. We also wish to thank those who worked on the media and ancillary materials: Lynn Pearlman, Senior Media Editor; Bridget O'Lavin; and Erin Grattan.

We wish to thank Barbara Murck, University of Toronto, co-author of *Visualizing Geology*, a book that was the principal source for topics covered in the first half of the book. We are grateful to Stacy Gold, Research Editor and Account Executive at the National Geographic Image Collection, for her valuable expertise in selecting NGS photos. Many other individuals at National Geographic offered their expertise and assistance in developing this book: Richard Easby, Executive Editor, National Geographic School Division; Mimi Dornack, Sales Manager, and Lori Franklin, Assistant Account Executive, National Geographic Image Collection; and Dierdre Bevington-Attardi, Project Manager, and Kevin Allen, Director of Map Services, National Geographic Maps.

ABOUT THE AUTHORS

Zeeya Merali has an undergraduate degree and a Master's in natural sciences from the University of Cambridge, and a PhD in theoretical cosmology from Brown University. She also holds a Master's degree in science communication from Imperial College, London. As a science writer, her work has appeared in *Scientific American* magazine, the journal *Nature*, and *New Scientist* magazine, and as a filmmaker, her work has been broadcast on The History Channel, UK.

Brian Skinner was born and raised in Australia, studied at the University of Adelaide in South Australia, worked in the mining industry in Tasmania, and in 1951 entered the Graduate School of Arts and Sciences, Harvard University, from which he obtained his PhD in 1954. Following a period as a research scientist in the United States Geological Survey in Washington, DC, he joined the faculty at Yale in 1966, where he continues his teaching and research as the Eugene Higgins Professor of Geology and Geophysics. Brian Skinner has been president of the Geochemical Society, the Geological Society of America, and the Society of Economic Geologists. He holds an honorary Doctor of Science from Toronto University and an honorary Doctor of Engineering from the Colorado School of Mines.

Alan Strahler earned his PhD in geography from Johns Hopkins in 1969, and is presently Professor of Geography at Boston University. He has published over 250 articles in the refereed scientific literature, largely on the theory of remote sensing of vegetation, and has also contributed to the fields of plant geography, forest ecology, and quantitative methods. In 1993, he was awarded the Association of American Geographers/Remote Sensing Specialty Group Medal for Outstanding Contributions to Remote Sensing. With Arthur Strahler, he is co-author of 7 textbook titles with 11 revised editions on physical geography and environmental science. He holds the honorary degree DSHC from the Université Catholique de Louvain, Belgium, and is a Fellow of the American Association for the Advancement of Science.

CONTENTS in Brief

Preface v

1	Introduction to Earth Science	2
2	Minerals: Earth's Building Blocks	34
3	Rocks: Keepers of Earth's History	62
4	Weathering, Soils, and Mass Wasting	98
5	Water On and Under the Ground	132
6	Extreme Climatic Regions: Deserts, Glaciers, and Ice Sheets	168
7	Plate Tectonics: Sculptor of Earth's Ever-Changing Landscape	200
8	Earthquakes and Earth's Interior	232
9	Volcanism and Other Igneous Processes	270
10	How Old Is Old? The Rock Record and Deep Time	302

11	A Brief History of Life on Earth	334
12	The Oceans	370
13	Where Ocean Meets Land	394
14	The Atmosphere: Composition, Structure, and Clouds	424
15	Global Circulation and Weather Systems	456
16	Global Climates Past and Present	488
17	Earth's Place in Space	516
Appendix	A Periodic Table of the Elements	553
	B Units and Their Conversions	554
	C Answers to Self-Tests	556

Glossary 557

Credits 563

Index 569

CONTENTS

1 Introduction to Earth Science 2

What Is Earth Science? 4
- Using the Scientific Method 4
- Earth System Science 7
  - ■ WHAT AN EARTH SCIENTIST SEES: ISLAND OR OPEN SYSTEM? 9

Earth in Space 12
- The Solar System 12
- The Terrestrial Planets 13
- What Makes Earth Unique? 15

Humans and Earth 19
- Renewable and Nonrenewable Resources 20
- Why Study Earth Science? 22

Amazing Places: Earth 26

2 Minerals: Earth's Building Blocks 34

Minerals, Elements, and Compounds 36
- Elements, Atoms, and Ions 36
- Compounds, Molecules, and Bonding 37

What Is a Mineral? 40
- Composition of Minerals 40
- Telling Minerals Apart 42
  - ■ CASE STUDY: MINERALS FOR ADORNMENT 47

Mineral Families 49
- Minerals of Earth's Crust 50
- Silicates: The Most Important Rock Formers 52

Mineral Resources 53
- ■ WHAT AN EARTH SCIENTIST SEES: HOW MANY MINERALS AND METALS DOES IT TAKE TO MAKE A LIGHT BULB? 54

Amazing Places: The Naica Mine, Chihuahua, Mexico 56

3 Rocks: Keepers of Earth's History 62

Rocks: A First Look 64
- The Three Rock Families 65
- The Rock Cycle 66

Igneous Rocks 67
- Rate of Cooling 67
  - ■ WHAT AN EARTH SCIENTIST SEES: PUTTING ROCKS UNDER A MICROSCOPE 68
- Chemical Composition 69

Sedimentary Rocks 72
- Clastic Sediments and Clastic Sedimentary Rocks 72
- Lithification of Clastic Sediment 72
- Chemical Sediments and Chemical Sedimentary Rock 75
  - ■ WHAT AN EARTH SCIENTIST SEES: A CHANGE IN THE ATMOSPHERE 76
- Biogenic Sediments and Biogenic Sedimentary Rock 77
- Rock Beds 77
- Interpreting Environmental Clues 77
- Sedimentary Facies 80
  - ■ CASE STUDY: SEDIMENTARY FACIES AND THE HISTORY OF HUMANKIND 81

Metamorphic Rocks 82
- The Limits of Metamorphism 83
- The Importance of Stress 84

Rocks as Resources 89
- Rocks as Construction Materials 89
- Mineral Resources in Igneous Rocks 90
- Mineral Resources in Chemical and Biogenic Sedimentary Rocks 90
- Gold in Metamorphic Rocks 90

Amazing Places: The Navajo Sandstone 91

4 Weathering, Soils, and Mass Wasting 98

Weathering: The Earth System at Work 100
- Weathering: How Rocks Disintegrate 100
- Mechanical Weathering: Breaking Rocks Apart 101
  - ■ WHAT AN EARTH SCIENTIST SEES: JOINT FORMATION 101
- Chemical Weathering: Breaking Rocks by Chemistry 102
- Factors Affecting Weathering 105

Soil: The Most Important Product of Weathering 109
- Soil 109
- Soil Profiles 110
- Factors that Influence Soil Formation 111
  - ■ CASE STUDY: BAD AND GOOD SOIL MANAGEMENT 116

Erosion and Mass Wasting: Gravity at Work 117
- Erosion by Water 117
- Erosion by Wind 118
- Erosion by Ice 119
- Gravity and Mass Wasting 119
- Tectonics and Mass Wasting 122

Resources Formed by Weathering and Erosion 124

Amazing Places: Monadnock—and Monadnocks 125

5 Water On and Under the Ground 132

The Hydrologic Cycle 134
   Water in the Earth System 135

How Water Affects Land 138
   Streams and Streamflow 138
   Stream Deposits 140
   Large-Scale Topography of Stream Systems 142
   ■ WHAT AN EARTH SCIENTIST SEES: DRAINAGE BASINS 143
   Lakes 144

Surface Water as a Hazard and a Resource 145
   Floods 145
   Surface Water Resources 149
   ■ CASE STUDY: MONO LAKE 152

Fresh Water Underground 153
   The Water Table 153
   How Groundwater Moves 154
   Where Groundwater Is Stored 156
   Groundwater Depletion and Contamination 158
   When Groundwater Dissolves Rocks 160

Amazing Places: Lechuguilla Cave 162

6 Extreme Climatic Regions: Deserts, Glaciers, and Ice Sheets 168

Deserts and Drylands 170
   Types of Deserts 171
   Wind Erosion 173
   Wind Deposits and Desert Landforms 175
   Stream Erosion and Deposition in Deserts 178
   Desertification 179

Glaciers and Ice Sheets 183
   Components of the Cryosphere 183
   Ever-Changing Glaciers 187
   ■ WHAT AN EARTH SCIENTIST SEES: PERIGLACIAL LANDFORMS 194

Amazing Places: Death Valley 195

7 Plate Tectonics: Sculptor of Earth's Ever-Changing Landscape 200

A Revolution in Earth Science 202
  Wegener's Hypothesis of Continental Drift 202
  The Puzzle-Piece Argument 202
  Matching the Rocks and Fossils 203
  Magnetic Poles that Seem to Wander 207

The Plate Tectonic Model 211
  Plate Tectonics in a Nutshell 211
  Types of Plate Margins 214
    ■ WHAT AN EARTH SCIENTIST SEES: THE RED SEA AND THE GULF OF ADEN 216
  Earthquakes and Plate Margins 217

The Search for a Mechanism 219
  Earth's Internal Heat 219
  Convection as a Driving Force 219
  Mantle Plumes 221
  The Tectonic Cycle 222
  The Supercontinent Cycle 223

Amazing Places: The Hawaiian Islands 224

8 Earthquakes and Earth's Interior 232

Earthquakes and Earthquake Hazards 234
  Earthquakes and Plate Motion 234
  Earthquake Hazards and Prediction 238
    ■ CASE STUDY: THE SUMATRA-ANDAMAN TSUNAMI OF 2004 240
  Designing for Earthquake Safety 243

The Science of Seismology 246
  Seismographs 246
  Seismic Waves 247
  Locating Earthquakes 248
  Measuring Earthquakes 250
    ■ WHAT AN EARTH SCIENTIST SEES: RICHTER MAGNITUDE: A LOGARITHMIC SCALE 251

Looking into Earth's Interior 252
  How Earth Scientists Look into Earth's Interior—Seismic Methods 253
  How Earth Scientists Look into Earth's Interior—Other Methods 256
    ■ WHAT AN EARTH SCIENTIST SEES: DIAMONDS: MESSENGERS FROM THE DEEP 257

How and Why Rock Breaks 260
  Stress and Strain 260
  Kinds of Deformation 260
  Kinds of Faults 262

Amazing Places: Loch Ness 264

9 Volcanism and Other Igneous Processes 270

Volcanoes and Volcanic Hazards 272
  Eruptions, Landforms, and Materials 273
  ■ WHAT AN EARTH SCIENTIST SEES: CRATER LAKE 276
  Hazards and Prediction 278
  ■ CASE STUDY: LAKES OF DEATH IN CAMEROON 279

Why and How Rocks Melt 285
  Heat and Pressure Inside Earth 285
  Fractional Melting 287
  Magma and Lava 287

Cooling and Crystallization of Magma 290
  Fractional Crystallization 290
  Bowen's Reaction Series 291

Plutons and Plutonism 293
  Batholiths and Stocks 293
  Dikes and Sills 294

Amazing Places: Mt. St. Helens 296

10 How Old Is Old? The Rock Record and Deep Time 302

Relative Age 304
  Stratigraphy 304
  ■ WHAT AN EARTH SCIENTIST SEES: THE PRINCIPLE OF STRATIGRAPHIC SUPERPOSITION 306
  Gaps in the Record 309
  Fossils and Correlation 310

The Geologic Column 312
  Eons and Eras 313
  Periods and Epochs 316

Numerical Age 317
  Early Attempts 317
  Radioactivity and Numerical Ages 319
  Magnetic Polarity Dating 322
  ■ CASE STUDY: DATING HUMAN ANCESTORS 324

The Age of Earth 325

Amazing Places: The Grand Canyon 328

11 A Brief History of Life on Earth 334

The Ever-Changing Earth 336
- Changes in the Atmosphere and Hydrosphere 337

Early Life 339
- Life in Three Not-So-Easy Steps 340
- Archean and Proterozoic Life 342

Evolution and the Fossil Record 345
- Evolution and Natural Selection 345
- How Fossils Form 347

Life in the Phanerozoic Eon 349
- The Paleozoic Era 349
  - ■ WHAT AN EARTH SCIENTIST SEES: A SAMPLE OF CAMBRIAN LIFE 349
- From Sea to Land 350
- The Mesozoic Era 354
- The Cenozoic Era 355
- Mass Extinctions 357

Amazing Places: The Burgess Shale 361

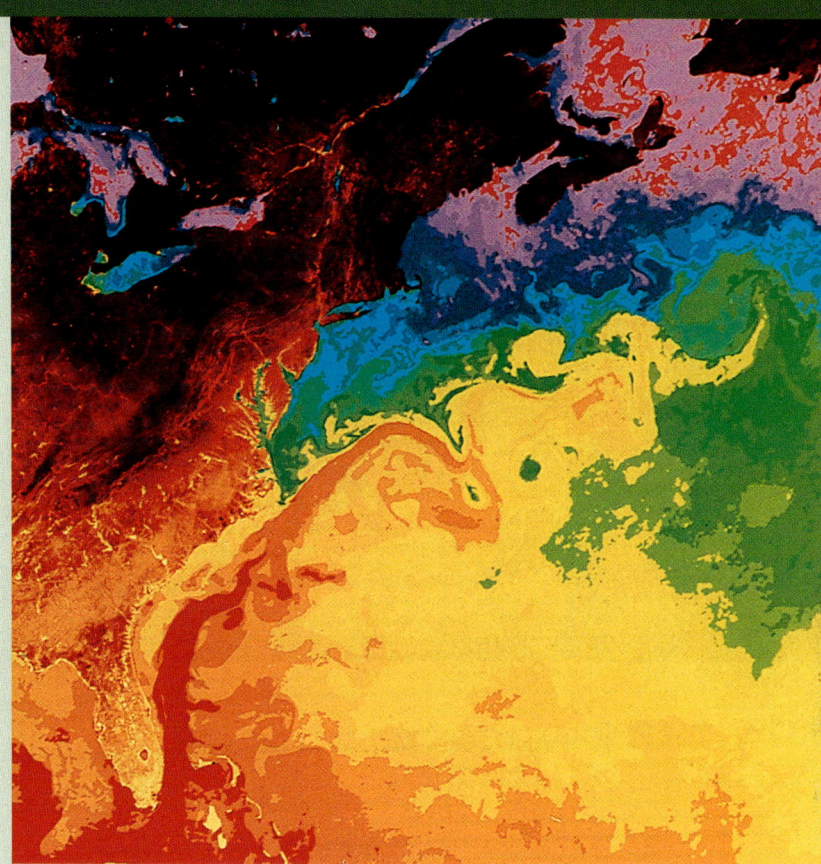

12 The Oceans 370

The Ocean Basins 372
- Ocean Geography 373
- Depth and Volume of the Oceans 373
  - ■ WHAT AN EARTH SCIENTIST SEES: UNDERSEA TOPOGRAPHY 375

The Composition of Seawater and the Movement of Sediment 376
- Turbidity Currents 378
- Biotic Zones and Deep-Sea Sediments 380

Ocean Water and Its Circulation 382
- Ocean Currents 382
- The Ocean Conveyor Belt 384
- How Oceans Regulate Climate 385

Amazing Places: Monterey Bay, California 389

14 The Atmosphere: Composition, Structure, and Clouds 424

Earth's Atmosphere 426
   The Thermal Structure of the Atmosphere 427
   ■ What an Earth Scientist Sees:
     The Bora Bora Sunset 427

Moisture in the Atmosphere 431
   Changes of State 431
   Humidity 432

The Global Energy System 434
   Solar Energy Losses in the Atmosphere 434
   Albedo 435
   Counterradiation and the Greenhouse Effect 436

Formation of Clouds 438
   The Adiabatic Principle 438
   Clouds 440

Precipitation 442
   Orographic Precipitation 445
   Convectional Precipitation 446
   Thunderstorms and Unstable Air 447
   Anatomy of a Thunderstorm 448

Amazing Places: Hole Punch Clouds in Mobile, Alabama 450

13 Where Ocean Meets Land 394

Changes in Sea Level 396
   Global Changes in Volume 397
   Tides 398

Waves 400
   ■ What an Earth Scientist Sees:
     Breaking Waves 401
   Wave Action Along Coastlines 401
   Erosion and Transport of Sediment by Waves 402
   Tidal Currents 404

Shorelines and Coastal Landforms 405
   Rocky Coasts 405
   Beaches and Barrier Islands 406
   Delta Coasts 408
   Coral Reefs 409

Humans versus the Sea 410
   Coastal Hazards 410
   Protection Against Shoreline Erosion 413
   Effects of Human Interference 416
   ■ Case Study: The Black Sea Coast 416

Amazing Places: The Florida Keys Reef 418

15 Global Circulation and Weather Systems 456

Atmospheric Pressure 458
  Air Pressure and Altitude 459

Why Air Moves 460
  Pressure Gradients 460
  Local Winds 460

Cyclones and Anticyclones 463
  Cyclones and Anticyclones 464

Global Wind Patterns 465
  Wind Systems 466

Winds Aloft 467
  The Geostrophic Wind 468
  Global Circulation at Upper Levels 468
  Rossby Waves, Jet Streams, and the Polar Front 469
  ■ WHAT AN EARTH SCIENTIST SEES: JET STREAM CLOUDS 470

Weather Systems 470
  Thunderstorms 474
  Wave Cyclones 474
  Tornadoes 476
  Tropical Cyclones 477

Amazing Places: New Orleans Before and After Katrina 482

16 Global Climates Past and Present 488

Global Climate Change in the Past 490
  What We Know 491
  How We Know It: The Temperature Record 493
  ■ WHAT AN EARTH SCIENTIST SEES: THE GUBBIO SEDIMENTS 494
  Causes of Climate Change 495
  Rapid Climate Change: The Younger Dryas Event 495

Global Climates Today 498
  Low-Latitude Climates 499
  Midlatitude Climates 501
  High-Latitude Climates 503

Present-Day Climate Change 505
  What We Know 505
  What We Think 506

Amazing Places: Barrow, Alaska 510

17 Earth's Place in Space 516

Astronomy and the Scientific Revolution 518
- Ideas from Antiquity 518
- Copernicus's Challenge 519
- Kepler and the New Astronomy 521
- Galileo and Newton 522

The Solar System 524
- ■ WHAT AN EARTH SCIENTIST SEES: THE CRAB NEBULA 525
- The Sun 526
- The Planets 530
- ■ WHAT AN EARTH SCIENTIST SEES: SATURN'S RINGS 533
- Other Solar Systems 537

Stars and Stellar Evolution 538
- The Hertzsprung-Russell Diagram 538
- Stellar Evolution: Birth to Death 539

The Universe and How It Came to Be 540
- Edwin Hubble and the Discovery of Galaxies 540
- The Big Bang 543
- ■ CASE STUDY: THE AFTERGLOW OF THE BIG BANG 544
- The Universe Today 545

Amazing Places: Mars 546

Appendix A Periodic Table of the Elements 553
B Units and Their Conversions 554
C Answers to Self-Tests 556

Glossary 557
Credits 563
Index 569

VISUALIZING FEATURES

Multipart visual presentations that focus on a key concept or topic in the chapter.

Chapter 1
Earth's Changing Face • Energy Consumption

Chapter 2
Four Types of Bonding • Atomic Structure of Minerals • Mineral Clevage

Chapter 3
The Three Rock Families • Tephra • From Clasts to Rocks • Metamorphic Rocks Derived from Shale

Chapter 4
Factors Affecting Rates of Weathering

Chapter 5
Reservoirs in the Hydrologic Cycle • Steam Valley Deposits • Principal Watersheds and Fresh-Water Resources of the World

Chapter 6
Dune Varieties • Glaciers and Ice Caps

Chapter 7
Evidence of Continental Drift • Our Tectonic Planet

Chapter 9
Volcanoes and Eruptions

Chapter 10
Strata and the Law of Horizontality • The Geologic Column • The Debate over Numerical Age

Chapter 11
The Three Steps of Life • The Effects of a Meteorite Impact • The Human Footprint on Earth's Surface

Chapter 12
Ocean Circulation and Sea-Surface Temperature

Chapter 13
Barrier Islands, Spits, and Lagoons

Chapter 14
Clouds

Chapter 15
Wave Cyclones and Tropical Cyclones

Chapter 16
Reconstructing Temperature Records • Future Impacts of Climate Change

Chapter 17
The Sun

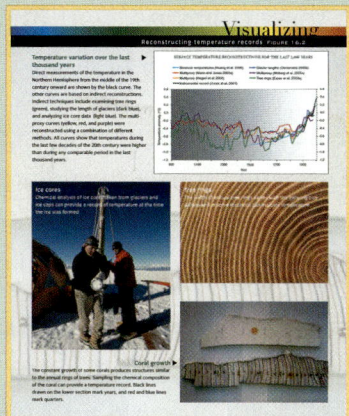

PROCESS DIAGRAMS

A series or combination of figures and photos that describe and depict a complex process.

Chapter 2
How Ions and Compounds Form

Chapter 3
Lithification • From Shale to Gneiss

Chapter 4
Ion Exchange

Chapter 5
The Hydrologic Cycle

Chapter 6
How Sand Dunes Form

Chapter 7
Seafloor Spreading

Chapter 8
How a Tsunami Was Unleashed by Motion on a Fault • Seismic Waves in Earth's Interior

Chapter 9
How a Stratovolcano Is Formed • Effect of Pressure and Temperature on Rocks

Chapter 10
Unconformities: Gaps in the Rock Record

Chapter 11
Darwin's Finches

Chapter 12
World Map of Surface Ocean Currents • The Ocean Conveyor Belt

Chapter 13
Littoral Drift

Chapter 14
How a Greenhouse Works • Cloud Formation and the Adiabatic Process

Chapter 15
Convection Loops • Life History of a Wave Cyclone "Low"

Chapter 16
Causes of the Younger Dryas Event

Chapter 17
The Seasons: Earth–Sun Relations through the Year • The Birth of Our Solar System

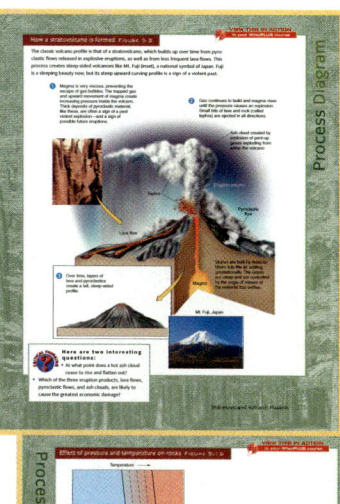
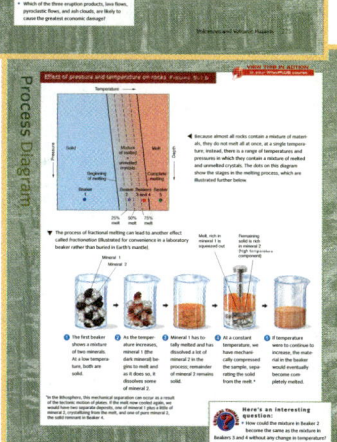

Achieve Positive Learning Outcomes

WILEY PLUS
www.wileyplus.com

WileyPLUS combines robust course management tools with interactive teaching and learning resources all in one easy-to-use system. It has helped over half a million students and instructors achieve positive learning outcomes in their courses.

*Wiley***PLUS** contains everything you and your students need – and nothing more – including:

- ⊕ The entire textbook online—with dynamic links from homework to relevant sections. Students can use the online text and save up to half the cost of buying a new printed book.
- ⊕ Automated assigning & grading of homework & quizzes.
- ⊕ An interactive variety of ways to teach and learn the material.
- ⊕ Instant feedback and help for students… available 24/7.

"WileyPLUS helped me become more prepared. There were more resources available using WileyPLUS than just using a regular [printed] textbook, which helped out significantly. Very helpful…and very easy to use."
– Student Victoria Cazorla,
Dutchess County Community College

See – and Try WileyPLUS in action! Details and Demo:
www.wileyplus.com

Wiley is committed to making your entire *WileyPLUS* experience productive & enjoyable by providing the help, resources, and personal support you & your students need, when you need it. It's all here: www.wileyplus.com –

TECHNICAL SUPPORT:
- A fully searchable knowledge base of FAQs and help documentation, available 24/7
- Live chat with a trained member of our support staff during business hours
- A form to fill out and submit online to ask any question and get a quick response
- **Instructor-only** phone line during business hours: 1.877.586.0192

FACULTY-LED TRAINING THROUGH THE WILEY FACULTY NETWORK:
Register online: www.wherefacultyconnect.com
Connect with your colleagues in a complimentary virtual seminar, with a personal mentor in your field, or at a live workshop to share best practices for teaching with technology.

1ST DAY OF CLASS...AND BEYOND!
Resources for You & Your Students Need to Get Started & Use WileyPLUS from the first day forward.

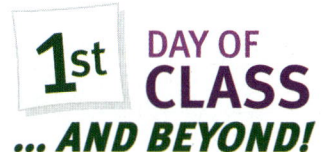

- 2-Minute Tutorials on how to set up & maintain your *WileyPLUS* course
- User guides, links to technical support & training options
- **WileyPLUS for Dummies**: Instructors' quick reference guide to using *WileyPLUS*
- Student tutorials & instruction on how to register, buy, and use *WileyPLUS*

YOUR WileyPLUS ACCOUNT MANAGER:
Your personal *WileyPLUS* connection for any assistance you need!

SET UP YOUR WileyPLUS COURSE IN MINUTES!
Select *WileyPLUS* courses with QuickStart contain pre-loaded assignments & presentations created by subject matter experts who are also experienced *WileyPLUS* users.

Interested? See–and Try *WileyPLUS* in action!
Details and Demo: www.wileyplus.com

Introduction to Earth Science

1

Not far from Alice Springs in central Australia, a ring of hills called Gosses Bluff (right) juts up from the endlessly flat landscape. This curious formation is the relic of a great meteorite that crashed to Earth 142 million years ago. The impact and explosion blasted out a crater 24 kilometers in diameter surrounded by a ring of debris. What we see today is the eroded remnant of the center of the impact site. You can see a ghostly, highly eroded remnant of the outer rim in a photograph from space (inset).

Our planet was formed about 4.56 billion years ago out of innumerable rocks very much like the Gosses Bluff meteorite. Though impacts of large meteorites are rare in the present era—fortunately—scars from past impacts remind us that we are not isolated in space but, rather, are part of an active solar system. We share energy and matter with the rest of the system.

The successes of the Space Age have made us more conscious of Earth's place in the larger solar system. We have learned that Earth is a system, too. Its components—rocks, water, air, living creatures—depend vitally on one another. Instead of studying various aspects of the planet in isolation, as we once did, scientists now try to look at the Earth system as a whole. In *Visualizing Earth Science* you will also learn how Earth has changed and is continuing to change today.

Global Locator

Outer rim of crater

Gosses Bluff

CHAPTER OUTLINE

■ **What Is Earth Science?** p. 4

■ **Earth in Space** p. 12

■ **Humans and Earth** p. 19

What Is Earth Science?

> **LEARNING OBJECTIVES**
>
> **Describe** the many areas of study embraced by Earth science.
>
> **Outline** the steps used in the scientific method.
>
> **Explain** what it means to take a systems approach to Earth science.
>
> **Identify** three types of systems.
>
> **Identify** the four major subsystems of the Earth system.

Earth is home to countless living things. No other planet so far discovered is known to support life, and no other planet has even been discovered to have the delicate balance of conditions—temperature, atmosphere, climate, liquid water—needed to maintain life. Earth may well be unique. **Earth science** investigates all aspects of our home planet, from its atmosphere and oceans, to its rocks, minerals, soils, mountains, deserts, and all living things. We all appreciate the beauty, the comfort, and the reliability of Earth, and we take a lot for granted. But science asks deeper questions: How do things work? How and why, for example, does an ocean current like the Gulf Stream flow where it does? Why have glaciers advanced and retreated naturally over time? What might happen if we inadvertently change something, like waterflow in a river system, or the composition of the atmosphere? Science is a special way of studying things, a way of observing, asking questions, and testing the answers. *To be scientific, answers must be testable.*

> **Earth science** The scientific study of all aspects of Earth.

The scientific study of Earth began in a systematic way in the 18th century, and for about 200 years all testable studies were confined to Earth. But with the dawn of the Space Age in the 1950s, Earth science expanded, first to study Earth from space, then other planets, and now beyond, as one of our spaceships has left the solar system. Unmanned spaceships now visit other planets, make tests and measurements, and send the results back to scientists on Earth. Astronauts have landed on the Moon, collected rocks and Moon dust, and returned the samples to Earth for detailed study and testing. Just as medical doctors sometimes study all members of a family to discover their genetic similarities and differences, so Earth scientists are being well rewarded by studies of the similarities and differences of Earth and its nearest neighbors.

USING THE SCIENTIFIC METHOD

Like all scientists, Earth scientists use a logical research strategy that has developed through trial and error over many years, called the **scientific method**. Although it varies in details, the scientific method includes these basic steps:

> **scientific method** The way a scientist approaches a problem; steps include observing, formulating a hypothesis, testing, and evaluating results.

Step 1. *Observe and gather data.* Scientists start with a question and acquire trustworthy evidence about it, especially measurements. In **FIGURE 1.1**, an Earth scientist asks the question, "How did this group of rocks form?" She observes and measures the sequence of layered rocks in question. She sees that the layers are *horizontal* and *parallel*—important clues. Further, each layer consists of innumerable *small* grains, and the size of the grains *varies* from layer to layer, but is approximately the same within each layer.

Step 2. *Formulate a hypothesis.* Scientists explain their observations by developing a **hypothesis**. In our case, our Earth scientist develops three hypotheses. She hypothesizes that the rocks were formed from material that was transported and deposited where she has found it; but how was it transported? Hypothesis 1 is that a *glacier* was the transporting agent. Hypothesis 2 is that *wind* did the transporting. Hypothesis 3 is that *water* did the transporting.

> **hypothesis** A plausible, but yet to be proved, explanation for how something happens.

Using the scientific method FIGURE 1.1

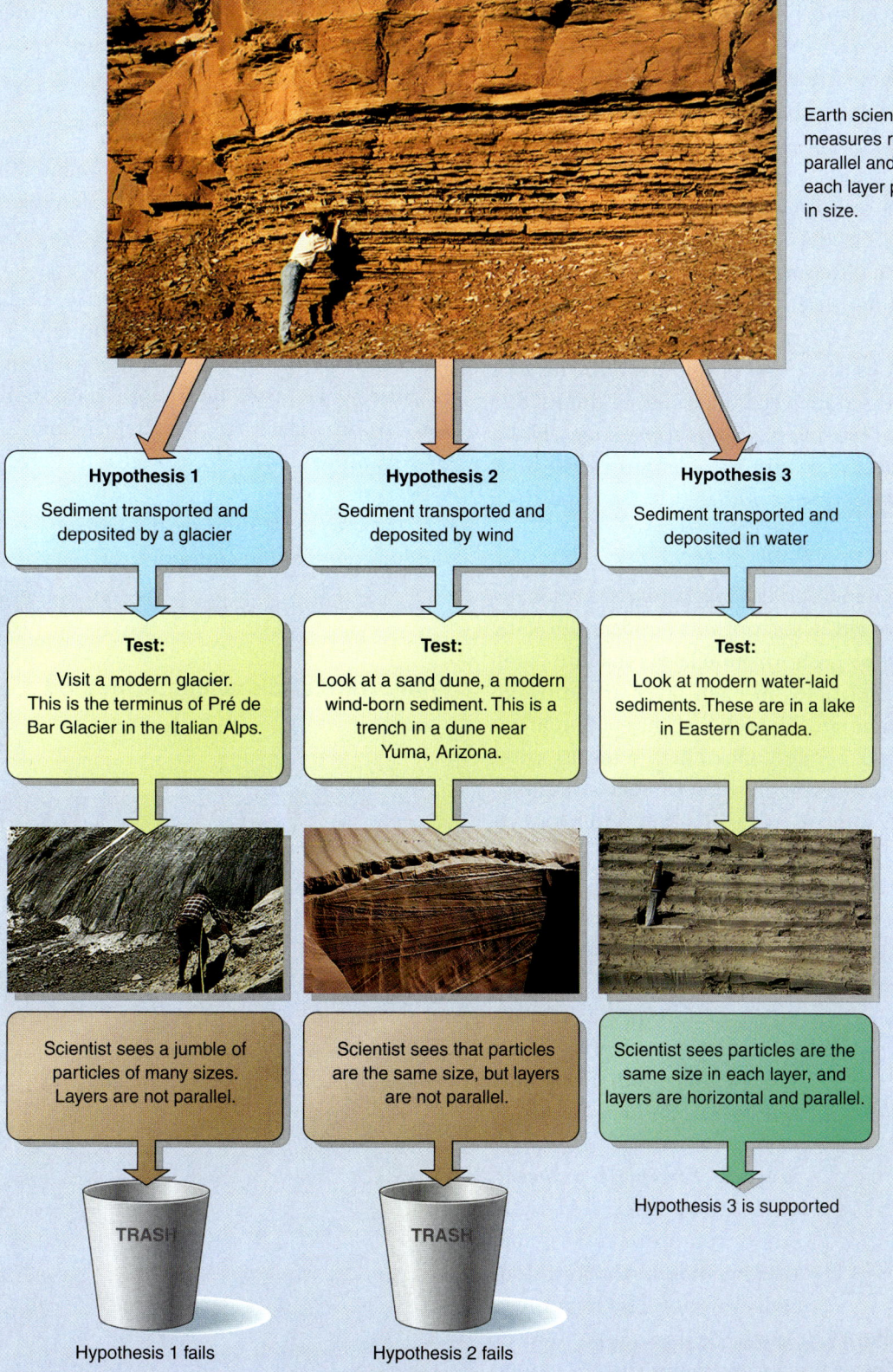

Step 3. *Test the hypotheses*. Scientists use a hypothesis—or in this case, multiple hypotheses—to make predictions and to develop tests. The tests may involve controlled experiments in a laboratory, further observations and measurements, and possibly the development of a mathematical model. Earth scientists in particular like to test their hypotheses against *real observations*.

- Our scientist travels to a modern glacier and studies the jumble of debris it deposits. She notes that the grains are different sizes, all mixed up, and not in neatly defined layers. So, Hypothesis 1 fails.

- Then she goes to a desert region where she sees wind-transported material deposited in dunes. She observes that particle sizes are approximately the same, but they aren't in parallel and horizontal layers—the layers are at odd angles. So, Hypothesis 2 fails.

- Finally, our scientist visits a lake and observes materials transported by a river and deposited in lake water. Now she sees horizontal layers that are parallel, and the particles in each layer are approximately the same size. Hypothesis 3 has potential but more testing is needed. Our Earth scientist notes that plants are growing in the lake. Like a responsible scientist, she goes a step further and hypothesizes that if the material that formed the rocks really was deposited in a lake, the remains of aquatic plants might still be present. If, on further observation, she finds fossilized fresh-water plant remains in the layered horizontal rocks, she would be even more confident that she was on the right track. This is how the scientific method tests and retests hypotheses.

theory A hypothesis that has been tested and is strongly supported by experimentation, observation, and scientific evidence.

Step 4. *Formulate a theory*. Once a hypothesis has withstood numerous tests, scientists become more confident in its validity. The hypothesis is elevated to a **theory**. It is not the final word, however, and a theory is always open to further testing. (Note: In everyday speech, people often misuse the term *theory* to mean *hypothesis* by saying, "Well, that's just a theory." What they really mean is, "That's just a hypothesis." In science, by the time a statement attains the stature of a theory, it is very substantial and must be taken seriously.)

Step 5. *Formulate a law or principle*. Ultimately, a theory or a group of theories may be formulated into a **law** or **principle**. Laws and principles are statements that some natural phenomenon invariably is observed to happen in the same way, and no deviations ever have been observed. For example, in Earth science the **Law of Original Horizontality** states that sediment deposited in water is always in horizontal layers (or nearly always so, because a lake or sea floor might have slight irregularities) and the layers are parallel to Earth's surface (or nearly always so). No exceptions have ever been observed.

FIGURE 1.2 sums up the key steps in the application of the scientific method.

Framework of the scientific method FIGURE 1.2

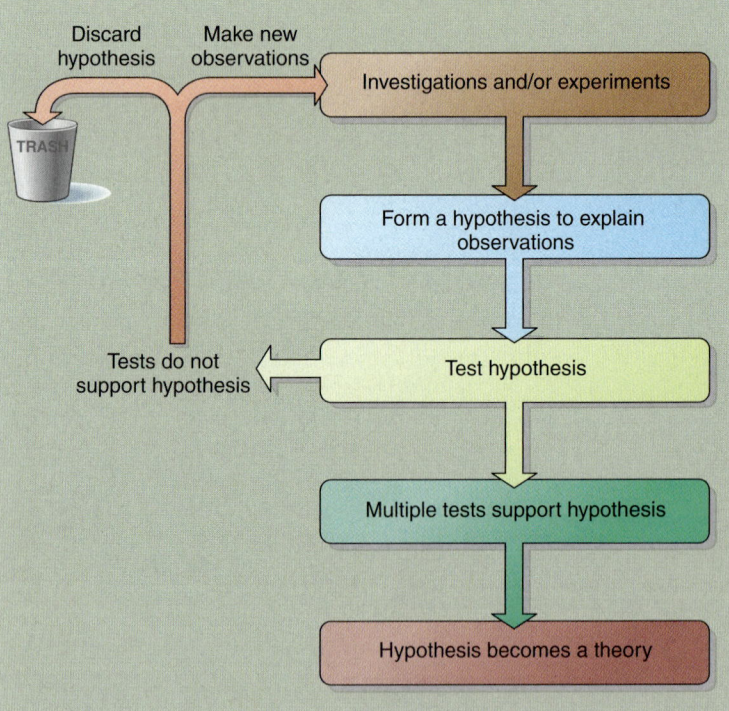

CHAPTER 1 Introduction to Earth Science

EARTH SYSTEM SCIENCE

Traditionally, scientists have studied Earth by focusing on separate units—meteorologists studied the atmosphere, oceanographers the ocean, geologists the solid Earth, botanists the plants, zoologists the animals—more or less in isolation from each other. However, the first photographs of Earth from space in the 1960s caused a dramatic rethinking of this traditional view (**FIGURE 1.3A**). For the first time, it was possible to see the planet in one sweeping view. We could see everything at a glance—the clouds, the oceans, polar ice caps, and the continents—all at the same time, and in their proper scale. The astronauts, like the rest of us, marveled at Earth's "overwhelming beauty . . . the stark contrast between bright colorful home and the stark black infinity" (Rusty Schweikart, *Apollo 9*). Yet from space it was also clear how small Earth is—just a dust speck compared to the vastness of the solar system and the universe. On such a small planet it no longer made sense to study all the pieces separately. Earth science changed from a collection of specialties to specialists studying the same complex object.

Instruments carried into space on satellites have now given us ways to study the relationships between all parts of Earth on a global scale, as we never could have before (**FIGURE 1.3B**). This new, more all-inclusive view of Earth has lead to the emergence of a new field of research called *Earth system science*.

system A portion of the universe that can be separated from the rest of the universe for the purpose of observing changes that happen in it.

The system concept

A **systems** approach is a helpful way to break down a large, complex problem into smaller pieces that are easier to study without losing sight of the connections between those pieces. A system may be large or small, simple or

Earth from orbit FIGURE 1.3

A Space flight and satellite photography have enabled us to see our planet whole, for the first time. Many of the most arresting images and surprising discoveries about our planet now come from satellites.

B Satellite images can reveal interactions within the Earth systems. In this photo, dust storms from the Sahara Desert blow far out into the Atlantic. In fact, Earth scientists have found African dust all the way across the Atlantic Ocean, and some think that it might contribute to the death of coral reefs off the coast of Florida. ▼

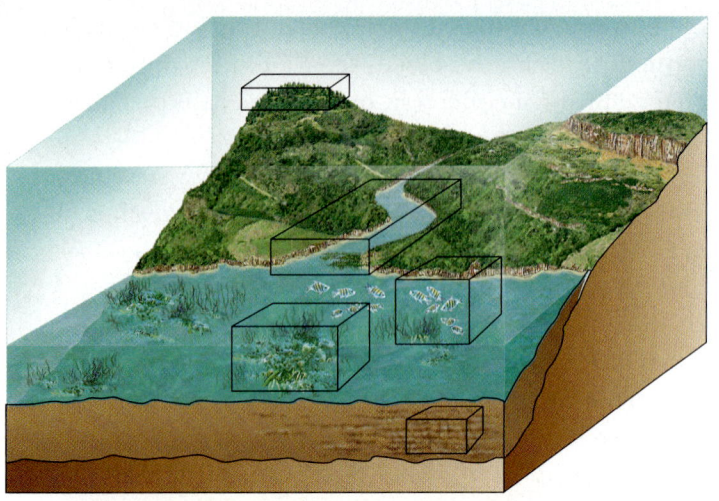

The system concept FIGURE 1.4

This figure shows a variety of systems. The entire diagram—mountains, river, lake—is one kind of system known as a *watershed*. The individual pieces enclosed by boxes, such as the river, are also systems. Even a small volume of water or lake sediment (foreground boxes) can be considered a system.

complex (FIGURE 1.4). It could be the contents of a beaker in a laboratory experiment or the contents of an ocean. A leaf is a system, but it is also part of a larger system (a tree), which is part of a still larger system (a forest).

The fact that we distinguish a system from the rest of the universe for specific study does not mean that we ignore its surroundings. In fact, the nature of a system's boundaries is one of its most important defining characteristics. FIGURE 1.5 illustrates the three basic kinds of systems. The easiest to understand is an *isolated system*. In an isolated system, the boundaries prevent the system from exchanging either matter or energy with its surroundings. However, no isolated systems exist in the real world. It is possible to have boundaries that do a pretty good job of preventing the passage of matter, but no boundary is so perfectly insulating that it prevents energy from entering or escaping.

A second type of system, and the nearest thing to an isolated system in the real world, is a *closed system*. Such a system has boundaries that permit the exchange of energy, but not matter, with its surroundings. An example of a closed system would be a perfectly sealed pressure cooker, which would allow the material inside to be heated but would not allow any of the cooking materials to escape. (Note that in real life, pressure cookers do allow some vapor to escape when the pressure gets too high, so they are not perfect examples of closed systems.)

The third kind of system, an *open system*, can exchange *both* matter and energy across its boundaries. Most of the systems we deal with are open systems. An island offers an example (see *What an Earth Scientist Sees*). The system concept can also be applied to either natural or artificial environments. For example, land-use planners sometimes use a systems approach in the study of cities. Enormous flows of energy and materials occur across city borders.

The Earth system Earth itself is a very close approximation to a closed system. Energy enters the Earth system as solar radiation. The Sun's energy is used in various life processes; it also causes winds, ocean currents, and numerous other events, and then departs in the form of heat radiated back into space. Very little matter crosses the boundaries of the Earth system. We do lose some hydrogen and helium atoms from the outer atmosphere, and we gain some matter in the form of meteorites, like the Gosses Bluff meteorite 142 million years ago. However, for most purposes, especially over the short term, the outflows and inflows are so tiny we can treat Earth as a closed system.

Three kinds of systems FIGURE 1.5

Scientists distinguish three kinds of systems: isolated, closed, and open. Most systems in Earth science are open.

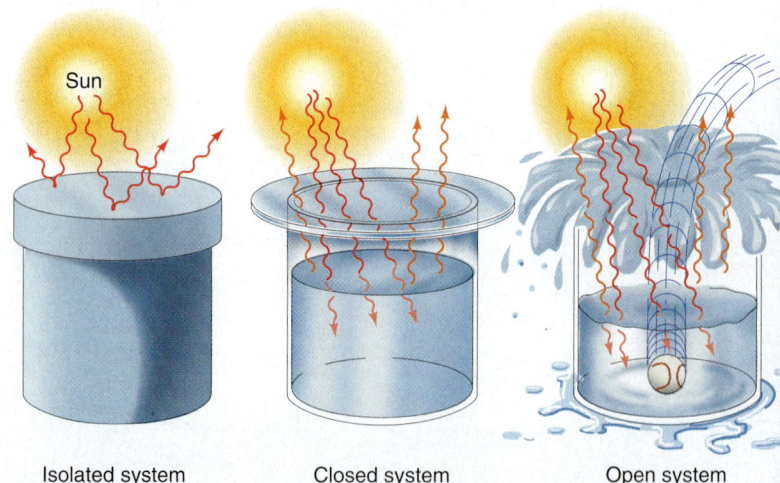

What an Earth Scientist Sees

Island or Open System?

The island of Bora Bora (above) is an exotic, though isolated, tourist destination. But how isolated is it, really?

Some islands may seem isolated to a tourist or to a shipwrecked sailor, but from the point of view of an Earth systems scientist, every island is an open system. (Remember from Figure 1.5 that an open system allows both matter and energy to cross its boundaries.)

Energy (in the form of sunlight) and matter (in the form of precipitation) reach the island from outside sources. The energy leaves the island as heat. The water either evaporates or drains into the sea. In the modern era, humans may also bring material into and out of the system, by importing and exporting resources.

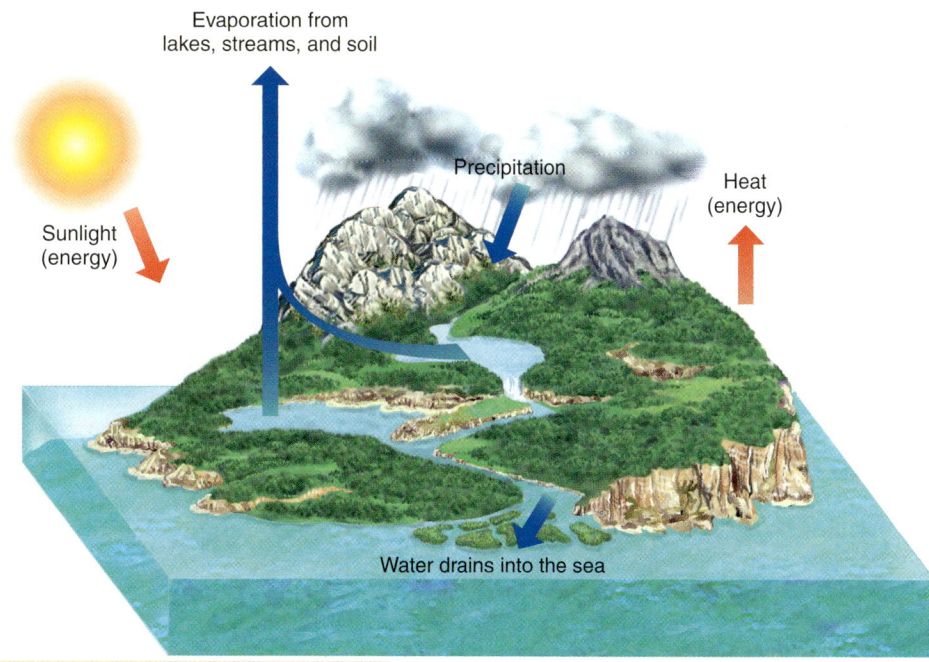

Here's an interesting question:
- Is Antarctica, which is covered by ice, an open system or a closed system?

What Is Earth Science? 9

A river no longer runs through it FIGURE 1.6

A The Colorado River and its tributaries provide drinking water to 25 million people in California, Nevada, Arizona, Utah, New Mexico, Colorado, and Wyoming. It also irrigates 3.5 million acres of fields. Because of the massive diversion of water away from the river in those states, not a drop of water reaches the river's historic terminus, the Gulf of California (or Sea of Cortez) in Mexico.

B All that remains of the river delta is the vast mud flat seen in this photo. The branched "stream" you see here is *not* the Colorado River: It is an inlet carved by the tides in the soft mud of the delta. It is almost as if the gulf were trying to recreate the river that no longer feeds it.

The fact that Earth is a closed system (on the scale of time that humans have existed) has extremely important implications. Here are a few examples and their implications:

- Any change in one part of a closed system will eventually affect other parts of the system. For instance, when we divert a river to provide drinking water for a city, we inevitably deplete the water resources somewhere else (**FIGURE 1.6**).

- The amount of matter in a closed system is fixed and finite. Therefore, the resources on Earth are *all we have* and, for the foreseeable future, *all we ever will have*. We must treat Earth resources with respect and use them wisely and cautiously.

- When we dispose of waste, we must remember that the waste will remain within the boundaries of the Earth system. As environmentalists sometimes say, "There is no away to throw things to."

The Earth system can be divided into four very large subsystems, which you can think of as Earth's principal reservoirs of materials and energy. These are the *lithosphere, biosphere, atmosphere,* and *hydrosphere* (**FIGURE 1.7**). We will define and discuss each term in detail throughout the book, starting with the lithosphere in this chapter. Each of these is an open system and they continually exchange matter and energy between them. For example, plants draw nutrients from the lithosphere and carbon dioxide from the atmosphere, and incorporate them into the biosphere. When plants die and decompose, some of the material they contain returns to the atmosphere as carbon dioxide, while other parts may fossilize and reenter the lithosphere. Rocks erode, and the minerals they contain become salts in the hydrosphere. Lakes evaporate and return their salts to the lithosphere. The exchanges of materials between the spheres never stop. Some Earth scientists refer to the collective interacting "whole" of these four open systems as the *geosphere*.

The place where the four reservoirs interact most intensively is the *life zone*. It is a region no greater than about 10 km above Earth's surface and 10 km below the surface (see **FIGURE 1.8**). In this narrow zone all known forms of life exist because it is only here that conditions favorable for life are created by interactions between the lithosphere, hydrosphere, atmosphere, and biosphere.

The four big open systems of the Earth system can, of course, be further divided into the many subsystems that are of interest to Earth scientists—all of the smaller subsystems are open systems. For example, the hydrosphere can be divided into oceans, glacial ice, streams, lakes, groundwater, raindrops, and so on. The important point to understand is that Earth is a closed system

Earth's four "spheres" FIGURE 1.7

This figure illustrates Earth's four principal subsystems: lithosphere, biosphere, atmosphere, and hydrosphere. Materials and energy cycle among these subsystems, as shown by the arrows, making them open systems.

Earth's climate provides an example of the interplay between positive and negative feedbacks. The Sun's heat causes water to evaporate from the sea and enter the atmosphere as water vapor. Water vapor increases the heat-trapping capacity of the atmosphere. As the temperature increases, the rate of evaporation also increases, so evaporation of seawater is a positive feedback. The negative feedback that balances the system is the formation of clouds (condensed droplets of water). As more water vapor accumulates in the atmosphere, more clouds form, and white clouds simply reflect the Sun's rays back into space before they reach the ocean. More cloud cover reduces the amount of evaporation.

The evaporation–condensation part of the climate system is in balance, but Earth's climate is currently warming, and one possible reason is that by burning oil and coal we have increased the carbon dioxide content of the atmosphere. Carbon dioxide, like water vapor, increases the heat-trapping capacity of the atmosphere—it is a positive feedback—but, unlike water vapor, does not form clouds, so there is not a counterbalancing negative feedback mechanism. Fortunately, other parts of the Earth

(or virtually so), and all of the subsystems within it, no matter how large or how small, are open systems.

feedback mechanisms Reactions that enhance (positive) or retard (negative) change in an open system.

Feedback mechanisms

If materials and energy move continually among Earth's many subsystems, it is appropriate to ask, Are there limits to how much things can change? The answer is, Yes, there are limits and they exist because of **feedback mechanisms**. Feedbacks keep the system stable. Mechanisms that enhance change are called *positive feedback mechanisms*. Mechanisms that tend to resist change and thereby to maintain or balance the system are called *negative feedback mechanisms*.

The life zone FIGURE 1.8

All life on Earth lives within a zone no wider than 20 km. It is the zone where interactions between the lithosphere, hydrosphere, and atmosphere create a habitable environment.

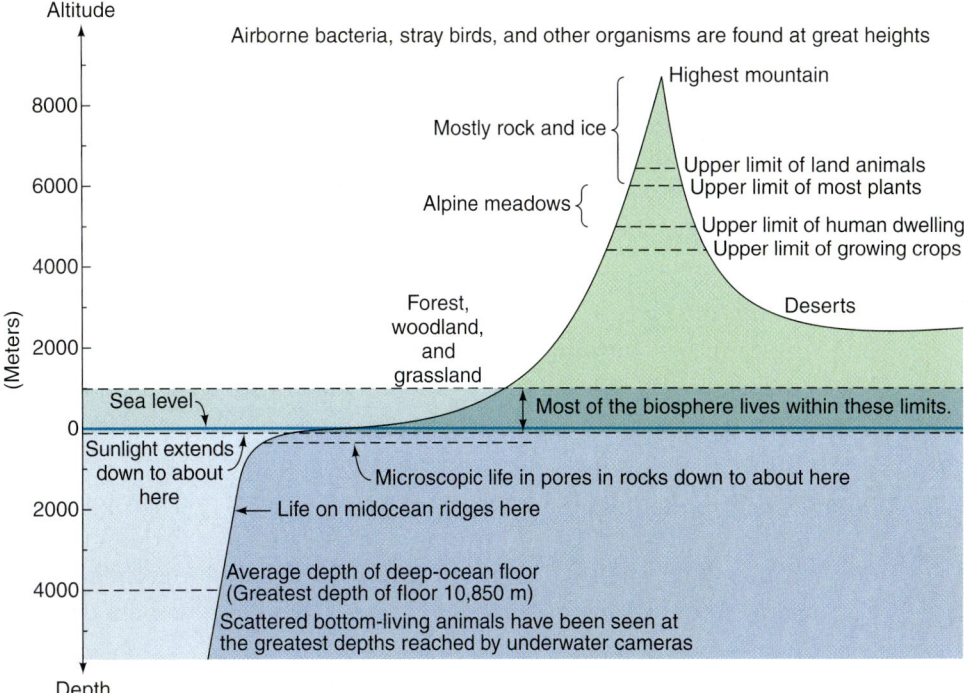

system have negative feedbacks. As mentioned above, as the temperature rises, cloud cover increases. Plants provide another negative feedback: Warmer temperatures promote the growth of plants, and plants remove carbon dioxide from the atmosphere. There must be a limit to the global temperature increase, where negative feedbacks balance positive feedbacks, but what that limit actually is, is a topic of intense research and raises issues that we will return to later in the book.

Earth science versus Earth system science

It is pertinent to ask a question at this point: How does Earth science differ from Earth system science? *Earth science*, as defined earlier in the chapter, *is the scientific study of all aspects of Earth*, and that means all of Earth's past history, as well as Earth as we know it today. **Earth system science** is that portion of Earth science dealing with the way Earth works today. Earth system science certainly draws evidence from the past, but its main focus is how the Earth works today and how, possibly, an Earth inhabited by humans is being changed by human activities.

Earth system science The study of Earth as a closed system composed of interacting open systems and how the open systems may be changed as a result of human activities.

CONCEPT CHECK STOP

What are some key questions investigated by Earth science?

Why is each of the steps in the scientific method important?

Why is the system concept a key part of Earth science?

How do feedback mechanisms work to keep open systems in balance?

Earth in Space

LEARNING OBJECTIVES

Describe the bodies that make up the solar system.

Discuss the internal structures of the terrestrial planets.

Identify the differences between compositional layers and rock strength layers in Earth's internal structure.

As Earth scientists we mainly study the materials and processes that occur on Earth, our home planet, in isolation from the rest of the solar system. But of course Earth is not isolated. For example, Earth gets energy from the largest of the objects in the solar system, the Sun; gravitational pulls from the Sun and Moon determine the tides in the ocean; and meteorites that originate within the solar system and sometimes crash into Earth can have massive consequences, as happened with the impact 65 million years ago that is hypothesized to have killed off the dinosaurs. Earth science requires that we broaden our perspective beyond Earth alone and ask some important questions: For example, why does Earth seem to be the only place in the solar system where life exists, and does the presence of an ocean of water make Earth operate differently from other nearby planets, such as Mars and Venus? The answers to such questions are not clear because they involve still-untested hypotheses; we return to some of those questions in Chapter 17. Next, we look at the structure of Earth and its neighbors in the solar system.

THE SOLAR SYSTEM

Earth is one of eight large objects traditionally called **planets** in our **solar system**, which consists of the Sun and the group of objects in orbit around it. In addition to the Sun and the planets, the solar system includes more than 140 moons (new ones seem to be discovered each time a spaceship visits a distant planet), a vast number of asteroids, millions of comets, and innumerable fragments of rock and dust called *meteoroids*. All of the objects in our solar system move through space in smooth, regular orbits, held in place by gravitational attraction. The planets, asteroids, comets, and meteoroids orbit the Sun, and the moons orbit the planets and some of the larger asteroids.

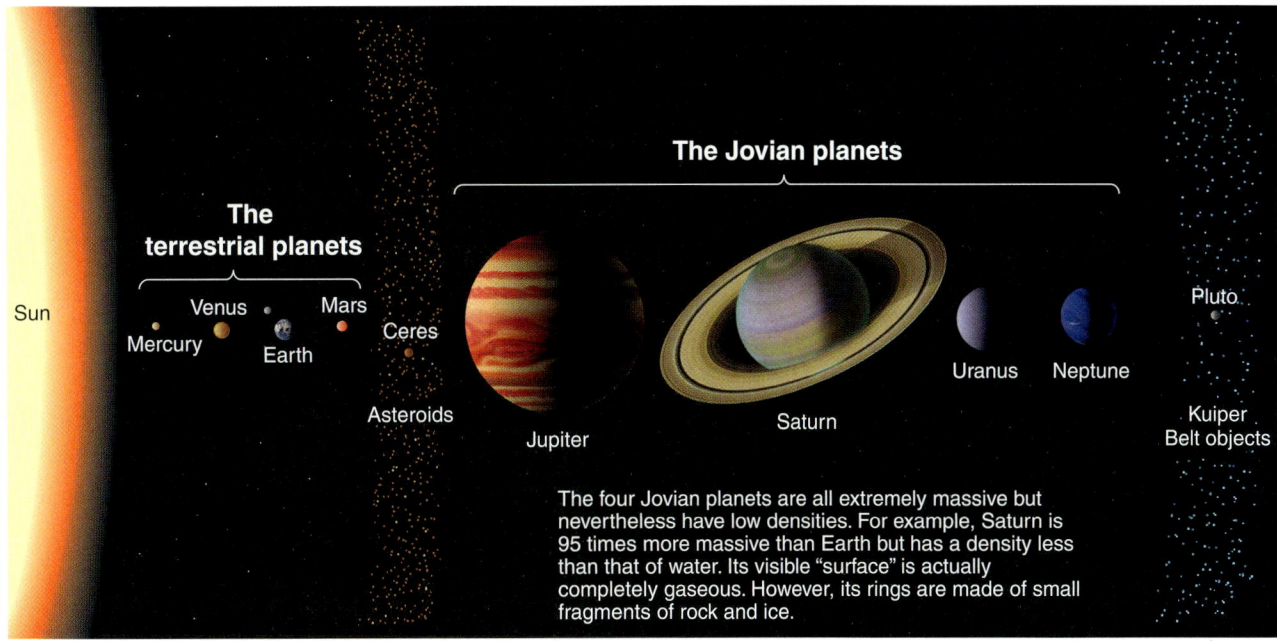

Family portrait of the solar system FIGURE 1.9

Our solar system's eight recognized planets, shown to scale against the Sun. (Note, however, that the distances between planets are much greater than shown, and the planets never line up neatly like this.) Between the terrestrial and Jovian planets lies the asteroid belt, consisting of more than 100,000 small pieces of rock that never coalesced into a planet. Beyond Neptune lies Pluto, formerly considered a ninth planet, and a large number of other icy objects, mostly like Pluto—some larger but mostly smaller—that collectively comprise the Kuiper Belt.

We can separate the planets into two groups on the basis of their physical characteristics and their distances from the Sun (FIGURE 1.9). The innermost planets—Mercury, Venus, Earth, and Mars—are small, rocky, and relatively dense. They are similar in size and chemical composition and are called *terrestrial planets* because they resemble Earth (*Terra* in Latin). The outermost planets are much larger and more massive than the terrestrial planets, yet much less dense. These *Jovian planets*—Jupiter, Saturn, Uranus, and Neptune—take their name from *Jove*, the name for Jupiter in Roman mythology. (Jupiter was the king of the gods, and the god of light and weather, among other things. In size, Jupiter is certainly the king of the planets.) The Jovian planets probably have small solid centers that may resemble the terrestrial planets, but most of their planetary mass is contained in thick atmospheres of hydrogen, helium, and other gases. The atmospheres are what we actually see when we observe these planets.

THE TERRESTRIAL PLANETS

Each of the terrestrial planets has a density between 3.9 and 5.5 g/cm^3. Common rocks have densities between 2.5 and 3.0 g/cm^3, so hidden inside the terrestrial planets there must be some heavy material that is denser than the rocks we find at the surface. Scientific research has discovered that at the center of each terrestrial planet there is a *core* that is very dense because it is largely metallic iron with an admixture of nickel and traces of other elements. Surrounding the metallic core is a layer of rock called a *mantle*. Indeed, research has shown that each of the terrestrial planets has the same basic layered structure, and the layering came about because early in the history of the solar system, each of the terrestrial planets underwent partial melting. As a result the dense material (iron) sank and the lighter material (silica-rich rocks) floated upward. This separation into layers of different composition as a result of melting is called *chemical differentiation*.

Layers of different composition The structure common to all of the terrestrial planets is most clearly demonstrated in the structure of Earth. At the center is the **core**, which, in Earth's case, has two parts, a solid inner core and a molten outer core. The two parts have essentially the same composition because the inner core has formed by the cooling and solidification of the outer core. Because the outer core is molten, we know that it must be at a very high temperature. Scientists have discovered that the temperature is about 5000°C, the same temperature as the surface of the Sun, which means that the core is a major source of Earth's internal (geothermal) heat. Motions in the fluid outer core due to Earth's rotation are also the source of Earth's magnetic field. Each of the terrestrial planets has a metallic core, but we still do not know whether they are all solid, all molten, or partly solid and partly molten (**FIGURE 1.10**). When satellites are sent to explore other planets and moons, one of the tests is for magnetism, which indicates the presence of a molten iron core.

The **mantles** of the terrestrial planets, like the cores, seem to be very similar. They

> **core** Earth's innermost compositional layer, where the magnetic field is generated and much geothermal energy resides.

> **mantle** The middle compositional layer of Earth, between the crust and the core.

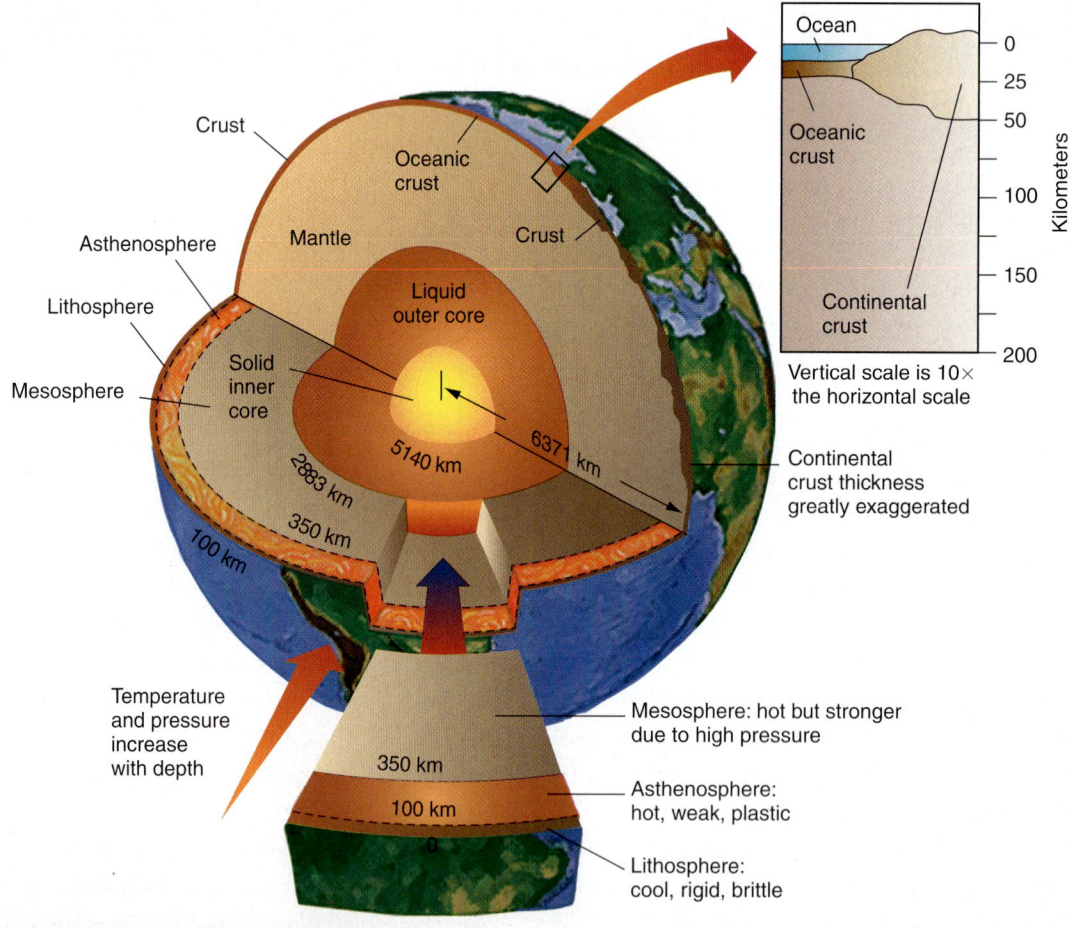

Inside view of Earth FIGURE 1.10

A sliced view of Earth reveals layers of different composition and zones of different rock strength. The compositional layers, starting from the inside, are the core, the mantle, and the crust. Note that the crust is thicker under the continents than under the oceans. Note, too, that boundaries between zones of different physical properties—*lithosphere* (outermost), *asthenosphere*, and *mesosphere*—do not coincide with compositional boundaries.

are much less dense than the cores and composed largely of compounds of silicon, oxygen, magnesium, iron, and calcium. Above the mantle lies the thinnest and outermost solid layer, the **crust**, which consists of rocky material that is less dense than mantle rock. There are considerable differences in detail between the crusts of the terrestrial planets.

> **crust** The outermost compositional layer of the solid Earth: part of the lithosphere.

Layers of different rock strength In addition to compositional layering, the rocky portion of Earth, that is, the crust and the mantle, can be divided into three layers based on differences in strength of the rock making up each layer: the *lithosphere, asthenosphere,* and the *mesosphere* (Figure 1.10).

The strength of a rock is controlled by both temperature and pressure. The outermost strength layer, the **lithosphere**, is approximately the outermost 100 km of the solid Earth. *Lithosphere* means "rock sphere," and rocks in the lithosphere, which includes all the crust and the upper part of the mantle, are strong and brittle, and can be broken or deformed only with difficulty. The great majority of earthquakes happen when rock in the lithosphere breaks.

> **lithosphere** Earth's outermost rocky layer, comprising the crust and the uppermost part of the mantle.
>
> **asthenosphere** A layer of weak, ductile rock in the mantle that is close to melting but not actually molten.

Within the upper mantle, from a depth of about 100 km to about 350 km, the balance between temperature and pressure is such that rocks have little strength; this region is called the **asthenosphere**, which means "weak sphere." Rock in the asthenosphere is ductile rather than brittle, and is easily deformed, like butter or warm tar. The rocky lithosphere floats on the ductile asthenosphere.

Beneath the asthenosphere, from a depth of 350 km to the core–mantle boundary at a depth of 2883 km, rock has considerable strength, even though the temperature is very high. This region of the mantle is the *mesosphere*, which means "middle, or intermediate, sphere."

As we will discuss in Chapter 7, Earth's interior strength properties play a major role in the formation and distribution of continents, ocean basins, mountain ranges, volcanoes, and other large-scale Earth structures.

WHAT MAKES EARTH UNIQUE?

Venus and Mars, Earth's nearest neighbors, are in some ways very similar to our planet. In terms of size, Venus is nearly Earth's twin. Yet there is no chance of mistaking either of them for Earth. Some obvious features make Earth unique.

Oxygen, water, and life As seen from space, Earth's blues, whites, and greens (Figure 1.3A) attest that it has three things neither Venus nor Mars, nor any other planet or moon in our solar system, possesses: an oxygen-rich *atmosphere* (blue); a *hydrosphere* that contains water as a solid, liquid, and vapor (white clouds); and a *biosphere* full of living organisms (green plants).

Soil The nature of Earth's solid surface is another special characteristic. Earth is covered by an irregular blanket of loose debris formed as a result of *weathering*—the chemical alteration and mechanical breakdown of rock caused by exposure to water, air, and living organisms. This layer is called *regolith* (from the Greek words for "blanket" and "stone"). It includes soil, river mud, desert sand, rock fragments, and all other unconsolidated debris. Other planets with rocky surfaces have regolith, too, but none of them has a property that is unique to Earth. Earth's regolith is unique because it teems with microscopic life. When material from the biosphere becomes incorporated with rock material, the result is a special type of regolith called *soil*.

Plate tectonics, mover of continents Another unique property of Earth is the nature and extent of its tectonic activity. Scientists have discovered that the outermost rocky layer of Earth, the lithosphere, approximately 100 km thick, is fractured into great slabs, or plates, that float on the asthenosphere and slide around the globe, slowly moving the continents (**FIGURE 1.11** on pages 16–17). **Plate tectonics** have shaped Earth's continents and oceans and govern, to a large extent,

> **plate tectonics** The movement and interactions of large fragments of Earth's lithosphere, called *plates*.

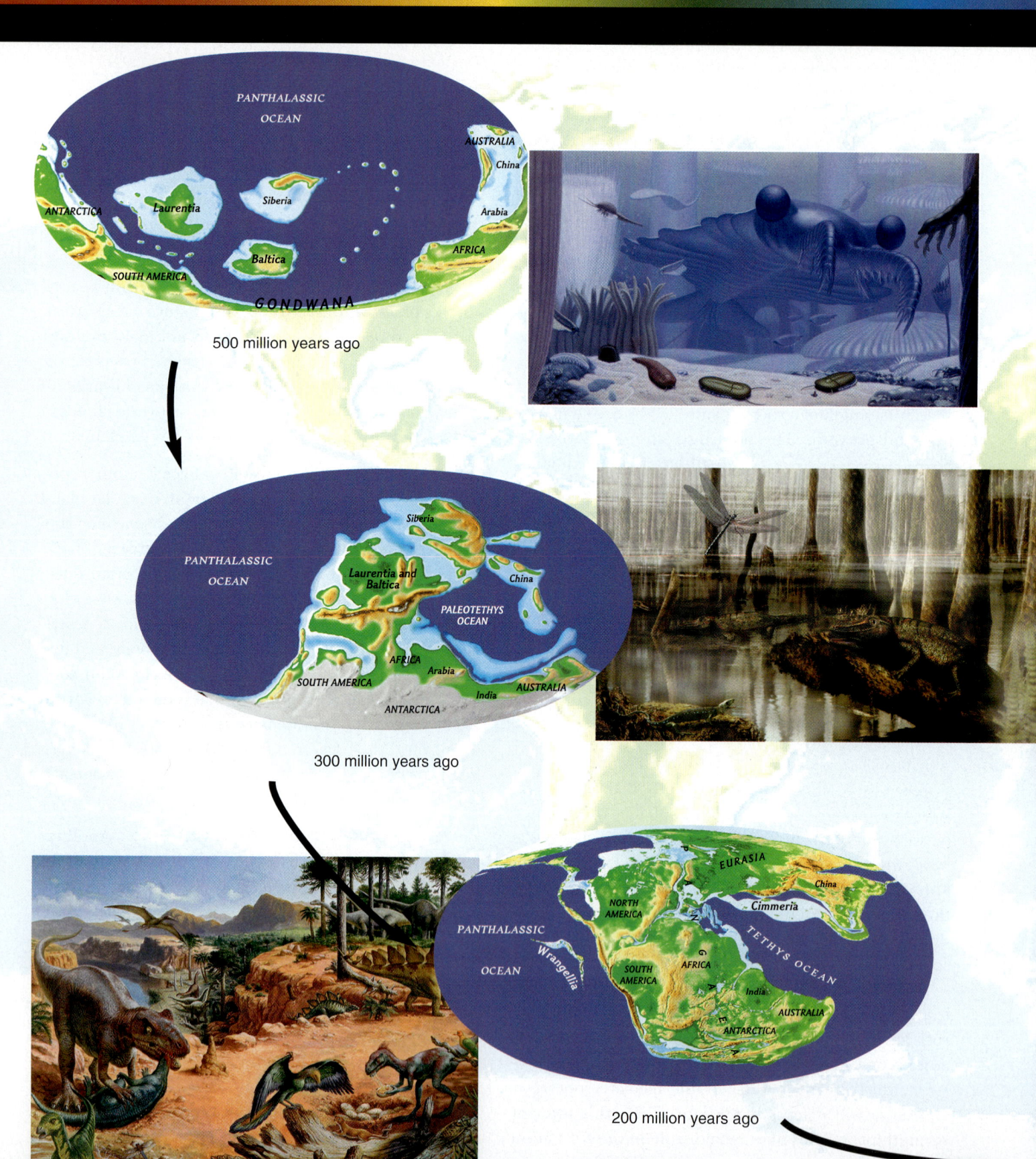

500 million years ago

300 million years ago

200 million years ago

Visualizing

Earth's changing face FIGURE 1.11

These figures show scientists' best reconstruction of the way Earth's landmasses have changed position over the last 500 million years. Note that Earth had one contiguous supercontinent from about 300 million years ago to about 200 million years ago. Life on Earth, revealed from fossils, is also shown in each time period.

Present Day

50 million years ago

100 million years ago

Earth in Space 17

Basalt: The most common volcanic rock in the solar system FIGURE 1.12

Basaltic lava erupts from Mauna Loa volcano in Hawaii. When it cools it will form the same kind of rock, called basalt, that forms most of the surface of Venus, Mars, and the "seas" (the dark-colored spots) of Earth's Moon. Though taken only a few years ago, this picture could represent what the surface of any of these planets could have looked like early in their lifetimes.

the location of mountain ranges and volcanoes and the occurrence of earthquakes. Because of plate tectonics, the map of Earth's surface is continually changing. Tectonic activity has given Earth two different kinds of crust—the thin, **oceanic crust** that is characterized by a common volcanic rock called *basalt*, and the thick, **continental crust**, in which an igneous rock called *granite* is predominant. We will discuss both basalt and granite in detail in later chapters. Granite seems to be unique to Earth; at least, it has not yet been found on Mars and Venus. But basalt is widespread and is a major kind of rock on each of the terrestrial planets, and also on the Moon; scientists currently think that it is the most common kind of volcanic rock in the solar system (**FIGURE 1.12**).

oceanic crust The thinner, denser, and younger part of Earth's crust, underlying the ocean basins.

continental crust The older, thicker, and less dense part of Earth's crust; the bulk of Earth's land masses.

Because the location of continents affects oceanic and atmospheric currents, plate tectonics exert a powerful influence on Earth's climate, which in turn affects the evolution of life. We will have much more to say about tectonics in later chapters.

CONCEPT CHECK STOP

How do the terrestrial planets differ from the Jovian planets?

What properties control the strengths of the lithosphere and the asthenosphere? Why are they different?

Why do the cores of the terrestrial planets have high densities?

Why do Venus and Mars lack a hydrosphere with liquid water?

How are Venus, Mars, and Earth similar, and how do they differ?

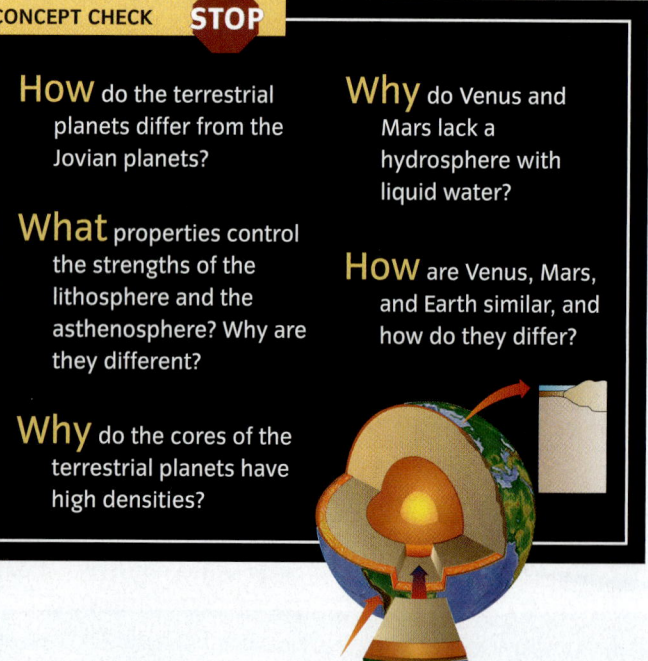

Humans and Earth

> **LEARNING OBJECTIVES**
>
> **Know** the workings and limits of the planet on which we live.
>
> **Understand** the issue of adequacy of food resources for a growing population.
>
> **Describe** how the loss of resources has affected past civilizations.
>
> **Explain** the difference between renewable and nonrenewable resources.
>
> **Identify** two kinds of nonrenewable resources that are crucial to modern society.

The fossils and organic debris trapped in sedimentary rocks provide an extensive record of life on Earth. The record is not perfect, but it is sufficient for us to be sure that never before in Earth's long history have so many large, warm-blooded creatures existed at the same time as exist today. Chief among those warm-blooded creatures are humans; there are now about 6.5 billion of us, and the number continues to rise: By 2020 there will be an estimated 8 billion people on Earth (**FIGURE 1.13**). Paul Ehrlich, a professor of population studies at Stanford University, views population growth with alarm; in his view, nothing in a billion years has posed a threat to terrestrial life comparable to that of human overpopulation. On a number of occasions Ehrlich has predicted that food supplies will fall short and famine will follow. So far, his gloomy predictions have been wrong. At the other end of the opinion scale stands Paul Waggoner, an agronomist and former director of the Connecticut Agricultural Research Station. Waggoner argues that by using current technology—nothing new required—farmers throughout the world could raise productivity to levels achieved in the United States, and that by doing so they could easily feed a population of 10 billion people using only half the land they now use.

How should Earth scientists respond to such disparate views about food? What about other resources—water, energy, minerals? Might such a large population destabilize the environment in which we live? Earth scientists are concerned with how all resources form, how they are distributed, and the consequences to the environment when we dig them up and use them. Earth scientists have a responsibility to use the scientific method to investigate and determine whether limits do exist, how to change the environment as little as possible, and how to ensure that the human population of the future can enjoy as secure a life as do those of us who are alive today.

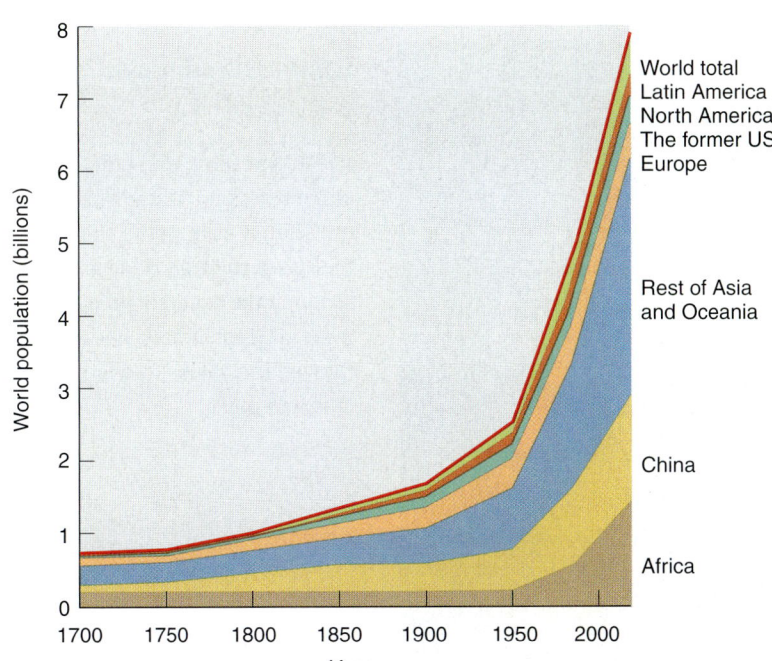

Cumulative growth of the human population FIGURE 1.13

The human population reached 1 billion in 1800, doubled to 2 billion by 1930, and rose dramatically through the rest of the 20th century and into the 21st century. Projected values extend measurements out to 2020.

RENEWABLE AND NONRENEWABLE RESOURCES

From the time our ancestors first picked up conveniently shaped stones and used them for hunting and skinning wild animals, down to the present day, humans have relied on a bountiful Earth to supply a seemingly endless flow of useful materials. Today it is hard to imagine life without an abundance of such resources making life easier and more convenient. But as you will learn, history suggests that there may be limits to this bounty.

Natural resources and ancient history Silent and inscrutable after 10 centuries, the giant stone heads of Easter Island (see **FIGURE 1.14**) are the principal remains of a civilization that once flourished on this remote outpost in the South Pacific. The Rapa Nui people transported hundreds of the megaliths, called *moai*, over hilly terrain, from the quarry where they were carved to their final locations. Archaeologists believe that the inhabitants used tree trunks and ropes to roll the stone figures across the land.

Just as interesting as the questions of how and why the Rapa Nui transported the moai is why they stopped. Scientists have found that palm trees grew on Easter Island until around 1500. Apparently the islanders cut down every last tree, leaving the ground bare and vulnerable to rapid erosion. In so doing, the Rapa Nui lost not only the ability to transport megaliths but also the ability to grow the food resources necessary to sustain their civilization, which was decimated by famine and warfare over the next century.

The fall of past civilizations is an endlessly fascinating and controversial topic. No civilization's decline can be attributed to a single cause. Nevertheless, a common thread runs through the stories of many societies that flourished and then collapsed. That thread is the depletion of **natural resources**. The Easter Islanders failed to conserve their trees, a *biologic resource*, and soil, an *Earth resource*.

> **natural resources**
> Useful materials obtained from the lithosphere, atmosphere, hydrosphere, or biosphere.

Another critical Earth resource is water. Halfway around the globe, in what is now Libya, the Garamantian culture arose in the Sahara Desert around the same time as the Roman Empire, and prospered for about a thousand years, in spite of the nearly complete lack of rainfall in the region. The Garamantes survived by tapping into immense underground aquifers and transporting the groundwater through a system of aqueducts called *foggara* (see **FIGURE 1.15**). Unfortunately, around 500 A.D. they apparently reached the limits of their technological ability to retrieve groundwater. The foggara had been depleted, and the Garamantian cities became ghost towns.

Like the Rapa Nui and the Garamantes, every human society depends on natural resources. Besides the most important resources—soil, water, and air—

Easter Island megaliths
FIGURE 1.14

The Polynesians who erected these stone heads on Easter Island were unwittingly sowing the seeds of their own culture's demise. They stripped the island of its palm trees, at least in part, because they needed the trunks to roll the megaliths into place.

A An ancient city in the Saharan desert, Garama was the center of the Garamantian culture from 500 B.C. to 500 A.D., and as many as 10,000 people may have lived here.

B Such a large population was supported by a vast underground network of foggara, or aqueducts, whose openings to the surface can still be seen.

Ghost town in the Sahara FIGURE 1.15

Earth's resources include building materials, metals, fertilizers, oil, coal, and gemstones. Biologic resources include crops, wild plants, and animals. Some of these resources are **renewable**. For example, even though we may consume a food crop each season, a new crop grows during the following season. A layer of soil that is lost to erosion will eventually be regenerated by the physical, chemical, and biologic processes of soil formation. Groundwater that is drawn from wells may eventually be replenished by rainwater. But what does "eventually" mean? Some resources take a very long time to regenerate—longer than humans are willing or able to wait. For example, the aquifers under the Sahara Desert formed tens of thousands of years ago, when the climate was moister than it is today, and they will not be replenished until the climate changes again, perhaps tens of thousands of years from now. For practical purposes, we consider these resources to be **nonrenewable**.

Resources such as coal, oil, copper, iron, gold, and fertilizers are mined from mineral deposits. Mineral deposits are known to be forming today, but the rate of formation is exceedingly slow. For example, it may take 600,000 years for a large copper deposit to be formed. From a human point of view, all mineral resources are one-crop resources, and Earth's supply of those "crops" is fixed.

Humans are the first and only species to routinely use and rely on nonrenewable resources. It does not seem to be in our nature to stop. However, history suggests that we should use such resources very

renewable resource A resource that can be replenished or regenerated on the scale of a human lifetime.

nonrenewable resource A resource that cannot be replenished or regenerated on the scale of a human lifetime.

Humans and Earth

judiciously and monitor how much we have left. That is what the Rapa Nui and the Garamantes failed to do. After the cultures that produced the moai and the foggara died out, their descendents had to adapt to the new conditions. The population of Easter Islanders collapsed, and their descendents developed new rituals that allocated their limited food resources. The inhabitants of the Sahara developed a nomadic lifestyle that was not dependent on the extraction of deeply buried groundwater. Like them, our descendents may have to make serious changes in their lifestyles as critical resources, such as oil, become scarce.

Resources and modern society Each of us uses—directly or indirectly—a very large amount of material derived from nonrenewable *mineral resources*. We may not be aware of our dependency on mineral resources, but think for a moment about all of the metals needed to build machines for manufacturing, transportation, and communications; they are all nonrenewable resources that are extracted from the ground. Without these resources, industry would collapse and living standards would deteriorate dramatically.

We are equally dependent on *energy resources*. Imagine what life would be like if we had to rely entirely on human muscle power. If a healthy adult rides an exercise bike that drives an electric generator connected to a light bulb, the best an average person can do in a nonstop eight-hour day of pedaling is to keep a 75-watt bulb burning. In North America, the same amount of electricity can be purchased from a power company for about 10 cents. Viewed in this way, we can see that human muscle power is puny. Over many thousands of years our ancestors found ways to supplement muscle power. At first they did this by domesticating beasts of burden such as horses, oxen, camels, elephants, and llamas. They later learned to make sails to use wind power, dams to use water power, and engines to convert the heat energy of wood, oil, and coal into mechanical power.

Today we use supplementary energy in every part of our lives, from food production and transportation to housing and recreation. North Americans are among the world's biggest energy consumers (see **FIGURE 1.16**). Whereas soil and water were the most critical nonrenewable resources for earlier civilizations, our society's weakest link may be its excessive dependence on nonrenewable energy sources. How and whether we can reduce this dependence is not just an Earth science problem, it is also a political and a social problem. Earth science plays an important role in three ways: identifying new supplies of traditional resources, allowing us to estimate how much we have left, and evaluating the environmental consequences of using new and unconventional resources.

WHY STUDY EARTH SCIENCE?

With this brief introduction to Earth science and the Earth system, you have probably deduced some of the reasons why it's important to study the subject. We need to understand Earth materials because we depend on them for all of our material resources—the mineral, rocks, and metals with which we construct our built environment; the energy with which we run it; the soil that supports agriculture and other plant life; and the air and water that sustains life itself. Many Earth resources are limited and require knowledgeable and thoughtful management. The materials of Earth also have physical and chemical properties that affect us, such as their tendency to flow or fail during a landslide, their capacity to hold or transmit fluids such as water or oil, or their ability to absorb waste or prevent it from migrating.

We have learned that Earth is essentially a closed system, which means that all materials remain within the system. Therefore, it is important to understand how materials move from one reservoir to another. It is also important to understand the time scales that govern these processes in order to get some perspective on the changes we see happening in the natural environment. Some Earth processes are hazardous—that is, damaging to human interests. These *natural hazards* include earthquakes, volcanic eruptions, landslides, floods, hurricanes, tornadoes, and even meteorite impacts. The more we know about these

Visualizing
Energy consumption FIGURE 1.16

An average American uses energy, directly or indirectly, at a rate equivalent to burning more than 150 75-watt light bulbs every minute of the day, every day of the year. (Canadians, with their cold winters, use even more energy.)

▲ A Frustrated drivers in heavy traffic on the Beltway around Washington, D.C.

United States

Watts/person

💡 = 75 watts of energy usage

World Average

Haiti

▲ B At the other extreme, Haiti uses the least energy per capita of any country in the Western Hemisphere—about 1.5 light bulbs per inhabitant. In the town of La Victoire, there are no cars in evidence and only one television set for the whole town. The photographer reported, "Sometimes it even works."

Humans and Earth 23

Natural hazards: predicted and not deadly, not predicted and deadly FIGURE 1.17

A Mt. Pinatubo in the Philippines erupted in 1991. Volcanologists predicted the eruption, making it possible to evacuate residents from the area and prevent thousands of deaths. When it erupted, the volcano sent this lethal cloud of searing, dust-laden gases rolling down its flanks, to spread rapidly across the surrounding plains. This particular car and driver escaped, but many houses, trees, and fields were smothered with volcanic ash.

B The tsunami generated by the Sumatra-Andaman earthquake of December 26, 2004, was not predicted. These people on vacation in Thailand were unprepared for the disaster as the giant wave swept ashore. An estimated 275,000 people in countries around the Indian Ocean died as a result of the tsunami.

hazardous processes, the more successful we will be in protecting ourselves from future natural disasters (Figure 1.17).

Finally, Earth is our home planet. The features that make Earth unique and the powerful natural processes that characterize the Earth system are a constant source of awe and fascination to those who study them. It makes sense to deepen and refine our understanding of the planet we live on.

From its beginnings a couple of centuries ago, Earth science has necessarily been an interdisciplinary science, because Earth operates through interactions of biologic, physical, and chemical processes. Yet we are discovering that the interactions are more complex and dynamic than we would have believed even a few decades ago. We are still learning about the complexities and interrelationships of subsystems such as climate, ocean currents, and shifting continents. We now appreciate more profoundly our own role in causing changes as well as the need to study the Earth system as a whole rather than in separate fragments.

Visualizing Earth Science starts your study of Earth. If you are planning to become a professional scientist, this book will be an introduction to some of the many fascinating possibilities that await you in your career. If you are taking this course out of personal interest or to fulfill a degree requirement, you will emerge more aware of the way our planet works, and better prepared to make informed decisions about the natural processes that affect your life on a daily basis.

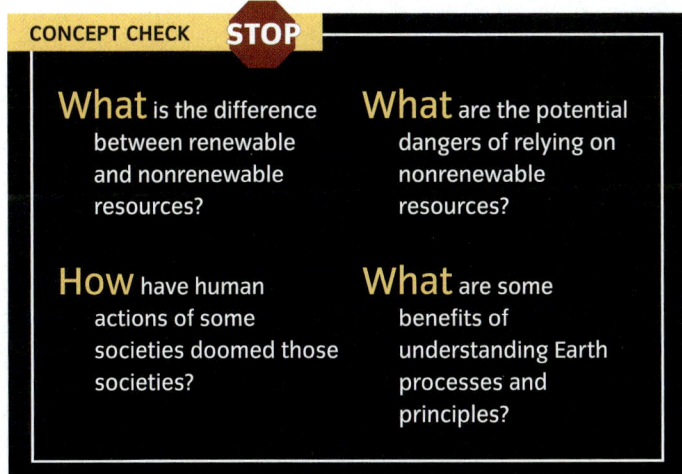

CONCEPT CHECK STOP

What is the difference between renewable and nonrenewable resources?

What are the potential dangers of relying on nonrenewable resources?

How have human actions of some societies doomed those societies?

What are some benefits of understanding Earth processes and principles?

Amazing Places

At the end of every chapter in this book, we will take you to an "Amazing Place" that is both beautiful and of interest to Earth scientists. Here is our itinerary:

Chapter 2: The Naica Mine, Chihuahua, Mexico, for the world's largest crystals.

Chapter 3: The Navajo Sandstone, Utah, for its beautiful sedimentary rock formations.

Chapter 4: Mt. Monadnock, New Hampshire and other monadnocks, for a look at the power of erosion.

Chapter 5: Lechuguilla Cave, New Mexico, for incredible shapes made by groundwater.

Chapter 6: Death Valley, California, for its extreme environment and desert landforms.

Chapter 7: The Hawaiian Islands, to see plate tectonics and volcanoes in action.

Chapter 8: Loch Ness, Scotland, for its faults (but not for its monster).

Chapter 9: Mt. Saint Helens, Washington, to witness the most famous eruption in U.S. history.

Chapter 10: The Grand Canyon, Arizona, for a look at geologic time.

Chapter 11: The Burgess Shale, British Columbia, Canada, for its fossil record of the first animals on Earth.

Chapter 12: Monterey Bay, California, to witness the exotic life that thrives in the deep.

Chapter 13: The Florida Keys Reef, for a geologic formation that is also alive.

Chapter 14: Hole Punch Clouds in Mobile, Alabama, for the remarkable patterns etched across the sky.

Chapter 15: New Orleans, Louisiana, before and after Katrina.

Chapter 16: Barrow, Alaska, a community on the frontline of climate change.

Chapter 17: Mars, to explore the rocky landscape of Earth's magnificent neighbor.

The most amazing place of all, however, is Earth itself (pages 26–27), the only world in the universe where we know that life exists.

Amazing Places: Earth

This is a computer rendering, compiled from many satellite images, of Earth at night. (In reality, the entire Earth is never dark at the same time.) Note how the lights are brightest in places with a high population density. Also, note how human settlement concentrates on coastlines. Why? What geologic features explain the dark areas? Can you spot where you live or identify any major cities?

NATIONAL GEOGRAPHIC

SUMMARY

1 What Is Earth Science?

1. **Earth science** is the study of all aspects of Earth. Earth scientists study the record contained in rocks of all that has happened in the past, the interactions between all parts of the Earth system today, and the probable future changes to the environment in which we live as a consequence of our collective human activities.

2. The **scientific method** is a research strategy that scientists use to study a problem by formulating a hypothesis and then testing it by performing an experiment. The steps include (1) observing and gathering data; (2) formulating a **hypothesis**; (3) testing the hypothesis; (4) formulating a **theory**; and (5) formulating a **law** or **principle**.

3. Earth scientists study Earth today using **Earth system science**. This concept comes from the discovery that Earth is an integrated **system** of interconnected and interdependent parts. Individual systems within the larger Earth system can be big or small, and can vary greatly in complexity, but regardless of size, each system operates within an identifiable boundary. There are three kinds of systems: *isolated, closed,* and *open*; the properties of the boundary determine the kind of system. In an isolated system, boundaries prevent the system from taking in or releasing any energy or matter. Because there is no perfect boundary against the passage of energy, isolated systems do not exist in the real world. A closed system has a boundary that permits the passage of energy, but not of matter, in and out of the system. The third kind of system, an *open system*, permits the exchange of both matter and energy across its boundary. Most environmental and geologic systems in the natural world are open systems.

4. Earth is considered a closed system, though some small amounts of matter do cross its boundary. The Earth system consists of four principal open subsystems, including the **atmosphere**, the envelope of gas that surrounds Earth; the **hydrosphere**, comprising of all the Earth's water; the **biosphere**, all of Earth's living organisms; and the **lithosphere**, Earth's rocky outer layer. Materials and energy are stored for varying lengths of time in each of these systems or reservoirs and can move among them via innumerable pathways and processes. Each of the four Earth subsystems can be further broken down into a vast number of still smaller subsystems, all of which are open.

5. An important component of Earth system science is the monitoring and study of the movement of materials among the subsystems. The system is kept in balance, or reaches a new balance following a change in some part of the system through feedback mechanisms. **Positive feedback mechanisms** work to change the system; **negative feedback mechanisms** work to resist change. Feedbacks are especially important in the **life zone**, the region between 10 km above and 10 km below sea level, where all of life on Earth is located.

2 Earth in Space

1. Earth is one of the eight bodies in the **solar system** recognized as planets. In addition to the Sun and *planets*, the solar system includes a vast number of moons, asteroids, comets, and fragments of rock called *meteoroids*. The four inner planets, or *terrestrial planets*, Mercury, Venus, Earth, and Mars, are similar in many ways. They are all small, rocky, and relatively dense, and they have similar sizes and chemical compositions. The four outer planets, or *Jovian planets*, in contrast, consist of huge gaseous atmospheres with small solid cores, giving them very low densities overall. The Jovian planets are Jupiter, Saturn, Uranus, and Neptune. Pluto, until recently considered to be the ninth planet, is not a Jovian planet because it is icy but not gassy. Instead, it is considered to be a "dwarf" planet and part of the Kuiper Belt, a region of the outer edge of the solar system that contains a large number of small, icy objects.

2. Early in its history, Earth underwent *differentiation* into a dense, metallic **core**, a rocky **mantle**, and a brittle, rocky outer **crust**. Because of the way in which temperature and pressure control the strength of rocks, the outermost 100 km or so of the solid Earth—that is the crust and upper part of the mantle—consists of rocks that are tough and resistant to breakage; this zone is called the **lithosphere**. Beneath the lithosphere, from a depth of about 100 km to a depth of about 350 km, is a zone where rocks are weak and easily deformed, and though not actually molten, so ductile they behave like very thick liquids. This zone is called the **asthenosphere**. The great bulk of the mantle, from a depth of 350 km to the boundary between the mantle and the core, is called the *mesosphere*, and it consists of rocks that are readily deformed but are not as ductile as the asthenosphere.

3. Earth is unique in the solar system in that it possesses an oxygen-rich atmosphere. Earth is also the only planet in the solar system with a hydrosphere in which water exists near the surface in solid, liquid, and gaseous forms, and a biosphere with living organisms. Finally, Earth is the only planet where true soil is formed from *regolith* by interactions among physical, chemical, and biologic processes, and where life as we know it could exist.

4. **Plate tectonics** is the motion and interaction of large segments of the lithosphere. It is because of plate tectonics that Earth has two fundamentally different types of crust: the relatively thin, dense **oceanic crust** of the volcanic rock *basalt* and the thicker, less dense **continental crust** comprised mainly of the igneous rock *granite*.

3 Humans and Earth

1. Study of human history reveals more than one example of societies that have collapsed because of a failure to use limited natural resources wisely. The Easter Islanders in the South Pacific and the Garamantes in North Africa are two examples.

2. **Natural resources**, which include all of the materials we take from the Earth system, can be divided into two families, the **renewable resources** and the **nonrenewable resources**. The renewables include materials such as water in streams and agricultural products, which are continually replaced or can be newly produced each growing season. Nonrenewables are those resources that cannot be regenerated on human timescales and so are one-crop materials.

3. The study of Earth science is important to human society for many reasons. Earth materials and processes affect our lives through our dependence on Earth resources: through geologic hazards such as volcanic eruptions, floods, and earthquakes; and through the physical properties of the natural environment.

KEY TERMS

- **Earth science** p. 4
- **scientific method** p. 4
- **hypothesis** p. 4
- **theory** p. 6
- **system** p. 7
- **feedback mechanisms** p. 11
- **Earth system science** p. 12
- **core** p. 14
- **mantle** p. 14
- **crust** p. 15
- **lithosphere** p. 15
- **asthenosphere** p. 15
- **plate tectonics** p. 15
- **oceanic crust** p. 18
- **continental crust** p. 18
- **natural resources** p. 20
- **renewable resource** p. 21
- **nonrenewable resource** p. 21

CRITICAL AND CREATIVE THINKING QUESTIONS

1. Do you think there may be life on a planet outside of our solar system? What would the atmosphere of that planet be like? Must it have a hydrosphere? Why or why not?

2. Why is the systems approach so useful in studying both natural and artificial processes? Can you think of examples of artificial (that is, human-built) systems other than those given in the text? Are they open systems or closed systems? (Think about the materials and energy in them.)

3. In this chapter we have suggested that Earth is a close approximation of a natural closed system, and we have hinted at some of the ways that living in a closed system affect each of us. Can you think of some other ways?

4. In what ways do Earth science processes affect your daily life?

5. Formation of clouds and removal of carbon from the atmosphere by growing plants are negative feedback mechanisms tending to cool Earth's atmosphere. Can you think of other negative feedback mechanisms that might involve the ocean or some other part of the Earth system?

6. How many things on which you rely for your daily activities require the use of nonrenewable resources? All nonrenewable resources in a closed system are limited; which ones do you think might have limits that will affect the long-term activities of the human population?

What is happening in this picture?

This rock, photographed in Saudi Arabia's Rhub al Khali (Empty Quarter), was discovered in 1965. It is believed to be the largest fragment of a meteorite that fell to Earth sometime before 1863 (when the first piece was discovered).

- How do you think these scientists can tell it is a meteorite?
- Why did it break up into pieces?
- Why is the desert a good place to look for meteorites?

(Hint: Think about what would have happened to this rock if it had fallen in a jungle or a mountain range.)

SELF-TEST

1. Earth science is the scientific study of _____.
   a. soils
   b. rocks and minerals
   c. all aspects of Earth
   d. all the terrestrial planets
   e. mines and oil fields

2. The scientific method is a way of investigating natural phenomena by _____.
   a. challenging entrenched beliefs
   b. using mathematics to confuse students
   c. defending personal ideas
   d. making observations and repeatedly testing conclusions
   e. taking photographs from space

3. On this illustration, label each of the following systems:

   isolated system
   closed system
   open system

4. The island depicted in the figure acts as a(n) _____.
   a. intermittent system
   b. closed system
   c. solar system
   d. open system
   e. isolated system

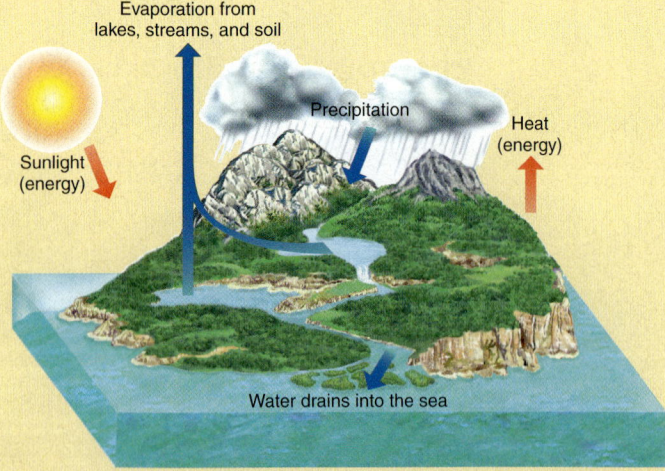

5. On the time scale of a human lifetime, Earth acts as a(n) _____.
   a. intermittent system
   b. closed system
   c. solar system
   d. open system
   e. isolated system

6. The _____ is a subset of the Earth system that comprises all of its bodies of water and ice, both on the surface and underground.
   a. atmosphere
   b. hydrosphere
   c. lithosphere
   d. ionosphere
   e. biosphere

7. Earth's climate is balanced by _____.
   a. the growth of forests acting as a negative feedback
   b. the absence of any negative feedbacks
   c. clouds of carbon dioxide acting as a negative feedback
   d. an interplay between positive and negative feedbacks
   e. the absence of any positive feedbacks

8. On this illustration label the following objects:

   Venus, Earth, Mars, Jupiter, the terrestrial planets, Neptune, asteroids, Pluto, Saturn, Kuiper Belt objects

9. The photograph is of a basalt flow on the island of Hawaii. Which one of the following statements is true?
   a. Basaltic lava flows have only occurred on Earth.
   b. These types of flows have been common only on Earth and in the early history of the Moon.
   c. Basalt is the most common volcanic rock known in our solar system.
   d. Basaltic lava flows would have been common in the early history of Earth, but in modern times they have largely ceased.
   e. None of the above statements is true.

10. Earth, Mars, and Venus all have _____.
    a. an oxygen- and nitrogen-rich atmosphere
    b. a core, mantle, and crust
    c. a hydrosphere with liquid water
    d. a biosphere
    e. All of the above statements are true.

11. On this illustration, label Earth's internal structure using the following terms:

    mantle   lithosphere   oceanic crust
    asthenosphere   outer core   inner core
    continental crust mesosphere

12. The asthenosphere is a layer whose distinctiveness from the rest of the mantle is based on its _____.
    a. difference in composition
    b. high strength
    c. low strength
    d. relatively low temperature
    e. increased brittleness

13. All known forms of life live in a restricted zone where the hydrosphere, atmosphere, and lithosphere interact to provide the right balance of physical conditions and nutrients; this life zone is _____.
    a. between the Arctic circle and the Antarctic circle
    b. from the shoreline to the top of the atmosphere
    c. between 20 km below and 20 km above sea level
    d. between 10 km below and 10 km above sea level
    e. between 5 km below and 5 km above sea level

14. Which one of the following statements is incorrect?
    a. The depletion of natural resources apparently led to the collapse of some ancient civilizations.
    b. Nonrenewable resources are never replenished.
    c. Renewable resources must be managed so they are not used at a rate that is greater than the rate of renewal or replenishment of the resource.
    d. Groundwater is, in principle, a renewable resource, but once depleted it may take a very long time to be replenished.
    e. Fossil fuels are nonrenewable resources, but metals are renewable resources.

15. The study of Earth science is important because _____.
    a. it helps us understand the processes that govern the Earth system
    b. it helps us assess the potential limitations of the supplies of natural resources on which civilization depends
    c. it helps us understand and mitigate the potential threats of natural hazards, such as floods, landslides, earthquakes, volcanic eruptions, and even meteorite impacts
    d. it makes us more aware of the uniqueness of this planet that we share with all other life-forms
    e. All of the above statements are true.

Minerals: Earth's Building Blocks 2

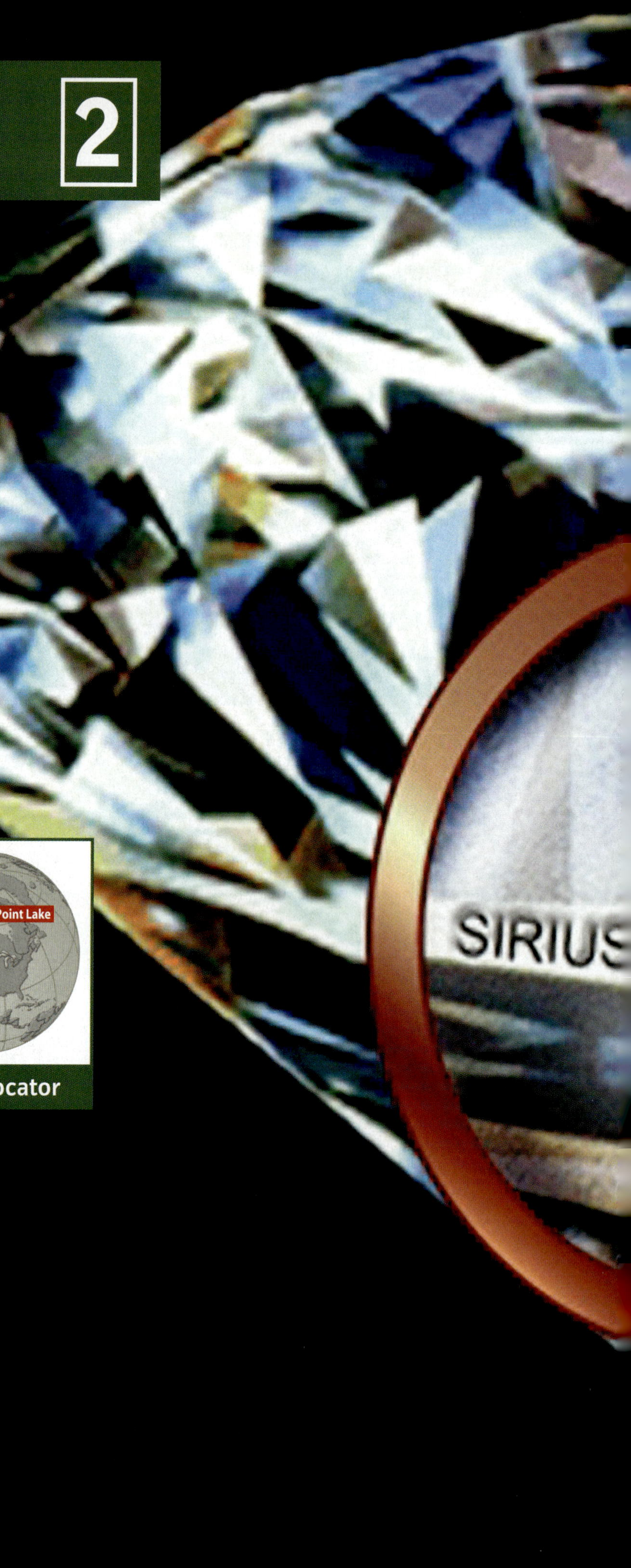

This diamond comes from Point Lake, Northwest Territories, Canada, where geologists Charles Fipke and Stewart Blusson discovered a rich diamond deposit in 1991. Fipke and Blusson hypothesized that diamonds found in Wisconsin had been carried there by glaciers during the last Ice Age. They were proved right. Since 1998, when the first mine opened, Canada has become the world's third largest exporter of diamonds. Every diamond produced in Canada is engraved with a tiny polar bear that attests to its origin.

Diamond is a remarkable mineral. It is pure elemental carbon, the same chemical found in graphite and charcoal. But, unlike graphite and charcoal, natural diamond forms only at extraordinarily high pressures and temperatures, deep under Earth's surface or at the center of meteorite impacts. It is the hardest mineral known and an excellent conductor of heat. Its properties have made it a valuable material for industrial purposes. If production costs can be brought down, artificial diamonds might one day replace silicon in our computers, which could then run at much higher temperatures.

In this chapter you will learn about minerals and why minerals of identical composition, such as diamond and graphite, have different properties. Some minerals are beautiful, some are economically important (but not beautiful), and some are vital as nutrients. A few minerals, such as certain varieties of asbestos, are potentially hazardous to human health. Earth scientists need to study minerals and their properties, so that we can learn to balance the positive and negative effects on our lives.

Global Locator

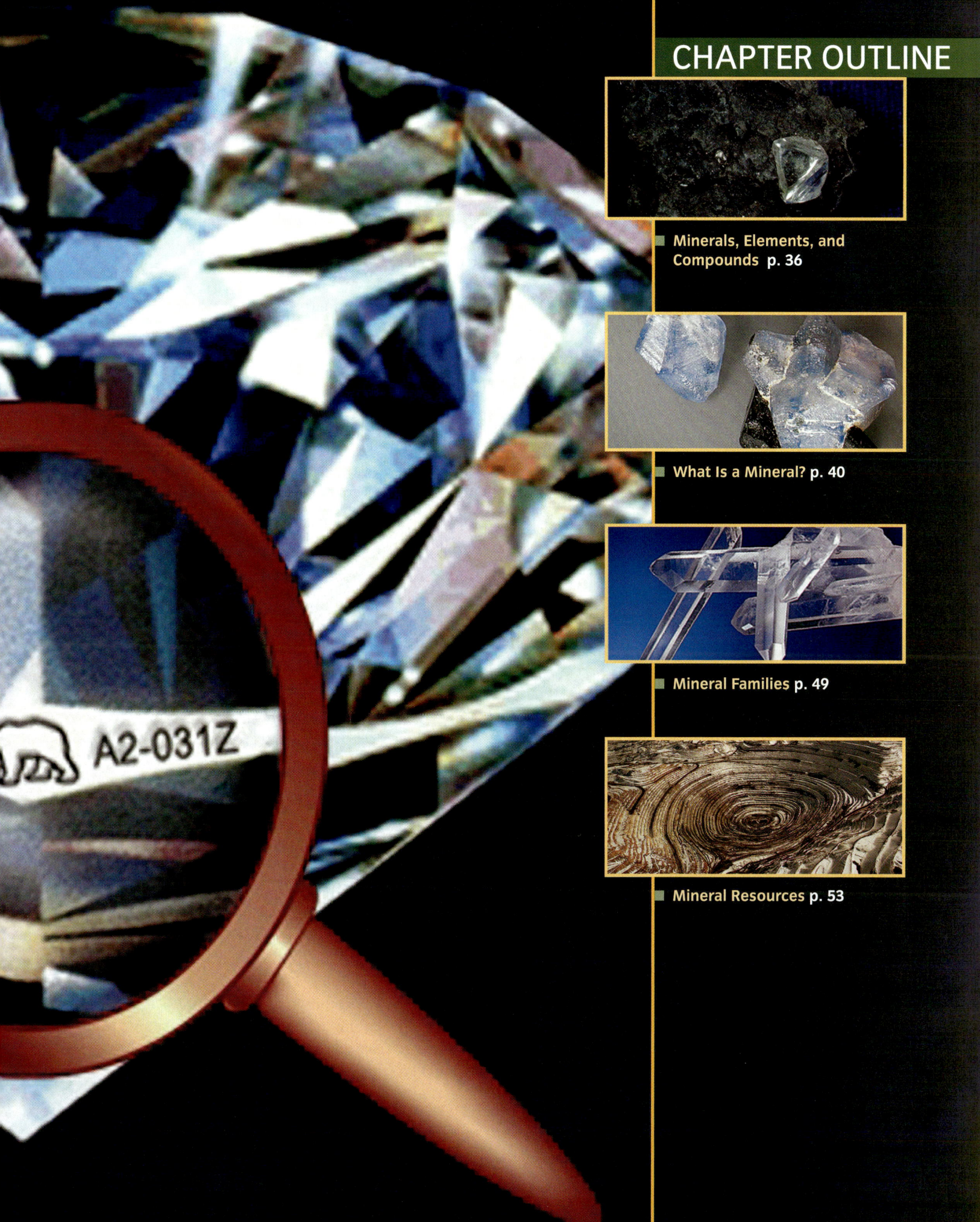

CHAPTER OUTLINE

- **Minerals, Elements, and Compounds** p. 36

- **What Is a Mineral?** p. 40

- **Mineral Families** p. 49

- **Mineral Resources** p. 53

Minerals, Elements, and Compounds

LEARNING OBJECTIVES

State the four requirements for a solid material to be classified as a mineral.

Define element, atom, compound, molecule, and ion.

Explain the difference between an atom and a molecule.

Describe the internal structure of an atom.

Identify four kinds of chemical bonding.

Explain how the kinds of bond in a material affect the properties of a mineral.

The word *mineral* has a specific connotation in Earth science. To be classified as a mineral, a substance must meet certain criteria. A **mineral** must

- Be a naturally occurring solid.
- Be formed by inorganic processes.
- Have a characteristic crystal structure.
- Have a specific chemical composition.

Each criterion is essential, and because minerals are solids, we will first discuss how solids form. Do not confuse a mineral with a rock. A **rock** is a solid aggregate of minerals, and in the next chapter we discuss how minerals combine to form rocks. In this chapter, we concentrate on minerals, and to do so we start with the fundamental particles that are present in all minerals—atoms.

- **mineral** A naturally formed, solid, inorganic substance with a characteristic crystal structure and a specific chemical composition.

- **rock** A naturally formed, coherent aggregate of minerals and possibly other nonmineral matter.

ELEMENTS, ATOMS, AND IONS

Chemical **elements** are the most fundamental substances into which matter can be separated and analyzed by ordinary chemical methods. All matter on Earth, including the page you are reading and the eyes you are reading it with, consists of one or more chemical elements. All of the chemical reactions that make life on Earth possible depend on the ways chemical elements interact. Ninety-two naturally occurring elements are known, and a number more have been synthesized by atomic scientists. Each element is identified by abbreviated symbols, such as H for hydrogen and Si for silicon. Some of the symbols come from other languages, such as Fe for iron, from the Latin *ferrum*, and Na for sodium, from the Latin *natrium*. Others are named in honor of famous scientists, such as element 99, Es, einsteinium. The periodic table of the elements is shown in Appendix A.

- **element** The most fundamental substance into which matter can be separated by chemical means.

- **atom** The smallest individual particle that retains the distinctive chemical properties of an element.

Even the tiniest grain of dust is made up of innumerable particles, called **atoms**, which are much too small to see (**FIGURE 2.1**). They are so tiny, about one-billionth of a millimeter, that they cannot even be seen at all with an optical microscope. Special microscopes that do not use light have succeeded in imaging atoms, but it would be more accurate to say that they "feel" the atoms rather than see them. It may seem strange that the properties of things as tiny as atoms should determine the properties of Earth, but they do.

Chemical reactions take place between atoms, and it is those reactions that produce the minerals, liquids, and gases that Earth scientists study. Atoms themselves are composed of even smaller particles, which have no independent chemical properties. The *nucleus* (plural *nuclei*) of an atom contains *protons*, with positive electric charges, and *neutrons*, which are electrically neutral. The number of protons in an atom—its *atomic number*—determines its chemical characteristics. Atomic numbers range from 1 for the lightest element, hydrogen, up to 92 for uranium, the heaviest naturally occurring element. Every element from atomic number 1 to 92

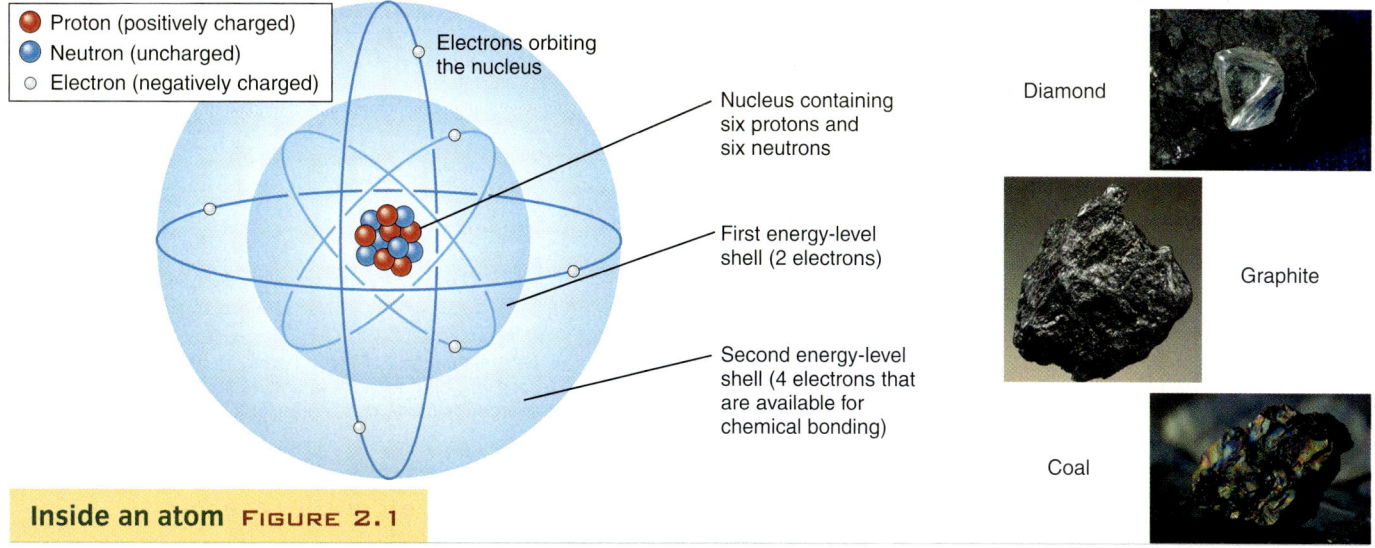

A Single Atom of Carbon-12 (Schematic Diagram)

Three things that contain carbon

Diamond

Graphite

Coal

Inside an atom FIGURE 2.1

As shown in the diagram of carbon-12 (left), six electrons orbit the nucleus in two complex paths called *orbitals,* rendered here (unrealistically) as circles. The orbitals arrange themselves in energy-level shells, which are more stable when completely filled—the first energy-level shell is filled when two electrons are present, the second shell can have eight electrons. There are two carbon isotopes, carbon-12 (the major component) and carbon-13, present in diamond, graphite, and coal. All living beings also contain carbon-12 and carbon-13, but in addition they contain trace amounts of carbon-14, a radioactive isotope.

has either been synthesized in the laboratory or found in nature, so there are no new elements to be discovered in that range. However, scientists are working on synthesizing heavier elements and have reached element 118 (ununoctium).

The number of protons plus the number of neutrons in the nucleus of an atom is the *mass number.* Atoms of a given element always have the same atomic number, but they can have different mass numbers. For example, there are three naturally occurring **isotopes** of carbon: carbon-12, carbon-13, and carbon-14. Each of the isotopes of carbon has 6 protons and thus an atomic number of 6. However, the three isotopes contain different numbers of neutrons: 6, 7, and 8 per atom, respectively (thus, different mass numbers: 12, 13, and 14).

> **isotopes** Atoms with the same atomic number and different mass numbers.

The third component of an atom is called an *electron.* Electron interactions help determine the makeup of ions and compounds. Electrons orbit the nucleus, as shown schematically in Figure 2.1 as circles of different sizes (the actual orbits are neither circles nor spheres; they are much more complex patterns). Electrons have a negative charge that is equal in magnitude but opposite in sign to the positive charge of the proton. In its ideal state, an atom has an equal number of protons and electrons and thus is electrically neutral. Under certain circumstances, however, an atom may gain or lose an electron during a chemical reaction. Although atoms can gain or lose electrons, they never gain or lose protons or neutrons as a result of chemical processes.

An atom that loses or gains one or more electrons has a net electric charge and is called an **ion**. If the charge is positive, meaning that the atom has lost one or more electrons, the ion is called a *cation*. If the charge is negative, meaning that the atom has gained one or more electrons, the ion is called an *anion*. A convenient way to indicate ionic charges is to record them as superscripts. For example, Na^+ is the symbol for an atom of sodium that has given up an electron; Cl^- is the symbol for an atom of chlorine that has accepted an electron; and Fe^{2+} is the symbol for an atom of iron that has given up two electrons.

COMPOUNDS, MOLECULES, AND BONDING

Chemical **compounds** form when atoms of one or more elements combine with atoms of another element in a specific

> **compound** A combination of atoms of one or more elements in a specific ratio.

Minerals, Elements, and Compounds

Process Diagram

How ions and compounds form FIGURE 2.2

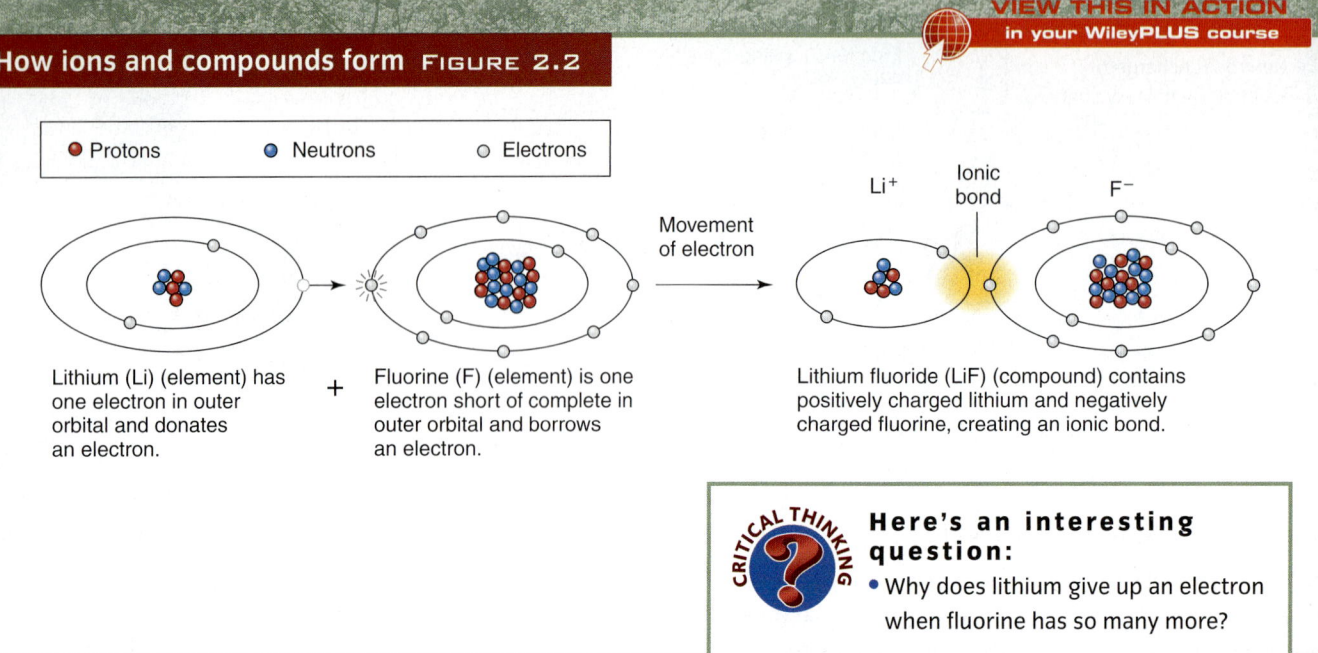

- Protons ● Neutrons ○ Electrons

Lithium (Li) (element) has one electron in outer orbital and donates an electron.

+ Fluorine (F) (element) is one electron short of complete in outer orbital and borrows an electron.

Movement of electron →

Lithium fluoride (LiF) (compound) contains positively charged lithium and negatively charged fluorine, creating an ionic bond.

CRITICAL THINKING Here's an interesting question:
- Why does lithium give up an electron when fluorine has so many more?

VIEW THIS IN ACTION in your WileyPLUS course

---

ratio. For example, sodium and chlorine combine to form sodium chloride (a mineral called *halite*, also known as table salt), which is written NaCl. For every Na atom in this compound, there is one Cl atom. The element that tends to form cations is written first, and the relative numbers of atoms are indicated by subscripts. For example, water forms when hydrogen (a cation, H^+) combines with oxygen (an anion, O^{2-}) in the ratio of two atoms of hydrogen to one atom of oxygen. Thus, for water, we write H_2O. **FIGURE 2.2** shows how lithium and fluorine combine to form an ionic compound, lithium fluoride (LiF) that is used in the ceramics industry and for making lenses used for ultraviolet devices.

Properties of compounds differ from the properties of their constituent elements. For example, hydrogen and oxygen are both gases at Earth's surface, whereas water is a liquid. Similarly, sodium and chlorine are both highly toxic elements, whereas their compound, salt, is essential for life.

The smallest unit that has the properties of a given compound is a **molecule**. Do not confuse a molecule and an atom; the definitions are similar, but a molecular compound always consists of two or more atoms. Molecules are held together by electromagnetic forces known as **bonds**.

■ **molecule** The smallest chemical unit that has all the properties of a particular compound.

Bonding involves the transfer of electrons from one atom to another or, in some cases, the sharing of electrons. The four principal kinds of bonds are illustrated in **FIGURE 2.3**. You will see that different bonding explains the difference in properties of diamond and graphite.

■ **bond** The force that holds the atoms together in a chemical compound.

Why have we spent all this time learning about elements, compounds, and bonding? Because minerals are chemical compounds (or, in a few cases, simply chemical elements), the chemical elements and kinds of bonds determine the properties of a mineral, and minerals are the main building blocks of the solid Earth. Now let's look more closely at the characteristics that define minerals and help us to identify them.

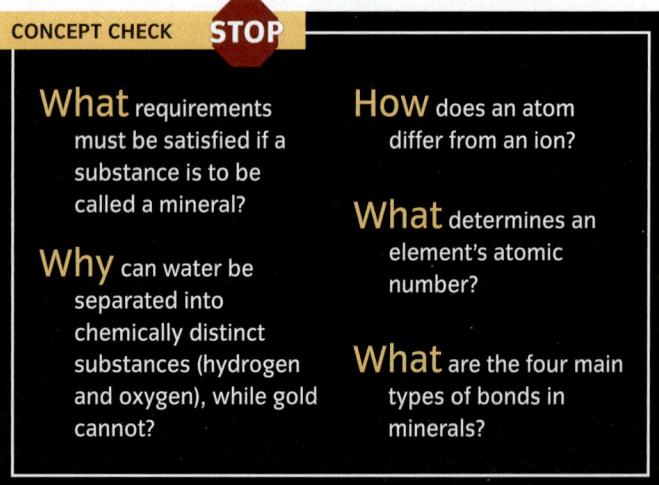

CONCEPT CHECK STOP

What requirements must be satisfied if a substance is to be called a mineral?

Why can water be separated into chemically distinct substances (hydrogen and oxygen), while gold cannot?

How does an atom differ from an ion?

What determines an element's atomic number?

What are the four main types of bonds in minerals?

Visualizing

Four types of bonding FIGURE 2.3

Ionic bonding: When one atom transfers an electron to another, as illustrated in Figure 2.2, an attractive force is set up that creates an ionic bond. In a crystal such as table salt (NaCl), the sodium ions (red) are attracted to all of the neighboring chlorine ions (gray), not just to one of them. Thus the ionic bonds form a cubic lattice. Compounds with ionic bonds tend to have moderate strength and hardness.

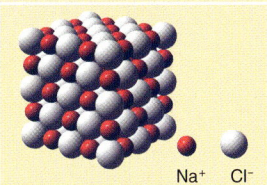
In table salt (sodium chloride, NaCl), each sodium cation is surrounded by chlorine anions.

Crystals of sodium chloride are rectangular, with straight edges.

Salt is a moderately hard solid that dissolves easily in water.

Covalent bonding: When electrons from different atoms "pair up," the force of this sharing is called a covalent bond. Note that electron sharing does not produce ions. These are the strongest chemical bonds, and elements and compounds with covalent bonds (such as diamond) tend to be strong and hard.

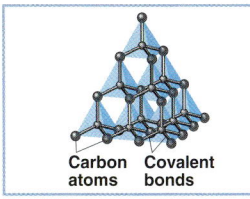
Diamond consists of carbon atoms connected in a network of covalent bonds. Each atom is connected to four others.

Diamond crystals appear in a rock called kimberlite. Covalent compounds are often strong and hard; diamond is one of the hardest substances known.

Cut and polished diamonds are prized gems. Tiny diamonds are used in industry for cutting and grinding instruments.

Metallic bonding: In metals, atoms are so tightly packed that electrons can be shared among several atoms. In fact, the outermost electrons are so loosely held that they can readily drift from one atom to another. This mobility of electrons explains why metals are good at conducting electricity and heat.

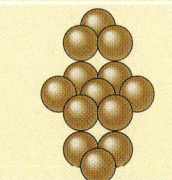
Atoms of gold are packed in the densest possible manner. Each atom is surrounded by, and in contact with, 12 other gold atoms

This nugget of gold was once embedded in rock, but weathering and erosion have removed most of the rock.

Gold is durable as well as malleable; it has been used as currency since ancient times. These are gold coins.

Van der Waals bonding: A weak attraction can occur between electrically neutral molecules that have an asymmetrical charge distribution. The positive end of one molecule will be attracted to the negative end of another molecule. For example, the carbon atoms in graphite form sheets in which each carbon atom has strong covalent bonds with three neighbors. The bonds between sheets are weak. This is why graphite feels slippery when you rub it between your fingers.

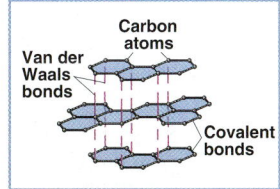
In graphite, carbon atoms form layers connected by covalent bonds. The layers are weakly held together by Van der Waals bonds.

Graphite is not a strong material and can be easily crumbled into small particles.

The "lead" in pencils is really graphite. When you write, the pressure of your hand breaks off a trail of carbon particles.

Minerals, Elements, and Compounds

What Is a Mineral?

LEARNING OBJECTIVES

Revisit the four requirements for a substance to be a mineral.

Explain the principle of atomic substitution.

Explain why crystals have flat faces with specific angles between them.

Identify at least six methods that Earth scientists use to tell minerals apart.

Explain why color is one of the least dependable ways to identify a mineral.

Chemists have been able to create millions of compounds in the laboratory, but there are only about 4000 compounds that qualify as minerals. It is important to keep the requirements for a substance to be a mineral clearly in mind. These requirements, as stated earlier, are that it be a *naturally formed solid*, be *formed inorganically*, and have a *specific chemical composition* and a *characteristic crystal structure*. Each of the items on this checklist is essential (**FIGURE 2.4**).

COMPOSITION OF MINERALS

An apparent confusion to the rule that a mineral must have a specific chemical composition is a phenomenon called *atomic substitution*. In some cases, two elements can be similar enough in size and in bonding properties that they can substitute for each other in a mineral. For example, magnesium and iron ions (Mg^{2+} and Fe^{2+}) are so similar in size, and have the same electrical charge, that one often takes the place of the other. The

Mineral or not a mineral? FIGURE 2.4

Ice is a mineral, though you may not usually think of it that way. It occurs in nature in the form of hexagonal crystals and has a specific chemical formula (H_2O).

Water is not a mineral, because it is not a solid. This criterion also means that such naturally occurring substances such as oil and natural gas cannot be considered minerals.

Bones are a tricky case. They do contain the same chemical compound found in a common mineral called apatite, but they are not minerals because they form by organic processes. Thus the bone in this modern crocodile skull is not a mineral.

However, this fossilized crocodile skull, from the Kenyan National Museum, is composed of minerals. Why? During fossilization, the original materials were replaced in an inorganic process called *mineralization*.

mineral olivine (an important component of Earth's mantle) can occur as pure Fe_2SiO_4 or pure Mg_2SiO_4 or an intermediate mixture in which some of the Fe^{2+} cations are replaced by Mg^{2+} cations. We show atomic substitution in the chemical formula by putting parentheses around the substituting elements and a comma between them. The formula for olivine, therefore, becomes $(Mg, Fe)_2SiO_4$, indicating that Mg and Fe can substitute for one another in this mineral. Note that the ratio of cations to anions is not changed by atomic substitution, so the specific composition rule is not violated.

The composition requirements for minerals specifically rule out a material whose composition varies so much that it cannot be expressed by an exact chemical formula. An example of such a material is glass, which is a mixture of many elements and can have a wide range of compositions.

Glass—even naturally formed volcanic glass—also fails the test of having a characteristic **crystal structure**. In a *crystal*, the atoms are arranged in regular, repetitive geometric patterns, as shown in FIGURE 2.5 (on the following page). By contrast, the atoms in a liquid or in an amorphous solid such as glass are mixed up or randomly jumbled. Sometimes amorphous solids are referred to as *mineraloids*; an example of a mineraloid is opal, a familiar stone that is often used in jewelry.

> **crystal structure**
> An arrangement of atoms or molecules into a regular geometric lattice. Materials that possess a crystal structure are said to be crystalline.

Coal fails the second of the four tests for a mineral because it is derived from the remains of plant material and was formed as a result of organic processes.

Steel (being made in the background) fails the first of the four tests for a mineral because it does not occur naturally. It is formed by extensive human processing of naturally occurring ores, which are minerals.

Quartz is an easily recognizable mineral. Its chemical formula is SiO_2. Note that some minerals have very complex formulas. For example, phlogopite, a form of mica, is $KMg_3AlSi_3O_{10}(OH)_2$. The important thing is that the elements combine in specific ratios.

Although **opals** are typically included in books about minerals, they are not true minerals because they do not have a specific composition and lack a crystalline structure. Opals are *mineraloids*.

Visualizing

Atomic structure of minerals FIGURE 2.5

▼ **A** The atoms of all crystalline materials are arranged in orderly lattices, like the cubical lattice illustrated here for a mineral called *galena* (PbS), the main source of lead. The atoms are so small that a cube of galena 1 cm on an edge would contain 10^{22} atoms (that's 1 followed by 22 zeros). The inset shows an exploded view of the packing arrangement of atoms in a galena crystal. The atoms are shown pulled apart along the black lines to demonstrate how they fit together. Compare the arrangement of atoms in NaCl (Figure 2.3); it is the same as the arrangement of lead and sulfur atoms in PbS.

All specimens of a given mineral have an identical crystal structure. Extremely sensitive scanning tunneling microscopes enable scientists to determine the crystal structures of minerals and actually detect the orderly arrangement of atoms in the mineral. As you can see in Figure 2.5, the atoms in a crystalline material resemble the regular, orderly rows in an egg carton.

▼ **B** Atoms are too small to see with an optical microscope, but a scanning tunneling microscope can detect the location of the atoms in a crystal. This is an image of what such a microscope detected in a galena crystal; the sulfur atoms look like large bumps and the lead atoms like small ones.

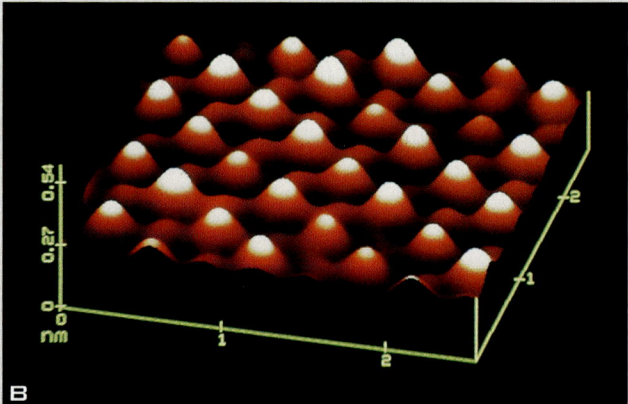

TELLING MINERALS APART

The compositions and crystal structures of minerals influence their physical properties and characteristics. If we have an unidentified mineral sample (such as **FIGURE 2.6**), we can apply a few simple tests to determine what mineral it is, without taking it to a laboratory or using expensive equipment. The properties most often used to identify minerals are the quality and intensity of light reflected from the mineral, crystal form and habit of the mineral, hardness, tendency to break in preferred directions, color, and specific gravity or density. Color, perhaps the most obvious characteristic, is often the least reliable identifier. Let's look at the properties that are used to identify different minerals.

Luster Suppose you collected a mineral sample like the one in Figure 2.6. What steps would you go through to identify it? One of the first things you would notice is

A "mystery" mineral FIGURE 2.6

Scientists have several effective, low-tech methods to identify minerals. We will use them to identify the mineral pictured here. First, note its metallic luster and cubic habit. (The answer is on page 48.)

42 CHAPTER 2 Minerals: Earth's Building Blocks

luster The quality and intensity of light that reflects from a mineral.

how shiny it is, or what scientists call its **luster**. Different minerals can reflect light in different ways as well as with different intensities. The luster of the mystery sample in Figure 2.6 is *metallic*, meaning that it looks like a polished metal surface. Some other kinds of *nonmetallic* lusters you might encounter are *vitreous*, like that of glass; *resinous*, like that of resin; *pearly*, like that of pearl (**FIGURE 2.7**); or *greasy*, as if the surface were covered by a film of oil. Two minerals with almost identical color can have quite different luster.

Crystal faces and mineral habits

The ancient Greeks were fascinated by ice. They were intrigued by the fact that needles of ice are six-sided and have smooth, planar surfaces. The Greeks called ice *krystallos*. Eventually the word *crystal* came to be applied to any solid body that has grown with flat or planar surfaces. The planar surfaces that bound a crystal are called *crystal faces*.

During the 17th century, scientists investigated crystal faces as a way to identify minerals. But the sizes of faces vary widely from one sample to another. Under some circumstances, a mineral species may grow a thin crystal; under others, the same mineral species may grow a fat crystal, as **FIGURE 2.8** shows. It is apparent from the figure that the overall crystal size and the relative sizes of crystal faces are not the same for these two crystals of quartz. In fact, crystal size and the relative sizes of crystal faces are not definitive for any mineral.

In 1669, a Danish physician, Nils Stensen (better known by his Latin name, Nicolaus Steno), unraveled the mystery of crystal faces. Steno demonstrated that the key property that identifies a given mineral is the angles between the faces. Steno's Law states that the angle between any corresponding pairs of crystal faces of a given mineral species is constant no matter what the overall shape or size of the crystal might be (see Figure 2.8).

Steno and other early scientists hypothesized that a mineral must have some kind of internal order that

Mineral luster FIGURE 2.7

Three examples of luster.

Vitreous
◀ Quartz (SiO_2) has a glassy luster.

Resinous
Sphalerite (ZnS, a source of zinc) has a resinous luster, like dried tree resin. ▶

Pearly
◀ Talc [$Mg_3Si_4O_{10}(OH)_2$] has a pearly luster.

Crystal faces and angles FIGURE 2.8

Crystals of the same mineral may differ widely in shape and size. However, the angles between faces will remain the same in all specimens. In the two quartz crystals pictured, numbers identify equivalent faces. According to Steno's Law, the angle between faces 1 and 3 (for example) is the same in both specimens.

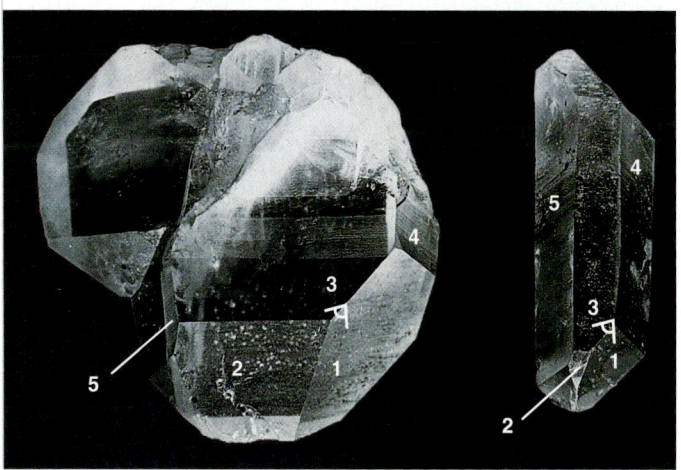

What Is a Mineral? 43

predisposes it to form crystals with constant interfacial angles. Support for their hypothesis finally arrived in 1912, when German scientist Max von Laue sent a beam of X-rays through a crystal and showed that the diffracted rays matched the patterns a geometric array would create. More recently, scanning tunneling microscopes have given more direct proof of atomic lattice structure, as shown in Figure 2.5.

Crystals develop planar faces most easily when mineral grains can grow freely in an open space. Because most mineral grains do not form in open, unobstructed spaces, nicely formed crystals are uncommon in nature, and very large crystals are very rare because large open spaces inside Earth are rare. Usually other mineral grains get in the way as minerals grow. As a result, most mineral grains have an irregular shape. However, in both a crystal and an irregularly shaped grain of the same mineral, all the atoms present are packed in the same strict geometric pattern. This is why we use the term *crystal structure*, rather than *crystal*, in the definition of a mineral.

Some minerals grow in such distinctive ways that their shape—called the mineral's **habit**—can be used as an identification tool. Our mystery sample in Figure 2.6 clearly has a *cubic* habit, because it looks like a collection of interlocked cubes. A very different example is the mineral chrysotile, shown in FIGURE 2.9,

Fibers of asbestos FIGURE 2.9

Some minerals have distinctive growth habits even though they do not develop well-formed crystal faces. The mineral chrysotile ($Mg_3Si_2O_5(OH)_4$) sometimes grows as fine, cotton-like threads that can be separated and woven into fireproof fabric. When the mineral occurs like this, it is said to have an *asbestiform* habit. Many different minerals can grow with asbestiform habits, and several are mined and commercially sold as asbestos.

which takes the form of fine fibers or threads. This *fibrous* habit is characteristic of asbestos minerals.

Hardness Hardness, like habit and crystal form, is governed by crystal structure and by the strength of the bonds between

habit The distinctive shape of a particular mineral.

hardness A mineral's resistance to scratching.

The Mohs Scale* of relative hardness of minerals TABLE 2.1

The 10 minerals of the Mohs scale are shown above, starting with the softest, talc, in the upper left-hand corner, and proceeding across two rows to diamond, the hardest mineral in the lower right-hand corner.

	Relative Hardness Number	Reference Mineral	Hardness of Common Objects
Softest	1	Talc	
	2	Gypsum	
	3	Calcite	Fingernail
	4	Fluorite	Copper penny
	5	Apatite	
	6	Potassium feldspar	Pocketknife; glass
	7	Quartz	
	8	Topaz	
	9	Corundum	
Hardest	10	Diamond	

*Named for Friedrich Mohs, a German mineralogist, who chose the 10 minerals of the scale.

Visualizing
Mineral cleavage FIGURE 2.10

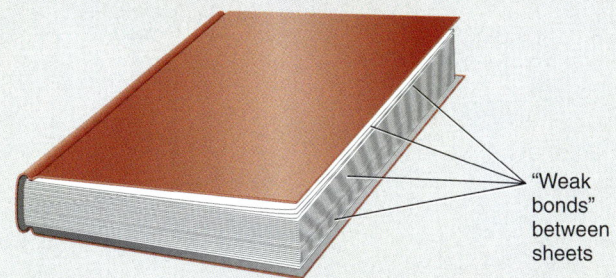

Muscovite (a potassium alumino-silicate mineral), a form of mica, cleaves so easily in one direction that it can be split by hand into flakes. The flakes suggest the leaves of a book, and Earth scientists will often refer to a "mica book."

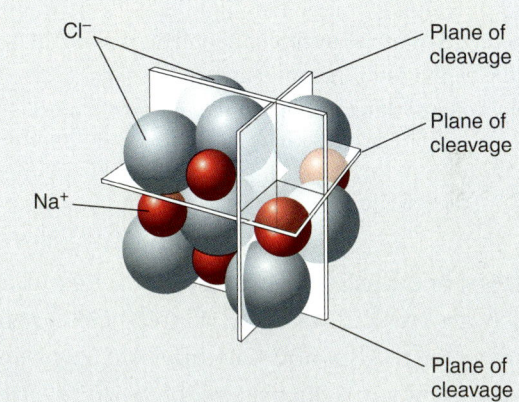

Halite, NaCl or table salt, has three distinct cleavage directions. No matter how small the bits you break it into, they will always have perpendicular faces.

atoms: The stronger the bonds, the harder the mineral. Relative hardness values can be assigned by determining whether one mineral will scratch another, using the *Mohs relative hardness scale*. The scale is divided into 10 steps, each marked by a common mineral listed in TABLE 2.1. Talc, the basic ingredient of most body ("talcum") powders, is the softest mineral known and therefore is assigned a value of 1 on the relative hardness scale. Diamond, the hardest mineral, has a value of 10. The 10 steps of the hardness scale do not represent equal intervals of hardness; the important thing is that any mineral on the scale will scratch all minerals below it. For convenience, we often test relative hardness by using a common object such as a penny or a pocketknife as the scratching instrument, or glass as the object to be scratched. The hardness values of these objects are also shown in Table 2.1.

The mystery mineral in Figure 2.6 is relatively soft, and it scratches easily with a copper penny or a piece of fluorite. However, it is barely scratched by a fingernail. This puts its hardness somewhere around 2.5 on the Mohs hardness scale.

Cleavage If you break a mineral with a hammer or drop it on the floor so that it shatters, some of the broken fragments will be bounded by surfaces that are smooth and planar, called **cleavage** surfaces. In FIGURE 2.10, the muscovite "book" cleaves easily into thin sheets (the "pages" of the book) but does not cleave at all in any other direction. In certain cases, such as the halite (NaCl) fragments shown in Figure 2.10, several cleavage directions are present, and all of the fragments are bounded by smooth, planar surfaces. Don't confuse crystal faces and cleavage

cleavage The tendency of a mineral to break in preferred directions along bright, reflective plane surfaces.

What Is a Mineral? 45

Same material, different color FIGURE 2.11

These cut gems are all synthetic sapphires, the mineral corundum. They are all the same type of material, but slight differences in chemical composition give them very different colors.

Uncut natural specimens of the same material, corundum (Al_2O_3), also have very different colors. The red crystals, rubies, are from Tanzania, and the blue crystals, sapphires, come from Newton, New Jersey. The ruby in the upper right-hand corner is about 2.5 cm (1 inch) across.

surfaces, even though the two often look alike. A cleavage surface is a breakage surface, whereas a crystal face is a growth surface. Also, note the difference between hardness and cleavage: a mineral might be quite hard—that is, resistant to scratching—but it may still cleave easily in one or more directions. Diamond, the hardest mineral, has four directions of cleavage.

The directions in which cleavage occurs are governed by the crystal structure. Cleavage takes place along planes where the bonds between atoms are relatively weak, as in the case of muscovite, or where there are fewer bonds per unit area, as in the case of diamond. Because cleavage directions are directly related to crystal structure, the angles between equivalent pairs of cleavage directions are the same for all grains of a given mineral. Thus, analogously to Steno's Law for crystals, the angles between cleavage planes are constant. A small hand lens is usually enough for an Earth scientist to spot these distinctive angles. As we remarked earlier, crystals and crystal faces are rare; but almost every mineral grain you see in a rock shows one or more breakage surfaces. That is why cleavage is such a useful aid in the identification of minerals.

The mystery mineral in Figure 2.6 has three directions of cleavage at right angles to each other.

Color and streak The color of a mineral, though often striking, is not a reliable means of identification (FIGURE 2.11). A mineral's color is determined by several factors, but the main determinant is chemical composition. Some elements can create strong color effects, even when they are present only as trace impurities. For example, the mineral corundum (Al_2O_3) is commonly white or grayish, but when small amounts of chromium are present as a result of atomic substitution of Cr^{3+} for Al^{3+}, corundum is blood red and is given the gem name *ruby*. Similarly, when small amounts of iron and titanium are present, the corundum is deep blue, producing another gem, *sapphire*. The *Case Study, Minerals for Adornment*, discusses a number of other gemstones. Many other colors such as green, gray, white, black, and pink are not useful in identifying minerals, because there are so many minerals that occur in these colors.

Color can be particularly confusing in opaque minerals that have metallic luster. This is because the color is partly a property of the size of the mineral grains. One way to reduce error is to prepare a **streak** by rubbing a

> **streak** A thin layer of powdered mineral made by rubbing a specimen on an unglazed fragment of porcelain.

46 CHAPTER 2 Minerals: Earth's Building Blocks

CASE STUDY
Minerals for Adornment

Humans started to use colored minerals for adornment in prehistoric times. Archeologists have found evidence of the practice in some of the most ancient settlements they have excavated. A mineral used for adornment is called a *gem*. The word means *precious stone* and comes from the Latin *gemma*.

The value and desirability of a gem depend on three properties. It must be beautiful so that the color and sparkle pleases the eye. It must be durable; a gem must be hard and tough enough so that it doesn't readily scratch or break. It must be rare. About 50 minerals are cut and polished for use as gems, but the most popular are diamond, ruby, sapphire, and emerald. Each satisfies the three criteria of rarity, durability, and beauty. Although diamonds are sometimes colored, as in the famous Hope Diamond (**A**), most are colorless and owe their beauty to their brilliant sparkle, a property that can be enhanced by cutting and polishing. Ruby and sapphire are colored forms of the same mineral, corundum (see Figure 2.11). Emerald is a green variety of the mineral beryl, as seen here in a necklace made for Catherine the Great of Russia in 1762 (**B**).

Gems that meet the criteria of beauty and durability, but are not so rare, include garnet, tourmaline, olivine (the gem name is *peridot*), quartz (gem names for different colors include *amethyst* and *citrine*), feldspar (gem form is called *moonstone*), spinel, topaz, and zircon. Gems in this less-rare group are sometimes said to be *semiprecious*, as opposed to *precious* for diamonds, rubies, sapphires, and emeralds. The terminology is confusing and meaningless, because an exceptional specimen of a topaz, or tourmaline, or some other less-rare gem can be far more valuable than an ordinary specimen of a diamond or emerald.

The term *gem* is also applied to certain mineraloids and organic materials that are used for adornment. Among the mineraloids are opals and natural volcanic glass. Organic materials include amber (fossil tree resin), and jet (a black, compact form of coal), coral, ivory, and pearls, displayed here in a necklace (**C**).

Mineral streak FIGURE 2.12

Although hematite (Fe_2O_3) is shiny, black or gray, and metallic looking, it makes a distinctive red-brown smear when rubbed on a porcelain streak plate.

specimen of a metallic luster mineral on an unglazed fragment of porcelain called a *streak plate*. The color of a streak is reliable because all the grains in the streak are very small and the effect of grain size is reduced. For example, hematite (Fe_2O_3), an important iron-bearing mineral, produces a reddish-brown streak, even though a specimen may look black and metallic (**FIGURE 2.12**). The streak of our mystery mineral from Figure 2.6 is gray.

Density

Another important physical property of a mineral is how light or heavy it feels. We are all familiar with the fact that two equal-sized baskets have different weights when one is filled with feathers and the other with rocks. This means the rocks have greater **density** than the feathers. Minerals that have a high density, such as gold, have closely packed atoms. Minerals with a low density, such as ice, have less closely packed atoms.

density The mass of material per unit volume.

It is not easy to measure the density of a mineral in a laboratory, because it requires dividing the mass by the volume, and the volume of an irregularly shaped grain is difficult to determine. This was the same problem that faced Archimedes more than 2000 years ago when he was asked to determine whether the gold in a crown was pure or alloyed with a less dense material. Earth scientists use a modern version of Archimedes' method. First, they measure the weight of the grain in air (W_A). Then they submerge it in water and measure the weight again (W_W). The weight in water is less than that in air, and the difference ($W_A - W_W$) is the weight of water displaced by the mineral grain. The ratio of W_A to ($W_A - W_W$) gives the *specific gravity* of the mineral. Because water has a density of 1.0 gram per cubic centimeter, the specific gravity is numerically equal to the density.

Many common minerals, such as quartz, have specific gravities in the range 2.5 to 3.0. Fortunately, fancy equipment for measuring specific gravity is usually not necessary. If a mineral is substantially lighter than 2.5 g/cm³, or substantially heavier than 3.0 g/cm³, an Earth scientist can immediately tell this by hefting it in his or her hand. Metallic minerals generally feel heavy, whereas minerals with vitreous luster tend to feel light. The mystery mineral in Figure 2.6 would feel quite a bit heavier than most minerals of the same size, and if we took it back to the lab, we would find that its specific gravity is 7.5.

The mystery mineral revealed

What is the identity of the mystery mineral in Figure 2.6? We have revealed several clues: It has metallic luster, a cubic habit, hardness around 2.5, gray color and streak, three perpendicular cleavage directions, and an unusually high specific gravity of 7.5. A quick look in a table of mineral properties would confirm an Earth scientist's suspicions: The mystery mineral is galena (PbS), a compound of lead and sulfur. This is the same mineral that is illustrated in Figure 2.5.

CONCEPT CHECK **STOP**

How can the hardness of a mineral be used for identification?

What is Steno's Law and why is it important?

What is the difference between a cleavage surface and a crystal face?

What is the difference between luster, color, and streak, and how are they related?

Why is color an unreliable way of identifying a mineral?

48 CHAPTER 2 Minerals: Earth's Building Blocks

Mineral Families

LEARNING OBJECTIVES

Identify the 12 most common chemical elements in Earth's crust.

Explain why Earth's crust contains many fewer minerals than one might expect and why the number of rock-forming minerals is even smaller.

Identify the two most common mineral families and four accessory mineral families.

Describe the various molecular structures of silicate minerals.

Scientists have identified approximately 4000 minerals. This number may seem large, but it is tiny compared with the number of synthetic materials, such as ceramics, concrete, drugs like aspirin, and solid chemical reagents. The reason for the disparity between the number of minerals and the millions of solids that have been synthesized in laboratories becomes clear when we consider the relative abundances of the chemical elements in nature. Out of every kilogram of material in Earth's continental crust, only 12 elements are present in quantities greater than one gram (see **FIGURE 2.13**). The abundant 12 account for 992.3 of the 1000 g; all common minerals have compositions based on one or more of these abundant elements. The remaining 80 elements, combined, make up less than 1% of the crust by weight and less than 2% by volume. Minerals made of the scarcer elements occur only in small amounts, and

Elements of the continental crust FIGURE 2.13

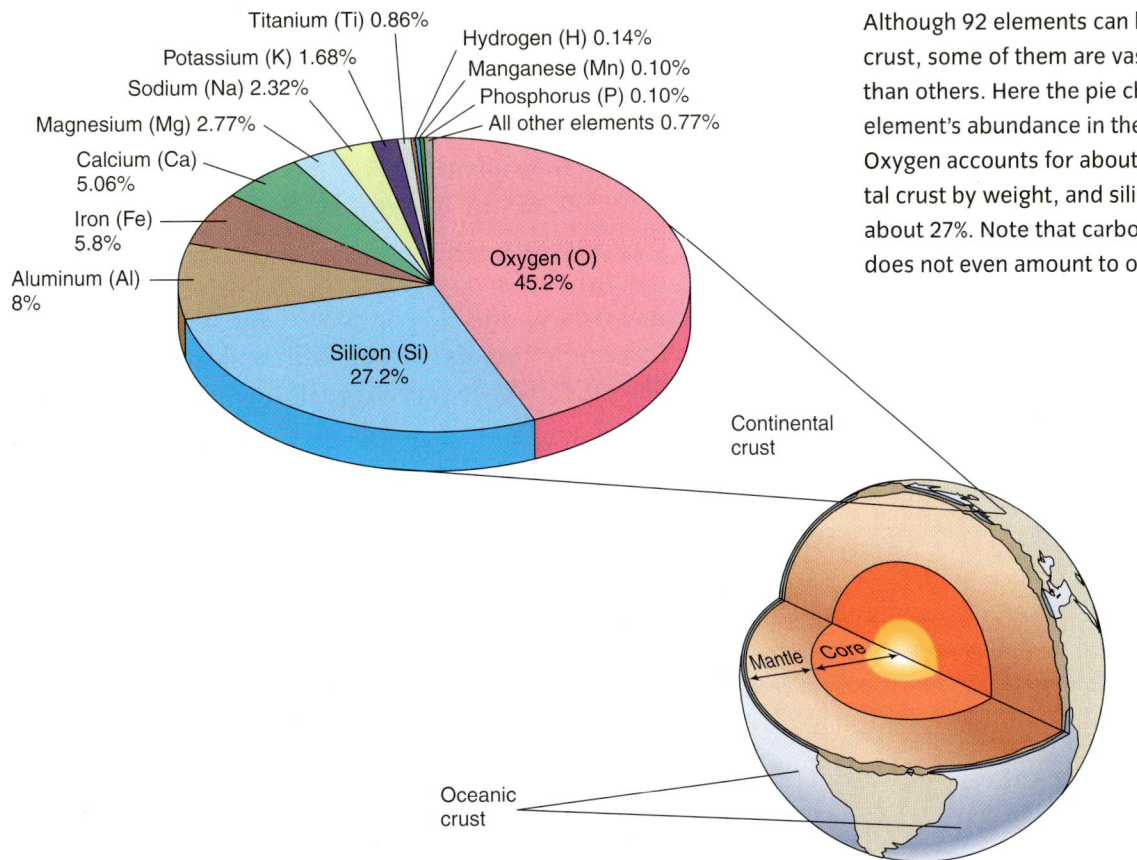

Although 92 elements can be found in Earth's crust, some of them are vastly more common than others. Here the pie chart illustrates each element's abundance in the continental crust. Oxygen accounts for about 45% of the continental crust by weight, and silicon accounts for about 27%. Note that carbon, though vital to life, does not even amount to one part per thousand.

Mineral Families 49

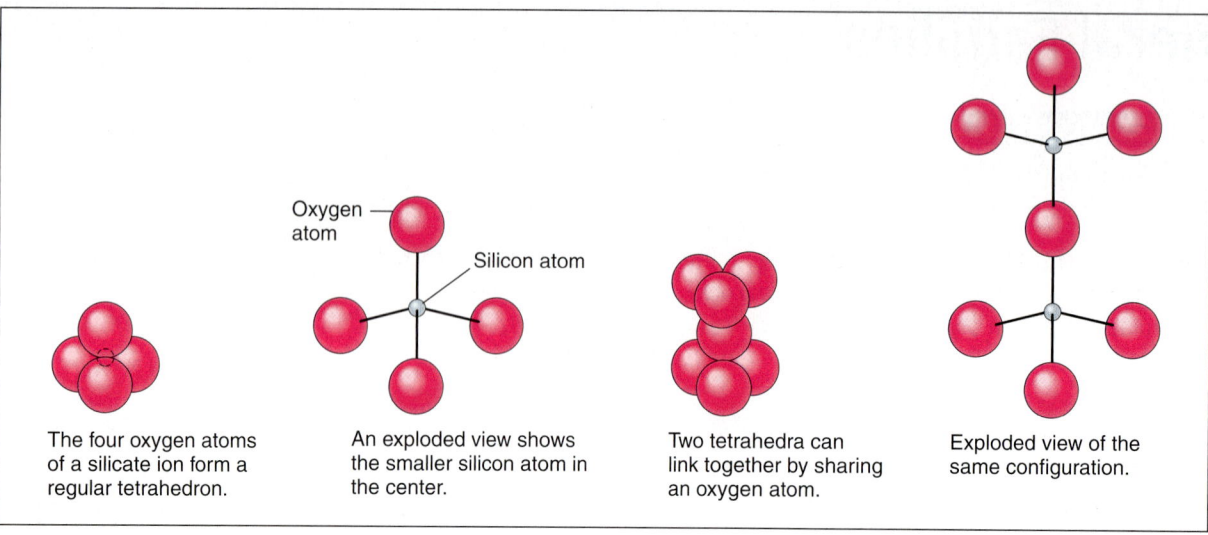

Silicate links FIGURE 2.14

Silicate minerals consist of $(SiO_4)^{4-}$ anionic groupings linked together.

ore deposits of scarce elements such as gold, uranium, and tin are rare and hard to find. However, these scarce elements can be extracted and used to synthesize a wide range of materials in the laboratory and in manufacturing for everyday use.

MINERALS OF EARTH'S CRUST

Mineral families are groups of minerals that are similar to one another in terms of chemistry or atomic structure or (more commonly) both. Two elements alone—oxygen and silicon—make up more than 80% of the atoms in Earth's crust and comprise more than 70% of its mass. Thus it should be no surprise that the great majority of minerals contain one or both of these elements. The most common family of minerals, the **silicate minerals**, contain a strongly bonded complex anion called the *silicate* anion, which contains both silicon and oxygen $(SiO_4)^{4-}$ (see FIGURE 2.14). The bonding in most silicate minerals is a mixture of ionic and covalent; as a result, silicates tend to be hard, tough minerals, and their properties, such as cleavage, reflect the way silicate anions are arranged in their crystal structures (TABLE 2.2). The next most abundant family is the **oxide minerals**, which contain the simple oxide anion O^{2-}.

Other mineral groups are based on different anions, some complex and some simple. For example, *carbonates* are based on the complex $(CO_3)^{2-}$ anion, *sulfates* are based on the complex $(SO_4)^{2-}$ anion, *phosphates* are based on the complex $(PO_4)^{3-}$ anion, and *sulfides* are based on the simple S^{2-} anion. A few silicate minerals and a few oxide minerals, together with calcium sulfate and calcium carbonate, comprise the bulk of Earth's crust—an estimated 99% by volume. These common minerals, of which there are about 30, are called the **rock-forming minerals**. Rock-forming minerals are everywhere—not only in rocks, but also in soils and sediments and even in the dust that we breathe. Rock-forming minerals are common and inexpensive but nevertheless economically vital; you rely on them every day when you drive on a paved road or enter a concrete building.

Less common minerals are called **accessory minerals**. These are widely present in common rocks but in such small amounts that they do not determine the properties of the rocks. Though they are less abundant than the silicates and oxides, many are economically important. Some are ore minerals. As mentioned, galena is the main source of lead, and chalcopyrite $(CuFeS_2)$ is the principal source of copper. Others have significant biological properties. The phosphate mineral apatite $[Ca_5(PO_4)_3(F, OH)]$ supplies the phosphorus for

Polymerization and the common silicate minerals TABLE 2.2

Silicate structure	Mineral/Formula	Cleavage	Example of a specimen
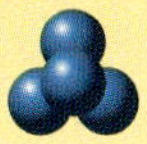 Single tetrahedron	Olivine Mg_2SiO_4	None	
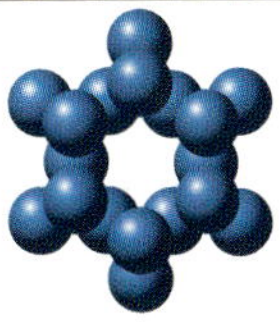 Hexagonal ring	Beryl (Gem form is emerald) $Be_3Al_2Si_6O_{18}$	One direction	
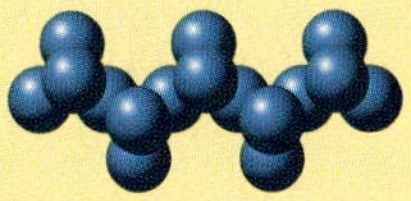 Single chain	Pyroxene group $CaMg(SiO_3)_2$ (variety: diopside)	Two directions at about 90°	
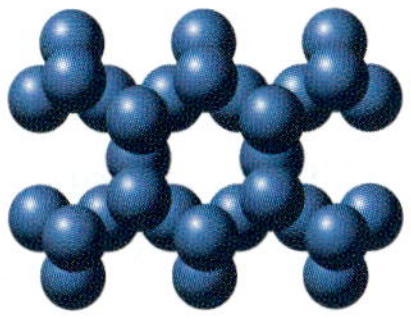 Double chain	Amphibole group $Ca_2Mg_5(Si_4O_{11})_2(OH)_2$ (variety: tremolite)	Two directions at about 120° and 60°	
 Sheet	Mica $KAl_2(AlSi_3O_{10})(OH)_2$ (variety: muscovite) $K(Mg,Fe)_3(AlSi_3O_{10})(OH)_2$ (variety: biotite)	One direction	
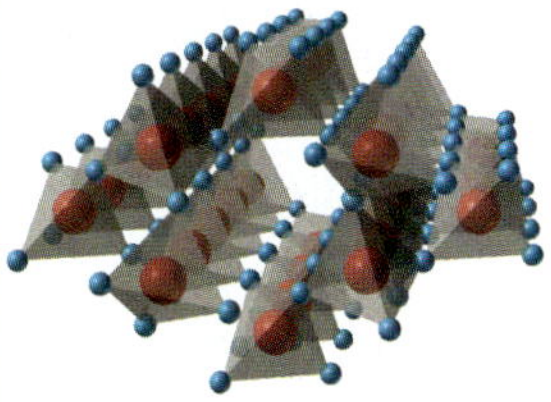 Network (quartz)	Feldspar $KAlSi_3O_8$ (variety: orthoclase)	Two directions at about 90°	
	Quartz SiO_2	None	

This chart summarizes the ways in which silica ions can polymerize to form minerals. Typical examples of each type are shown in the photographs. There are many other silicate materials in each category, but all of the principal categories that occur in nature are illustrated here.

Mineral Families 51

agricultural fertilizers and is also the fundamental compound in our bones and teeth.

Sometimes two minerals can have the same chemical formula but different crystal structures. A very common example is calcium carbonate ($CaCO_3$), which forms two very different minerals called *calcite* (the principal ingredient in limestone and marble) and *aragonite* (found in the shells of many organisms, such as mollusks). These different forms are called *polymorphs*. Likewise, graphite and diamond are both polymorphs of carbon, as explained in Figure 2.3. Remarkably, water ice has 14 different polymorphs that have been discovered so far, but only one of them occurs naturally on Earth—the others have been made in the laboratory. Some of the other polymorphs of ice are likely to turn up on the ice-covered moons of other planets, such as Ganymede, a moon of Jupiter.

Note that some materials occur in nature as native elements; that is, they are not combined in compounds with other elements. Minerals that can occur in this form include some metals, such as gold (Au) and silver (Ag), and some nonmetals, such as sulfur (S) and graphite and diamond (C).

SILICATES: THE MOST IMPORTANT ROCK FORMERS

Not only are silicates the most common minerals, they also have an unusual diversity of atomic structures. To explain this phenomenon, we begin by looking at the silicate anion itself, in which four oxygen atoms are tightly bonded to the single silicon atom. As shown in Figure 2.14, if we draw a stick figure joining the centers of the oxygen atoms with lines, we would get a regular tetrahedron (i.e., a pyramid with a triangular base). The small ionically bonded silicon atom occupies the space at the center of the tetrahedron, and the oxygen atoms occupy the corners.

Two silica tetrahedra can bond by sharing an oxygen atom, in a process called *polymerization*. Through polymerization, even larger complex ions can form, consisting of rings, chains, sheets, or three-dimensional frameworks of tetrahedra. Different common cations, such as calcium (Ca^{2+}), aluminum (Al^{3+}), magnesium (Mg^{2+}), iron (Fe^{2+}), sodium (Na^+), and others can fit into the spaces or *interstices* between the polymerized tetrahedra. The identity of a silicate mineral is determined by three factors: how the silica tetrahedra occur within the mineral; which cations are present; and how the cations are distributed throughout the structure. Table 2.2 shows the common silicate mineral families. For example, polymerization in mica produces a sheet, and the pronounced cleavage in mica is parallel to the sheets. In quartz, the three-dimensional framework, shown in Table 2.2, is equally strong in all directions, and so quartz is a hard, tough mineral that lacks cleavage.

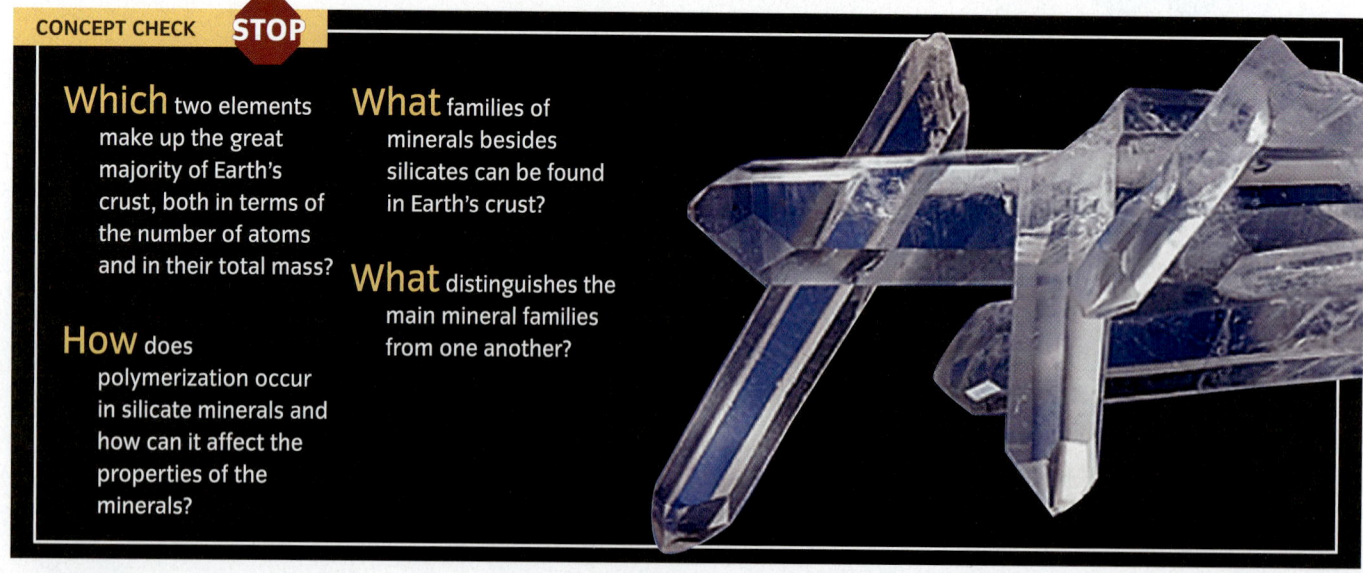

CONCEPT CHECK STOP

Which two elements make up the great majority of Earth's crust, both in terms of the number of atoms and in their total mass?

How does polymerization occur in silicate minerals and how can it affect the properties of the minerals?

What families of minerals besides silicates can be found in Earth's crust?

What distinguishes the main mineral families from one another?

Mineral Resources

LEARNING OBJECTIVES

Explain why no nation today is able to supply all its mineral needs.

Explain the difference between renewable and nonrenewable resources.

Identify environmental problems caused by mining.

Can you imagine a world without machines? Our modern world, with its 6.5 billion inhabitants (and the number continues to grow), could not operate without machines. We use machines to produce our food, make our clothes, transport us around, and help us communicate. The metals that are needed to build machines all come from minerals dug from Earth. So great has our dependence on minerals become that today we have industrial uses for almost all the naturally occurring chemical elements, and more than 200 kinds of minerals are mined and used.

Recall from Figure 2.13 that only 12 chemical elements make up a little more than 99.2% of the mass of Earth's crust. Six of the 12 are useful metals—silicon, aluminum, iron, magnesium, titanium, and manganese. All the other useful metals, such as copper, lead, zinc, tin, platinum, gold, silver, and tungsten, are present in Earth's crust in tiny amounts. In order to mine such scarce materials, it is necessary to find a place where some natural process has produced a localized concentration of a scarce element; such a localized concentration is called an **ore deposit**.

> **ore deposit** A localized concentration in the crust from which one or more minerals can be profitably extracted.

Ore deposits have four distinctive aspects:

1. They are limited in abundance and distinctly localized in the crust. As a result, no nation is self-sufficient in mineral supplies, and supplies must be searched out around the globe (see *What an Earth Scientist Sees* on page 54).

2. The quantity of a given mineral resource in any one country is rarely known with accuracy, and the likelihood that new deposits will be found is hard to assess. A country that is an exporter of a mineral resource today may be an importer tomorrow. A century and a half ago, Britain was a great mining nation, producing tin, copper, tungsten, lead, and iron. Today the known deposits have been exhausted. Even the commonest of objects on which we rely requires an efficient global trade system.

3. Unlike plants and animals, which are *renewable resources* because they can be harvested seasonally and replenished by growth, ore deposits are *nonrenewable resources* because they are depleted by mining and eventually exhausted. This disadvantage can be offset only by finding new deposits or by using the same material repeatedly—that is, by recycling. Depletion raises a question: "Are Earth's supplies of nonrenewable resources large enough to meet the needs of the world's growing population?" The world's population will find out the answer at some unknown time in the future.

Mining disturbs the land surface FIGURE 2.15

The Bingham Canyon Copper Mine in Utah is the largest human excavation on Earth. Located near Salt Lake City, it is open to the public. It is currently four kilometers wide and nearly a kilometer deep.

What an Earth Scientist Sees

How Many Minerals and Metals Does It Take to Make a Light Bulb?

Gas — Usually a mixture of *nitrogen* and *argon* to retard evaporation of the filament.

Bulb — Soft glass is generally used, made from *silica, trona (soda ash), lime, coal,* and *salt*. Hard glass, made from the same minerals, is used for some lamps to withstand higher temperatures and for protection against breakage.

Filament — Usually is made of *tungsten*. The filament may be a straight wire, a coil, or a coiled-coil.

Button & Button Rod — Glass, made from the same materials listed for the bulb (plus *lead*), is used to support and to hold the tie wires placed in it.

Lead-in-wires — Made of *copper* and *nickel* to carry the current to and from the filament.

Base — Made of *brass (copper* and *zinc)* or *aluminum*. One lead-in wire is soldered to the center contact and the other soldered to the base.

Stem Press — The wires in the glass are made of a combination of *nickel-iron* alloy core and a *copper* sleeve.

Fuse — Protects the lamp and circuit if the filament arcs. Made of *nickel, manganese, copper,* and/or *silicon* alloys.

Even a simple appliance like a light bulb contains materials that must be extracted from different minerals mined in different countries around the world. Many of these materials can be recycled, but none of them can be replenished.

CANADA: Copper, Molybdenum, Nickel, Zinc

UNITED STATES: Argon, Coal, Copper, Lead, Limestone, Molybdenum, Nitrogen, Salt, Silica (sand), Trona (soda ash), Tungsten

JAMAICA: Aluminum

GUINEA: Aluminum

RUSSIA: Coal, Copper, Lead, Manganese, Nickel, Salt, Tungsten, Zinc

CHINA: Coal, Manganese, Salt, Tungsten

ZAMBIA: Copper

BRAZIL: Manganese

CHILE: Copper

SOUTH AFRICA: Manganese

AUSTRALIA: Aluminum, Lead, Nickel, Zinc

Here's an interesting question:
- Fluorescent light bulbs are more efficient than filament bulbs. What do fluorescent bulbs contain and what countries supply the materials? (You may need to do some research to answer this question.)

CHAPTER 2 Minerals: Earth's Building Blocks

Destruction of an environment FIGURE 2.16

Shown here is small-scale gold mining in Sulawesi, Indonesia. First, the large boulders and gravel are removed and discarded. Then the bottom layers of sediment, where the gold is concentrated, are pumped out through the white hoses and treated in the devices visible in the upper right side of the image. When mining is completed, the miners walk away and leave the mess.

4. Mining disturbs Earth's surface in many ways (FIGURE 2.15). The search for ore deposits has increasingly taken miners to places of topographic and climatic extremes, and the result has been destruction of some environmentally sensitive areas. Small-time miners, in particular, tend to mine an area and leave it, without any effort at reclamation (FIGURE 2.16). The United Nations estimates that there are between 10 and 15 million small-scale miners, including 4 million women and 1 million children. The problems are increasing as population and the rate of mineral consumption grow. In many countries mining is under increasingly strict control, but major damage to the environment is nevertheless widespread. For example, the United Nations estimates that as much as 1000 tons of mercury are released to the environment each year by small-scale miners who ignore the need for conservation.

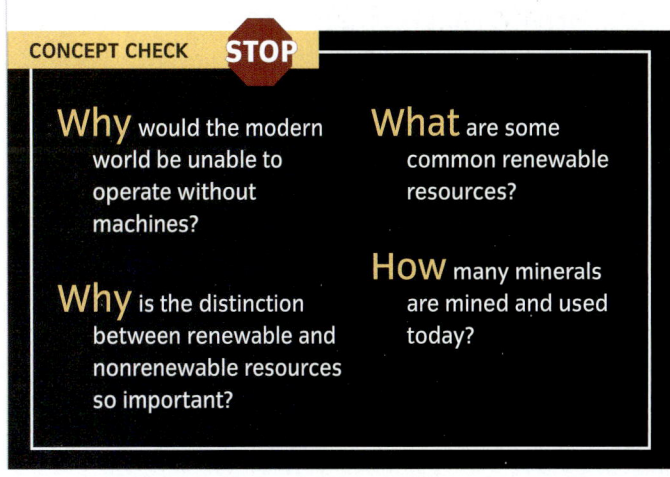

CONCEPT CHECK STOP

Why would the modern world be unable to operate without machines?

Why is the distinction between renewable and nonrenewable resources so important?

What are some common renewable resources?

How many minerals are mined and used today?

Mineral Resources

Amazing Places: The Naica Mine, Chihuahua, Mexico

In 1912, miners seeking extensions to lead and zinc ore bodies in the Naica mining district of Mexico broke through into a cavernous opening that contained giant crystals of gypsum (calcium sulfate) up to 2 m in length. The cave, which is known as the Cave of Swords (A), is 120 m below the surface. In 2000, an even more astonishing discovery was made in the same mine at a depth of 290 m—the largest crystals of any kind discovered anywhere in the world (B and C).

The crystals grew from saline waters rich in calcium sulfate at a temperature just below 58°C. Scientists who have studied the crystals estimate that it took hundreds of thousands of years for the giant crystals to grow.

A The Cave of Swords, discovered in 1912, is still in its original condition because it has been protected from vandalism by the mining company.

B The Cave of Crystals, discovered in 2000, contains the most striking collection of giant crystals ever discovered. The sizes of the crystals can be judged from the size of the miner.

C The largest crystals in the Cave of Crystals can be seen in the foreground. They exceed 11 m (36 ft) in length and weigh an estimated 55 tons.

SUMMARY

1 Minerals, Elements, and Compounds

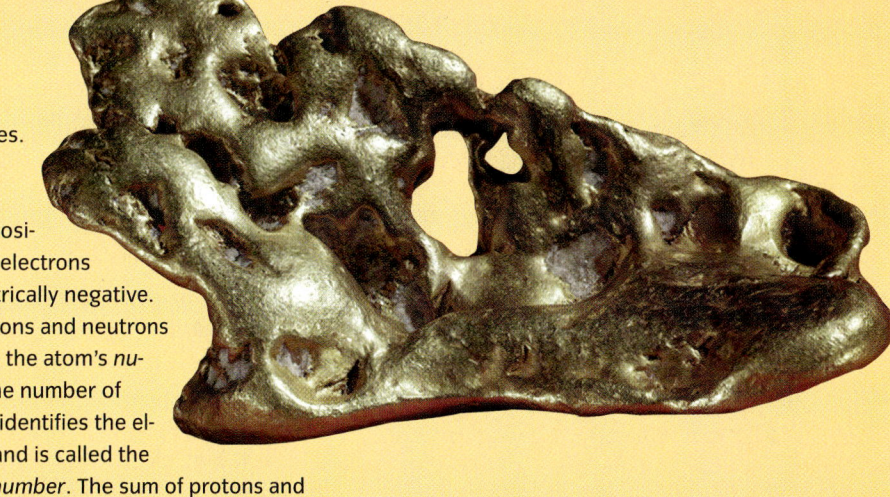

1. Never confuse **rocks** and **minerals**. Minerals are to rocks as letters are to words. All specimens of a given mineral have the same basic composition. Rocks have variable compositions because they are aggregates of one or more kinds of minerals and may also contain fragments of other rocks, amorphous substances, and organic matter.

2. **Elements** are the most fundamental of all naturally occurring substances because they cannot be separated into chemically distinct materials. The minute particles that make up all matter, including elements, are **atoms**. All atoms of a particular element are the same.

3. Atoms are made of smaller particles called *protons*, *neutrons*, and *electrons*, which have no independent chemical properties. Protons are electrically positive and electrons are electrically negative. The protons and neutrons reside in the atom's *nucleus*. The number of protons identifies the element, and is called the *atomic number*. The sum of protons and neutrons is the *mass number*. Most elements have several different **isotopes**, which differ in the number of neutrons. Ordinarily an atom has equal numbers of protons and electrons, but it may gain or lose electrons, in which case it becomes electrically charged and is called an **ion**.

4. **Compounds** consist of multiple elements, and therefore multiple types of atoms. The smallest unit that has the properties of a given compound is a **molecule**. Molecules in a compound are held together by **bonds**. The most common form of bonding is *ionic bonding*, caused by the electrostatic attraction of two oppositely charged ions. Other forms of bonding are *covalent*, *metallic*, and *Van der Waals bonding*.

2 What Is a Mineral?

1. **Minerals** are *naturally occurring solids*, each with a unique **crystal structure**. Minerals are formed by *inorganic processes* and must have a *fixed chemical composition*. One extension of the last rule is *atomic substitution*, in which other atoms of like size and ionic charge may be substituted for specific atoms in a mineral without causing the crystal structure to change.

2. The *crystal faces* that bound a mineral are a direct consequence of its atomic lattice. In some cases, a mineral will not have enough room to grow identifiable crystals, but the underlying geometric lattice remains the same.

3. Several properties can be used to tell minerals apart. One of the most easily visible is a mineral's **luster**, or the quality and intensity of light that reflects from it. Measuring the angles between adjacent faces is a useful way to identify a crystal, for though crystal size and overall shape in a mineral may vary, *Steno's Law* states that the angle between corresponding faces remains constant. Other factors that can aid in mineral identification include the distinct external shape, or **habit**; a mineral's resistance to scratching, known as its **hardness**; how a mineral behaves when broken, known as **cleavage**; and its **density**, or *specific gravity*. Color may also be useful but is often misleading. However, when a mineral is rubbed on a *streak plate*, it produces a thin layer of powdered mineral, known as its **streak**, which is more reliable than color for identification.

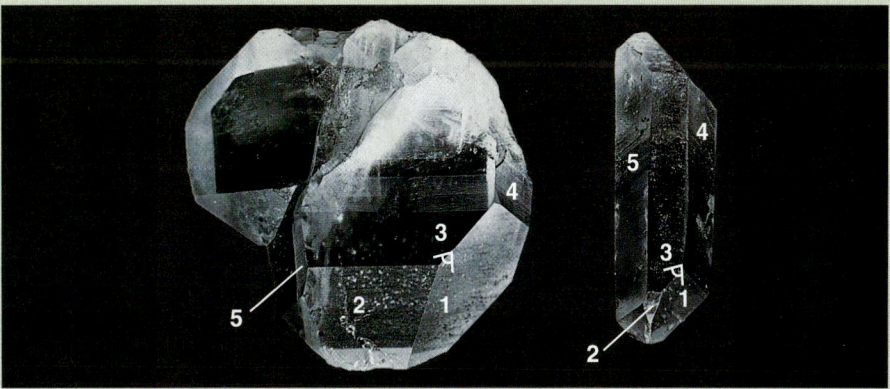

3 Mineral Families

1. The distribution of elements in Earth's crust is far from even. Only 12 elements are present at a level of more than one part in a thousand by mass, and of these, oxygen and silicon dominate. As a result, the number of naturally occurring minerals is relatively small, and the number of important rock-forming minerals is even smaller—only about 30 or so.

2. **Silicate minerals**, based on the $(SiO_4)^{4-}$ anion, are the most abundant family of **rock-forming minerals**. They can adopt a variety of crystal structures because of the ability of the *silicate anions* to link together. These structures include chains, sheets, and three-dimensional lattices. **Oxide minerals** are the next most abundant family. Other important mineral families include *sulfides* and *sulfates*, *carbonates*, and *phosphates*. Less abundant minerals are called **accessory** minerals, and they usually do not affect the properties of the rock they are found in. Nevertheless, these minerals are frequently of great economic importance as sources of metal ore.

3. Two minerals can have the same chemical formula but different crystal structures. These different forms are *polymorphs*. Other minerals occur in nature as *native elements* that are uncombined with other elements.

4. Silicates are the most common minerals. Large, complex ions can be formed by *polymerization*, a process of bonding based on the sharing of an oxygen atom. Three factors affect the identity of a silicate mineral: whether the silicate tetrahedra are single or polymerized, which cations are present, and how the cations are distributed.

4 Mineral Resources

1. All of the metals used in industry and elsewhere come from minerals mined on Earth. The distribution of **ore deposits** around the world is very uneven, and as a result no country is self-sufficient in all the mineral resources needed.

2. Mining disturbs the ground surface. Responsible mining restores the surface and/or mitigates the impact after mining is finished, but small-scale mining, which is widespread, does not restore the surface and, in addition, is all too often the source of releases of toxic chemicals, such as mercury, into the environment.

KEY TERMS

- **mineral** p. 36
- **rock** p. 36
- **element** p. 36
- **atom** p. 36
- **isotopes** p. 37
- **compound** p. 37
- **molecule** p. 38
- **bond** p. 38
- **crystal structure** p. 41
- **luster** p. 43
- **habit** p. 44
- **hardness** p. 44
- **cleavage** p. 45
- **streak** p. 46
- **density** p. 48
- **ore deposit** p. 53

CRITICAL AND CREATIVE THINKING QUESTIONS

1. When astronauts brought back rock samples from the Moon, the minerals present were mostly the same as those found on Earth. Can you think of reasons why this might be so? Would you expect minerals on Mars or Venus to be the same as, or at least very similar to, those on Earth?

2. Which of the following materials are minerals, and why (or why not)?: water; beach sand; diamond; wood; vitamin pill; gold nugget; fishbone; emerald.

3. The minerals calcite and aragonite have the same chemical formula ($CaCO_3$) but different crystal structures. Are they polymorphs? The materials halite (NaCl) and galena (PbS) have the same geometric patterns in their crystal structures but different compositions. Are they polymorphs?

4. Do some research to find out whether any valuable ores are mined in your area. What are the minerals involved and what commodity is recovered from the ore?

5. Almost everything we use is made from materials dug or pumped from the ground, but all resource production disrupts the land and has the potential to cause major impacts to the environment. Do some research and address the question, "How would you reduce the environmental impact of resource production, or, if you advocate stopping production, how would you help society to do without some materials?"

What is happening in this picture?

The Earth scientist is part of a team studying the feasibility of establishing a platinum mine near Stillwater, Montana. He is looking at a core sample through a hand lens, one of the simplest and most useful tools in any Earth scientist's backpack.

- What features might he be looking for?

- The woodlands in the background suggest a pristine, natural environment. What questions should the team be thinking about if they discover ore and a mine is opened?

SELF-TEST

1. A(n) _____ is an atom that has gained or lost one or more electrons and has a net electric charge.
   a. molecule
   b. isotope
   c. element
   d. ion
   e. compound

2. On this illustration, locate and label the following parts of the atom:

   proton

   electron

   neutron

   nucleus

   first-energy-level electron shell

   second-energy-level electron shell

   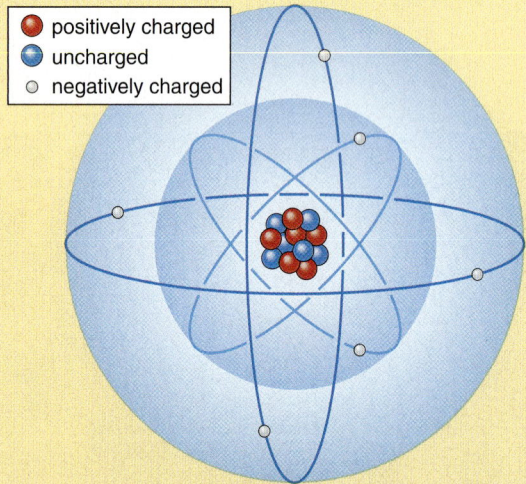

3. In _____, electrons from different atoms are shared and the force of this sharing forms the strongest of chemical bonds.
   a. ionic bonding
   b. covalent bonding
   c. metallic bonding
   d. Van der Waals bonding

4. In _____, the mobility of electrons in the outermost shell allows materials with these types of bonds to act as good conductors of electricity and heat.
   a. ionic bonding
   b. covalent bonding
   c. metallic bonding
   d. Van der Waals bonding

5. To be considered a mineral, a substance must _____.
   a. have a specific chemical composition
   b. be formed by inorganic processes
   c. be a naturally formed solid
   d. have a characteristic crystal structure
   e. have all of the characteristics listed above

6. Atomic substitution is a process whereby one element can substitute for another in a mineral provided that _____.
   a. the elements have essentially the same size
   b. the elements have similar bonding properties
   c. positive ionic charges remain balanced with the negative charges
   d. All of the above statements are true.

7. Volcanic glass is not considered a mineral because it _____.
   a. is amorphous (that is, it lacks a crystal structure)
   b. is not naturally occurring
   c. does not have enough silicon or oxygen in its chemical composition
   d. All of the above statements are correct.

8. The photograph shows natural samples of the mineral corundum. What best explains the striking difference in color between the red (ruby) and the blue (sapphire) samples?
   a. polymorphism
   b. small variations in composition
   c. differences in crystalline structure
   d. polymerization

60 CHAPTER 2 Minerals: Earth's Building Blocks

9. Which of the following minerals is the hardest?
   a. quartz
   b. calcite
   c. muscovite
   d. feldspar

10. Which of the following physical properties is the least useful in identifying many varieties of common minerals?
    a. cleavage or fracture
    b. hardness
    c. color
    d. density

11. _____ is the most abundant element by weight in Earth's continental crust.
    a. Silicon
    b. Iron
    c. Calcium
    d. Aluminum
    e. Oxygen

12. Earth's crust is mostly composed of a small number of rock-forming minerals because _____.
    a. polymorphs are common
    b. of the overwhelming abundance of oxygen and silicon
    c. of a lack of carbon in igneous rocks
    d. All of the above statements are true.

13. Gold is an example of a(n) _____, quartz is a(n) _____, and pyrite is a(n) _____.
    a. native element/silicate mineral/sulfide mineral
    b. oxide mineral/sulfide mineral/phosphate mineral
    c. phosphate mineral/sulfide mineral/carbonate mineral
    d. oxide mineral/silicate mineral/phosphate mineral

14. On the illustration, label each silicate structure with its proper name. For each structure give an example of a mineral (name and chemical formula) and indicate the prominent cleavage.

Structure 1

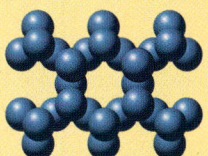

Structure 2

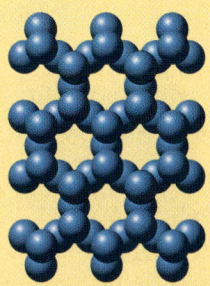

Structure 3

15. No nation today is self-sufficient in mineral supplies because _____.
    a. environmental laws prohibit mining
    b. mining is too expensive
    c. the big ore deposits have all been exhausted
    d. ore deposits are limited in number and unevenly distributed
    e. economic agreements favor less-developed countries

Rocks: Keepers of Earth's History

Aboriginal people call it *Uluru*. Australians of European descent named it *Ayers Rock*. Now called *Uluru/Ayers Rock*, it is a World Heritage site and one of Earth's most striking features. This aerial photograph reveals vertical layers of sedimentary rock. About 550 million years ago, those layers were horizontal layers of sediment deposited by streams flowing off the now deeply eroded Musgrave and Petermann Ranges.

Global Locator

Most sandstone contains only quartz; the presence of feldspar in Uluru/Ayers Rock indicates something about the environment in which the rocks formed. Quartz is resistant, but feldspar quickly breaks down to clay under the combined chemical attack of rainwater and the atmosphere. Weathering must have been so rapid that feldspar did not have time to break down before it was transported and buried. This means the ancient mountains must have been high, steep, and subject to rapid erosion. It also suggests that the climate was rather arid, so there was not much rainfall to promote the breakdown of feldspar.

Based on evidence from elsewhere in central Australia, it has been shown that tectonic forces tilted the originally horizontal layers of sediment to their present vertical position between 300 and 400 million years ago. Because the rocks are tougher and more resistant to erosion than other sedimentary rocks, Uluru/Ayers Rock has remained standing as erosion has worn down the surrounding landscape.

Uluru/Ayers Rock is located in the Kata Tjuta National Park, which is overseen by the local aboriginal community, for whom it has many tribal and spiritual connections.

CHAPTER OUTLINE

- Rocks: A First Look p. 64

- Igneous Rocks p. 67

- Sedimentary Rocks p. 72

- Metamorphic Rocks p. 82

- Rocks as Resources p. 89

Rocks: A First Look

LEARNING OBJECTIVES

Explain the difference between a rock and a mineral.

Identify the three major families of rocks.

Describe the rock cycle.

Rocks, as we saw in Chapter 2, are different from minerals. The important distinction is that rocks are *aggregates*. This means that rocks are collections of mineral particles (and sometimes other types of particles such as organic debris or bits of volcanic glass) stuck together or intergrown to make a coherent mass. Rocks usually consist of several types of minerals, but sometimes they are made of just one common mineral, such as quartz or calcite. In any case, a rock will

◀ **Igneous rock**

This rounded granite boulder on Mt. Desert Island in Acadia National Park, Maine, was transported to its current location by a glacier. The boulder is sitting on a thick platform of granite that was once part of an enormous chamber full of molten rock—magma—underlying a volcano. The volcano is now gone and its top has eroded away, leaving the solidified remnants of the magma chamber exposed. The weathered surfaces of these granites make them appear buff-colored.

▲ Granite is an igneous rock. It forms from magma that cools and solidifies slowly, deep in the crust. The minerals in this sample are quartz (gray), potassium feldspar (light pink), plagioclase feldspar (white), and biotite (black). The same minerals are present in the sample of gneiss, shown on the facing page, but the overall appearance of the two rocks is quite different.

always contain many grains of the constituent mineral or minerals.

The kind of rock that forms depends on the environment in which it forms, and for that reason, rocks are the record keepers of Earth's long history. Rocks are the words that tell the story; minerals are the letters that form the words.

THE THREE ROCK FAMILIES

Rocks are grouped into three large families, according to the processes that formed them (see **FIGURE 3.1**).

Within each of these families, called *igneous*, *sedimentary*, and *metamorphic* rock, there is a range of possible *mineral assemblages*—the types and relative proportions of the minerals that constitute the rock—and a range of possible *textures*—the overall appearance of a rock because of the size, shape, and arrangement of mineral grains. Mineral assemblages and textures help identify rocks (for example, whether it is granite or basalt), textures help separate the rock families (for example, whether a sample is igneous or metamorphic), and the two together reveal much about the particular environment in which a given rock formed.

Visualizing

The three rock families FIGURE 3.1

Landscapes reflect the kind of rock that underlies them, just as mineral assemblages and textures serve to define rock types (see insets next to scenery photographs).

◢ Metamorphic rock ▶

The beautiful Lauterbrunnen Valley in Switzerland lies within a great mountain range—the Alps—where the rocks have been uplifted and chemically and physically altered by enormous tectonic forces. In this photo, the Lauterbrunnen Falls cascade down a steep cliff of metamorphic rock.

◀ Sedimentary rock

This remarkable landscape is part of Bryce Canyon National Park in Utah. The horizontal rock layers are mainly sandstones and limestones. The tall spires, called *hoodoos*, and the deeply incised crevices between them, are the result of erosion over many, many years. Bryce Canyon today is a desert, but the grains of sediment in these rocks were originally deposited in an environment that was alternately hot and dry (sandstone) or covered by a shallow sea (limestone).

◀ This photo shows a sample of gneiss (pronounced "nice"), a metamorphic rock similar to the rocks that form many of the cliffs in the Swiss Alps. This rock may have started as either a sedimentary or igneous rock, but it has been altered by heat and pressure so that a new set of minerals has developed, along with its banded appearance. The minerals present are potassium feldspar (light pink), plagioclase (white), quartz (gray), and biotite (black). The minerals are similar to those in the igneous rock granite, shown on the facing page, but the overall appearance is quite different.

◀ Sandstone is a sedimentary rock that consists largely of grains of quartz, with traces of other minerals. The grains became rounded as they were transported by running water to the place where they were deposited. After being deposited in layers, the grains eventually became cemented together and turned into sandstone.

Actual inset samples are 7 cm wide and 4 cm high.

THE ROCK CYCLE

Earth's surface is a meeting place. It is where the activities of Earth's internally driven processes—the movement of continents as a result of plate tectonics, earthquakes, elevation of mountains, eruptions of volcanoes—confront the quicker-paced activity of Earth's surface layers, the atmosphere, hydrosphere, and biosphere, which constantly break down and erode away the surface rocks. The surface layers are the most dynamic parts of the Earth system. The external forces of wind, water, ice, and life constantly modify the surface, breaking down and cutting away material here, depositing material there, and sculpting the landscapes that surround us. These external forces are the most obvious ones of the Earth system; they are the forces that confront us every day in our daily lives. We are less aware of Earth's internal forces because they act more slowly and to a large extent are neither seen nor felt. But the internal forces play an important role in the Earth system because they cause changes such as the elevation of mountain ranges and the slow shifting of continents, and those changes in turn cause climates to change and influence the places where plants and animals can live.

All of this activity, both fast and slow, is called the **rock cycle**, one of the great cycles that drive the Earth system, as shown in FIGURE 3.2. The Earth system has three great cycles: the *rock cycle*, the *hydrologic cycle*, and the *tectonic cycle*. The three cycles are interconnected, and they influence everything that happens on and in Earth. Earth would not be a habitable planet without the three great cycles. We will discuss the cycles and the ways they interact at many places in later chapters.

The rock cycle has no beginning or ending: instead, it is an endless process, powered by Earth's internal heat energy and by the incoming energy from the Sun. We will discuss specific details of the rock cycle at several places in later chapters. For example, in Chapter 4, we will discuss the processes involved in **weathering**, by which bedrock breaks into smaller rock and mineral fragments, and **erosion**, the group of related processes by which the products of weathering are moved around on Earth's surface. Weathering and erosion are critical steps in the rock cycle; they are powered by energy from the Sun. Earth's internal heat energy powers the forces that thrust up mountains and cause volcanoes to erupt, processes that expose new rocks to the forces of weathering and erosion.

> **rock cycle** The set of crustal processes that form new rock, modify it, transport it, and break it down.

> **weathering** The chemical and physical breakdown of rock exposed to air, moisture, and living organisms.

> **erosion** The wearing away of bedrock and transport of loosened particles by a fluid, such as water.

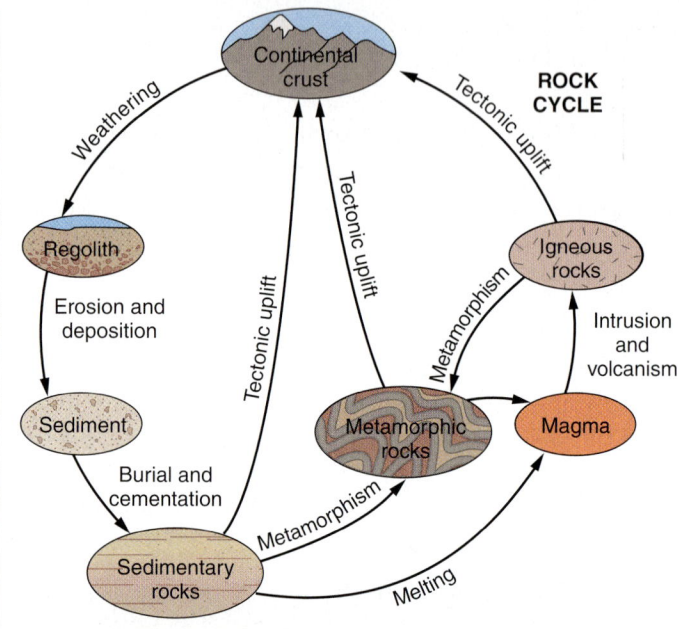

The rock cycle FIGURE 3.2

The cycle of rock change has been active since our planet became solid and internally stable. It continually forms and re-forms rocks of the three major families. Not even the most ancient igneous and metamorphic rocks that we have found are the original rocks of Earth's crust, for they were recycled eons ago.

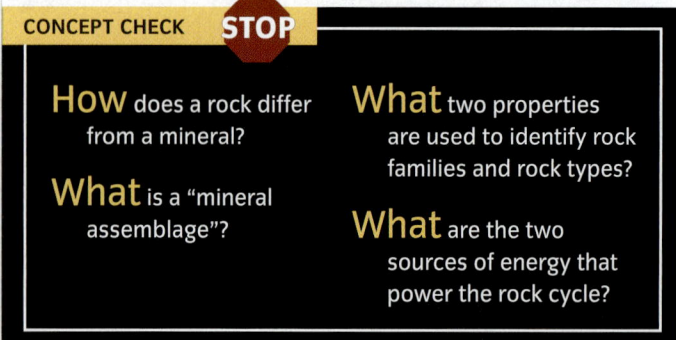

CONCEPT CHECK STOP

How does a rock differ from a mineral?

What is a "mineral assemblage"?

What two properties are used to identify rock families and rock types?

What are the two sources of energy that power the rock cycle?

Igneous Rocks

LEARNING OBJECTIVES

Distinguish between volcanic and plutonic igneous rocks.

Identify different textures of igneous rocks and explain the physical processes that produce them.

Identify the three main types of volcanic rock and their plutonic equivalents.

Igneous rocks (named from the Latin *ignis*, meaning *fire*) form by the cooling and *crystallization* of **magma**. When magma cools, mineral grains begin to crystallize, just as ice crystals form in water as it cools. The physical properties of the rock will differ, depending on whether the cooling process is slow or fast. The *rate of cooling* determines how large the individual mineral grains in the rock will grow—slow cooling yields large grains; fast cooling yields tiny grains. Grain size determines the appearance or *texture* of a rock. The *composition* of the magma determines the final mineral assemblage in the solidified rock. Let's examine each of these factors more closely because mineral assemblage and texture are the properties by which igneous rocks are classified and named.

igneous rocks Rocks that form by cooling and solidification of molten rock.

magma Molten rock.

RATE OF COOLING

Even with a cursory look at igneous rocks, you can see that some contain easily visible mineral grains and others do not. Rate of cooling is the main factor that distinguishes the two groups. **Volcanic rocks** (also called *extrusive igneous rocks*) form at Earth's surface, where lava contacts the much cooler air and water. They solidify so quickly that large mineral grains have no time to form. **Plutonic rocks** (also called *intrusive igneous rocks*) form when magma crystallizes deep underground. This process is much slower and therefore gives the mineral grains time to grow larger.

volcanic rock An igneous rock formed from lava.

plutonic rock An igneous rock formed underground from magma.

Rapid cooling: Volcanic rocks and their textures
Sometimes lava cools so rapidly that mineral grains do not have a chance to form at all. The resulting volcanic rock is not crystalline but *glassy* (**FIGURE 3.3A**).

Volcanic rock textures FIGURE 3.3

A *Glassy texture.* This photograph shows obsidian obelisks from a Mayan grave in Guatemala. Though these were sculpted by humans, a shiny, curvy appearance is typical for volcanic glass.

B *Aphanitic texture.* In this fine-grained rock, individual mineral grains cannot be discerned with the naked eye. Note also the vesicles (due to trapped volcanic gas). This sample of basalt is from Hawaii, and each of the largest vesicles is about the size of a small pea. Such a bubble-rich rock is called *vesicular* basalt.

C *Porphyritic texture.* This volcanic rock from Nevada contains large mineral grains, or *phenocrysts*, suspended in an aphanitic material called the *groundmass*. The largest grains visible in the photo are approximately 6 mm in length.

What an Earth Scientist Sees

Putting Rocks Under a Microscope

Earth scientists can identify the minerals in an aphanitic rock by studying it under a microscope.

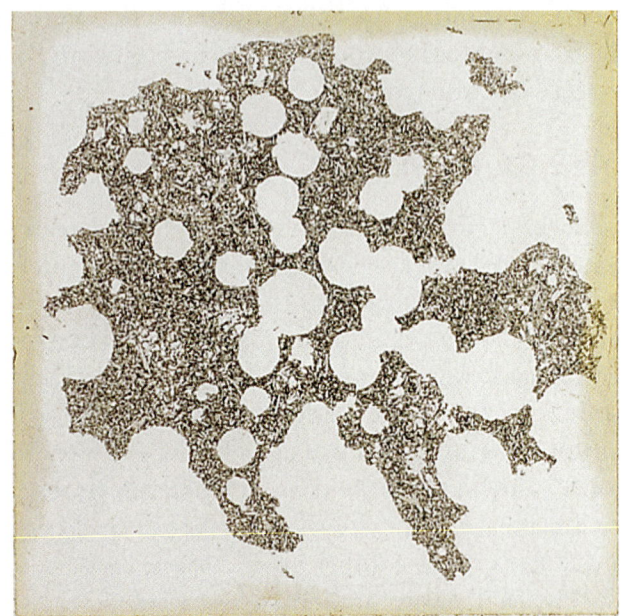

A The specimen of aphanitic volcanic rock from Figure 3.3 has been cut and polished into a wafer that is thin enough to allow light to pass through. Earth scientists call these rock slices *thin sections*. This magnified thin section is about 0.03 mm thick, as thin as a piece of tissue paper, and about 3.5 cm across.

B Using polarized light, Earth scientists can see individual mineral grains and identify them by optical and other properties. Using a microscope to study thin sections is a common technique in Earth science to analyze rocks of all types for many purposes.

 **Here's an interesting question:**
- There are two obviously different minerals visible in photo B. One is olivine, the other is feldspar. One appears yellowish, the other white. Which is which?

More often, mineral grains do form in solidifying lavas, but they are extremely small, and can be seen only under magnification (see *What an Earth Scientist Sees*). Rocks with this very fine-grained texture are said to be *aphanitic* (FIGURE 3.3B).

Most igneous rocks contain mineral grains of approximately the same size—that is, they are unimodal. But some volcanic rocks are bimodal in grain size and have a texture known as *porphyritic* (FIGURE 3.3C), which means they consist of large mineral grains embedded in an aphanitic matrix. This happens when magma starts to crystallize slowly at depth and grows large grains before it erupts, and the suspended large grains erupt along with the liquid lava. Once erupted, the remaining magma cools rapidly to form an aphanitic matrix in which are embedded the large mineral grains that formed earlier.

Dissolved gases, too, can affect the texture of volcanic rock. Erupting lava may froth and bubble as dissolved gases are released; if the froth is blasted into the air and cools quickly, it forms *pumice*, which is a glassy mass of bubbles. As lava cools, the viscosity increases and it becomes increasingly difficult for gas bubbles to escape. When the lava finally solidifies into rock, the last bubbles to form may become trapped; the bubble holes are called *vesicles*. In basaltic lava, this process can create a volcanic rock with lots of bubble holes that looks like Swiss cheese (see Figure 3.3B).

Slow cooling: Plutonic rocks and their textures
Unlike the minerals in volcanic rocks, those in plutonic rocks usually have time to form mineral grains that can be readily seen by the unaided eye. This coarse-grained texture is said to be *phaneritic* (**Figure 3.4**). Exceptionally large mineral grains (sometimes up to several meters!) typically form in the last stage of crystallization of a plutonic rock, when gases build up in the remaining magma. The vapor facilitates the growth of large crystals, because chemicals can migrate quickly to the growing faces. A coarse-grained plutonic rock with mineral grains larger than 2 cm in diameter is called a *pegmatite*.

The process that produces porphyritic texture in volcanic rocks also occurs in plutonic rocks. Magma starts to slowly crystallize deep underground, forming big crystals; then the remaining magma, with large crystals suspended within it, is intruded higher in the crust—though not extruded as a lava. In the new position, cooling is more rapid, so the remaining magma crystallizes to smaller crystals. The resulting texture is porphyritic even though the matrix is phaneritic.

CHEMICAL COMPOSITION

Scientists subdivide the most common igneous rocks into three broad categories based on their silica contents. Rocks that contain large amounts of silica (about 70% SiO_2 by weight) are usually light-colored. They are said to be *felsic* (a word formed from *feldspar* and *silica*), because feldspar and quartz are the most common minerals found in them. At the other end of the scale are rocks called *mafic* (a word formed from *magnesium* and *ferrous*, or iron-rich), which contain large amounts of dark-colored minerals, such as olivine and pyroxene, rich in magnesium and iron. They are usually lower in silica content (about 50% SiO_2 by weight). Mafic rocks that contain even less silica, and consist almost entirely of olivine and pyroxene, are said to be *ultramafic*. Igneous rocks with silica contents about 60% SiO_2 by weight, halfway between felsic and mafic, are said to be *intermediate* in composition. The origins of different magma types, and the way magmas crystallize into different kinds of igneous rocks through processes such as Bowen's Reaction Series, are discussed in Chapter 9.

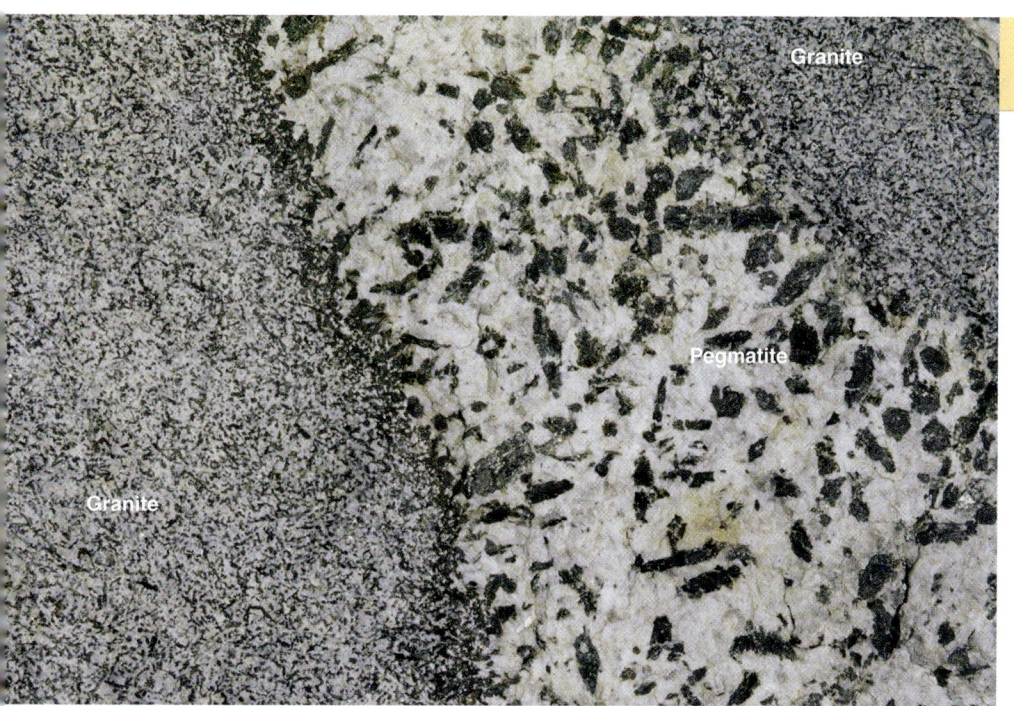

Plutonic rock and textures
Figure 3.4

Two distinct grain sizes of plutonic rock can be seen in this granite specimen from California's Sierra Nevada. Both textures are phaneritic; in the two outside layers there are small but visible grains of plagioclase, potassium feldspar, quartz (white), and biotite (black). Sandwiched between them is a vein of pegmatite, which contains the same minerals but in much larger grains.

Scientists organize igneous rock types on the basis of both grain size and silica content. The results are shown in **TABLE 3.1**. The rocks on the left are volcanic (aphanitic and fine-grained), and the ones on the right are plutonic (phaneritic and coarse-grained). The ones on the top contain the most silica and are the lightest in color (that is, they are felsic), while those on the bottom contain the least silica and are darkest in color (they are mafic).

Volcanic and plutonic equivalents TABLE 3.1

Grain Size →

Silica Content of Magma	Resulting Volcanic Rocks	Resulting Plutonic Rocks
High (= 70%–75%)	**Rhyolite** lies at the felsic, high-silica end of the scale, and consists largely of quartz and feldspars. It is usually pale, ranging from nearly white to shades of gray, yellow, red, or lavender.	 **Granite**, the plutonic equivalent of rhyolite, is common because felsic magmas usually crystallize before they reach the surface. It is found most often in the continental crust, especially the cores of mountain ranges. 
Intermediate (= 60%)	**Andesite** is an intermediate-silica rock, with lots of feldspar mixed with darker mafic minerals such as amphibole or pyroxene. It is usually light to dark gray, purple, or green.	 **Diorite** is the plutonic equivalent of andesite, an intermediate-silica rock. 
Low (= 45%–50%)	**Basalt**, a mafic rock, is dominant in oceanic crust, and the most common igneous rock on Earth. Large, low-viscosity lava flows from shield volcanos and fissures are usually basaltic. Dark-colored pyroxene and olivine give it a dark gray, dark green, or black color.	 **Gabbro** is the plutonic equivalent of basalt, a low-silica rock. 

(↑ *Silica Content*)

Visualizing
Tephra FIGURE 3.5

A Large volcanic *bombs*, fist-sized and larger, erupt from Mt. Etna in Sicily.

Volcanic bombs

C Volcanic ash, the smallest tephra, blankets a farm in Oregon after the eruption of Mt. St. Helens in 1980. Though we call it "ash," it is really quite different from what you find in your fireplace, because it consists of microscopic pieces of volcanic glass.

B Intermediate-sized tephra, called *lapilli*, cover the Kau Desert in Hawaii.

Lava flows versus pyroclastic eruptions The texture of an extrusive igneous rock is closely related to the silica content of the parent magma. The silica content controls the *viscosity* of the magma. If the magma has low silica content (low viscosity), the gas bubbles escape fairly easily. Basaltic magma, for example, flows readily and basaltic lavas are often vesicular. Rhyolitic magma has a high silica content, flows less freely (high viscosity), and it is difficult for gas bubbles to form and escape. When gas does escape from high-viscosity magma, it often does so explosively, shattering the magma into a mass of tiny, hot fragments.

A fragment of shattered magma ejected during a volcanic eruption is called a *pyroclast* (from Greek words meaning *fire* and *broken*). Collectively, all the ejecta from a volcanic eruption are known as *tephra,* and they range from car-sized rocks to ultrafine *volcanic ash* whose individual fragments can be seen only under a microscope (**FIGURE 3.5**).

Loose pyroclasts are often welded together during an eruption, or cemented together afterward, forming *pyroclastic rocks*. These rocks are called *agglomerates* when the tephra particles are large and *tuff* when the particles are small.

CONCEPT CHECK STOP

What is the relationship between grain size and the rate of crystallization of an igneous rock?

How do igneous rocks with a bimodal texture form? What are such rocks called?

How does silica content influence the viscosity of a magma?

How do igneous rocks with high silica content differ in appearance from those with low silica content?

What are pyroclasts and how do they form?

Sedimentary Rocks

> **LEARNING OBJECTIVES**
>
> **Identify** three types of sediment.
>
> **Explain** the three processes that lead to lithification of sediments.
>
> **Describe** clastic sediment in terms of size, sorting, and roundness.
>
> **Explain** where and how chemical and biogenic sediments form.
>
> **Explain** how features like ripple marks, cracks, and fossils can indicate the kind of environment in which a sedimentary rock originated.

The next major rock family forms when regolith, which consists of mineral and rock particles, is transported by water, wind, or ice and then deposited. This fragmented, transported, and deposited material is called *sediment*. We will discuss sediment-forming processes in more detail in Chapter 4. As sediment accumulates layer by layer, increasing temperature and pressure transform the sediment into **sedimentary rock**.

Nearly every geologic process leaves its mark in the sedimentary record. Tectonic forces form mountains and basins, which in turn control where the source material for sediment comes from (erosion of mountains) as well as the sites of **deposition** (basins). To understand the complicated record these events leave, first we have to understand the different kinds of sediment. Earth scientists separate sediments into three broad categories: *clastic*, *chemical*, and *biogenic*.

CLASTIC SEDIMENTS AND CLASTIC SEDIMENTARY ROCKS

Clastic sediment derives its name from *clasts*, individual grains of mineral or fragments of rock. Clasts range in size from large boulders down to clay particles finer than flour. In fact, the size of clasts is the primary basis for classifying clastic sediment and the sedimentary rocks they eventually form (see **FIGURE 3.6**). There are four basic classes of clastic sedimentary rocks: **Conglomerates** have many clasts larger than 2 mm in diameter; medium-sized grains in **sandstone** range from 0.05 to 2 mm in size; *mudstone* and *siltstone* consist of silt- or clay-sized particles; and **shale** is a special kind of mudstone that is *fissile*, which means that it splits into sheet-like fragments.

LITHIFICATION OF CLASTIC SEDIMENT

Lithification is a step in the rock cycle. Sediment originates from the erosion of pre-existing rocks that have been broken down through weathering (Chapter 4). When the sediment is deposited and buried, it lithifies to become new rock. Each kind of sediment becomes lithified, but because clastic sediment is the most abundant kind, it is convenient to discuss lithification here.

> **sedimentary rocks** Rocks that form from sediment under conditions of low pressure and low temperature near the surface.
>
> **deposition** The laying down of sediment.
>
> **clastic sediment** Sediment formed from fragmented rock and mineral debris produced by weathering and erosion.
>
> **conglomerate** A clastic sedimentary rock with large fragments in a finer-grained matrix.
>
> **sandstone** A medium-grained clastic sedimentary rock in which the clasts are typically, but not necessarily, dominated by quartz grains.
>
> **shale** A very fine-grained fissile or laminated sedimentary rock, consisting primarily of clay-sized particles.
>
> **lithification** The group of processes by which sediment is transformed into sedimentary rock.

Visualizing

From clasts to rocks FIGURE 3.6

SEDIMENT

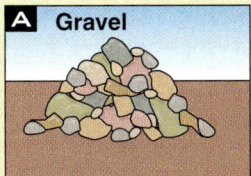 **A** Gravel

 **B** Sand

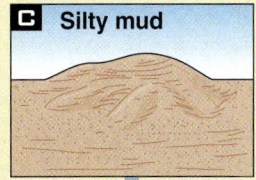 **C** Silty mud

 **D** Clayey mud

...WITH COMPRESSION AND TIME, CAN BECOME...

A sediment with pea-sized or larger particles is called *gravel*. When gravel is cemented, the rock so formed is a **conglomerate**.

Sand consists of somewhat smaller particles, about the size of a pinhead. When compacted and cemented, sand becomes **sandstone**.

Sediment with even finer particles, the size of a grain of table salt, is called *silt*. The corresponding rock type is **siltstone**.

The finest sedimentary particles, the size of flour or smaller, are called *clay*.* The corresponding rock type is **shale** or **mudstone**.

ROCK

This shows pebble-rich strata of conglomerates interbedded with sandstone; the area photographed is about 1 m wide.

This sandstone is about 7.5 cm across. The colors of depositional layers vary because of different iron contents.

This siltstone sample is about 8 cm across. Note the impressions of fossil marine shells.

This shale sample is about 10 cm across. The colors are caused by differing contents of organic matter.

*Note that "clay" in this context refers only to particle size, not composition.

Volcaniclastic sediments are another kind of clastic sediment. What makes them different is that all of the clasts are volcanic in origin. As discussed in Chapter 9, explosive volcanic eruptions blast out large quantities of fragments. An old saying explains the uniqueness of volcaniclastic sediments: "They are igneous on the way up but sedimentary on the way down." All of the clasts in volcaniclastic sediment are pyroclasts.

In order for newly deposited, loose sediment to be lithified and turn into rock, the individual particles must somehow be bound together into a cohesive unit. After a layer of sediment is buried, either by accumulation of more sediment or by tectonic processes, it is placed under higher pressure, leading to *compaction*—the reduction of pore space. Compaction is generally accompanied by *cementation*, which is the deposition of substances dissolved in the pore water. Deposition and cementation can happen in various ways, such as the evaporation of groundwater under desert conditions. As the water evaporates, chemicals such as silica, calcium carbonate, and iron hydroxide precipitate and cement the grains of sediment together. Another method of lithification is *recrystallization*, in which new crystalline grains form from old ones. Recrystallization is especially common in sediment consisting of grains of calcium carbonate. Carbonate grains in coral reefs, for example, change from their original form (aragonite) to a more stable mineral, calcite. In the process of recrystallizing, mineral grains that were separate can grow together.

Sedimentary Rocks

From **Figure 3.7**, we can see that pressures caused by sediment accumulation or tectonic force initiate the lithification process. Although the pressures involved in lithification may be high by everyday human standards, they are low pressures by geologic standards, and low in comparison with the pressures that induce metamorphism. All of the low-temperature, low-pressure changes that happen to sediment after deposition are collectively called *diagenesis*. They include lithification, as well as processes involving chemical reactions or microbial activity that are not part of lithification.

Process Diagram

Lithification FIGURE 3.7

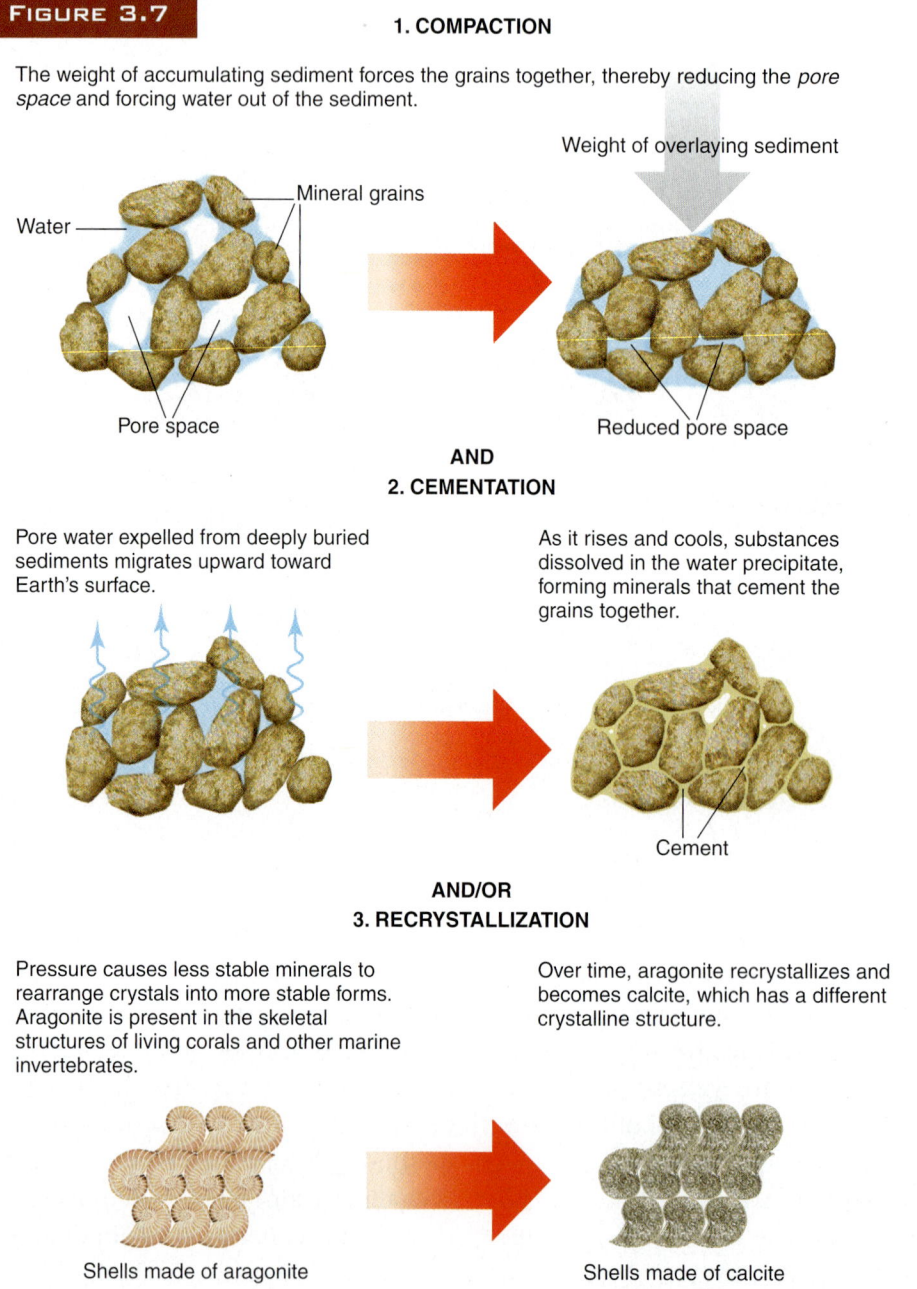

1. COMPACTION

The weight of accumulating sediment forces the grains together, thereby reducing the *pore space* and forcing water out of the sediment.

AND

2. CEMENTATION

Pore water expelled from deeply buried sediments migrates upward toward Earth's surface.

As it rises and cools, substances dissolved in the water precipitate, forming minerals that cement the grains together.

AND/OR

3. RECRYSTALLIZATION

Pressure causes less stable minerals to rearrange crystals into more stable forms. Aragonite is present in the skeletal structures of living corals and other marine invertebrates.

Over time, aragonite recrystallizes and becomes calcite, which has a different crystalline structure.

VIEW THIS IN ACTION in your WileyPLUS course

CRITICAL THINKING — Here's an interesting question:
- Consider the chemical composition of Earth's crust. What do you predict is the most common mineral cement in clastic sedimentary rocks? (You might find it helpful to refer to Figure 2.13.)

 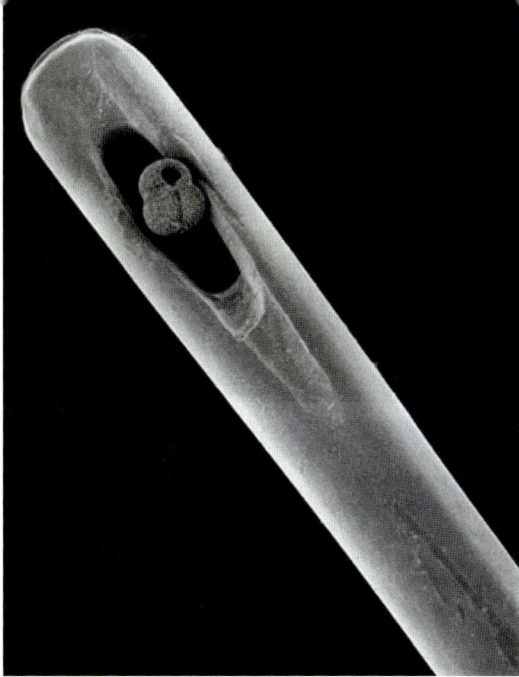

A Utah's Bonneville Salt Flats is one of the most desolate landscapes on Earth. These chemical sediments, which contain magnesium and potassium chloride in addition to ordinary salt, have been formed over the last 14,000 years by the evaporation of the prehistoric Lake Bonneville. The Great Salt Lake, a remnant of Lake Bonneville, covers less than one-tenth of the former lake's area.

B Inside the eye of this needle is the shell of a foraminifer, a one-celled plankton that is abundant in the ocean. The shells of foraminifera cover roughly one-third to one-half of the ocean floor. Over time these biogenic sediment deposits are buried and converted into limestone or chalk.

Chemical and biogenic sediments FIGURE 3.8

CHEMICAL SEDIMENTS AND CHEMICAL SEDIMENTARY ROCKS

All surface water and groundwater contain dissolved chemicals that eventually find their way to a lake or the sea. No natural water on or in Earth is completely free from dissolved matter. When dissolved matter precipitates from sea or lake waters, **chemical sediment** is the result.

> **chemical sediment** Sediment formed by the precipitation of minerals dissolved in lake, river, or seawater.

This precipitation can happen in two ways. First, plants and animals living in the water can alter its chemical balance. For example, increasing the amount of carbon dioxide dissolved in the water will cause calcium carbonate to precipitate. Many limestones are formed this way.

Second, if an inland sea is subjected to an increasingly warm and dry climate, or if the inflow of fresh water is restricted for some reason, evaporation may exceed the input of fresh water. The sea may then become so shallow and saline that salts that were dissolved in the water will begin to precipitate as solids. Modern examples of this process are found in the Aral Sea (Uzbekistan), Mono Lake, California, and Utah's Great Salt Lake (see FIGURE 3.8A).

Chemical sedimentary rocks often form as **evaporites**. Calcite, gypsum, and halite typically form from the evaporation of seawater, while the evaporation of lake water may yield more exotic minerals, such as sodium carbonate and borax. Many evaporite minerals are mined because they have industrial uses, such as gypsum for plasterboard. In addition, most of the salt we eat comes from evaporites.

> **evaporite** A rock formed by evaporation of lake water or seawater, followed by lithification of the resulting salt deposit.

Sedimentary Rocks 75

An unusual but important kind of chemical sedimentary rock is a *banded iron formation*. Such rocks are the source of most of the iron mined today. Not only are they valuable for their ore, they also tell the story of a critical period in Earth's history. Almost all of the banded iron formations are about the same age—1.8 to 2.5 billion years old. This strongly suggests that unique conditions existed on Earth during that period (see *What an Earth Scientist Sees: A Change in the Atmosphere*).

What an Earth Scientist Sees

A Change in the Atmosphere

Banded iron formations, such as this 2.5-billion-year-old stratum in the Hamersley Range in Australia, formed when iron that was dissolved in seawater precipitated as chemical sediment. Today, seawater contains only slight traces of iron, because the oxygen in the atmosphere reacts with it to form insoluble iron compounds. If the ocean was once rich in dissolved iron, there must have been very little oxygen in the atmosphere at that time.

How did Earth make the transformation from an oxygen-poor atmosphere 2.5 billion years ago to the oxygen-rich atmosphere of today? In some 2.5-billion-year-old rocks there are microscopic fossils of cyanobacteria. These bacteria are thought to be the first organisms on Earth to extract energy from sunlight by photosynthesis, a chemical process that releases oxygen. Scientists hypothesize that cyanobacteria might have oxygenated Earth's atmosphere very rapidly—and possibly more than once, as they proliferated and then poisoned themselves by producing too much oxygen. Each time the oxygen concentration changed, an iron mineral would precipitate out of the seawater, creating a new band of iron-rich sediments. Eventually, about 2.0 to 1.8 billion years ago, the oxygen level of the atmosphere reached a point where the ocean could no longer retain much iron, and banded iron formations could no longer form.

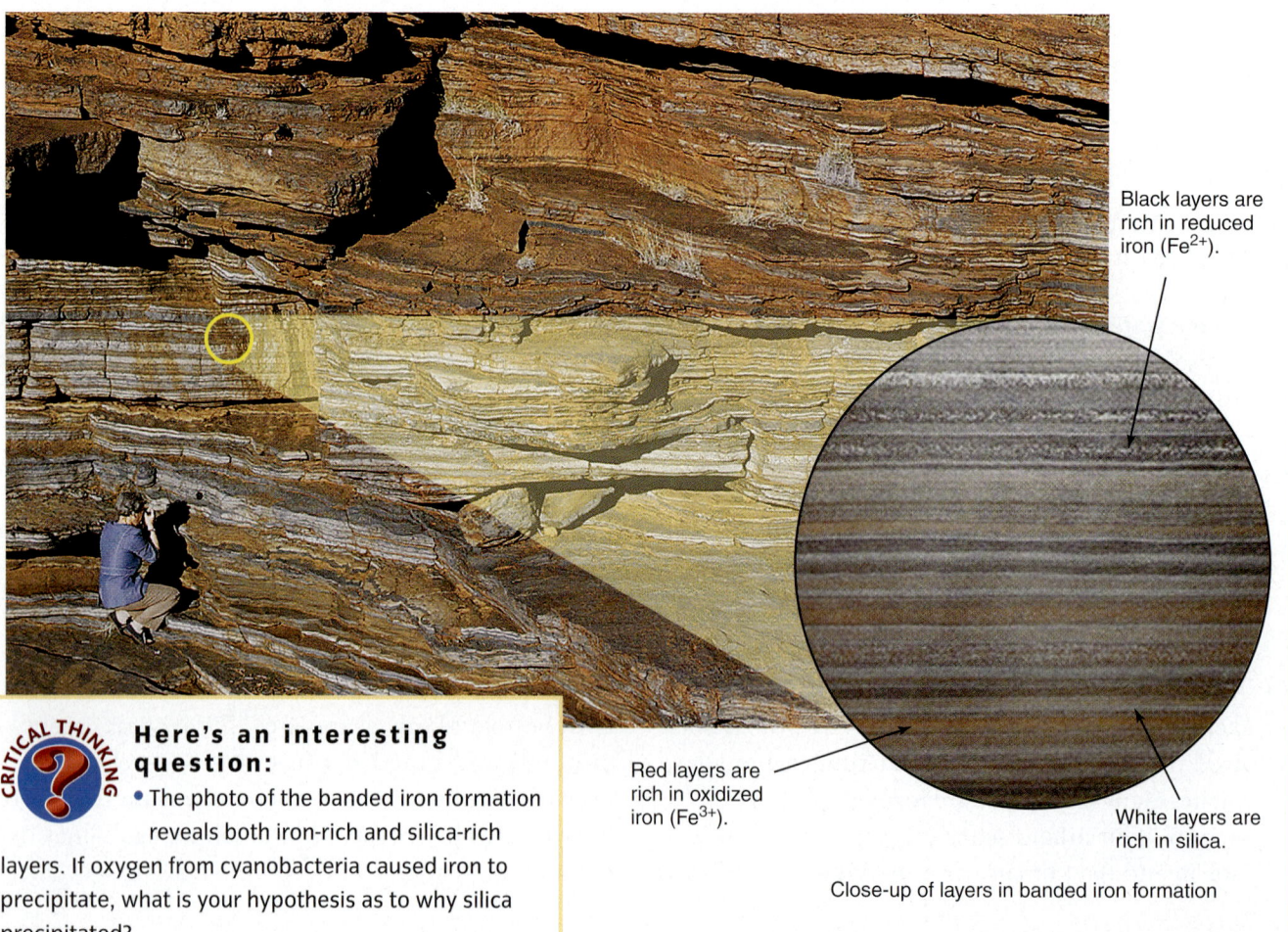

Black layers are rich in reduced iron (Fe^{2+}).

Red layers are rich in oxidized iron (Fe^{3+}).

White layers are rich in silica.

Close-up of layers in banded iron formation

CRITICAL THINKING

Here's an interesting question:
- The photo of the banded iron formation reveals both iron-rich and silica-rich layers. If oxygen from cyanobacteria caused iron to precipitate, what is your hypothesis as to why silica precipitated?

BIOGENIC SEDIMENTS AND BIOGENIC SEDIMENTARY ROCK

Biogenic sediment is composed of the remains of plants and animals. This includes the hard parts of large animals, such as shells, bones, and teeth, as well as fragments of plant matter, such as wood, roots, and leaves (see **FIGURE 3.8B**). Two of the most common rock types that come from biogenic sediments are limestone, formed of the calcium carbonate skeletons of marine invertebrates, and coal, formed from partially decomposed terrestrial plant material. (Note that limestone can be either biogenic or chemical in origin.)

Limestone is the most abundant biogenic sedimentary rock. It is formed from the lithified shells and other skeletal material from marine organisms. Some of these organisms build their shells or skeletons from calcite, but most construct them from aragonite, which, like calcite, is calcium carbonate. During diagenesis the aragonite is transformed to calcite, which becomes the main mineral of limestone.

Calcite is sometimes replaced by the mineral dolomite (a carbonate mineral containing both magnesium and calcium); the resulting rock is called *dolostone*. Another kind of biogenic rock, which consists of extremely tiny particles of quartz, is called *chert*. The quartz in chert does not come from sand, but from the shells of microscopic sea animals.

An important class of biogenic sediment consists of the accumulated remains of terrestrial plants. Over time, and with pressure, this mass gradually becomes **peat**. Eventually, given enough time and pressure, peat may lithify and become **coal**. Lithification (also called *coalification*) involves further compaction, release of water, and slow chemical changes that weld the plant fragments together, thereby making the coal relatively lower in water and richer in carbon than the original peat.

> **biogenic sediment** Sediment that is composed primarily of plant and animal remains, or precipitates as a result of biologic processes.
>
> **limestone** A sedimentary rock that consists primarily of the mineral calcite.
>
> **peat** A biogenic sediment formed from the accumulation and compaction of plant remains.
>
> **coal** A combustible rock formed from the lithification of plant-rich sediment.

ROCK BEDS

When you look at an outcrop of sedimentary rock, such as the one shown in **FIGURE 3.9**, one of the first things you will notice is the **bedding**. The banded appearance comes from the fact that sedimentary particles are laid down in distinct *beds*, or *strata*. Over time, the mineral composition of the sediments in a particular location may change, or they may be transported or deposited in different ways. This will cause the adjacent strata to look different. The boundary between adjacent strata is called a *bedding surface*. It is the presence of bedding and bedding surfaces that indicates that a rock was once sediment.

> **bedding** The layered arrangement of strata in a body of sediment or sedimentary rock.

INTERPRETING ENVIRONMENTAL CLUES

Just as history books record the changing patterns of civilization, sedimentary rocks record the environmental history of our planet. Layers of sedimentary rock, like pages in a book, record how environmental conditions have changed throughout Earth history. Earth scientists can "read" this story by interpreting the evidence in the rocks.

Layers of rock: Bedding FIGURE 3.9

The Bungle Bungle Range in northwestern Australia derives its unique coloration from layers of sandstone that have different permeabilities. Algae grow in the more permeable strata, tinting the rock black.

Patterns formed by currents of water or air moving across sediment can be preserved and later exposed on bedding surfaces. For example, bodies of sand that are being moved by wind, streams, or coastal waves are often rippled; these may be preserved in sandstone as *ripple marks* (see **FIGURE 3.10A** and **B**). Similarly, *mud cracks* (**FIGURE 3.10C** and **D**), fossil tracks (see **FIGURE 3.11**), and even raindrop impacts can be recorded on bedding surfaces, attesting to moist surface conditions at the time they were formed.

Ancient and modern features compared FIGURE 3.10

▲ **A** Ripples are forming in shallow water near the shore of Ocracoke Island, North Carolina.

▲ **B** Almost identical ripples are exposed on a bedding surface of sandstone at Artists Point, Colorado National Park, Colorado.

▲ **C** Mud cracks formed on this modern riverbed as the river dried up.

▲ **D** Similarly shaped mud cracks are preserved on the surface of shale exposed at Ausable Chasm, New York. We can infer that this rock formation was deposited in an intermittently wet environment, such as a seasonal lakebed or a tidal flat. Note that in both (**B**) and (**D**) the present-day environment bears no relation to the environment in which the sedimentary rocks were deposited.

▲ **A** Seagull footprints are imprinted in sandbars along the Alsek River in Glacier Bay National Park, Alaska.

B Fossilized footprints are preserved in sandstone in the Painted Desert, near Cameron, Arizona. The animal that left these prints may have been hunting for prey stranded by the falling tide, just like the present-day seagulls in Alaska. ▶

Footprints in the sand FIGURE 3.11

Fossils also provide significant clues about former environments. Some animals inhabit warm, moist climates, whereas others can live only in cold, dry climates. Using the climatic ranges of modern plants and animals as guides, we can infer the general character of the climate in which similar ancestral forms lived (see **FIGURE 3.12**). Even microscopic fossils are important: The shells of foraminifera can tell us about former temperatures and salinity conditions in the oceans. Fossils are also the basis for determining the relative ages of strata. In fact, fossils have played an essential role in efforts to reconstruct the past 635 million years of Earth's history.

The color of fresh, unaltered sedimentary rock is determined by the colors of the minerals, rock fragments, and organic matter of which it is composed. Iron sulfides and organic detritus buried with sediment are responsible for most of the dark colors in sedimentary rocks. The presence of these materials implies that the sediment was deposited in an oxygen-poor (reducing) environment. Reddish and brownish colors result mainly from the presence of iron oxides, occurring either as coatings on mineral grains or as very fine particles. These minerals point to oxygen-rich (oxidizing) conditions in the environment.

Plants and climate FIGURE 3.12

This fossilized seed fern, found in Greenland, dates from the Triassic Period more than 200 million years ago. Fossils of tropical, moisture-loving plants, such as cycads and palms, can be found in Triassic-aged sedimentary rocks all over the world. The particular fern species in this photograph became extinct after the Triassic, so the fossil can be used to date the sediment in which it was found. And despite the fact that Greenland's modern climate is cooler, the presence of this fossil shows this was not always the case.

Sedimentary Rocks 79

Each depositional environment leaves its own kind of sedimentary record, which may change over time. This section shows a variety of depositional environments in which distinctive facies are deposited. On the land surface, they lie side by side. In a vertical section, they lie one above another. Notice the seaward dip of the boundaries between adjacent facies. This indicates that, over time, the boundaries have migrated in the landward direction, due to a rise in sea level.

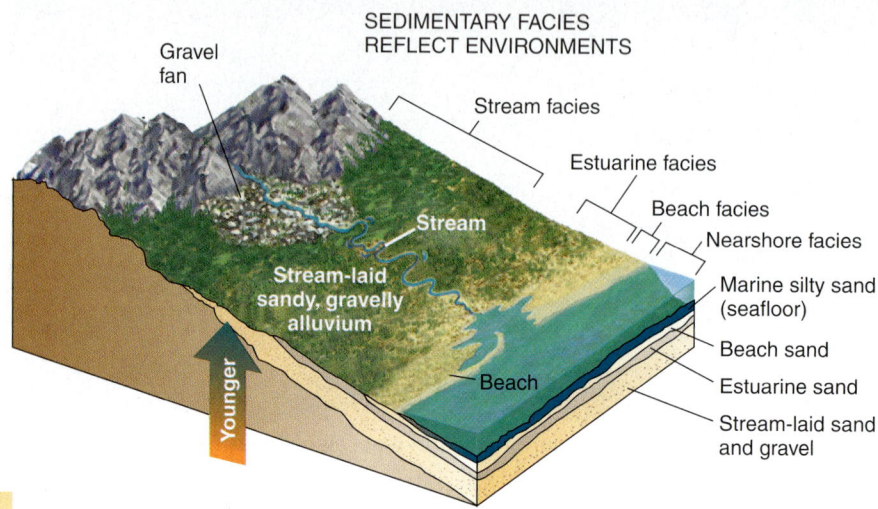

Sedimentary facies FIGURE 3.13

SEDIMENTARY FACIES

If you examine a vertical sequence of exposed sedimentary rocks, you may notice differences as you move upward from one bed to the next. You may note changes in the sedimentary rock type, color, fossil content, and thickness. The differences indicate that the environmental conditions in that location changed over time. If you trace a single bed laterally for a few kilometers, you may also notice changes, indicating that at any given time during deposition of the sediment, conditions differed from one place to another.

The changes in the character of sediment from one environment to another are referred to as changes of *sedimentary facies* (pronounced *fay-sheez*). One facies may be distinguished from another by differences in grain size, grain shape, stratification, color, chemical composition, depositional structure, or fossils. Adjacent facies can merge into each other either gradually or abruptly (see FIGURE 3.13). For example, coarse sand and gravel on a beach may gradually pass into finer sand, silt, and clay on the floor of the sea or a lake. Coarse, boulder-like glacial sediment, on the other hand, may end abruptly at the margin of a glacier.

By studying relationships among different sedimentary facies and using these characteristics to identify original depositional settings, we can reconstruct a picture of the environmental conditions that prevailed during past geologic times (see the *Case Study*).

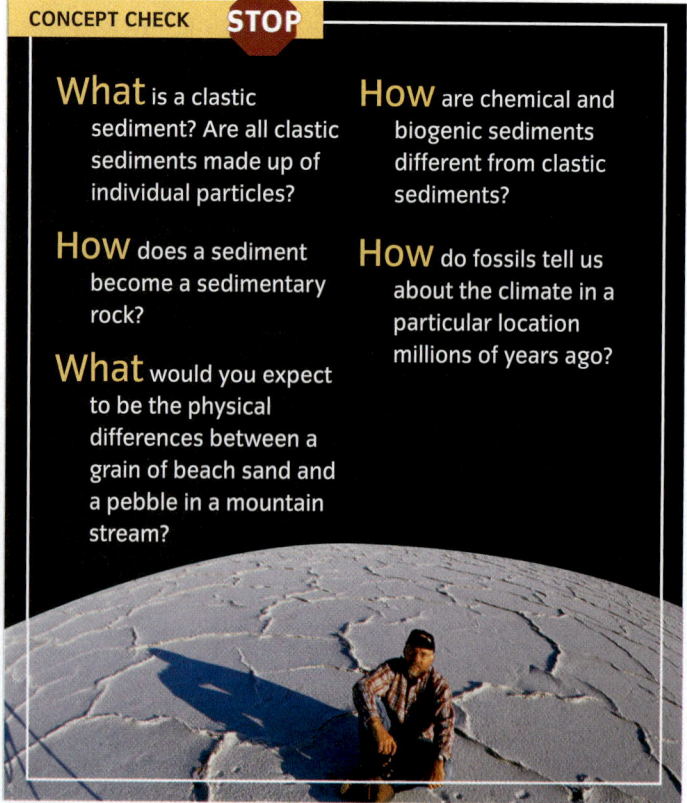

CONCEPT CHECK STOP

What is a clastic sediment? Are all clastic sediments made up of individual particles?

How does a sediment become a sedimentary rock?

What would you expect to be the physical differences between a grain of beach sand and a pebble in a mountain stream?

How are chemical and biogenic sediments different from clastic sediments?

How do fossils tell us about the climate in a particular location millions of years ago?

CASE STUDY
Sedimentary Facies and the History of Humankind

Once upon a time, 3.3 million years ago, a little girl wandered near a stream in Africa. Around her were open grasslands and shaded woodlands, home to elephants, hippos, rhinoceroses, and antelopes. Though the tiny three-year-old walked upright on two feet, she had strong muscles and gorilla-like shoulders and could climb a nearby tree to escape from predators, take shelter, or forage for fruit.

Today we know this part of the world as Ethiopia. The rocky hills, south of the Awash River near a place called Dikika, are now dry, hot, and seemingly barren (A). But the tiny set of bones that emerged from these rocks is the oldest and most complete fossil of a human-like (*hominid*) child ever found (B). The paleontologists who found the girl's remains named her Selam—"peace" in the Amharic language. The strata of the Hadar Formation that held her bones reveal much about how Selam lived—and died.

The facies reveal that Selam and others of her species, *Australopithecus afarensis* (including the famous fossil "Lucy" found not far away in 1974), lived in an environment where deltas formed rapidly in the ephemeral lakes. Flooding was a common occurrence in this constantly shifting terrain. Perhaps the little girl died by falling into a swiftly flowing stream—her bones were found encased in river channel deposits of gravel and sand, with clear evidence of rapid sedimentation. Her small body would have been buried quickly by sediment-laden water, protected from predators and the elements for millennia.

A Here, members of the team sift loose sediment, looking for small fossils. Sedimentary strata of the Ethiopian badlands are in the background.

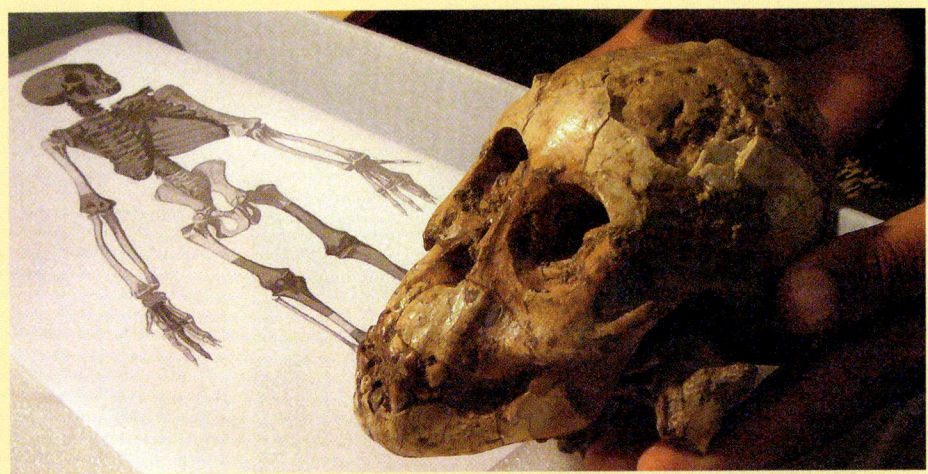

B Dr. Zeresenay Alemseged, the paleontologist who heads the Dikika research team, holds the baby's fossilized skull, which fits easily into the palm of his hand.

NATIONAL GEOGRAPHIC

Metamorphic Rocks

LEARNING OBJECTIVES

Describe the temperature and pressure conditions required for metamorphism.

Describe how pore fluids and stress affect the results of metamorphism.

Define the grade of metamorphism.

Define foliation.

Describe the common metamorphic products of shale.

The third major rock family is **metamorphic rock**, rock whose original sedimentary or igneous form and mineral assemblage have been changed as a result of exposure to high temperature, high pressure, or both. The term *metamorphic* comes from the Greek *meta*, meaning *change*, and *morphe*, meaning *form*—hence, change of form. Metamorphism occurs in a variety of tectonic and geologic environments. It can occur as an end result of the processes of burial and compaction through which sediments become transformed into rocks. It can also occur when rocks are subjected to the extreme pressures and temperatures associated with mountain building. Temperature-induced metamorphism can occur when rocks are heated—essentially baked—by nearby magmas. How and why metamorphism occurs, the kinds of metamorphic rocks, and their characteristic mineral assemblages are important areas of research.

Perhaps the best everyday analogy for **metamorphism** is cooking—a process that takes, for instance, flour, sugar, salt, yeast, and water and transforms them into a completely new substance called bread. As in cooking, metamorphism of a mass of rock is accompanied by chemical reactions that change both the mineral assemblage and the texture. But one important kind of change does *not* take place during metamorphism: Rocks do not melt. That is one of the things that make metamorphic rocks so interesting. They may have been squeezed, stretched, heated, and altered in complex ways (like the gneiss in **FIGURE 3.14**), but metamorphic rocks remain solid at all times and thus preserve a record of all the heatings, squeezings, and stretchings they have suffered since they were formed. When we encounter an area with many metamorphic rocks, we can be sure it has been tectonically busy.

Metamorphism and cooking FIGURE 3.14

High temperatures and pressures deep in the crust long ago converted the ingredients in this mix of minerals (dark-colored clay and light-colored quartz) into a new metamorphic rock called *gneiss*. This rock outcropping was photographed near Isua, Greenland.

> **metamorphic rocks** Rocks that have been altered by exposure to high temperature, high pressure, or both.

> **metamorphism** The mineralogical, textural, chemical, and structural changes that occur in rocks as a result of exposure to elevated temperatures and pressures.

Quartz vein FIGURE 3.15

This sample of gneiss contains many quartz *veins* (white) that criss-cross the rock. The quartz precipitated out of pore fluid that was expelled from the rock during metamorphism.

THE LIMITS OF METAMORPHISM

The three most important factors in metamorphism are heat, pressure, and the amount of *pore fluid*—the watery fluid present in *pores*, the tiny open spaces present in all rocks. The heat that causes metamorphic reactions is Earth's internal heat. We know from drilling deep gas and oil wells, and from deep gold mines, that temperature in the continental crust increases with depth at a rate of about 30°C/km. At a depth of about 5 km, the temperature is about 150°C. This is the temperature at which some rocks start to show distinct signs of metamorphic change. Below 150°C, sediments are lithified to sedimentary rocks; above 150°C, metamorphism commences.

The upper temperature limit of metamorphism is about 800°C. Above this temperature rocks begin to melt, and metamorphic processes give way to magmatic processes. Pressure also influences metamorphic changes. The air pressure at sea level is defined as one *atmosphere*—about half the pressure of the air inside an automobile tire. Due to the weight of overlying rock, pressure in the crust increases with depth. The rate of increase is about 300 atmospheres (atm) per kilometer, or about 30 megapascals (MPa) per kilometer. Therefore, the pressure 5 km below the surface is 1500 times greater than the surface atmospheric pressure—that is, 1500 atm, or 150 MPa. This is the depth at which both temperature and pressure are high enough for recrystallization and growth of new minerals to start.

The third important factor in metamorphism is the presence of *pore* fluid. The watery or aqueous fluid is never just pure water—it has small amounts of gases and salts dissolved in it, along with traces of the mineral constituents present in the enclosing rock. Pore fluids enhance metamorphism in two ways. They permit material to dissolve in one place, move quickly via the fluid, and be precipitated in another place. In this way pore fluids speed up recrystallization. Pore fluids also enhance metamorphism by acting as a reservoir during growth of new minerals. Pore fluids speed up chemical reactions in the same way that water in a stew pot speeds up the cooking of a tough piece of meat.

As pressure and temperature increase and metamorphism proceeds, the amount of pore space decreases and pore fluids are driven out of the rock. The escaping pore fluids carry with them small amounts of dissolved mineral matter. As the fluid flows through fractures in the rock, some of the dissolved mineral matter may precipitate, creating *veins* (see FIGURE 3.15). Veins are quite common in metamorphic rocks.

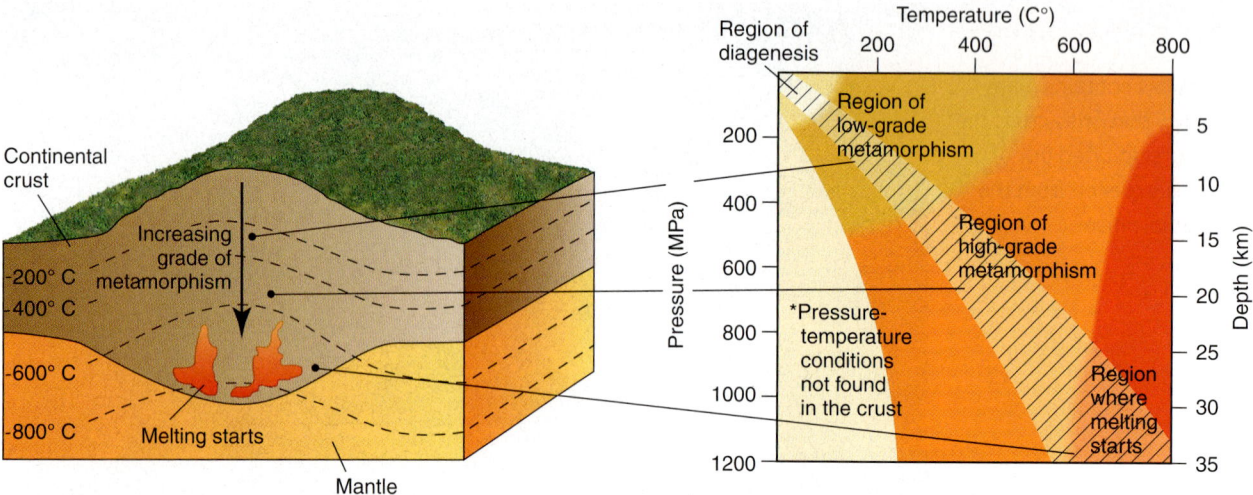

A This sketch shows schematically where in the crust metamorphism and melting occur.

B Four colored bands illustrate the temperature and pressure conditions for diagenesis, low-grade and high-grade metamorphism, and melting. Note that the vertical scale is given in units of both pressure and depth beneath the surface. The shaded diagonal band indicates the pressures and temperatures most commonly found in continental crust.

Temperature and pressure conditions for metamorphism FIGURE 3.16

FIGURE 3.16 shows the range of temperatures and pressures over which metamorphism occurs. In FIGURE 3.16B, pressure is represented on the vertical axis, increasing with depth. Temperature is represented on the horizontal axis, increasing from left to right. The pressure and temperature conditions under which sediments are formed and changed into sedimentary rocks during lithification and diagenesis occupy the upper left-hand corner of the diagram. Below this, and up to a depth of about 15 km, where the pressure reaches 400 MPa and the temperature is about 400°C, metamorphism occurs but is not very intense. In these *low-grade* metamorphic rocks, the minerals have changed, but the appearance is still similar to that of sedimentary rocks. At higher temperatures and pressures, extending to the onset of melting, *high-grade* metamorphic rocks are formed. In the lower right-hand corner of the diagram, representing the hottest temperatures and the highest pressures, rocks may begin to melt and would no longer be considered metamorphic. The changes in mineral assemblage and the rock types produced as shale moves successively from diagenesis to high-grade metamorphism are shown in FIGURE 3.17.

THE IMPORTANCE OF STRESS

In discussing the way rock deforms, we use the word *stress* rather than *pressure*. These two words are related in meaning but different in connotation. Both are defined as the force acting on a surface per unit area. The term *pressure*, as used in Earth science, implies that forces on a rock are essentially uniform in all directions, like the forces on a body of rock immersed in a liquid such as water or magma. Rocks, however, are solids: Unlike liquids or gases, solids can resist different pressures in different directions at the same time. For this reason, *stress* is a more versatile term for discussing rock deformation because it does not imply that forces are necessarily the same in all directions. To be even more precise, we sometimes use the term *differential stress* when the force is greater from one direction than from another.

From shale to gneiss FIGURE 3.17

As shale is subjected to higher and higher temperatures and pressures, it develops into a sequence of metamorphic rocks that have different mineral assemblages. The photos in the diagram were all taken under a microscope, with a 3 mm field of view.

Diagenesis

Shale (sedimentary)

1. Shale is a sedimentary rock made of clay particles and quartz grains.

← Increasing temperature and pressure (metamorphism)

Low-Grade

Slate (metamorphic)

2. Low-grade slate develops from shale. It contains quartz, chlorite, muscovite mica, and feldspars, but no clay.

Medium-Grade

Phyllite (metamorphic)

3. As temperature continues to increase, chlorite disappears and is replaced by biotite mica.

High-Grade

Schist (mica-rich)
Gneiss (mica-poor)
(metamorphic)

4. At even higher temperatures, schist or gneiss is formed. Minerals such as garnet, kyanite, and sillimanite appear.

Mineral stability bars: Clay | Chlorite | Muscovite (white) mica | Biotite (dark) mica | Garnet | Kyanite* | Sillimanite* | Feldspar | Quartz

*Sillimanite and kyanite both have the same formula—Al_2SiO_5—but different crystal structures.

CRITICAL THINKING — Here's an interesting question:
- Look at the gneiss photo in the lower right-hand corner of the figure. What is the direction of maximum stress under which the foliation in the rock developed?

Effects of uniform and differential stress FIGURE 3.18

The two rocks in these photos have similar mineral assemblages but look very different because of their different stress histories.

A This granite consists of quartz (glassy), feldspar (white), and biotite (dark), which crystallized from magma (a liquid) under conditions of uniform stress. Note that the biotite grains are randomly oriented.

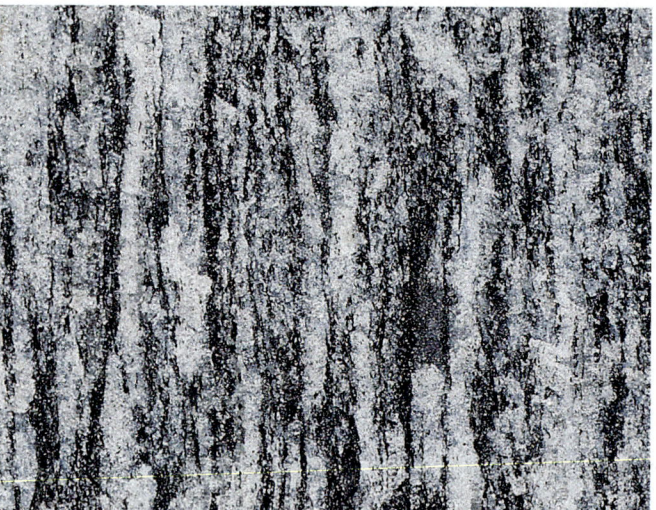

B This gneiss, a high-grade metamorphic rock, contains the same minerals as the granite, but they developed entirely in the solid state and under differential stress. The biotite grains are aligned, giving the rock a pronounced layered texture.

Rocks buried deep underground and undergoing metamorphism are typically subjected to differential stresses and as a result develop a distinctive layered or planar texture called **foliation** (see FIGURE 3.18).

Metamorphic rocks with foliation

Low-grade metamorphic rocks tend to be so fine-grained that the new mineral grains can be seen only under a microscope. Low-grade metamorphism of shale produces **slate**, and the foliation in slate causes a distinctive style of fracture called *slaty cleavage*. The orientation of the cleavage planes says a great deal about conditions under which metamorphism occurred, as explained in FIGURE 3.19. Another kind of foliation, *schistosity*, forms under conditions of high-grade metamorphism, which causes minerals to grow large enough to be seen with the naked eye. Schistose rocks tend to break along wavy or distorted surfaces, while slaty cleavage is strictly planar.

There are some subtleties in terminology in the sequence of metamorphic rocks derived from shale (see Figure 3.19). The names *slate* and *phyllite* describe textures. They are usually used without adding mineral names as adjectives because their mineral assemblages are not easy to see. On the other hand, the mineral grains in **schist** and **gneiss** are large enough to be identified, so we usually add the mineral assemblage to the name, as in quartz-plagioclase-biotite-garnet gneiss.

- **foliation** A planar arrangement of textural features in a metamorphic rock, which give the rock a layered or finely banded appearance.

- **slate** A very fine-grained, low-grade, metamorphic rock with slaty cleavage; the product of metamorphism of shale.

- **schist** A high-grade metamorphic rock with pronounced schistosity, in which individual mineral grains are large enough to be visible.

- **gneiss** A coarse-grained, high-grade metamorphic rock.

Visualizing

Metamorphic rocks derived from shale FIGURE 3.19

SHALE
A sedimentary rock

A The source rock, *shale*, consists primarily of clay minerals and quartz.

SLATE
A low-grade metamorphic rock

B Under conditions of low-grade metamorphism, muscovite and/or chlorite forms from the clay minerals and pore fluids, and the rock becomes a *slate*. The tiny new mineral grains (too small to be seen without a microscope) produce slaty cleavage.

PHYLLITE
A medium-grade metamorphic rock

C Continued low-grade metamorphism of a slate produces larger grains of mica and a changing mineral assemblage. In a *phyllite* the grains of mica are just large enough to be visible.

SCHIST
A mica-rich high-grade metamorphic rock with pronounced *schistosity*.

GNEISS
A mica-poor high-grade metamorphic rock with a coarsely banded (*gneissose*) texture.

D Still further metamorphism of phyllite leads to *schist*. Though the most obvious change is the size of the grains, there are also changes in the mineral assemblage. At high grades of metamorphism, minerals may begin to segregate into separate bands. In *gneiss* (pronounced *nice*), the dark bands contain mica, and the light bands consist of quartz or feldspar.

A *Marble* is a metamorphic rock composed mainly of calcite. Pure marble is snow white. The pink color of this sample (from Tate, Georgia) come from impurities in the marble. In this case, the pink color is due to minute amounts of manganese (Mn^{2+}) and iron (Fe^{2+}) replacing calcium (Ca^{2+}) in the calcite by atomic substitution (Chapter 2).

B Quartz-rich sandstone becomes *quartzite* when its pore spaces are filled with silica and the entire mass recrystallizes as a result of metamorphism. The specimen shown here is from Minnesota.

Nonfoliated metamorphic rocks FIGURE 3.20

Metamorphic rocks without foliation Two kinds of sedimentary rock consist almost entirely of a single mineral species and therefore are said to be *monomineralic*. The first is sandstone, a clastic sedimentary rock that is usually dominated by quartz grains. The second is limestone, a chemical or biochemical sedimentary rock, whose only essential mineral is calcite. Neither of these rocks contains the ingredients (mainly aluminum, potassium, and iron) necessary to form micas and other minerals that might impart foliation to a metamorphic rock. As a result, **quartzite** and **marble**—the metamorphic rocks derived from quartz-rich sandstone and limestone respectively—usually lack foliation (see **FIGURE 3.20**).

quartzite The product of metamorphism formed by recrystallization of sandstone.

marble The product of metamorphism formed by the recrystallization of limestone.

CONCEPT CHECK STOP

How is temperature involved in metamorphism?

What does the presence of a vein in a rock reveal about the way the rock formed?

What is schistosity, and how does it differ from slaty cleavage?

Why are some metamorphic rocks foliated, while others are not?

What is the rock sequence that develops from shale under increasingly higher grades of metamorphism?

Rocks as Resources

LEARNING OBJECTIVES

Discuss rocks as construction materials.

Identify mineral resources found in igneous rocks.

Identify resources found in chemical and biogenic rocks.

Discuss resources in metamorphic veins.

When we say *resources*, people immediately tend to think of gold, diamonds, and other things of high value. But the most valuable resources are rather mundane things such as crushed rock for road construction and gypsum for making sheetrock walls. In the following short section we briefly review some of the ways we use rocks in our society.

ROCKS AS CONSTRUCTION MATERIALS

Almost all kinds of rocks find a use somewhere in the construction industry, but some rocks have more desirable properties than others and so are more widely used. Most of us are familiar with beautiful stone countertops in kitchens and bathrooms—the materials used tend to be limestone, marble, or phaneritic igneous rock, and they are popular because they have striking patterns and colors. But the rocks differ in durability. The main mineral in limestone and marble is calcite, which is soft and scratches easily, and calcite reacts with many common household liquids, so staining and marking are problems. Igneous rocks contain silicate minerals such as quartz, feldspar, and mica, which are physically durable and chemically resistant and do not get easily scratched or stained.

Limestones, marbles, and igneous rocks—especially granites—are widely used as cut stone for building, although the cost tends to restrict their use to publicly funded structures. Granite is far more durable than either limestone or marble because calcite is soluble, albeit slowly, in rainwater (see **FIGURE 3.21**). High-grade metamorphic rocks have many of the same physical properties as granites and can be used for the same purposes in the construction industry. Low-grade metamorphic rocks, especially slates, which break into flat sheets due to slaty cleavage, are widely used for durable—though expensive—roof coverings.

The amount of rock used for countertops and buildings pales to insignificance compared to the use of crushed rock for road construction and for aggregate in concrete and asphalt. The most suitable types of rock for crushed rock are basalt, limestone, and gneiss; their use amounts to nearly 10 tons per year for every person living in the United States.

Effects of rainwater on limestone FIGURE 3.21

Statues carved from limestone several hundred years ago, on the façade of the Bayeux Cathedral, Normandy, France, show the disfiguring effects caused by the acidity of rainwater.

Hydrothermal gold deposit FIGURE 3.22

Delicate sheets of native gold were deposited in the center of a quartz vein in Burgin Hill Mine, California.

MINERAL RESOURCES IN IGNEOUS ROCKS

Many valuable ore minerals occur as accessory minerals in igneous rocks. On occasion, the amount of a mineral present is sufficient to warrant mining. Granites, for example, and especially pegmatitic granites, sometimes contain beryl, one of the sources of beryllium, a metal important in the nuclear power industry. Other minerals found in granites are the source for the metals tantalum, tin, and niobium, each of which can be added to metals such as copper and iron to improve strength and lithium, which is used in ceramics and in batteries for watches and hearing aids.

MINERAL RESOURCES IN CHEMICAL AND BIOGENIC SEDIMENTARY ROCKS

Chemical and biogenic sedimentary rocks, because of the way they form, are concentrations of substances precipitated from sea or lake water. Many of the precipitates have valuable chemical properties. For example, the fertilizer element, potassium, is recovered from marine evaporites, and phosphorus, the most important of all fertilizer elements, is found in rocks of marine biogenic origin. Other examples are halite (table salt) and gypsum (for sheetrock), both of which are found in marine evaporites. One especially important chemical sedimentary rock is banded iron formation, discussed earlier in this chapter (see *What an Earth Scientist Sees: A Change in the Atmosphere*, page 76).

GOLD IN METAMORPHIC ROCKS

Metamorphic rocks start out as sedimentary or igneous rocks, so any mineral deposit present in those rocks will be preserved as a metamorphosed mineral deposit. There are many such metamorphosed deposits in terrains of geologically older rocks that have been subjected to metamorphism.

As discussed earlier in this chapter, metamorphism is accompanied by the expulsion of aqueous fluids, and in many cases the fluids flow through fractures and form veins that are predominantly quartz-rich (see Figure 3.15). Metamorphic fluids are never pure water—small amounts of many minerals are present in solution and these may be deposited with the quartz. When gold is among the minerals deposited, valuable ore deposits can be the result. Some of the gold produced in California is thought to have its origins in metamorphic vein deposits (FIGURE 3.22).

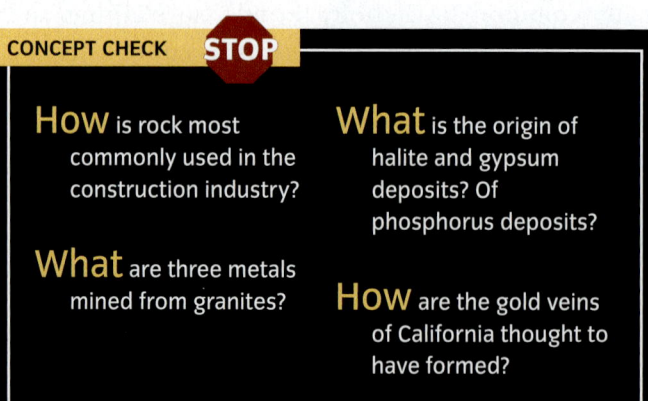

CONCEPT CHECK STOP

How is rock most commonly used in the construction industry?

What are three metals mined from granites?

What is the origin of halite and gypsum deposits? Of phosphorus deposits?

How are the gold veins of California thought to have formed?

Amazing Places: The Navajo Sandstone

NATIONAL GEOGRAPHIC

Global Locator

Many of the rock formations in Utah's Zion National Park resemble sand dunes—because that is what they originally were. The imposing cliffs (**A**), the roadside hill (**B**), and the stunningly etched ground seen from above (**C**) were all once part of a vast sea of sand, larger and thicker than the Sahara Desert is today. The dunes lithified into sandstone and subsequently eroded into the formations you see today. The sedimentary formation they belong to, called the *Navajo Sandstone,* extends over several states in the American Southwest and attains a thickness of 700 m in Zion National Park.

But if all of this rock was once sand, that poses a puzzle: Where did all the sand come from? The dune patterns indicate that the prevailing winds came from the north, but no mountain ranges of suitable size or age can be found there. The mountain range we call the Rockies did not yet exist in the Lower Jurassic Period, 190 million years ago, when the Navajo sand was being deposited.

However, the Appalachian Mountains did exist then. Lifted up by a plate tectonic collision that began 500 million years ago, they were once as tall as the Himalaya are today and extended all the way into what is now Texas. By the Lower Jurassic the Appalachians had eroded considerably—and scientists now think that it was their sediment that formed the Navajo Sandstone. Rivers flowing westward carried the erosional sediment toward what is today the center of the continent (**D**). As the climate became arid, winds transported the now-dried river sands southward, where they became one of the largest sand seas that has ever existed on Earth.

\\\ Appalachians   ⧘ Grenville Mountains

SUMMARY

1 Rocks: A First Look

1. Rocks are coherent aggregates, or complex *assemblages of minerals*, that are sometimes mixed with other materials, such as volcanic glass and organic matter.

2. Rocks come in three major families. **Igneous rocks** are formed by the solidification of **magma**. **Sedimentary rocks** are formed by the **deposition** of many layers of sediment. **Metamorphic rocks** start as either sedimentary or igneous rocks but change their form and mineral assemblage as a result of high temperature, high pressure, or both.

3. The **rock cycle** consists of all the processes whereby rocks within and on top of Earth's crust are **weathered**, and the products of weathering are **eroded**, transported, deposited as sediment, converted into new rock, metamorphosed, melted, uplifted, and exposed again to weathering.

2 Igneous Rocks

1. The process of solidification of cooling magma is called *crystallization*. The process of crystallization of magma influences the final properties of igneous rock, such as its *texture* and grain size. The crystallization process and the rock's final *mineral assemblage* in turn depend on such factors as the magma composition and the rate of cooling. A rock formed by rapid cooling of magma contains very small mineral grains.

2. Igneous rocks can be classified into **volcanic** and **plutonic** rocks. Volcanic rocks crystallize from lava; plutonic rocks crystallize underground from magma. Because lavas cool rapidly, mineral grains in volcanic rocks are small, microscopic, or even absent. These rocks are known as *aphanitic*. Plutonic rocks, on the other hand, are *phaneritic*, and have easily recognized mineral grains.

3. *Felsic* rocks are typically light in color, have high silica contents, and consist largely of quartz and feldspar. *Mafic* rocks are typically darker in color, with lower silica content, and consist largely of iron-containing minerals such as olivine and pyroxene. The most common volcanic rock is *basalt*, a dark, low-silica, mafic rock. *Granite* is a light-colored, high-silica, felsic rock that is a dominant constituent of the continental crust.

3 Sedimentary Rocks

1. Sediments and sedimentary rocks are our best record keepers of how climates and environments have changed throughout Earth's history. Earth scientists use three broad categories to distinguish types of sediment—*clastic*, *chemical*, and *biogenic*.

2. **Clastic sediment** consists of fragmented rock and mineral debris produced by weathering, together with broken remains of organisms. The fragments, called *clasts*, are classified on the basis of size. Clasts may become *sorted* during transport by water or wind but typically remain unsorted when transported by glaciers. Volcaniclastic sediments, also called *pyroclasts*, are volcanic in origin.

3. Clastic sedimentary rocks, like sediments, are classified mainly on the basis of clast size. **Conglomerate**, **sandstone**, and *mudstone* are the common rock equivalents of gravel, sand, and mud respectively. **Shale** is a mudstone that readily splits into thin layers.

4. **Lithification** is the group of processes that transform loose sediment into sedimentary rock. The principal processes are *compaction* of sediment, *cementation*, and *recrystallization* of mineral grains.

5. **Chemical sediment** is formed when substances carried in solution in lake water or seawater are precipitated, generally as a result of evaporation or other process that concentrates dissolved substances.

6. **Biogenic sediment** is composed of the accumulated remains of organisms. Plants, animals, and microscopic life-forms may all contribute skeletal and/or organic material to sediments.

7. When sediment is turned into sedimentary rock, the **bedding**, or layered arrangement of beds, is generally preserved. The presence of bedding and *bedding surfaces*, the boundaries between adjacent beds, indicates that the rock was once sediment.

8. Chemical sedimentary rocks formed by evaporation are **evaporites**. *Banded iron formations*, though uncommon, are significant both economically and for understanding the history of our planet. **Limestone**, **peat**, and **coal** are important kinds of biogenic sedimentary rocks.

9. A *sedimentary facies* is a single unit consisting of sedimentary rocks with the same composition, deposited at more or less the same time. Earth scientists learn about the history of a region by studying the way that different sedimentary facies adjoin and grade into each other in space and time.

4 Metamorphic Rocks

1. New rock textures and new mineral assemblages develop when rocks are subjected to elevated temperatures and stresses. **Metamorphism** is a term that describes all such processes that occur at higher temperatures and stresses.

2. Metamorphism commences at a temperature of about 150°C, which typically occurs at a depth of about 5 km. Between 5 and 15 km depth, rocks are subjected to *low-grade* metamorphism. The region of *high-grade* metamorphism lies from 15 km to the depth where melting commences, typically between 30 and 40 km.

3. *Pore* fluids enhance metamorphism by permitting material to dissolve, move around, and be precipitated somewhere else in the rock. *Veins* in metamorphic rocks mark the passageways through which pore fluids escaped from a rock undergoing metamorphism.

4. *Differential stress* during metamorphism produces a distinctive texture known as **foliation** marked by parallel cleavage planes. It is particularly noticeable in rocks containing minerals of the *mica* family. Foliation developed in low-grade metamorphic rocks is known as *slaty cleavage*. In higher-grade metamorphic rocks foliation is called *schistosity*.

5. The names of metamorphic rocks are based partly on texture and partly on composition. Foliated metamorphic rocks developed from shale are **slate**, *phyllite*, **schist**, and **gneiss** in order of increasing metamorphic grade. **Marble**, the metamorphic product of limestone, and **quartzite**, the metamorphic product of sandstone, are two common kinds of metamorphic rock that lack cleavage. Both rocks are *monomineralic*, or very nearly so.

5 Rocks as Resources

1. Essentially all kinds of rocks can be used somewhere in the construction industry. Limestone, marble, and granite are widely used as cut stones for building. The principal materials used as crushed rock for road building and concrete aggregate are basalt, limestone, and gneiss.

2. Igneous rocks are the source of many valuable ore minerals. Examples of metals recovered from igneous rocks are beryllium, tantalum, and lithium.

3. Chemical sedimentary rocks are the source of much of the world's potassium for fertilizers, table salt, and gypsum for sheetrock. Biogenic rocks are the source of phosphorus minerals used for fertilizers.

KEY TERMS

- rock cycle p. 66
- weathering p. 66
- erosion p. 66
- igneous rocks p. 67
- magma p. 67
- volcanic rock p. 67
- plutonic rock p. 67
- sedimentary rocks p. 72
- deposition p. 72
- clastic sediment p. 72
- conglomerate p. 72
- sandstone p. 72
- shale p. 72
- lithification p. 72
- chemical sediment p. 75
- evaporite p. 75
- biogenic sediment p. 77
- limestone p. 77
- peat p. 77
- coal p. 77
- bedding p. 77
- metamorphic rocks p. 82
- metamorphism p. 82
- foliation p. 86
- slate p. 86
- schist p. 86
- gneiss p. 86
- quartzite p. 88
- marble p. 88

CRITICAL AND CREATIVE THINKING QUESTIONS

1. When a volcano erupts, spewing forth a column of hot volcanic particles, the particles are tiny fragments of solidified magma that slowly fall down to Earth's surface, forming a layer of sediment. Would the rock formed from cemented volcanic particles be igneous or sedimentary? Can you think of other circumstances that might form rocks that are intermediate between two of the major rock families?

2. Volcanic rock is sometimes called *extrusive* and plutonic rock is sometimes called *intrusive*. Why do you think that Earth scientists describe them this way? (You may wish to look up these words in a dictionary.)

3. Do any sedimentary rocks outcrop in the area where you live? If so, see if you can recognize the kinds of rocks present and identify the environment in which the sediments were deposited.

4. Exploration for oil has led to the discovery of about 14 km of sedimentary rock on the continental shelf of eastern North America. The oldest beds were deposited in the Jurassic Period, 150 million years ago. The youngest are still being deposited today. What is the average rate of deposition?

5. Examining the texture of a rock is an important rule-of-thumb for Earth scientists because the texture can reveal whether the rock has been metamorphosed and under what conditions. Explain why. Would this rule work for both foliated and nonfoliated rocks? Why (or why not)?

What is happening in this picture?

This roof in Ireland was made from a commonly occurring planar metamorphic rock.

- What kind of rock do you think it is?
- What properties of this rock would make it very suitable for use as a roof tiling material?
- Do you think this rock is used in modern roofing? Why, or why not?

SELF-TEST

1. On this illustration, locate and label the following processes of the rock cycle:

   weathering
   tectonic uplift
   burial and cementation
   intrusion and volcanism
   melting
   metamorphism
   erosion and deposition

2. The three major rock families are _____.
   a. igneous, sedentary, metamorphic
   b. sedimentary, metamorphosis, igneous
   c. sedimentary, igneous, metamorphic
   d. metamorphic, igneous, metamorphosis

3. These two photographs are closeup views of two igneous rocks. Label each appropriately with the following terms:
   porphyritic-texture volcanic rock
   phaneritic-texture plutonic rock

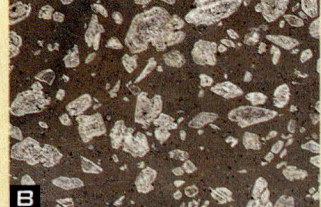

4. Which of the samples depicted in the photographs in question 3 records two distinct phases of cooling—slow followed by rapid?
   a. The sample labeled "A."
   b. The sample labeled "B."
   c. Neither. Both rocks cooled at the same rate.

5. Which of the following are plutonic igneous rocks correctly paired with their compositionally equivalent volcanic igneous rocks:
   a. granite/rhyolite; gabbro/andesite; diorite/basalt
   b. granite/basalt; gabbro/rhyolite; diorite/andesite
   c. diorite/basalt; granite/rhyolite; gabbro/andesite
   d. granite/rhyolite; diorite/andesite; gabbro/basalt

6. _____ sediment forms from loose rock and mineral debris produced by weathering and erosion.
   a. Clastic   b. Biogenic   c. Chemical

7. Which of the three rock samples in these photographs is not a sedimentary rock?
   a. Sample A
   b. Sample B
   c. Sample C
   d. All of the rock samples displayed are sedimentary rocks.

8. Which of the following sedimentary rocks is typically composed entirely of calcite?
   a. sandstone
   b. limestone
   c. shale
   d. conglomerate
   e. evaporite

9. Which of the following is a medium-grained sedimentary rock of clastic origin?
   a. sandstone
   b. limestone
   c. shale
   d. conglomerate
   e. evaporite

10. Which of the following rocks is most likely to have formed in a seasonally dry desert lake?
    a. sandstone
    b. limestone
    c. shale
    d. conglomerate
    e. evaporite

11. Metamorphism will occur under relatively _____ conditions.
    a. high temperature and high pressure
    b. high temperature and low pressure
    c. low temperature and high pressure
    d. All of the above statements are correct.

12. Pore fluids enhance metamorphism by _____.
    a. permitting material to dissolve
    b. permitting chemical components to move around and be precipitated somewhere else
    c. speeding up some chemical reactions
    d. All of the above statements are correct.

13. Foliated metamorphic rocks derived from shales are _____, in order of increasing metamorphic grade.
    a. phyllite, slate, and schist or gneiss
    b. phyllite, gneiss, and schist or slate
    c. slate, phyllite, and schist or gneiss
    d. slate, gneiss, and phyllite or schist

14. The two rock samples shown in these photographs are nonfoliated metamorphic rocks. Sample A is composed of calcite, and Sample B was formed through the metamorphism of sandstone. Which of the following correctly identifies the two samples?
    a. Sample A is quartzite and Sample B is schist.
    b. Sample A is quartzite and Sample B is marble.
    c. Sample A is marble and Sample B is quartzite.

15. Limestone is not the most desirable stone for buildings because _____.
    a. it is fissile and tends to splinter
    b. it is too hard and cannot be cut easily
    c. it slowly dissolves in rainwater
    d. All of the above statements are true.

Weathering, Soils, and Mass Wasting 4

In May 2003, New Hampshire's beloved landmark, the Old Man of the Mountain, fell victim to the forces of nature. The Old Man was granitic rock, sculpted in part by glacial activity during the last ice age, that bore a conspicuous resemblance to a human face jutting out from a cliff—a sort of Mt. Rushmore sculpted by nature. It had been commemorated just three years earlier on the New Hampshire state quarter (see inset). Caretaker Niels Nielson and his son are making a regular inspection of the monument.

The Old Man's demise can be attributed to the inexorable forces of weathering. Mere rock cannot resist the combined assault of water, ice, chemical changes, and gravity. Deep inside the Old Man's granite features, water repeatedly froze and thawed; each freeze widened and extended numerous minute cracks inside the rock, while at the same time chemical changes chiseled away at the outer edges. After hundreds of years of this wear and tear, the five granite ledges that created the man's profile broke apart and tumbled to the bottom of the hill.

New Englanders were dismayed at the loss of the seemingly timeless monument. Plans are underway to create a museum with high-tech imagery of the formation before its fall.

In reality, rocks exposed at Earth's surface are never timeless or immutable. They are constantly broken down by weathering and erosion; without these processes, we would not have soil or beaches. At the same time, new rock is constantly forming deep in Earth's crust. This constant recycling of material is the rock cycle, discussed at the beginning of Chapter 3.

CHAPTER OUTLINE

- Weathering: The Earth System at Work p. 100

- Soil: The Most Important Product of Weathering p. 109

- Erosion and Mass Wasting: Gravity at Work p. 117

- Resources Formed by Weathering and Erosion p. 124

Weathering: The Earth System at Work

LEARNING OBJECTIVES

Describe the two major kinds of rock weathering.

Explain the three types of chemical reactions involved in chemical weathering.

Describe the influences of climate, topography, and rock composition on mechanical and chemical weathering.

We met the terms **weathering** and **erosion** in Chapter 3; they are essential steps in the *rock cycle*. *Weathering* is the group of processes that break bedrock into smaller rock and mineral fragments, and eventually into soils. *Erosion* is the collection of processes by which the products of weathering are transported from one place to another (weathering happens in place).

> **weathering** The chemical and physical breakdown of rock exposed to air, moisture, and living organisms.

> **erosion** The wearing away of bedrock and transport of loosened particles by a fluid, such as water.

WEATHERING: HOW ROCKS DISINTEGRATE

Weathering takes place throughout the zone in which the materials of the lithosphere, hydrosphere, atmosphere, and biosphere coexist; it is the zone in which the Earth system creates the conditions within which life can exist. This zone extends downward below Earth's surface as far as air, water, and microscopic organisms can readily penetrate, and it ranges from one meter to hundreds of meters in depth. Rock in the weathering zone usually contains numerous fractures and pores through which water, air, and tiny organisms can enter. Given enough time, they produce major changes in the rock (see **FIGURE 4.1**). The zone of weathering is the most intensely active part of the Earth system.

The product of weathering is called **regolith**, from the Greek words meaning "blanket" and "stone." Fragments in the regolith range in size from microscopic to many meters across, but all of them have been formed by the physical and chemical breakdown of bedrock. As the particles get smaller and smaller, plants become capable of growing roots into the regolith and extracting mineral nutrients from it. At this point, the regolith becomes **soil**.

The breakdown processes involved in weathering fall into two categories. In **mechanical weathering**, the

> **regolith** A loose layer of broken rock and mineral fragments that covers most of Earth's surface.

> **soil** The uppermost layer of regolith, which can support rooted plants.

> **mechanical weathering** The breakdown of rock into solid fragments by physical processes that do not change the rock's chemical composition.

From rock to soil FIGURE 4.1

In this photograph, taken in South Africa, rock has broken down by weathering through disintegration of a layer of sedimentary rock. Water and air penetrate through the fractures in the rock and react with the minerals there. Rock near the surface is more heavily weathered than the lower layers because it has been exposed more to water, air, and microorganisms. At the surface, the rock has been completely changed to soil. Note that weathering does not have to be accompanied by erosion. Here, the products of weathering have remained in the place where they were formed.

chemical weathering The decomposition of rocks and minerals by chemical and biochemical reactions.

rock physically breaks down into small pieces, but there is no change in the mineral content. **Chemical weathering** involves dissolving of minerals or chemical reactions with air and water that replace the original minerals with new minerals that are stable at Earth's surface. Although mechanical weathering is distinct from chemical weathering, the two processes almost always occur together, and their effects are sometimes difficult to separate.

MECHANICAL WEATHERING: BREAKING ROCKS APART

Rocks in the upper half of the crust are brittle, and like any other brittle material they break when they are twisted, squeezed, or stretched by tectonic forces. Though we cannot always determine the timing of the breaks or the origin of the forces the caused them, we can see the results in the form of **joints**. *What an Earth Scientist Sees* shows a common way by which joints form.

joint A fracture in a rock, along which no appreciable movement has occurred.

What an Earth Scientist Sees

Joint Formation

A These heavily jointed rocks are found in the Joshua Tree National Monument in California. Weathering has widened some of the joints to the point of detaching stones from the main rock body.

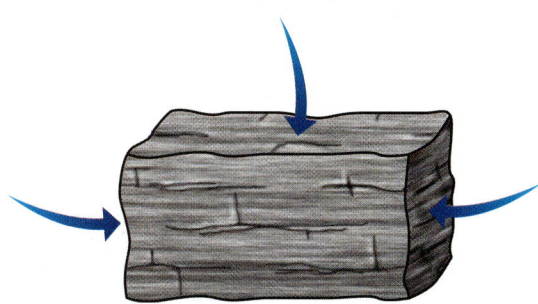

B The rock originally formed underground, where it was subject to great pressure from the overlying and surrounding rocks. Squeezing and twisting by tectonic forces cause the rock to fracture and form joints.

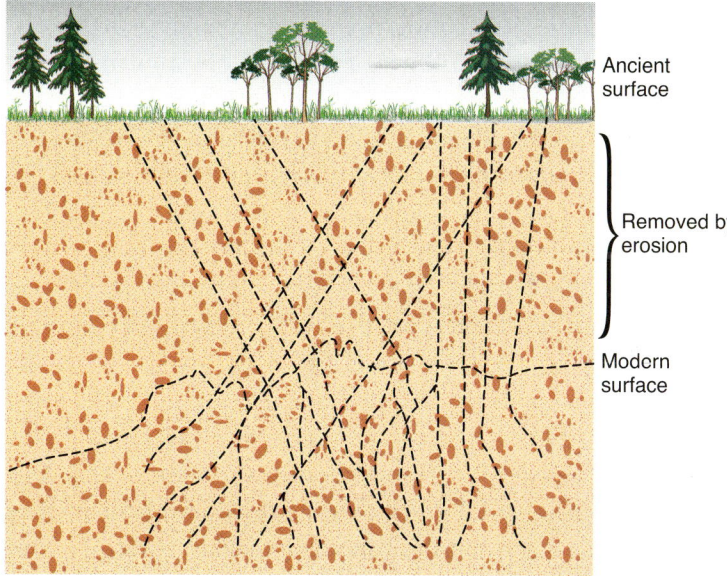

C As the rock rose to the surface and the overlying rock was eroded, the pressure decreased, causing the rock to expand and crack, creating additional joints. Later, erosion rounded off and widened the joints even further.

CRITICAL THINKING — Here are two interesting questions:
- Why are some granite bodies extensively jointed, while others are essentially joint free?
- How might the presence of joints influence the rate of weathering?

Weathering: The Earth System at Work

Joints are the main passageways through which rainwater, air, and small organisms enter the rock and lead to mechanical and chemical weathering. They differ from *faults* (Chapter 7), another kind of rock fracture, because there is slippage along faults, but no noticeable slippage along joints—they are simply cracks in a previously solid body of rock.

Joints do not always have to be straight. In a process called *sheet jointing* or *exfoliation* (see **FIGURE 4.2A**), large, curved slabs of rock peel off from the surface of a uniformly textured igneous rock. As with other types of joints, sheet jointing may be due to pressure release or a combination of forces that contribute to mechanical weathering.

Mechanical weathering takes place in four main ways—through freezing of water, precipitation of salt crystals from groundwater, penetration by plant roots, and abrasion. By far the most widespread type of mechanical weathering involves the freezing of water.

Water is an unusual substance. Most liquids contract when they freeze, and the volume of the resulting solid is smaller than the volume of the liquid. However, when water freezes it expands, increasing in volume by about 9%. If you put a full, capped bottle of water in the freezer, the bottle will burst when the water freezes, because it cannot contain the larger volume of ice. Wherever temperatures fluctuate around the freezing point for part of the year, water in the ground will alternately freeze and thaw. If the water gets inside a joint in the rock, the freeze-thaw cycles act like a lever prying the rock apart, and eventually the rock shatters. This process is known as **frost wedging** (see **FIGURES 4.2B** and **4.2C**).

The formation of salt crystals can also cause mechanical weathering. Water moving slowly through rock fractures will dissolve soluble material, which may later precipitate out of solution to form salt crystals of various kinds. The force exerted by growing crystals, either within rock cavities or along grain boundaries, can cause rocks to fall apart. Mechanical weathering by crystal growth occurs mostly in hot desert regions, where compounds such as calcium carbonate (calcite) and calcium sulfate (gypsum) are precipitated from groundwater as a result of evaporation; the process can also happen in near-seashore environments, where sodium chloride (halite) can precipitate from windborne sea spray.

Another type of mechanical weathering is caused by penetration by plant roots. Trees are very resourceful and can grow where there seems to be hardly any soil to sustain them. A tree may become rooted in a crack in the bedrock, eventually widening the crack and wedging apart the bedrock, as shown in **FIGURE 4.2D**. Large trees swaying in the wind can also cause fractures to widen. When trees are blown over, they can cause additional fracturing of rock. Although it is difficult to measure, the total amount of rock breakage caused by trees and other plants must be very large. The final method of mechanical weathering, **abrasion**, is the wearing away of bedrock by loose particles transported by water, wind, or ice. Abrasion is a form of mechanical weathering because it reduces the size of rocks by a physical (rather than chemical) process.

CHEMICAL WEATHERING: BREAKING ROCKS BY CHEMISTRY

Chemical weathering is primarily caused by water that is slightly acidic. As raindrops form and fall through the air, they dissolve atmospheric carbon dioxide:

$$H_2O + CO_2 \longrightarrow H_2CO_3$$

Rainwater thus is a weak solution of carbonic acid (H_2CO_3)—a weak version of soda water. When weakly acidified rainwater sinks into the regolith and becomes soil water or groundwater, it may dissolve additional carbon dioxide from decaying organic matter, becoming more strongly acidified. Another way that rainwater can become acidified is by interacting with *anthropogenic* (human-generated) sulfur and nitrogen compounds released into the atmosphere. This produces a phenomenon called *acid rain*. Human-caused acid rain is stronger than natural acid rain and causes accelerated weathering.

dissolution The separation of a material into ions in solution by a solvent, such as water or acid.

Through **dissolution**, some minerals can be dissolved and completely removed without leaving a residue. Some common rock-forming minerals, such as

Mechanical weathering FIGURE 4.2

Smooth, rounded surface where sheets have exfoliated (peeled off) in the past.

▲ **A Sheet Jointing**
Sheet jointing results in curved domes, like the famous Half Dome in Yosemite National Park, California.

Frost wedging has widened this joint in the rock.

▲ **B Frost Wedging**
This granite boulder in the San Andres Mountains of New Mexico has been split apart by repeated freezing and thawing of water that penetrated along the joints.

▲ **C** An Earth science student takes notes in Victoria Land, Antarctica, on a steep slope covered with a blanket of loose angular fragments called a *talus slope*. All of the loose fragments in this photograph were separated from the bedrock by frost wedging.

▲ **D Root Wedging**
This Ponderosa pine began growing into a crack in an outcrop of granite. Eventually, a large flake of rock broke away, exposing the tree's root system.

calcite (calcium carbonate) and dolomite (calcium magnesium carbonate), dissolve in slightly acidified water (see **FIGURE 4.3**). The products of dissolution, calcium cations and bicarbonate anions, are removed in solution:

$$CaCO_3 + H_2CO_3 = Ca^{2+} + 2(HCO_3)^-$$
Calcite + Carbonic acid = Calcium cation + Bicarbonate anion

Water can also alter the mineral content of rock without dissolving all of it. One reaction of special importance in chemical weathering is *ion exchange*. Ions, which are atoms with a positive or negative electric charge, exist both in solution and in minerals. They form when atoms give up or accept electrons (see Chapter 2). The difference is that ions in minerals are tightly bonded and fixed in a crystal lattice, whereas ions in solutions can move about randomly and cause chemical reactions. In ion exchange, hydrogen ions (H^+) from acidic water enter and alter a mineral by displacing larger, positively charged ions such as potassium (K^+), sodium (Na^+), and magnesium (Mg^{2+}), which go off in solution (see **FIGURE 4.4**).

The following chemical equations show how one of Earth's most common minerals, potassium feldspar ($KAlSi_3O_8$), breaks down to a clay mineral called *kaolinite*. First, carbonic acid dissociates in the following manner, to form hydrogen cations (H^+) and bicarbonate anions (HCO_3)$^-$:

$$H_2CO_3 = H^+ + (HCO_3)^-$$
Carbonic acid = Hydrogen cation + Bicarbonate anion

Hydrogen cations and water then react with feldspar to form kaolinite, potassium cations, and silica in the following reaction:

$$4KAlSi_3O_8 + 4H^+ + 2H_2O = Al_4Si_4O_{10}(OH)_8 + 8SiO_2 + 4K^+$$
Feldspar   Kaolinite

Where do the potassium and other ions go after they are replaced by hydrogen ions? Some remain in the groundwater, accounting for the taste of "mineral water" that some people find pleasant and others do not, and some flow out to sea and form part of the ocean's reserve of dissolved salts. On the other hand, if the water evaporates, the dissolved materials can precipitate out again as solid evaporites, such as halite and gypsum.

Dissolution of calcite FIGURE 4.3

A This marble tombstone, which has stood in a New England cemetery since the early nineteenth century, has been gradually dissolved by rainwater. The once sharply chiseled inscriptions have become rounded and less legible. Marble, which contains the soluble mineral calcite, is vulnerable to dissolution.

B In the same cemetery, and a few meters away from the tombstone in A, there is a granite tombstone from the nineteenth century. Note how sharp the letters are and how fresh and unaltered the rock appears. Feldspar and quartz, the main minerals in the granite, react with rainwater much more slowly than calcite does.

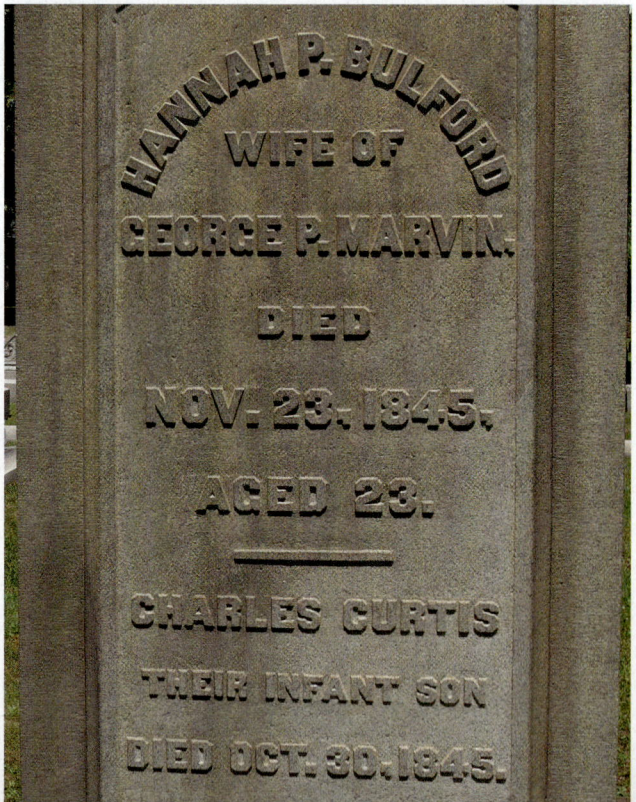

Ion exchange FIGURE 4.4

This photo, taken through a microscope, shows where a feldspar grain has been altered by ion exchange. The clay residue has been removed to make the pattern of alteration more visible.

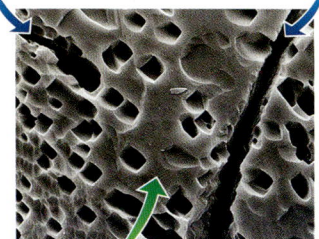

Clay residue forms along cleavage planes.

Unaltered feldspar

1 Acidified water containing hydrogen ions (H^+) enters feldspar crystal along existing fractures.

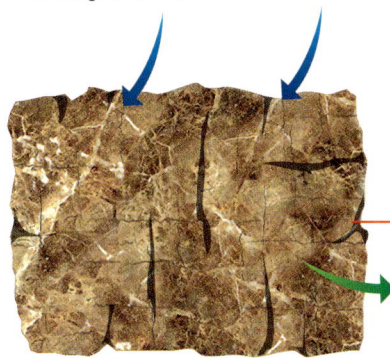

Potassium ion (K^+) leaves in solution.

Unaltered feldspar

Alteration products (clay)

2 Where potassium has washed away, an insoluble residue of clay remains.

CRITICAL THINKING — Here's an interesting question:
- On Earth, clay minerals are the most common products of weathering. Samples from the surface of the Moon brought back by astronauts do not contain any clay minerals. Why?

Another very important process of chemical weathering is *oxidation*, a reaction between minerals and oxygen dissolved in water. Iron and manganese, in particular, are present in many rock-forming minerals. When such minerals undergo chemical weathering, the iron and manganese are released and are immediately oxidized (see **FIGURE 4.5**). Oxidized iron commonly forms an insoluble yellowish hydrous material called limonite, and manganese forms an insoluble black mineral called pyrolusite.

FACTORS AFFECTING WEATHERING

As **FIGURE 4.6** on pages 106–107 shows, many factors influence the susceptibility of a rock to chemical and mechanical weathering. The most important factors are tectonic setting, composition of the rock, rock structure (the abundance of openings such as joints), topography, amount of vegetation and biologic activity, and climate (especially temperature and rainfall).

Oxidation FIGURE 4.5

The yellow rocks are limonite, an oxidized mineral of iron that contains water. When the water is removed, limonite becomes hematite (red rocks), a common ingredient in red tropical soils. The green rocks are a less common iron ore called siderite, an iron carbonate. The rocks were mined in Egypt.

◀ **Tectonic setting:** Young, rising mountain ranges, such as the Himalaya, weather very rapidly. This view of the Indus River in Pakistan shows many boulders that have separated from the bedrock because of mechanical weathering.

▼ **Rock structure:** The pace of weathering is also strongly affected by the closeness of joints. Sugarloaf Mountain in Rio de Janeiro, Brazil, is a large, unjointed mass of granite, and it stands out against an otherwise deeply eroded landscape.

Unjointed rock weathers slowly.

Steep slope weathers quickly.

Topography: Weathering ▶ proceeds more quickly on a steep slope than a gentle one. This rockslide on the island of Madeira in the Atlantic Ocean will expose new bedrock to weathering.

Visualizing

Factors affecting rates of weathering FIGURE 4.6

◀ **Biologic activity:** Animals—even microorganisms—contribute significantly to the breakdown of rocks. Here, a student lifts a mat of algae. The algae feed on a bacterium, *Thiobacillus ferrooxidans*, that thrives in acidic runoff from abandoned sulfur- and iron-bearing mines.

Bacterial activity can promote chemical weathering.

Composition: Different minerals weather at different rates. Calcite weathers quickly by dissolution, and feldspar weathers at an intermediate rate by ion exchange. However, quartz is very resistant to weathering, because it dissolves very slowly and is not affected by ion exchange. The knob atop Pilot Mountain, in North Carolina, is made of quartzite that has weathered much more slowly than the surrounding sedimentary rock. ▶

Resistant quartzite knob weathers slowly.

Deforestation accelerates weathering and erosion.

◀ **Vegetation:** Plants contribute to both mechanical and chemical weathering. They tend to hold a deeper regolith in place, which promotes weathering because it retains water. However, the removal of plants through slash-and-burn agriculture, clear-cut logging, or natural landslides can also accelerate erosion, as seen here in a photo of Madagascar.

Climate and weathering FIGURE 4.7

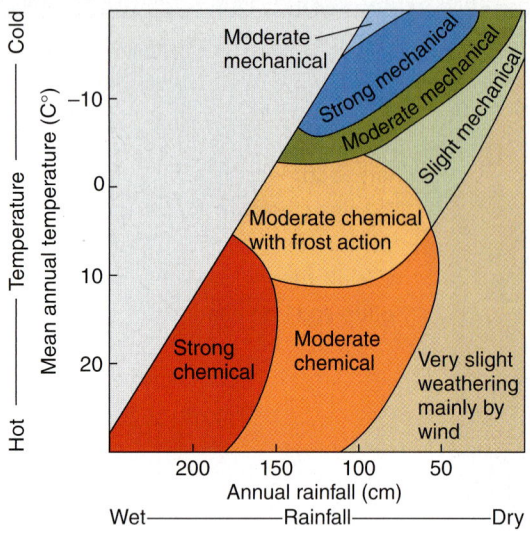

A Different climates cause rocks to weather at different rates and by different types of weathering.

B This map of North and South America illustrates some locations where the climate corresponds to the different zones of weathering shown in A.

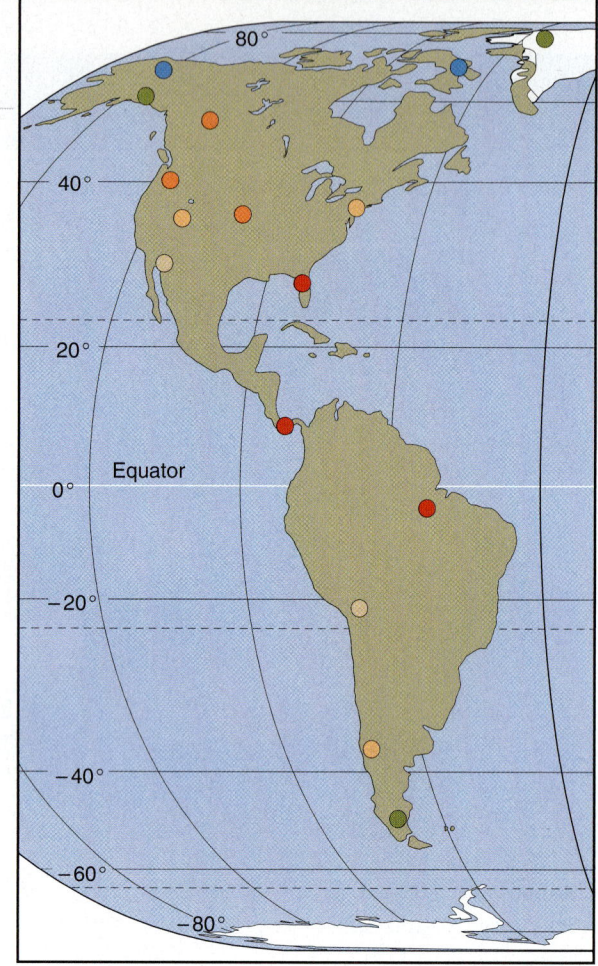

From the perspective of a human lifespan, the chemical weathering of rocks happens very slowly. For example, granite and other hard bedrock surfaces in New England, Canada, and Scandinavia still display polish and fine grooves made by the scraping of glaciers of the last ice age, which ended more than 10,000 years ago. In such regions, which have plentiful rainfall but cool climates, it takes hundreds of thousands of years for a regolith like that shown in Figure 4.1 to develop. In warmer regions, chemical weathering occurs more quickly at the surface and extends to depths of many tens of meters. Through the use of various dating techniques, it has been estimated that deep tropical weathering of 500 m or more requires many millions and possibly tens of millions of years.

Climate affects the rate of weathering in two different ways. As noted earlier, chemical weathering is more intense and extends to greater depths in a warm, wet, tropical climate than in a cold, dry, Arctic climate. In cold, dry climates like Greenland or Antarctica, chemical weathering proceeds very slowly. However, mechanical weathering is fairly rapid in these harsh environments. The only environment where both kinds of weathering proceed very slowly is a hot, dry climate (see FIGURE 4.7).

CONCEPT CHECK STOP

What are the main processes involved in mechanical weathering?

How do joints contribute to mechanical weathering?

What are three processes that contribute to chemical weathering?

Why is mechanical weathering most effective in cold regions and chemical weathering most effective in warm, moist regions?

Soil: The Most Important Product of Weathering

LEARNING OBJECTIVES

Identify three end products of weathering.

Describe the soil horizons found in most soils.

Explain why soils from different climates have different profiles.

Describe the link between human activity and soil erosion.

ll rocks break down under the chemical attack of the hydrosphere, atmosphere, and biosphere. With prolonged exposure to chemical weathering, minerals that are stable at higher temperatures and pressures begin to decompose, while new minerals, stable at the conditions of Earth's surface, are formed. For example, as we discussed earlier in the chapter, feldspar breaks down to form kaolinite, or other **clay** minerals. Clay minerals are one of the most stable kinds of mineral on Earth's surface. They are a major component of mud, both on the land and in the sea. Particles of

> **clay** A family of hydrous aluminosilicate minerals. The term is also used for tiny mineral particles of any kind that have physical properties like those of the clay minerals.

clay minerals tend to be tiny (typically less than 0.002 mm in diameter) and because tiny particles are hard to identify, scientists who study soils classify all particles less than 0.002 mm in size, regardless of mineral type, as *clay*. The very finest of the clay-sized particles are so small they can remain suspended in water indefinitely; they are called *colloids*.

Some minerals, such as quartz, are more resistant to weathering than feldspar, and hence do not break down into grains as small as those found in clay. Instead, quartz usually ends up as **sand**, which has grain sizes as coarse as 1 to 2 mm, roughly 100 to 1000 times larger than the grains in clay. Sediment with grain sizes between those of sand and clay is called *silt*

> **sand** A sediment made of relatively coarse mineral grains.

(see **FIGURE 4.8**). The grain sizes of particles weathered from bedrock decrease with the distance traveled from the source. Some of the finest sand is found at the beach, because it has had an especially long journey from the mountains.

SOIL

The most complex product of weathering is also the most familiar: soil. It consists of matter in three states—solid, liquid, and gas. Solid matter consists of a mixture of mineral grains ranging in size from clay to silt to sand, plus material of biologic origin. The term *soil texture* refers to the proportion of particles that fall into each of the three size ranges. A *loam* is a soil containing substantial proportions of each of the three size ranges of minerals. Loam is classified as sandy, clay rich, or silty, when one of the three size ranges is dominant. If you look at soil under a magnifying glass, you will find, in addition to mineral grains, fragments of **humus** and possibly some tiny insects and worms. With a very strong microscope, you will see bacteria,

> **humus** Partially decayed organic matter in soil.

fungi, and other microorganisms living on the humus.

Both air and water are present in soil. Water wets mineral grains and humus fragments and is critical to

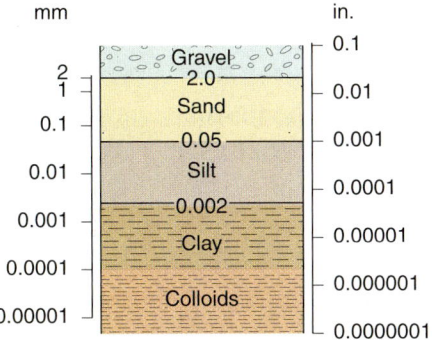

Mineral particle sizes FIGURE 4.8

Size grades are named *sand*, *silt*, and *clay* (which includes colloids). Gravel is not included when discussing soil texture. Size grades are defined using the metric system, and each unit on the scale represents a power of 10. English equivalents are also shown.

Earth and lunar "soil"—not the same! FIGURE 4.9

This microscopic view of Earth soil (**A**) and lunar regolith (**B**) shows some significant differences. Most importantly, Earth soil contains organic material and hydrous minerals such as clay, while lunar regolith contains none of either.

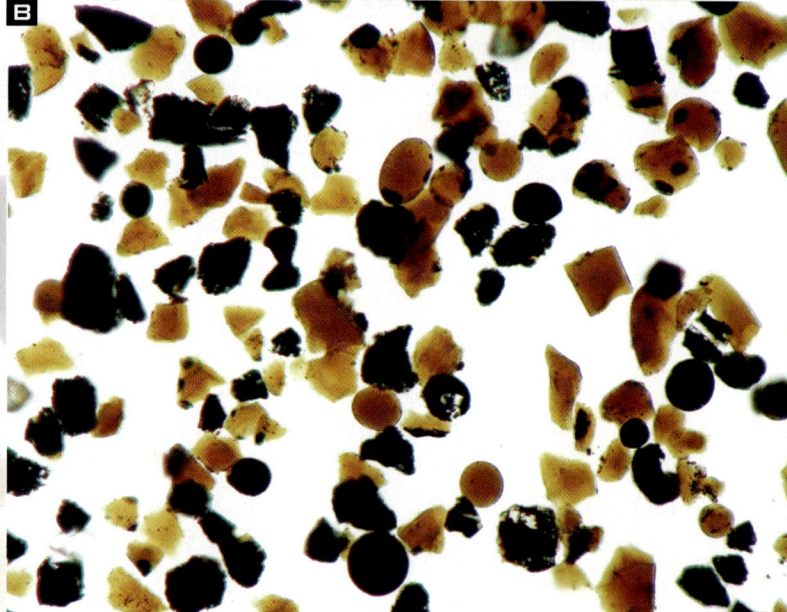

soil because plants need water to grow. Soil moisture tends to contain high levels of dissolved substances, which are the nutrients needed for plant growth. Air fills all the open spaces in soil and tends to contain high levels of carbon dioxide and methane, and low levels of oxygen.

Soil is a complex medium in which all parts interact and play essential roles. Humus retains some of the chemical nutrients released by decaying organisms and by the chemical weathering of minerals. Humus is critical to soil fertility, which is the ability of soils to provide nutrients such as phosphorus, nitrogen, and potassium needed by growing plants. All of the processes that involve living organisms and other soil constituents produce a continuous cycling of plant materials between the regolith and the biosphere. Earth is the only planet in the solar system that has true soil, but soil is not present everywhere. For example, dunes of moving sand, bare rock surfaces of deserts and high mountains, and surfaces of fresh lava do not have soil layers. The other rocky bodies in the solar system have blankets of loose, rocky material (*regolith*) that have sometimes been pulverized to a very fine texture, but their regoliths lack humus and microorganisms (see FIGURE 4.9).

SOIL PROFILES

Soil evolves gradually, from the top down. As erosion removes the top layer, weathering of the underlying material continually creates new soil. When fully developed, soil consists of **soil horizons**, each of which has distinct physical, chemical, and biologic characteristics.

Soil profiles (see FIGURE 4.10) vary considerably, being influenced by such factors as climate and the parent material from which the soil is developed. However, certain kinds of horizons are common to many profiles.

The uppermost horizon in many soil profiles, the *O horizon*, is an accumulation of organic matter. Below it

> **soil horizon** One of a succession of zones or layers within a soil profile, each with distinct physical, chemical, and biologic characteristics.
>
> **soil profile** The sequence of soil horizons from the surface down to the underlying bedrock.

lies the *A horizon*, which is typically dark in color because of the humus present. An *E horizon*, which is sometimes present below A, is typically grayish in color because it contains little humus and the mineral grains do not have dark coatings of iron and manganese hydroxides. Both the A and E horizons have had the soluble minerals leached out of them. E horizons are most common in the acidic soils of evergreen forests.

The *B horizon* underlies the A horizon (or E, if one is present). B horizons are brownish or reddish in color because of the presence of iron hydroxides that have been transported downward from the horizons above. The B horizon is a zone of accumulation, where materials that were leached from the A horizon are redeposited. Clays are usually abundant in the B horizon. The *C horizon* (commonly known as the *subsoil*) is deepest, consisting of parent rock material in various stages of weathering. Oxidation of iron in the parent rock gives the C horizon a yellowish or rusty color.

Beneath the C horizon lays the unweathered bedrock. Scientists studying the rocks of a region search for samples of "fresh," unweathered bedrock in order to form a true picture of a rock's original properties. Such samples can be found in natural outcrops along stream banks, steep sides of mountains, cliff faces where the soils are thin, or at artificial exposures such as highway road cuts, quarries, or mines.

FACTORS THAT INFLUENCE SOIL FORMATION

As we have emphasized repeatedly in this and previous chapters, Earth's four large open systems interact at Earth's surface, and as a result rock is broken down to regolith and regolith eventually forms soil. The five important soil-forming factors are parent material, climate, living organisms, topography, and time.

Parent material Soil develops from regolith, and regolith is of two kinds: *residual regolith* and *transported regolith*. Residual regolith forms when the underlying rock weathers in place; transported regolith develops on sediment that has been transported and deposited elsewhere. Soils developed on residual regolith are called *residual soils*, and they develop slowly because the

Soil profile FIGURE 4.10

This is a typical sequence of soil horizons that would commonly develop in moist, temperate climates. The A horizon, which lies within reach of plant roots, is commonly called the *topsoil*.

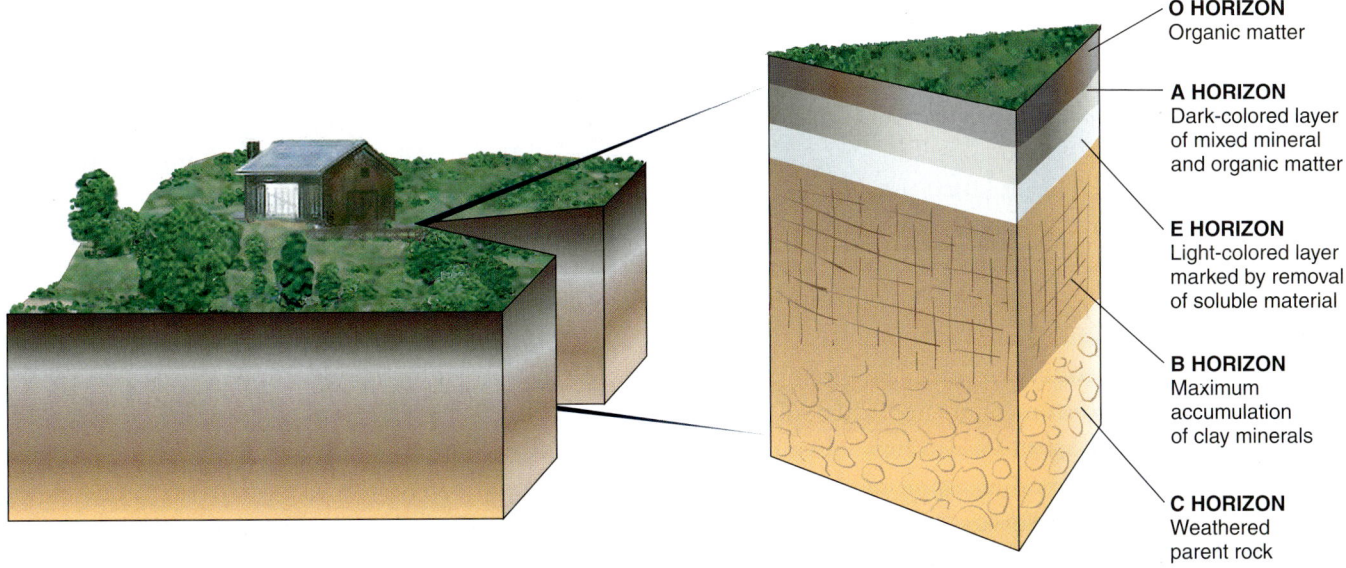

O HORIZON
Organic matter

A HORIZON
Dark-colored layer of mixed mineral and organic matter

E HORIZON
Light-colored layer marked by removal of soluble material

B HORIZON
Maximum accumulation of clay minerals

C HORIZON
Weathered parent rock

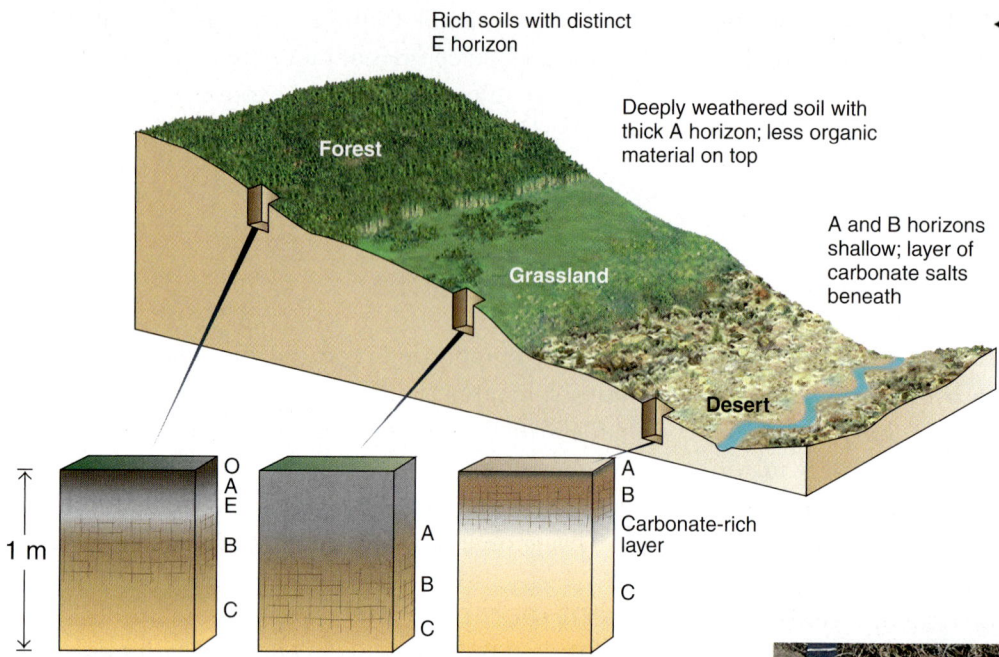

A Soil horizons are strongly influenced by the climatic zone in which they form. For example, a layer of carbonate minerals forms in many desert soils because the hot, dry conditions cause groundwater to evaporate and precipitate the dissolved minerals. The precipitates form a hard layer, sometimes called *hardpan* or *caliche*. If the hard layer is near the surface, plant roots may not be able to grow to their normal depth.

B The photographs show two similar-looking soils from different locations. The "Steedman profile," left, from Kansas, has a dark A horizon whose bottom is indicated by the white arrow. The lighter B horizon extends to the bottom of the picture, and the C horizon (not seen here) is gray and starts just below this photo. The "Windsor profile," right, from Connecticut, also has a dark A horizon and a brown-colored B horizon. The yellowish brown C horizon is the sandy parent material that was deposited by glacial meltwaters.

Climate and soils FIGURE 4.11

weathering processes that break down rocks to form regolith are slow. Rock and mineral particles in transported regolith are already separate small grains, so that the processes that form *transported soils* proceed more rapidly.

Climate You might not think that the climate above the ground would have anything to do with the soil profile underground, but in fact they are closely related. As FIGURE 4.11 shows, the soil profile in an arid region is quite different from a region that gets plenty of rainfall. Recognizing the variability of soil texture and humus content, and the important role played by climate, soils of the world have been classified according to a system developed by scientists of the U.S. Soil Conservation Service, in cooperation with soil scientists of many other nations. There are 12 *soil orders* (see TABLE 4.1) and many suborders. The importance of climate in the development of soil is apparent in FIGURE 4.12 on pages 114–115. For example, oxisol soils develop in the tropics and are found near the Equator, while organic-rich spodosols form in cold, boreal forests, and are found in the high latitudes of Asia and North America.

Living organisms Plants, animals, and a multitude of microorganisms reside in soil, and their importance in soil formation cannot be overstated. Plants are the main source of the organic matter in soil. Microorganisms such as bacteria and fungi break the organic matter down to humus, and animals such as worms, mice, and moles burrow in the soil, mixing the components, and their burrows provide passageways for water and air to enter.

Soil orders* TABLE 4.1

Group I
Soils with well-developed horizons or with fully weathered minerals, resulting from long-continued adjustment to prevailing soil temperature and soil-water conditions.

Oxisols	Very old, highly weathered soils of low latitudes, with a subsurface horizon of accumulation of mineral oxides and very low base status.
Ultisols	Soils of equatorial, tropical, and subtropical latitude zones, with a subsurface horizon of clay accumulation and low base status.
Vertisols	Soils of subtropical and tropical zones with high clay content and high base status. Vertisols develop deep, wide cracks when dry, and the soil blocks formed by cracking move with respect to each other.
Alfisols	Soils of humid and subhumid climates with a subsurface horizon of clay accumulation and high base status. Alfisols range from equatorial to subarctic latitude zones.
Spodosols	Soils of cold, moist climates, with a well-developed B horizon of illuviation and low base status.
Mollisols	Soils of semiarid and subhumid midlatitude grasslands, with a dark, humus-rich epipedon and very high base status.
Aridisols	Soils of dry climates, low in organic matter, and often having subsurface horizons of accumulation of carbonate minerals or soluble salts.

Group II
Soils with a large proportion of organic matter.

Histosols	Soils with a thick upper layer very rich in organic matter.

Group III
Soils with poorly developed horizons or no horizons, and capable of further mineral alteration.

Entisols	Soils lacking horizons, usually because their parent material has accumulated only recently.
Inceptisols	Soils with weakly developed horizons, having minerals capable of further alteration by weathering processes.
Gelisols	Soils underlain by permafrost, with organic and mineral materials churned by frost action.
Andisols	Soils with weakly developed horizons, having a high proportion of glassy volcanic parent material produced by erupting volcanoes.

Base status refers to degree of leaching. *Low* means highly leached; *high* means minimal leaching.

Topography The important topographic variable connected with soil is the slope of the land surface. Slope influences the ability of water to soak into the regolith, and the steeper the slope the faster the rate of erosion. Soils are thin and poorly developed on steep slopes because rainwater runs off rapidly instead of soaking in; as a result, plant life tends to be poorly developed on steep slopes. Soil that does form on a sloping surface tends to mass-waste downslope under the pull of gravity and to be washed off by surface waterflow, so deep soil profiles do not develop on steep slopes.

Time The last of the five important soil-forming variables is time. Many Earth processes are slow and soil formation is no exception. The longer soil-forming processes have been operating, the more mature the soil. Of course, the actual time needed to develop a mature soil profile depends on all the soil-forming factors, but even in the most favorable circumstances, it takes thousands of years for a mature soil to develop.

Soil erosion Because soil formation is part of the never-ending rock cycle, soil is not static. Soil can be formed, and it can be depleted by natural processes, but natural processes tend to work slowly compared to human activities. Soil can be very strongly affected by human activities, and this makes proper soil management a vital issue (see the *Case Study* on page 116).

Soil: The Most Important Product of Weathering

World soils and soil classification system FIGURE 4.12

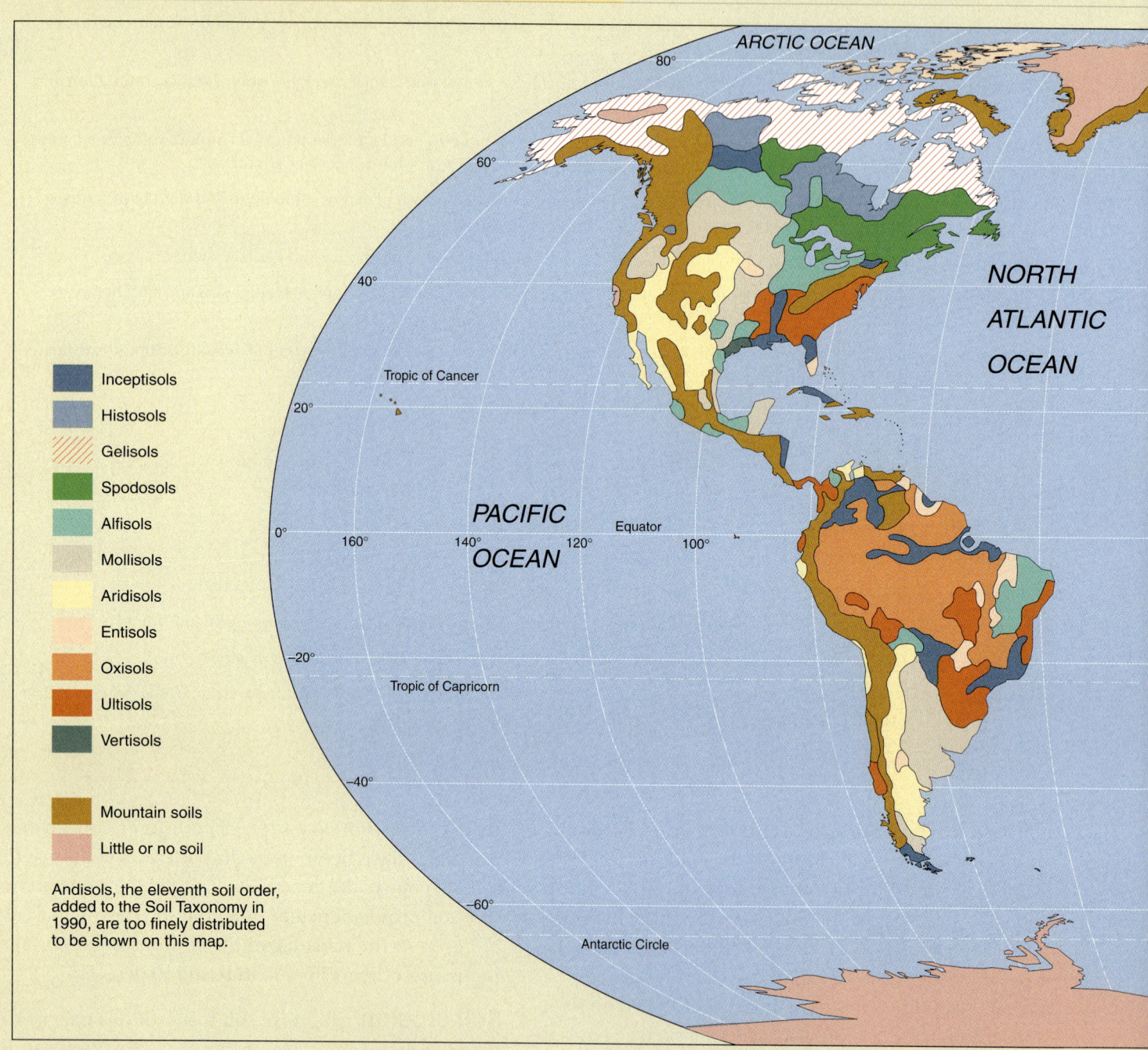

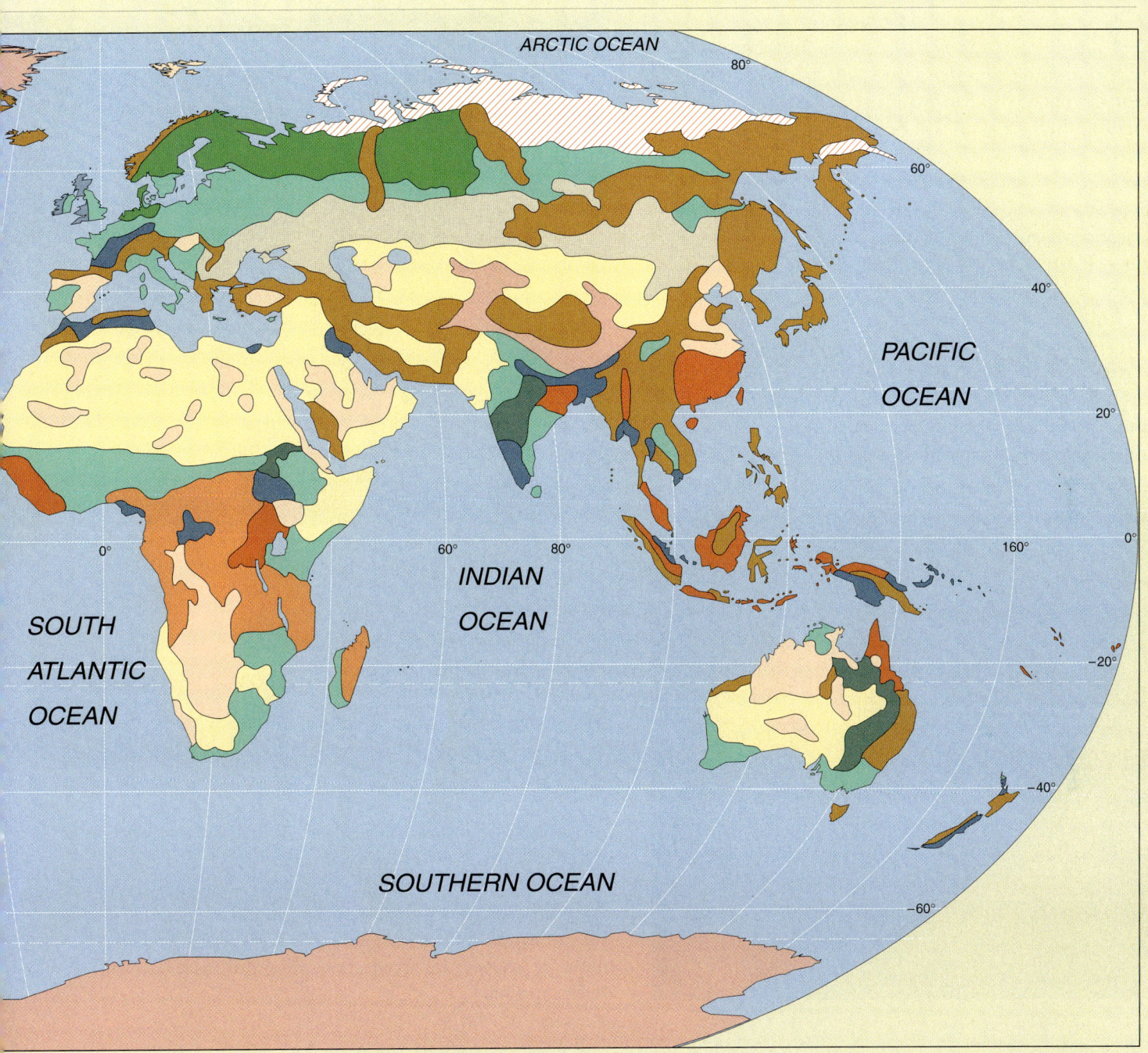

Soil: The Most Important Product of Weathering 115

CASE STUDY

Bad and Good Soil Management

Providence Canyon in Georgia is a gorgeous example of deeply weathered soil, but it is also a dreadful example of poor soil management. In photograph **A**, you can readily spot the dark brown A horizon, the bright red B horizon that is full of clay, and the paler E horizon. Some people have called Providence Canyon the "Little Grand Canyon," but in reality the two are very different. While the layered appearance of the Grand Canyon is due to the strata of bedrock, the appearance of Providence Canyon is caused by soil alone.

Another difference is age. Providence Canyon, believe it or not, is less than 200 years old. There was no canyon here when settlers from Europe arrived in the early 1800s. But the farmers plowed straight up and down the hills, and the furrows rapidly developed into gullies. By 1850, the gullies were 1 to 2 m deep. The farmers had to abandon their fields, but by then

Global Locator

erosion in the gullies was running amok. The canyon is now more than 50 m deep. Unfortunately, there are many such locations in North America.

To fight erosion, modern farmers use contour plowing (**B**). Instead of going in straight lines, the furrows follow the contour of the land, slowing runoff and inhibiting the formation of gullies.

Even so, the erosion of farmland soil is a massive worldwide problem. In the United States, the amount of agricultural soil eroded each year exceeds the amount of replenished soil by about a billion tons. For every kilogram of food we eat, the land loses 6 kg of soil. Although there is a small *sustainable farming* movement, we are very far from consuming only as much as we can put back.

CONCEPT CHECK STOP

What are the primary differences between clay, sand, and soil?

Why is Earth the only planet with true soil?

What are the soil horizons normally observed in temperate climates?

What factors control the formation of soil?

How do good and bad soil management practices affect the landscape? (See the *Case Study*.)

Erosion and Mass Wasting: Gravity at Work

> **LEARNING OBJECTIVES**
>
> **Explain** the distinction between weathering, erosion, and mass wasting.
>
> **Define** turbulent and laminar flow.
>
> **Describe** how ice, water, and air transport regolith across Earth's surface.
>
> **Define** and give examples of mass wasting by slope failure.
>
> **Define** and give examples of mass wasting by sediment flow.

Be careful not to confuse the terms *weathering* and *erosion*. Weathering is the breakdown of rock in place to form regolith; erosion is a term that describes the transport of regolith from one place to another. Both processes can, of course, happen at the same time; for instance, a rock can be abraded (a weathering process) and the particles that break off the bedrock can be transported elsewhere by the wind (an erosion process).

Erosion requires a natural fluid to pick up and transport the weathered material. That requirement differentiates erosion from mass wasting, when regolith moves downslope under the pull of gravity, with no transporting medium required. (We will discuss mass wasting later in this section.)

The fluids that cause the most erosion on Earth are water, wind, and ice. (Even though we usually think of ice as a solid, it does flow when it forms a glacier or ice sheet.) As discussed in Chapter 3, different fluids display differing degrees of runniness, or *viscosity*. Ice behaves as an extremely viscous fluid, but it flows so slowly that its motion cannot be seen by the human eye. Water is much less viscous and flows freely, and air is the least viscous. A fluid's viscosity in turn partly determines whether its flow is *laminar* or *turbulent*. In laminar flow, all fluid particles travel in parallel layers. Turbulent flow is erratic and complex, full of swirls and eddies. Turbulent flow is more effective at picking particles up off the ground. Airflow is almost always turbulent; waterflow is usually turbulent except when the velocity of flow is very low; the flow of glacial ice is laminar.

EROSION BY WATER

Erosion by water begins even before a distinct stream has formed on a slope. This happens in two ways: by impact, which occurs when raindrops hit the ground and dislodge small particles of soil, and by overland flow, which occurs during heavy rains. Overland flow involves water moving as sheets over the ground, not in channels. When the water starts flowing in a channel, particles are moved in several ways: The largest particles, which form the **bed load** (boulders, cobbles, and pebbles), roll or slide along the stream bed due to the force of the flowing water. Smaller (sand-sized) particles move along the stream bed by **saltation** (see **FIGURE 4.13** on the following page).

The particles in the **suspended load**—silt and clay—are small and do not sink to the bottom as long as the water is flowing. If you took a sample of water from a muddy stream and let it stand, the silt and clay would settle to the bottom as mud. But in the turbulent environment of a stream, the upward-moving currents keep them from sinking. Thus, mud deposits form only where velocity decreases and turbulence ceases, as in a lake, the sea, or a reservoir.

> **bed load** Sediment that is moved along the bottom of a stream.
>
> **saltation** Sediment transport in which particles move forward in a series of short jumps along arc-shaped paths.
>
> **suspended load** Sediment that is carried in suspension by a flowing stream of water or wind.

Bed load and suspended load FIGURE 4.13

A A stream's bed load consists of the particles that are too heavy to stay suspended in the water. Pebbles creep or roll along the bottom, while sand-sized particles move in small jumps, called *saltation*. Very fine silt and clay particles form the suspended load and give the water a muddy appearance.

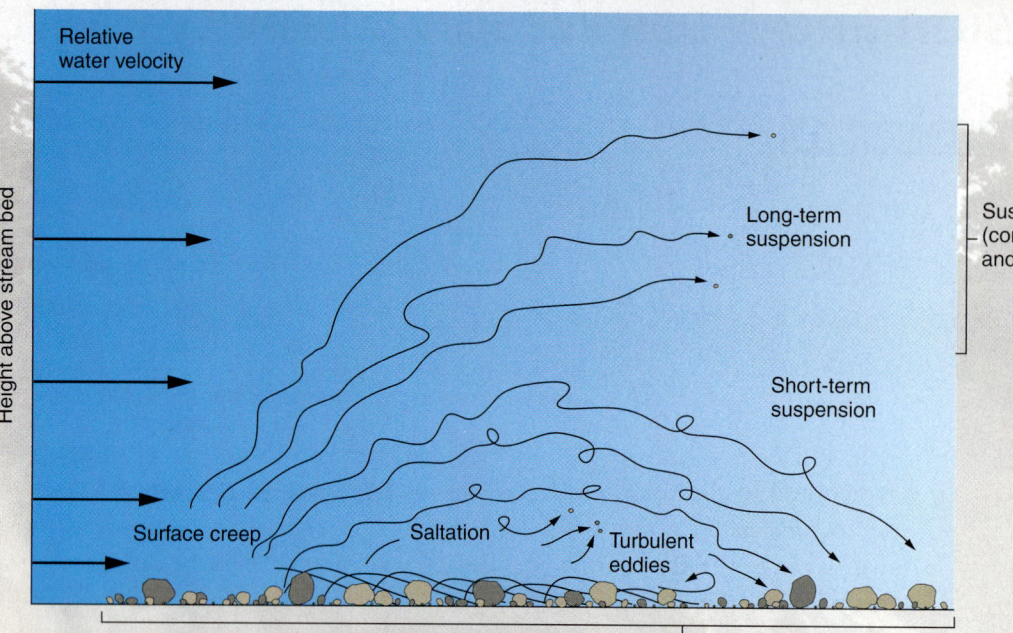

B The turbulent flow of this river in Gabon, Africa, enables it to carry a large suspended load.

Streams also carry a *dissolved load* of soluble materials released by chemical weathering. The dissolved load may also contain organic matter, which accounts for the infamous "black water" found in rivers that drain swamps.

EROSION BY WIND

Because the density of air is about 800 times less than that of water, air cannot move the large particles that water flowing at the same velocity can move. In exceptional cases such as hurricanes and tornadoes, winds can reach speeds of 300 km/h and sweep up coarse rock particles several centimeters in diameter. In most regions, however, wind speeds rarely exceed 50 km/h, a velocity that is described as a strong wind. As a result, most (at least 75%) of the sediment transported by wind occurs through saltation of sand grains. Only the finest particles of dust remain aloft long enough to be moved by suspension. Even so, space-based imaging has recently given Earth scientists a greater appreciation for the massive amounts of material that can be moved by wind (see **FIGURE 4.14**).

EROSION BY ICE

Ice is a solid. However, it does flow under the influence of gravity, albeit very slowly, in parts of the world where there is enough year-round ice to form a **glacier**. Compared with water and air, ice is extremely viscous. Glacial ice therefore moves only by laminar flow (see Figure 6.16).

glacier A semipermanent or perennially frozen body of ice, consisting largely of recrystallized snow, that moves under the pull of gravity.

Glaciers play a three-part role in erosion and transport: They act as a plow, a file, and a sled. As a plow, a glacier scrapes up weathered rock and soil and plucks out blocks of bedrock (see **FIGURE 4.15A** on the following page). As a file, the load of sediment rasps away and polishes the bedrock (see **FIGURE 4.15B**). As a sled, a glacier carries away the load of sediment acquired by plowing and filing, as well as additional debris that falls onto it from adjacent slopes (see **FIGURE 4.15C**).

GRAVITY AND MASS WASTING

Landscapes may seem fixed and unchanging, but if you made a time-lapse movie of almost any hillside for a few years you would see that the slope changes constantly as a result of **mass wasting**. There is no such thing as a static hillside—a lesson learned all too often by people who live at the top or the bottom, or on a steep slope. Exactly how movement happens and how fast it happens is controlled by

mass wasting The downslope movement of regolith and/or bedrock masses due to the pull of gravity.

Storms from Africa FIGURE 4.14

In this satellite photo, winds blowing from east to west lift a colossal cloud of dust from the Sahara Desert into the air. The dust can travel all the way across the Atlantic Ocean to America, and it has been implicated in the bleaching of coral reefs off the coast of Florida.

Erosion and Mass Wasting: Gravity at Work 119

Glacial erosion FIGURE 4.15

▲ **A** This rocky debris was deposited at the edge of Matunuska Glacier, Alaska. Note the extreme range in size of the rock fragments, from boulders much larger than the person in the photo down to tiny pebbles.

▲ **B** This polished and grooved rock surface was produced by the Findelen Glacier in the Swiss Alps. The glacier has retreated in modern times, exposing the weathered rock beneath. A famous mountain, the Matterhorn, is in the background.

C Two ice streams, bearing debris that has fallen from adjacent mountain slopes, merge to form the Kaskawulsh Glacier in the Yukon, Canada. The smooth, parallel streams are a hallmark of laminar flow.
◄

slope failure The falling, slumping, or sliding of relatively coherent masses of rock.

flow Any mass-wasting process that involves a flowing motion of regolith containing water and/or air within its pores.

the composition and texture of the regolith and bedrock, the amount of air and water in the regolith, and the steepness of the slope. For convenience, we divide mass wasting into two categories: **slope failure** and **flow**. We illustrate several types of slope failures and sediment flows in Figures 4.16 and 4.17.

Slope failures occur as one of three basic types (**FIGURE 4.16**). A **fall** is a sudden vertical, or nearly vertical, drop of rock fragments or debris. Rockfalls and debris falls are sudden and usually very dangerous. **Slides** involve rapid displacement of a mass of rock or sediment in a straight line down a steep or slippery slope. A **slump** involves *rotational* movement of rock and regolith—that is, downward and outward movement along a curved surface. Slumps often result from poor engineering practices, for example, when slopes have been oversteepened for construction of buildings or roads.

Slope failures FIGURE 4.16

A Kaibito Canyon in Arizona has been the site of repeated rockfalls, as you can see both from the debris at the base of the cliff and from the scars on the cliff face where rocks have detached themselves in the recent past. ▼

C A slump is a slower kind of failure in which the debris moves rotationally (as shown by the curved arrow). This slumping failure occurred in central California.

B In this rockslide in the Andes mountains in Argentina, the rocks moved in a roughly straight line from the point of detachment to the valley floor.

Erosion and Mass Wasting: Gravity at Work   121

Flowing regolith can be either wet or dry (**FIGURE 4.17**). *Slurry flows* occur when the regolith is saturated with water; they can occur either rapidly or slowly. Rapid slurry flows can move at speeds up to 160 km/h and are very dangerous. Slow slurry flow, a process known as *solifluction*, is common in areas with high rainfall, where soil is thin over bedrock, and in areas of permafrost, where the ground is frozen at depth. Flowing regolith that is not water saturated is called a *granular flow*. Like slurry flows, granular flows can be either slow or fast. The most common kind of granular flow (and the most common kind of mass wasting) is called **creep**. Movement is aided by expansion and contraction of regolith due to freezing and thawing, or to wetting and drying.

> **creep** The imperceptibly slow downslope granular flow of regolith.

Expansion is perpendicular to the slope of the ground and contraction is vertically down, so each cycle moves particles a tiny step downslope. More rapid granular flows include *debris avalanches*, which are rare, spectacular, and extremely dangerous. Debris avalanches often start as a rockfall or slide but gain speed when the material pulverizes and begins to flow downslope like a fluid. Debris avalanches also can be triggered by earthquakes or volcanic eruptions.

TECTONICS AND MASS WASTING

The locations of the world's major historic and prehistoric landslides tend to cluster along belts that lie close to the boundaries between converging plates of lithosphere. They do so for two main reasons.

Flows FIGURE 4.17

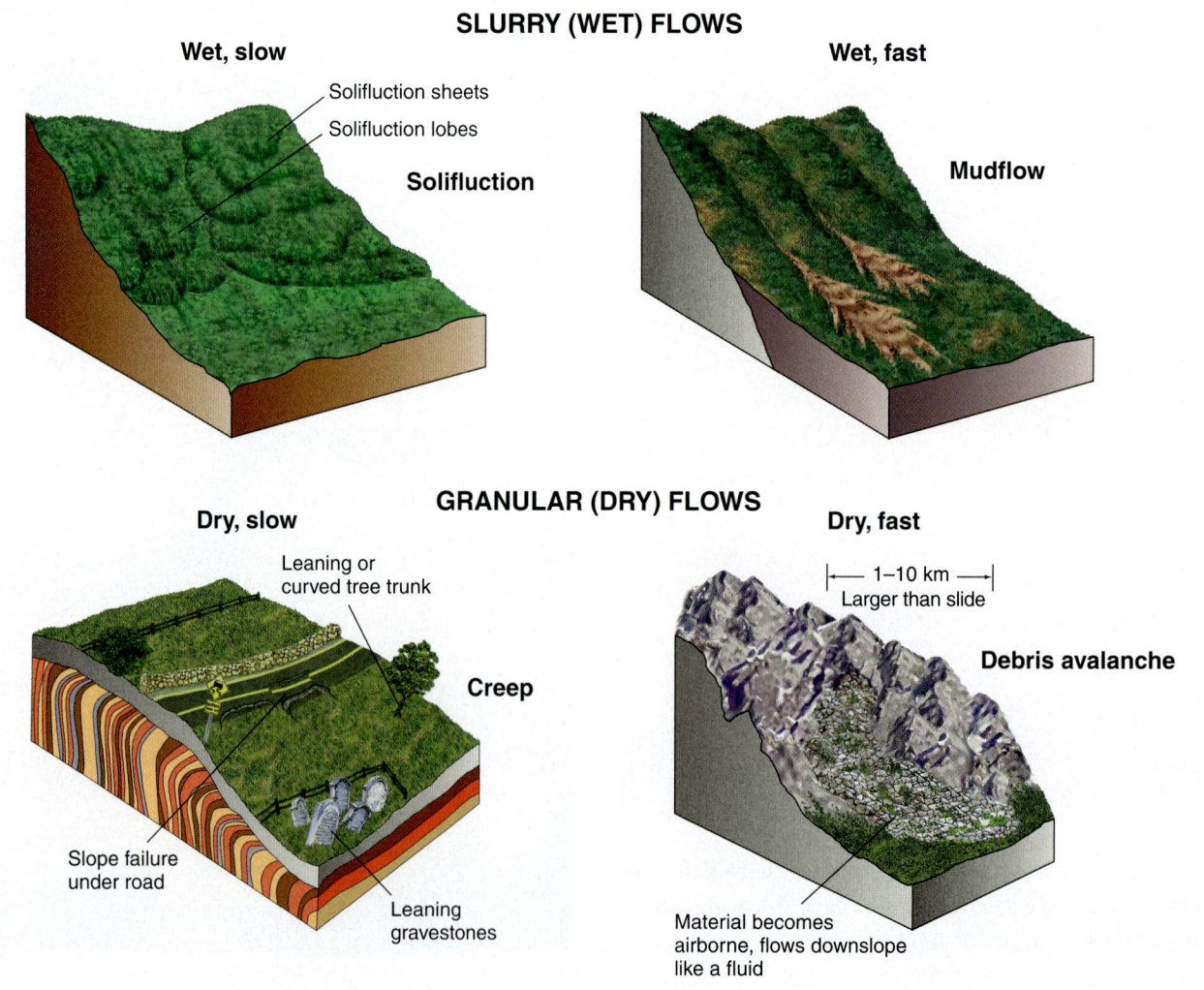

Why landslides occur near plate boundaries FIGURE 4.18

Earthquakes, particularly those close to plate edges, often trigger landslides. The subduction-related quake of March 28, 1964, the Great Alaska Earthquake, caused many slides.

A In downtown Anchorage, the Turnagain Heights district broke apart and slid toward the sea.

B Farther inland, part of the Chugach Mountains collapsed and caused a flow that covered part of the Sherman Glacier.

First, the world's highest mountain chains lie at or near plate boundaries, and the rocks of many mountain ranges consist of well-jointed strata that have been strongly fractured and deformed as they were uplifted. Both the joint planes and the bedding surfaces are potential zones of failure. Volcanoes found in the same regions also tend to have steep slopes conducive to landslides.

Second, most large earthquakes occur along the boundaries between plates where plate margins slide past or over one another. Earthquakes often trigger landslides in areas where the regolith is unstable. Several historic landslides were directly related to major earthquakes (see FIGURE 4.18).

It may seem as if landslides and mass wasting should ultimately level all the world's mountains and leave the continents as flat, featureless plains. That will never happen, because uplift is always taking place at the same time. For example, at Nanga Parbat, a Himalayan mountain that lies at the boundary between the Indian and Eurasian plates, uplift rates are as high as 5 mm/yr. At this rate, the mountain should increase in altitude by 5000 m every million years. However, high mountains also mean high erosion rates, and erosion is tearing down the mountain almost as rapidly as it is rising. Much of this destruction is the result of mass wasting.

CONCEPT CHECK STOP

Which three fluids are responsible for most of the erosion on Earth?

What are the differences between a stream's bed load, its suspended load, and its dissolved load?

What is the primary agent that causes mass wasting?

Which kind of mass wasting is most common, and which kinds are most dangerous?

Why is mass wasting most rapid in regions of active tectonic uplift?

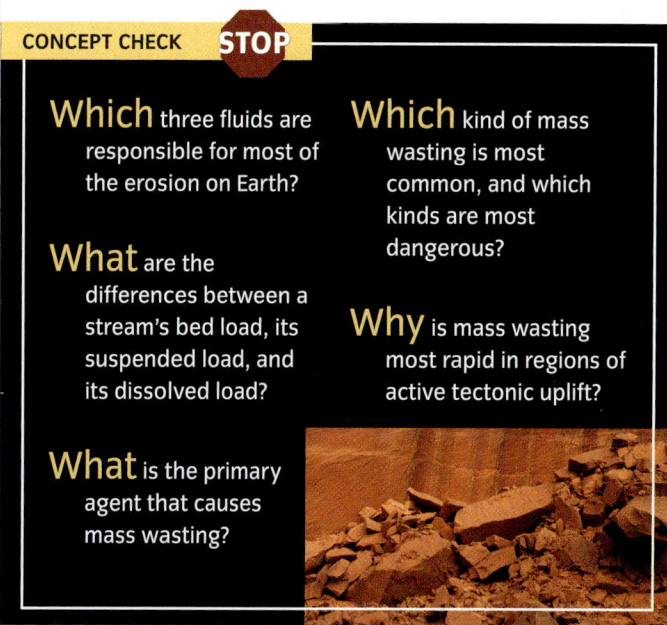

Erosion and Mass Wasting: Gravity at Work

Resources Formed by Weathering and Erosion

LEARNING OBJECTIVES

Explain how chemical weathering can concentrate minerals.

Explain why aluminum and manganese are mined chiefly in the tropics.

Discuss the influence of density on the concentration of minerals.

Chemical weathering involves the chemical breakdown of minerals under the combined attack of air, water, and living organisms. Some of the breakdown products go into solution and are removed, but nonsoluble materials form new minerals and remain in place. Some of the nonsoluble materials are valuable resources.

As discussed earlier in the chapter, feldspar, the most abundant mineral family in the crust, breaks down to form clay minerals. But with continued chemical weathering under the warm and high-rainfall conditions found in the tropics, clay minerals can break down still further, forming the mineral *gibbsite*, principal constituent of *bauxite* according to the following reaction:

$$Al_4Si_4O_{10}(OH)_8 + 2H_2O = 4Al(OH)_3 + 4SiO_2$$
Kaolinite + Water = Gibbsite + Silica

Gibbsite remains as insoluble residue and the silica is removed in solution (see **FIGURE 4.19**). Gibbsite is actually the aluminum-rich member of a family of insoluble residues called *laterites*. Most laterites are iron-rich because the nonsoluble residue is the reddish-colored iron hydroxide mineral, limonite. Another chemical element that is concentrated by chemical weathering under tropical conditions is manganese. Manganese, like iron, forms nonsoluble oxide and hydroxide minerals, and it is these compounds that are found in manganese-rich laterites. Manganese is an important alloying element in steels, and certain of the naturally occurring manganese hydroxide minerals are the essential components of common manganese batteries.

Flowing water can separate heavy particles such as gold from light particles such as quartz (**FIGURE 4.20**). Concentration of high-density particles are called *placer deposits*.

CONCEPT CHECK STOP

How do laterites form?

What important mineral resources are formed by chemical weathering?

How does gold become concentrated in placer deposits?

Residual mineral deposits FIGURE 4.19

Chemical weathering of rock can concentrate insoluble minerals by removing the more soluble ones. The most important mineral formed this way is bauxite, an aluminum ore. In this bauxite sample from Queensland, Australia, rounded masses of aluminum hydroxide (gibbsite) are imbedded in a matrix of iron and aluminum hydroxides.

Gold deposits FIGURE 4.20

The world's largest gold deposits, the Witwatersrand deposits in South Africa, are placer deposits formed about 2.7 billion years ago. The pebbles and associated gold particles were eroded and washed downstream from a source that has never been discovered.

Amazing Places: Monadnock—and Monadnocks

Mt. Monadnock (**A**), a 1156-m peak in New Hampshire, is one of the world's most frequently climbed mountains—no doubt because it is easy to climb and yet rewards the climber with a beautiful view of all six New England states. Its name, which comes from an Algonquin Indian phrase meaning "mountain standing alone," has actually become a generic term for a mountain that rises out of a surrounding plain. (A synonym used more often by Earth scientists is *inselberg*.) Many of the world's most scenic and best-loved peaks are of this form.

Monadnocks (or inselbergs) are isolated by erosion, either because they are unjointed or because they were made of more resistant material than the surrounding land mass. They are often domed because of exfoliation, like Stone Mountain in Georgia (**B**). They can be made of any of the three rock types. Mt. Monadnock is made of schist (a metamorphic rock). Ayers Rock in Australia, seen in the opening photo for Chapter 3, is an inselberg made of sandstone, a sedimentary rock. Another famous inselberg is Enchanted Rock in Texas (**C**). Here a climber tests his skills on the granite face of the rock.

As the map shows, there are lots of excellent examples of inselbergs in the United States.

SUMMARY

1 Weathering: The Earth System at Work

1. The **rock cycle** is the continuous cycle of processes by which rock is formed, modified, transported, decomposed, and reformed. Most of Earth's surface is covered by a blanket of weathered rock, which we call **regolith**. Regolith fragments can range in size from many meters to microscopic. When small regolith particles are altered by biologic processes, the result is **soil**, a material from which rooted plants can extract nutrients.

2. When rocks are exposed at Earth's surface, they are constantly subjected to **weathering**, the process by which air, water, and microbes break down bedrock into smaller rock and mineral fragments. Weathering extends as far down as air, water, and living organisms can readily penetrate Earth's crust. The two categories of weathering are **mechanical**, the physical breakage of rock, and **chemical**, which involves the removal of some minerals in solution and the transformation of others into new minerals that are stable at Earth's surface. The agents of weathering, such as water from rain or melted snow, enter bedrock along fractures and via *pores*.

3. **Joints** are fractures along which no appreciable movement has occurred; they are the main passageways through which agents of weathering enter the rock and lead to weathering. Mechanical weathering takes place in four main ways: by **frost wedging**, or the freezing of water; by the growth of salt crystals in confined spaces; by the prying action of roots; or by **abrasion**, the direct wearing away of bedrock by loose particles transported by moving water, wind, or ice.

4. Chemical weathering involves the removal of some minerals in solution and the transformation of others into new

minerals that are stable at Earth's surface. This type of weathering is caused primarily by water that is slightly acidic. *Acid rain* can be naturally occurring or human-generated. Human-generated acid rain, which is created when rainwater interacts with *anthropogenic* sulfur and nitrogen compounds, is stronger than natural acid rain and causes accelerated weathering.

5. There are three principal processes of chemical weathering—**dissolution**, when minerals dissolve in acidified water; *ion exchange,* in which hydrogen ions from acidic water enter and alter a mineral by displacing larger, positively charged ions; and *oxidation,* a reaction between minerals and oxygen dissolved in water.

6. From a human perspective, chemical and mechanical weathering occur very slowly, over many thousands of years. The effectiveness of weathering depends on the type and structure of rock, the tectonic setting, the steepness of the slope, the local climate, and the amount of biological activity. Chemical weathering is most active in moist, warm climates, whereas mechanical weathering is more active in cold, dry climates.

126   CHAPTER 4   Weathering, Soils, and Mass Wasting

2 Soil: The Most Important Product of Weathering

1. Given enough time, weathering eventually breaks rock down into very fine particles that can be classified, primarily by size, as **clay**, silt, or **sand**. *Clay* refers to a family of hydrous aluminosilicate minerals, but it is more commonly used to describe any tiny mineral particles that have physical properties similar to clay minerals. Silt particles are larger than clay, and sand is larger than silt.

2. Soil is developed on regolith; it is supplemented by organic matter and small organisms that can support rooted plants. Earth's soil is different from the "soil" of other planets because it contains **humus**, partially decayed organic matter. Humus retains some chemical nutrients released by decaying organisms and the chemical weathering of minerals. It is critical for soil fertility, because it provides nutrients that plants need to grow, such as phosphorus, nitrogen, and potassium.

3. Soils weather from top to bottom, and develop distinctive **soil horizons** whose properties are a function of the duration, intensity, and nature of the weathering process. In a typical **soil profile**, the *O horizon* is the topmost layer of accumulated organic matter. The *A horizon* is rich in humus; the A and *E horizons* are layers from which soluble material, especially iron and aluminum, has been lost through leaching. Farther down, clay minerals accumulate in the *B horizon*. Deeper still, the *C horizon* consists of slightly weathered parent rock. Soil profiles can vary greatly from location to location, as they are strongly affected by climate.

4. The factors controlling the formation of soil from regolith are the nature of the parent regolith (whether residual or transported); the climate; the presence of living organisms such as plants, worms, bacteria, and fungi; the slope of the ground; and the time over which the soil profile has developed.

3 Erosion and Mass Wasting: Gravity at Work

1. **Erosion** involves the removal and transport of regolith through the combined actions of ice, water, wind, and gravity. This is different from weathering, which happens in place, though both processes can occur at the same time. Most processes of erosion involve a fluid that picks up and transports material.

2. Both air and water move particles by the process of **saltation**, a mechanism of sediment transport in which particles move forward in a series of short hops along arc-shaped paths. Only the smaller, sand-sized particles move along the bottom of a stream by saltation, whereas the larger particles move by rolling or sliding. These larger particles are known as the **bed load**; lighter particles carried in suspension form the **suspended load**; and dissolved ions released by chemical weathering form the *dissolved load*. The dissolved load may also contain organic matter.

3. Air and water carry suspended particles most effectively through *turbulent* flow. Turbulent flow is dynamic, nonlinear, and generally more effective at picking particles up off the ground. A *glacier* is a permanent body of ice consisting largely of recrystallized snow. Glaciers have very high viscosities and particles in a glacier are moved by *laminar* flow. All fluid particles travel in parallel layers in laminar flow. Glaciers play a significant role in erosion and transport by acting in three ways: as plows, as files, and as sleds.

(continued)

4. **Mass wasting** is the *en masse* downslope movement of rock or regolith under the pull of gravity. In contrast to other types of erosion, the materials moved in mass wasting do not need to be transported by a fluid. **Slope failures** involve downslope movement of relatively coherent masses of rock or regolith. Slope failures can be one of three types: a **fall**, or a sudden vertical drop of rock fragment or debris; a **slide**, which involves rapid displacement of a mass of rock or sediment in a straight line down a steep slope; or a **slump**, a slower type of slope failure that involves the *rotational* movement of rock.

5. **Flows** are mixtures of regolith and water or air. Flowing regolith can be wet or dry. Wet flows include *slurry flows* that can occur rapidly or slowly—a slow slurry flow is also known as *solifluction*. *Granular flows* can be rapid, as in debris avalanches, or slow. The most common type of granular flow, which is also the least noticeable because of its slow motion, is called **creep**.

6. Both weathering and erosion are controlled by climate and topography. Mass-wasting processes—especially landslides—tend to be particularly frequent along plate boundaries, where earthquakes commonly act as triggering mechanisms.

4 Resources Formed by Weathering and Erosion

1. Chemical weathering can lead to a concentration of valuable resources by the chemical breakdown of common minerals into a soluble component that is removed, leaving a nonsoluble residue behind.

2. *Laterites* is the general term for nonsoluble residues left from chemical weathering. Most laterites are iron-rich, but manganese and aluminum residues are also found. Aluminum-rich residues are called *bauxites*.

3. Clastic sediment particles get separated and sorted in flowing water by density and mass. High-density particles of gold concentrated in streams have yielded more than half of the gold ever mined. The largest deposits are in South Africa.

KEY TERMS

- **weathering** p. 100
- **erosion** p. 100
- **regolith** p. 100
- **soil** p. 100
- **mechanical weathering** p. 100
- **chemical weathering** p. 101
- **joint** p. 101
- **dissolution** p. 102
- **clay** p. 109
- **sand** p. 109
- **humus** p. 109
- **soil horizon** p. 110
- **soil profile** p. 110
- **bed load** p. 117
- **saltation** p. 117
- **suspended load** p. 117
- **glacier** p. 119
- **mass wasting** p. 119
- **slope failure** p. 121
- **flow** p. 121
- **creep** p. 122

CRITICAL AND CREATIVE THINKING QUESTIONS

1. Look around for evidence of mechanical and chemical weathering. How might you determine their relative importance in your area?

2. Many features have been recently discovered on Mars that are suggestive of erosion by water. What is the evidence that Mars once had a hydrosphere? How long ago was this? Where did the water go?

3. The Moon lacks an atmosphere, a hydrosphere, and a biosphere, but when the first astronauts landed, they discovered that the Moon has a deep regolith. How might the regolith have been formed, and how does it differ from Earth's regolith?

4. What kinds of mass-wasting processes occur where you live? Can you identify any evidence that would suggest how rapidly or how slowly mass wasting is moving regolith downslope? Look especially for signs of creep, which occurs almost everywhere. Some clues are bent tree trunks, curved fences, lobes of soil on grassy slopes, and tilted gravestones.

5. Keep an eye out for the structures in your town used to stabilize slopes or protect property from mass wasting. Are the slopes in your area engineered, or have they been left more or less in their natural state? Where you find retaining walls, do they appear to have stabilized the slope as intended?

6. Do some research and find out where bauxite is mined today. Now check where aluminum smelters are located. Can you offer explanations why the two are rarely close together?

What is happening in this picture ?

- Mudflows are a particularly rapid and dangerous form of mass wasting. In January 2005, 400 thousand tons of mud cascaded down on the California town of La Conchita, killing 10 people.

- From an Earth scientist's point of view, a town should never have been built in this location. Why?

- What do you think might have caused the mudslide?

SELF-TEST

1. In _____, rock breaks down into solid fragments by physical processes that do not change the rock's chemical composition.
   a. chemical weathering
   b. mechanical weathering
   c. mass wasting
   d. erosion

2. Joints are fractures in rocks that form by _____.
   a. prolonged weathering under deep regolith
   b. intrusion of granite magma into cold rocks
   c. tectonic forces deep underground
   d. meteorite impacts
   e. all of the above

3. This illustration shows the chemical weathering of a common feldspar mineral. The process shown depicts what kind of chemical reaction?
   a. strong chemical
   b. moderate chemical
   c. dissolution
   d. ion exchange

Unaltered feldspar

Alteration products (clay)

4. Death Valley in California is one of the hottest and driest spots in North America. Summer air temperatures commonly reach 50°C and rainfall averages less than 5 cm a year. In this desert environment, what type of weathering would you expect to find?
   a. strong chemical weathering
   b. strong mechanical weathering
   c. moderate chemical weathering
   d. moderate mechanical weathering
   e. very slight weathering, mostly by wind

5. Sediments found on Earth's surface _____.
   a. are one product of the mechanical weathering process
   b. are one product of the chemical weathering process
   c. are the result of a combination of mechanical and chemical weathering processes
   d. None of the above statements is correct.

6. A dark-colored layer of mixed mineral and organic matter defines the _____.
   a. O soil horizon
   b. A soil horizon
   c. E soil horizon
   d. B soil horizon
   e. C soil horizon

7. This illustration shows three climatic zones and accompanying soil profiles. Draw a line on the illustration linking soil profiles to the correct climatic zone.

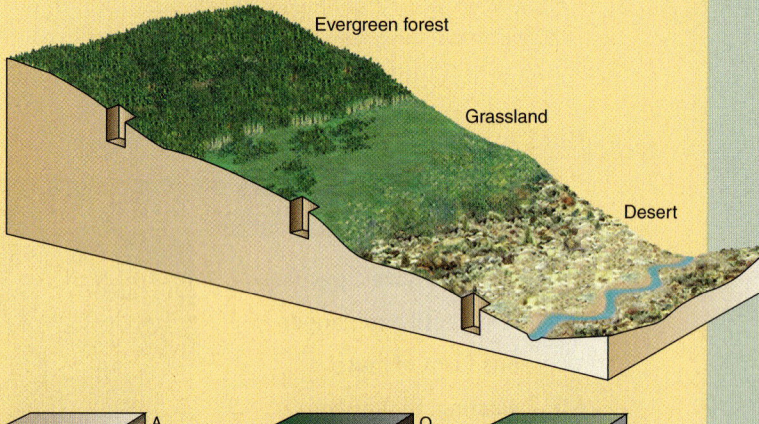

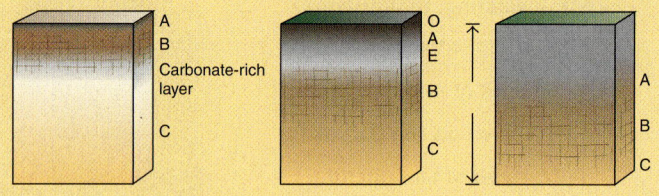

8. _____ involves the removal and transport of regolith through the combined actions of ice, water, wind, and gravity. This is different from _____, which happens only in place, though both processes can occur at the same time. The process of _____ further alters Earth's surface by the downslope displacement of regolith due to the pull of gravity.
   a. Erosion/mass wasting/weathering
   b. Weathering/erosion/mass wasting
   c. Mass wasting/weathering/erosion
   d. Erosion/weathering/mass wasting
   e. Mass wasting/erosion/weathering

9. _____ is dynamic, nonlinear, and generally more effective at picking particles up off the ground than is _____, in which particles travel in parallel layers.
   a. Viscous flow/laminar flow
   b. Viscous flow/turbulent flow
   c. Laminar flow/turbulent flow
   d. Turbulent flow/laminar flow

10. Air and water tend to carry suspended particles most effectively through _____.
    a. a combination of viscous and laminar flow
    b. turbulent flow
    c. laminar flow
    d. a combination of turbulent and laminar flow

11. _____ transport regolith through laminar flow.
    a. Streams   c. Debris flows
    b. Winds   d. Glaciers

12. _____ are a form of slope failure involving rapid displacement of a mass of rock or sediment in a straight path down a steep or slippery slope.
    a. Rockfalls   c. Slides
    b. Slumps   d. Slurries

13. _____ involve rotational movement of rock or regolith.
    a. Rockfalls   c. Slides
    b. Slumps   d. Slurries

14. This illustration shows block diagrams of mass wasting of hill slopes through sediment flow. Identify each block diagram based on processes related to rate and degree of wetness from the following list:

    wet, slow   dry, slow
    wet, fast   dry, fast

15. Gold becomes concentrated in placer deposits because of its _____.
    a. color   c. density
    b. chemical resistance   d. all of the above

Self-Test 131

Water On and Under the Ground

"Mosi-oa-Tunya" or "the smoke that thunders," as it was dubbed by the local Makololo tribe, is the largest single sheet of falling water in the world—over 100 meters tall and 1.5 kilometers wide.

Today, we know this curtain of water as Victoria Falls. The falls are formed as the Zambezi River flows across a basaltic lava plain, then plummets into a chasm about 120 m wide, carved by its waters along a fracture in the basalt. The river is the border between Zambia on the north (to the right in the photograph) and Zimbabwe to the south. The waterfall converts the calm river into a ferocious torrent. The roar, as almost 9.1 million cubic meters (321 million cubic feet) of water per minute thunder over the edge of the cliff, can be heard from 40 km (25 miles) away.

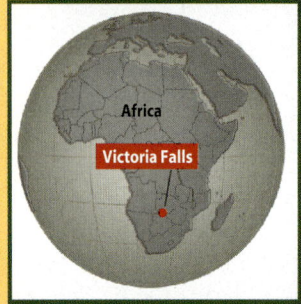

Global Locator

Humans have learned to harness the power of waterfalls, here and around the world. The first power station was set up to generate hydroelectric power in the third gorge below Victoria Falls in 1938. In North America, the Niagara Power project exploits the 52-m (171-ft) drop over Niagara Falls by withdrawing water upstream of the waterfalls and carrying it through tunnels to generating plants located about 6 km (4 mi) downstream.

It's easy to imagine how the immense force of these waterfalls can mold the land below. Over the course of millennia, even gentle flows of running water have sculpted the landforms around us by eroding land and depositing sediment.

Author Skinner (dark helmet) views the Falls from above.

CHAPTER OUTLINE

- The Hydrologic Cycle p. 134

- How Water Affects Land p. 138

- Surface Water as a Hazard and a Resource p. 145

- Fresh Water Underground p. 153

The Hydrologic Cycle

LEARNING OBJECTIVES

Define and describe the hydrologic cycle.

Identify the main pathways in the hydrologic cycle.

Identify the main reservoirs in the hydrologic cycle.

Four great reservoirs make up the Earth system: the lithosphere, hydrosphere, atmosphere, and biosphere. Water moves—and helps move materials—among all four spheres. Water evaporates from the ocean, and water vapor enters the atmosphere. Water is a constituent of many common minerals (such as micas and clays) in the lithosphere, where it is tightly bonded in their crystal structures. And, of course, water is a fundamental component of living things in the biosphere. The **hydrologic cycle**, also called the *water cycle*, describes how water moves among these four reservoirs (**FIGURE 5.1**), and the scientific study of water is called *hydrology*.

hydrologic cycle A model that describes the movement of water through the reservoirs of the Earth system; the *water cycle*.

Process Diagram

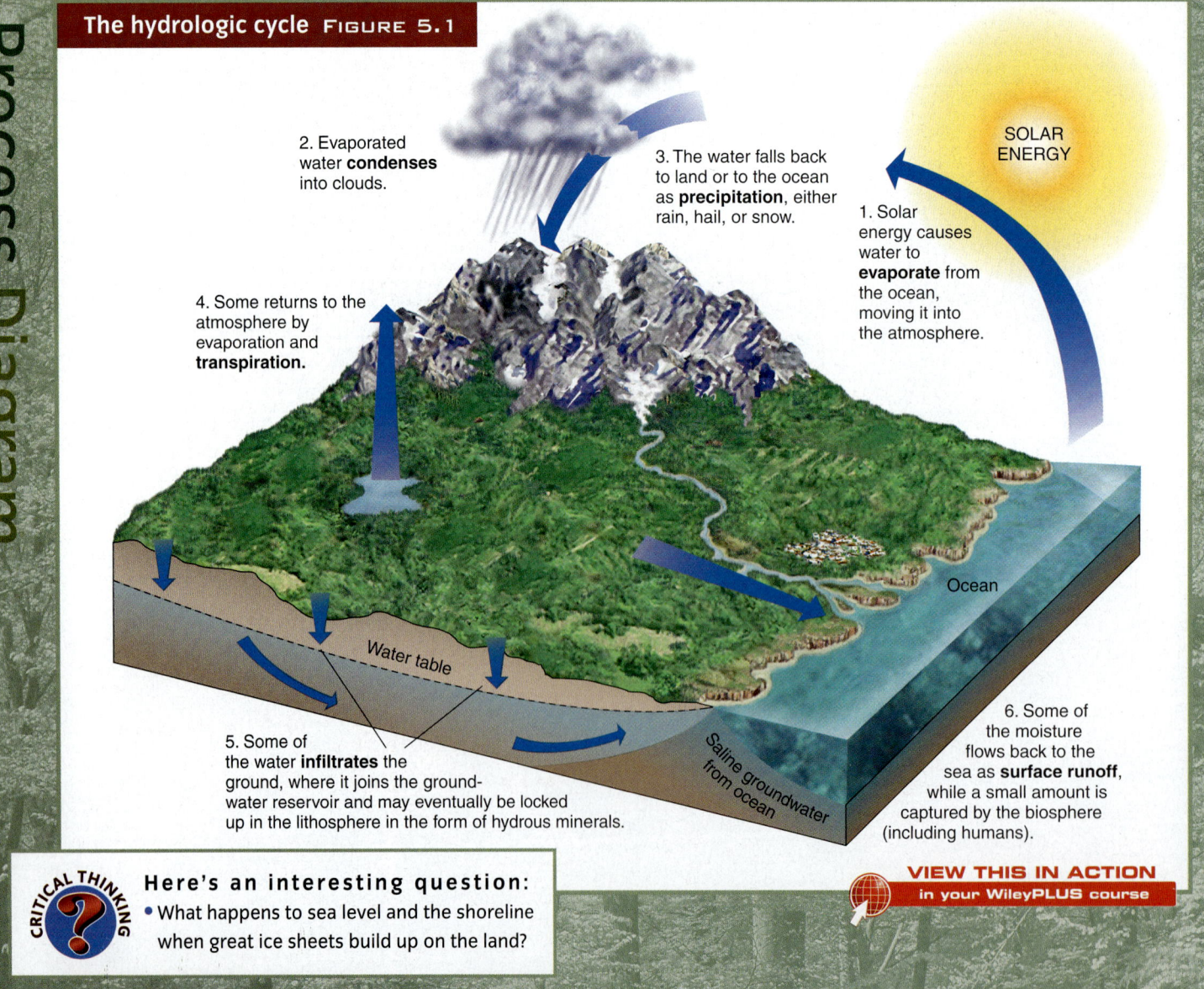

The hydrologic cycle FIGURE 5.1

1. Solar energy causes water to **evaporate** from the ocean, moving it into the atmosphere.
2. Evaporated water **condenses** into clouds.
3. The water falls back to land or to the ocean as **precipitation**, either rain, hail, or snow.
4. Some returns to the atmosphere by evaporation and **transpiration.**
5. Some of the water **infiltrates** the ground, where it joins the groundwater reservoir and may eventually be locked up in the lithosphere in the form of hydrous minerals.
6. Some of the moisture flows back to the sea as **surface runoff**, while a small amount is captured by the biosphere (including humans).

CRITICAL THINKING

Here's an interesting question:
- What happens to sea level and the shoreline when great ice sheets build up on the land?

VIEW THIS IN ACTION in your WileyPLUS course

The hydrologic cycle as part of the Earth system FIGURE 5.2

Diagram of the hydrologic cycle, showing how water moves between the reservoirs to maintain a balanced cycle. Because the amount of water in the cycle is fixed, when any part of the cycle changes, the rest of the cycle adjusts so that a new balance is reached.

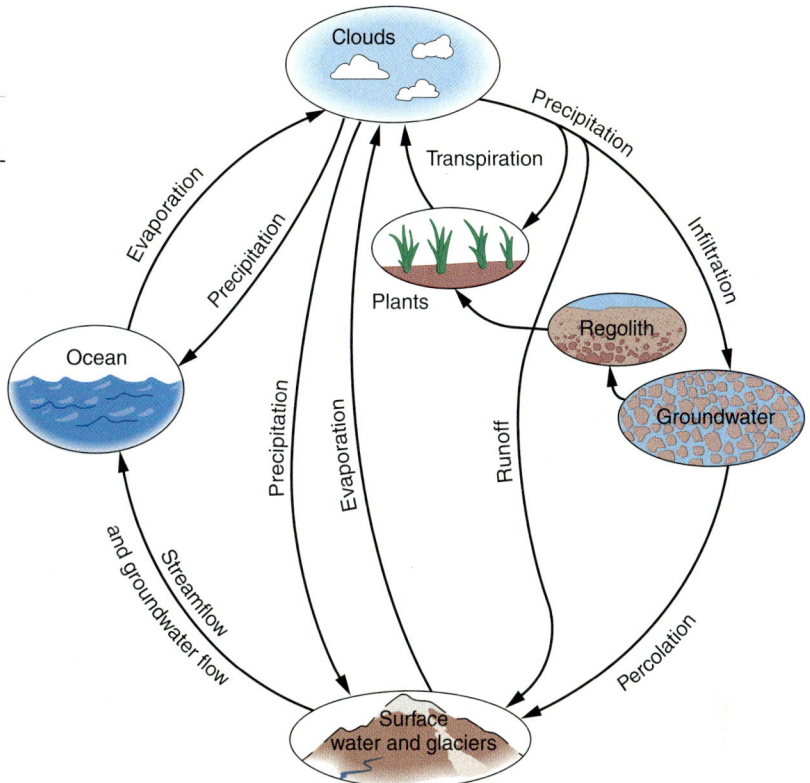

HYDROLOGIC CYCLE

■ **evaporation** The process by which water changes from a liquid to a vapor.

■ **transpiration** The process by which water taken up by plants passes directly into the atmosphere.

■ **condensation** The process by which water changes from vapor into a liquid.

■ **deposition** The process by which water changes from a vapor into a solid.

Water moves through the hydrologic cycle along numerous pathways and processes. These include **evaporation** and **transpiration**, both of which are powered by energy from the Sun. Depending on local conditions of temperature, pressure, and humidity, some of the water vapor in the atmosphere will undergo **condensation**, changing to a liquid, or **deposition**, changing to a solid and falling back to the land or ocean as rain, snow, or hail via the process of **precipitation**. Some of this precipitation becomes **surface runoff**, whereas some trickles directly into the ground via **infiltration**.

WATER IN THE EARTH SYSTEM

The schematic representation of the hydrologic cycle (see FIGURE 5.2) is a *closed cycle* of *open systems*. Because it is a closed cycle, the total amount of water is fixed. However, all of the local reservoirs within the cycle, such as rivers and trees, are free to gain or lose water. They sometimes do so quite dramatically, as during a flood or drought.

The water cycle is easily observable and readily studied. We can measure the amount of global precipitation; using satellite monitoring, we can even measure the amount of evaporation. With these

■ **precipitation** The process by which water that has condensed in the atmosphere falls back to the surface as rain, snow, or hail.

■ **surface runoff** Precipitation that drains over the land or in stream channels.

■ **infiltration** The process by which water works its way into the ground through small openings in the soil.

The Hydrologic Cycle

Visualizing
Reservoirs in the hydrologic cycle FIGURE 5.3

The world's water resources (in proportion):
97.5 liters saltwater (A)

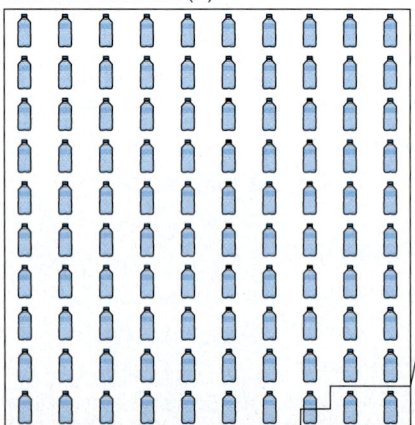

1.85 liters frozen water (B)

0.64 liters groundwater (C)

10 milliliters surface water (D)*
(= 2 teaspoons)

*includes water in biosphere and atmosphere

The vast majority of Earth's water is salty (**A**), frozen (**B**), or underground (**C**). The most visible everyday sources of fresh water, such as rivers, lakes, and the atmosphere (**D**), together comprise less than one hundredth of a percent of Earth's water budget.

Living on the water's edge FIGURE 5.4

Many of the world's great cities are built on riverbanks or coastlines. St. Louis has long benefited from its proximity to the Mississippi River. In the foreground, a towboat pushes barges upstream; in the background, the Mississippi Queen riverboat brings tourists to the Gateway Arch.

measurements, along with the overall mass balance of the water cycle, Earth scientists can roughly deduce how much water is exchanged along each of the pathways shown in Figure 5.2.

It is also instructive to compare the sizes of each of the reservoirs in the water cycle (see **FIGURE 5.3**). The largest reservoir—by far—is the world ocean, which holds 97.5% of Earth's water. Thus, the vast majority of Earth's water is saline (salty), not fresh. This has important consequences for humans, because we depend on fresh water as a resource for drinking, agriculture, and industrial use.

The other numbers in Figure 5.3 may surprise you. Most of Earth's fresh water (almost 74%) is locked up in polar ice sheets, while the majority of the unfrozen fresh water (98.5%) lies underground. Historically, human settlement has concentrated along lakes or rivers, where the other 1.5% of the world's fresh water—namely, surface water—is readily available (see **FIGURE 5.4**). But in most rural areas and quite a few urban areas as well, underground water sources are much more plentiful.

Although water is continuously cycling from one reservoir to another, the total volume of water in each reservoir is approximately constant over short time intervals. However, the volume of water in each reservoir can change dramatically over longer intervals. During glacial ages, for example, vast quantities of water evaporate from the ocean and are precipitated on land as snow. The snow slowly accumulates to build ice sheets that are thousands of meters thick and cover vast areas. At such times, the amount of water removed from the ocean is so large that the global sea level can fall by many meters, and the expanded glaciers increase the ice-covered area of Earth.

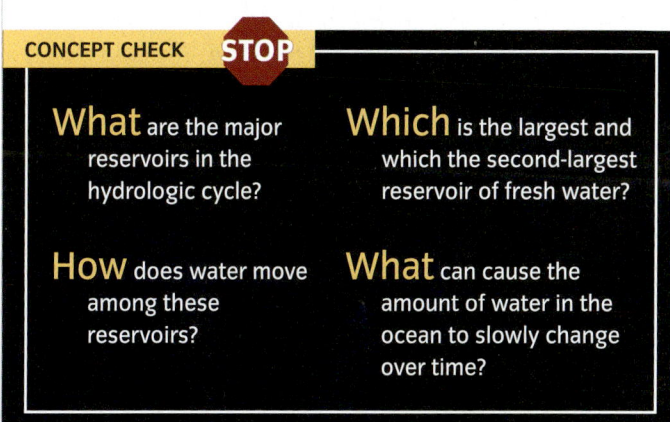

CONCEPT CHECK STOP

What are the major reservoirs in the hydrologic cycle?

How does water move among these reservoirs?

Which is the largest and which the second-largest reservoir of fresh water?

What can cause the amount of water in the ocean to slowly change over time?

The Hydrologic Cycle

How Water Affects Land

> **LEARNING OBJECTIVES**
>
> **Identify** the basic characteristics of streams.
>
> **Explain** how straight, braided, and meandering channels form.
>
> **Identify** three common land formations made by stream deposits.
>
> **Define** a drainage basin.
>
> **Describe** how lakes form and disappear.

If you stand outside during a heavy rain, you can see that, initially, water tends to move downhill in a process called *overland flow* (or *sheet flow*, because the flowing water often takes the form of a thin, broad sheet). After traveling a short distance, overland flow begins to be concentrated into well-defined channels, thereby becoming *streamflow*. Overland flow and streamflow together constitute surface runoff, one of the pathways in the water cycle (Figure 5.2). Let's look more closely at streams, streamflow, and their interactions with the land.

STREAMS AND STREAMFLOW

Every **stream** or *river* has a **channel**. Several factors affect the shape of a channel and the types of landforms it creates. The most important are the channel's **gradient**, **discharge**, and **load**.

Gradient, discharge, and load are interrelated. For example, if the gradient of a stream becomes steeper along a particular stretch of channel, the velocity of the flow is likely to increase as well. If the velocity is high, a greater load can be carried. If the discharge increases, it means that the channel must handle more water in a given period; as a result, both the velocity of flow and the depth of the water in the stream will increase. (Note that this is the first of two different meanings of *discharge* used in this chapter.) In some cases, the channel itself will increase in width and depth as the water scours the banks and the bottom. This scouring in turn adds sediment to the load. When the velocity of the water eventually decreases, the sediment settles out, filling in the channel and allowing it to return to its original size.

stream A body of water that flows downslope along a clearly defined natural pathway.

channel The clearly defined natural passageway through which a stream flows.

gradient The steepness of a stream channel.

discharge The amount of water passing by a point on the channel's bank during a unit of time.

load The suspended and dissolved sediment carried by a stream.

Types of channels Streams create landforms through two processes: *erosion* (see Chapter 4) and *deposition* (see Chapter 3). Both processes go on throughout a stream's existence and along its length, but one or the other may predominate at a particular location or during a particular time, depending on a variety of factors.

The gradient, discharge, and load are all important determinants of a channel's size and shape. This means that topography (which determines gradient), climate (which determines the amount of precipitation), and the general character of the landscape through which the stream flows are all important factors, too. The physical and chemical characteristics of the underlying rocks are also important. For example, a stream may suddenly bend or its gradient may increase when it passes from erosion-resistant rock into rock that is easily eroded. Because these factors interact in different ways, no two channels are exactly alike. Nevertheless, we can classify them into three broad categories: *straight, meandering,* and *braided* (see **FIGURE 5.5**).

Unlike engineered aqueducts and canals, natural streams are never really straight from start to finish.

Three kinds of streams FIGURE 5.5

A Straight channels, such as this stream that drains a glacier in Alaska, usually occur only in relatively short stretches. They are often found where streams have a high gradient (near the stream's headwaters), and they generally have a classic V-shaped valley shaped by mass wasting and overland flow.

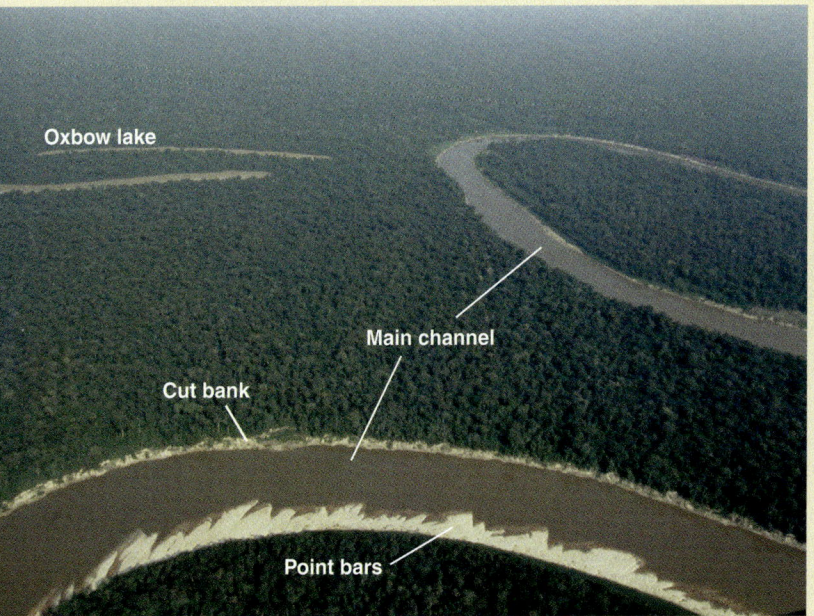

B Rio Itaquai, a tributary of the Amazon River in Brazil, shows several features typical of a meandering river: a low gradient, light-colored sandy bars on the inside edge of the bends, and nearby oxbow lakes marking previous watercourses that have been cut off.

C This stream flows north across the Arctic National Wildlife Refuge in Alaska. It has the classic profile of a braided stream, with a low gradient and a large and variable load of sediment—produced in this case by the action of glaciers.

How Water Affects Land

Straight channels may occur over short distances, particularly in *upstream* areas (that is, near the *headwaters* or source of the stream), where the gradient is high and the channel deeply incised. Even in a "straight" channel, close examination will show that the deepest part of the channel oscillates from side to side.

Meandering streams tend to develop where the stream gradient is low, typically in the lower or *downstream* parts of a stream system (close to the *mouth*, where the stream empties into a lake or the sea). The erosion in a meandering stream concentrates along the sides of the channel rather than the bottom. As water sweeps around a bend, it flows more rapidly along the outer bank, undercutting and steepening it to form a *cut bank*. Meanwhile, along the inner side of each meander, where the water is shallow and velocity is low, sediment accumulates to form a *point bar*, as shown in Figure 5.5. Thus, meanders slowly change shape and shift their position along a valley as the stream erodes material from one bank and deposits sediment on the other. Sometimes the water finds a shorter route downstream, bypassing a meander by cutting across the narrow part of the loop. As sediment is deposited along the banks of the new channel route, the former meander is cut off and converted into a curved *oxbow lake* (**Figure 5.6A**).

Braided channels arise when a stream's ability to move its sediment load varies over time. At times of high flow, a stream can carry more sediment. If the discharge decreases but the load does not, the stream deposits the excess sediment in its own channel as bars or islands. These variations in flow over time cause the channel to repeatedly divide and reunite, as shown in Figure 5.5. Braided patterns tend to form in streams with a highly variable discharge and large load of coarse sediment. Braided patterns can form in streams with discharge that varies seasonally (e.g., following rapid snowmelt) and in streams with easily eroded banks.

STREAM DEPOSITS

Stream deposits form along channel margins, valley floors, mountain fronts, and at the stream's mouth where it opens into the ocean or a lake. These are all places where the stream loses energy, and therefore its ability to carry a load. A point bar is one example of a stream deposit. Others include *floodplains, alluvial fans,* and *deltas.*

When a stream rises during a flood, water overflows the banks and inundates the **floodplain** (see **Figure 5.6**). As sediment-laden water flows out of the channel, its depth, velocity, and turbulence decrease abruptly at the margins of the channel. The result is a sudden, rapid deposition of the coarser part of the load along the margins, which builds up a broad, low ridge of **alluvium** atop each bank, called a *natural levee*. Farther away, the finer particles settle out in the quiet water covering the valley. This creates the broad, flat, fertile land that is typical of floodplains.

Another kind of alluvial structure develops where a stream draining a steep upland region suddenly emerges onto the floor of a much broader lowland valley. The stream will slow down and lose some of its ability to carry sediment. It will deposit the coarser part of its load (the part that can no longer be transported) in an *alluvial fan* (see **Figure 5.6B**). Such fans are typical of semiarid conditions where vegetation is sparse and infrequent rainfall creates streams that are heavily laden with sediment.

A similar situation occurs when a stream flows into a standing body of water, such as an ocean or lake. It quickly loses velocity and fans out, dropping its sediment load (the heaviest particles first and the finer particles farther seaward). Over time, the sediment builds up a deposit called a *delta*. Most of the world's great rivers, including the Nile, Ganges-Brahmaputra, Huang He, Amazon, and Mississippi, have built massive deltas (see **Figure 5.6C**).

> **floodplain** The relatively flat valley floor adjacent to a stream channel, which is inundated when the stream overflows its banks.
>
> **alluvium** Unconsolidated sediment deposited by a stream.

Visualizing
Stream valley deposits FIGURE 5.6

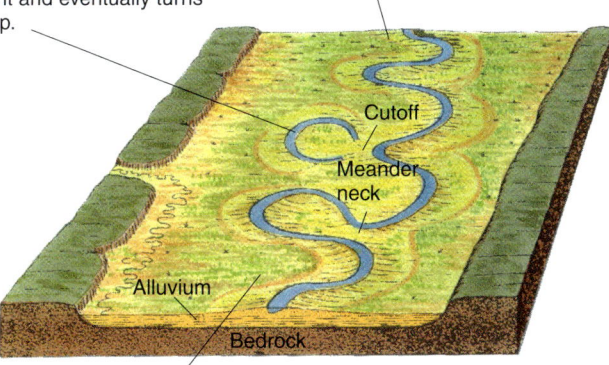

Oxbow lake After the cutoff, silt and sand seal the abandoned channel, producing a lake. The lake fills with fine sediment and eventually turns into a swamp.

Floodplain The meandering river channel dominates the floodplain. Parts of the channel were abandoned after cutoff events, when the river cut across a meander loop to create a shorter, more direct path.

Natural levees are created during flooding, when sand and silt are deposited next to the channel creating belts of higher land on either side of the channel.

▲ **A** Some of the landforms created by sediment in stream valleys are floodplains, oxbow lakes, natural levees, and alluvial fans.

B Where this stream emerges from the mountains into Death Valley, California, it abruptly slows and deposits its sediment load. This has created a symmetrical alluvial fan, covered by a braided system of channels. The stream was dry at the time the photograph was taken.

C Where the Nile River empties into the Mediterranean Sea, the sediment it deposits has formed a fan-shaped delta that supports the green vegetation seen in this satellite image. Its triangular shape, similar to the Greek letter delta (δ), gave rise to the term *delta* that we use today.

How Water Affects Land

DECONSTRUCTING A COAST

1839

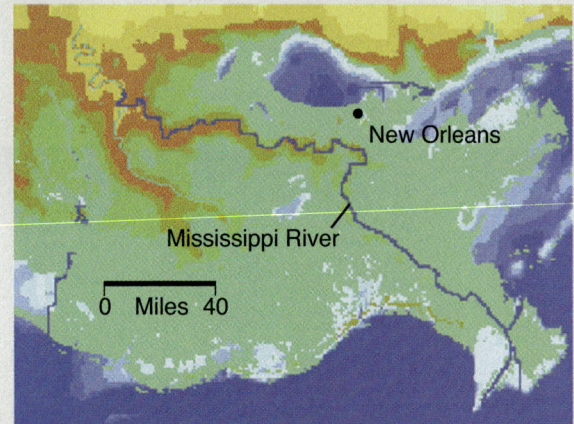

Mississippi River
0 Miles 40
New Orleans

1993 ☐ Land loss on coast

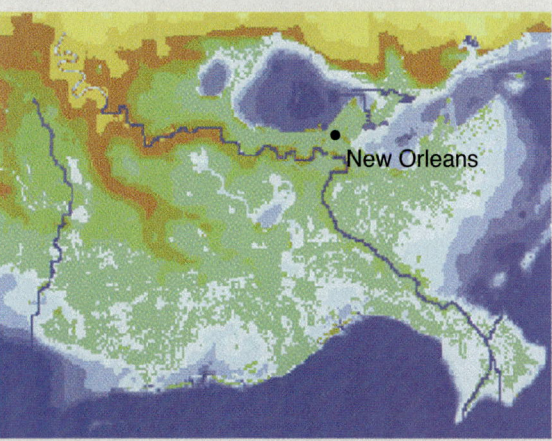

New Orleans

2090 *Projection*

New Orleans

Disappearing coastline FIGURE 5.7

The Mississippi River has been changed by levees and channels, slowing deposition of new sediment in its delta. As a result, the Mississippi Delta has been shrinking. Marshes give way to open water and barrier islands shrink. The top image shows the delta in 1839, the middle image in 1993, and the bottom image shows a projection in 2090. The degradation of the delta is partly responsible for the vulnerability of New Orleans to hurricane damage.

Prominent deltas do not form in places where strong wave, current, or tidal action redistributes sediment as quickly as it reaches the coast. However, if the rate of sediment supply exceeds the rate of coastal erosion, a delta will form. The converse holds, too: If the rate of deposition slows down, the delta will disappear (see **FIGURE 5.7**).

LARGE-SCALE TOPOGRAPHY OF STREAM SYSTEMS

Streams are governed by a simple principle: Water flows downhill. Rainwater that falls on the land surface will move from higher to lower elevations under the influence of gravity. A stream's headwater region is the area of relatively higher elevation from which streams have their source. Small, high-gradient *tributary streams* carry water downslope from the headwater region, combining their flow to form a larger stream. The gradient gradually decreases toward the low-lying region of the stream's mouth.

Every stream is surrounded by its **drainage basin** (sometimes called a *catchment* or a *watershed*). Drainage basins range in size from less than a square kilometer to areas the size of subcontinents. In general, the greater a stream's annual discharge, the larger its drainage basin. The vast drainage basin of the Mississippi River encompasses more than 40% of the total area of the contiguous United States (see **FIGURE 5.8**). From an environmental perspective, a drainage basin is a more natural geographic entity than a country or a state, because issues of water supply, pollution,

> **drainage basin**
> The total area from which water flows into a stream.

The Mississippi River drainage basin FIGURE 5.8

The drainage basin of the Mississippi River encompasses most of the midwestern United States and extends into southern Canada. In this diagram, the widths of the rivers are exaggerated to represent the discharge in cubic meters per second.

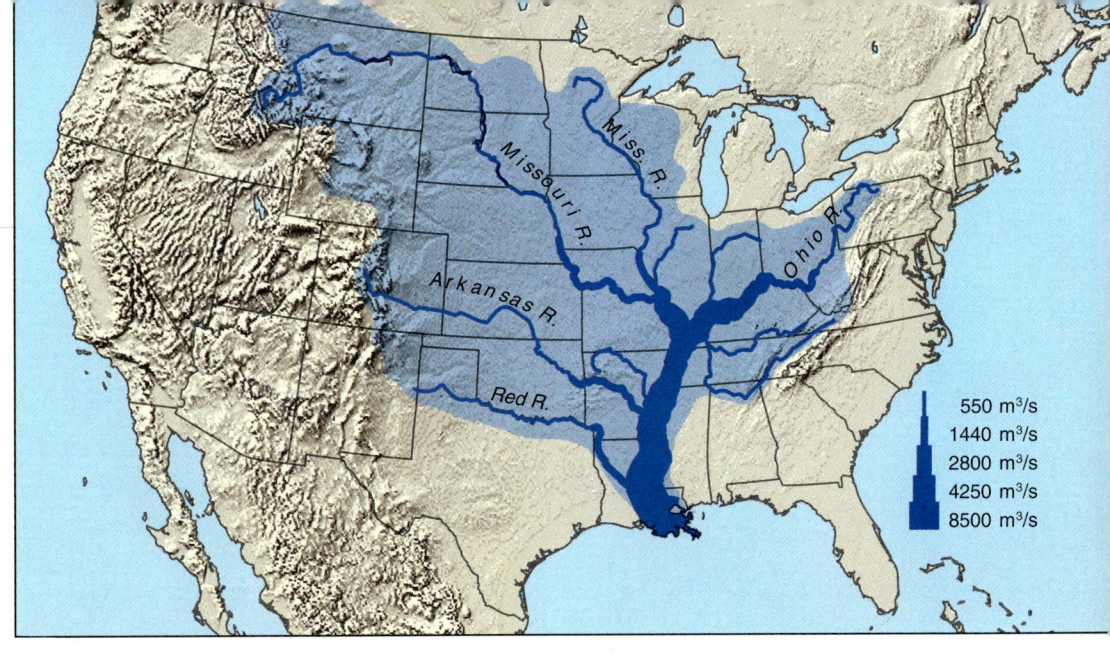

or wildlife management that affect one part of a watershed are likely to affect it all. (See *What an Earth Scientist Sees*.)

If you have driven across North America, you may have seen highway signs marking the *continental divide*. This **divide** separates streams that drain toward one side of the continent from streams that drain toward the other side. The continental divide of western North America lies along the length of the Rocky Mountains. In general, any two adjacent watersheds are separated by a divide, even if they ultimately flow into the same ocean.

divide A topographic high that separates adjacent drainage basins.

What an Earth Scientist Sees

Drainage Basins

The drainage basins of many rivers are hard to see from a satellite because they are covered by vegetation. However, the basin of this river, Wadi Al Masilah in South Yemen, adjacent to the Rubh-al-Khali, a desert, is very easy to spot. We have outlined the boundary—that is, the divide—that surrounds and defines the drainage basin of one tributary here; note that this basin would itself be a part of the drainage basin of the larger river at the top of the photograph.

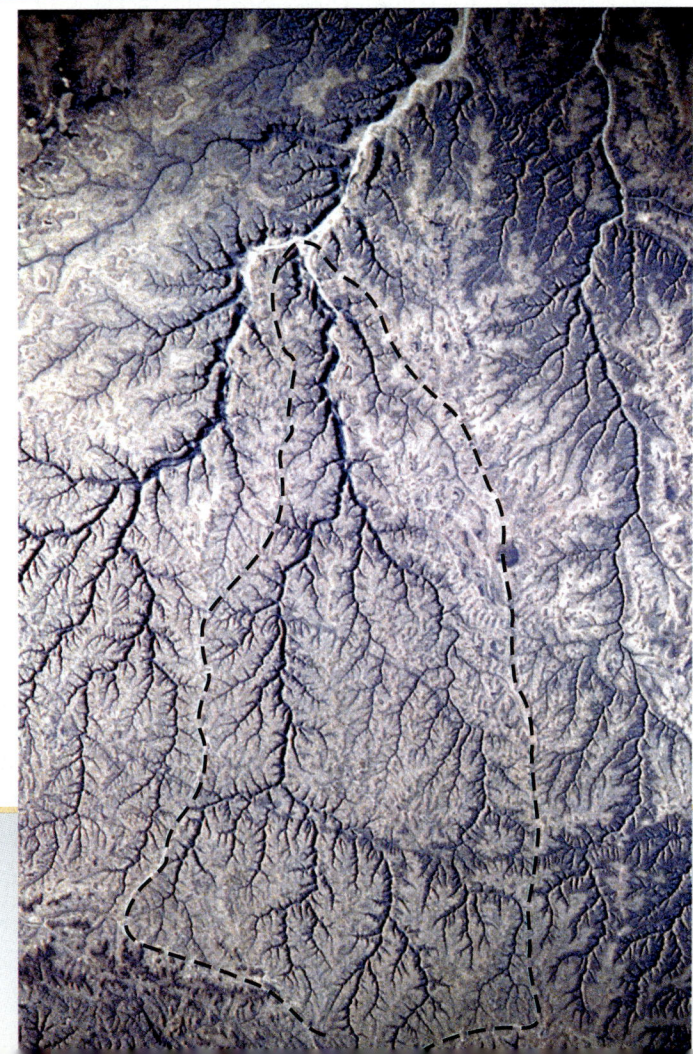

 **Here's an interesting question:**
- The stream channels in the drainage basin outlined seem to be completely dry. In fact, all of the small channels seem to be dry. But look closely and you can see that the main river channel, visible at the top of the photo, has water in it. Where did the water come from?

A dying lake FIGURE 5.9

This lake in Florida has turned green and mucky-looking because of the growth of algae stimulated by excessive nutrients (probably from sewage or fertilizer). Eventually, the algae will use up all the oxygen in the water, making it impossible for other life to survive in the lake. This is called *eutrophication*.

LAKES

Lakes are standing bodies of water with open surfaces. Water enters lakes from streams, overland flow, and groundwater, and exits either by evaporation or by flowing through an outlet. All fresh-water lakes have outlets; however, some saline lakes lack an outlet and therefore lose water only through evaporation, which inevitably leads to a buildup of salt. The Great Salt Lake in Utah is an important example of an inland saline lake.

Lakes are important to us as sources of fresh water and food. The majority of lakes in the United States have been altered, and in some cases even created, by humans. Some changes are deliberate, as when a *reservoir* lake is created by a dam, often for hydroelectric power, or a wetland is drained for farming or building. In other cases, the human effects are inadvertent. For example, runoff of sewage or fertilizer into a lake can cause *eutrophication*, which can kill most of the life in the lake (see **FIGURE 5.9**). Eutrophication can occur as a natural part of the process of swamp formation.

Lakes can form as a result of several different Earth processes. Crustal faulting creates many large, deep lakes. Lava flows and landslides often form dams in river valleys, causing water to back up as lakes. Throughout formerly glaciated regions of North America and Europe, plains of glacial sand and gravel contain natural pits and hollows left by the melting of stagnant ice masses that were buried in the sand and gravel deposits. Kettle lakes, a common landform in New England, form when these pits fill with water.

An important characteristic of lakes is that they are short-lived features in Earth's history. They disappear by one of two processes, or a combination of both. First, lakes that have stream outlets will be gradually drained as the outlets are eroded to lower levels. Second, lakes accumulate inorganic sediment carried by streams entering the lake, and organic matter produced by plants within the lake. Eventually, they fill up, forming a boggy *wetland* with little or no free water surface.

In arid climates, many lakebeds are either dry or only intermittently filled with shallow water. Streams bring dissolved salts to these *ephemeral* lakes. Since evaporation removes only pure water, the salts remain behind and salinity levels increase. Eventually the salts may be precipitated as solid evaporites.

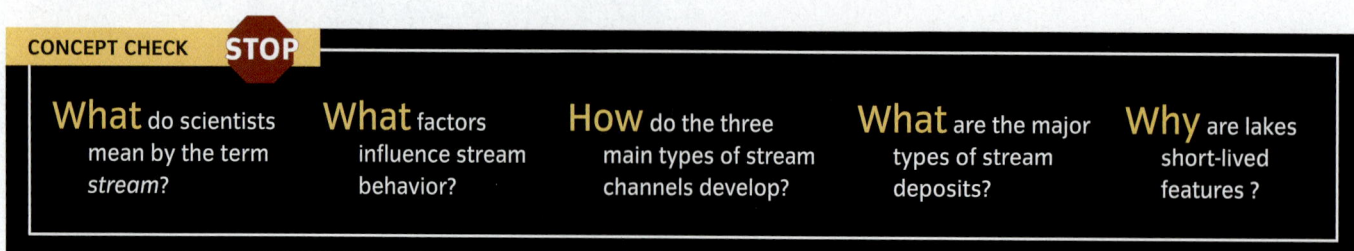

CONCEPT CHECK STOP

What do scientists mean by the term *stream*?

What factors influence stream behavior?

How do the three main types of stream channels develop?

What are the major types of stream deposits?

Why are lakes short-lived features?

Surface Water as a Hazard and a Resource

LEARNING OBJECTIVES

Describe how floods occur and what factors may make them worse.

Define recurrence interval and show how it is used to predict floods.

Explain why flood prevention efforts sometimes have the opposite effect.

Explain why interbasin transfer of water can be a flawed solution to water scarcity.

hough water is a vital resource, it can also be a dangerous force. Uneven distribution of rainfall through the year causes some water bodies to dry up, but others rise and overflow their banks, creating hazardous circumstances for people who live in the area.

FLOODS

Under natural circumstances, all water bodies undergo changes in the volume of water they hold or transport. From time to time, when the discharge or water level becomes too much to handle, the water body will **flood** (see FIGURE 5.10).

flood An event in which a water body overflows its banks.

During a stream flood, the "extra" water flowing in the channel, contributed by excess precipitation, is called *storm runoff*. The

Mississippi River flood FIGURE 5.10

A pair of satellite images shows the region where the Missouri River joins the Mississippi River at St. Louis, Missouri. **A** This photo shows a dry summer with low flows (July 1988). **B** This photo shows the same region in July 1993. Weeks of rain hundreds of kilometers away caused the rivers to overflow their levees. Numerous towns, along with 44,000 km² of farmland in nine states, were flooded by an estimated 3 km³ of floodwater.

peak discharge of a flood usually comes well after the rains that produced it. Hydrologists record the development of the flood with a *hydrograph*, which shows the stream's discharge as a function of time. **FIGURE 5.11** shows an example in which a passing storm generated a brief interval of intense rainfall. As the runoff moved into the stream channel, the discharge quickly rose. The *crest* of the resulting flood—the time when the peak flow passed the *hydrologic station* where the measurements were made—occurred about two and a half hours after the storm. It took another eight hours for the flood runoff to pass through the channel, for the water level to drop, and for the discharge to return to normal.

A hydrograph of stream discharge FIGURE 5.11

This diagram illustrates the hydrograph of a stream after a brief, intense storm.

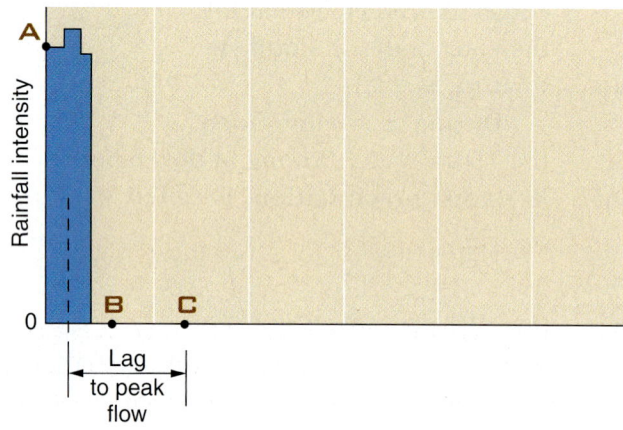

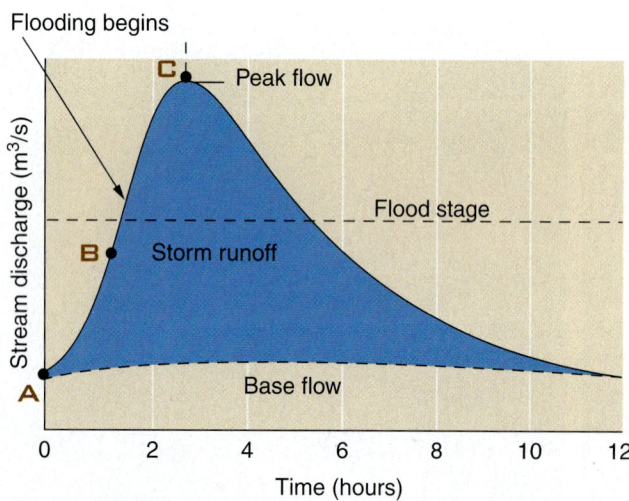

*Base flow is the "normal" flow of water in a stream, contributed by groundwater.

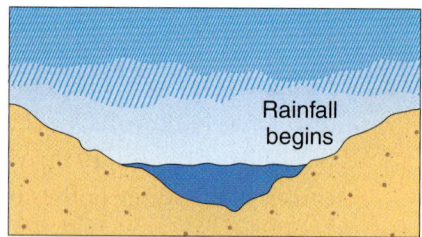

A Onset of storm (0 hours): The peak discharge is delayed as the runoff collects and runs down the stream channel.

B 1 hour: One hour after the cloudburst, the stream can still contain the increased volume.

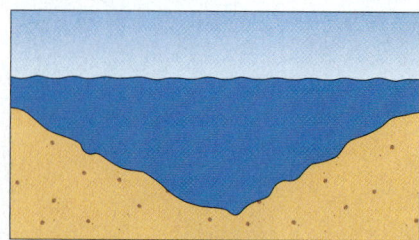

C 2 hours: After two hours, the stream reaches its peak flow and cannot be contained by its banks any more.

Coastal flood FIGURE 5.12

As Hurricane Katrina moved toward New Orleans from the Gulf of Mexico in August 2005, high waters breached the city's protective levees. The subdivision shown here in St. Bernard Parish (*inset*) was constructed on sand dredged from the Mississippi River and was protected by a levee, but the levee was not sufficient to withstand the storm surge. Ten days after the storm, when the larger photo was taken, the subdivision was still under almost 2 m of water.

Lakes also flood, as do oceanic coastal zones. In the case of coastal flooding, it may be the inflow of water from the ocean, rather than the runoff of water from the land, that does most of the damage. The *storm surge* associated with Hurricane Katrina in 2005 temporarily raised the water level near New Orleans by 6 m or more and breached the levees even before the main force of the storm hit the city (see **FIGURE 5.12**).

Though both coastal flooding and stream flooding often take people by surprise, Earth scientists view them as normal and inevitable events. The record shows that floods have been occurring throughout Earth's history, for as long as there has been a hydrosphere. Even though flooding is a natural Earth process, it can quickly become a human catastrophe when it affects population centers. The Huang He in China, sometimes called the Yellow River because of its heavy load of yellowish-brown silt, has a long history of catastrophic floods. In 1887, the river inundated 130,000 km^2 and swept away many villages in the heavily populated floodplain. In 1931, another Huang He flood killed a staggering 3.7 million people.

Surface Water as a Hazard and a Resource

Yet these same floods help replenish the soil in the floodplain, which explains why people keep moving back to the area.

Human activity sometimes increases the chance that flooding will occur (instead of decreasing it). Urban development can exacerbate the problem of flooding in a variety of ways. Urban construction on compressible sediments, often accompanied by withdrawal of groundwater, can lead to **subsidence** (a drop in the ground surface), which increases the danger of flooding (as it did in New Orleans). The impermeable ground cover associated with urbanization can add substantially to surface runoff in urban areas. Storm sewers can contribute to flooding because they allow the runoff from paved areas to reach the river channel more quickly. Floods in urbanized basins often have higher peak discharges and reach their peaks more quickly than floods in undeveloped basins. This quicker and higher crest means that people living in a flood-prone area must be able to move very quickly or be able to predict when a flood might strike.

Flood prediction and prevention Because floods can be so damaging, predicting and preparing for them is essential (see **FIGURE 5.13**). To do this, the frequency of past floods of different sizes is plotted on a graph, producing a flood-frequency curve. The average time interval between two floods of the same magnitude is called the *recurrence interval*. For example, a "10-year flood" has a recurrence interval of 10 years, which means that there is a 1-in-10 (or 10%) chance that such a flood will occur in any given year. A flood with an even greater discharge having a recurrence interval of 50 years would be termed a "50-year flood" for this particular stream, and have a 1-in-50 (2%) chance of occurring in any given year. Regional planners should (and do) keep these intervals in mind when planning development on or near a floodplain.

Predicting floods FIGURE 5.13

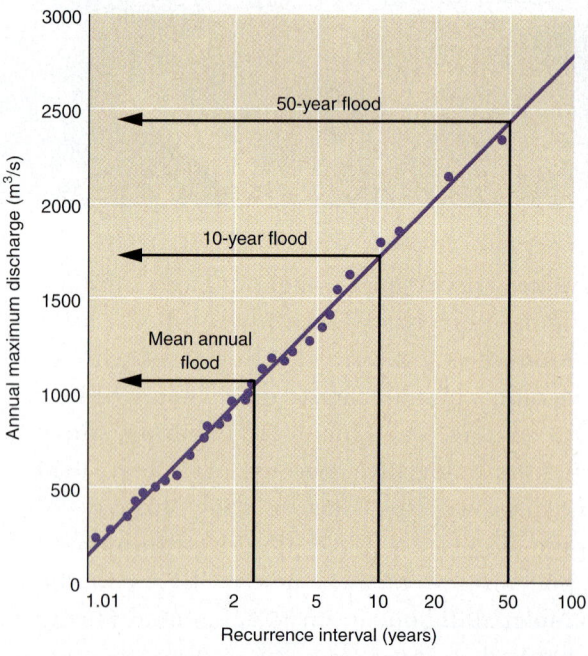

The graph below shows the frequency of floods of different sizes on the Skykomish River at Gold Bar, Washington. A flood with a discharge of 1750 m³/s has a recurrence interval of 10 years, and hence a 1-in-10 chance of occurring in any given year.

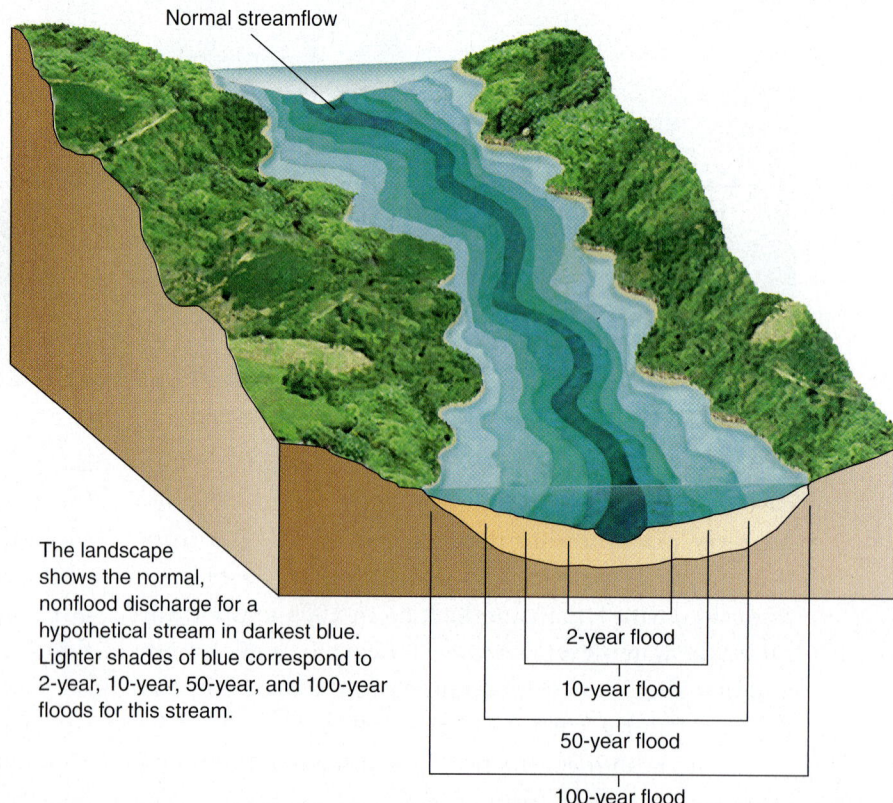

The landscape shows the normal, nonflood discharge for a hypothetical stream in darkest blue. Lighter shades of blue correspond to 2-year, 10-year, 50-year, and 100-year floods for this stream.

148   CHAPTER 5   Water On and Under the Ground

Another aspect of flood prediction is the real-time monitoring of storms and water levels. Scientists can combine information about the weather with their knowledge of a river basin's characteristics and topography to forecast the peak height of a flood and the time when the crest will pass a particular location. Such forecasts, which are often made with the aid of computer models and Geographic Information Systems (GISs), can be very useful for planning evacuation or defensive measures.

Understandably, many people throughout history have been unsatisfied with simply predicting floods and have attempted to prevent them. River channels are often modified or engineered for the purpose of flood control and protection as well as to increase access to floodplain lands, facilitate transport, enhance drainage, and control erosion. The modifications usually consist of some combination of widening, deepening, straightening, clearing, or lining of the natural channel. All of these approaches are collectively called *channelization*.

Like dams, channelization projects can contribute to the economic well-being of a community, but at a price. Channel modifications interfere with natural habitats and ecosystems. The aesthetic value of the river can be degraded, and water pollution aggravated. Projects sometimes control flooding in the immediate area but contribute to more intense flooding downstream. Perhaps most importantly, any modification of a channel's course or cross-section renders invalid the hydrologic data collected there in the past. During the Mississippi River floods of 1973 and 1993, experts could not account for water levels that were higher than predicted by the historical data; the likely cause was extensive upstream modifications of the river channel by humans.

SURFACE WATER RESOURCES

A reliable water supply is critical—not only for human survival and health, but also for the role it plays in industry, agriculture, and other economic activities (see FIGURE 5.14 on the following page). Twenty-six countries worldwide, with a total population of almost 250 million people, are today designated as water-scarce. The lack of water in these countries places serious constraints on agricultural production, economic development, health, and environmental protection.

Globally, crop irrigation accounts for about 73% of the demand for water, industry for about 21%, and domestic use for the remaining 6%, though the proportions vary from one region to another. Demand in each of these sectors has more than quadrupled since 1950. Population growth is partly responsible for the increasing demand, but improvements in standards of living around the world have also contributed to the large increase in water use per capita over the past few decades. The total amount of water being withdrawn (that is, diverted from rivers, lakes, and groundwater) for worldwide human use is now about eight times the annual streamflow of the Mississippi River.

Sometimes, because of population growth and development, regions with the greatest demand for water do not have an abundant and readily available supply of surface water. For this reason, surface water is often transferred from one drainage basin to another, sometimes over long distances. Besides raising political issues related to water rights, such *interbasin transfer* can have negative environmental impacts (see the *Case Study* on page 152).

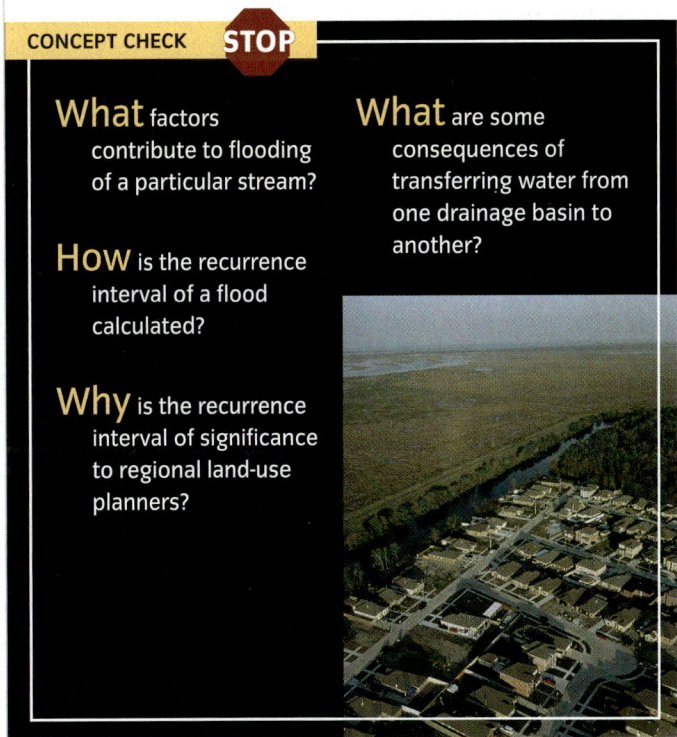

CONCEPT CHECK STOP

What factors contribute to flooding of a particular stream?

How is the recurrence interval of a flood calculated?

Why is the recurrence interval of significance to regional land-use planners?

What are some consequences of transferring water from one drainage basin to another?

Surface Water as a Hazard and a Resource

WATER ACCESS AND USE

In arid and poorly developed regions, large proportions of inhabitants lack clean water supplies. Most water is consumed by agriculture for irrigation.

▲ Water gatherers
Women wait to fill water jugs from an irrigation canal in southern Ethiopia. In this drought-plagued area, more than 80% of rural inhabitants lack access to clean drinking water.

Primary watersheds
Annual renewable water, 2000 (cubic meters per person)
- More than 100,000
- 10,000 to 100,000
- 4,001 to 10,000
- 1,701 to 4,000
- 1,001 to 1,700
- Less than 1,000
- No data
- ▲ Water related conflict in the last 100 years
- − Large dam—volume (in thousands) greater than 38,000 cu m (50,000 cu yds)

Access to fresh water ▶
In many regions, drinkable water is becoming scarce because of increasing demand and decreasing quality. Pollution of surface and ground water with contaminants and organisms is a major threat. Contamination of aquifers by pesticides and heavy metals is especially worrisome.

Percent of total population using improved drinking water sources, 2000
- More than 90%
- 76%–90%
- 51%–75%
- 25%–50%
- Less than 25%
- No data available

Visualizing

Principal watersheds and fresh-water resources of the world FIGURE 5.14

WATER AVAILABILITY

Within a watershed, water availability depends on both precipitation and the number of people the water must support.

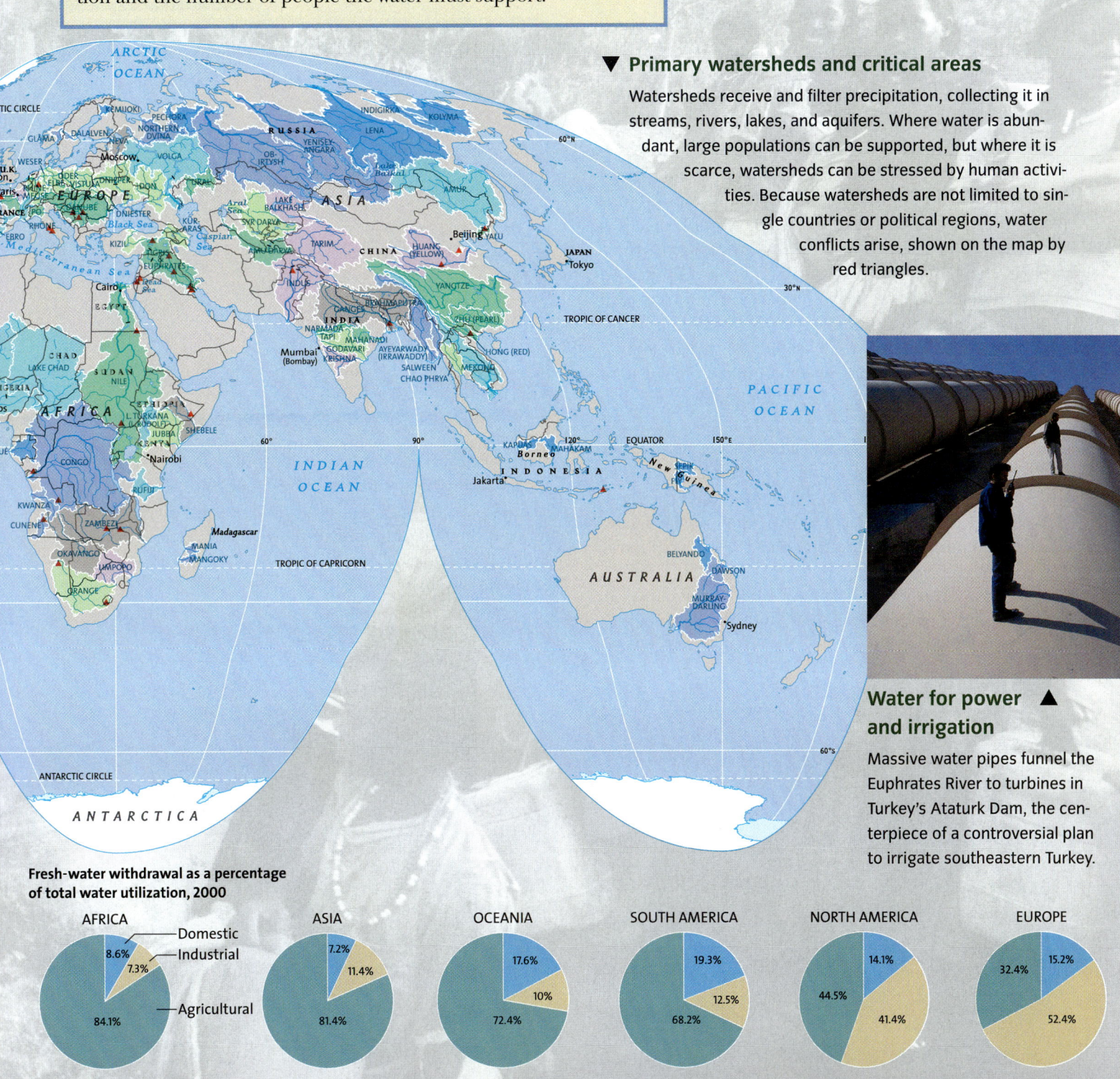

▼ Primary watersheds and critical areas

Watersheds receive and filter precipitation, collecting it in streams, rivers, lakes, and aquifers. Where water is abundant, large populations can be supported, but where it is scarce, watersheds can be stressed by human activities. Because watersheds are not limited to single countries or political regions, water conflicts arise, shown on the map by red triangles.

Water for power and irrigation ▲

Massive water pipes funnel the Euphrates River to turbines in Turkey's Ataturk Dam, the centerpiece of a controversial plan to irrigate southeastern Turkey.

Fresh-water withdrawal as a percentage of total water utilization, 2000

Region	Domestic	Industrial	Agricultural
AFRICA	8.6%	7.3%	84.1%
ASIA	7.2%	11.4%	81.4%
OCEANIA	17.6%	10%	72.4%
SOUTH AMERICA	19.3%	12.5%	68.2%
NORTH AMERICA	14.1%	41.4%	44.5%
EUROPE	15.2%	52.4%	32.4%

Surface Water as a Hazard and a Resource

CASE STUDY

Global Locator

Mono Lake

California's Mono Lake, in the Sierra Nevada Mountains, has been the site of a collision between human water needs and the needs of a unique habitat and its wildlife.

These calcium carbonate spires (see **A**) were once underwater. But in 1941, the Los Angeles Aqueduct began diverting water from four of the six streams that empty into Mono Lake. With insufficient input to make up for its evaporation losses, the lake shrank to half its original volume and doubled in salinity. Migratory birds, such as the nation's second largest colony of California gulls, were placed in jeopardy. Their food supply (brine shrimp) was dying because of the high salinity of the water, and the islands on which they nested were in danger of being connected to the mainland because of the retreating water, making the birds vulnerable to predators.

By the late 1980s, the lake and its ecosystem were nearing collapse. In 1994, after a court battle between the City of Los Angeles and environmental groups, both sides agreed to a plan to raise the water level by 5 m and compensate Los Angeles for the loss of part of its water supply.

Water levels in the lake have risen about halfway to the target set by the agreement, and the lake is expected to reach its target level in 15 to 20 years. Dams that once diverted streams into the aqueduct (see **B**) have been reengineered to do the opposite: They maintain a steady flow into Mono Lake and divert water into the aqueduct only if there is an overflow. For the California gulls (see **C**), the change may have come just in time.

152 CHAPTER 5 Water On and Under the Ground

Fresh Water Underground

LEARNING OBJECTIVES

Define water table.

Explain how porosity and permeability of rocks affect the motion of groundwater.

Identify two types of aquifers.

Explain why some wells require pumping, while others flow unaided.

Describe how subsidence relates to groundwater.

Describe how a cave forms.

Less than 1% of all water in the hydrosphere cycle is **groundwater**. Although this sounds small, the volume of groundwater is 40 times greater than that of all the water in fresh-water lakes and streams. Water can be found everywhere beneath the land surface, even beneath parched deserts. About half of it is near the surface, no more than 750 m below ground. At greater depths, the pressure exerted by overlying rocks reduces the pore space, making it difficult for water to flow freely. In addition, water that occurs at great depths tends to be *briny*, or rich in dissolved mineral salts, and not well suited for human use. Therefore, from a practical perspective, we can think of groundwater as the water found between the land surface and a depth of about 750 m, even though an equally large amount of water is present at greater depths.

groundwater Subsurface water contained in pore spaces in regolith and bedrock.

water table The top surface of the saturated zone.

THE WATER TABLE

Much of what we know about groundwater has been learned from the accumulated experience of generations of people who have dug or drilled millions of wells. This experience tells us that a hole penetrating the ground ordinarily first encounters a zone in which the spaces between the grains in regolith or bedrock are filled mainly with air, although the material may be moist to the touch. This is the *zone of aeration*, also known as the *vadose zone* (see **FIGURE 5.15**).

After passing through the zone of aeration, the hole reaches the **water table** and enters the *saturated zone*,

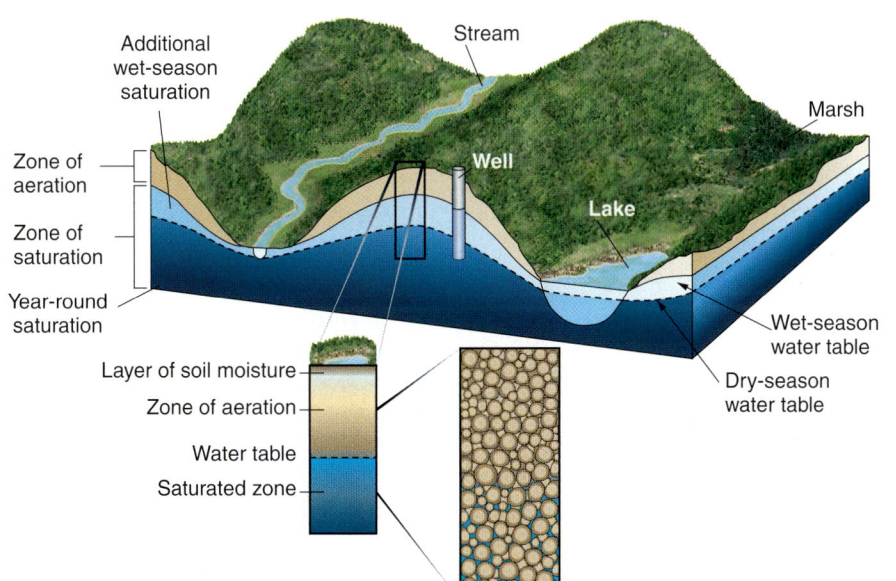

Water under the ground
FIGURE 5.15

A well first passes through the zone of aeration, where pores in the soil are filled with both air and water. Eventually, it reaches the water table, where the pore spaces are completely filled with water. Underground water exists everywhere, and surface water occurs wherever the ground intersects the water table.

Fresh Water Underground 153

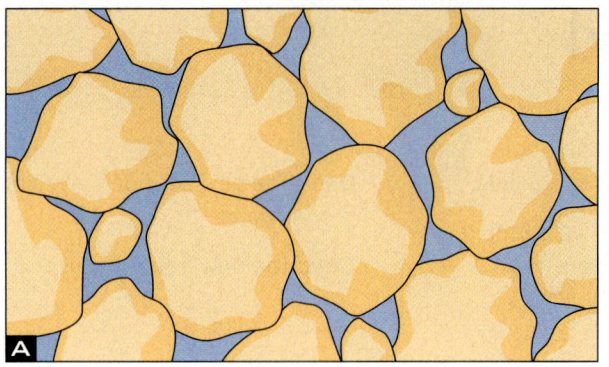
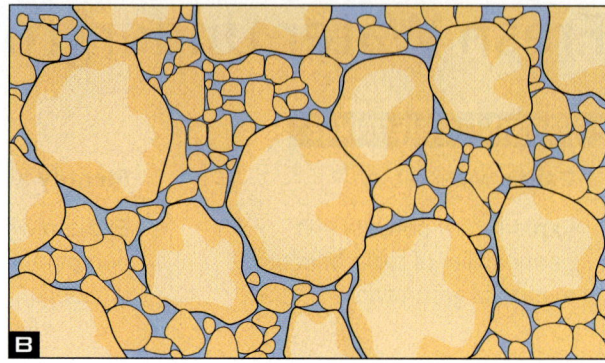

Porosity in sediments and rocks
FIGURE 5.16

In these examples, all of the pore spaces are filled with water, as they would be in the saturated zone. **A** The porosity is about 30% in this sediment, with particles of uniform size. **B** This sediment, in which fine grains fill the space between larger grains, has a lower porosity, around 15%. **C** In sedimentary rock, the porosity may be reduced by cement that binds the grains together and fills the pores.

also known as the *phreatic zone,* in which all openings are filled with water. The water table is high beneath hills and low beneath valleys. This may seem surprising, because the surface of a glass of water or a lake is always level. But water underground flows very slowly and is strongly influenced by surface topography. If all rainfall were to cease, the water table would slowly flatten. Seepage of water into the ground would diminish and then stop entirely, and streams would dry up as the water table fell. During droughts, the depression of the water table is evident from the drying up of springs, streambeds, and wells. Repeated rainfall, which soaks the ground with fresh supplies of water, maintains the water table at a normal level and keeps surface water bodies replenished.

Whether it is deep or shallow, the water table marks the upper limit of readily usable groundwater. For this reason, a major aim of groundwater specialists and well drillers is to determine the depth and shape of the water table. To do this they must first understand how groundwater moves and what forces control its distribution underground.

HOW GROUNDWATER MOVES

Most groundwater is in motion. Unlike the swift flow of rivers, however, which is measured in kilometers per hour, the movement of groundwater is so slow that it is measured in centimeters per day or meters per year. The reason is simple: Whereas the water of a stream flows through an open channel, groundwater must move through small, constricted passages. Therefore, the rate of groundwater flow is dependent on the nature of the rock or sediment through which the water moves, especially its porosity and permeability.

Porosity and permeability Porosity determines the amount of fluid a sediment or rock can contain. The porosity of sediment is affected by the size and shape of the particles and the compactness of their arrangement (see FIGURE 5.16A and B) The porosity of a sedimentary rock

> **porosity** The percentage of the total volume of a body of rock or regolith that consists of open space.

is also affected by the extent to which the pores have been filled with cement (see **FIGURE 5.16C**). Plutonic igneous rocks and metamorphic rocks, which consist of many closely interlocked crystals, generally have lower porosities than do sediment and sedimentary rocks. However, joints and fractures may increase their porosity.

> **permeability** A measure of how easily a solid allows fluids to pass through it.
>
> **percolation** The process by which groundwater seeps downward and flows under the influence of gravity.

A rock with low porosity is likely also to have low **permeability**. However, high porosity does not necessarily mean high permeability, because both the sizes and the continuity of the pores (that is, the extent to which the pores are interconnected) influence the ability of fluids to flow through the material.

Percolation After water from a rain shower soaks into the ground, or infiltrates, some of it evaporates, while some is taken up by plants. The remaining water continues to **percolate** under the influence of gravity until it reaches the water table. The "perc test" that must be carried out when a new septic system is being installed is a measure of percolation. The movement of groundwater in the saturated zone is similar to the flow of water that occurs when you gently squeeze a water-soaked sponge. Water moves slowly through very small pores along threadlike paths. The water flows from areas where the water table is high toward areas where it is lower. In other words, it generally flows toward surface streams or lakes (see **FIGURE 5.17**). Some of the flow paths turn upward and enter the stream or lake from beneath, seemingly defying gravity. This upward flow occurs because groundwater is under greater pressure beneath a hill than beneath a stream or lake. Because water tends to flow toward points where pressure is low, it flows toward bodies of water at the surface.

Recharge and discharge

Recharge of groundwater occurs when rainfall and snowmelt infiltrate the ground and percolate downward to the saturated zone (see Figure 5.17). The water then moves slowly along its flow path toward zones where **discharge** occurs. (Note that earlier in the chapter we used the term *discharge* for a somewhat different concept—the flow of water along a stream channel.) In discharge zones, subsurface water either

> **recharge** Replenishment of groundwater.
>
> **discharge** The process by which subsurface water leaves the saturated zone and becomes surface water.

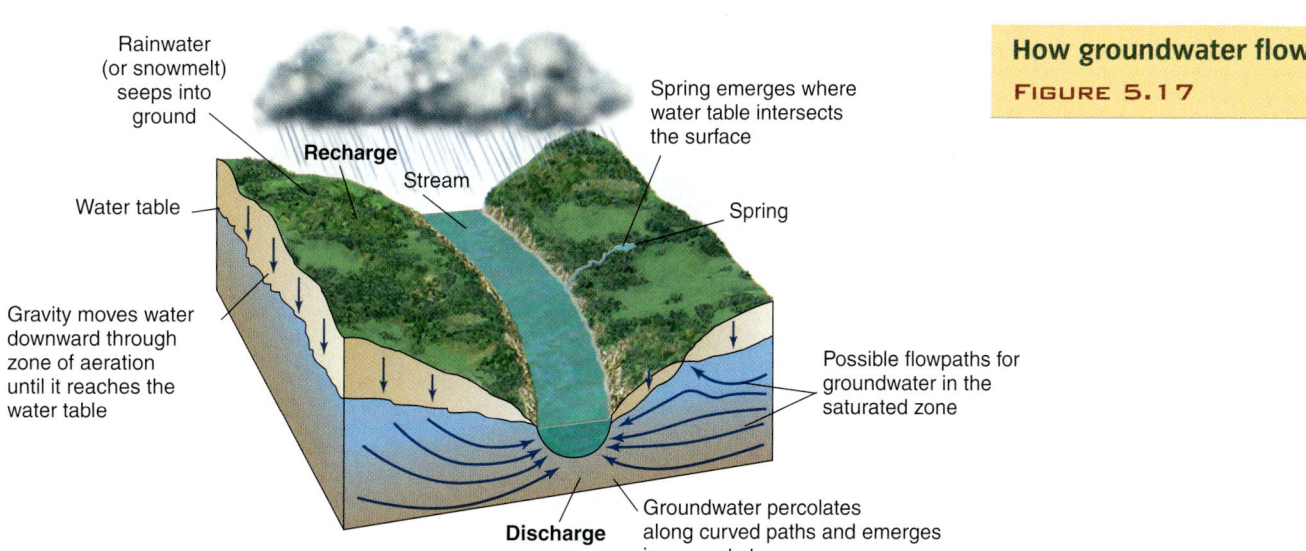

How groundwater flows
FIGURE 5.17

spring Where the water table intersects the land surface, allowing groundwater to flow out.

flows out onto the ground surface as a **spring** or joins bodies of water such as streams, lakes, ponds, swamps, or the ocean. The discharge of groundwater is what maintains the base flow of a stream. Pumping groundwater from a well also creates a point of discharge. The amount of time water takes to move through the ground to a discharge area depends on distance and rate of flow. It may take as little as a few days or as long as thousands of years.

WHERE GROUNDWATER IS STORED

When we wish to find a reliable supply of groundwater, we search for an **aquifer** (Latin for "water carrier"). An aquifer is not a body of water—it is a body of rock or regolith that is water-saturated, as well as porous and permeable. Gravel and sand generally make good aquifers; sandstone is often a good aquifer, as is fractured or cavernous limestone or granite. An aquifer in which the water is free to rise to its natural level is called an *unconfined aquifer* (see **FIGURE 5.18**). In a well drilled into an unconfined aquifer, the water will rise to the level of the surrounding water table. To bring it to the surface one would need a pump or a bucket.

A *confined aquifer* is overlain by impermeable rock units, called *confining layers,* or **aquicludes** (see **FIGURE 5.19**). Common aquicludes are shale and clay layers. The water in a confined aquifer is held in place by the overlying aquiclude, and its recharge zone may be many kilometers away at a higher elevation. If a well is drilled into the aquifer, the high water pressure due to the elevation of the recharge zone will cause the water to rise or even flow out of the well without having to

aquifer A body of rock or regolith that is water-saturated, porous, and permeable.

aquiclude A layer of impermeable rock.

Aquifers, confined and unconfined FIGURE 5.18

An unconfined aquifer is open to the atmosphere through pores in the rock and soil above the aquifer. In contrast, the water in a confined aquifer is trapped between impermeable rock layers.

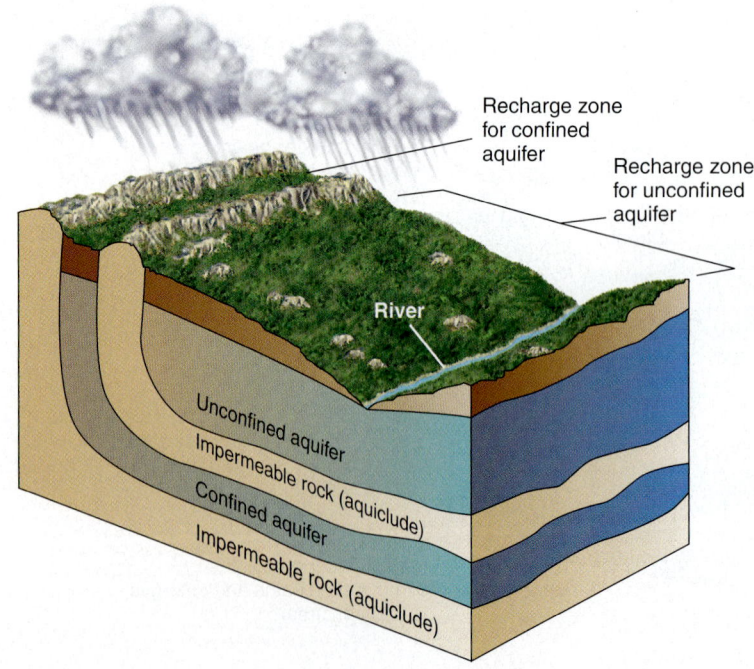

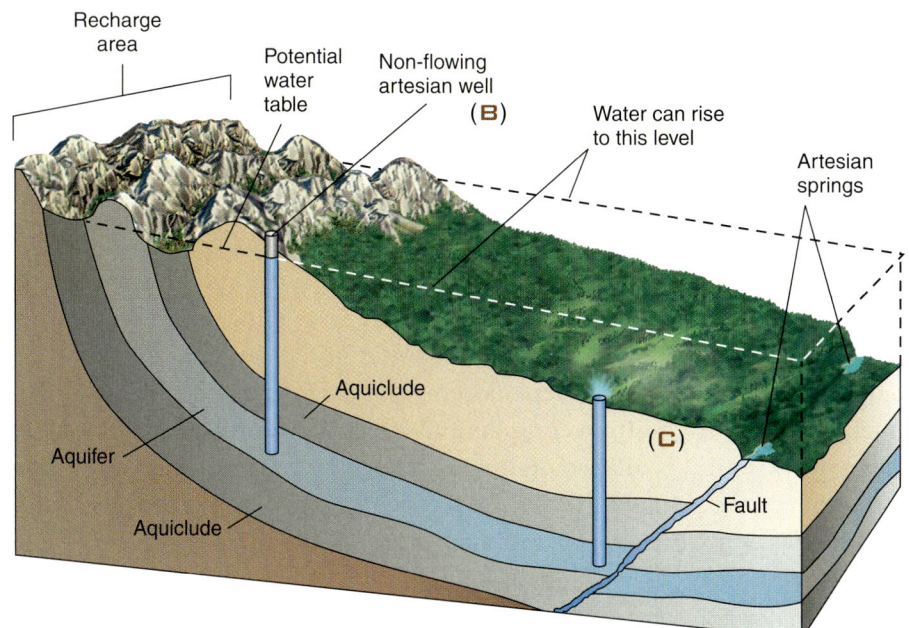

Artesian water
FIGURE 5.19

This diagram illustrates the natural conditions that produce artesian wells or springs. Note that the water enters the aquifer at a higher elevation than the well. As a result, the water flows or gushes out of the well with positive water pressure.

be pumped. This is called an *artesian well*. A fault can also serve as a natural conduit for artesian water.

A change in permeability of the rocks at ground level often gives rise to a spring, which is a natural analogue of an artesian well. Such a change may be due to the presence of an aquiclude (see **FIGURE 5.20**) or it may happen along the trace of a fault (see Figure 5.19).

What causes springs? FIGURE 5.20

A This spring in the Grand Canyon is fed by water from the porous Redwall and Muav Limestone. These cavernous limestones are the water source for many springs. The impermeable shale unit beneath them, the aquiclude, is the Bright Angel Shale.

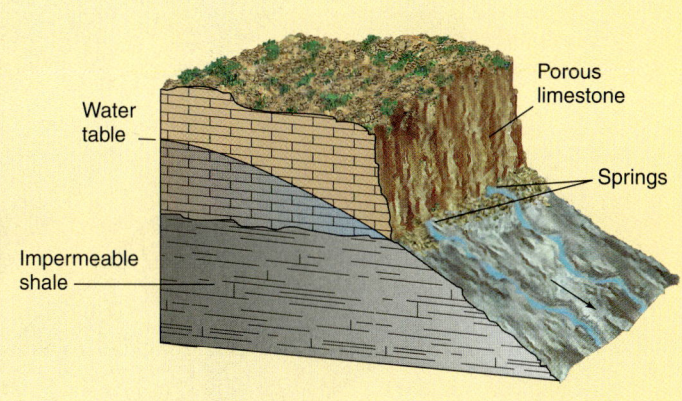

B Water flows from a spring in a limestone aquifer underlain by an impermeable shale aquiclude.

Fresh Water Underground

GROUNDWATER DEPLETION AND CONTAMINATION

A well will supply water if it is deep enough to intersect the water table. As shown in **FIGURE 5.21**, a shallow well may become dry during periods when the water table is low, whereas a deeper well may yield water throughout the year. When water is pumped from a well, a *cone of depression* (a cone-shaped dip in the water table) will form around the well. In most small domestic wells, the cone of depression is hardly discernible. Wells pumped for irrigation and industrial uses, however, sometimes withdraw so much water that the cone may become very wide and steep and can lower the water levels in surrounding wells. When large cones of depression from pumped wells overlap, the result is regional depression of the water table.

If the rate of withdrawal of groundwater regularly exceeds the rate of natural recharge, the volume of stored water steadily decreases; this is called *groundwater mining*. It may take hundreds or even thousands of years for a depleted aquifer to be replenished. The results of excessive withdrawal include lowering of the water table; drying up of springs and streams; *compaction* of the aquifer; and *subsidence*, a decline in land surface elevation. Sometimes it is possible to recharge an aquifer by pumping water into it. In other cases, the effects of depletion may be permanent. When an aquifer suffers compaction—that is, when its mineral grains collapse on one another because the pore water that held them apart has been removed—it is permanently damaged and may never be able to hold as much water as it originally held.

Wells: Year-round and seasonal FIGURE 5.21

Seasonal changes affect the height of the water table. A well will produce year-round only if it extends into the year-round zone of saturation.

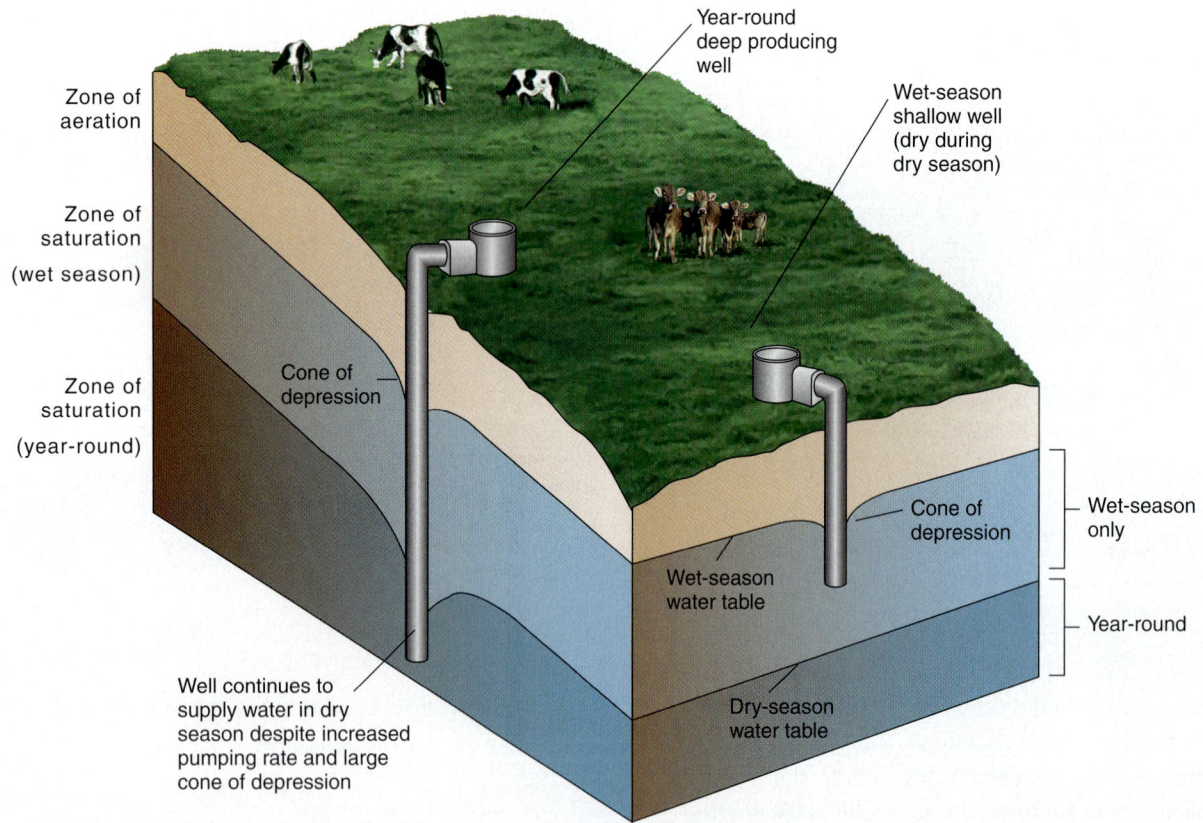

CHAPTER 5 Water On and Under the Ground

A Newtown Creek runs for about 5 km through New York City, where it forms part of the border between Queens (at right) and Brooklyn (at left).

B The surface of the creek is constantly fouled with oil, from an underground spill that took place more than half a century ago. According to a Coast Guard study, in 1948 approximately 65 million liters of oil leaked from a refinery owned by the Standard Oil Company of New York (later renamed Mobil). (For comparison, this is more than the amount of oil spilled in the much more famous *Exxon Valdez* accident in 1989.) The successor company, ExxonMobil Company, has cleaned up part of the mess, but most of the spill continues to contaminate the groundwater and ooze to the surface in Newtown Creek.

The forgotten oil spill FIGURE 5.22

Laws and policies relating to water rights are very complicated, and the application to groundwater is even more complicated than for surface water. Because groundwater is hidden from view, it is difficult to monitor its flow and regulate its use. If you drill a well into an aquifer underlying your property, are you entitled to withdraw as much water as you need from that well? Should you withdraw water only for your own purposes, or should you be permitted to withdraw the water and sell it elsewhere? What happens if withdrawing the groundwater depletes the aquifer and your neighbor's well runs dry?

Many of the types and sources of contaminants that affect surface water also cause groundwater contamination. Because of its hidden nature, however, groundwater contamination can be much more difficult to detect, control, and clean up than surface water contamination. The most common source of water pollution in wells and springs is untreated sewage. Agricultural pesticides and fertilizers, significant sources of surface water pollution, are also common contaminants of groundwater. Harmful chemicals leaking from waste disposal facilities can also infiltrate into groundwater reservoirs and contaminate them. Probably the most serious groundwater contamination problem in North America is caused by leaking underground storage tanks at gas stations, refineries, and other industrial settings (see **FIGURE 5.22**).

Fresh Water Underground

WHEN GROUNDWATER DISSOLVES ROCKS

As discussed in Chapter 4, all rainwater and most groundwater is slightly acid due to solution of carbon dioxide. In regions underlain by rocks that are highly susceptible to chemical weathering by acid waters, groundwater creates extensive systems of underground caverns. In such areas, a distinctive landscape forms on the surface. *Karst topography*, named for the Karst region of the former Yugoslavia (now Slovenia), is characterized by many *sinkholes* (small, closed basins) and disrupted drainage patterns (see **Figures 5.23 and 5.24**). Streams disappear into the ground and join the groundwater. Large springs form where the water table intersects the land surface, or along favorable routes of escape, such as faults. Karst is most typical of regions underlain by soluble carbonate rocks (limestone and dolostone), although it can also occur in regions with extensive evaporite (salt) deposits. In carbonate terrains with karst topography, the rate of dissolution is faster than the average rate of erosion of surface materials by streams and mass wasting.

Caves and sinkholes

Caves and caverns are formed when circulating groundwater at or below the water table dissolves carbonate rock. The process begins with dissolution along interconnected fractures and bedding planes. A cave passage then develops along the most favorable flow route.

> **cave** and **cavern**
> Underground open space; a cavern is a system of connected caves.

The development of a continuous passage by slowly moving groundwater may take up to 10,000 years, and the further enlargement of the passage by more rapidly flowing groundwater needed to create a fully developed cave system may take an additional 10,000 to 1 million years. Finally, the cave may become accessible to humans after the water table drops below the floor of at least some of the chambers.

In the parts of the cave that lie above the water table (in the vadose zone), groundwater continues to percolate downward, dripping from the ceiling to the floor. Calcium carbonate dissolved in the water precipitates out of solution and builds up beautiful icicle-like decorations on the cave walls, ceilings, and floor. These include stalactites (hanging from the ceiling) and sta-

Evolution of a karst landscape FIGURE 5.23

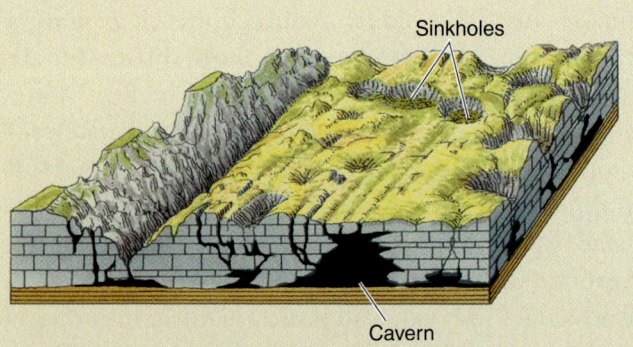

A Over time, rainwater dissolves limestone, producing caverns and sinkholes. In warm, humid climates, solution of pure limestone can form towers (left side of diagrams).

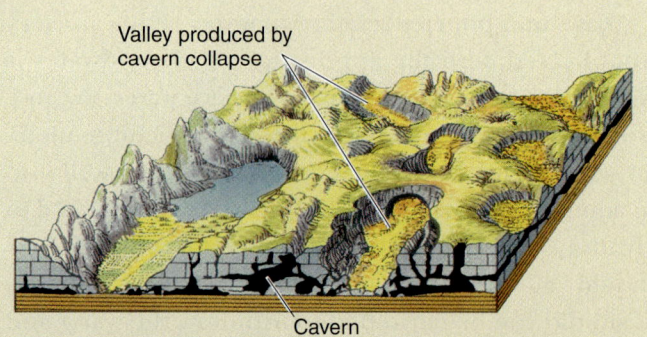

B Eventually, the caverns collapse, leaving open, flat-floored valleys. Surface streams flow on shale beds beneath the limestone. Some parts of the flat-floored valleys can be cultivated.

A These sinkholes in Florida are part of a karst terrain. Most of Missouri and large parts of Kentucky, Tennessee, Texas, and New Mexico also have carbonate karst landscapes. In Arecibo, Puerto Rico, the world's largest radio telescope was built in a sinkhole, taking advantage of its naturally circular shape.

B This is a karst terrain near Guilin, China. White limestone is visible in the pillars that remain after cavern collapse created the valley in which the village is located. The pillars are riddled with caves and passageways.

Karst FIGURE 5.24

lagmites (projecting upward from the floor), as well as columns, draperies, and flowstones. In the saturated zone, below the water table, water movement is largely horizontal, creating tubular passages.

Caves are dissolution cavities that are closed to the surface or have only a small opening. In contrast, a *sinkhole* is a dissolution cavity that is open to the sky. Some sinkholes are formed very abruptly when the roofs of caves collapse. Most sinkholes, though, develop much more slowly and less catastrophically, simply growing wider over time as the carbonate bedrock slowly dissolves.

CONCEPT CHECK STOP

What is the difference between an unconfined and a confined aquifer?

What is a water table? What lies above and below the water table?

What are some of the problems associated with the use of groundwater resources?

How and why do caves and sinkholes develop in certain regions?

Fresh Water Underground

Amazing Places: Lechuguilla Cave

In 1986, cavers (also known as *spelunkers*) discovered the deepest known cave in the United States, called Lechuguilla Cave, in Carlsbad Caverns National Park. Its entrance had been known for decades, but it had been considered a dead end until cavers dug through the floor to a huge network of passages on the other side.

Lechuguilla Cave has now been explored to a depth of 475 m, and has almost 160 km of mapped passages. It is as spectacular as it is deep, but is closed to the public to preserve its unusual formations.

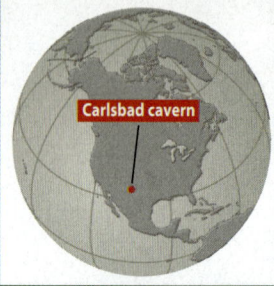

Global Locator

◀ **A** A "bush" made of fragile aragonite pokes out of a stalagmite made of calcite.

B "Soda straws" reach down from the ceiling in this small chamber. Water flows down through the center of the straw. If water starts flowing down the outside, it will build a stalactite. ▶

◀ **C** The origin of this rare formation, called "pool fingers," is a mystery, perhaps related to bacterial activity. The "fingers" crystallized in a pool of water and were left behind when the water retreated.

D Gypsum crystals provide a clue to this cave's unusual history. Unlike most limestone caves, which are formed by carbonic acid in rainwater, Lechuguilla formed from the bottom up. Hydrogen sulfide from lower-lying oil deposits percolated up into the groundwater and formed sulfuric acid, which dissolved away the rock. As the water table dropped, gypsum deposits precipitated out of the acidic water. ▶

SUMMARY

1 The Hydrologic Cycle

1. The **water cycle**, or **hydrologic cycle**, describes the movement of water from one reservoir to another in the hydrosphere. The ocean is the largest reservoir, followed by the polar ice sheets. The largest reservoir of unfrozen fresh water is groundwater. Surface water bodies, the atmosphere, and the biosphere are much smaller water reservoirs.

2. The pathways or processes by which water moves from one reservoir to another include **evaporation**, **condensation**, **precipitation**, **transpiration**, **infiltration**, and **surface runoff**.

3. Like the tectonic and rock cycles, the hydrologic cycle is a *closed cycle* of *open systems*. Because it is closed, the global hydrologic cycle maintains a mass balance.

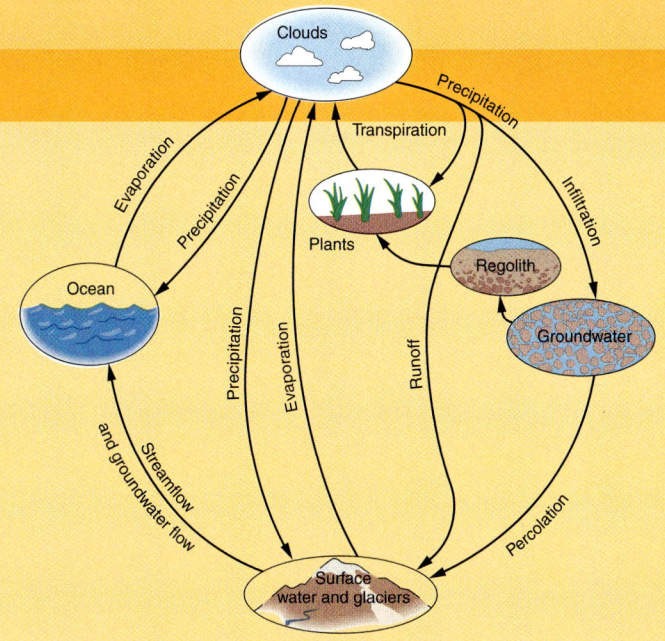

HYDROLOGIC CYCLE

2 How Water Affects Land

1. **Streams** and **rivers** flow downslope along a clearly defined natural passageway, the **channel**. Interrelated factors that influence the behavior of a stream include the **gradient**, **discharge**, **load**, and velocity of the water.

2. Streams create landforms through *erosion* and *deposition*. Straight, braided, and *meandering* channels and *oxbow lakes* are erosional landforms. *Point bars, alluvial fans, deltas,* and **floodplains** are depositional landforms made

of recently deposited sediment, or **alluvium**.

3. Every stream is surrounded by its **drainage basin**, the total area from which water flows into the stream. The topographic "high" that separates adjacent drainage basins is a divide. Streams originate in a *headwater* region, where they collect water from a large number of smaller *tributary streams*. Water exits a stream system from the stream's *mouth*, which may empty into either a larger river, a lake, or an ocean.

4. *Lakes* form in topographic basins created by faults, glacial debris, landslides, and lava flows, as well as in areas of poor drainage. Lakes are usually short-lived features, because they tend to disappear by erosion of the outlet or by silt deposition.

3 Surface Water as a Hazard and a Resource

1. A **flood** occurs when a stream's discharge becomes so great that it exceeds the capacity of the channel, causing the stream to overflow its banks. The risk of floods is increased by *subsidence,* and flooding can be exacerbated by human activity.

2. Prediction of flooding is based on analysis of the frequency of occurrence of past events and on real-time monitoring of storms using a hydrograph. River channel modifications made for purposes of flood control and protection, as well as for navigation and other purposes, are collectively known as *channelization*.

3. Agriculture is by far the largest consumer of fresh water. Surface water can be transported from one drainage basin to another but only at the risk of long-term environmental changes in the affected basins.

4 Fresh Water Underground

1. **Groundwater** is subsurface water contained in spaces within bedrock and regolith. In the *zone of aeration* (*unsaturated* or *vadose zone*), water is present, but does not completely saturate the ground. In the *phreatic* or *saturated zone,* all openings are filled with water. The top of the saturated zone is the **water table**.

2. The rate of groundwater flow is dependent on the characteristics of the rock or sediment through which the water must move. **Porosity** is the percentage of the total volume of a body of rock or regolith that consists of open spaces. **Permeability** is a measure of how easily a solid allows fluids to pass through it. Groundwater in the saturated zone moves slowly by **percolation** through very small pores from areas where the water table is high to where it is lower.

3. Groundwater **recharge** occurs when rainfall and snowmelt infiltrate and percolate downward to the saturated zone. **Discharge** occurs where subsurface water leaves the saturated zone and becomes surface water in a stream, lake, or **spring**.

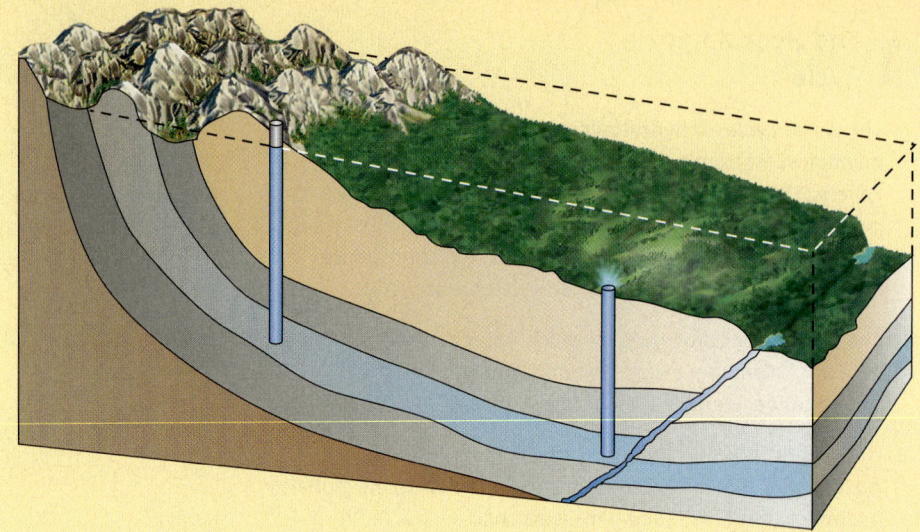

4. An **aquifer** is a body of permeable rock or regolith in the zone of saturation. An *unconfined aquifer* is in contact with the atmosphere through the pore spaces of overlying rock or regolith, while a *confined aquifer* lies between layers of impermeable rock, called **aquicludes**. The water pressure in a confined aquifer may be high enough to force the water partway or all the way to the surface when a well is drilled into it. Such a well is called *artesian*. Excessive withdrawal from an aquifer can cause the water table to drop, springs and streams to dry up, and regolith to compact and subside.

5. Many of the types and sources of contaminants that affect surface water also cause groundwater contamination. These include untreated sewage, agricultural pesticides and fertilizers, and leaks or spills of chemicals from commercial and industrial sites.

6. Caves, caverns, sinkholes, and karst topography are formed when rocks—most commonly carbonate rocks—are dissolved by circulating groundwater, creating underground cavities. Stalactites and stalagmites are built up from calcium carbonate precipitated from percolating groundwater.

KEY TERMS

- hydrologic cycle p. 134
- evaporation p. 135
- transpiration p. 135
- condensation p. 135
- deposition p. 135
- precipitation p. 135
- surface runoff p. 135
- infiltration p. 135
- stream p. 138
- channel p. 138
- gradient p. 138
- discharge (1) p. 138
- load p. 138
- floodplain p. 140
- alluvium p. 140
- drainage basin p. 142
- divide p. 143
- flood p. 145
- groundwater p. 153
- water table p. 153
- porosity p. 154
- permeability p. 155
- percolation p. 155
- recharge p. 155
- discharge (2) p. 155
- spring p. 156
- aquifer p. 156
- aquiclude p. 156
- cave p. 160
- cavern p. 160

CRITICAL AND CREATIVE THINKING QUESTIONS

1. List as many ways as you can in which we depend on the availability of fresh water in our daily lives.

2. Investigate the ways in which water shortages, floods, or other processes associated with water (such as erosion and deposition of sediment) have affected human history. How did societies respond, and what effects did those responses have?

3. The concept of residence time is very important in Earth science. It applies not only to water but also to any substance that moves from one reservoir to another in the Earth system. What other substances, besides water, move around in the Earth system? Why would we want to monitor these substances and keep track of their residence times?

4. What evidence has been gathered for water-shaped landscapes on Mars or elsewhere in the solar system? How are these landforms similar to or different from their analogues on Earth?

5. Where does your community obtain its water supply? Is it a groundwater or a surface fresh-water source? Is either the quantity or the quality of the water threatened?

6. Visit a stream before and after an intense rainfall. Observe the gradient, discharge, load, and velocity of the streamflow. What changes do you notice?

What is happening in this picture?

This 100-m-wide sinkhole opened up one day in Winter Park, Florida, and grew to the point where it eventually swallowed up part of a house, six commercial buildings, and the municipal swimming pool.

- What could have caused this to happen?
- What kind of rock is prone to this sort of collapse?

SELF-TEST

1. On this illustration, label each stage of the hydrologic cycle (1 through 6) using the following terms:

   precipitation
   surface runoff
   infiltration
   cloud formation through condensation
   surface evaporation and transpiration
   evaporation from the ocean

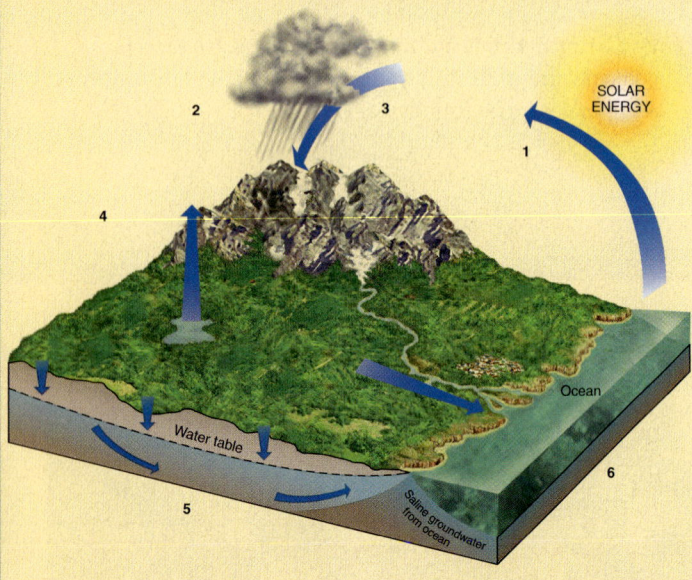

2. The illustration above depicts Earth's hydrologic cycle. Which reservoir holds most of the fresh-water resources of the planet?
   a. oceans
   b. ice sheets
   c. lakes
   d. groundwater

3. The _____ of a stream will have a large influence over channel development and the evolution of associated landforms.
   a. discharge
   b. gradient
   c. sediment load
   d. All of the above statements are correct.

4. _____ will form in streams when there is a low slope gradient and large and variable sediment load.
   a. Straight channels
   b. Meandering channels
   c. Braided channels

5. Lakes are ephemeral features that disappear by _____.
   a. the accumulation of inorganic sediment carried in by streams
   b. the accumulation of organic matter produced by plants within the lake
   c. gradually being drained by stream outlets eroding to lower levels
   d. All of the above statements are correct.

6. Oxbow lakes form when _____.
   a. levees collapse and cause floods
   b. meanders form cut-offs
   c. meanders fill with sediment
   d. point bars collapse

7. A(n) _____ is the total area from which water flows into a stream.
   a. alluvial fan
   b. braided channel
   c. water cycle
   d. drainage basin

8. Urban development can lead to increased risk of flooding by _____.
   a. compressing underlying sediments causing subsidence
   b. increasing surface runoff
   c. channeling runoff more quickly to rivers through storm drains
   d. All of the above are correct.

9. There is a _____ chance of a 50-year flood occurring in any given year.
   a. 1%
   b. 2%
   c. 5%
   d. 10%
   e. 50%

10. Though the proportions vary from one region to another, globally what accounts for the greatest demand for water?
    a. industry
    b. domestic use
    c. crop irrigation

11. What factors control the porosity of an aquifer?
    a. size of the particles in the material
    b. shape and uniformity of the material
    c. amount of cementing agent in the pore space of the material
    d. All of the above are correct.

12. A(n) _____ is an underground reservoir of water that is overlain by impermeable rock units.
    a. aquiclude
    b. aquifer
    c. confined aquifer
    d. unconfined aquifer

13. In a well drilled into a(n) _____, the water will rise to the level of the surrounding water table.
    a. aquiclude
    b. aquifer
    c. confined aquifer
    d. unconfined aquifer

14. Placing a well in a position where Earth's surface is below the water table will result in a well that _____.
    a. requires pumping to produce water
    b. does not require pumping to produce water

15. On this illustration, label each groundwater feature using the following terms:

    dry-season water table
    wet-season water table
    zone of saturation
    additional wet season saturation
    year-round saturation

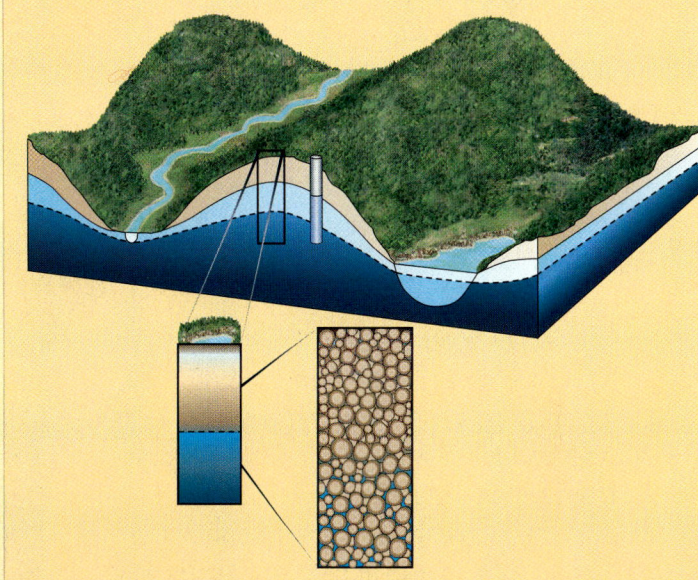

Self-Test 167

Extreme Climatic Regions: Deserts, Glaciers, and Ice Sheets 6

The Mandara Lakes in Libya excite the senses. Flat as a mirror, the water reflects a turquoise sky. The palm trees are dwarfed by the dunes of the Sahara Desert, which creep up to within a few feet of them. It is easy to see why explorers have always loved the desert: Its vastness and the beauty of its oases make human creations seem trivial by comparison.

Explorers of another stripe, like the mountaineer exulting at the summit of Aurora Peak in Alaska (inset), have always loved the world's icy wildernesses—in this case, the mountaineer is looking directly at Black Rapids Glacier. Almost every description of the Sahara Desert could also apply to this landscape: abstract, beautiful, immense, remote . . . and vulnerable. For Earth scientists, deserts and glaciers are linked by more than beauty; they are joined by extreme climates. Erosional forces such as ice and wind, which play lesser roles elsewhere, are dominant in these regions.

Deserts and glaciers are important as sensitive indicators of climate change. Glaciers and ice caps are retreating, a trend that many climatologists believe will continue throughout the 21st century. It is a trend that many people think humans may be partially responsible for because of our production of "greenhouse gases." Likewise, deserts grow or shift, in part because of changes in rainfall, and in part because of the way humans manage soil and water resources. Every one of us has good reason to pay attention to deserts and glaciers, even if we never set foot in the Sahara sand or on an Alaskan mountaintop.

Global Locator

NATIONAL GEOGRAPHIC

CHAPTER OUTLINE

- Deserts and Drylands p. 170

- Glaciers and Ice Sheets p. 183

Deserts and Drylands

> **LEARNING OBJECTIVES**
>
> **Identify** five different geographic settings for deserts.
>
> **Explain** how the wind shapes desert landforms through abrasion and deflation.
>
> **Describe** the structure of a sand dune.
>
> **Identify** five types of sand dunes.
>
> **Explain** why human activity sometimes causes desertification and what can be done about it.

Convection in the atmosphere (see Chapter 15) creates huge belts or cells of rising and falling air masses (see FIGURE 6.1). This results in three global belts of high rainfall and four belts of low rainfall. The high-rainfall belts are regions of *convergence* where huge, moist air masses meet and rise upward. These belts lie in the equatorial region and along the two polar fronts, at approximately 50°N and S latitudes, resulting in the warm-humid (tropical) and cold-humid (polar front)

The world's deserts FIGURE 6.1

A This map shows the distribution of arid and semiarid climates and the major deserts associated with them. Many of the world's great deserts are located where belts of dry air descend along the 30°N and 30°S latitudes. Notice also that regions of cold, descending air also surround both poles.

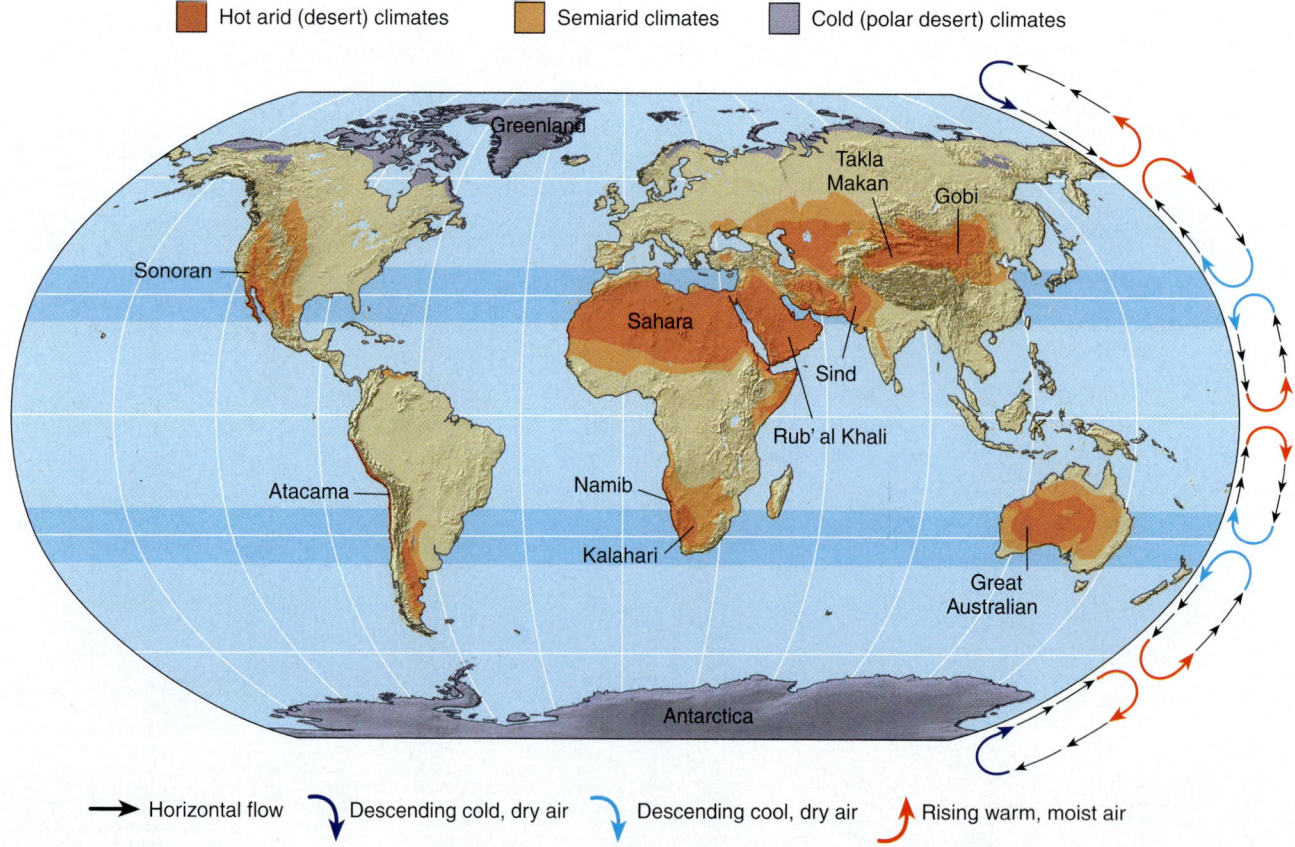

CHAPTER 6 Extreme Climatic Regions: Deserts, Glaciers, and Ice Sheets

climate zones. The four global belts of low rainfall are regions of *divergence*, where cool, dry air masses move downward and apart. These belts lie in the two polar regions and in the subtropical regions along the 30°N and S latitudes. The dry air masses and resulting low annual precipitation in these regions create two belts of dry climate in the subtropical regions, as well as the dry, cold climates of the polar regions.

TYPES OF DESERTS

The word *desert* literally means a deserted (that is, almost uninhabited) region that is nearly devoid of vegetation. However, in recent years, irrigation based on groundwater has changed the meaning of the word by making many desert regions suitable for agriculture and therefore habitable. As a result, the term **desert** is now defined in terms of the annual precipitation: less than 250 millimeters per year. Desert lands total about 25% of the land area of the world outside the polar regions. In addition, there is a smaller percentage of *semiarid* land in which the annual rainfall ranges between 250 and 500 mm.

Deserts can be separated into five categories, of which the most extensive, *subtropical deserts*, are associated with the two belts of low rainfall near 30°N and S latitudes. These include the Sahara, Kalahari, Namib, and Great Australian deserts. *Polar deserts* receive as little precipitation as subtropical deserts, but because it comes in the form of snow that never gets a chance to melt, these deserts gradually build up a thick ice sheet.

> **desert** An arid land that receives less than 250 mm of rainfall or snow equivalent per year, and is sparsely vegetated unless it is irrigated.

B The Namib Desert of, Namibia is located at 30°S latitude, and is one of the hottest and driest places on Earth.

C Antarctica, a cold desert, is the world's largest desert. This is Queen Maude Land in East Antarctica. Despite being covered by ice, the polar regions receive little precipitation and are considered to be frozen deserts.

Hot deserts FIGURE 6.2

A The Sahara is the greatest of the world's subtropical deserts. Here, a camel caravan crosses the desert in Libya.

B Mongolian nomads transport their belongings through the Altai Mountains, which border on the Gobi desert. This is a typical continental interior desert, which (even though it lies outside the subtropical latitudes) receives little rain because it is so far from the ocean.

C Rainshadow deserts form when a mountain range creates a barrier to the flow of moist air, causing a zone of low precipitation to form on the downwind side of the range. Mountains do not completely block air, but they do effectively remove most of its moisture. Death Valley, seen here, lies just east of the Sierra Nevada, the tallest mountain range in the continental United States.

D Baja California, the long, narrow peninsula in Mexico just south of its border with the western United States, consists mostly of coastal desert. Coastal deserts occur locally along the western margins of continents, where cold, upwelling seawater cools and stabilizes maritime air flowing onshore, decreasing its ability to form precipitation.

172 CHAPTER 6 Extreme Climatic Regions: Deserts, Glaciers, and Ice Sheets

Dust storm FIGURE 6.3

Driven by strong winds, a dust storm approaches an American facility in Iraq. The dust reached an altitude of 1000 meters and the storm lasted for 45 minutes, leaving a thick layer of dust in its wake.

The other three categories are less related to global air circulation patterns and more related to local geography. They are *continental interior deserts*, *rainshadow deserts*, and *coastal deserts* (see **FIGURE 6.2**).

Later in the chapter, we will take a closer look at the unique environment of the polar desert. For now, let's focus on the processes that characterize the world's hot deserts.

WIND EROSION

As we discussed in Chapter 4, wind is an important agent of erosion and transport. Processes related to wind—*eolian* processes—are particularly effective in arid and semiarid regions.

Windblown sediment Sediment carried by the wind tends to be finer than that moved by water or ice. Because the density of air is far less than that of water, air cannot move as large a particle as water flowing at the same velocity. In most regions, the largest particles that can be lifted in the airstream are grains of sand.

> **surface creep** Sediment transport in which the wind causes particles to roll along the ground.

Though they are smaller in size, windblown particles are similar to waterborne sediments in the way they travel. The largest grains are transported through **surface creep**. As wind speed increases, smaller grains may be bumped or lifted into the air, where they experience **saltation** (see Figure 4.13). Finer, dust-sized particles may be carried aloft to heights of a kilometer or so, where they can travel along in **suspension** as long as the wind keeps blowing. Such particles can remain suspended for many hours and cause great dust storms (**FIGURE 6.3**).

Mechanisms of wind erosion Flowing air erodes the land surface in two ways. The first, **abrasion** (a term used in Chapter 4 in the discussion of mechanical weathering), results from the impact of wind-driven grains of sand. Abraded rocks acquire distinctive, curved shapes and a surface polish. A bedrock surface or stone that has been abraded and shaped by windblown sediment is a *ventifact* (wind artifact). When preserved in sedimentary strata, ventifacts can indicate the direction of prevailing winds in the past. Other landforms characteristic of desert regions, such as steep-sided but flat-topped buttes, also result, at

> **saltation** Sediment transport in which particles move forward in a series of short jumps along arc-shaped paths.
>
> **suspension** Sediment transport in which the wind carries very fine particles over long distances and periods of time.
>
> **abrasion** Wind erosion in which airborne particles chip small fragments off rocks that protrude above the surface.

Deserts and Drylands 173

least in part, from wind erosion of bedrock (see **Figure 6.4**).

The second erosional process is called **deflation** (see **Figure 6.5A**). Deflation on a large scale takes place only where there is little or no vegetation and loose particles are fine enough to be picked up by the wind. It is especially severe in deserts but can occur elsewhere during times

> **deflation** Wind erosion in which loose particles of sand and dust are removed by the wind, leaving coarser particles behind.

of drought when no vegetation or moisture is present to hold soil particles together. Continued deflation sometimes leads to the development of desert pavement; most of the fine particles are removed, leaving a desert pavement covered by coarse particles (see **Figure 6.5B**).

Another way desert pavements are thought to form is by accretion. A layer of gravel at the surface

Abrasion by wind FIGURE 6.4

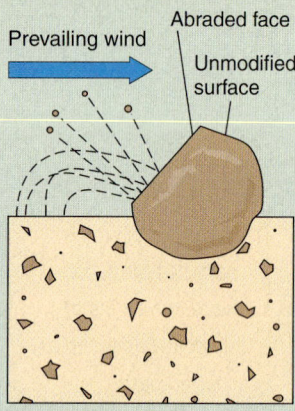

A Windblown sand peppers the upwind side of an exposed rock, eventually abrading it to a smooth, inclined surface.

B Ventifacts, each with at least one smooth, abraded surface facing upwind, litter the ground near Lake Vida in Victoria Valley, Antarctica.

C The Finger of God, a pillar-like butte of sandstone in the Kalahari Desert of South Africa, rests precariously on a pyramid of shale. In North America, rock pillars like these are known as *hoodoos*. They form when a resistant stratum lies atop a stratum that is less resistant to weathering. Note the sandstone mesa in the background, which has formed in the same way.

CHAPTER 6 Extreme Climatic Regions: Deserts, Glaciers, and Ice Sheets

From deflation to desert pavement FIGURE 6.5

These drawings show how progressive removal of sand from sediment with different-sized particles can lead to the formation of desert pavement.

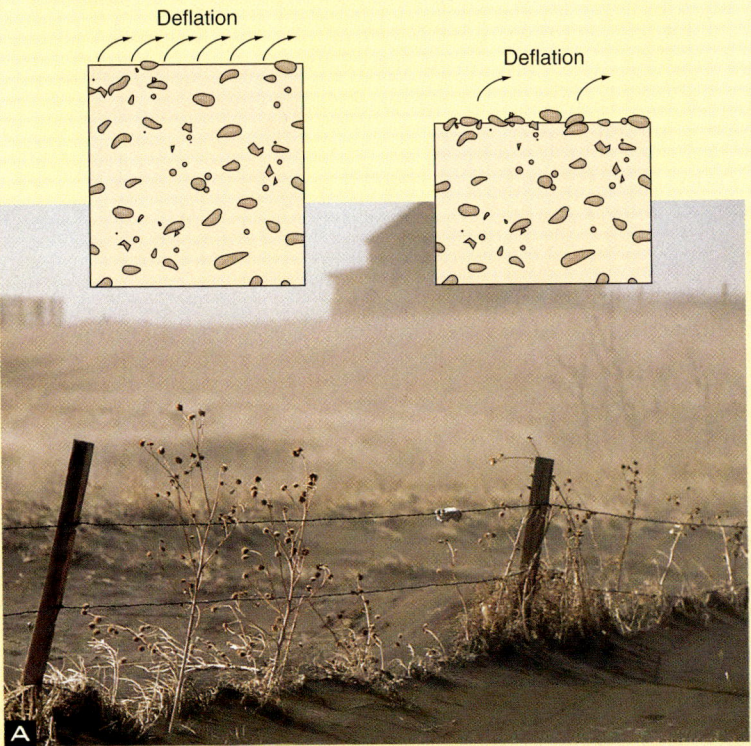

A Deflation is in progress in this plowed field in eastern Colorado. If the deflation continues long enough, and if the soil contains mixed particle sizes, including gravel, the result can be what you see here.

B This photo shows desert pavement on the floor of Searles Valley in California. One hypothesis is that the gravel is too coarse for the wind to move, and the pavement formed by deflation of fine-grained material. A second hypothesis is that a layer of gravel on the surface trapped wind-transported dust.

traps fine, windblown dust. Microorganisms live in the dust and their actions force gravel particles upward. The end result is a gravel layer covering a fine silt layer.

WIND DEPOSITS AND DESERT LANDFORMS

Any sediment that the wind removes from one place must eventually be deposited somewhere else—sometimes very far away. For example, distinctive reddish dust particles from the Sahara Desert have been identified in the soils of Caribbean islands, in the ice of Alpine glaciers, in deep-sea sediments, and in the tropical rainforests of Brazil. However, the most distinctive and characteristic eolian deposits in the deserts themselves are *dunes*.

Dunes

Although little is known about how **dunes** begin, it is likely that they start where some minor surface irregularity or obstacle distorts the flow of air. On encountering an impediment, the wind sweeps over and around it but leaves a pocket of slower-moving air immediately downwind. In this pocket of low wind velocity, sand grains moving with the wind drop out and begin to form a mound. The mound in turn influences the flow of air over and around it and may continue to grow into a dune.

dune A hill or ridge of sand deposited by the wind.

Deserts and Drylands

A typical dune is asymmetric, with a gentle windward slope (the side facing toward the wind) and a steep leeward face (the side facing away from the wind). Pushed by the wind, sand moves by surface creep and saltation up the gentle windward slope (see **FIGURE 6.6**). When it reaches the top, the sand cascades down the steep leeward slope, also called the *slip face*. The slip face is always on the leeward side, so we can tell which way the wind was blowing from the asymmetrical form of the dune. Crisscrossed strata within the dune, called *cross-beds*, are former slip faces.

The sliding sand on the slip face comes to rest at the angle of repose, the steepest angle at which loose particles will come to rest. The angle of repose varies for different materials, depending on factors such as the size and angularity of the particles. For dry, medium-sized sand particles it is about 33–34°; the angle is generally steeper for coarser materials such as gravel and gentler for finer materials such as silt.

Three factors control the shape and behavior of a dune: the wind conditions, the amount of vegetation cover, and the characteristics and quantity of sand available. These three factors are shown schematically in the triangle in **FIGURE 6.7**, along with photos of common types of dunes. Clearly, if there is no sand, or if there is plenty of vegetative cover to anchor sediments in place, then dunes will not form. Some types of dunes tend to remain stationary. However, others may migrate over long distances, and they can cause severe degradation and loss of agricultural productivity when they invade nondesert lands. This is part of the process of *desertification*.

Process Diagram

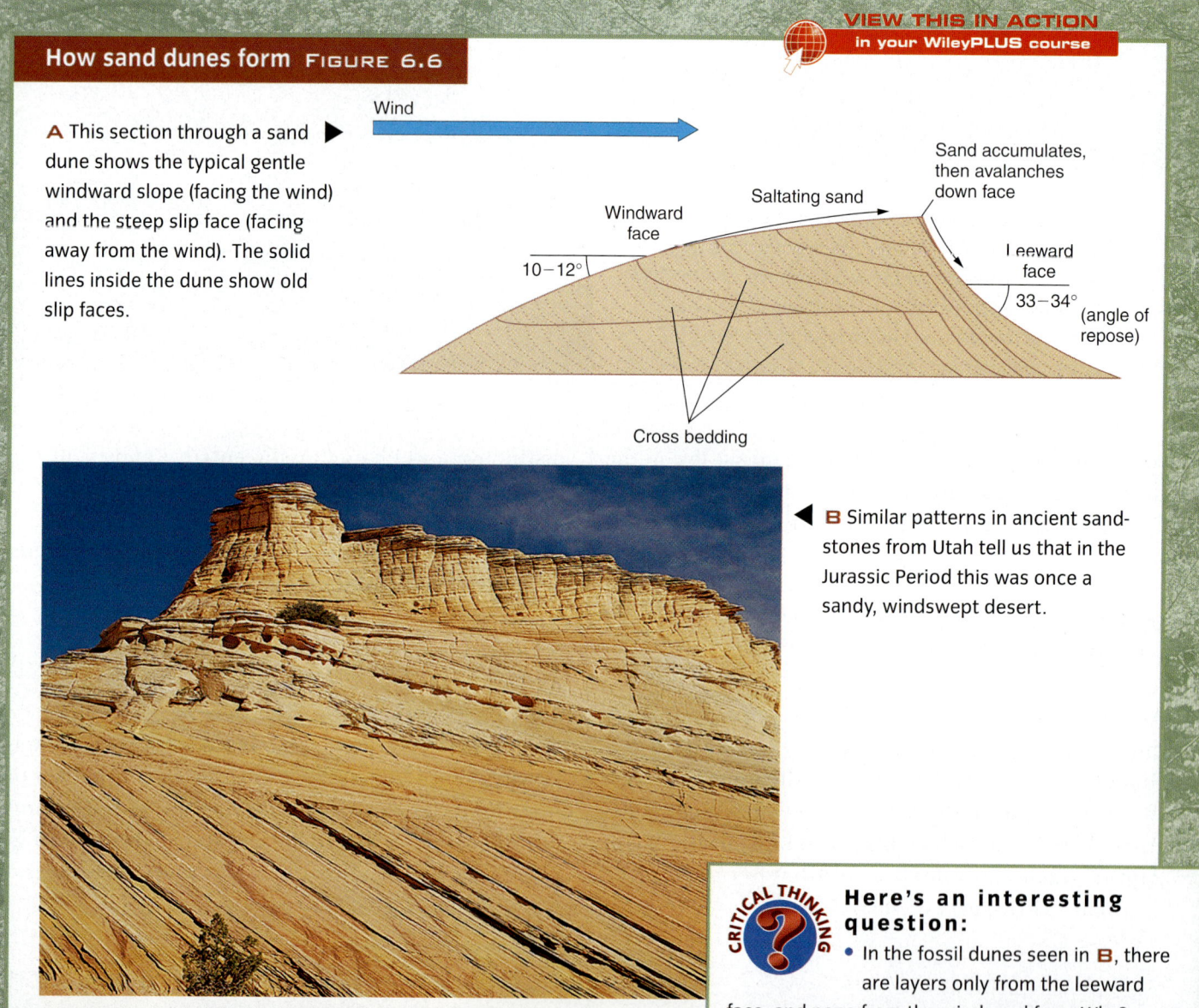

How sand dunes form FIGURE 6.6

A This section through a sand dune shows the typical gentle windward slope (facing the wind) and the steep slip face (facing away from the wind). The solid lines inside the dune show old slip faces.

B Similar patterns in ancient sandstones from Utah tell us that in the Jurassic Period this was once a sandy, windswept desert.

CRITICAL THINKING — Here's an interesting question:
- In the fossil dunes seen in **B**, there are layers only from the leeward face, and none from the windward face. Why?

Visualizing

Dune varieties FIGURE 6.7

▲ **A** Crescent-shaped dunes, called *barchans*, are very mobile. These form when the wind blows predominantly in one direction (the "horns" of the crescent point downwind). The dunes shown here are crossing a dry drainage system in Namibia, southwestern Africa.

▲ **B** When there is a copious sand supply, barchan dunes can merge and form *transverse dunes*, such as these dunes in the Empty Quarter of Saudi Arabia. They are oriented perpendicular to the prevailing winds.

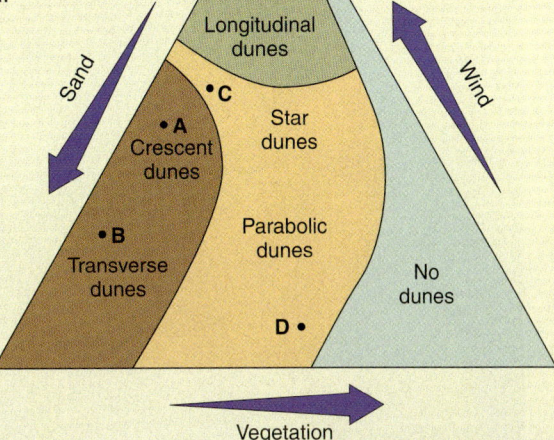

▼ **C** If the wind regularly blows in several different directions, it piles the sand up into stationary *star dunes*, such as the ones seen here in the Empty Quarter.

D Coastal regions, where the moist wind off the ocean allows vegetation to grow, form a typical environment for *parabolic dunes*. These are oriented in the opposite direction from barchan dunes: The arms, stabilized by vegetation, point upwind.

▲ **E** *Longitudinal dunes*, like these dunes in Death Valley, California, run parallel to the prevailing winds. They form in deserts with a meager sand supply. They can also form in areas with bidirectional winds, which first push on one side of the dune and then on the other.

177

Sand dunes on Mars FIGURE 6.8

A Shown here is a field of transverse dunes in Victoria Crater. The crater is an ancient meteorite impact site about 750 m across and 70 m deep. The walls of the crater have been modified by mass wasting and wind erosion. The crater has been the focus of a long study by Mars exploration rover, Opportunity.

B Giant barchans in the Hellespontus region of Mars were photographed by NASA's Mars Orbiter.

Sand dunes are not restricted to Earth. One of the striking features about Mars is that it is a desert-covered planet that has frequent, massive dust storms. Spacecraft orbiting Mars have seen and recorded a number of great dune fields. In Victoria Crater (**FIGURE 6.8A**), a field of transverse dunes is present; in the Hellespontus region, there are huge barchans (**FIGURE 6.8B**). Although we do not yet understand why, the Martian barchans reach heights of 300 m, almost 10 times larger than barchans on Earth.

STREAM EROSION AND DEPOSITION IN DESERTS

Contrary to popular belief (and many Hollywood movies), most deserts do not consist of endless expanses of sand dunes. Only a third of the Arabian Peninsula, the sandiest of all dry regions, and only a ninth of the Sahara Desert are covered with sand dunes. The remaining land area is either crossed by systems of stream valleys or covered by alluvial fans and alluvial plains. Running water, therefore, is important in the erosion of deserts, just as it is in rainy regions. However, instead of acting slowly and steadily, running water in deserts acts suddenly and in short bursts.

Rainfall in a desert region typically occurs during intense downpours that occur during a brief rainy season. The rapid runoff erodes steep-sided canyons called *arroyos* into the landscape (see **FIGURE 6.9**). These are likely to be dry most of the year but are subject to *flash floods* during the wet season. When arroyos draining an upland region meet the flat desert floor below, they lose their ability to transport sediment and drop their load of sand and gravel in an *alluvial fan* (see Figure 5.6B). If canyons are closely spaced along the base of a mountain range, the alluvial fans will sometimes coalesce into a broad alluvial apron called a *bajada*.

DESERTIFICATION

In the region south of the Sahara lies a drought-prone belt of dry grassland known as the Sahel. There the annual rainfall is normally only 100 to 300 mm (4 to 12 in), most of it falling during a single brief rainy season. In the early 1970s, the Sahel experienced the worst drought of the century. For several years in a row the annual rains failed to appear, causing the adjacent desert to spread southward by as much as 150 km. The drought extended from the Atlantic to the Indian Ocean, and affected a population of at least 20 million.

The fringes of deserts naturally migrate back and forth as a result of climate changes, and such was the case in the Sahel. However, the results of the drought were intensified by the fact that between about 1935 and 1970 the human population of the region had

Flash flood FIGURE 6.9

After a storm, a flash flood thunders down this arroyo on the Navajo reservation in Arizona. As the floodwater subsides, sediments are deposited across the alluvial floor of the canyon.

Desertification FIGURE 6.10

A In the Sahel region of Niger, a herd of goats grazes on pasture at the edge of the desert. As the goats consume the remaining grass and bushes, the dunes of the desert will inevitably advance.

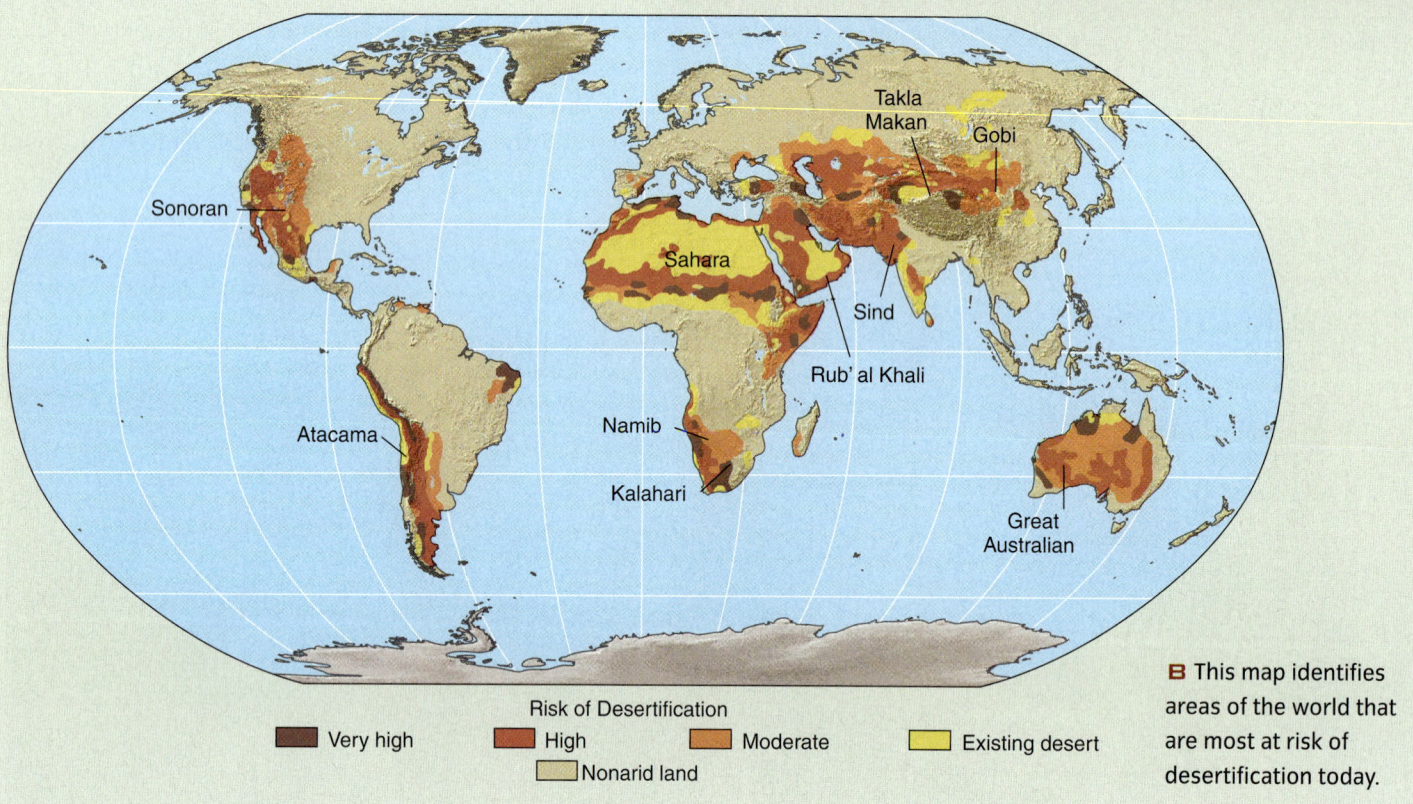

B This map identifies areas of the world that are most at risk of desertification today.

Risk of Desertification: Very high | High | Moderate | Existing desert | Nonarid land

doubled and the number of domestic livestock had also increased dramatically. This resulted in severe overgrazing, so the grass cover was devastated by the drought. Millions of people suffered from thirst and starvation. The overgrazing is continuing today (see **FIGURE 6.10A**), leaving the Sahel at risk for another devastating famine if the dry weather should ever return.

desertification
Invasion of desert conditions into non-desert areas.

Desertification can result from natural environmental changes or from human activities, or a mixture of both. The major signs of desertification include lower water tables, higher levels of salt in water and topsoil, reduction in surface water supplies, unusually high rates of soil erosion, and destruction of vegetation. Areas that are most susceptible to desertification, whether by nat-

180 **CHAPTER 6** Extreme Climatic Regions: Deserts, Glaciers, and Ice Sheets

ural or human-made causes, are shown in **FIGURE 6.10B**. Most are semiarid fringe lands adjacent to the world's great deserts. In many of these places, humans have lived successfully in a semiarid environment for centuries; what has changed is an explosion in the size of the population and the adoption of agricultural practices that may not be suited to that part of the world.

One of the best-known examples of desertification occurred in the United States during the mid-1930s, when huge dust storms swept across the Great Plains and drove many farm families off their land. John Steinbeck, in his award-winning novel, *The Grapes of Wrath*, described this resettlement, the largest forced migration in the country's history. The southern plains came to be called the *Dust Bowl*, and historians refer to that period as the "Dust Bowl years" (see **FIGURE 6.11**).

The Dust Bowl had both natural and human causes. Like the Sahel famine, it was triggered by a multiyear drought. However, the effects of the drought were exacerbated by decades of poor land-use practices. The grasses that originally grew on the prairies protected the rich topsoil from wind erosion. However, settlers gradually replaced these tall grasses with plowed fields and seasonal grain crops, which left the ground

The Dust Bowl FIGURE 6.11

On April 18, 1935, a massive dust storm closes in on Stratford, Texas. Within a few minutes, the town would be enveloped in pitch darkness, and it would be impossible even to see the house in the foreground.

Preventing the next Dust Bowl FIGURE 6.12

One of the biggest changes in agriculture in the Great Plains since the 1930s is the extensive use of groundwater for irrigation. These fields in Kansas use central-pivot irrigation, a method that minimizes evaporative loss of water and gives the fields a distinctive circular shape. In June, when this satellite photo was taken, wheat fields are bright yellow. Corn fields, in dark green, are growing vigorously, and the sorghum crop, light green, is just starting to come up. Irrigation helps stabilize the topsoil and thus prevents another Dust Bowl, but at a cost: It is slowly depleting the High Plains Aquifer.

bare and vulnerable for part of the year. Today, improved farming and irrigation practices have greatly reduced the risk of similar catastrophes (see FIGURE 6.12).

How can desertification be halted or even reversed? The answer lies largely in understanding the Earth science principles involved and in the application of measures designed to reestablish a natural balance in the affected areas. Soil management techniques are widely known and readily available; they simply need to be applied more aggressively. These techniques include crop rotation, terracing of steep slopes, and reforestation of vulnerable lands. Earth scientists can identify and map soils that are unsuitable for agriculture. Land-use planners can eliminate the incentives to exploit arid and semiarid lands beyond their capacity. The preservation of productive lands is essential to maintaining the world's food production capacity at the level needed for an increasing population.

CONCEPT CHECK STOP

Why is a desert more likely to develop in the subtropics than in the tropics?

How do subtropical deserts differ from rainshadow deserts, coastal deserts, and continental interior deserts? Discuss location, causes of aridity, and dune patterns.

Why is water more important than wind in sculpting many desert landscapes?

What are the main causes of desertification?

Glaciers and Ice Sheets

LEARNING OBJECTIVES

Distinguish between several different kinds of glaciers and ice formations.

Explain how temperate and polar glaciers differ.

Describe how ice in a glacier changes form, accumulates, ablates, and moves.

Identify several kinds of landforms created by glacial sediments.

Desertification can be an expression of climatic change. Natural processes involving changes in both precipitation and temperature have caused the Sahara Desert to advance and retreat many times over the past 10,000 years, independent of recent human activities in the region.

Climate changes in hot, sandy deserts and polar regions are interconnected; another climate battle is played out in the vast deserts of the polar ice sheets. The expansion and shrinking of glaciers and ice sheets, both in the polar regions and in more temperate alpine settings, is an expression of the complex interplay between temperature and precipitation in the global climate system (Chapter 14). The existence of **glaciers** and **ice sheets** is linked to the interaction of several parts of the Earth system: tectonic forces that produce high, mountainous areas; the ocean, a source of moisture; and the atmosphere, which delivers the moisture to the land in the form of snow. We now turn our attention to the great polar deserts and other parts of the **cryosphere**.

glacier A semipermanent or perennially frozen body of ice, consisting largely of recrystallized snow, that moves under the pull of gravity.

ice sheet The largest type of glacier on Earth, a continent-sized mass of ice that covers all or nearly all the land within its margins.

cryosphere The perennially frozen part of the hydrosphere.

COMPONENTS OF THE CRYOSPHERE

Annual snowfall is generally very low in polar regions because the air is too cold to hold much moisture. The small amount of snow that does fall doesn't usually melt, because summer temperatures stay very low. In areas where more snow falls each winter than melts during the following summer, the covering of snow gradually grows thicker. As the snow accumulates, its increasing weight causes the snow at the bottom to compact into a solid mass of ice. When the accumulating snow and ice become so thick that the pull of gravity causes the frozen mass to move, a glacier is born. The five main types of glaciers are illustrated in **FIGURE 6.13** on pages 184–185.

Glaciers are cold because they consist primarily of ice and snow. However, scientists have found by drilling holes through glaciers that interior temperatures are not all the same. Some glaciers are warmer than others, and the difference influences the behavior and movement of the ice. In one kind of glacier, the ice is near its melting point throughout the interior. These glaciers, called *temperate glaciers*, form in low and middle latitudes. Meltwater and ice can exist together at equilibrium in temperate glaciers. At high latitudes and altitudes, where the mean annual temperature is below freezing, the temperature in a glacier remains low and little or no seasonal melting occurs. Such a cold glacier is commonly called a *polar glacier*.

Ice sheets, which occur at high latitudes, are worth special mention. At present, ice sheets are found only in Greenland and Antarctica, though they

▲ **A** A *cirque glacier*, such as this one in Montana's Glacier National Park, occupies a bowl-shaped depression on a mountainside and often serves as the source for a valley glacier.

◀ **B** This *valley glacier* is in Alaska's Wrangell Saint Elias National Park.

NATIONAL GEOGRAPHIC

Visualizing

Glaciers and ice caps FIGURE 6.13

C An *ice cap* covers a mountaintop (or low-lying land in the polar regions) completely and usually displays a radial flow pattern. In this aerial photograph, an ice cap in Greenland surrounds the Nunatak Mountains.

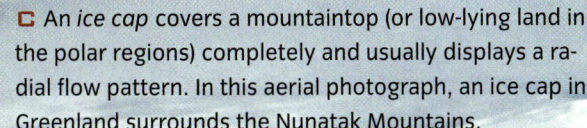

D When a glacial valley is partly filled by an arm of the sea, the valley is called a *fjord* and the glacier is a *fjord glacier*. Such glaciers often give rise to icebergs that break off and float away.

E When a glacier flows all the way out of the mountains and onto the surrounding lowlands, it is called a *piedmont glacier*. The Columbia Glacier in Alaska starts as a valley glacier and then spreads out as a piedmont glacier.

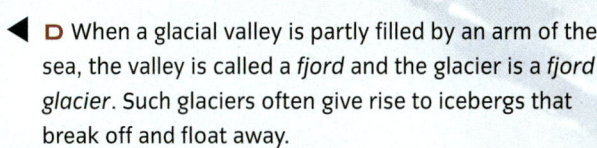

Glaciers and Ice Sheets 185

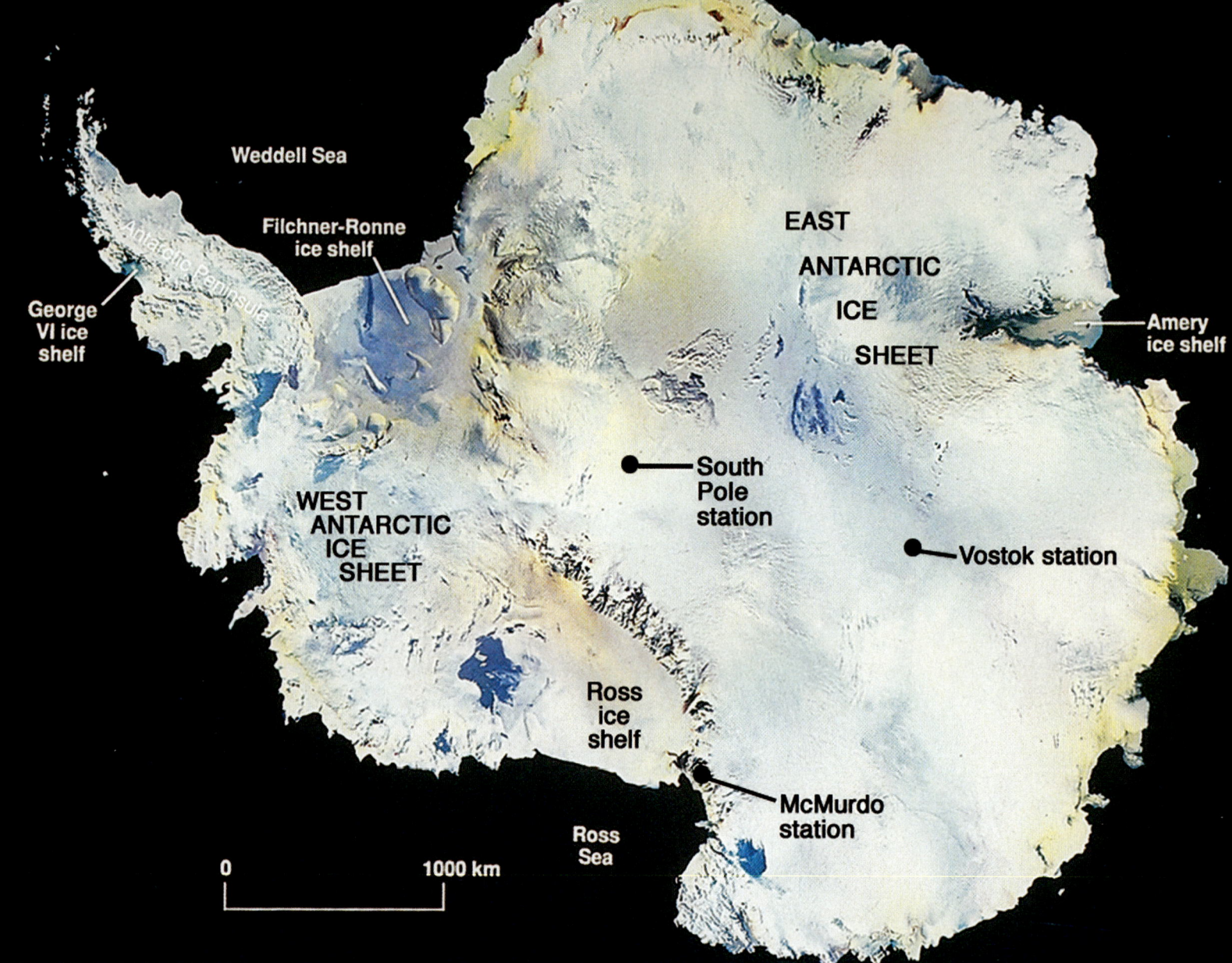

Welcome to Antarctica FIGURE 6.14

The East Antarctic Ice Sheet covers most of the continent of Antarctica, whereas the West Antarctic Ice Sheet overlies a volcanic island arc and the surrounding seafloor. In this satellite image, you can also see four ice shelves that occupy large bays. Glaciers that flow down from the mainland feed these shelves.

have been much more extensive in the past. They contain 95% of the world's glacial ice (and 70% of the world's fresh water). They are so thick and heavy that some of the land underneath Antarctica has actually been pushed below sea level (see FIGURE 6.14).

Ice shelves are sheets of floating ice hundreds of meters thick that adjoin glaciers on land (see Figure 6.14). They are constantly replenished by land-based glaciers, but also lose ice when large pieces called *icebergs* calve, or break off, from them.

Finally, *sea ice* never touches land at all but forms by the direct freezing of seawater. Antarctica is surrounded by sea ice, which in wintertime roughly doubles the apparent area of the continent. Most of the Arctic Ocean is covered by sea ice year-round.

EVER-CHANGING GLACIERS

Although we define a glacier as a semipermanent or perennially frozen body of ice, glaciers are constantly changing in several ways. For example, the snow that falls on the surface of glaciers gradually changes to ice. Glaciers also shrink and grow in response to seasonal changes in temperature and precipitation. The ice in a glacier moves, slowly but surely, under the influence of gravity, and changes in climatic conditions cause the margins of glaciers to advance or retreat. Let's take a closer look at some of these changes.

How glaciers form Newly fallen snow is very porous and easily penetrated by air. The presence of air in the pore spaces allows the delicate points of each snowflake to sublimate (change from solid to vapor without melting). The resulting water vapor crystallizes in tiny spaces in the snowflakes, eventually filling them. In this way, the ice crystals in the snow pack slowly become smaller, rounder, and denser, until the pore spaces between them disappear (see **FIGURE 6.15**). Snow that survives for a year or more becomes more compact as it is buried by successive snowfalls. As the years go by, the snow gradually becomes denser and denser until it is no longer penetrable by air and becomes *glacier ice*. This process may take from decades in the case of temperate glaciers, to millennia in the case of polar ice sheets.

Further changes take place as the glacial ice is buried deeper and deeper. As snowfall adds to the glacier's thickness, the increasing pressure causes the small grains of glacier ice to grow. This increase in size is similar to what happens when a fine-grained rock recrystallizes as a result of metamorphism in Earth's crust (Chapter 2). Ice is in fact a mineral (Chapter 3), and therefore, glacier ice is technically a rock. However, the properties of this rock are very different from any other naturally occurring rock because of its very low melting temperature and its unusually low density. Ice floats in water because it is only nine-tenths as dense as water.

How glaciers grow and shrink The mass of a glacier constantly changes as the weather varies from season to season and, over time, as local and global climates change (see Chapter 14). In a way, a glacier is like a checking account. Instead of being measured in terms of money, the balance of a glacier's account is measured in terms of the amount of snow deposited, mainly through snowfall in the winter, and the amount

From snow to ice FIGURE 6.15

As a new snowflake is slowly converted into a granule of ice, it loses its delicate points, through evaporation and recrystallization, and becomes much more compact.

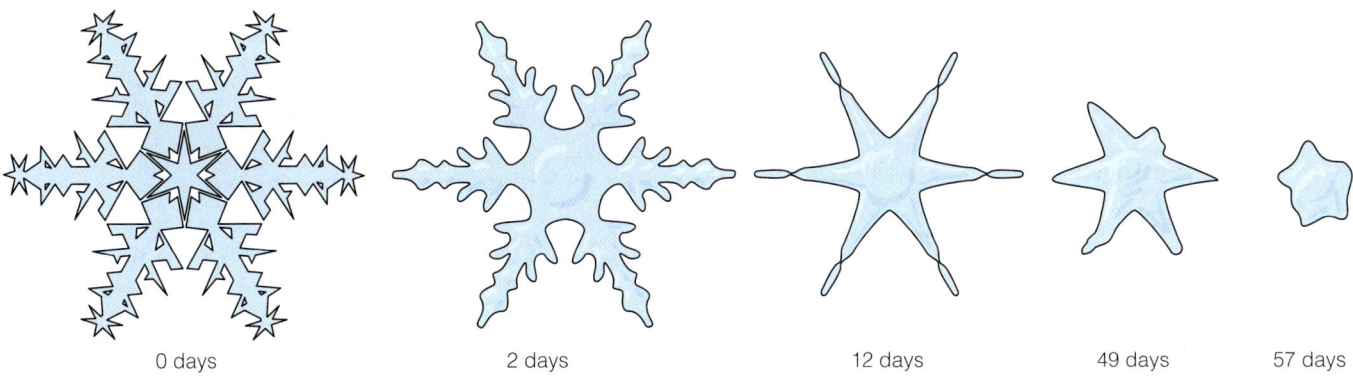

0 days   2 days   12 days   49 days   57 days

Glaciers and Ice Sheets 187

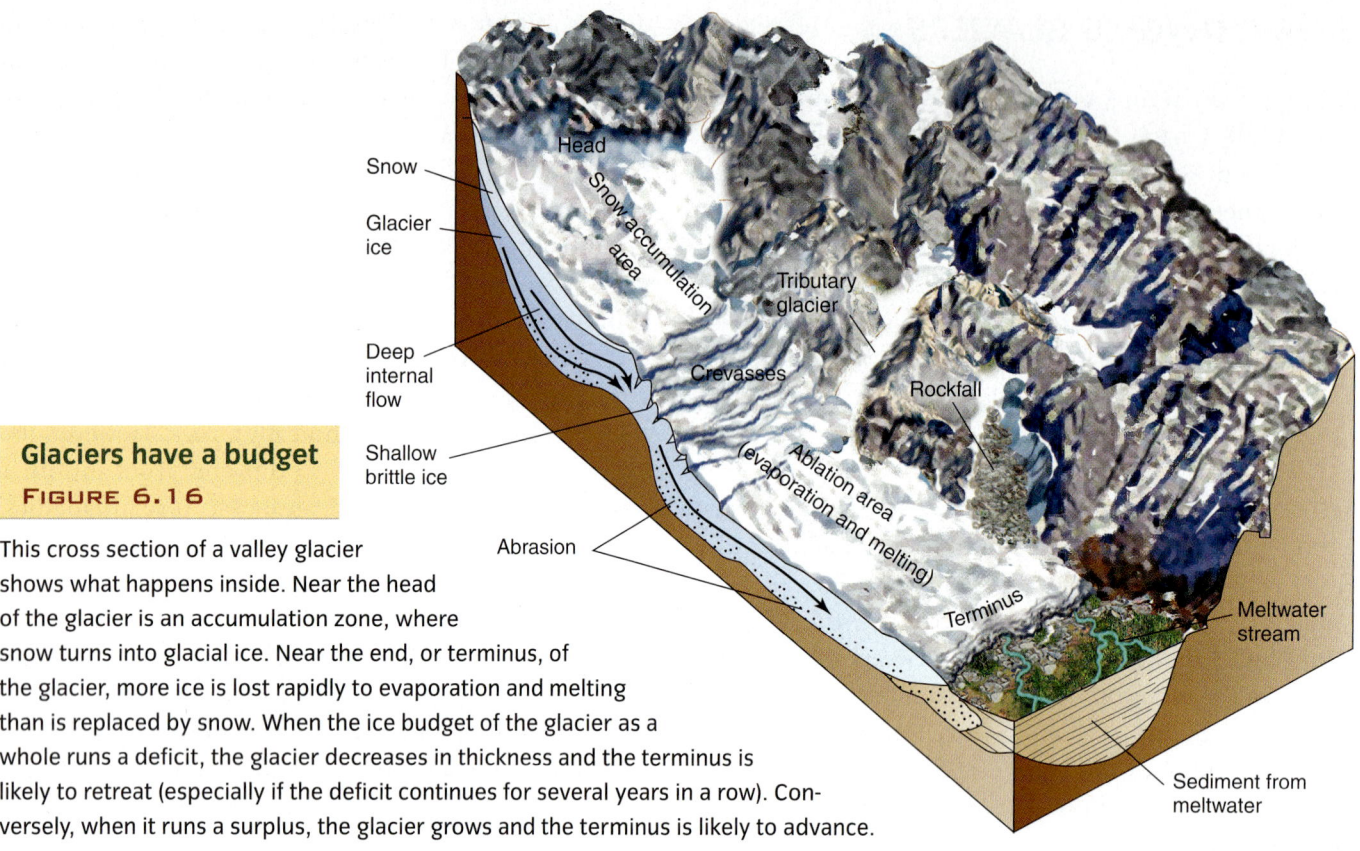

Glaciers have a budget
FIGURE 6.16

This cross section of a valley glacier shows what happens inside. Near the head of the glacier is an accumulation zone, where snow turns into glacial ice. Near the end, or terminus, of the glacier, more ice is lost rapidly to evaporation and melting than is replaced by snow. When the ice budget of the glacier as a whole runs a deficit, the glacier decreases in thickness and the terminus is likely to retreat (especially if the deficit continues for several years in a row). Conversely, when it runs a surplus, the glacier grows and the terminus is likely to advance.

of snow (and ice) withdrawn, mainly through melting during the summer. The additions are collectively called **accumulation** and the losses **ablation** (see **FIGURE 6.16**). The total added to the account at the end of a year—the difference between accumulation and ablation—is a measure of the glacier's *mass balance*. The account may have a surplus (a positive balance) or a deficit (a negative balance), or it may hold the same amount at the end of the year as it did at the beginning.

How glaciers move Part of the definition of a glacier is that it moves because of the pull of gravity. How can we detect this movement? One method is to carefully measure the position of a boulder on its surface relative to a fixed point beyond the glacier's edge. If you measure the boulder's position again a year later, you will find that it has moved "downstream," usually by several meters. Actually, it is the ice that has moved, carrying the boulder along.

Measurements of velocity show that the ice in the central part of the glacier moves faster than the ice at the sides, and the uppermost layer moves faster than the lower layers. This is similar to what happens to water flowing in a stream (Chapter 5). In most glaciers, flow velocities range from a few centimeters to a few meters a day. It may take hundreds of years for an ice crystal that fell as a snowflake at the head of a glacier to reach the terminus and melt. The glacial ice moves in two basic ways: by internal flow and by basal sliding across the underlying rock or sediment.

Internal flow As the weight of overlying snow and ice in a glacier increases, individual ice crystals are subjected to higher and higher stress. Under this stress, ice crystals deep within the glacier creep along internal crystal planes (see **FIGURE 6.17**). As the compacted, frozen mass moves, the crystal axes of the individual ice crystals are forced into the same orientation and end

Deforming ice FIGURE 6.17

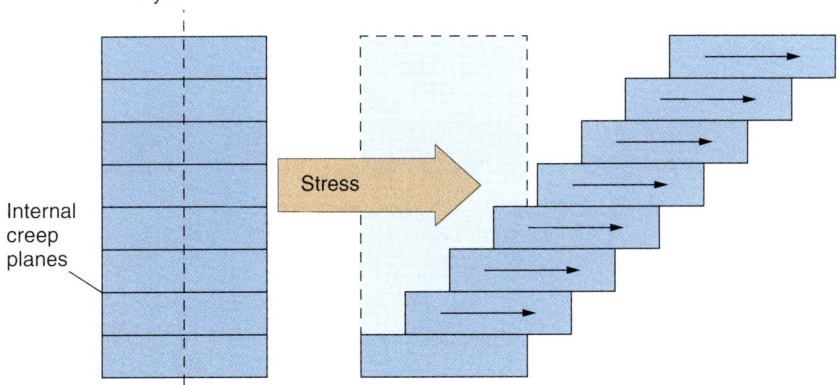

A Randomly oriented ice crystals are reorganized by stress so that their internal crystal planes are parallel.

B The ice creeps along its internal planes in layers, in a process very similar to playing cards in a deck of cards sliding past one another.

up with their internal crystal planes oriented in the same direction.

In contrast to the deep parts of a glacier, where ice flows by internal creep, the surface portion has relatively little weight on it and is brittle. When a glacier passes over a change in slope, such as a cliff, the surface ice cracks as tension pulls it apart. When the crack opens up, it forms a *crevasse*, a deep, gaping fissure in the upper surface of a glacier (see **FIGURE 6.18**). Thus, ice moves in a glacier through a combination of ductile deformation at depth and brittle deformation at the surface.

Basal sliding

Sometimes ice at the bottom of a glacier slides across its *bed* (the rock or sediment on which the glacier rests). This is called *basal sliding*. In temperate glaciers, meltwater at the base can act as a lubricant. Basal sliding may account for up to 90% of total observed movement in such a glacier, with the remaining 10% being internal flow. By contrast, polar glaciers are so cold that they are frozen to their bed; they seldom move by basal sliding, so all movement is by internal flow.

Crevasses FIGURE 6.18

Deep fissures in the ice, called *crevasses*, open up as a result of stresses in the brittle surface layer of a glacier. The glacier flows in a direction perpendicular to the crevasse.

Glacial flow →

On infrequent occasions, a glacier seems to go berserk (see FIGURE 6.19). Ice in one part of the glacier begins to move rapidly downslope, producing a chaos of crevasses and broken pinnacles. Rates of movement have been observed that are up to 100 times those of ordinary glaciers. These episodes are called *surges,* and their causes are not fully understood. Glacial specialists believe that, in many cases, a buildup of water pressure at the base of the glacier reduces friction and permits very rapid basal sliding. The surge stops when the water finds an exit.

The glacial landscape

As glaciers move, they change the landscape by eroding and scraping away material as well as by transporting and depositing material at their ends and along their margins. In changing the surface of the land over which it moves, a glacier acts like a file, a plow, and a sled. As a file, it rasps away firm rock. As a plow, it scrapes up weathered rock and soil and plucks out blocks of bedrock. As a sled, it carries away the load of sediment acquired by plowing and filing, along with rock debris that falls onto it from adjacent slopes. Let's look more closely at the landforms that result from these processes.

The "galloping glacier" FIGURE 6.19

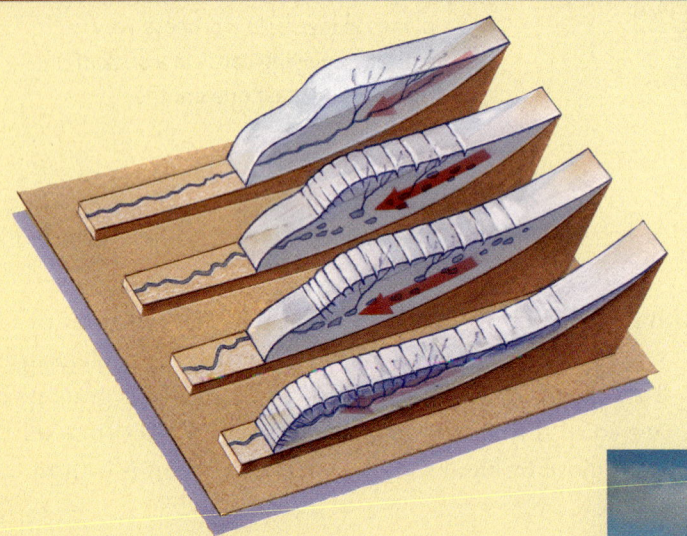

A According to one theory, a glacial surge gets started when water at the base of the glacier gets blocked from flowing out. The buildup of pressure lubricates the base and allows the glacier to flow very rapidly, until the water finds a way out again.

B In 1986, Hubbard Glacier in Alaska surged across the mouth of Russell Fjord, damming it up and creating a freshwater lake. This picture was taken soon after the dam broke and reopened the fjord.

C Here is a view of the ice dam before it broke. The helicopter hovering in front should give you an idea of the size of this wall of ice.

Glacial erosion The base of a glacier is studded with rock fragments of various sizes that are all carried along with the moving ice. When basal sliding occurs, small fragments of rock embedded in the basal ice scrape away at the underlying bedrock and produce long, nearly parallel scratches called *glacial striations*. Larger particles gouge out deeper *glacial grooves* (see **Figure 6.20A** on the following page). Because glacial striations and grooves are aligned parallel to the direction of ice flow, they help Earth scientists reconstruct the flow paths of former glaciers.

Mountain glaciers produce a variety of distinctive landforms. Bowl-shaped cirques are found at a glacier's head. Two cirques on opposite sides of a mountain can meet to form a sharp-crested ridge called an *arête*. Cirques developing on all sides of a mountain may carve its peak into a prominent horn (like Mt. Everest, seen in **Figure 6.20B**).

When glacial ice moves downward from a cirque, it scours a valley channel with a distinctive U-shaped cross section and a floor that usually lies well below the level of tributary valleys (see **Figure 6.20C**). Continental ice sheets can gouge the bedrock to form lakes; some examples of very large glacially formed lakes are the Great Lakes, Lake Winnipeg, and Great Bear Lake. Large glaciers and ice sheets are more effective agents of erosion, and carve deeper valleys and lakes, than small tributary glaciers. At the intersection of a smaller and larger glacier, there will usually be an abrupt change in elevation of the valley floor due to the different depths of erosion.

Glacial deposition Like streams, glaciers carry a load of sediment particles of various sizes. Unlike a stream, however, a glacier can carry part of its load at its sides and even on its surface. A glacier can carry very large rocks and small fragments side by side. When deposited by a glacier, the load of mixed rocky fragments (called glacial **till**) is not sorted, rounded, or stratified the way stream deposits usually are. In most cases, the boulders and rock fragments in a till are different from the underlying bedrock (see **Figure 6.21A** on page 193).

> **till** A heterogeneous mixture of crushed rock, sand, pebbles, cobbles, and boulders deposited by a glacier.

The boulders, rock fragments, and other sediment carried by the glacier may be deposited along its margins or at its terminus. These form ridges called **moraines**; specifically, *lateral moraines* form along the edges and a *terminal moraine* forms at the terminus (see **Figure 6.21B**), and *recessional* or *end moraines* form as a glacier melts and recedes. If two glaciers converge, they may trap lateral moraines between them, forming a ridge of material that rides along the middle of the ice stream, called a *medial moraine* (see **Figure 6.21C**). Earth scientists have used the locations of glacial moraines in the United States and Canada to determine how far the glacial ice cover extended over North America during the last ice age.

> **moraine** A ridge or pile of debris that has been, or is being, transported by a glacier.

The sinuous deposit shown in **Figure 6.21D** may look perplexing at first: It seems like an upside-down streambed embossed upon the landscape. What could create such a feature? We have already mentioned that the bottoms of some temperate glaciers contain meltwater. This water may actually form a stream that tunnels through the glacier. (These streams can sometimes be seen emerging from the terminus of an active glacier.) Like any other stream, it deposits sediment. If the glacier subsequently retreats, that sediment is left behind in a raised bed called an *esker*, like the one shown.

The retreat of a glacier can leave behind a terrain full of pits and pockmarks, due to abandoned blocks of ice embedded in the glacial debris. These subsequently melt and the depressions left behind are called *kettles*. Many kettles fill with water to form *kettle ponds* and *kettle lakes*. One famous example of a small kettle pond is Walden Pond, in Massachusetts, immortalized by the writer Henry David Thoreau.

Periglacial landforms Regions subjected to **periglacial** conditions, such as areas near glacial ice, also have a distinctive set of landforms, resulting from intense frost action and a large annual range in temperatures. The most common type of environment in present-day periglacial regions is tundra, a

> **periglacial** Conditions that are near glacial.

Glacial sculpting FIGURE 6.20

A These *glacial grooves* in Ohio were cut into limestone by the Wisconsin Glacier during the most recent ice age, about 25,000 years ago.

B The Western Cwm, a deep cirque on the west side of Mt. Everest, is flanked by sharp-crested *arêtes*.

C The gorgeous Lauterbrunnen Valley in Switzerland has the classic U-shape of a glacial valley. The glacier that formed it no longer exists.

Glacial deposits FIGURE 6.21

◀ **A** Glacial till can sometimes include very large boulders, such as these boulders in Yellowstone National Park. When they are different from the bedrock, such boulders are called *erratics*.

▲ **B** This *terminal moraine* near Mt. Robson in British Columbia marks the farthest advance of the glacier at left in the 19th and 20th centuries.

Medial moraine

▲ **C** The dark stripes running down the center of Kaskawulsh Glacier, in the Yukon, are a *medial moraine*.

▲ **D** The curving ridge of sand and gravel in this photo is an *esker* in Kettle Moraine State Forest in Wisconsin.

Glaciers and Ice Sheets   193

What an Earth Scientist Sees

Periglacial Landforms

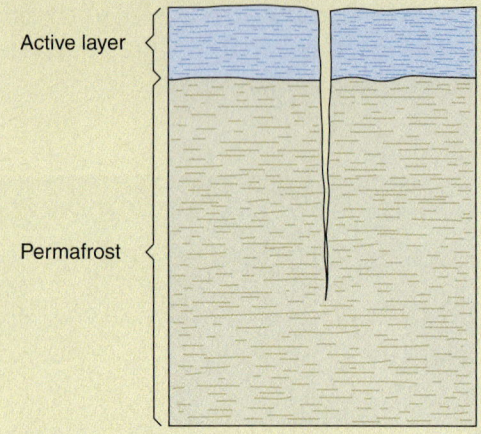

A An ice wedge forms when water seeps into an open crack in the ground and freezes.

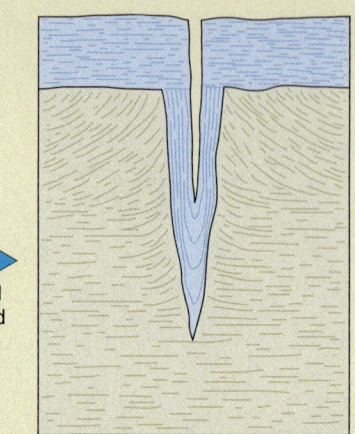

B In summer the crack opens or partially melts, allowing more water to enter. In winter the ice freezes again. The ice wedge continues to grow as the melting and refreezing cycle repeats itself hundreds of times. Such wedges can grow as wide as 3 m and as deep as 30 m.

C In Beacon Valley, Antarctica, ice wedges have grown and interconnected to form what Earth scientists call *patterned ground*.

CRITICAL THINKING

Here's an interesting question:
- If an Earth scientist finds fossil mud cracks in ancient strata, how might she or he tell whether they were the result of periglacial conditions?

treeless landscape with long winters, very short summers, poorly developed soils, and low, scrubby vegetation. Tundra regions often lie on top of a layer of **permafrost**. During the short summer, the ice melts only in a thin layer near the surface, called the *active layer*. The freeze-thaw cycle perennially occurring below produces characteristic landscape formations called *ice wedges* and *patterned ground* (see *What an Earth Scientist Sees*).

permafrost
Ground that is perennially below the freezing point of water.

CONCEPT CHECK STOP

How do the glaciers of temperate regions differ from those of polar regions?

What has to happen (in terms of glacial budget) for a glacier to advance?

How does glacial erosion differ between mountain glaciers and continental ice sheets?

What is a moraine? What is significant about a terminal moraine?

Amazing Places: Death Valley

A place of extremes, Death Valley in California boasts the lowest point in North America, a mostly dry lakebed called Badwater that is 86 m below sea level (**A**). From nearby Dante's Peak, on a clear day, you can see both the lowest and the highest point (Mt. Whitney, at 4418 m) in the continental United States. Death Valley is also notorious for its heat. The hottest temperature ever recorded in the Western Hemisphere (57°C) was measured here, at Furnace Creek, in 1913. With 4.8 cm of rain per year, Death Valley also ranks as one of the country's driest places. These extremes result in part from the topography of Death Valley: It is a classic rainshadow desert, lying just east of the Sierra Nevada mountain range. Its low elevation and oppressive heat come from its location in a natural basin formed by extensional stresses caused by plate tectonic motions.

Though Death Valley is dry today, during the last ice age it was almost entirely underwater. You can still see signs of Lake Manly, which once covered this area. Shoreline Butte (**B**) was once an island in the lake, and the successive levels of the water are etched into the rock. As the lake dried up, it left large evaporite deposits, such as the salt pans at Badwater (**C**).

Glaciers and Ice Sheets 195

SUMMARY

1 Deserts and Drylands

1. The term **desert** refers to arid lands where annual rainfall is less than 250 mm. Five types of deserts have been identified: *subtropical, continental interior, rainshadow, coastal,* and *polar*. Subtropical and polar deserts result from the global wind patterns that create dry high-pressure air masses around 30°N and S and at the poles. The other three kinds of deserts result from local topographic conditions.

2. *Eolian* (wind) erosion is particularly effective in arid and *semiarid* regions. Wind moves particles through **surface creep**, **saltation**, and **suspension**. Flowing air erodes the land surface through the processes of **abrasion** and **deflation**.

3. Dunes are hills or ridges of sand deposited by winds. They are asymmetrical, with a gentle slope facing the wind and a steeper *slip face* on the leeward side. Common types of dunes are *barchan, transverse, star, parabolic,* and *longitudinal*. The types that will form in a given place depend on the amount of sand, the wind conditions, and the amount of vegetation.

4. Contrary to the popular image of a desert, the majority of desert lands are not covered by sand. Water erosion is an important natural process in deserts. Flash floods carve deep canyons called *arroyos* and create depositional landforms such as *alluvial fans*.

2 Glaciers and Ice Sheets

1. The perennially frozen part of the hydrosphere is called the **cryosphere**. It includes **glaciers** on land as well as sea ice. Glaciers come in several varieties, and can form either in polar regions or at high altitudes in temperate regions. Most of the world's fresh water is locked in vast **ice sheet**s in Antarctica and the North Pole region.

2. *Glacial ice* is formed by the compaction of grains of snow and recrystallization of small ice crystals into larger ones—processes that are quite similar to lithification and metamorphism in ordinary rocks. The mass of a glacier can change from season to season and year to year through **accumulation** and **ablation**. Accumulation (by new snowfall) predominates at the head of the glacier, and ablation (by melting) predominates at the foot.

3. Just like ordinary rocks, glaciers can move by either ductile or brittle deformation. The internal movement is ductile, and aligns the ice crystals in the direction of flow. Brittle deformation occurs when a glacier goes over a change in slope, opening up fissures called *crevasses*. At the bottom of the glacier, where *basal sliding* occurs, small fragments of rock embedded in the ice scrape away at the underlying bedrock, producing glacial striations and grooves.

4. In mountainous regions, glaciers produce a variety of distinctive erosional landforms, such as cirques, arêtes, and U-shaped valleys. Common periglacial features include permafrost and patterned ground, which are formed by the repeated freezing and thawing of groundwater.

5. Glacial deposits often consist of unsorted **till**. A *moraine* is a ridge or pile of debris being carried along by a glacier or deposited along its edge or terminus. When a glacier retreats, the moraine is left behind. Such deposits are useful for identifying the previous extent of glaciers that have retreated or disappeared. Other features often left behind by retreating glaciers are *kettle lakes* and elevated *eskers*.

KEY TERMS

- desert p. 171
- surface creep p. 173
- saltation p. 173
- suspension p. 173
- abrasion p. 173
- deflation p. 174
- dune p. 175
- desertification p. 180
- glacier p. 183
- ice sheet p. 183
- cryosphere p. 183
- till p. 191
- moraine p. 191
- periglacial p. 191
- permafrost p. 194

CRITICAL AND CREATIVE THINKING QUESTIONS

1. Look at the photograph of sandstones shown in Figure 6.6. Can you tell which way the wind was blowing when it produced these dune deposits?

2. Investigate the current status of drought and land degradation in the Sahel or elsewhere. Can you find any information about soil erosion control techniques or other methods that are being used to combat desertification?

3. Do some research on the most recent ice age. Do you live in an area that was formerly covered by ice? How thick was the ice? Is there any evidence in the landforms around you that indicates the area was formerly glaciated?

4. At the height of the most recent ice age, vegetation in North America south of the ice front must have been different from the vegetation today. Do some research and find out what is known of vegetation changes in your area over the past 20,000 years.

5. There is evidence that glaciers in Alaska and elsewhere are melting. Do some research on the extent of the glaciers feeding into Glacier Bay in Alaska and see whether you can determine how far they have melted back in the last hundred years.

What is happening in this picture?

This house was built on top of permafrost in the Canadian Arctic.

- How might warming temperatures affect the permafrost?
- Why would this cause the surface to buckle and subside?

SELF-TEST

1. _____ and _____ deserts result from the global wind patterns that create dry high-pressure air masses around 30°N and S and at the poles.
   a. Subtropical/continental interior
   b. Subtropical/polar
   c. Subtropical/rainshadow
   d. Continental interior/rainshadow
   e. Continental interior/polar

2. Ventifacts are the result of _____.
   a. scouring at the bottom of a glacier
   b. intermittent floods in desert stream channels
   c. abrasion by wind-carried sand particles
   d. All of the above statements are correct.

3. Deflation on a large scale takes place only _____.
   a. where there is little or no vegetation
   b. where loose particles are fine enough to be picked up by the wind
   c. in desert environments
   d. Both a and b are correct.
   e. Both b and c are correct.

4. A typical sand dune _____.
   a. is asymmetrical
   b. has a gentle windward slope
   c. has a steep leeward face
   d. All of the above statements are correct.

5. The illustration below depicts dune formation as a function of wind, sand supply, and vegetation cover. Label the ternary diagram with the following terms:

   longitudinal dunes   transverse dunes
   crescent dunes   parabolic dunes
   star dunes

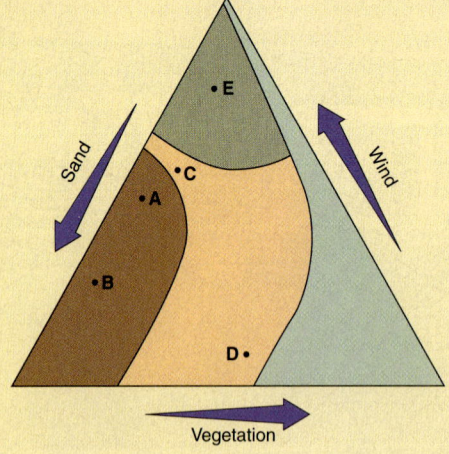

6. The major signs of desertification include all of the following except _____.
   a. lower water tables and a reduction of surface waters
   b. increased surface temperatures in summer months
   c. higher levels of salt in water and topsoil
   d. unusually high rates of soil erosion
   e. destruction of vegetation

7. The Dust Bowl of the American Southwest was a result of _____.
   a. prolonged drought conditions
   b. destruction of natural desert grasses
   c. poor farming practices
   d. All of the above statements are correct.

8. Glaciers form by _____.
   a. freezing of stream waters
   b. compaction of snow
   c. a combination of compaction of snow and freezing rain
   d. All of the above statements are correct.

9. This illustration is a block diagram showing various glaciers at Earth's surface. Label the illustration with the appropriate terms for the individual glaciers A through E:

   cirque glacier   fjord glacier
   piedmont glacier   ice cap
   valley glacier

198   CHAPTER 6   Extreme Climatic Regions: Deserts, Glaciers, and Ice Sheets

10. In _____, meltwater and ice can exist together in equilibrium.
    a. temperate glaciers
    b. polar glaciers
    c. ice caps

11. In _____, little or no seasonal melting occurs.
    a. temperate glaciers
    b. polar glaciers
    c. ice caps

12. Glacial ice moves under the influence of gravity through _____.
    a. basal sliding
    b. internal flow
    c. Both a and b are correct.

13. Internal flow in a glacier happens when _____.
    a. randomly oriented ice grains become reoriented by stress
    b. internal planes of atoms in ice crystals slide past each other (creep)
    c. Both a and b are correct.

14. This photograph is of a glacial landscape. What is the curving ridge of sand and gravel that dominates the picture called?
    a. a drumlin   c. a kettle
    b. an esker   d. an end moraine

15. Periglacial landforms are the result of _____.
    a. freezing of the ground under an ice sheet
    b. intense frost action and a large annual temperature range
    c. cracking of permanently frozen ground
    d. stranded ice masses as an ice sheet melts back
    e. All of the above statements are correct.

Plate Tectonics: Sculptor of Earth's Ever-Changing Landscape 7

The peak of Ama Dablam, in the Himalaya of Nepal, basks in the late afternoon sunshine. For mountain climbers and explorers, the Himalaya is the top of the world. Many of the peaks here are more than 7000 meters tall. The tallest of them all, of course, is Mt. Everest, which has become a favorite, and sometimes fatal, challenge for thousands of mountaineers since Edmund Hillary and Tenzing Norgay first scaled it in 1953.

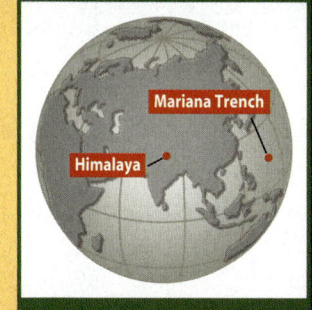

Global Locator

Only two people—a Swiss engineer named Jacques Piccard and a Navy lieutenant, Donald Walsh—have ever been to the lowest point on Earth's surface. The Mariana Trench, near Guam, bottoms out at about 11,000 m below sea level—deep enough that it could swallow up Mt. Everest. To get there, Piccard had to build a special spherical submarine (inset; the submarine is the little sphere hanging below the large buoyancy chamber) designed to resist the intense pressure of the deep ocean.

Amazingly, both the Himalaya and the Mariana Trench formed as a result of the same tectonic process—the interactions between two plates of lithosphere. In one case, the plates compressed each other and thrust up the world's highest mountain range. In the other case, one plate sank beneath the other, dragging the seafloor down with it.

The imperceptibly slow motion of plates, known as *plate tectonics*, is constantly reshaping our world. Over millions of years, it has restructured and relocated continents, built mountain ranges, opened up new oceans where none were before, and closed oceans that were once the size of today's Atlantic Ocean. Plate tectonics unites all of Earth's topographic features into a single system.

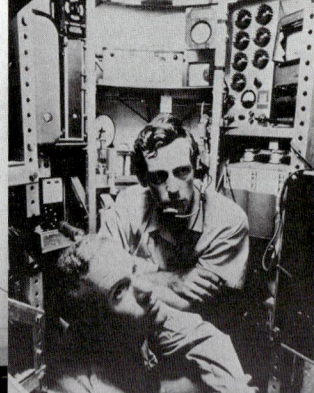

CHAPTER OUTLINE

- **A Revolution in Earth Science** p. 202

- **The Plate Tectonic Model** p. 211

- **The Search for a Mechanism** p. 219

A Revolution in Earth Science

> **LEARNING OBJECTIVES**
>
> **Describe** the supercontinent called Pangaea.
>
> **Identify** the early arguments for and against Wegener's hypothesis of continental drift.
>
> **Explain** how paleomagnetism provided the definitive evidence for continental drift.
>
> **Define** seafloor spreading.

Scientific revolutions challenge us to look at the world in a new way. They turn accepted ideas upside down. However, they don't happen overnight. For example, for millennia, people thought that Earth was the center of the universe, with the Sun and all the planets revolving around it. In 1543, Nicolaus Copernicus offered a hypothesis that argued it was the other way around—that the planets revolve around the Sun. It took decades of debate, new inventions like the telescope, many tests of the hypothesis, and finally the persuasive writing of such scientists as Galileo and Newton to convince astronomers that Copernicus was right. Today the Copernicus hypothesis has become a law; the planets orbit the Sun and the moons orbit the planets. They are a system of bodies called the *solar system*, all held in their orbits by gravitational attraction.

Earth science underwent a revolution in the 1960s, when new evidence was discovered that supported some aspects of a 50-year-old hypothesis called *continental drift*. As with Copernicus's hypothesis, the suggestion that continents move provoked controversy and was rejected by most scientists. However, in recent decades, the hypothesis of continental drift has reemerged and grown into the *theory of plate tectonics*. Plate tectonics has changed our view of the planet. It has encouraged us to think of Earth as a system in constant flux, a dynamic world whose appearance has changed considerably over the eons, and that continues to change before our eyes. Early Earth scientists gathered information about Earth and its processes painstakingly, one piece at a time. Plate tectonics shows us how the pieces fit. Testing of the continental drift hypothesis and the emergence of the plate tectonic theory is a nice example of the scientific method at work.

WEGENER'S HYPOTHESIS OF CONTINENTAL DRIFT

In 1912, a German meteorologist named Alfred Wegener proposed a hypothesis called **continental drift**. Wegener's hypothesis was that the continents had once been joined together in a single "supercontinent," which he called *Pangaea* (pronounced Pan-JEE-ah), meaning "all lands" (see **FIGURE 7.1**). He suggested that Pangaea had split into fragments like pieces of ice floating on a pond and that the fragments, today's continents, had slowly drifted to their present locations.

> ■ **continental drift**
> The slow, lateral movement of continents across Earth's surface.

Wegener's proposal created a storm of protest in the scientific community, one that continued for decades. Some of the criticisms were well-founded. Contemporary scientists could not envision any reasonable mechanism by which continents could be moved. Remember that the continents are anchored in solid rock, and a drifting continent would have to plow through or across a seafloor also made of solid rock. Even Wegener could not explain how this could happen.

Let's put this problem aside for a moment and look at the evidence as the scientists of the time did. No single piece of evidence is conclusive on its own. It took all of these arguments (and more) to convince scientists several decades later that Wegener was right.

THE PUZZLE-PIECE ARGUMENT

It's easy to see from a map that the Atlantic coastlines of Africa and South America seem to match, almost like puzzle pieces. But is this an accident, or does it truly support the hypothesis that the continents were once joined together?

To decide whether the continents really do match, we must first note that there is more to a continent than

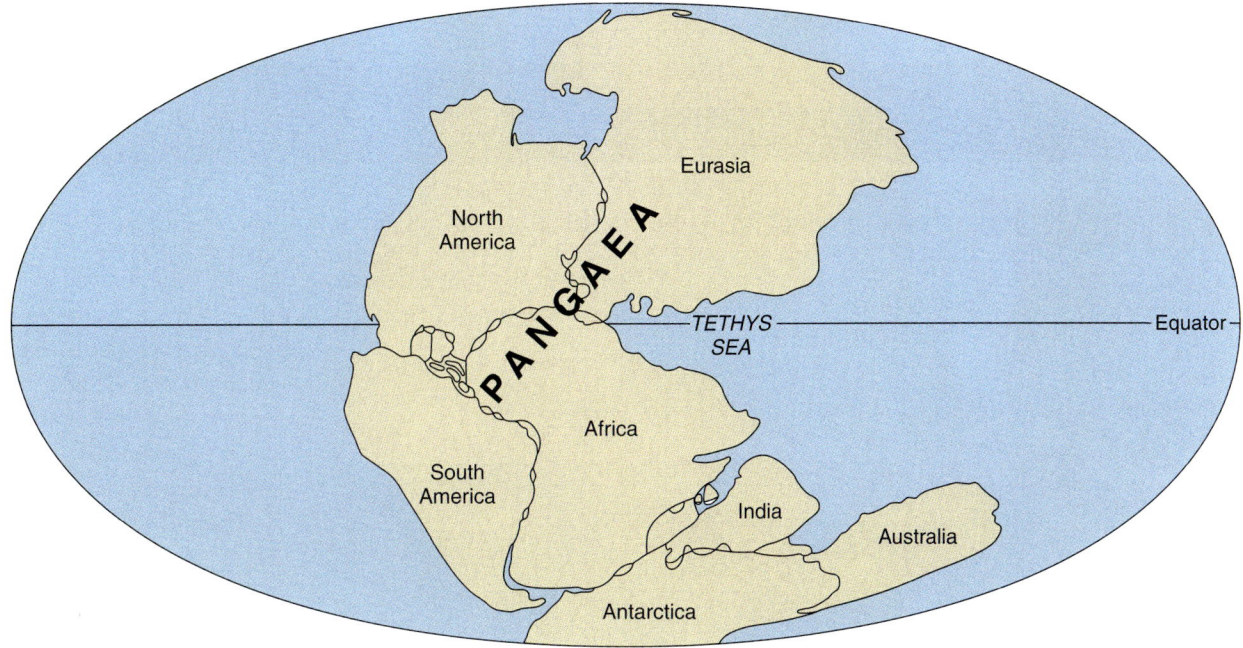

Pangaea FIGURE 7.1

In 1915, Alfred Wegener drew a map much like this one, showing the distribution of the continents about 300 million years ago in the Carboniferous Period. Wegener proposed that the continents at that time were joined in one "supercontinent," which he named Pangaea. The apparent close fit of the coastlines of the modern continents, particularly Africa and South America, had been observed by Leonardo da Vinci and Francis Bacon as early as the 1500s, but the hypothesis of a supercontinent was Wegener's. Modern reconstructions of Pangaea differ from Wegener's version in some respects but his basic idea was correct.

meets the eye. The Atlantic coasts of South America and Africa, like those of many continents, do not terminate abruptly at the shoreline but slope gently seaward. This gently sloping land, part above sea level and part below, is called the **continental shelf**. At a water depth of about 100 m, there is an abrupt change of slope called the *shelf break*, below which lies the steeper **continental slope** (see FIGURE 7.2A on the next page). The edge of a continent, where continental crust meets oceanic crust, is about halfway down the continental slope.

As FIGURE 7.2B shows, when we fit together South America and Africa along the true edges of the continents, we get an even better match than we might expect. The most significant overlaps consist of sedimentary or volcanic rock that was added *after* the continents are thought to have split apart.

MATCHING THE ROCKS AND FOSSILS

The close fit between Africa and South America suggests they were once joined. But a jigsaw puzzle requires both the shapes and the designs to match. If the continents were once joined together, we should find similar features in rocks on both sides of the join. However, matching the rocks on opposite sides of an ocean is more difficult than you might imagine. Erosion and rock formation since the breakup of Pangaea may have destroyed or covered up some of the evidence, so we will be putting together a puzzle with some pieces missing and others defaced.

Matching rocks A starting point is to see whether the ages of similar igneous and metamorphic rocks match up across the ocean. In Wegener's time, the technique of radiometric dating (Chapter 10) was just being developed, so it was not easy to determine the exact age of a rock. But now we know that there is, indeed, some similarity in the ages of rocks and correlation between rock sequences on both sides of the ocean. As shown in FIGURE 7.3A on page 205, the match is particularly good between rocks about 550 million years old in northeast Brazil and West Africa.

We can also check for the continuity of old mountain chains. FIGURE 7.3B shows a reconstruction of the northern part of the supercontinent Pangaea. Notice, again, how mountain chains of similar ages seem to line up when the continents are moved back into this position. The oldest portions of the Appalachian Mountains, extending from the northeastern United States

A Revolution in Earth Science

What is the "true" edge of a continent? FIGURE 7.2A

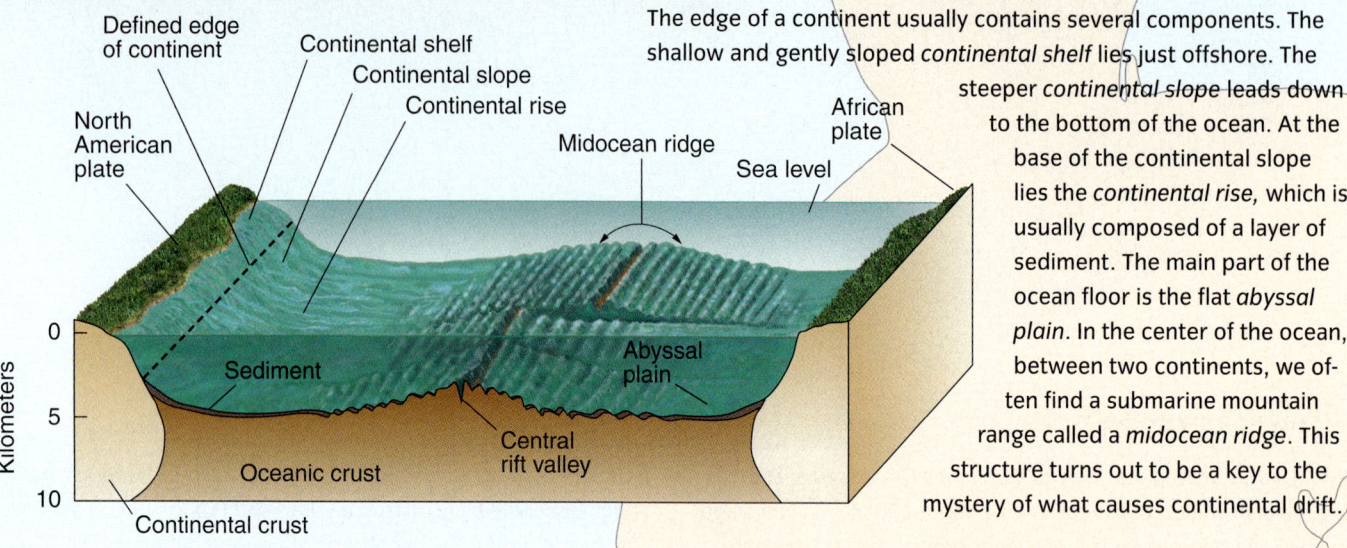

The edge of a continent usually contains several components. The shallow and gently sloped *continental shelf* lies just offshore. The steeper *continental slope* leads down to the bottom of the ocean. At the base of the continental slope lies the *continental rise,* which is usually composed of a layer of sediment. The main part of the ocean floor is the flat *abyssal plain.* In the center of the ocean, between two continents, we often find a submarine mountain range called a *midocean ridge.* This structure turns out to be a key to the mystery of what causes continental drift.

How well do the continents fit? FIGURE 7.2B

This map reconstructs the fit of South America and Africa along the *true* edges of the continents—not along the shoreline but along the submerged edge of the continental crust. The darkly shaded areas show overlap. Note that the inclusion of the shallow off-coastal areas makes the fit better, not worse. Wegener did not have access to this information.

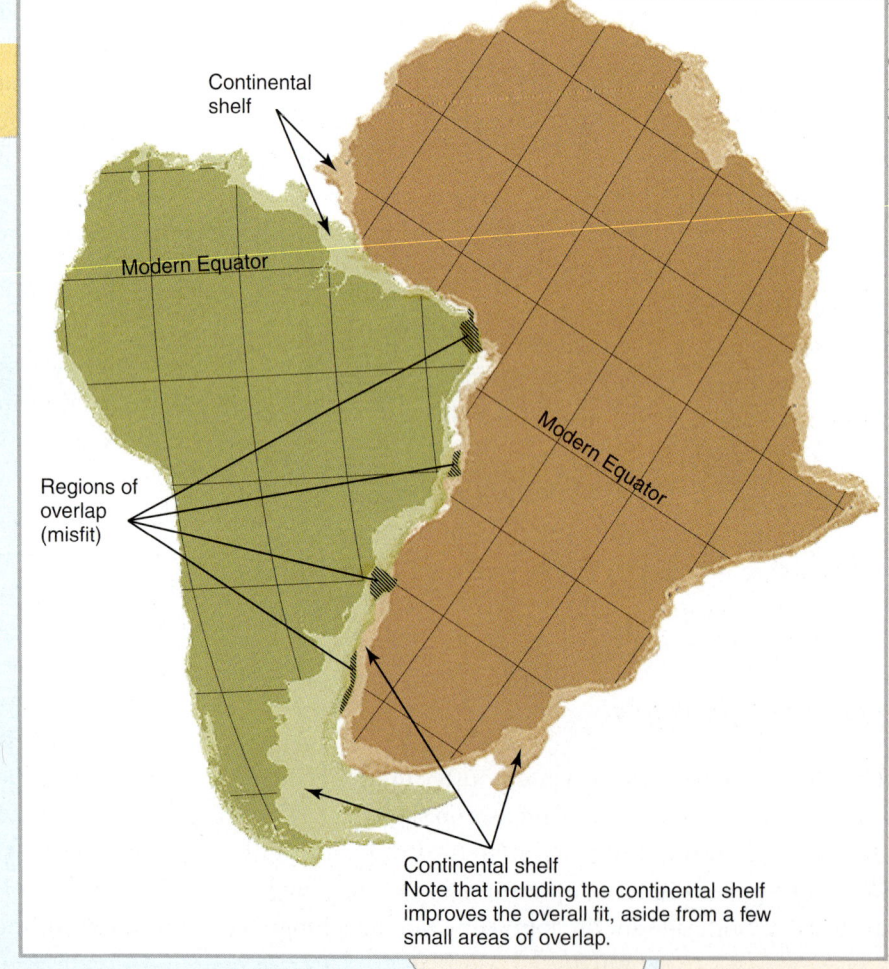

Continental shelf
Note that including the continental shelf improves the overall fit, aside from a few small areas of overlap.

Visualizing
Evidence of continental drift

How well do the ages match? FIGURE 7.3A–B

- Older than 550 million years
- About 550 million years old
- Younger than 550 million years

Close match in rock ages

Africa

Modern Brazil

South America

A When the continents are rotated back together, the ages of rock units generally match, particularly in the regions of northeast Brazil and West Africa.

B In this reconstruction of the northern part of Pangaea, mountain belts of similar ages match up.

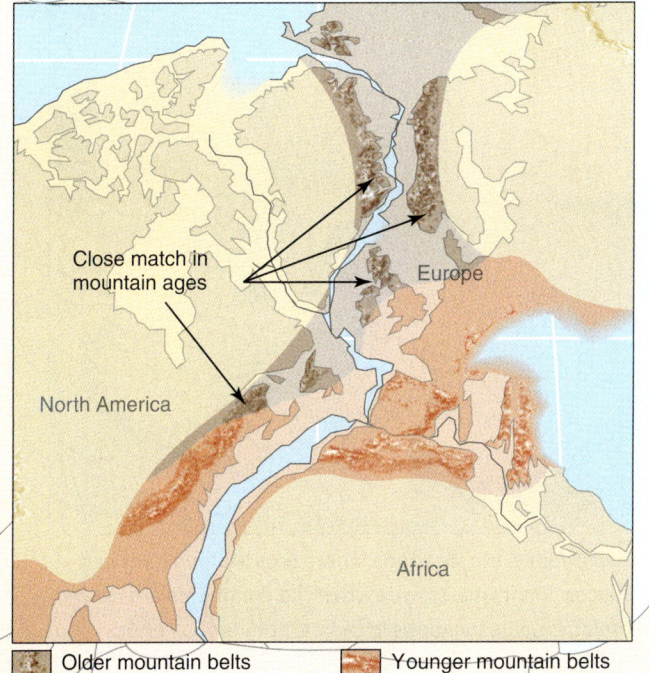

Close match in mountain ages

North America

Europe

Africa

- Older mountain belts
- Younger mountain belts

What evidence did the glaciers leave? FIGURE 7.3C–D

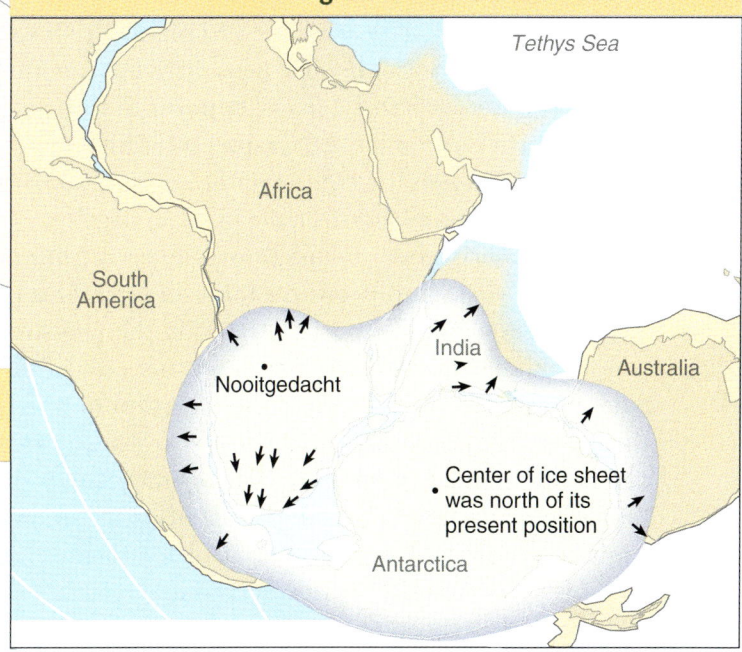

Tethys Sea

Africa

South America

Nooitgedacht

India

Australia

Center of ice sheet was north of its present position

Antarctica

C In this reconstruction of the southern continents of Pangaea, we have shown the rough extent of glacial deposits. Small arrows indicate the direction in which the ice was moving during the glaciation, as deduced from evidence such as grooves and scratches in the bedrock.

D As glaciers advance, the sediment and rocks contained in them grind and abrade the bedrock, producing a polished and grooved surface. This polished surface, at Nooitgedacht in South Africa (see map above), was created by the Pangaean ice sheet. The grooves (and the pen, upper right) indicate the direction of ice movement.

A Revolution in Earth Science

through eastern Canada, match up with the Caledonides of Ireland, Britain, Greenland, and Scandinavia. A younger part of the Appalachians lines up with a belt of similar age in Africa and Europe.

The deposits left by ancient continental ice sheets also match up across continental joins. In South America and Africa there are thick glacial deposits of the same age (Permian-Carboniferous; Chapter 10), which match almost exactly when the continents are moved back together (see **FIGURE 7.3C** on the previous page).

As glacial ice moves, it cuts grooves and scratches in underlying rocks and produces folds and wrinkles in soft sediments (see **FIGURE 7.3D** on the previous page). These features provide evidence not only of the extent of glaciation but also of the direction in which the ice was moving during the glaciation. When Africa and South America are moved back together, the direction of ice movement on both continents is consistent, radiating outward from the center of the former ice sheet. It's hard to imagine how such similar glacial features could have been created if the continents had not been joined together. The glacial deposits also suggest not only that Africa and South America were once joined, but that they were closer to the South Pole at the time, and that they had cooler climates.

Matching fossils If Africa and South America were really joined at one time, with the same climate and matching rock units, we should expect that in ice-free intervals they would have been inhabited by the same plants and animals. To check this hypothesis, Wegener looked at fossils. He found that some communities of plants and animals apparently evolved together until the time that Pangaea split apart, and that after that time they evolved separately.

For example, Wegener found fossils of an ancient tree, *Glossopteris,* in matching areas of southern Africa, South America, Australia, India, and Antarctica (see **FIGURE 7.4A** and **B**). Could the seeds of this plant have been carried by wind or water from one continent to another? Probably not; the seeds of *Glossopteris* were large and heavy, and probably would not travel far on the wind or water currents. Not only that, *Glossopteris* flourished in a cold climate; it would not have thrived in the warm present-day regions where its fossil remains are found. This, too, is consistent with the idea that

How well do the fossil records match? FIGURE 7.4

▲ **A** The tongue-shaped fossil leaves shown here came from a tree of Carboniferous age called *Glossopteris* (named after the Greek word for "tongue"). Similar fossils have turned up in Africa, India, Australia, Antarctica, and South America, providing strong evidence that these areas were once contiguous.

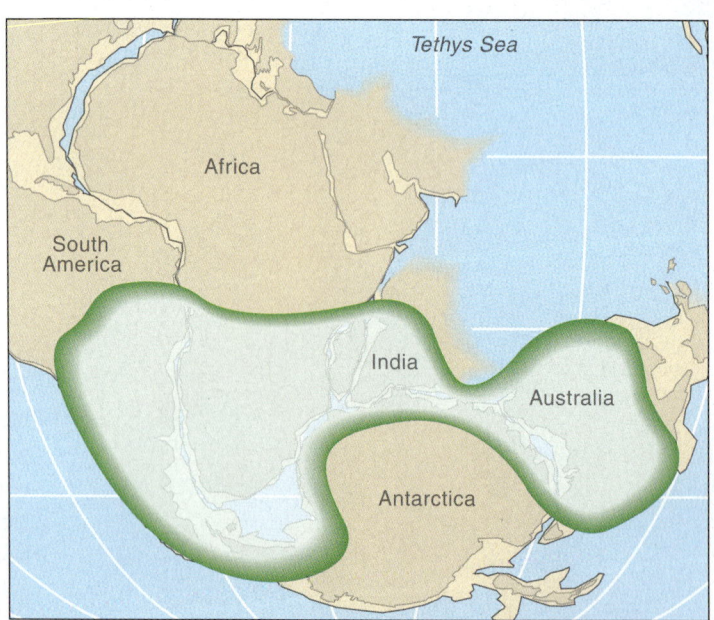

▲ **B** This map outlines the locations where *Glossopteris* has been found. The continents match well when the continents are moved back to their probable locations in the Carboniferous Period.

these continents were once joined together in a more southerly location.

Certain animal fossils, too, match up well. The fossil remains of *Mesosaurus,* a small reptile from the Permian Period, are found both in southern Brazil and in South Africa. The types of rocks in which the fossils are found are very similar. *Mesosaurus* did swim but was too small (about half a meter long) to swim all the way across the ocean. Fossil remains of certain types of earthworms also occur in areas that are now widely separated. Since an earthworm could not have hopped across a wide ocean, the landmasses in which they lived must once have been connected.

Rejection of the hypothesis
Despite all the evidence that seemed to support the former existence of Pangaea, scientists of Wegener's day could not advance a reasonable hypothesis to explain *how* it had broken apart and how the pieces had moved. A few Earth scientists, particularly in the Southern Hemisphere where a lot of the best evidence was located, supported Wegener's hypothesis, but most rejected the concept and sought other reasons to explain the evidence Wegener used to support his hypothesis. Wegener died in 1930, and that seemed to be the end of continental drift, until the 1950s when new and unexpected evidence revived the concept.

MAGNETIC POLES THAT SEEM TO WANDER

In the 1950s, **paleomagnetism** emerged as a tool for studying Earth's history, and it was destined to break the intellectual logjam. When magma cools and solidifies into rock, grains of iron-bearing minerals such as magnetite become magnetized and record a locked-in-rock record of the magnetic field at the instant of magnetization. That is, the rock magnetism records both the prevailing *polarity* (north–south orientation) of Earth's magnetic field at that time and the direction of Earth's magnetic poles.

> **paleomagnetism**
> The study of rock magnetism in order to determine the intensity and direction of Earth's magnetic field in the past.

For reasons that are not yet well understood, Earth's magnetic poles have reversed direction at various times

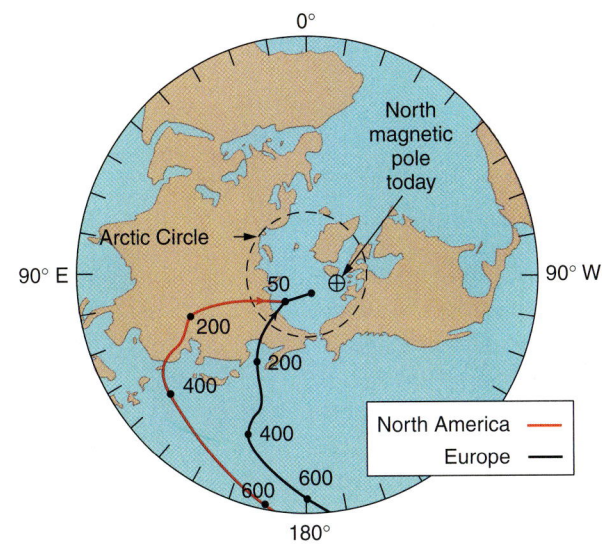

Wandering poles? FIGURE 7.5

This map traces the *apparent* path of the north magnetic pole over the last 600 million years. The numbers indicate millions of years before the present. Rocks in North America point to the apparent magnetic poles shown on the red curve, and rocks in Europe point to the apparent poles shown on the blue curve. The *actual* magnetic pole probably never wandered far from the geographic North Pole (center). Its apparent motion is created by the motion of the continents, which carried magnetically oriented rocks along with them.

in Earth's history. That is, the magnetic compass that today points to the north magnetic pole (called *normal polarity*) might start pointing south (called *reverse polarity*). Nothing happens to the compass; it is Earth's magnetic field that reverses. The last time a reversal of polarity happened was about 700,000 years ago. Reversal happens quickly—over a period of a few thousands of years—in relation to Earth's long history, which goes back several billion years. But despite reversals of polarity, the actual positions of the magnetic pole remain very close to Earth's geographic North and South Poles (the poles marking the axis of rotation). Paleomagnetism allows us to reconstruct both the locations of Earth's past magnetic poles and also the reversals. Exactly when the reversals happened has been determined by radiometric dating of volcanic rocks (Chapter 10).

Apparent polar wandering
In the 1950s, geophysicists studying paleomagnetic pole positions found evidence suggesting that Earth's magnetic poles had wandered all over the globe over the last several hundred million years. They plotted the pathways of the poles on maps such as FIGURE 7.5 and called the

A Revolution in Earth Science 207

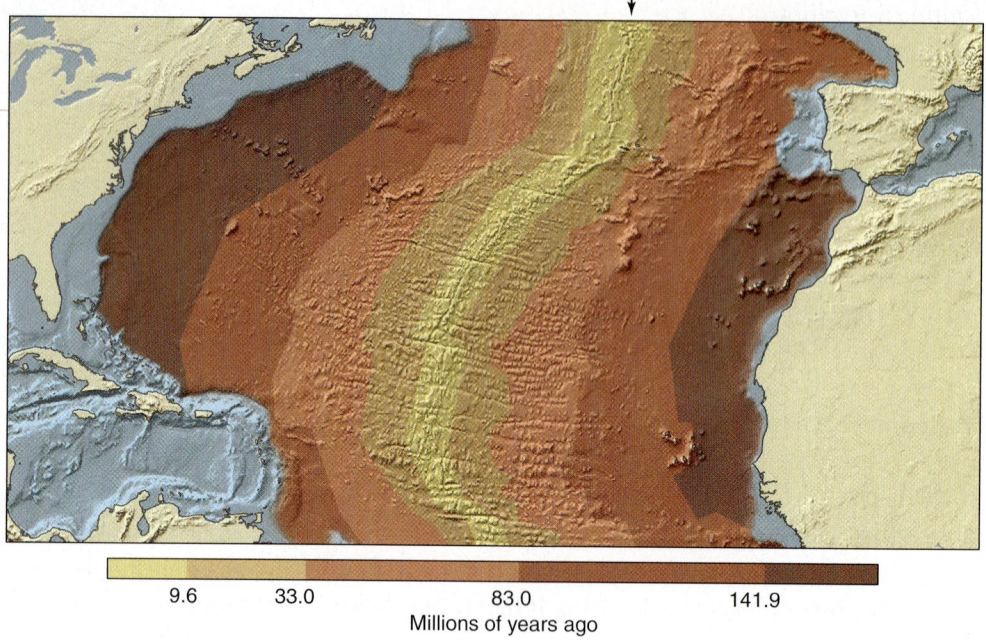

Banded rocks on the seafloor FIGURE 7.6

This map shows the ages of magnetically banded rocks on either side of the Mid-Atlantic Ridge. The numbers indicate the ages of the rocks in millions of years. The youngest bands lie along the midocean ridge, and the oldest are far away from the ridge, indicating that the seafloor has been spreading over time.

phenomenon *apparent polar wandering*. Geophysicists found this puzzling because they knew that the magnetic poles could not actually wander very far from Earth's axis of rotation. Even more puzzling, geophysicists found that the path of apparent polar wandering measured from North American rocks differed from that of European rocks. Earth certainly cannot have two different north magnetic poles at the same time! The only solution is that the magnetic pole is fixed and the continents themselves moved, carrying the rocks with them. In that case, the apparent motion of the poles would be an illusion, like the apparent motion of trees when you drive by them. In fact, the apparent polar wandering path of a continent, determined from rocks of various ages, provides a historical record of the position of that continent relative to the magnetic poles.

Notice also that the apparent wandering paths for Europe and North America actually look quite similar from 600 million years ago to 200 million years ago. In fact, if you rotated Europe and America so that they were side by side, the two paths would overlap each other exactly. This indicates, once again, that Europe and North America were moving as a single continent during that time. So paleomagnetism suggested that Wegener had been right—continents do move. But how they move remained a mystery.

The missing clue: Seafloor spreading By the early 1960s, many clues had been amassed in support of continental drift. The hypothesis made testable predictions, such as matching geology and fossils, and those predictions had been confirmed. The hypothesis was also consistent with evidence that was not known during Wegener's lifetime—the polar wandering paths. Even so, many scientists were still skeptical. So far, all the evidence was circumstantial. The ripping apart and relocation of an entire continent should have left signs that would still be visible today.

Then, three decades after Wegener's death, scientists found the evidence they needed—at the bottom of the sea. The missing clue turned up when oceanographers discovered that the seafloor consists of magnetized volcanic rocks with alternating bands of normal and reversed polarities (see Chapter 10). The bands are hundreds of kilometers long. More important, they are symmetrical on either side of a centerline that coincides with the ridge running down the middle of the Atlantic Ocean. In other words, if you could fold the seafloor in half along the midocean ridge, the bands on either side would match. As seen in FIGURE 7.6, later evidence would show that not only the paleomagnetic polarities but also the ages of the rocks were symmetrically banded, and mirrored on either side of the ridge,

seafloor spreading The processes by which the seafloor splits and moves apart along a midocean ridge and new oceanic crust forms along the ridge.

with the youngest rocks closest to the ridge itself.

The symmetrical pattern of magnetic bands provided powerful support for another hypothesis, first proposed in 1960 and called **seafloor spreading**. According to this hypothesis, the midocean ridge is a place where the seafloor has split apart and the rocks are moving away from one another (see **FIGURE 7.7**). When magma from below wells up into the crack, it solidifies into new volcanic rock on the seafloor, and as it cools and crystallizes, it records the polarity of the magnetic field. Over time, the expanding seafloor operates as a conveyer belt, carrying the newly magnetized bands of rock away from the ridge in either direction.

Seafloor spreading FIGURE 7.7

Lava extruding along a midocean ridge forms new oceanic crust. As the lava cools, it becomes magnetized with the polarity of Earth's magnetic field at the time. As the plates on either side of the midocean ridge move apart from one another, successive bands of oceanic crust have alternating normal and reversed polarities. The resulting magnetic bands are symmetrical on either side of the midocean ridge.

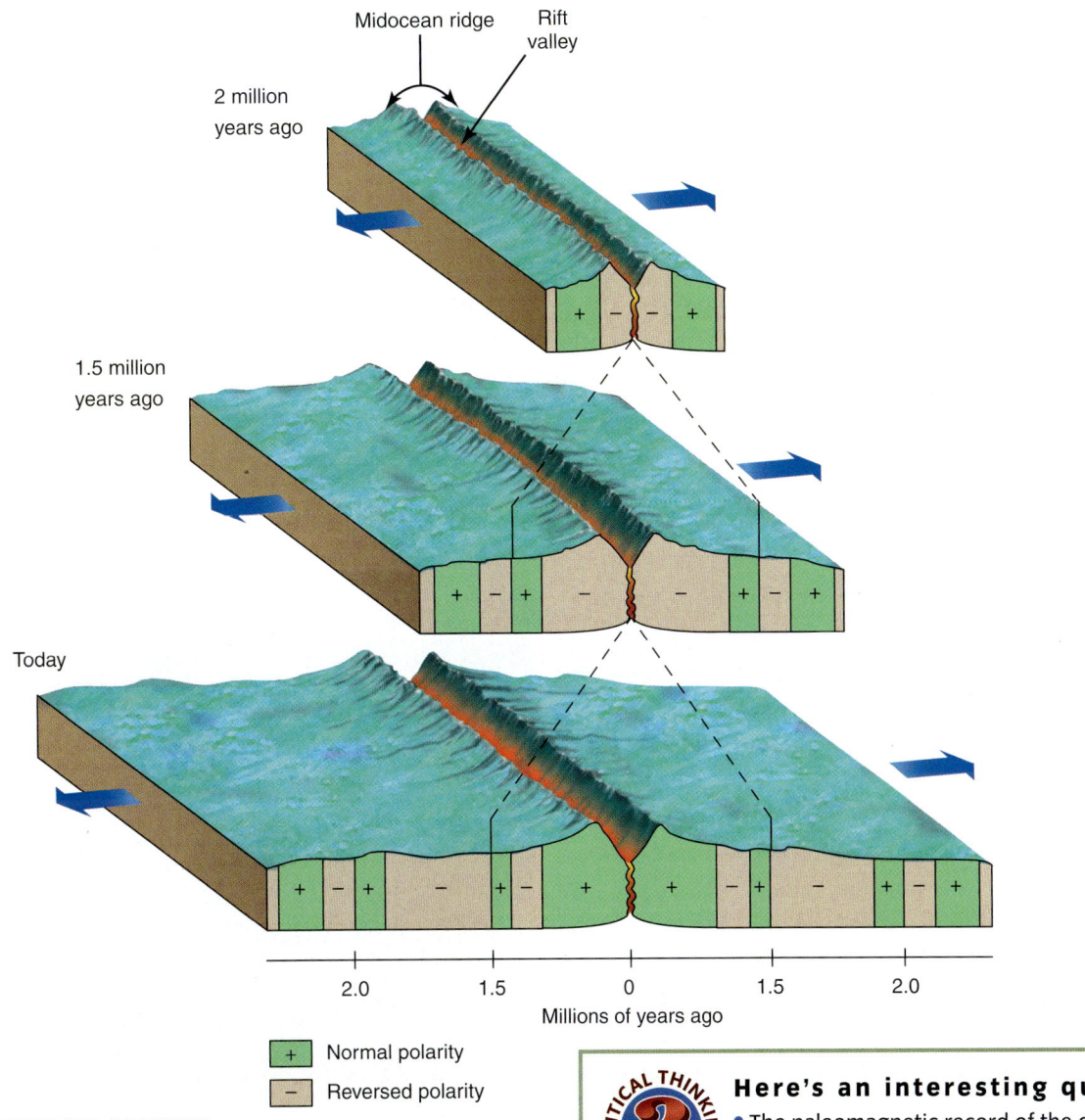

Process Diagram

VIEW THIS IN ACTION in your WileyPLUS course

Here's an interesting question:
- The paleomagnetic record of the ocean floor shows that the most ancient oceanic crust in today's oceans is only about 200 million years old. Why might this be so?

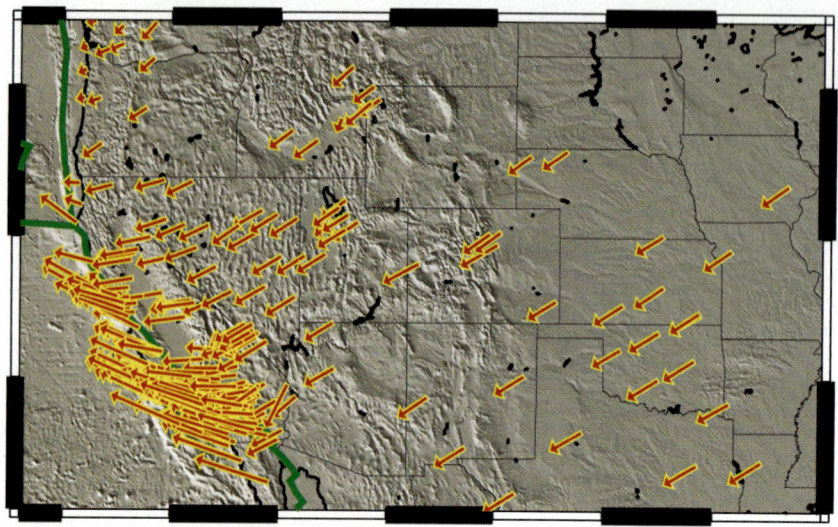

Measuring plate motion using GPS FIGURE 7.8

The arrows show surface motions from a continuous series of GPS measurements. Note that points on the same plate are generally moving in the same direction, confirming that the plates are rigid. In this figure, GPS stations on the North American plate are moving slowly southwest, while those on the Pacific plate are moving more rapidly northwest. The speeds are too slow for humans to be aware of (1 to 10 cm per year) but are easily detectable by GPS sensors.

The decisive piece of evidence for seafloor spreading is that the ages of seafloor rocks increase with distance from the ridge. The youngest rocks are found along the center of the ridge, where new molten material wells up (as shown in Figure 7.7). By combining the hypothesis of seafloor spreading with the hypothesis of continental drift, scientists made a conceptual leap forward. The continents are not trundling along on top of a static ocean floor, which would be physically impossible. Instead, they are embedded in oceanic crust and are conveyed in opposite directions by a dynamic ocean floor that is constantly spreading and replenishing itself. Ironically, it was the geophysicists—the group that had most vigorously opposed Wegener's ideas—who found the final piece of evidence that proved Wegener correct. The many investigations seeking to test the continental drift hypothesis provide a good example of the scientific method at work.

Recently, an even more direct piece of evidence for continental plate motion became available. In the 1980s, the U.S. Department of Defense declassified the data from a network of satellites that form the Global Positioning System (GPS). GPS receivers help motorists, hikers, boaters, and pilots find their exact location quickly and easily, and they now can be used to track continental drift. Earth scientists have attached GPS receivers permanently to the ground, so that they do not move unless the ground moves. This allows them to plot maps of Earth's surface velocities (see FIGURE 7.8). The GPS data confirm that the continents have been moving—at least for the last 20 years! Of course, the other evidence presented in this chapter shows that they have been moving for at least 600 million years.

CONCEPT CHECK STOP

What were the arguments that Wegener gave in support of his hypothesis of continental drift?

What further evidence for continental drift was discovered after Wegener's death?

How does seafloor spreading create paleomagnetic bands?

How quickly do continents move, and how do we know this?

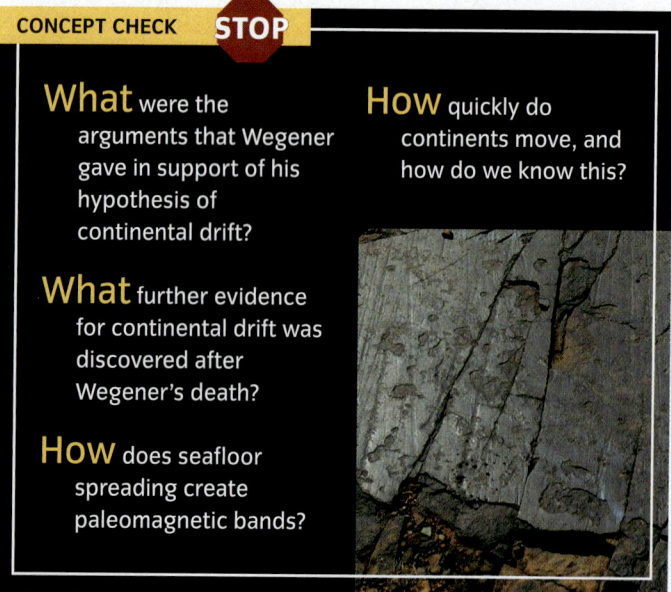

The Plate Tectonic Model

LEARNING OBJECTIVES

Define the theory of plate tectonics.

Describe three different types of plate margins.

Explain why different kinds of plate margins are often associated with earthquakes.

Relate Earth's large-scale topographic features to plate margins.

Once the realities of continental drift and seafloor spreading were established, many previously puzzling phenomena in Earth science suddenly began to make sense. For instance, scientists could now explain the location of mountain ranges and deep ocean trenches. They could explain why, in some places (such as the Tibetan Plateau), Earth's crust seems to be squeezed together, while in other places (such as the East African Rift Valleys) it seems to be pulling apart. They could also understand the distribution of earthquakes and volcanic activity around the planet, which is far from uniform (see **FIGURE 7.9** on the following page). Other chapters in this book discuss each of these developments. The unifying theory that emerged from all the research is called **plate tectonics**. Note that we call it a *theory* rather than a *hypothesis*. As we discussed in Chapter 1, a hypothesis is a tentative explanation. Before it can become a theory, it must be supported by extensive experimentation and observation. A theory is an explanatory model that is supported—in this case, quite strongly—by a lot of scientific evidence.

> **plate tectonics** The movement and interactions of large fragments of Earth's lithosphere, called *plates*.

PLATE TECTONICS IN A NUTSHELL

Earth's *lithosphere*, or rocky outer layer, is very thin relative to Earth as a whole. The solid rock that makes up the lithosphere is strong, but it lies on top of a vast mantle of hotter, weaker material that is constantly in motion (albeit very slow motion). The layer directly below the lithosphere, called the *asthenosphere*, is especially weak because it is close to the temperature at which melting begins. (The lithosphere and asthenosphere were introduced in Chapter 1 and will be discussed in more detail in Chapter 8.) The relationship between these two layers is a condition called *isostasy*, which means that the lithosphere is essentially "floating" on the asthenosphere, like a sheet of ice floating on water.

If you place a very thin, cool, hard shell on top of hot, ductile material that is moving around, what would you expect to happen? It's almost a certainty that the shell will crack, and that is exactly what has happened to Earth's lithosphere. It has broken into a set of enormous rocky fragments we call **plates**. Today there are seven large plates—the North American, South American, African, Eurasian, Australian, Antarctic, and Pacific plates—each extending for thousands of kilometers, and many smaller ones (see Figure 7.9).

Driven in part by the underlying motion of the asthenosphere, these plates collide, split apart, and slide past one another. This generates a lot of Earth-changing activity, such as earthquakes and mountain building. Much of this occurs along the boundaries at which plates interact with one another, typically marked by huge fractures, or **faults**.

> **fault** A fracture in the lithosphere along which movement has occurred.

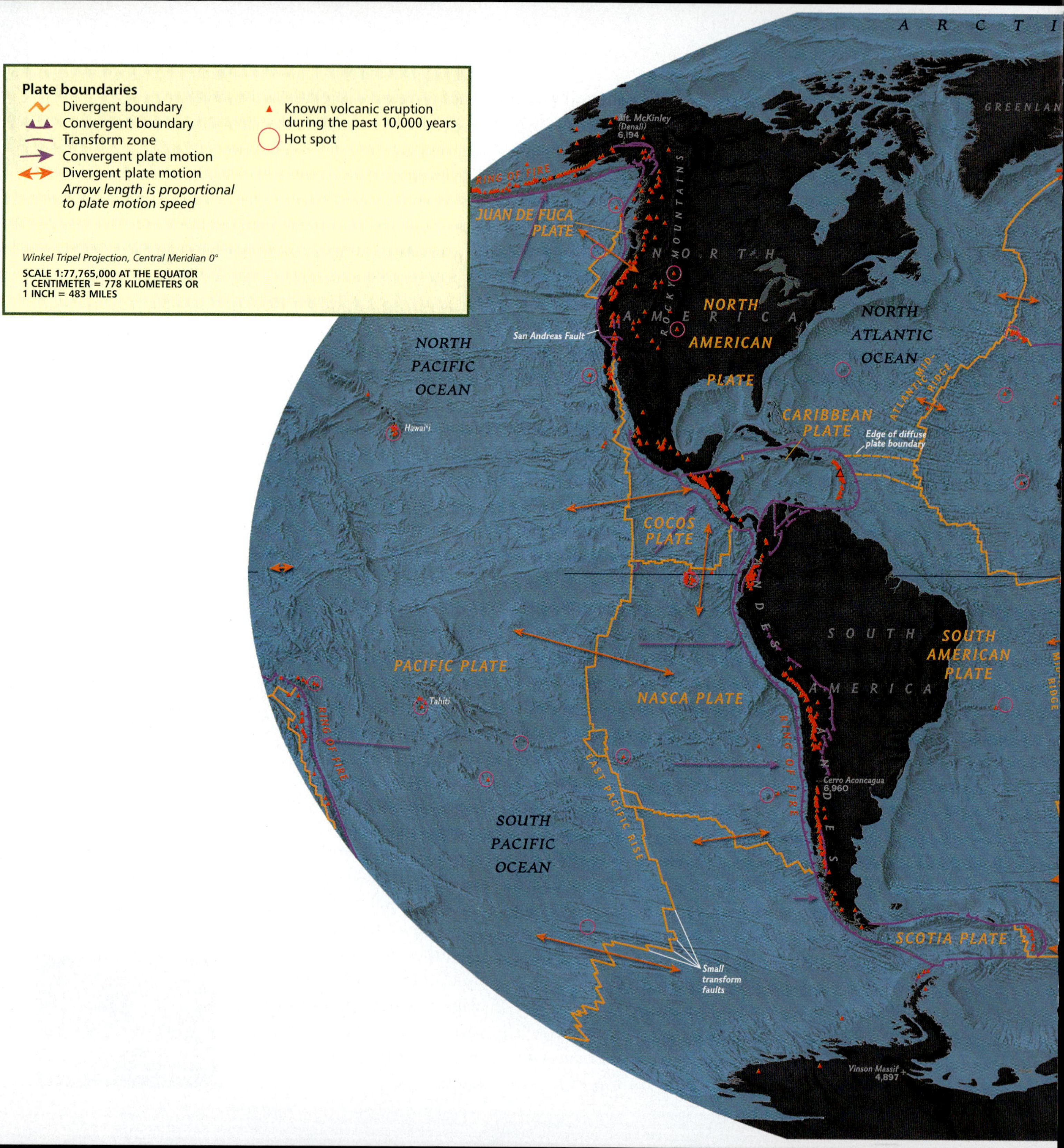

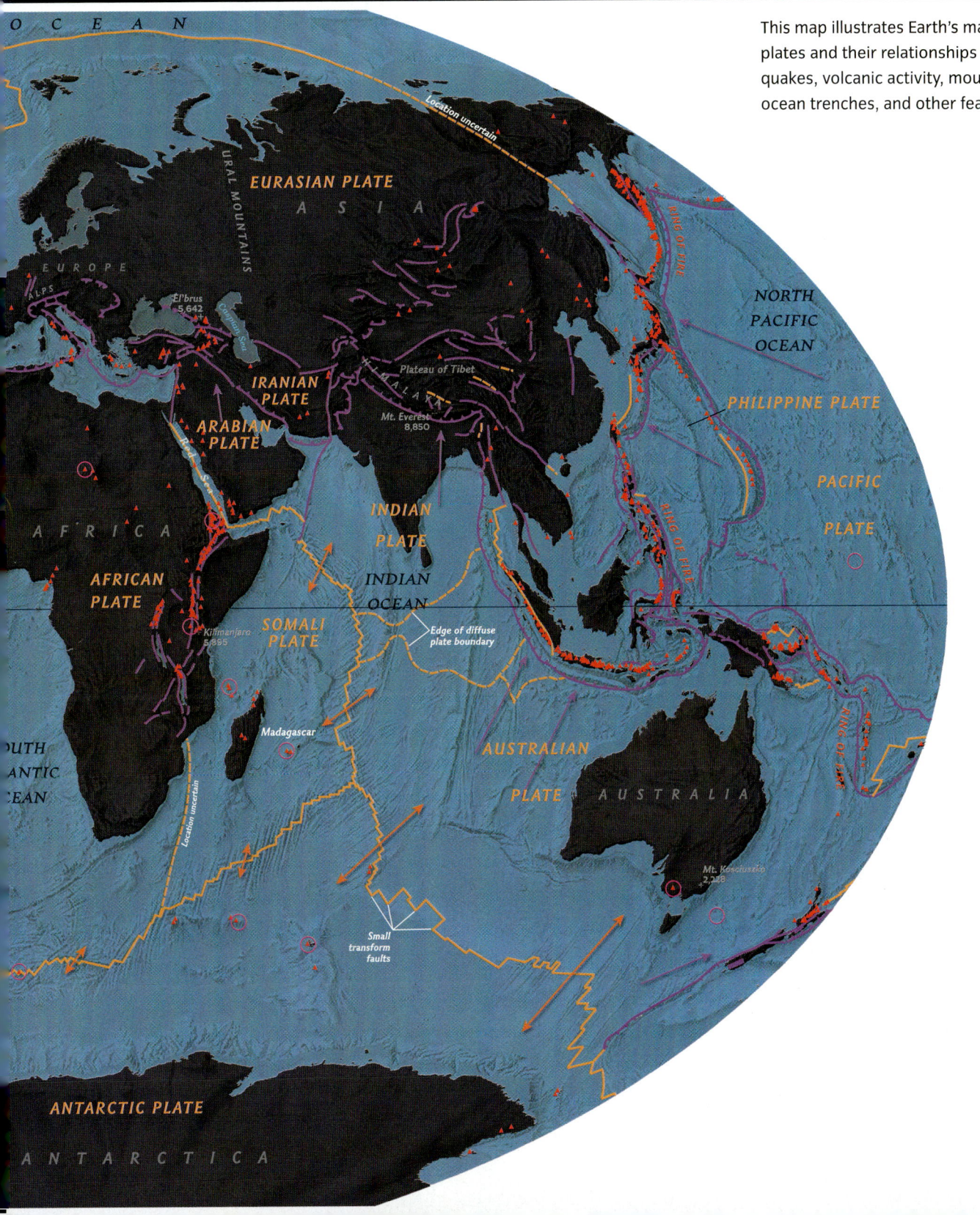

Visualizing

Our tectonic planet FIGURE 7.9

This map illustrates Earth's major and minor plates and their relationships to earthquakes, volcanic activity, mountain ranges, ocean trenches, and other features.

TYPES OF PLATE MARGINS

Interactions between plates of lithosphere occur mainly along their edges. There are three fundamentally different ways in which they can interact. They can move away from each other (*diverge*); they can move toward each other (*converge*); or they can slide past each other along a long fracture. We will take a look at examples of each of these types of margins.

Divergent margins (see **FIGURE 7.10A**), also called *rifting* or *spreading centers*, occur where two plates are moving apart. They can occur either in continental or oceanic crust. In East Africa, for example, continental lithosphere is being stretched and torn apart, creating long *rift valleys* that separate the African plate from the Somali plate. Lake Victoria and the other great east African lakes are in the African Rift Valley. Eventually a rift in the continental crust may widen sufficiently for the ocean to enter; a modern example of the sea entering a young rift is the Red Sea, which was created by Africa splitting apart from Arabia (see *What an Earth Scientist Sees* on page 216). When rifting progresses far enough for the formation of a new ocean basin, as in the case of the Atlantic Ocean, the rifting center is marked by a *midocean ridge*.

■ **divergent margin** A boundary along which two plates move apart from one another.

Convergent margins occur where two plates move toward each other. This leads to different types of margins, depending on whether the boundary is between two oceanic plates, two continental plates, or one of each.

■ **convergent margin** A boundary along which two plates move toward one another.

A convergent margin begins when a plate of oceanic lithosphere fractures and one of the fractured edges starts to sink. Fracturing creates two plates. When both plates are oceanic, the sinking plate will slide beneath the other plate, plunging into the asthenosphere, where water released from the wet rocks of the sinking plate promotes the formation of magma (this process is discussed in more detail in Chapter 9). **Subduction zones** are marked by very deep *oceanic trenches*—the deepest points in the ocean—and, on the surface, by lines of volcanoes formed as a result of melting in the mantle, generated by water released from the subducting plate (see **FIGURE 7.10C**). The Aleutian Islands, a chain of volcanoes west of Alaska, are an example of convergence between two plates of oceanic lithosphere.

■ **subduction zone** A boundary along which one plate of lithosphere descends into the mantle beneath another plate.

When oceanic lithosphere sinks beneath continental lithosphere, the chain of volcanoes formed as a result of subduction forms on the overriding continental plate. The great chain of volcanoes along the western edge of South America is an example.

Continental lithosphere is too light to sink into the asthenosphere. As a result, when one continent meets another continent along a convergent margin, they crumple upwards and downwards as the lithosphere thickens, in a *collision zone*. The Himalaya (see **FIGURE 7.10B**), Earth's highest mountain range, was thrust up in just this way by an ongoing, violent collision over the last 50 million years between the Indian plate and the Eurasian plate.

Transform faults occur where two plates slide past each other, grinding and abrading their edges as they do so. The San Andreas Fault in California, shown in **FIGURE 7.10D**, is a well-known example: It separates the northwesterly moving Pacific plate from the adjacent North American plate, which is moving toward the southwest.

■ **transform fault** An approximately vertical fracture in the lithosphere along which two plates slide past each other.

Transform faults are much more common in the ocean floor, where hundreds of them run perpendicular to the midocean ridges (see Figure 7.9). This causes the plates to have complicated, jagged boundaries in which spreading centers alternate with transform faults.

214 CHAPTER 7 Plate Tectonics: Sculptor of Earth's Ever-Changing Landscape

Types of plate margins FIGURE 7.10

A Divergent margin
The only place in the world where a divergent plate margin in oceanic lithosphere rises to the surface is in Iceland. The rift valley seen here lies directly atop the Mid-Atlantic Ridge, a divergent plate margin where the North American and Eurasian plates are gradually moving apart. The valley is getting wider at a rate of about 2 cm per year, and part of it has filled with water, creating Lake Thingvallaran (at right). The rift itself extends in width off the right side of the photo.

Rift valley

B Convergent margin

Collision zone

▲ **B** In this photograph, the snow-covered Himalaya is on the right, the lower-lying Indian peninsula is off to the left, and the convergent margin lies roughly on the border between them. This is an example of a convergent margin between two plates of continental lithosphere.

▼ **C** The string of volcanoes marching down the spine of the island of Java—and continuing in the islands to the east—is typical of an ocean–ocean convergent margin. Some of the more famous volcanoes are named.

D The constant grinding of the Pacific plate and the North American plate has made the area near the San Andreas Fault, a transform fault margin, highly unstable. There have been several notable earthquakes related to this fault, including the 1906 earthquake that destroyed much of San Francisco. (For more on this earthquake and others, see Chapter 8.)
▼

D Transform fault margin

Transform fault margin

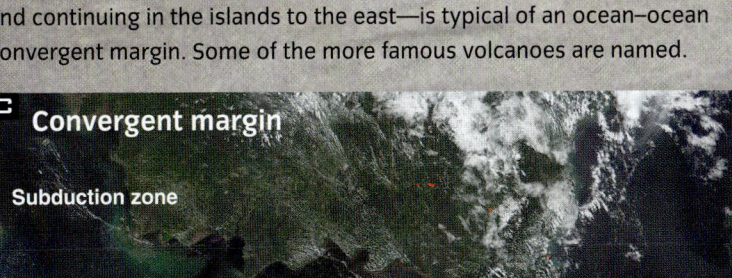

C Convergent margin

Subduction zone

Subduction zone

Krakatau
Galunggung
Merapi
Semeru
Agung
Tambora

The Plate Tectonic Model 215

What an Earth Scientist Sees

The Red Sea and the Gulf of Aden

This spectacular photo, taken by astronauts on the Gemini 11 mission, shows the southern end of the Red Sea (left) and the Gulf of Aden (right) separating the southern tip of the Arabian Peninsula from the northeastern corner of Africa. The Earth scientist sees Arabia splitting off from Africa. A spreading edge runs down the center of the Gulf of Aden and joins another spreading edge that runs up the center of the Red Sea (both are underwater). The pieces of land on either side of the ocean fit together like jigsaw pieces.

There is a third spreading edge that is splitting off a portion of Africa (see inset). The spreading edge is marked by frequent earthquakes and numerous volcanoes in the African Rift Valley.

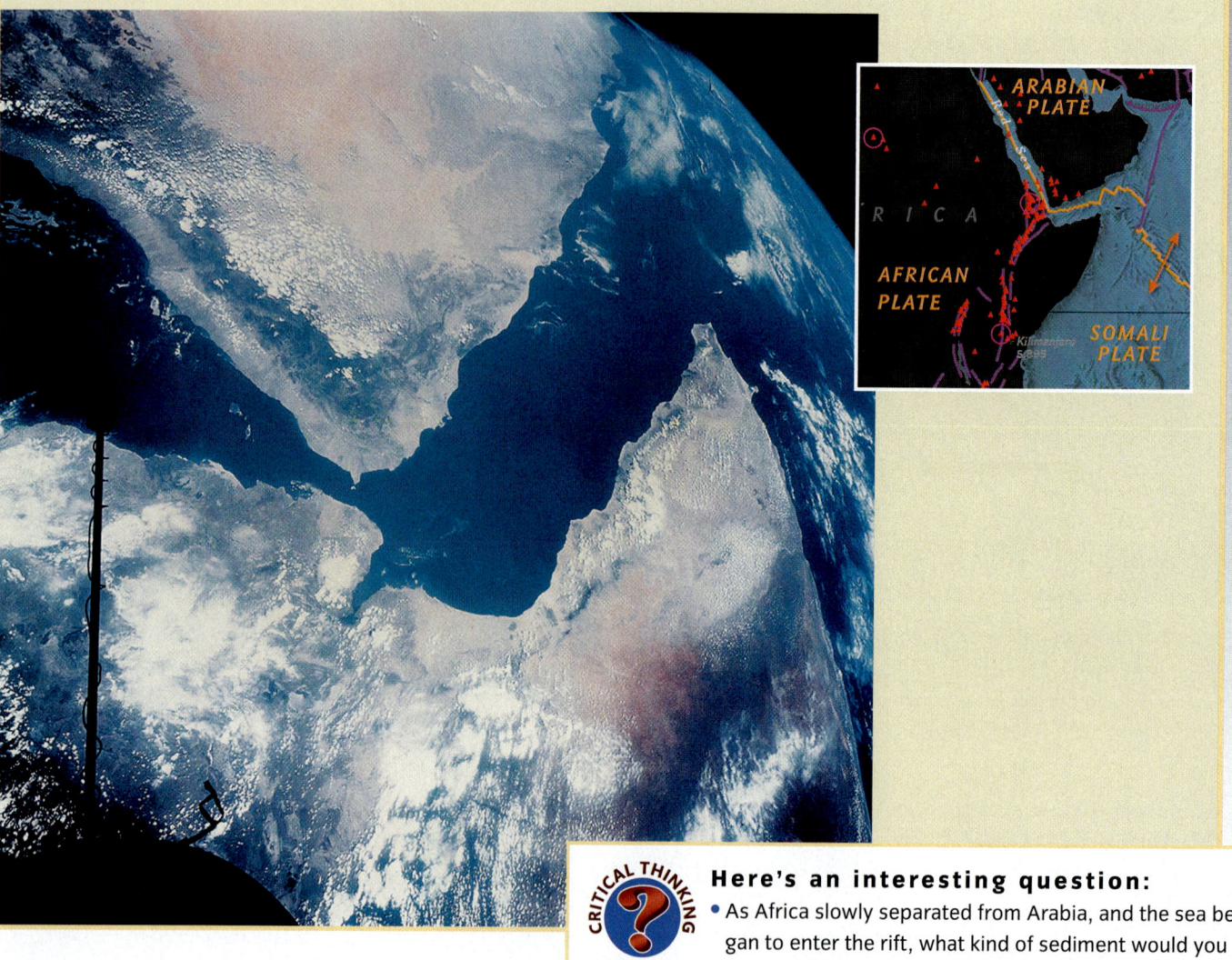

CRITICAL THINKING — Here's an interesting question:
- As Africa slowly separated from Arabia, and the sea began to enter the rift, what kind of sediment would you expect to have been deposited?

There is no solid evidence to suggest that Earth has either grown or shrunk significantly since its formation. This means that the total area of Earth's crust stays constant. For every square kilometer of crust that is created at a midocean ridge, therefore, somewhere else a square kilometer is consumed by subduction at a convergent margin. Overall, there is a balance between creation and destruction in the tectonic cycle. Convergent and divergent plate margins are summarized in **FIGURE 7.11**.

EARTHQUAKES AND PLATE MARGINS

Earthquakes and volcanic eruptions, which occur primarily along plate margins, are the most obvious manifestation of active plate interactions (see Figure 7.9).

Earthquakes occur along faults, where huge blocks of rock are grinding past each other (this topic is discussed in greater detail in Chapter 8). Tectonic motions produce directional pressure, which causes rocks on either side of a large fracture to move past each other.

Plate margins: A summary figure FIGURE 7.11

Several different types of plate margins are illustrated schematically in this diagram. In reality, these different types of margins would not be so close to one another. At middle left and middle right are divergent margins, which may occur in continental or oceanic crust. At far left, center, and far right are three types of convergent margins. Continent–continent convergent margins thicken the lithosphere and cause high mountain chains to form. At ocean–continent and ocean–ocean margins, one plate is subducted underneath the other. The subducting oceanic plate, which typically carries ocean floor sediments, has a relatively low melting temperature because of its high water content (Chapter 9). Therefore, it will begin to melt at a relatively shallow depth. The melt rises to the surface, creating a line of volcanoes called a *volcanic arc* along the edge of the overriding plate.

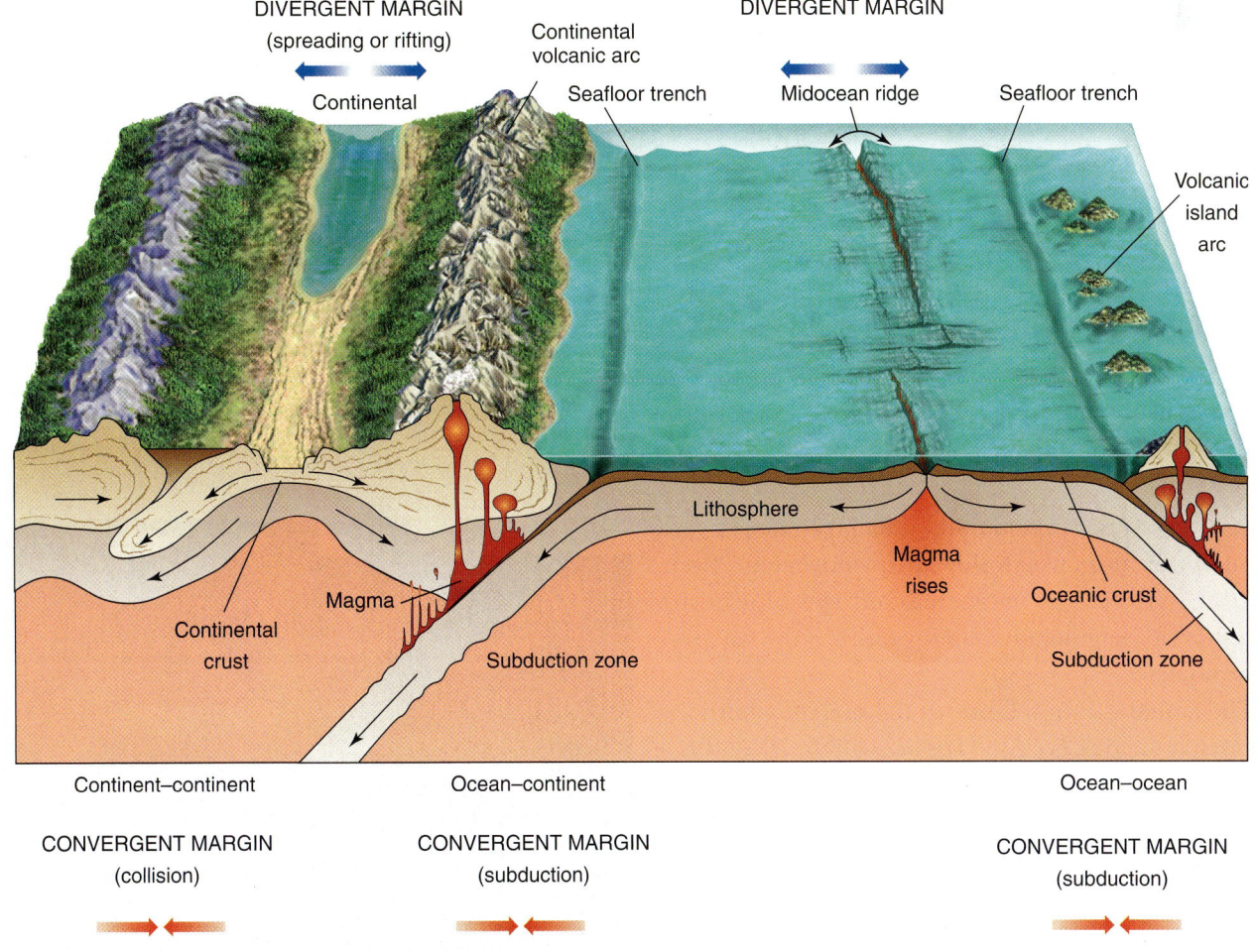

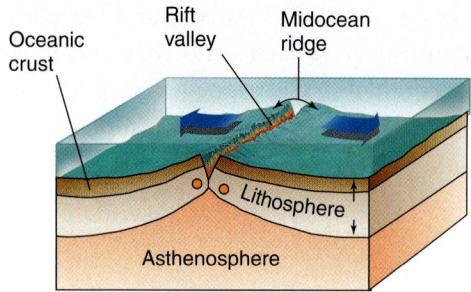

DIVERGENT BOUNDARY

At divergent margins, earthquakes tend to be fairly weak and shallow. Earthquakes can only occur in rock that is cold and brittle enough to break; at a midocean ridge, this means they cannot be very deep.

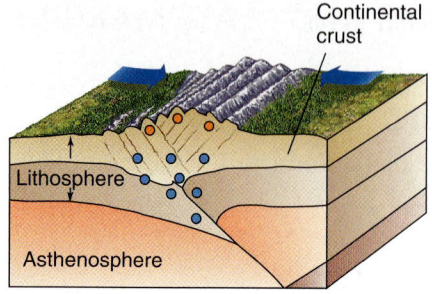

CONTINENTAL COLLISION BOUNDARY

In collision zones the earthquakes can be deep and also very powerful.

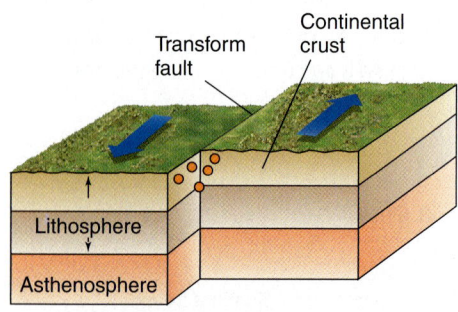

TRANSFORM FAULT BOUNDARY

Transform fault margins have shallow earthquakes, but they can be very powerful.

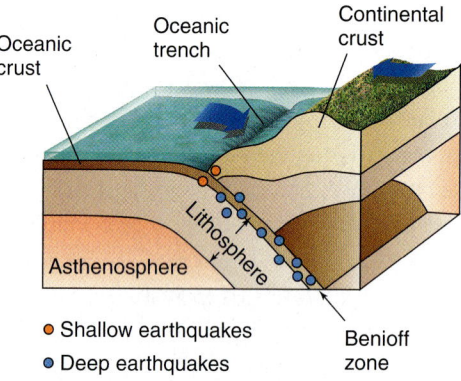

- Shallow earthquakes
- Deep earthquakes

SUBDUCTION ZONE BOUNDARY

The deepest and most powerful quakes occur in subduction zones. Here, an oceanic plate moves downward relative to a continental plate. The earthquake foci are shallow near the oceanic trench but become deeper along the descending edge of the subducting plate. These zones of shallow- and deep-focus earthquakes, called *Benioff zones*, first alerted scientists to the phenomenon of *subduction*.

Earthquake depths FIGURE 7.12

Earthquakes occur at depths from a few kilometers to several hundred kilometers below the ground surface. The depth of the quake tells Earth scientists a lot about the characteristics of the plate margin.

The movement is rarely smooth; usually the blocks stick because of friction, which slows their movement. Eventually, the friction is overcome and the blocks slip abruptly, releasing pent-up energy with a huge "snap"— an **earthquake**.

The actual location beneath the surface where the earthquake begins is called the *focus*. This should not be confused with the better-known *epicenter*, which is the point on Earth's surface that lies directly over the focus. The depths of foci provide useful information about the characteristics of the plate margin (see FIGURE 7.12).

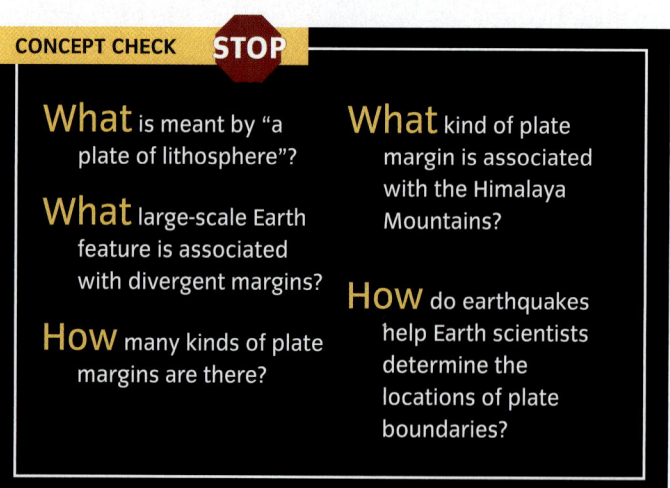

CONCEPT CHECK STOP

What is meant by "a plate of lithosphere"?

What large-scale Earth feature is associated with divergent margins?

How many kinds of plate margins are there?

What kind of plate margin is associated with the Himalaya Mountains?

How do earthquakes help Earth scientists determine the locations of plate boundaries?

The Search for a Mechanism

LEARNING OBJECTIVES

Identify the two mechanisms by which heat moves inside Earth.

Explain the role of mantle convection in plate tectonics.

Define the tectonic cycle.

Discuss the possible existence of mantle plumes.

Although virtually all Earth scientists accept the basic theory of plate tectonics, some questions remain. What, exactly, drives plate motion? How does the mantle interact with the crust? What initiates subduction? Scientists have a basic understanding of these processes, but the details have not been completely worked out. Thermal motion in the mantle is at least partially responsible for the motion of plates. This thermal motion in turn results from the release of heat from Earth's interior. Let's take a closer look at some of the complexities of Earth's heat-releasing processes.

EARTH'S INTERNAL HEAT

Earth gives off heat for two main reasons. First, it is slowly cooling off from its initial formation processes, including formation of a molten iron core. Second, heat is *constantly being generated by the decay of radioactive elements in the interior, primarily uranium, potassium, and thorium.* If Earth did not release heat into outer space, the entire interior would eventually melt.

■ **conduction** The process by which heat moves through a solid body without deforming it.

Some of Earth's heat is released through **conduction**. This is a gentle and slow process similar to what you can feel when you hold a cup of hot coffee in your hands. The heat moves through the wall of the cup by conduction, a gradual transfer of energy from atom to atom. Conduction is heat energy on the move, but in a manner that does not involve the movement of hot material from one place to another.

We know that molten rock erupts from volcanoes, so Earth's heat clearly can move with its hot material in a second process, called **convection**. When you boil water in a pot on a stove, you will see the water churning around in big circles called *convection cells*. A mass of hot water at the bottom is slightly less dense than the cooler water at the top, and hence it will rise. When it reaches the surface, it sheds its heat, moves sideways as it cools, and then sinks back down to the bottom, where it is reheated. This mechanism of heat transfer is more efficient than conduction. The convection cells act like couriers, carrying heat directly from the burners to the top of the pan instead of passing the heat along from atom to atom.

■ **convection** A form of heat transfer in which hot material circulates from hotter to colder regions, loses its heat, and then repeats the cycle.

CONVECTION AS A DRIVING FORCE

Even though Earth's mantle is composed mostly of solid rock, it, too, releases heat through convection. Solid rock, if it is hot enough, can behave like a flowing viscous fluid, just as the solid ice in a glacier does. Rock deep in the mantle heats up and expands, becoming buoyant. Very, very slowly it moves upward, in huge convection cells. Just like the water in the boiling pot, the hot rock moves laterally near the surface as it sheds its heat. This lateral movement

Mantle convection FIGURE 7.13

A An everyday example of convection can be seen when you boil a pot of water. The water closest to the burners is hotter than the rest of the water. As it heats up, it becomes less dense and rises to the top. At the surface, it cools down and moves sideways to make room for the hot water rising beneath it. As the water at the surface cools, it becomes denser and sinks.

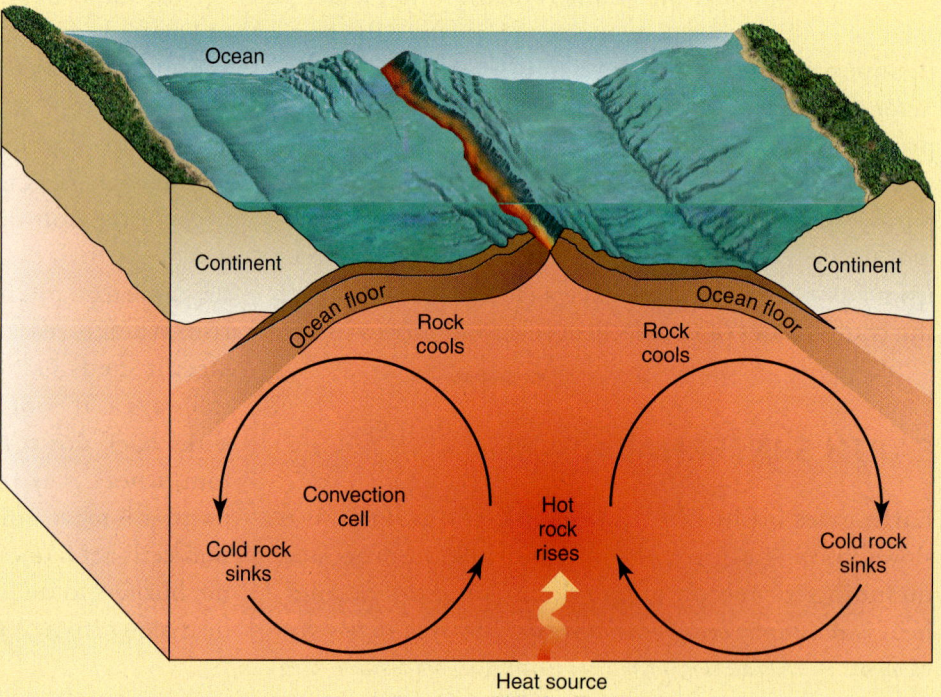

B The same process happens in Earth's mantle on a much grander scale and over a much longer time. Hot rock rises slowly and plastically from deep inside Earth, then cools, flows sideways, and sinks. The relation between convection cells and lithospheric plates is far more complex than what we see, for example, in a pot of boiling water.

of rock in the asthenosphere is believed to be one of the causes of the motion of plates of lithosphere (see **FIGURE 7.13**).

Convection in the mantle is complex and incompletely understood. Many challenging questions remain unanswered. For example, does the whole mantle convect as a unit, or does the top part convect separately from the bottom, creating rolls upon rolls? In subduction zones, are plates of lithosphere pushed down or pulled down, or do they sink into the asthenosphere under their own weight? (Plates with subduction edges move faster than plates without subduction edges, which suggests that the sinking plate edge is pulling the rest of the plate.) How do midocean ridges start, and what role do they play in plate motion? There appears to be a force that pushes plates apart, away from the midocean ridge. What are the shapes and distribution of convection cells?

220 CHAPTER 7 Plate Tectonics: Sculptor of Earth's Ever-Changing Landscape

MANTLE PLUMES

One of the most puzzling features of convection in Earth's mantle is evidence that hot rocks do not travel only in neatly packaged convection cells; some of the hot rock from deep in the mantle seems to rise in long, thin blobs called *plumes*. The plumes give rise to long-lived sources of magma deep in the mantle. We know this must be the case because plumes create long lines of volcanoes, which means that the plate on which a volcano forms moves and eventually loses contact with the source of magma, so a new volcano is formed (see *Amazing Places* on pages 224–225). Do the plumes originate in the middle of the mantle, or perhaps even deeper, where the core meets the mantle? Earth scientists continue to actively research questions such as these. In the process of researching plumes, they have identified about 50 long-lived centers of volcanism on Earth, each of which is thought to sit above a plume (**Figure 7.14**).

Volcanism and hot spots Figure 7.14

Most volcanism on Earth occurs underwater at the midocean ridges. Most volcanoes above sea level are associated with subduction zones. In addition, hot-spot volcanism associated with rising plumes of hot rock in the mantle (yellow circles) has been identified at many locations. Some hot spots, such as Iceland, are associated with plate boundaries. Other hot spots, such as Hawaii, lie within plate interiors.

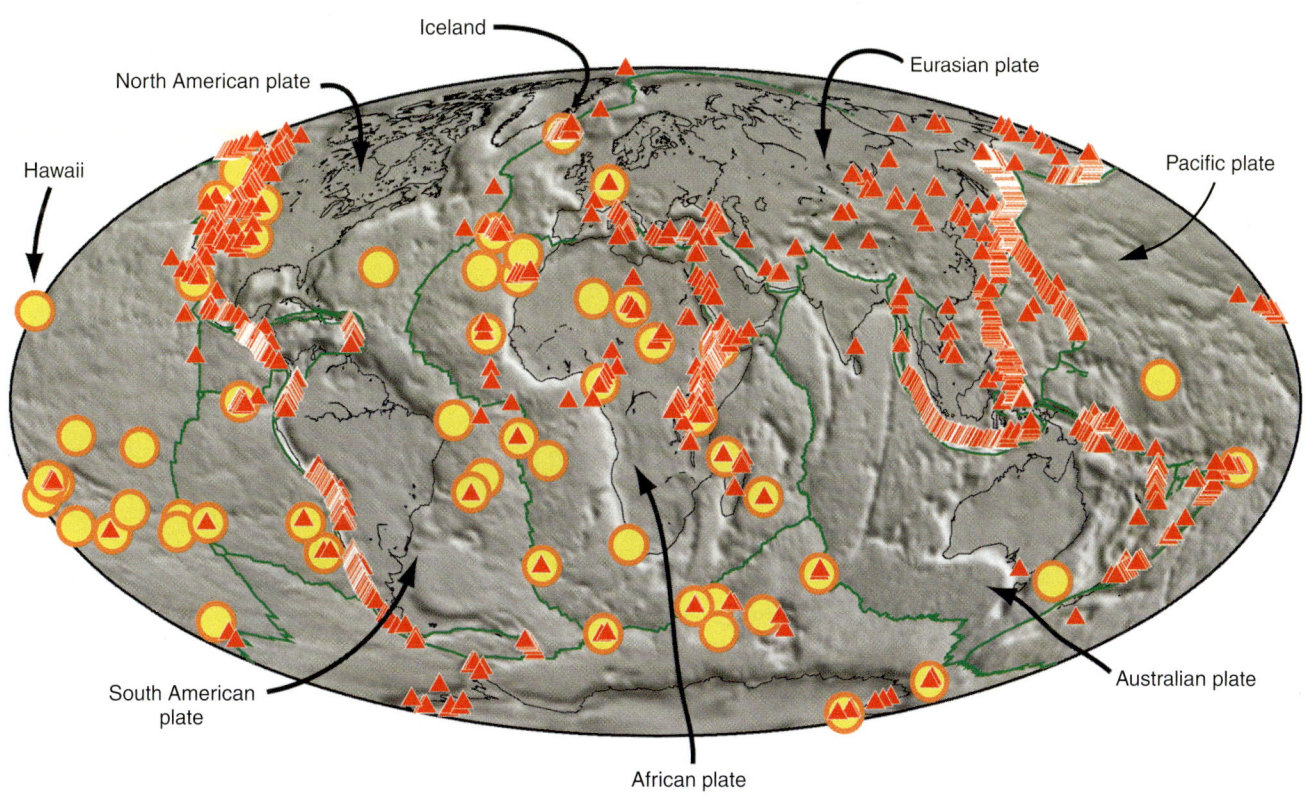

The Search for a Mechanism

THE TECTONIC CYCLE

Recall that in Chapter 1 we introduced the concept of interacting cycles in the Earth system. In this chapter, we introduce you to the **tectonic cycle** (FIGURE 7.15). New material is constantly added to oceanic crust by volcanism along divergent plate boundaries, while subduction consumes an equal amount of oceanic crustal material at convergent margins. Because of these processes, the seafloor renews itself on the order of every 200 million years.

> **tectonic cycle**
> Movements and interactions of the lithosphere by which rocks are cycled from the mantle to the crust and back; includes earthquakes, volcanism, and plate motion, driven by convection in the mantle.

Continental crust lasts much longer than oceanic crust. Because of its lower density (greater buoyancy), it cannot easily be subducted. The recycling of continental crust involves plate tectonics, erosion (discussed in Chapter 4), and magmatic activity (discussed in Chapter 9).

Plate tectonics affects all life on Earth, sometimes in subtle ways. It influences climate through the distribution of continents and ocean basins, for example. These changes are slow but profound; they can lead to ice ages or periods of unusual warmth, both of which strongly affect the evolution of species. Other effects are more obvious, such as major earthquakes or volcanic eruptions.

At the beginning of this chapter, we described plate tectonics as a "unifying" theory of how the Earth system works. This is because the plate-tectonic model brings together many diverse observations of Earth's major features and unifies them into a single, reasonably straightforward story. As a model, plate tectonics is truly representative of the Earth system science approach because it illustrates how internal Earth processes are integrated with all other parts of the Earth system.

The tectonic cycle FIGURE 7.15

The continuous formation of new oceanic lithosphere and removal of an equal amount of old oceanic lithosphere cause the plate motions that drive the tectonic processes that shape Earth's surface features.

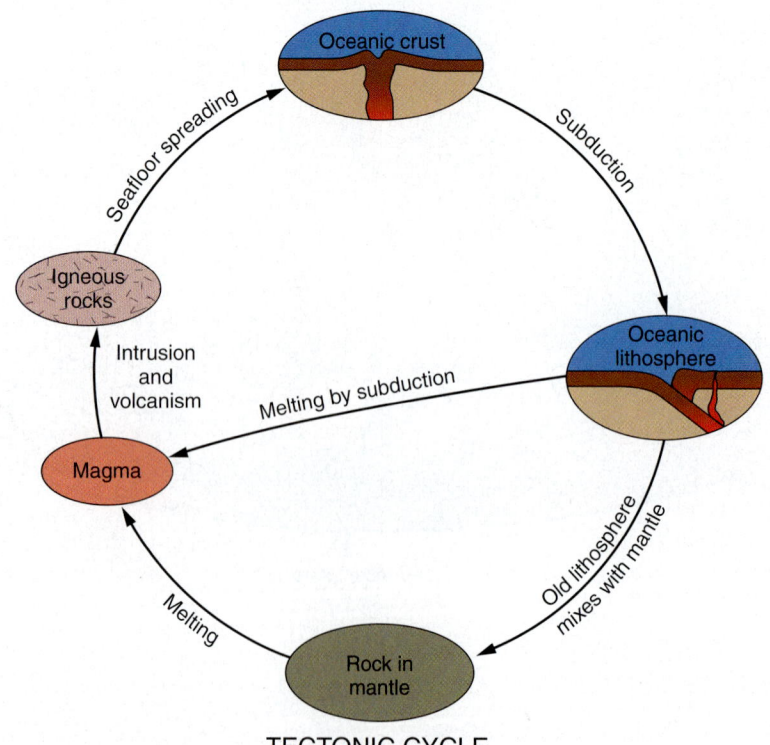

222 CHAPTER 7 Plate Tectonics: Sculptor of Earth's Ever-Changing Landscape

Rodinia: An ancient supercontinent FIGURE 7.16

A proposed arrangement of continental fragments in the 1100-million-year-old supercontinent Rodinia. The dark red strip marks the location of a now deeply eroded mountain chain that formed as a result of collisions as Rodinia was assembled but that is now dispersed among Earth's present-day continents.

THE SUPERCONTINENT CYCLE

So far, we have presented the plate tectonics model in terms of the breakup and dispersal of the continental fragments of Pangaea. However, the breakup of Pangaea began about 200 million years ago, so it is pertinent to ask whether the plate tectonics model can explain activities still earlier in Earth's history.

There is convincing evidence that fragments of continental crust collided together to form Pangaea between 350 and 450 million years ago. The collisions formed mountain ranges, such as the Appalachians in North America and the Urals in Russia. The eroded remnants of those collisions remain as evidence of the assembly of Pangaea. There is also convincing evidence that a supercontinent existed about 1100 million years ago (FIGURE 7.16). The evidence is in the paleomagnetic record and the existence of now deeply eroded ancient mountain ranges. Given the name *Rodinia*, the late Proterozoic supercontinent continues to be the subject of much study. Fragmentary evidence suggests that prior to Rodinia other ancient supercontinents formed, existed for a few hundred million years, and then broke apart. The formation and breakup of a supercontinent is called a *Wilson Cycle*, named for Professor J. Tuzo Wilson of Canada, one of the major contributors to the plate tectonic revolution.

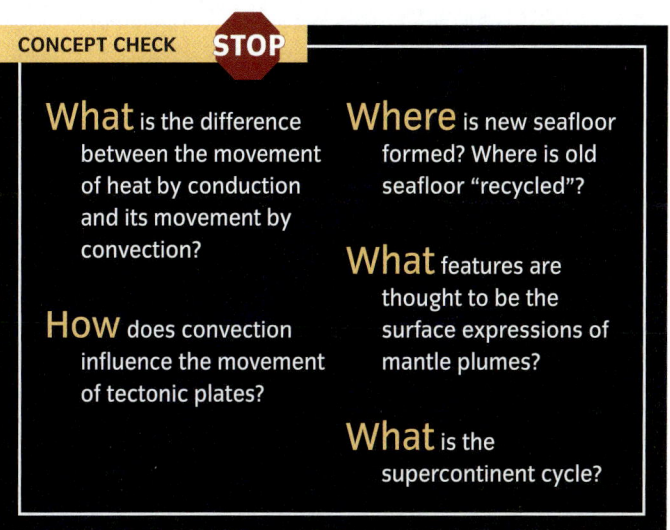

CONCEPT CHECK **STOP**

What is the difference between the movement of heat by conduction and its movement by convection?

How does convection influence the movement of tectonic plates?

Where is new seafloor formed? Where is old seafloor "recycled"?

What features are thought to be the surface expressions of mantle plumes?

What is the supercontinent cycle?

The Search for a Mechanism 223

Amazing Places: The Hawaiian Islands

A GROWING ISLAND CHAIN

One of the most striking features of the Hawaiian Islands, first noted by geologist and explorer James Dwight Dana on an expedition to the islands in 1840–1841, is the increasing ages of the islands from southeast to northwest. The Na Pali (**A**), on the north shore of Kauai, the northernmost of the major islands, shows highly eroded volcanic formations. By contrast, on the south end of the big island of Hawaii (**B**), an ongoing eruption of Kilauea Volcano has added new land to the island almost ceaselessly from 1983 to the present day. Just 15 km south of Hawaii, the next Hawaiian island is already being born. Loihi seamount (**C**), an underwater volcano that appears as a raised feature in this topographical map of the seafloor, already has a peak that is only 1 km below the ocean surface (shown in red), and it is still growing. Loihi erupted as recently as 1996.

Even though Hawaii is thousands of kilometers from the nearest plate boundary, the processes that have formed the islands, as well as their differing ages, have a great deal to do with plate tectonics. The big island of Hawaii is currently sitting above a long, thin plume of hot material, called a *hot spot*, rising from deep in the mantle. The plume itself is stationary, but the lithospheric plate above it is moving to the northwest. The plume provides magma for the currently active volcanoes of Mauna Loa and Kilauea, and the undersea volcano of Loihi. The older islands, such as Kauai, once lay above the hot spot but moved off it as the Pacific plate moved northwest. With no new volcanic additions to their land mass, they will eventually erode into the sea.

A The oldest of the large Hawaiian islands

B **The youngest Hawaiian Island**

C **The next Hawaiian island?**

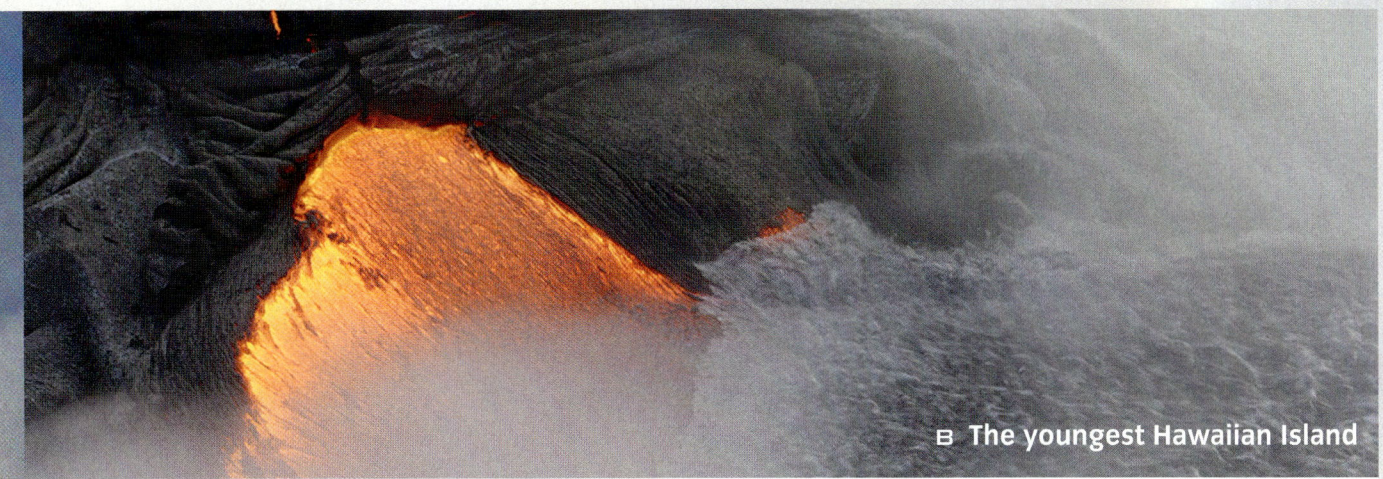

Approximate ages of islands in millions of years

Kauai, 3–5.5 — A
Oahu, 2.25–3.25 — B
Molokai, 1.25–2 — C, D
Maui, 0.5–1 — E
Hawaii, present–0.8

Direction of plate movement

Loihi — Hot spot

Mantle plume

D **Hot spot**

The Search for a Mechanism 225

SUMMARY

1 A Revolution in Earth Science

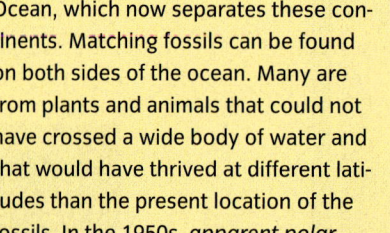
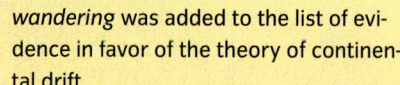

1. A revolution in earth science began almost 100 years ago with the hypothesis that the continents have not always been in their present positions but moved into their present-day positions after the breakup of a "supercontinent" known as *Pangaea*. This became known as the **continental drift** hypothesis. At first, it was quite controversial because scientists could not envision a satisfactory physical mechanism that could move the continents through solid rock.

2. The evidence for continental drift included the fit between the coastlines of continents. The evidence is stronger when the fit is made along the "true" edge of the continent—the **continental shelf**. In addition, geologic features such as mountain ranges, ancient glacial deposits, and rock types match very closely on both sides of the Atlantic Ocean, which now separates these continents. Matching fossils can be found on both sides of the ocean. Many are from plants and animals that could not have crossed a wide body of water and that would have thrived at different latitudes than the present location of the fossils. In the 1950s, *apparent polar wandering* was added to the list of evidence in favor of the theory of continental drift.

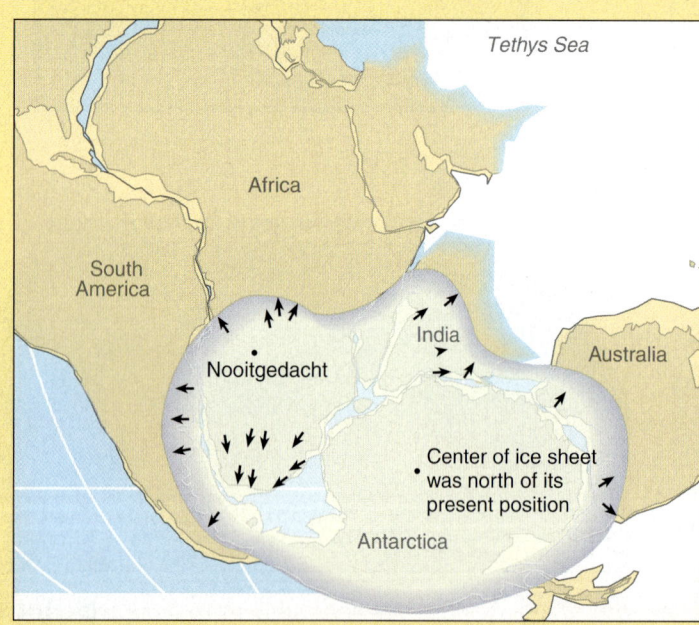

3. The most significant piece of evidence in support of continental drift was the discovery of bands of magnetized rock on the seafloor with alternating normal and reversed polarities, aligned symmetrically on either side of the *midocean ridges*. The only plausible explanation for this discovery was **seafloor spreading**. It was quickly realized that continents and adjacent seafloor move together. This evidence was decisive because it finally led to a mechanism that could account for the continents' movement. At midocean ridges, the ocean floor is constantly replenished, pushing older rocks apart. At deep-sea trenches, old ocean floor sinks back into the mantle.

4. Global positioning satellites (GPS) can now measure the speed and direction of continental drift directly and precisely, which shows that the continents are still moving today at the rate of several centimeters per year.

2 The Plate Tectonic Model

1. According to **plate tectonics**, the lithosphere has fragmented into several large **plates**. These plates essentially float on an underlying layer of hot, ductile rock called the *asthenosphere*. The asthenosphere is in slow, but constant, motion and this forces the more rigid *lithospheric* plates to move around, collide, split apart, or slide past each other. At present, there are six major lithospheric plates and many smaller ones.

2. Plate tectonics predicts three different kinds of interactions along plate margins. At **divergent margins**, two plates move apart. If the boundary occurs in oceanic crust, a divergent margin coincides with a *midocean ridge*. When the divergent boundary occurs in continental crust, it produces long and relatively straight *rift valleys*. Eventually, if the divergence continues for a long enough time, the rift valley will become wide and deep enough to form a sea or an ocean.

3. At **convergent margins**, two plates move toward each other. When an oceanic plate collides with another oceanic plate or with a continental plate, one oceanic plate slides beneath the other plate. This creates a **subduction zone**, which is marked by a deep *oceanic trench* and a great deal of volcanic activity. A second kind of convergent margin occurs where two continental plates collide over an ancient subduction zone, forming a *collision zone*. Continental crust is generally too buoyant to subduct, so instead the lithosphere crumples and thickens, building up giant mountain ranges.

4. At **transform fault margins**, which are fractures in the lithosphere, two plates slide past each other. Transform faults are common on the ocean floor, where they run perpendicular to midocean ridges.

5. Many **earthquakes** and active volcanoes are located along plate margins. Studies of the locations and depth of earthquakes enable scientists to determine the shapes of lithospheric plates and the type of margin between the plates. Earthquakes occur as a result of the motion of rocks along a **fault**. The **focus**, where the motion starts, is underground; the **epicenter** is the place on the surface that lies directly above the focus.

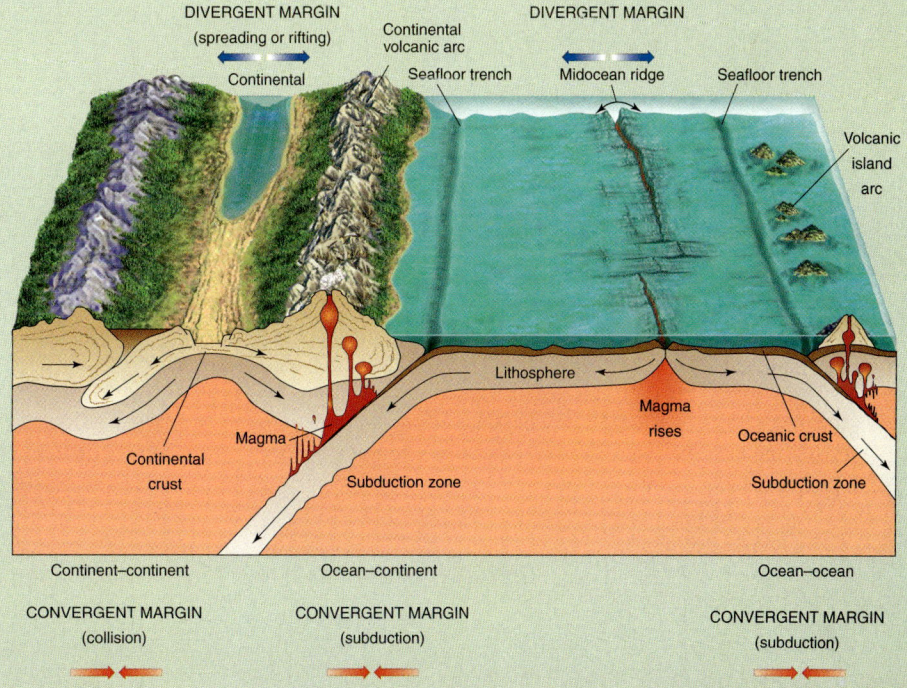

3 The Search for a Mechanism

1. Heat is released from Earth by both conduction and convection, but convection is the more important of the two mechanisms.

2. The release of heat from Earth's interior creates huge **convection** cells, which are at least partly responsible for driving plate motion. Convection brings hot rock up from deep in the mantle and recycles cold rock back into the mantle. Many unanswered questions remain about *convection cells*.

3. Hot rock rises from deep in the mantle, possibly from the core–mantle boundary, in *plumes* that do not seem to be related to the convection cells. Plumes are marked at Earth's surface by very long-lived centers of volcanism.

4. Plate tectonics unifies the processes of continental drift, seafloor spreading, mountain building, faulting, earthquakes, and volcanism into the **tectonic cycle**. The tectonic cycle creates and recycles oceanic crust on a time scale of roughly 200 million years. Continental crust lasts much longer than oceanic crust.

KEY TERMS

- **continental drift** p. 202
- **paleomagnetism** p. 207
- **seafloor spreading** p. 209
- **plate tectonics** p. 211
- **fault** p. 211
- **divergent margin** p. 214
- **convergent margin** p. 214
- **subduction zone** p. 214
- **transform fault** p. 214
- **conduction** p. 219
- **convection** p. 219
- **tectonic cycle** p. 222

CRITICAL AND CREATIVE THINKING QUESTIONS

1. Why was the discovery of paleomagnetic bands on the Atlantic Ocean floor such an important turning point in the acceptance of the theory of plate tectonics? Suppose you could rewrite history so that the satellite measurements of continental drift came first. Would the discovery of seafloor spreading still have been important?

2. What are some of the important questions about plate tectonics that remain unanswered today? How might you proceed in trying to resolve them?

3. We have called plate tectonics a scientific revolution. What other scientific revolutions do you know about? How is the theory of plate tectonics similar to, or different from, other revolutions, such as the Copernican revolution in astronomy?

4. Why do Earth scientists call plate tectonics a *unifying theory*?

5. What is the tectonic environment of the place where you live or attend school? Do you live near a plate margin (what type?) or in the middle of a plate (which one?)?

What is happening in this picture?

The location is the Rift Valley in Kenya, in Africa, and the view is to the north. What we see is a beautiful illustration of rifting due to continental crust being split apart by a spreading center. The steep slopes to the right appear to be surface expressions of faults that mark the edge of the Somali plate.

■ Are they normal faults or reverse faults?

Land to the left of the scarps has dropped down, forming the flat-bottomed farming valley. Still farther to the left, and out of view, is the eastern edge of the African plate. The rift valley ranges in width from 30 to 60 km.

■ What will happen if spreading continues and the valley widens?

An Earth scientist would also make note of the way local agriculture exploits the stepped landscape at the edge of the Rift Valley with cultivated fields laid out in the flat tops of terraces.

The flat faces of the slopes on the fault scarps between gullies are likely to be remnants of the original fault planes.

Critical and Creative Thinking Questions

SELF-TEST

1. The work of Earth scientists over the years has supported Wegener's contention that the current continental masses were assembled into a single supercontinent, which Wegener called _____.
   a. Pangaea
   b. Transantarctica
   c. Gondwana
   d. Tethys
   e. Laurasia

2. Which of the following lines of evidence supporting continental drift was not used by Wegener when he first proposed his hypothesis?
   a. The apparent fit of the continental margins of Africa and South America
   b. Ancient glacial deposits of the Southern Hemisphere
   c. The apparent polar wandering of the north magnetic pole
   d. Close match of ancient geology between West Africa and Brazil
   e. Close match of ancient fossils on continents separated by ocean basins

3. Analysis of apparent polar wandering paths led geophysicists to conclude _____.
   a. that Earth's magnetic poles have wandered all over the globe in the last several hundred million years
   b. that the continents had moved because it is known that the magnetic poles themselves are essentially fixed
   c. that the apparent wandering path of a continent provides a historical record of the position of that continent over time
   d. Both b and c are correct.

4. _____ is the process by which oceanic crust splits and moves apart along a midocean ridge and new oceanic crust forms.
   a. Continental drift
   b. Paleomagnetism
   c. Seafloor spreading
   d. Continental rifting

5. This illustration is a map showing the age of the seafloor, across the northern extent of the Atlantic Ocean. The Mid-Atlantic Ridge can be seen stretching roughly north–south (in the yellow band) down the middle of the map. Yellow through red colors show rocks of similar age; number them from 1 (oldest) through 5 (youngest).

6. _____ technology has allowed scientists to measure the movement of continental crust.
   a. Global Positioning System (GPS)
   b. Seismic recording
   c. Magnetometer
   d. Gravity meter

7. This map shows radiometric ages for the Hawaiian island chain in the middle of the Pacific plate. These islands formed over a hot spot in Earth's mantle. Draw an arrow on the map showing the direction of movement of the Pacific plate over this hot spot as indicated by the ages of the islands in the Hawaiian chain.

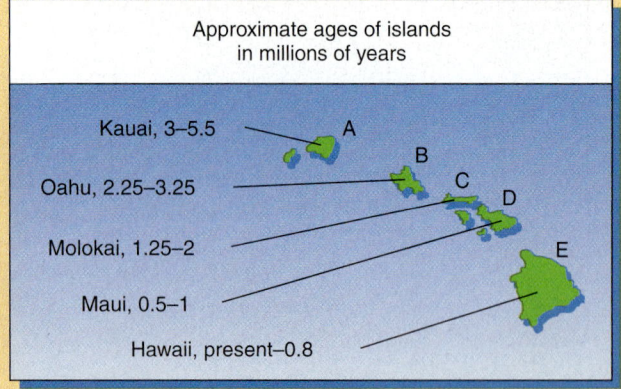

8. At a _____, two lithospheric plates slide past one another horizontally.
   a. divergent boundary
   b. transform fault boundary
   c. subduction zone boundary
   d. continental collision boundary

9. At a _____, oceanic crust is consumed back into the asthenosphere.
   a. divergent boundary
   b. transform fault boundary
   c. subduction zone boundary
   d. continental collision boundary

10. At a _____, new oceanic crust forms along midocean ridges.
    a. divergent boundary
    b. transform fault boundary
    c. subduction zone boundary
    d. continental collision boundary

11. A _____ is a convergent margin along which subduction is no longer active and high mountain ranges are formed.
    a. divergent boundary
    b. transform fault boundary
    c. subduction zone boundary
    d. continental collision boundary

12. Heat from the solid mantle is released through a process of _____.
    a. polar wandering
    b. paleomagnetism
    c. convection
    d. magnetic reversal

13. This illustration shows different types of plate boundaries. For each block diagram, choose the appropriate label from the following list:

    Divergent boundary   Continental collision boundary
    Transform fault boundary   Subduction zone boundary

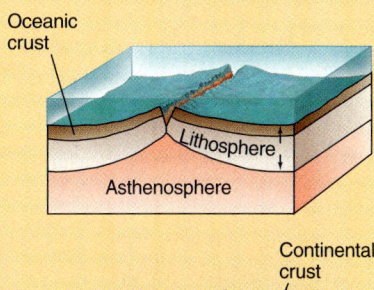
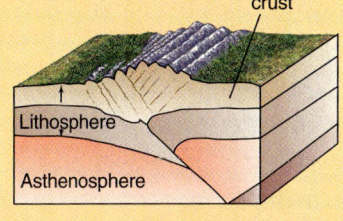
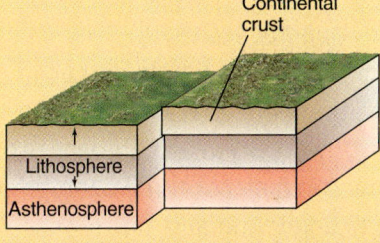
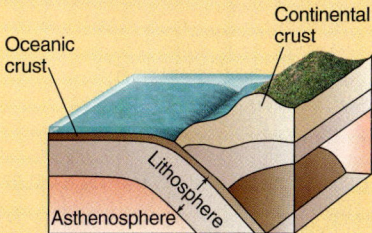

14. On the block diagrams in question 13, indicate the locations of earthquakes for each type of plate boundary. Use a red dot to show shallow-focus earthquakes and a blue dot to show the location of deep-focus earthquakes.

15. A Wilson Cycle is _____.
    a. eruption and eventual subduction of basaltic lava on the seafloor
    b. formation of mountains by plate collisions and subsequent erosion of the mountains
    c. formation and subsequent breakup of supercontinents
    d. the cycle of water through the hydrosphere

Earthquakes and Earth's Interior   8

On December 26, 2004, a powerful earthquake shook the Indian Ocean floor about 160 kilometers off the island of Sumatra. This earthquake—the Sumatra-Andaman earthquake of 2004—caused the largest and deadliest tsunami in history.

The quake began when part of the Indian plate, which is slipping under the Eurasian plate, suddenly slipped downward about 15 meters. The motion pushed the seafloor up as much as 5 m on the Eurasian side. Unlike most earthquakes, which are over in seconds, the slippage continued for 10 minutes as the interface between the plates slipped, section by section, for 1200 km to the north. On the surface of the ocean, waves swept toward Indonesia, Thailand, Sri Lanka, and India. As they approached the shore and the ocean shallowed, the waves grew in height; when they reached the shore, they had grown to heights of 20 or 30 m. Sweeping inland, the waves obliterated everything in their path. The photograph here shows destruction in Sumatra, and the satellite views (inset) show houses, roads, and bridges in Banda Aceh, on the island of Sumatra, before and after the tsunami.

Although earthquakes and tsunamis are common in the Indian Ocean, no warning system was in place when this disaster occurred. The resulting devastation caused hundreds of billions of dollars in damages and at least 227,000 deaths.

Earthquakes give Earth scientists a window into Earth's internal structure. We cannot prevent earthquakes, but scientists are getting better at understanding and preparing for their after effects, including devastating tsunamis. The knowledge we gain from such events allows us to prepare more effectively for future disasters.

Global Locator

BEFORE

AFTER

NATIONAL GEOGRAPHIC

CHAPTER OUTLINE

- Earthquakes and Earthquake Hazards p. 234
- The Science of Seismology p. 246
- Looking into Earth's Interior p. 252
- How and Why Rock Breaks p. 260

Earthquakes and Earthquake Hazards

LEARNING OBJECTIVES

Explain the connection between earthquakes and plate tectonics.

Describe the theory of elastic rebound.

Identify several earthquake-related hazards.

Compare short-term prediction and long-term forecasting of earthquakes.

The Sumatra-Andaman **earthquake** occurred in a subduction zone, where the oceanic lithosphere of the Indian plate is being subducted beneath the Eurasian plate (see **FIGURE 8.1**). It was the fifth giant subduction zone earthquake since 1950. The others were in Kamchatka, Russia, in 1952; the Aleutians in 1957; southern Chile in 1960; and Prince William Sound, Alaska, in 1964.

earthquake A sudden motion in Earth caused by the abrupt release of slowly accumulated energy.

This association of subduction zones and large earthquakes suggests that the motion of the plates is somehow responsible for earthquakes. However, plate motion is very gradual, typically on the order of a few centimeters per year. Why, then, should earthquakes be so sudden, and the big ones so catastrophic? The science of **seismology** works to answer this and related questions.

seismology The scientific study of earthquakes and seismic waves.

EARTHQUAKES AND PLATE MOTION

Most earthquakes are due to stresses that cause sudden movement along a fracture in the lithosphere called a *fault*. If the rocks could slide past one another smoothly, like the parts of a well-oiled engine, big quakes like the Sumatra-Andaman quake would not happen. In the real world, smooth sliding is rare; friction between the huge blocks of rock causes them to seize up, bringing the motion along that part of the fault to a temporary stop. While the fault remains locked by friction, energy

Megathrust earthquakes FIGURE 8.1

Major earthquakes at the boundary between a subducting plate and an overriding plate are sometimes called *megathrust* earthquakes. The world's five most powerful earthquakes since 1950 (shown on this map) have been of this variety. Compare with Figure 7.9 (pages 212–213), where the locations of subduction plate boundaries are plotted.

234 CHAPTER 8 Earthquakes and Earth's Interior

A When these orange trees were planted on land that lies over the San Andreas Fault in southern California, the rows were straight. In 1938, earthquake motion along the fault displaced the trees significantly. Arrows show the direction of movement of the plates.

B The second most powerful earthquake on record, the Alaska "Good Friday" earthquake, struck Anchorage on March 27, 1964. The vertical motion along the fault amounted to several meters in some places. In this photograph, officials examine damage in the Turnagain Heights area.

Evidence of lateral and vertical motion on faults FIGURE 8.2

continues to build up as a result of the plate motion, causing rocks adjacent to the jammed section to bend and buckle. Finally, the stress becomes great enough to overcome the friction along the fault. All at once, the blocks slip, and the pent-up energy in the rocks is released as the violent tremors of an earthquake.

This cycle of slow buildup of energy followed by abrupt movement along a fault repeats itself many times, so earthquakes tend to happen again and again on the same fault. Although movement along a large fault may eventually total many kilometers, this distance is the sum of numerous smaller slips happening over many millennia. In some places, these small slips (collectively called *seismic creep*) are frequent, though imperceptible to humans. Nevertheless, they can create visible distortion of surface features, as shown in FIGURE 8.2A. Earthquakes can also cause vertical dislocations of the ground surface (see FIGURE 8.2B). The largest abrupt vertical displacement on record occurred in 1899, at Yakutat Bay, Alaska, when a long stretch of the Alaskan shore was suddenly lifted 15 m above sea level during a major earthquake.

The elastic rebound theory
The initial vertical or horizontal motion, dramatic as it may appear, often is not what does the most damage during an earthquake. It is the sustained shaking of the ground that destroys buildings, bridges, and cities, sometimes many kilometers away from the location of the quake. In 1910, Harry Fielding Reid, a member of a commission appointed to investigate the infamous 1906 San Francisco earthquake that destroyed much of the city, proposed a now widely accepted explanation for the shaking. Reid's **elastic rebound theory** suggests that rocks, like all solids, are *elastic* (within limits). This means that they will stretch or bend when subjected to stress, and then snap back when the stress is removed. The snap back happens when the two blocks on either side of a fault manage to overcome friction and slip past one another. However, the strained rocks on either side of the fault don't just snap back and stop. Like a guitar string after it is plucked, they continue vibrating. These vibrations are called **seismic waves**. Like sound waves from a guitar string,

> **elastic rebound theory** The theory that continuing stress along a fault results in a buildup of elastic energy in the rocks, which is abruptly released when an earthquake occurs.
>
> **seismic wave** An elastic shock wave that travels outward in all directions from an earthquake's source.

Earthquakes and Earthquake Hazards

they can travel a long distance from their place of origin. (We will investigate the scientific concepts of *stress* and *strain* in greater depth later in this chapter.)

The first evidence to support Reid's elastic rebound hypothesis came from studies of the San Andreas Fault, a large, complex fault in California that generated both the 1906 quake and the 1989 "World Series" quake in Oakland. Beginning in 1874, scientists from the U.S. Coast and Geodetic Survey had been measuring the precise positions of many points both adjacent to and distant from the fault. As time passed, movement of the points revealed that at some places the two sides of the fault were smoothly slipping in opposite directions. Near San Francisco, however, the fault appeared to be locked by friction and did not reveal any slip. Then, on April 18, 1906, the two sides of this locked section of fault shifted abruptly (see **FIGURE 8.3**). The elastically stored energy in the rocks was released as the crust

Horizontal motion along fault blocks FIGURE 8.3

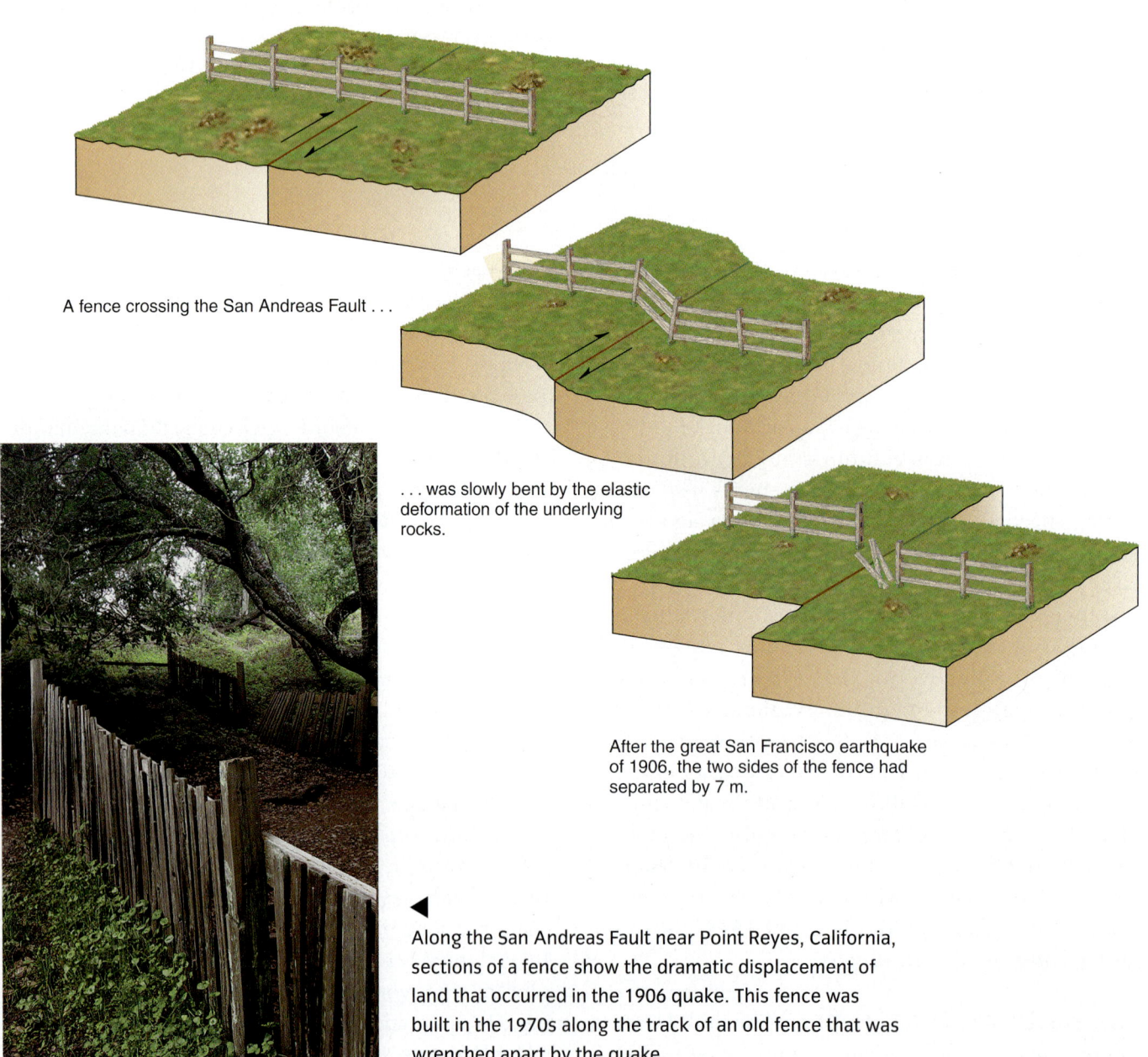

A fence crossing the San Andreas Fault . . .

. . . was slowly bent by the elastic deformation of the underlying rocks.

After the great San Francisco earthquake of 1906, the two sides of the fence had separated by 7 m.

◀ Along the San Andreas Fault near Point Reyes, California, sections of a fence show the dramatic displacement of land that occurred in the 1906 quake. This fence was built in the 1970s along the track of an old fence that was wrenched apart by the quake.

How a tsunami was unleashed by motion on a fault FIGURE 8.4

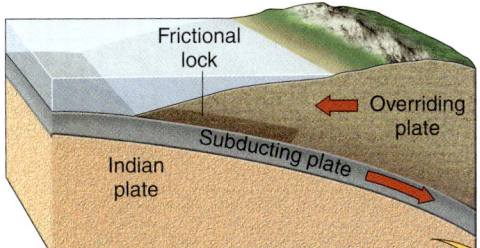

1 The subducting Indian plate sticks to the overriding Eurasian plate as a result of friction.

2 Stress builds and the seafloor gradually deforms as the Eurasian plate is bent downwards.

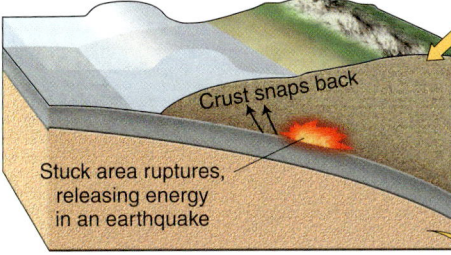

Earthquake occurs as the crust snaps back, displacing water and initiating the tsunami

3 A sudden rupture frees the stuck section of the Eurasian plate. It rebounds elastically and releases energy in the form of seismic waves. The seafloor along the fault is lifted, causing the displacement of a very large volume of water.

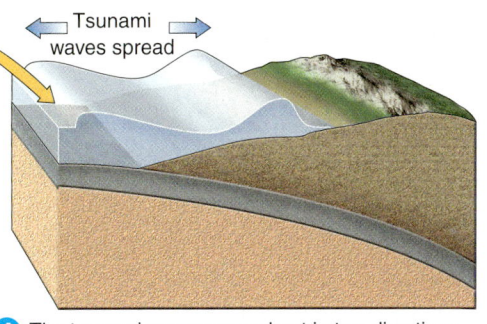

4 The tsunami waves spread out in two directions from the uplifted region, traveling mainly to the east and west at right angles to the subduction zone.

CRITICAL THINKING

Here's an interesting question:
- Notice in the lower right-hand drawing that tsunami waves are spreading in two directions. Why would this happen? Why would they not spread in just one direction?

VIEW THIS IN ACTION in your WileyPLUS course

---

snapped to its new position, creating a violent earthquake. Reid's measurements after the quake revealed that the bending, or strain, stored in the crust had disappeared.

The elastic rebound theory also explains why the Sumatra-Andaman earthquake of 2004 lifted the seafloor and caused a giant tsunami (see **FIGURE 8.4**).

Causes of earthquake-like vibrations Earthquakes are caused by the sudden release of a locked fault, but any sudden release of energy, such as the impact of a large meteorite or a volcanic eruption, can cause tremors that are very much like earthquakes. Blasting operations are routinely monitored for tremors, especially if the quarry or mine is close

Earthquakes and Earthquake Hazards 237

to habitation. Very large explosions, such as the underground tests of atom bombs, cause tremors that can be detected around the world. The tremors are like those caused by earthquakes but sufficiently different that they can be used to monitor underground explosions. Building collapses can also create tremors—the collapse of the Twin Towers of the World Trade Center in Manhattan on September 11, 2001, caused tremors that were detected in many places around New York City.

EARTHQUAKE HAZARDS AND PREDICTION

Each year more than a million earthquakes occur around the world. Fortunately, only a few are large enough, or close enough to major population centers, to cause much damage or loss of life. A great deal of research focuses on earthquake prediction and hazard assessment. Scientists are working hard to improve their forecasting ability to the point where effective and accurate early warnings can be issued. Let's look briefly at the hazards associated with earthquakes and at efforts to predict them.

Earthquake hazards Earthquakes can cause total devastation in a matter of seconds. The most disastrous quake in history occurred in Shaanxi Province, China, in 1556, killing an estimated 830,000 people. The earthquake caused the caves in which most of the population lived to collapse. In all, 20 earthquakes in history have caused 50,000 or more deaths apiece (**TABLE 8.1**).

Ground motion, with the resulting collapse of buildings, bridges, and other structures, is usually the most significant *primary hazard* to cause damage during an earthquake. In the most intense quakes, the surface of the ground can be observed moving in waves. Sometimes large cracks and fissures open in the ground. Where a fault breaks the ground surface, buildings can be split, roads disrupted, and anything that lies on or across the fault broken apart. To make matters worse, movement on one part of a fault can cause stress along another part of the fault, which in turn slips, generating another earthquake, called an *aftershock*. Aftershocks triggered by large earthquakes tend to be on the same fault system as the original quake, though they

Earthquakes during the past 800 years that have caused 50,000 or more deaths TABLE 8.1

Place	Year	Estimated number of deaths
Silicia, Turkey	1268	60,000
Gulf of Chili, China	1290	100,000
Naples, Italy	1456	60,000
Shaanxi, China	1556	830,000
Shemaka, Russia	1667	80,000
Catania, Italy	1693	60,000
Beijing, China	1731	100,000
Calcutta, India	1737	300,000
Lisbon, Portugal	1755	60,000
Calabria, Italy	1783	50,000
Messina, Italy	1908	160,000
Gansu, China	1920	180,000
Tokyo and Yokohama, Japan	1923	143,000
Gansu, China	1932	70,000
Quetta, Pakistan	1935	60,000
T'ang Shan, China	1976	240,000
Iran	1990	52,000
Sumatra-Andaman	2004	227,000
Kashmir, Pakistan	2005	86,000
Sichuan, China	2008	69,000

may be quite far from the original location, causing the damage to be spread more widely as time passes. The 1992 Landers earthquake, near Los Angeles, triggered major aftershocks at 14 locations, some of them hundreds of kilometers away.

Ground motion is not the only source of damage in an earthquake. Sometimes the aftereffects, or *secondary hazards* related to an earthquake, can cause even more damage than the original quake. Examples of secondary hazards that can be initiated by earthquakes include landslides, fires, ground liquefaction, and tsunamis. (See **FIGURE 8.5** and the *Case Study* on pages 240–241.)

Earthquake-related secondary hazards FIGURE 8.5

◀ Landslide
Huascaran, Peru

◀ Open fissure
Golcuk, Turkey

▼ Fire
San Francisco, California

▼ Tsunami, Kalutara, Sri Lanka ▼

CASE STUDY

The Sumatra-Andaman Tsunami of 2004

Tsunamis are uncommon compared with other secondary hazards associated with earthquakes, but in the final week of 2004, the world witnessed their devastating potential as never before.

The progress of the tsunami from 0 to 3 hours after the Sumatra-Andaman earthquake is shown in **A**. In the open sea, tsunamis are barely noticeable; the tsunami wave velocity may be hundreds of kilometers per hour and the wavelength many kilometers, but the peak wave height is typically half a meter or less. Upon approaching the shore, the wave begins to "hit bottom," slows down, and may build up to colossal heights; wave heights of 20 to 30 m were reported in Sumatra.

The frequency of tsunamis is greatest around the Pacific Ocean, where an extensive circum-Pacific

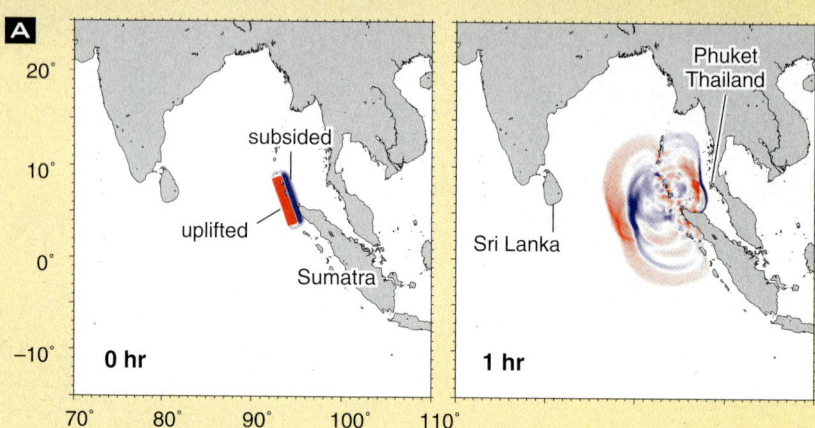

tsunami early warning system has been effective in reducing fatalities. As a result of the tsunami generated by the Sumatra-Andaman earthquake, research is being directed to the installation of early warning systems for the Indian Ocean. Research is also proceeding on ways to reduce the damage caused by monster waves.

Earthquake prediction

Charles Richter, the inventor of the Richter scale for quantifying the severity of earthquakes, once said, "Only fools, charlatans, and liars predict earthquakes." Today, unfortunately, this is still more or less correct: No one can predict the exact magnitude and time of occurrence of an earthquake. However, scientists' understanding of seismic mechanisms and the tectonic settings in which earthquakes occur has improved greatly since Richter's time, and advances in modern seismology may yet prove him wrong.

There are two aspects to the problem of earthquake prediction. *Short-term prediction* would identify the precise time, magnitude, and location of an earthquake in advance of the actual event, providing an opportunity for authorities to issue an early warning. *Long-term forecasting* involves the prediction of a large earthquake years or even decades in advance of its occurrence.

Short-term prediction and early warning Unfortunately, the short-term prediction of earthquakes has not been very successful to date. Attempts at short-term prediction are based on observations of anomalous *precursor phenomena*—that is, unusual activity preceding and leading up to the occurrence of an earthquake. For example, the magnetic or electrical properties of the rock could change, the level of well water could drop, or the amount of radon gas in the groundwater could rise in advance of an earthquake, any of which may indicate unusual activity in the underlying rock. Strange animal behavior, glowing auras, and unusual radio waves have also been reported as precursors near the sites of large earthquakes; there are plausible scientific explanations

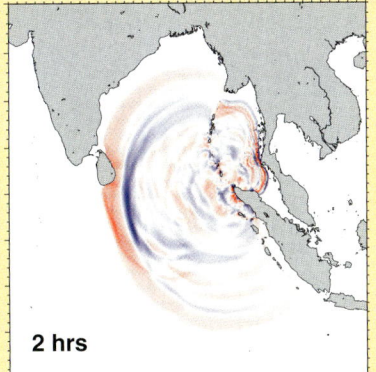
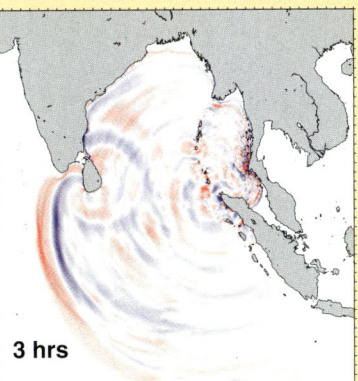

The height of the wave approaching the shore in B can be judged from this photo of a boat mooring in Thailand as it is about to be devastated by the 2004 tsunami.

for these. Small cracks and fractures can develop in severely strained rock and cause swarms of tiny earthquakes—*foreshocks*—that may presage a big quake.

The most famous successful earthquake prediction, made by Chinese scientists in 1975, was based on slow tilting of the land surface, fluctuations in the magnetic field, and numerous foreshocks that preceded a large quake that struck the town of Haicheng. Half the city was destroyed, but because authorities had evacuated more than a million people beforehand, only a few hundred were killed.

However, less than two years after the prediction of the Haicheng earthquake, the devastating 1976 T'ang Shan earthquake struck with no apparent precursory activity. With an official death toll of 240,000 (and unofficial reports suggesting many more deaths), it rivaled the Shaanxi earthquake of 1556 as the most disastrous in history. In 1976, and still today, short-term prediction and early warning of earthquakes remain elusive goals for seismologists.

Long-term forecasting Long-term earthquake forecasting is based mainly on our understanding of the tectonic cycle, tectonic settings in which earthquakes occur, and the hypothesis that earthquakes are repetitive. The hypothesis is supported by observation in places where earthquakes are known to occur repeatedly, such as along plate boundaries; in such places, seismologists have detected patterns in the recurrence intervals of large quakes. Because historical records seldom go back as far as seismologists would like, they also use the information provided by **paleoseismology**.

paleoseismology
The study of prehistoric earthquakes.

Earthquakes and Earthquake Hazards

Prehistoric quakes leave evidence in the sedimentary-rock record, such as vertical displacement of sedimentary layers, indications of liquefaction, or horizontal offset of natural features (see **FIGURE 8.6**). If the pattern of recurrence suggests regular intervals of, say, a century between major quakes, it may be possible to predict within a decade or two when a large quake is due to happen next in that location.

By studying ancient earthquakes, scientists have identified a number of *seismic gaps* around the Pacific Rim. A seismic gap is a place along a fault where a large earthquake has not occurred for a long time, even though tectonic movement is still active and stress is building. Some earthquake experts consider seismic gaps to be the places most likely to experience large earthquakes.

Long-term forecasting has met with reasonable success. Seismologists know where most (but not all) hazardous areas are. They can calculate the probability that a large earthquake will occur in a particular area within a given period. They have a theory of earthquake generation that successfully unites their predictions and observations in the context of plate tectonic theory. Forecasting helps people who live in seismically active areas to plan and prepare well in advance of a major event. If short-term prediction could advance as much as forecasting has done, many lives could be saved.

Evidence of ancient quakes   FIGURE 8.6

Layers of sediment were offset by ancient earthquakes. The carbon-rich layers of sediment have yielded carbon-14 dates, which help scientists pinpoint the ages of ancient quakes. (Carbon-14 dates are discussed further in Chapter 10.)

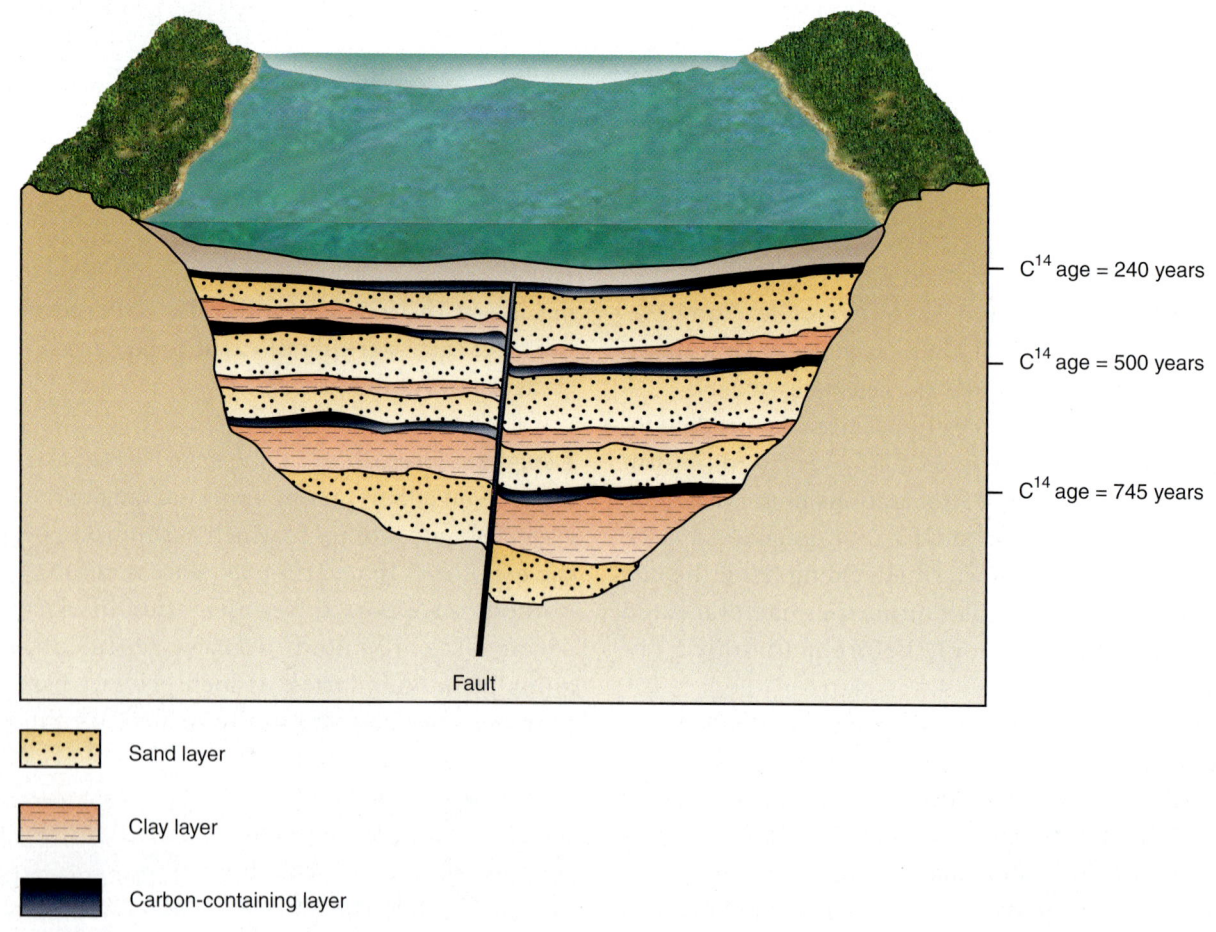

Earthquake disaster FIGURE 8.7

Some of the 5,000,000 people made homeless by the Great Sichuan Earthquake of 2008 managed to escape with a few belongings.

DESIGNING FOR EARTHQUAKE SAFETY

It is important to realize that the most powerful quakes are not necessarily the deadliest. The death toll depends to a great extent on the population of the affected region and how well prepared they are for a major quake. The Great Sichuan Earthquake, May 12, 2008, in China, was a much greater disaster than it need have been, because many buildings, especially schools, had not been constructed to withstand earthquakes, even though the region was known to be prone to quakes (FIGURE 8.7). Every earthquake hazard, from fires to tsunamis, can be reduced in severity (though not eliminated) by proper planning and preparedness (see FIGURE 8.8 on the following page). For example, skyscrapers and bridges can be built with reinforced concrete and large counterweights to help them resist shaking.

Some of the tragic loss of life in the tsunami of 2004 was especially preventable because one to two hours elapsed between the time of the earthquake off Sumatra and the arrival of the waves in Sri Lanka and India. Early warning systems would probably not have helped Sumatra or Thailand, because they were so close to the tsunami's source, but an early warning system could have given people in more distant locations the opportunity to move to higher ground.

CONCEPT CHECK **STOP**

How does the elastic rebound theory explain the violent tremors that occur during earthquakes?

How do scientists predict earthquakes?

What are some of the primary and secondary hazards associated with earthquakes?

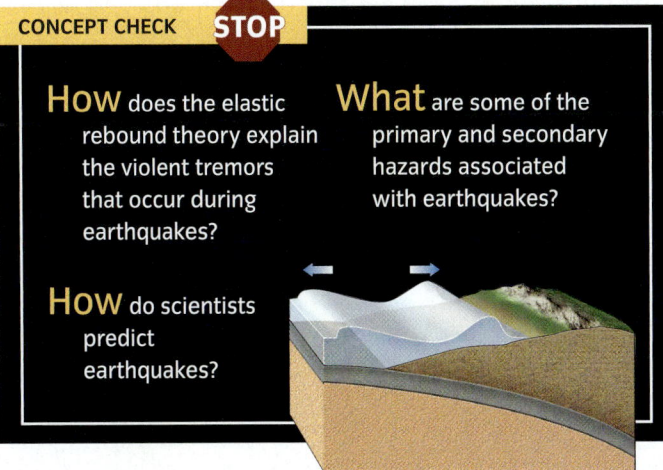

Earthquakes and Earthquake Hazards

UTILITY LINES RUPTURE
Major quakes can cut unprotected electric power lines, water mains, and gas lines, adding fire and flooding to a city's miseries.

HOUSES FALL
Older wood-frame houses are shaken off their foundations, sending chimneys, roofing, and unsecured items toppling onto residents.

MID-RISES COLLAPSE
The most vulnerable structures are often unreinforced masonry mid-rises, whose rigid walls tend to crack and crumble rather than sway.

HIGHWAYS BUCKLE
Roads and bridges give way under relentless shaking and swaying. Even where decks don't collapse, damage can slow rescue efforts.

Building for protection FIGURE 8.8

A Engineers and scientists are collaborating to find ways to make the built environment safer during earthquakes.

Building for Protection

Earthquake-resistant methods and materials have consistently improved since serious study of earthquakes began in the early 1900s. In many of the world's at-risk urban centers, high-rises and homes now regularly weather seismic events and stand firm as the ground rolls around them. In cities that enforce strict construction standards, new structures of all kinds—bridges, tunnels, stadiums—are designed from the start to withstand at least some shaking. But experts still worry about those built decades ago, before quake resistance became common. For such structures, engineers have developed a host of innovations that help hold things together when quakes strike.

1 HOUSES
Several retrofits can help homeowners outlast an earthquake. Wood-frame houses should be bolted to their foundations; in masonry homes, walls should be reinforced with fiber mesh and tied to each other and the roof. Water heaters and other appliances can be strapped or bolted down to prevent tipping, and cabinet doors can be latched shut to keep their contents from spilling out.

2 SUBWAY LINES
Tunnels are most vulnerable where they run from one ground material into another, say from rock to soil. Flexible joints let tunnels bend with quake motion and resist breaking.

3 MID-RISES
Engineers can reinforce the masonry walls of mid-rises with sprayed concrete. They can also isolate the structure from its foundation, hoist it onto steel and rubber pads, and insert dampers to absorb shock and steady the building.

4 UTILITY LINES
Plunging underground like concrete veins, common utility ducts (CUDs) carry lifelines like water pipes and electricity cables. CUDs move with the ground during a quake, reducing damage, and allow for quicker repairs.

5 HIGHWAYS AND BRIDGES
Driving micropiles—long, pipe-like anchors—through bridge foundations boosts stability. Bridge columns are strengthened by encasing them in steel jackets or fiber mesh. Roads can be preserved by anchoring loose earth to rock on shoulders, reducing slides.

6 HIGH-RISES
High-rises are buttressed with braces and shock absorbers bolted to inner steel skeletons. This allows movement but prevents catastrophic swaying.

HIGH-RISES RATTLE Designed to sway, skyscrapers generally don't collapse but do sustain damage. Windows explode, girder welds crack, fires erupt.

BUILDINGS SINK When soil is saturated, quakes can turn solid ground into a molasses-like mix that causes buildings to lean or even topple.

4 PROTECTED UTILITY LINES

5 REINFORCED COLUMNS

B Japanese schoolchildren practice what to do in case of an earthquake.

C Apartment buildings in Niigata, Japan, fell over after an earthquake caused water-saturated sediment to shake and lose strength, a process known as *liquefaction*.

D A major quake in Kobe, Japan, caused highways and bridges to buckle and collapse.

The Science of Seismology

LEARNING OBJECTIVES

Explain how a seismograph works.

Define body waves and surface waves.

Identify the two kinds of body waves and explain how they differ.

Explain how seismologists locate the epicenter of an earthquake.

Describe the Richter and moment magnitude scales.

Seismologists can quickly locate an earthquake anywhere on Earth and tell how strong it is. They are also very good at telling the difference between earthquakes caused by natural movements and other seismic disturbances, such as explosions and landslides. The device used to study all aspects of earthquakes is a *seismograph*.

SEISMOGRAPHS

The earliest known **seismographs** (also called *seismometers*) were invented in China in the second century (see **FIGURE 8.9A**). The first seismographs in Europe were invented much later, in the 19th century. Modern seismographs provide a printed or digital record of seismic waves, called a **seismogram** (see **FIGURE 8.9B**).

The most advanced seismographs measure the ground's motion optically and amplify the signal electronically. Vibrations as tiny as one hundred-millionth (10^{-8}) of a centimeter can be detected. Indeed, many instruments are so sensitive they can sense vibrations caused by a moving automobile many blocks away.

> **seismograph** An instrument that detects, measures, and records vibrations of Earth's surface.

> **seismogram** The record made by a seismograph.

Ancient and modern seismographs FIGURE 8.9

▲ **A** Ancient Chinese seismograph.

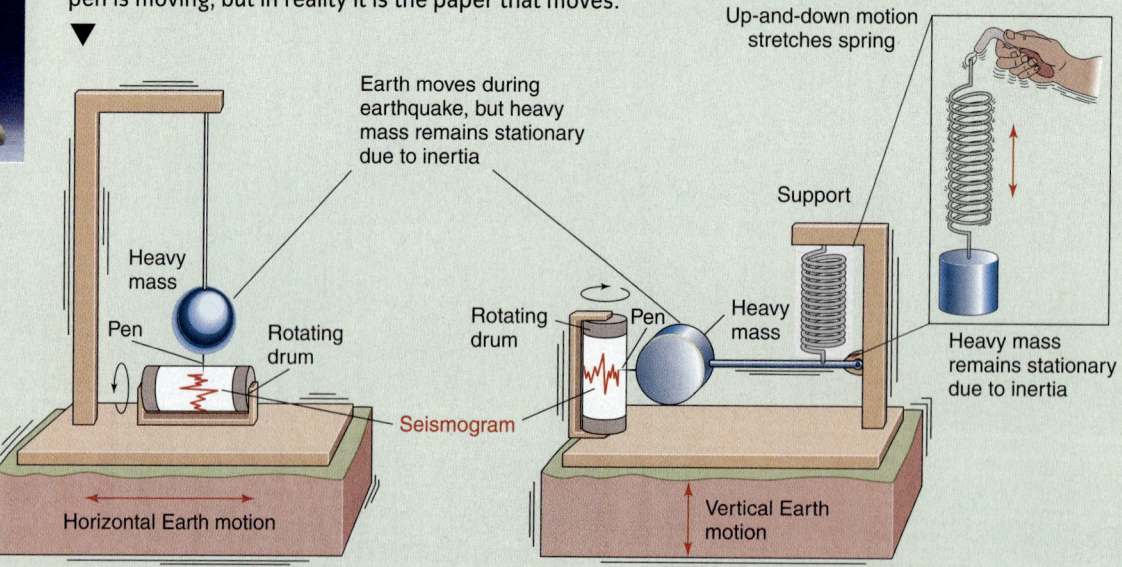

B Seismographs use the principle of *inertia*—the resistance of a mass to motion. In this schematic diagram, seismic waves cause the support post and the roll of paper to vibrate back and forth. However, the large mass attached to the pendulum, and the pen attached to it, barely move at all. It looks to an observer as if the pen is moving, but in reality it is the paper that moves.

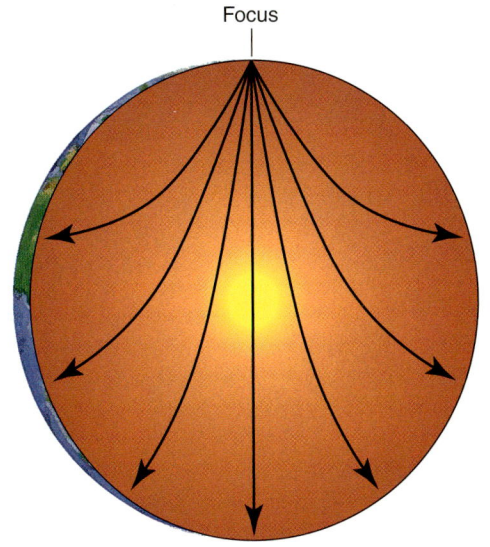

Travel paths of seismic body waves FIGURE 8.10

The energy released during an earthquake travels through Earth from its source (the focus). If Earth were of uniform density throughout, the waves would travel in straight lines. However, rock density increases with depth as a result of increasing pressure. Seismic waves travel faster through denser rocks; hence, they travel more quickly at greater depths. This increase in velocity with depth causes seismic wave paths to be curved, rather than straight. (This diagram is not completely accurate, because the increase in rock density and seismic velocity with depth is not smooth; you will see a more detailed diagram later in the chapter.)

SEISMIC WAVES

body wave A seismic wave that travels through Earth's interior.

surface wave A seismic wave that travels along Earth's surface.

focus The location where rupture commences and an earthquake's energy is first released.

P wave The first, or primary, wave to be detected by a seismograph.

compressional wave A seismic body wave consisting of alternate pulses of compression and expansion in the direction of travel; P wave or primary wave.

S wave The second kind of body wave to be detected by a seismograph.

shear wave A seismic body wave in which rock is subjected to side-to-side or up-and-down forces, perpendicular to the direction of travel; S wave or secondary wave.

The energy released by an earthquake is transmitted to other parts of Earth in the form of seismic waves. The waves move as elastic deformations of the rocks; they leave no record behind them once they have passed, so they must be detected while they pass. The waves, which include both **body waves** and **surface waves**, travel outward in all directions from the earthquake's **focus** (see FIGURE 8.10).

Body waves Following an earthquake, a seismograph will record two kinds of *body waves*; they travel at different speeds so there is a time gap between their arrivals. The first set of waves to arrive and be detected by a seismograph are called **P waves** (or primary waves). P waves are **compressional waves** (see FIGURE 8.11A on page 248); they are like sound waves and can pass through solids, liquids, and gases. They have the highest velocity of all seismic waves—typically 6 km/s in the uppermost portion of the crust.

The second waves to reach and be recorded by a seismograph after earthquakes are called **S waves** (or secondary waves). They travel through solid materials by an undulating, or shearing motion (see FIGURE 8.11B). Solids tend to resist a shear force and bounce back to their original shape afterward, whereas liquids and gases do not. Without this elastic rebound, there can be no wave. Therefore, **shear waves** cannot be transmitted through liquids or gases. This has important consequences for the interpretation of seismic waves, as you will soon see. Shear waves travel only 3.5 km/s—not as fast as compression waves.

Surface waves In addition to body waves that travel through Earth, earthquakes generate surface waves that travel along or near Earth's surface, like waves along the surface of the ocean. They travel more slowly than P and S waves, and they pass around Earth rather than through it. Thus, surface waves are the last to be

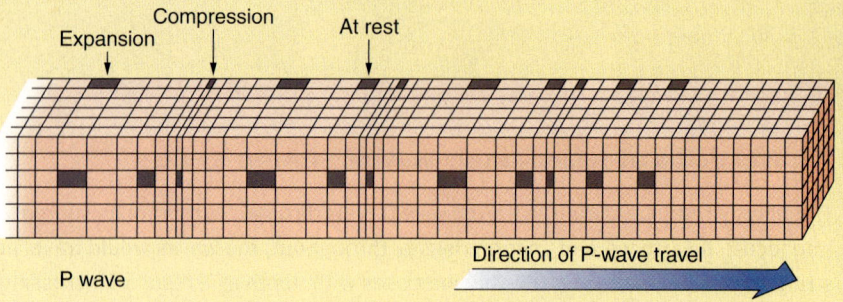

A A compressional wave alternately squeezes and stretches the rock as it passes through. The grid is intended to help you visualize how the rock responds. All the divisions in the grid start out square, but the wave alternately squeezes them down to narrow rectangles and then stretches them to long rectangles. Sound waves travel through air by the same means—alternate compressions and expansions of air.

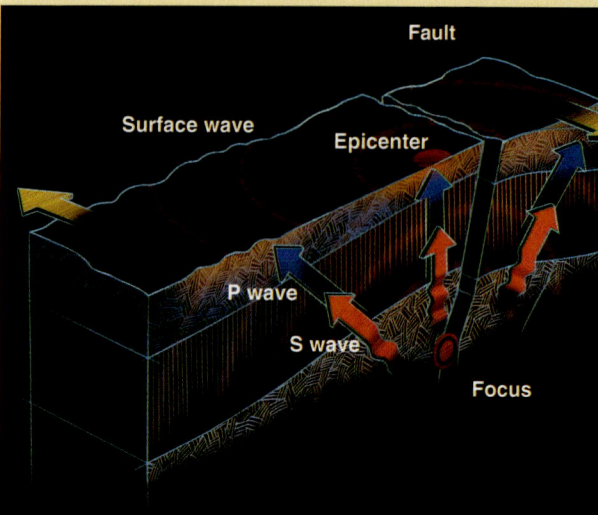

C P and S waves travel outward from the focus, generating waves that travel along the surface.

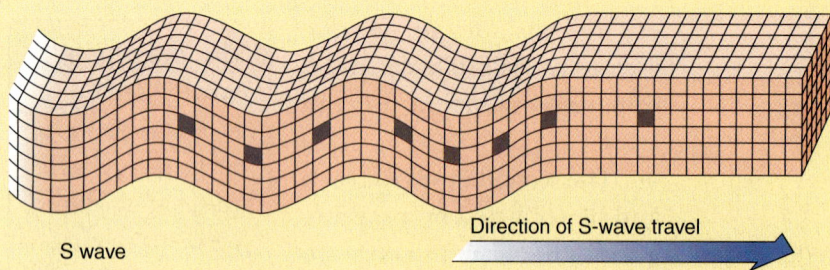

B A shear wave causes the rock to vibrate up and down, like a rope whose end is being shaken. In this case, the squares do not expand or contract but do get distorted, changing shape alternately from a square to a parallelogram and back to a square again.

Seismic body waves FIGURE 8.11

detected by a seismograph. FIGURE 8.12A shows a typical seismogram, in which the P waves' arrival is first, followed by the arrival of the S waves, and finally by that of the surface waves. There are two kinds of surface waves, and both are named for the English mathematicians who first predicted their existence. Rayleigh waves, named for Lord Rayleigh, cause Earth's surface to move up and down, like a wave on the ocean. Love waves, named for A. E. H. Love, cause the surface to shake in a sideways motion, but do not cause any vertical motion. Surface waves are responsible for much ground shaking and structural damage during major earthquakes.

LOCATING EARTHQUAKES

The **epicenter**, or surface location, of an earthquake can be determined through simple calculations using travel-time curves from at least three seismographs that have recorded the quake. The first step is to find out how far each seismograph is from the source of the earthquake. The greater the distance traveled by the seismic waves, the more the S waves will lag behind the P waves. Thus, the lag time between the P and S waves on a seismogram (see FIGURE 8.12B) provides seismologists with the necessary distance information.

After determining the distance from each seismograph to the source of the earthquake, the seismologist draws a circle on a map, with the seismic station at the center of the circle. The radius of the circle is the distance from the seismograph to the focus. It is a circle because the seismologist knows only the distance, not the direction. When this information is calculated and plotted for three or more seismographs, the unique point on the map where the three circles intersect is the location of the epicenter (see FIGURE 8.12C). This process is called *triangulation*.

epicenter The point on Earth's surface that is directly above an earthquake's focus.

Using seismograms to locate an earthquake   FIGURE 8.12

A Seismogram of a typical earthquake.

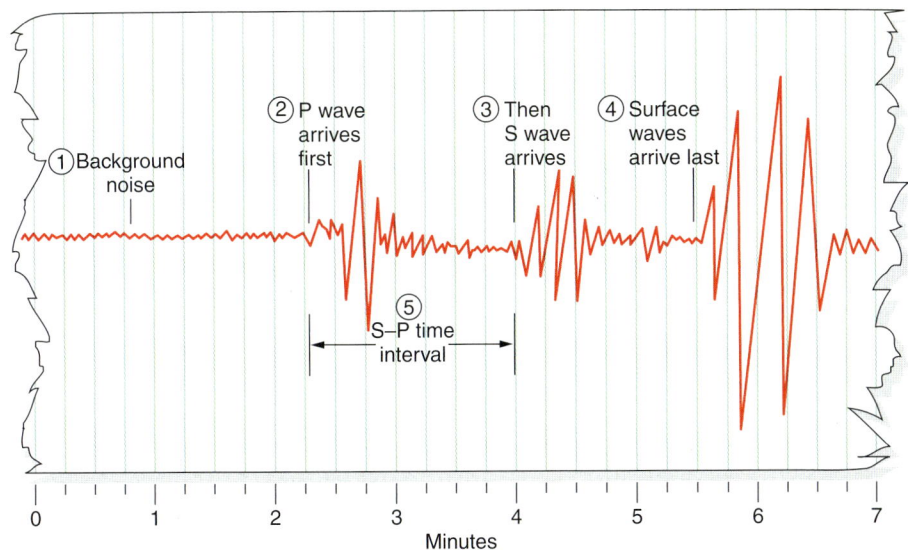

① The earthquake happens at time 0.

② The first P waves arrive a little over 2 minutes later.

③ The first S waves arrive 4 minutes later.

④ The surface waves, which travel the long way around Earth's surface, arrive last.

⑤ The S–P interval, here slightly less than 2 minutes, tells the seismologist how far away the earthquake was.

▼ **B** P and S waves leave the focus of an earthquake at the same instant. The fast-moving P waves reach a seismograph first, and sometime later the slower-moving S waves arrive. The delay in arrival time increases with distance traveled. Average travel-time curves are used to locate an epicenter. For example, when seismologists at a station measure the S–P time interval to be 13.7 min – 7.4 min = 6.3 min, they know the epicenter is 4000 km away from the station.

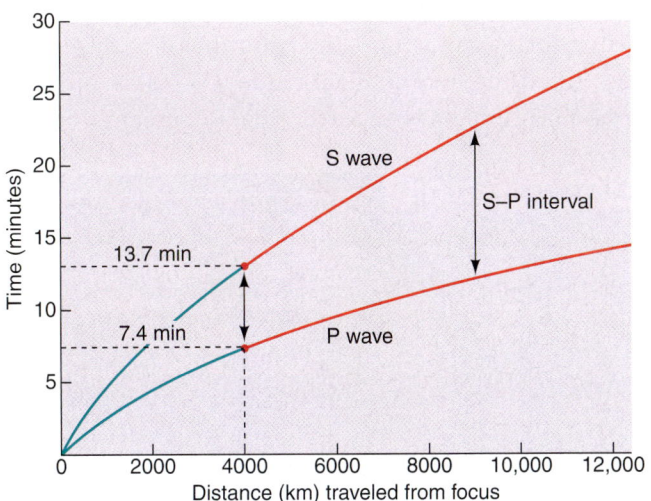

▲ **C** The method of triangulation. If three seismic stations—shown here in Stockholm, Honolulu, and Manila—record an earthquake, each one can independently determine its own distance from the focus of the quake, thus generating a circle on which the epicenter of the earthquake must lie. The three circles have a unique intersection point, which is the location of the epicenter: in this case Kobe, Japan (the site of a major earthquake in 1995). The black arrows indicate the radius of the circle on which each station estimated the epicenter must lie.

MEASURING EARTHQUAKES

Earth scientists use several different scales to quantify the strength or *magnitude* of an earthquake, by which we mean the amount of energy released during the quake. The earliest scale, developed by an Italian scientist in 1902 and later modified, is called the *Modified Mercalli Intensity Scale*. The scale is based on descriptions of vibrations that people felt, saw, and heard and on the extent of damage to buildings. The scale ranges from I (not felt, except under favorable circumstances) to XII (waves visible, practically all buildings destroyed). The Mercalli intensity of an earthquake varies with distance from the epicenter—an earthquake could have an intensity of X near the epicenter whereas a hundred kilometers away the intensity could be only II. The Modified Mercalli Scale is most useful for studying the earthquakes that happened before the development of modern seismic equipment.

Richter magnitude scale A scale of earthquake intensity based on the heights, or amplitudes, of the seismic waves recorded on a seismograph.

The most familiar of the modern intensity scales is the **Richter magnitude scale**. It has been superseded for most purposes by the *moment magnitude*, which uses the same scale but is computed in a somewhat different way.

The Richter magnitude scale

Charles Richter developed his famous magnitude scale in 1935. Though it was not the first earthquake intensity scale, it was an important advance because it was the first to use data from seismographs rather than subjective estimates of damage. Also, it compensated for the distance between the seismograph and the focus. This means that each seismic station will (in principle) calculate the same magnitude for a given earthquake, no matter how far from the epicenter it may be located. As discussed above, a scale such as the Modified Mercalli Scale, based on estimates of the damages sustained, yields the highest magnitude closest to the epicenter, where the damage is greatest.

The Richter scale is *logarithmic*, which means that each unit increase on the scale corresponds to a 10-fold increase in the amplitude of the wave signal. Thus, a magnitude 6 earthquake has an amplitude 10 times larger than that of a magnitude 5 quake. A magnitude 7 earthquake has an amplitude 100 times larger (10×10) than that of a magnitude 5 quake. However, even this comparison understates the difference, because the amount of damage done by an earthquake is more closely related to the amount of energy released in the quake. Each step in the Richter scale corresponds roughly to a 32-fold increase in energy (see *What an Earth Scientist Sees: Richter Magnitude: A Logarithmic Scale*). The actual damage done by a quake will also depend, of course, on local conditions, such as how densely populated the area is.

Moment magnitude

Seismologists today determine magnitudes using both the Richter scale and **moment magnitude**, which are calculated using different starting assumptions. Richter scale calculations are based on the assumption that an earthquake focus is a point. Therefore, the Richter scale is best suited for earthquakes in which energy is released from a relatively small area of a locked fault. In contrast, the calculation of seismic moment takes account of the fact that energy may be released over a large area. A classic example was the Sumatra-Andaman earthquake of 2004, when a 1200-km length of fault moved. Though the method of calculation is different, the scales are the same because they measure the same thing—the amount of energy released. In either system, magnitude 9 is catastrophic, whereas magnitude 3 is imperceptible to humans.

moment magnitude A measure of earthquake strength based on the rupture size, rock properties, and amount of displacement on the fault surface.

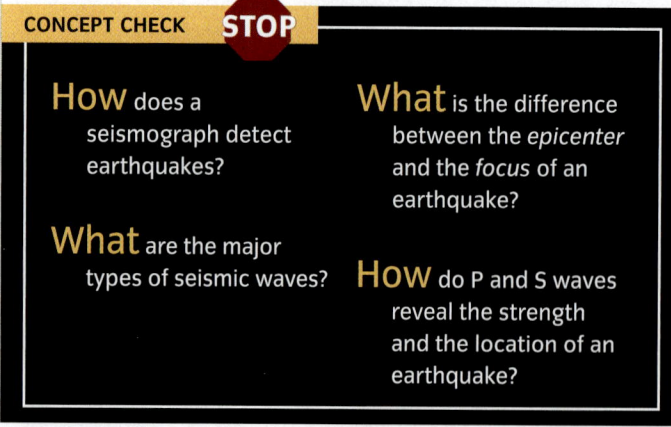

CONCEPT CHECK STOP

How does a seismograph detect earthquakes?

What are the major types of seismic waves?

What is the difference between the *epicenter* and the *focus* of an earthquake?

How do P and S waves reveal the strength and the location of an earthquake?

Richter Magnitude: A Logarithmic Scale

The energy released in an earthquake increases exponentially with its magnitude. A magnitude 6 earthquake releases as much energy as the atomic bomb dropped on Hiroshima, the largest ever used in combat. A magnitude 7 quake would be equivalent in energy to about 32 Hiroshima bombs, and a magnitude 8 quake would be equivalent to 32×32, or about 1000, of them. A magnitude 9 quake, such as the Sumatra-Andaman quake, is equivalent to $32 \times 32 \times 32$, or about 32,000 bombs. An Earth scientist would confirm the magnitude of an earthquake determined from a seismogram by the degree of damage close to the epicenter.

◀ **Richter magnitude 6**
Damage on surface close to the epicenter: small objects broken, sleepers awakened (Mercalli intensity ≈VII)
Energy released: about the same as one atomic bomb

Parkfield, CA, 2004

Richter magnitude 7 ▶
Damage on surface close to the epicenter: some walls fall, general panic (Mercalli intensity ≈IX)
Energy released: about the same as 32 atomic bombs

San Francisco, CA, 1906

Kobe, Japan, 1995

◀ **Richter magnitude 8**
Damage on surface close to the epicenter: wide destruction, thousands dead (Mercalli intensity ≈XI)
Energy released: about the same as 1000 atomic bombs

Here's an interesting question:
- What might cause there to be a limit to the largest possible magnitude for an earthquake, or is there no limit to the magnitude of an earthquake?

What an Earth Scientist Sees

Looking into Earth's Interior

LEARNING OBJECTIVES

Explain how the materials in Earth's interior affect seismic waves.

Explain why seismic data point to the existence of a liquid core.

Discuss the limitations of direct sampling of Earth's interior.

Identify several ways in which scientists can study Earth's interior indirectly or remotely.

Although earthquakes are significant to society because of the damage they cause, they also have benefits from a scientific perspective. They provide us with some of our most detailed information about Earth's interior—including parts that we can never hope to observe directly.

When scientists cannot study something by direct sampling, a second method comes to the forefront: indirect study or *remote sensing*. Some familiar objects—including the human eye—are actually remote-sensing devices. A camera, for instance, is a remote sensing instrument that collects information about how an object reflects light. Medical techniques such as X-rays allow doctors to study the inside of the body remotely without opening it up surgically.

The seismic waves from an earthquake are much like X-rays, in the sense that they enter Earth near the surface, travel all the way through it, and emerge on the other side. They travel along different paths depending on the different kinds of materials they encounter. We will first discuss what earthquakes reveal about Earth's structure and then describe other sources of information about Earth's interior.

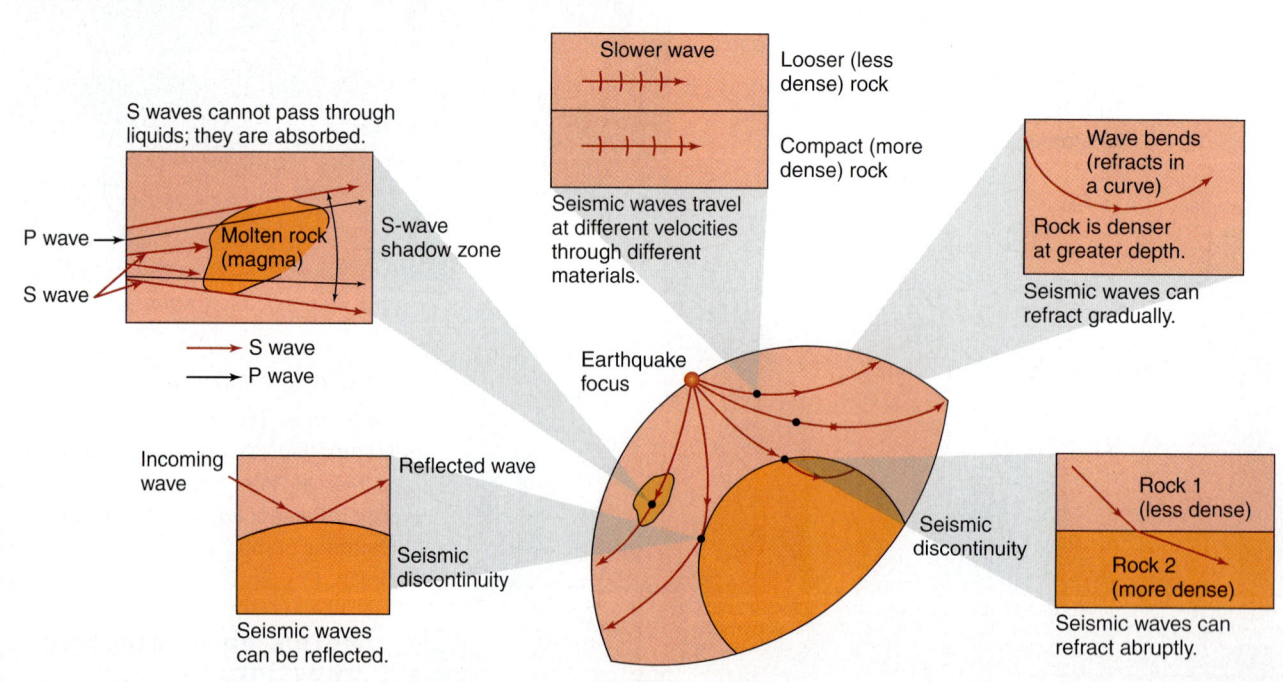

Seismic waves in Earth's interior FIGURE 8.13

Process Diagram

VIEW THIS IN ACTION in your WileyPLUS course

252 CHAPTER 8 Earthquakes and Earth's Interior

HOW EARTH SCIENTISTS LOOK INTO EARTH'S INTERIOR—SEISMIC METHODS

Before 1906, scientists' understanding of seismic waves was limited. However, in that year, British scientist Richard Dixon Oldham first identified the difference between P waves and S waves and then suggested an explanation for the complicated patterns recorded after an earthquake: Underneath thousands of kilometers of solid rock, he postulated, Earth has a liquid core. Oldham's theory, now universally accepted by Earth scientists, is illustrated in **FIGURE 8.13**.

When a major earthquake strikes, the tremors are measured at seismic stations around the world, and the arrival times of P waves and S waves can be used to analyze the kinds of rock that the seismic waves passed through. Even a crude diagram of Earth's structure shows that the pattern of arrival times is quite complex (as shown in Figure 8.13).

How do seismic waves tell us what is inside Earth? Because Earth's interior is not homogeneous, there are boundaries between different materials. Seismic waves behave differently, depending on the properties of the materials they pass through—most importantly, whether the material is a liquid or a solid. Three distinct things can happen to seismic waves when they meet such a boundary, called a **seismic discontinuity** (see Figure 8.13):

> **seismic discontinuity**
> A boundary inside Earth where the velocities of seismic waves change abruptly.

1. They can be refracted, or bent, as they pass from one material into another. This is the same thing that happens to light waves when they pass from air to water.

2. They can be reflected, which means that all or part of the wave energy bounces back, like light from a mirror.

3. They can be absorbed, which means that all or part of the wave energy is blocked.

The combination of the various behaviors (left) creates a complex pattern of arrival times of seismic waves at distant locations following an earthquake. Note especially the *shadow zones*, where P waves and S waves are not observed.

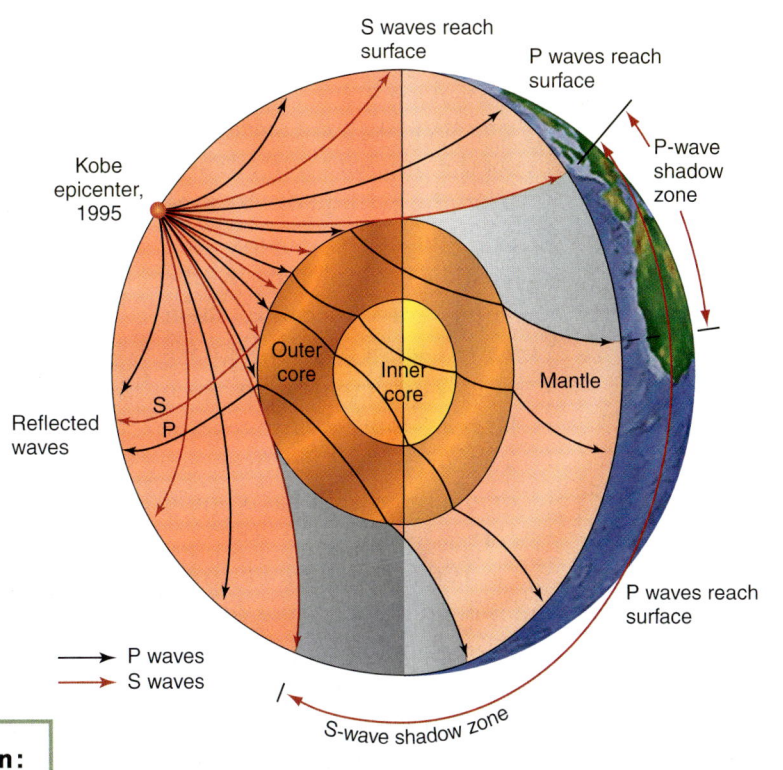

CRITICAL THINKING — Here's an interesting question:
- A seismometer located in the P-wave shadow zone would actually record a P wave, but it would have a delayed arrival. Why?

Beneath the surface FIGURE 8.14

Seismologists don't always have to wait for earthquakes to be able to study Earth's interior. For shallow, near-surface studies, they can use explosive charges (or large "thumper trucks") as the source for seismic waves, and then map out the underground rock layers with seismic tomography. Here, two oil-prospecting seismologists in Texas are looking at seismic images made from data collected in Alaska.

Refraction, **reflection**, and *absorption* all play a role in Oldham's model (see Figure 8.13). P waves are bent dramatically when they pass from the mantle to the outer core. This bending creates a ring-shaped *P-wave shadow zone* on the opposite side of Earth from the earthquake. S waves, on the other hand, are blocked completely by the outer core, because shear waves cannot pass through liquid. This creates an even larger *S-wave shadow zone*, as well as providing firm evidence that Earth has a liquid core.

- **refraction** The bending of a wave as it passes from one material into another material, through which it travels at a different speed.

- **reflection** The bouncing back of a wave from an interface between two different materials.

If this picture looks complicated to you, imagine how confusing it was to Oldham and his contemporaries, who were figuring it out for the first time; the results are a nice example of the scientific method at work. We have presented the story *deductively*—that is, proceeding from a model to a conclusion. However, Oldham proceeded *inductively*—that is, he began with observed effects and inferred a hypothesis that explained them. This is much more difficult to do, and justifiably conferred fame on him when he succeeded.

Seismic tomography

As more sophisticated seismic equipment was developed and scientists began to make more detailed observations, they discovered more boundaries and layers within Earth. Today, seismologists use seismic waves to probe Earth's interior in much the same way that a doctor uses X-rays and CAT scans to probe the interior of a human body. In CAT (*computer-aided tomography*) scanning, a series of X-rays along successive planes can be used to create a three-dimensional picture of the inside of the body. Similarly, *seismic tomography* allows seismologists to superimpose many two-dimensional seismic snapshots to create a three-dimensional image of the inside of Earth. These techniques have also helped us to understand more about how plate tectonics works. They also allow scientists to map the locations of seismic discontinuities, the distribution of hot and cold masses, and the distribution of dense and less dense materials inside Earth (see **FIGURE 8.14**).

Seismic discontinuities Oldham's discovery of the boundary between the liquid outer core and the mantle was the first seismic discontinuity to be identified and explained. The boundary that separates the crust from the mantle was the next discontinuity to be discovered. It was named the *Mohorovičić discontinuity* after the seismologist who discovered it in 1909, but it is usually called the *Moho* for short. Mantle rocks, being denser and compositionally different from crustal rocks, transmit P waves much more quickly. The Moho is thus an example of a boundary between two layers of rock that have different compositions and densities but similar physical characteristics (rigidity).

The mantle extends from the Moho to the core and accounts for about 80% of Earth's volume. All the available evidence suggests that its composition is uniform throughout. Nevertheless, studies have revealed several seismic boundaries within the mantle. Extending from a depth of 100 to about 350 km below the surface is a region of distinctly low seismic velocities (see **FIGURE 8.15**). The low-velocity zone coincides with the *asthenosphere,* a region where temperature and pressure are such that rock is ductile and has little strength. The upper boundary of the asthenosphere is the base of the *lithosphere,* which is marked by a distinct change in rock properties, from weak and ductile in the asthenosphere, to strong and brittle in the lithosphere.

The *mesosphere* is the portion of the mantle that stretches from 350 km to the core–mantle boundary. Although temperatures in the mesosphere are very high, the rocks are a bit stronger than in the asthenosphere because they are so highly compressed. Additional seismic discontinuities exist within the

Seismic discontinuities in the mantle FIGURE 8.15

Earth's mantle has several seismic boundaries within it. We know the boundaries exist because P and S waves slow down or speed up abruptly and are refracted or reflected at these boundaries. The boundaries do not seem to be caused by compositional layering, but the exact nature of each boundary is still not completely understood.

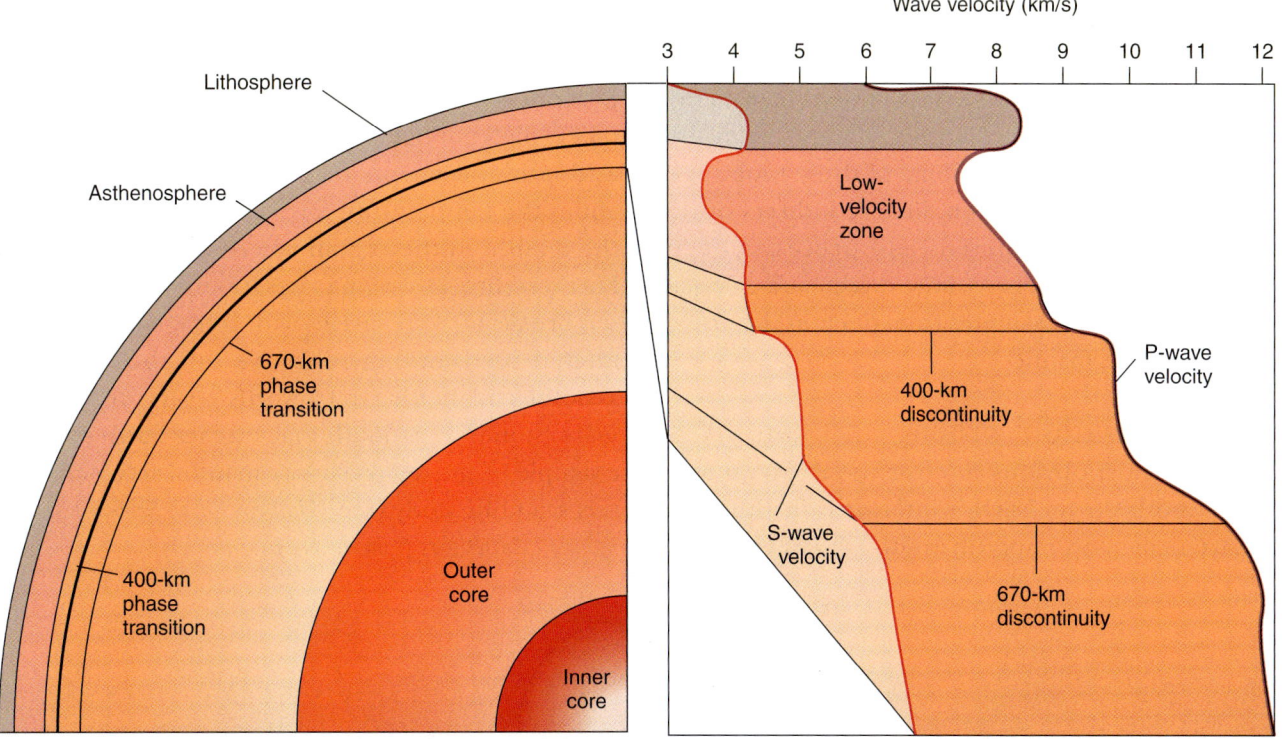

Looking into Earth's Interior

mesosphere, with transitions at about 400 km and at about 670 km below the surface (see Figure 8.15). These discontinuities are not well understood; they seem to result from changes in physical properties. For example, when the mineral olivine is squeezed at a pressure equal to that found at a depth of 400 km, the atoms rearrange themselves into a more compact structure or *polymorph* of olivine (see Chapter 2). Perhaps this change to a more compact form causes the 400-km seismic discontinuity. It is important to remember that the mantle is solid rock, because both P waves and S waves can travel through it. Nevertheless, pressures and temperatures deep within Earth are so high that even solid rock can flow, in very, very slow convection currents, as described in Chapter 7. The mechanism of flow is similar to that of ice flowing in a glacier, and the rate of flow is roughly the same as the rate of growth of a human fingernail. Seismologists have recently detected hot regions in the mantle that may coincide with the rising limbs of convection cells that help drive plate motion.

HOW EARTH SCIENTISTS LOOK INTO EARTH'S INTERIOR— OTHER METHODS

Earthquakes have provided a great deal of information about Earth's interior, but Earth scientists have many other tools and techniques that allow them to study the deepest parts of our planet. Some of these tools, like the use of seismic information, involve indirect or remote observation of materials and processes deep within the planet. Others are more direct, and give scientists access to actual samples from deep in the crust, and even from the mantle.

Direct observation: drilling and xenoliths

Perhaps the most obvious tool for the retrieval and study of samples from Earth's interior is drilling. To date, Earth's deepest mine (in South Africa) is 3.6 km deep, and the deepest hole ever drilled (in the Kola Peninsula of Russia) reached a depth of just over 12 km. Recall from Chapter 1 that Earth's crust varies from an average thickness of 8 km for oceanic crust to an average of 45 km for continental crust. Therefore, a 12-km hole sounds just about right for sampling the top part of the mantle—or does it?

The problem with drilling is that areas where the crust is thin tend to have high heat flow. In other words, if you try to drill a hole through thin oceanic crust all the way to the mantle, you will quickly encounter temperatures that could destroy your drilling equipment. Another problem with oceanic crust is that it's deep underwater, which makes drilling more difficult. Thus, the only place where the rocks are both accessible and cool enough for drilling to great depth is precisely where the crust is very thick—on the continents. The hole in the Kola Peninsula went through more than 12 km of thick continental crust and never even came close to reaching the mantle. Thus, although drilling has yielded much interesting and useful information about the composition and properties of the crust, it hasn't even reached Earth's shallowest seismic discontinuity.

If we can't obtain samples from deep within Earth by reaching in to retrieve them, perhaps we can wait for them to come to us. This does happen, in two different ways. Molten rock, or *magma*, is formed in the upper portions of Earth's mantle, in areas where the temperature is high enough. By studying magma that originates at depth and erupts to the surface, scientists learn about the temperature, pressure, and composition of the mantle in the region where the magma formed. Furthermore, as magma rises toward the surface, it often breaks off and carries with it fragments of the unmelted surrounding rock. We call these fragments **xenoliths**, from the Greek words *xenos* (foreigner) and *lithos* (stone). A xenolith that reaches Earth's surface is a sample of the deep crust or mantle, accessible for direct scientific study (see *What an Earth Scientist Sees: Diamonds: Messengers from the Deep*).

What an Earth Scientist Sees

Diamonds: Messengers from the Deep

A Diamonds form at the extremely high pressures found at depths of 100 to 300 km. The different colors in this diamond—nicknamed the "Picasso" diamond—show various zones of growth. The diamond is approximately 1 millimeter across, and the colors are revealed by a special type of photography that highlights small variations in composition. An observant Earth scientist would discover that within a diamond there are tiny inclusions of other minerals, such as olivine, garnet, and graphite, that were trapped as the diamond grew. These tiny inclusions are pristine samples from deep in the mantle. ▼

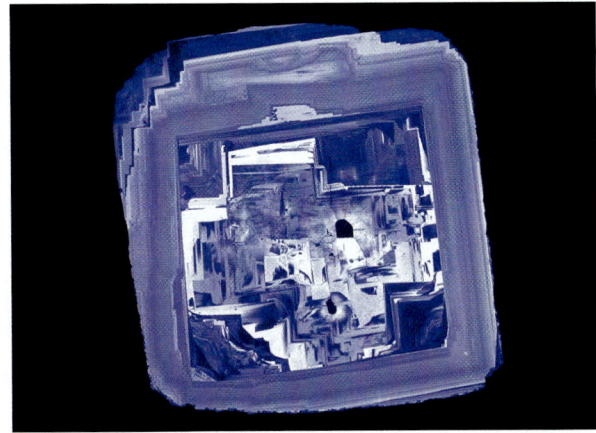

B The beautiful uncut yellow diamond is the 253.7-carat Oppenheimer Diamond from the Smithsonian Institution; it was discovered in South Africa in 1964. ▼

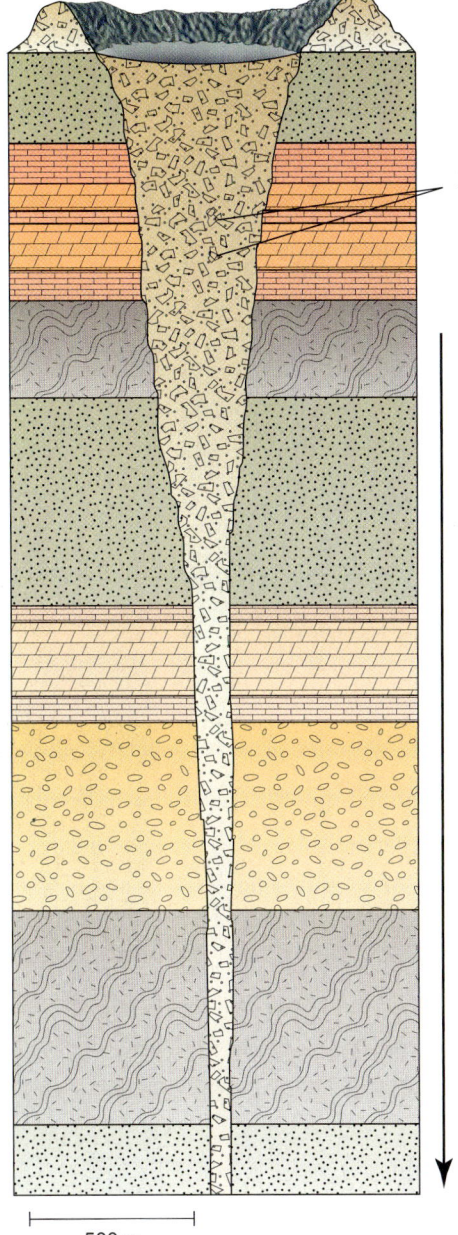

Magma vent is circular when viewed from above.

Xenoliths of mantle rock

Pipe extends 150–200 km down into mantle

500 m

C To reach the surface from such great depths, diamonds must be carried by an eruption of unusual ferocity. These eruptions leave behind a long, cone-shaped tube of solidified magma, called a *kimberlite pipe*. Although most people treasure diamonds for their beauty and luster, Earth scientists treasure them also as "messengers"—samples from an otherwise inaccessible region of Earth's interior.

Here's an interesting question:
- How might you distinguish a natural diamond from a synthetic one grown in the laboratory?

A Earth is surrounded by a magnetic field, which causes a compass needle to point north. More precisely, the needle is aligned along the field lines that lead to the north and south magnetic poles, which are almost—but not exactly—aligned with Earth's north (N) and south (S) geographic poles.

B This is a photograph of the aurora borealis, or northern lights, as seen from Fairbanks, Alaska. This phenomenon is caused by charged particles from the Sun entering Earth's atmosphere at high latitudes along magnetic field lines.

Earth's magnetic field FIGURE 8.16

Indirect observation: methods from physics, astronomy, and chemistry The availability of indirect or remote techniques for the study of Earth's interior has increased considerably since the dawn of the Space Age. Many of these techniques are also used to study other planets in the solar system. We have already discussed the single most important method for the study of Earth's interior: the study of seismic waves generated by earthquakes. Another important geophysical method relies on the study of Earth's magnetic field.

Magnetism is a force created either by permanent magnets (ferromagnets) or by moving electrical charges. We can try thinking of Earth as having a huge dipole bar magnet with north and south poles at its center, offset slightly from the geographic North and South Poles. The problem with this analogy is that solids, including bar magnets, lose their magnetism at temperatures above a critical transition temperature, called the *Curie point*, which is specific to each material. The Curie point for iron is about 770°C, but we know that the temperature in the core is *much* higher than this—at least 5000°C.

Since there can't be a giant bar magnet inside the planet, moving electrical charges must be responsible for generating Earth's magnetic field (see **FIGURE 8.16**). Physicists have shown that the movement of an electrically conducting liquid inside a planet could generate a self-sustaining magnetic field, much like a rotating coil of wire in an electric motor. This is consistent with the observation from seismology that at least the outer part of Earth's core is liquid. However, molten rock is not a good enough electrical conductor to generate a magnetic field in this manner; for this and other reasons, Earth scientists believe that the liquid outer core is made of molten iron and nickel. This is consistent with evidence from meteorites, discussed later.

We can also gain a certain amount of information about any planet's interior—including Earth's—from astronomical observations. The first step is to determine the planet's *mass*. This can be deduced from the planet's gravitational influence on other planets and satellites. Second, we need to know the *diameter* of the planet. Knowing the dimensions of the planet and its shape (in the case of Earth, a very slightly flattened sphere), it is a simple matter to figure out its volume and average density (mass divided by volume).

What do these kinds of measurements reveal about Earth's interior? For one thing, we can determine whether material is distributed evenly throughout the planet. The rocks at Earth's surface are very light (low-density) compared to the planet as a whole. Surface rocks have an average density of about 2.8 g/cm^3, whereas Earth's overall density is 5.5 g/cm^3. (For comparison, water has a density of 1 g/cm^3 at 4°C.) For the planet as a whole to have such a high density, with such low-density rocks at the surface, there must be a concentration of denser material somewhere inside the planet. Seismic evidence indicates that the density of the mantle is less than 4.0 g/cm^3, so the missing mass must be in the core and the density of the core must be greater than 10 g/cm^3, which is consistent with the densities of iron meteorites.

A third way to study Earth's interior is to analyze the building blocks that formed it. Planetary scientists have discovered that most (though not all) meteorites were formed at about the same time and in the same part of the solar system as Earth. Some of these meteorites are *primitive*—that is, they have remained unaffected by melting and other processes since the beginning of the solar system. These meteorites give scientists an idea of the overall composition of the solar system and its constituent bodies. Other meteorites—the *irons*, *stony-irons*, and some kinds of *stony meteorites*—show signs of melting and differentiation, and may be more representative of Earth's core and mantle. It is highly significant that a core with the composition of a typical iron meteorite (mostly iron and nickel) would bring Earth's overall density up to the observed value of 5.5 g/cm^3.

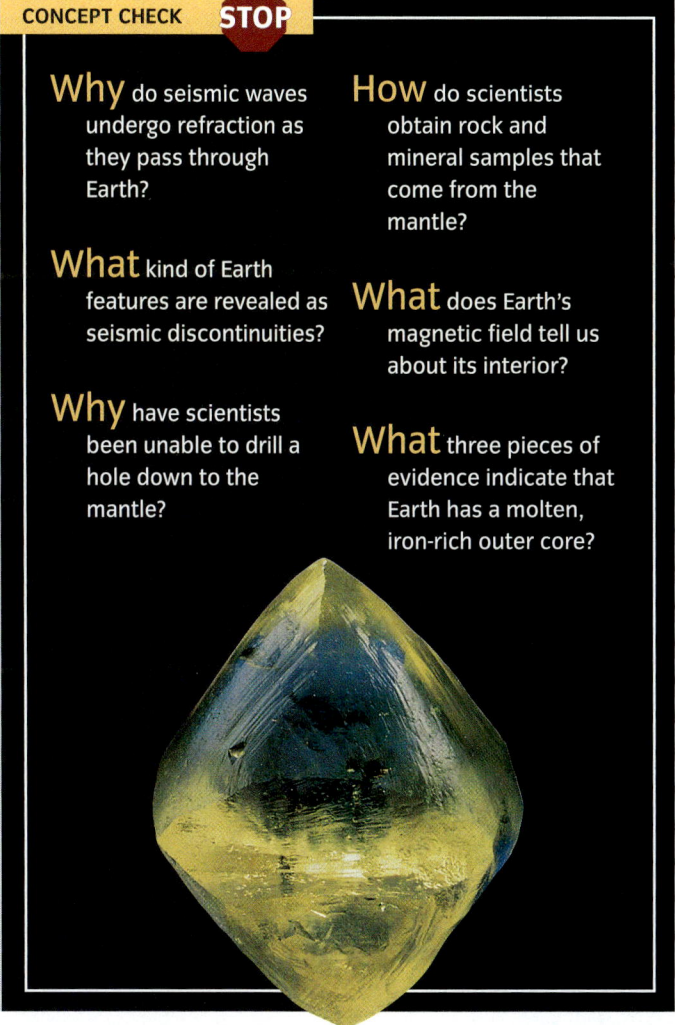

CONCEPT CHECK **STOP**

Why do seismic waves undergo refraction as they pass through Earth?

What kind of Earth features are revealed as seismic discontinuities?

Why have scientists been unable to drill a hole down to the mantle?

How do scientists obtain rock and mineral samples that come from the mantle?

What does Earth's magnetic field tell us about its interior?

What three pieces of evidence indicate that Earth has a molten, iron-rich outer core?

How and Why Rock Breaks

> **LEARNING OBJECTIVES**
>
> **Explain** the difference between stress and pressure.
>
> **Define** the three kinds of stress.
>
> **Identify** the kinds of stress associated with various types of faults.
>
> **Relate** the severity of earthquakes to different types of faults.

The elastic rebound theory says that an earthquake is caused by the storage of energy in rocks that have been bent and buckled because friction stopped motion along a fault. Faults are not all alike, so it is helpful briefly to review the different kinds of faults and the earthquakes they cause.

STRESS AND STRAIN

> **stress** The force acting on a surface, per unit area, which may be greater in certain directions than others.
>
> **pressure** A particular kind of stress in which forces acting on a body are the same in all directions.

In discussing deformation and fracture of rocks, we use the word **stress** rather than **pressure**. These two words are related in meaning but different in connotation. Both are defined as the force acting on a surface per unit area. The term *pressure*, as used in Earth science, implies that the forces on a body of rock are essentially uniform in all directions. Sometimes this is called *uniform stress* or *confining stress*. These are appropriate terms to describe, for instance, the stress on a small body immersed in a liquid such as water or magma.

Rocks, however, are solids; unlike liquids and gases, solids can resist different pressures in different directions at the same time. For this reason, *stress* is a more versatile term for discussing rock deformation, because it does not imply that the forces are necessarily the same in all directions. To be even more precise, we sometimes use the term *differential stress* when the force is greater from one direction than from another. The stresses that cause rock to fracture or change shape are differential. They can be classified into three different kinds, as illustrated in **FIGURE 8.17**: **tension**, **compression**, and **shear**.

In response to stress, a rock will experience **strain**. Uniform stress causes a change in volume only, while differential stress may cause a change in shape. For example, if a rock is subjected to uniform stress by being buried deep in Earth, its volume will decrease; that is, the rock will be compressed but it is not likely to fracture.

> **tension** A stress that acts in a direction perpendicular to and *away* from a surface.
>
> **compression** A stress that acts in a direction perpendicular and *toward* a surface.
>
> **shear** A stress that acts in a direction *parallel* to a surface.
>
> **strain** A change in the shape or volume of a rock in response to stress.

KINDS OF DEFORMATION

The way a rock responds to differential stress depends not only on the amount and kind of stress but also on the nature of the rock itself. For example, a rock may stretch like a metal spring and then return to its original shape when the stress is removed. Such a nonper-

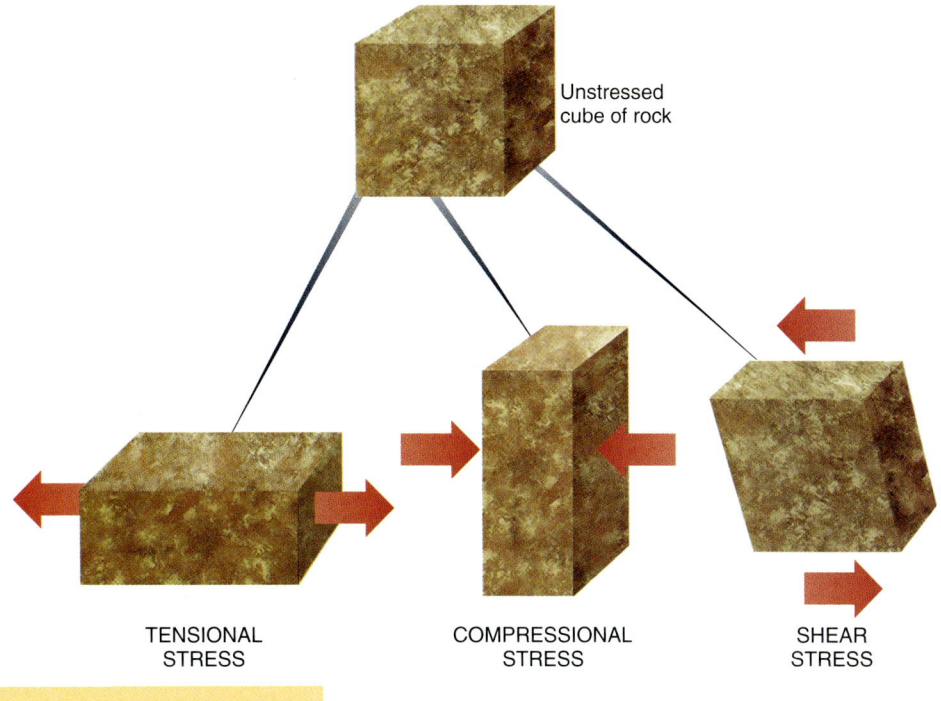

Three types of stress FIGURE 8.17

The shape of a cube of rock changes, depending on the type of stress applied to it. The arrows indicate tensional, compressional, and shear stress. Rocks that are subjected to differential stress—stress that is stronger in one direction than another—typically respond by changing their shape, as shown by these blocks.

manent change is called **elastic deformation**. For most solids, including rocks, there is a degree of stress—called the *elastic limit*—beyond which the material is permanently deformed. If the rock is subjected to more stress than this, it will not return to its original size and shape when the stress is removed.

When rocks are stretched past their elastic limit, they can deform in two different ways. **Ductile deformation**, also called *plastic deformation*, is one type of permanent deformation in a rock (or other solid) that has been stressed beyond its elastic limit. Alternatively, the rock may undergo **brittle deformation**. A brittle material deforms by fracturing, whereas a ductile material deforms by changing its shape. Drop a piece of chalk on the floor and it will break. Drop a piece of play dough, and it will bend or flatten instead of breaking. Under the conditions of room temperature and atmospheric pressure, chalk is brittle, and play dough is ductile. Rock in the lithosphere is brittle, like chalk, and so deformation in the lithosphere is commonly by fracture. Rock in the asthenosphere and mesosphere is very hot and, like play dough, has little strength, so deformation in these regions is ductile. Faults and earthquakes are confined to the lithosphere.

■ **elastic deformation** A temporary change in the shape or volume from which a material rebounds after the deforming stress is removed.

■ **ductile deformation** A permanent but gradual change in the shape or volume of a material, caused by flowing or bending.

■ **brittle deformation** A permanent change in shape or volume, in which a material breaks or cracks.

How and Why Rock Breaks

Faults and associated thrusts FIGURE 8.18

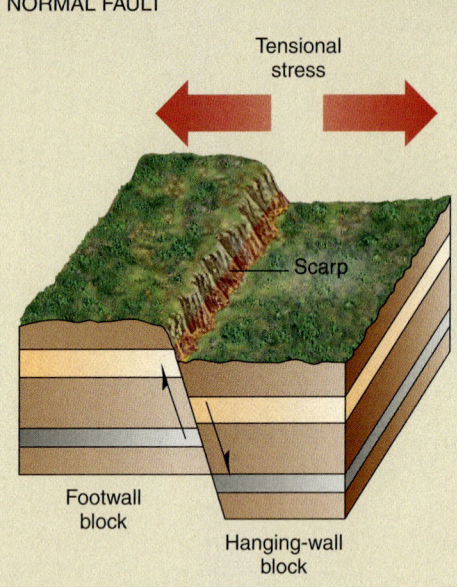

NORMAL FAULT

A When the crust is stretched by tension, normal faults occur. The block on the overhanging part of the fault moves down relative to the block underneath the fault.

REVERSE FAULT

B In a reverse fault, compression pushes the overhanging block up and over the one underneath.

KINDS OF FAULTS

A *fault*, as defined in Chapter 7, is a fracture along which movement has occurred. Most faults are small, only meters long, but others are very large, and can be 100 km or more in length (see *Amazing Places: Loch Ness* on page 264).

Different kinds of faults are caused by different kinds of stress (see FIGURE 8.18). Tensional (or *extensional*) stress pulls the crust apart and causes **normal faults** (FIGURE 8.18A). Rocks are weak when subjected to tensional forces—that is, they deform very little before they break. As a consequence, earthquakes associated with normal faults tend to be of low magnitude and have shallow foci. Such earthquakes are common along spreading centers between two plates.

Rocks are strong when compressed—that is, they can store a lot of elastic energy before they break. Compressional stress is responsible for **reverse** and **thrust faults**, and for Earth's largest magnitude earthquakes (FIGURES 8.18B and 8.18C). In reverse faults, one side of the fault rides up over the other side. Thrust faults are reverse faults that have gentle slopes. Reverse faults and thrust faults are common along convergent plate boundaries, and earthquakes associated with them tend to be of large magnitude and often have deep foci.

■ **normal fault** A fault in which the block above the fault surface moves down relative to the block below.

■ **reverse fault** A fault in which the block on top of the fault moves up and over the block on the bottom.

■ **thrust fault** A reverse fault that cuts Earth's surface at a shallow angle.

THRUST FAULT

TRANSFORM FAULT

C A reverse fault that cuts the surface at a shallow angle is called a *thrust fault*. On a map, an Earth scientist would indicate a thrust fault with a row of triangles pointing toward the overhanging block.

D In a transform or strike-slip fault, motion is mostly horizontal and parallel to the trace of the fault on Earth's surface.

> **transform fault**
> An approximately vertical fracture in the lithosphere along which two plates slide past each other.

Rocks subjected to shear stress are intermediate in strength between those subjected to tensional and compressional stresses. Shear stress typically creates **transform faults** (also called *strike-slip faults*) along which adjacent blocks are displaced horizontally to one another (**FIGURE 8.18D**). Because transform faults pass down through the lithosphere, earthquake foci can be either deep or shallow, and earthquake magnitudes can be almost as large as those associated with thrust faults. Transform faults are associated with transform fault plate margins.

CONCEPT CHECK STOP

Why is a rock buried deep underground more likely to undergo ductile deformation than brittle deformation?

Which kind of stress tends to stretch rocks? Which kind of stress shortens rocks?

What kinds of faults are often associated with plate convergence margins? What kinds of earthquakes are associated with convergent margins?

What is the difference between a reverse fault and a thrust fault?

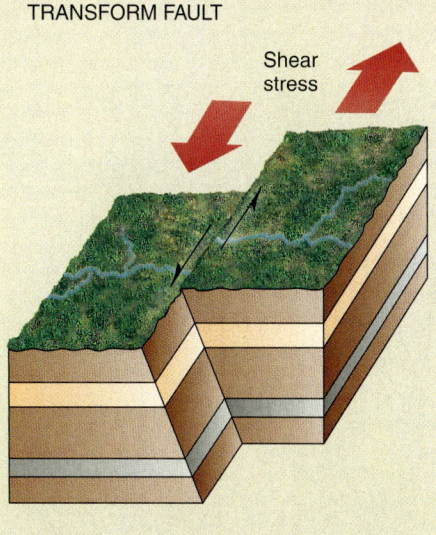

How and Why Rock Breaks 263

Amazing Places: Loch Ness

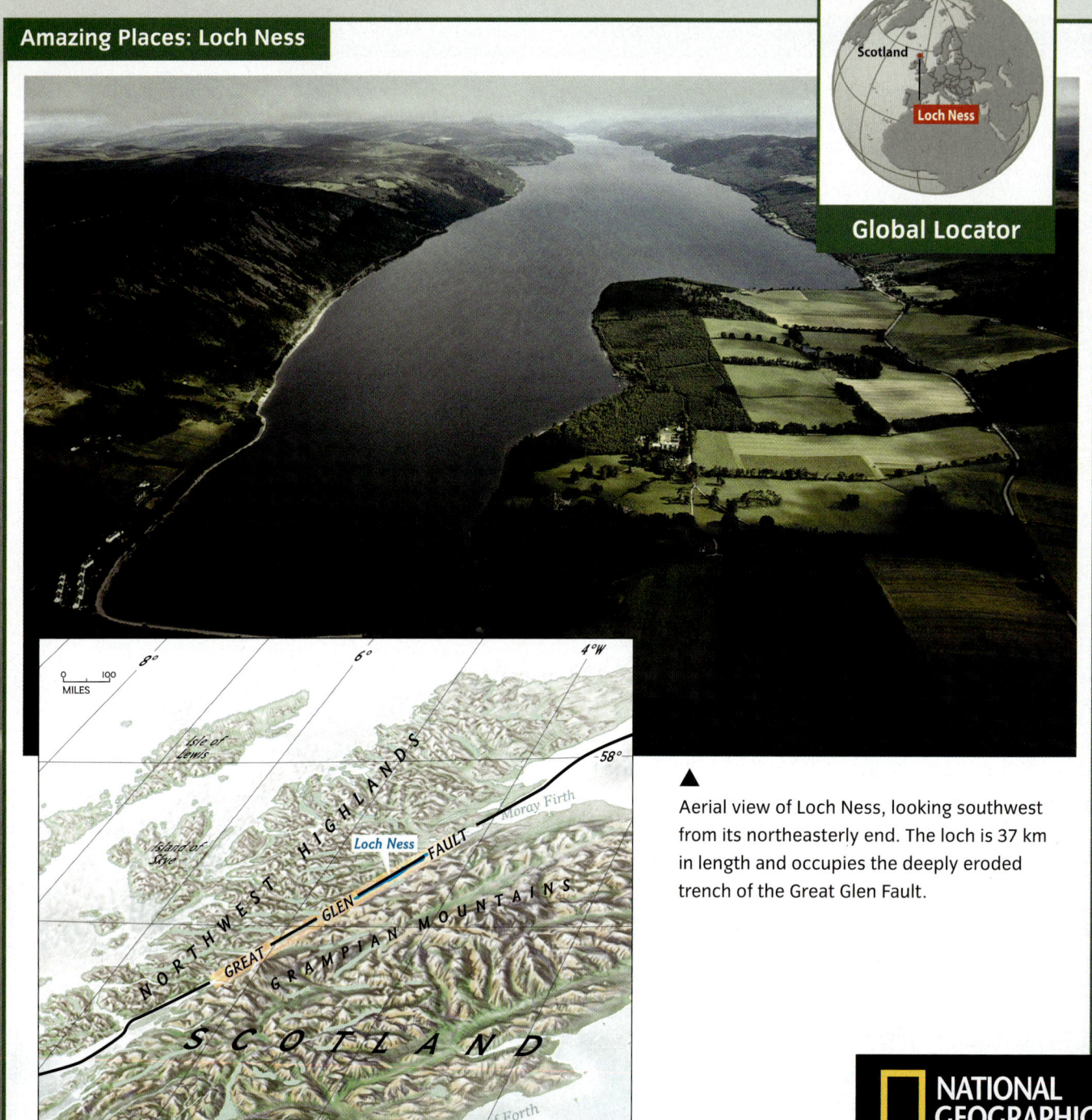

▲ Aerial view of Loch Ness, looking southwest from its northeasterly end. The loch is 37 km in length and occupies the deeply eroded trench of the Great Glen Fault.

Though it is most famous for its fabled inhabitant, the "Loch Ness Monster," this Scottish lake has a fascinating history. It lies on the Great Glen Fault, which 300 million years ago was part of a plate margin and was as active as the San Andreas Fault is today. The Great Glen Fault, like the San Andreas, is a transform fault. Over time, the rocks on the north side of the fault moved at least 100 km to the left (see inset). Movement on the fault crushed rocks along its margins. These rocks were particularly susceptible to erosion during the most recent ice age. Thus, they have eroded into a long and remarkably straight valley that runs along the fault. Today this valley is occupied by several lakes, including Loch Ness.

264 CHAPTER 8 Earthquakes and Earth's Interior

SUMMARY

1 Earthquakes and Earthquake Hazards

1. The subject of **seismology** relates **earthquakes** to the processes of plate tectonics. Although the motion of tectonic plates is very gradual, friction causes the rocks in the crust to jam together for long periods and then to break suddenly and lurch forward, causing an earthquake to occur. Earthquakes can cause large vertical or horizontal displacements of the ground, but much of the damage they cause comes as a result of the violent shaking that accompanies the displacement.

2. The shaking motion experienced during an earthquake can be explained by the **elastic rebound theory**, which says that the energy stored in bent and deformed rocks is released as **seismic waves**. After the earthquake, the rocks return to their previous state.

3. In many cases, the destructiveness of earthquakes is magnified by *secondary hazards*, such as fires, landslides, soil liquefaction, and tsunamis. Proper building design and earthquake preparedness can greatly reduce the loss of life from earthquakes and secondary hazards.

4. *Short-term forecasting* of earthquakes is still very unreliable. Scientists have concentrated their efforts on finding *precursor phenomena*, such as *foreshocks*, changes in water level, release of gases, and even unusual animal behavior—but with limited success. However, *long-term forecasting* can provide a good idea of which regions are at risk. One of the main tools of long-term forecasting is **paleoseismology**, which reveals when past earthquakes occurred in a given region, as well as the periodicity and magnitudes of past earthquakes.

2 The Science of Seismology

1. **Seismographs** produce recordings of seismic waves that are called **seismograms**. In a basic seismograph, a pen is attached to a heavy suspended mass. Seismic waves cause the paper to shake while the pen stays still and traces a wavy line on the vibrating paper.

2. Earthquakes produce three main types of seismic waves: **compressional** or **P waves** (*primary waves*), **shear** or **S waves** (*secondary waves*), and a variety of **surface waves**. Compressional and shear waves are called **body waves** because they travel through Earth's interior.

3. Compressional waves travel faster than shear waves, and hence arrive at seismographs first. The difference in arrival times between the P and S waves allows seismologists to compute the distance, but not the direction, to the **focus** of the earthquake. To determine the precise location of the **epicenter**, seismologists need measurements from three separate seismic stations. They can then determine the location by *triangulation*.

4. The **Richter** and **moment magnitudes** are measures of earthquake intensity that can be determined regardless of the distance to the earthquake or the amount of damage done. Both are *logarithmic* scales, in which each unit of magnitude corresponds roughly to a 10-fold increase in the amplitudes of seismic waves, but a 32-fold increase in the amount of energy released by the earthquake. The Richter magnitude assumes all of the energy is released from a point source—an assumption that is seldom valid for large quakes. In general, therefore, seismologists prefer to use the moment magnitude to describe the sizes of earthquakes.

3 Looking into Earth's Interior

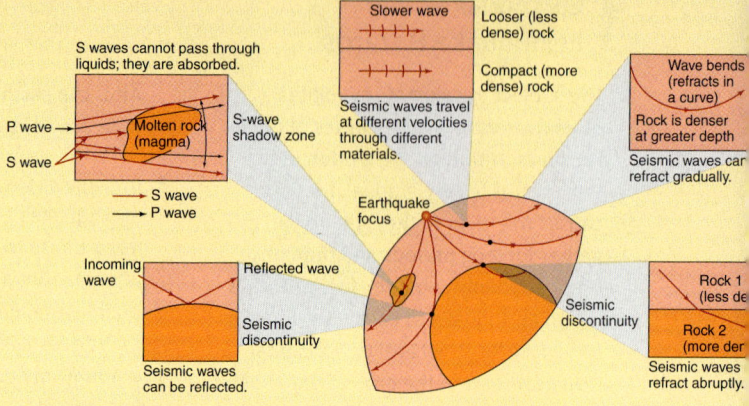

1. After an earthquake, seismic waves travel downward into Earth's interior as well as upward and along the surface. Seismic waves travel at different velocities through different materials, and they change velocity and direction when they pass from one material to another material with different physical and/or compositional properties. This understanding of seismic waves has allowed seismologists to identify many **seismic discontinuities** in Earth's interior.

2. Seismic discontinuities can result from either a change in composition or a change in physical properties of the material. They may **refract**, **reflect**, or even block seismic waves. P waves are strongly refracted, or bent, when they pass from the mantle to the core. S waves, on the other hand, are completely blocked by the core. This discovery provides evidence that Earth has a liquid outer core.

3. The mantle contains several discontinuities that are caused by differences in physical property rather than composition. The most important discontinuities are the boundaries between the *lithosphere*, the *asthenosphere*, and the *mesosphere*. The lithosphere is about 100 km thick; the asthenosphere begins about 100 km beneath the surface and ends at a depth of about 350 km; and the mesosphere extends from a depth of 350 to 2883 km.

4. Earth scientists use a variety of remote or indirect techniques to understand Earth's interior, in addition to the study of naturally occurring earthquakes. They can set off explosive charges at the surface and use *seismic tomography*, analogous to medical tomography, to detect seismic discontinuities underground.

5. Other sources of information about Earth's interior include drilling, the *magnetic field*, the *mass* and *diameter* of Earth, and *meteorite* studies. Evidence points to the likelihood that Earth has a dense core that consists mainly of iron and nickel.

6. So far, scientists have been unable to drill deep enough to sample the mantle directly. However, some mineral samples from the mantle come to the surface as *xenoliths*, carried along by magma that rises to the surface.

4 How and Why Rock Breaks

1. In response to **stress**, a rock may undergo **strain**; that is, it may change its shape, volume, or both. A nonpermanent change is **elastic deformation**. A permanent change that involves folding or flowing is **ductile** (or *plastic*) **deformation**. A permanent change that involves fracturing is **brittle deformation**.

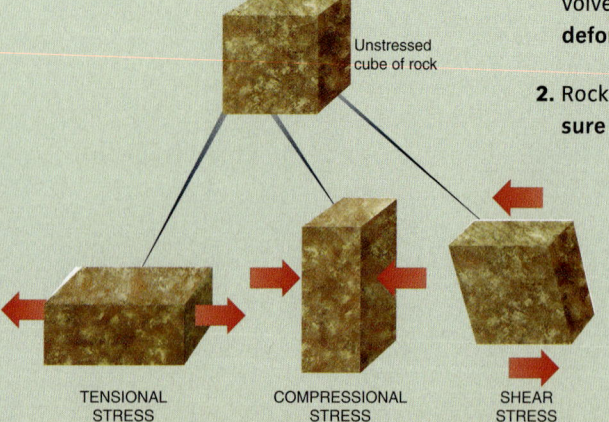

2. Rock deformation results from **pressure** or stress placed on rocks by the movements and interactions of lithospheric plates.

3. Stress can be *uniform* (the same in all directions) or *differential* (stronger in one direction than in another). **Compression** results from forces that squeeze a rock. **Tension** results from forces that stretch a rock or pull it apart. **Shear** stress causes a body of rock to be twisted and to change shape.

4. **Normal faults**, in which the upper block moves down relative to the lower block, are caused by tensional (pull-apart) stress. In **reverse faults**, caused by compressional (squeezing) stress, the upper block moves up and over the lower block. Shallowly sloping reverse faults are called **thrust faults**. In **transform faults** (also called *strike-slip faults*), caused by shear stress, the movement is mainly horizontal and parallel to the trace of the fault on Earth's surface.

266 CHAPTER 8 Earthquakes and Earth's Interior

KEY TERMS

- earthquake p. 234
- seismology p. 234
- elastic rebound theory p. 235
- seismic wave p. 235
- paleoseismology p. 241
- seismograph p. 246
- seismogram p. 246
- body wave p. 247
- surface wave p. 247
- focus p. 247
- P wave p. 247
- compressional wave p. 247
- S wave p. 247
- shear wave p. 247
- epicenter p. 248
- Richter magnitude scale p. 250
- moment magnitude p. 250
- seismic discontinuity p. 253
- refraction p. 254
- reflection p. 254
- stress p. 260
- pressure p. 260
- tension p. 260
- compression p. 260
- shear p. 260
- strain p. 260
- elastic deformation p. 261
- ductile deformation p. 261
- brittle deformation p. 261
- normal fault p. 262
- reverse fault p. 262
- thrust fault p. 262
- transform fault p. 263

CRITICAL AND CREATIVE THINKING QUESTIONS

1. Use the elastic rebound theory to describe what happens to rocks at the focus just before, during, and after an earthquake.
2. Why is short-term prediction of earthquakes so much less successful than long-term prediction? Why do you think seismologists are extremely cautious about making predictions? Do you think it will ever be possible to predict earthquakes accurately? Research your answer.
3. If you were asked to determine the exact shape and size of Earth, how would you go about it? What would you do differently if you were not allowed to use Space Age technology such as satellite photographs and orbital data?
4. Some of the boundaries inside Earth represent transitions between layers with differing compositions, whereas others represent transitions between layers with different physical states. Find out more about these different layers, and draw a detailed diagram to show the layering.
5. Which of the techniques used to study Earth's interior could also be used to study other planets? Which ones cannot, and why? Scientists know more about the surface of the Sun than about the interior of our own planet; why do you think this is so?
6. Find real examples of plate boundaries along which each of the following types of stress predominates: (a) compression, (b) tension, and (c) shearing. Try to find different examples than those used in the text.

What is happening in this picture ?

This photograph shows a stream in Carrizo Plains, California. The stream makes an abrupt turn to the right and then a 90-degree turn to the left.

What reason can you suggest for this stream's strange behavior?

SELF-TEST

1. At which type of plate boundary have the largest recorded earthquakes occurred?
   a. divergent boundaries
   b. transform fault boundaries
   c. subduction zone boundaries
   d. continental collision boundaries

2. According to the elastic rebound model, earthquakes are caused by _____.
   a. the slow release of gases from the athenosphere
   b. the sudden release of energy stored in rocks through continuing stress
   c. the sudden movement of otherwise stable tectonic plates
   d. the rapid release of gases from the asthenosphere

3. _____ and the resulting collapse of buildings, bridges, and other structures are usually the most significant primary hazards to cause damage during an earthquake.
   a. Fire
   b. Tsunami
   c. Ground liquefaction
   d. Ground shaking

4. _____ can provide a good idea of which regions are at risk for severe earthquakes.
   a. Short-term forecasting
   b. Long-term forecasting
   c. Unusual animal behavior studies
   d. Studies of groundwater levels

5. Body waves _____.
   a. move through Earth's interior
   b. cannot penetrate Earth's liquid outer core
   c. move along Earth's surface, causing great destruction
   d. Both b and c are correct.

6. Illustrations **A** and **B** depict two different types of seismic waves. Which of the following statements can be made about these two seismic waves?
   a. The wave depicted in A is a P wave and has a greater velocity through Earth's crust than other types of seismic waves.
   b. The wave depicted in A is an S wave and has a greater velocity through Earth's crust than other types of seismic waves.
   c. The wave depicted in B is a P wave and has a greater velocity through Earth's crust than other types of seismic waves.
   d. The wave depicted in B is an S wave and has a greater velocity through Earth's crust than other types of seismic waves.

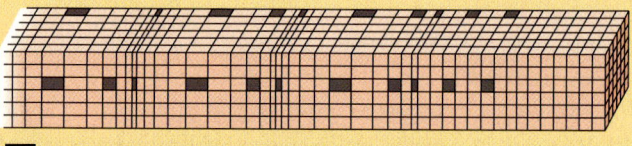

A

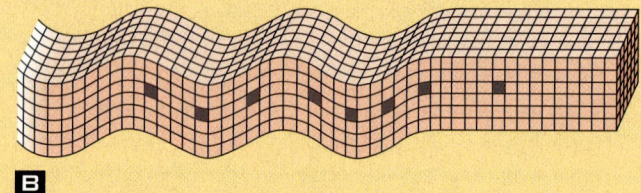

B

7. This illustration shows a seismogram of a hypothetical earthquake. On the seismogram, label the following:
   S–P interval   First arrival of P wave
   First arrival of S wave   Background noise
   First arrival of surface waves

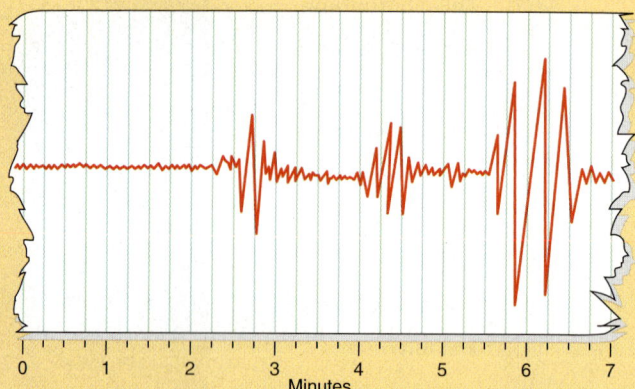

8. Using seismograms from three different seismic recording stations, A, B, and C, you determine the epicenter of an earthquake. Stations A and B both had an S–P interval of 3 seconds while C had an S–P interval of 11 seconds. Which of the following statements most accurately depicts the location of the epicenter?
   a. The epicenter is closest to Station A and equally far from B and C.
   b. The epicenter is closest to Station B and equally far from A and C.
   c. The epicenter is closest to Station C and equally far from A and B.
   d. The epicenter is equally close to A and B and farthest from Station C.

9. For the earthquake mentioned above, which seismic recording station would have recorded the P wave first?
   a. Station A
   b. Station B
   c. Station C
   d. Both stations A and B would have recorded the P wave before station C.

10. A magnitude 8 earthquake releases approximately _____ times more energy than a magnitude 7 event.
    a. 2   c. 20   e. 31.5
    b. 10   d. 21.5

11. Moment magnitude differs from Richter magnitude _____.
    a. because the Richter magnitude assumes earthquakes are generated at a point source, whereas moment magnitude takes into account that earthquakes can be generated over a large area of rupture
    b. because the moment magnitude assumes earthquakes are generated at a point source, whereas Richter magnitude takes into account that earthquakes can be generated over a large area of rupture
    c. in that moment magnitude uses Roman numerals to designate strength of an earthquake
    d. in that Richter magnitude uses Roman numerals to designate strength of an earthquake

12. When seismic waves reach a discontinuity inside Earth's interior, _____.
    a. they can be refracted, or bent, as they pass from the first material into the second
    b. they can be reflected, which means that all or part of the wave energy bounces back
    c. they can be absorbed, which means that all or part of the wave energy is blocked by the second material
    d. All of the above statements are correct.

13. In _____, rocks will bend as long as stress is applied to the crust, but resume their original shape if the stress is released.
    a. elastic deformation
    b. brittle deformation
    c. ductile deformation

14. This diagram shows a faulted block of Earth's crust. What type of fault is depicted in the diagram?
    a. normal fault
    b. thrust or reverse fault
    c. strike-slip fault

15. The fault depicted in the diagram in question 14 must have formed in response to what kind of stress?
    a. tension
    b. compression
    c. shear

Volcanism and Other Igneous Processes   9

Global Locator

The eruption of Mt. Pinatubo, on the island of Luzon in the Philippines, on June 15, 1991, was the second-largest volcanic eruption of the 20th century and the largest in a densely populated area. The top of the mountain exploded (main photo), blasting a gaping hole 2.5 kilometers in diameter and propelling volcanic ash and sulfurous gases more than 30 km into the atmosphere. The cloud lingered in the stratosphere and lowered worldwide temperatures for the next year by half a degree.

By unhappy coincidence, a typhoon was bearing down on the island at the time of the eruption. The rain-soaked ash caused many roofs to collapse. It formed loose, unstable mud that continued to flow and slide downhill for months, burying towns, wiping out bridges, and ultimately causing more damage than the eruption itself. The two inset photos show the town of Bamban, 30 km away from Mt. Pinatubo, one month (top) and three months after the eruption (bottom).

Despite the damage, the Mt. Pinatubo eruption killed relatively few people. Thanks to early warnings from Earth scientists, most of the area around the volcano had been evacuated. Although 847 people died from the effects of the eruption (including the mudslides afterward), scientists estimate that 5,000 to 20,000 lives were saved by the timely evacuation.

This chapter will explore the processes that lead to volcanic eruptions, the kinds of rocks they form, the ways in which scientists can tell when an eruption might be coming, and the reasons some volcanoes erupt violently and others do not.

CHAPTER OUTLINE

- Volcanoes and Volcanic Hazards p. 272

- Why and How Rocks Melt p. 285

- Cooling and Crystallization of Magma p. 290

- Plutons and Plutonism p. 293

Volcanoes and Volcanic Hazards

LEARNING OBJECTIVES

Identify several different categories of volcanoes.

Explain why stratovolcanoes tend to erupt explosively whereas shield volcanoes tend to erupt nonexplosively.

Describe how volcanic features such as calderas, geysers, and fumaroles arise.

Identify the hazards of volcanoes and the ways in which they can have beneficial effects.

Describe how scientists monitor volcanic activity.

For many people, the thought of a **volcano** conjures up visions of fountains of **lava** spurting up into the air and pouring out over the landscape (**FIGURE 9.1**). Although it's true that most volcanoes produce at least some liquid lava, many other types of materials can emerge from volcanoes as well, such as fragments of rock, glassy volcanic ash, and gases. Stored-up gases can cause a volcano to explode, covering the surrounding area with a catastrophic shower of volcanic ash and broken rock (**FIGURES 9.1A** and **9.1B**). Or gases can seep out silently and poison a whole town overnight, as you will discover in the *Case Study*. The different kinds of eruptions and the volcanoes they build have much to do with the physical properties of **magma** that lies at their source. We will begin our discussion by taking a look at some of the different kinds of volcanoes.

> **volcano** A vent through which magma, rock debris, volcanic ash, and gases erupt from Earth's crust to its surface.
>
> **lava** Molten rock that reaches Earth's surface.
>
> **magma** Molten rock, with any suspended mineral grains and dissolved gases, that forms when melting occurs in the crust or mantle.

Different eruption styles, different hazards FIGURE 9.1

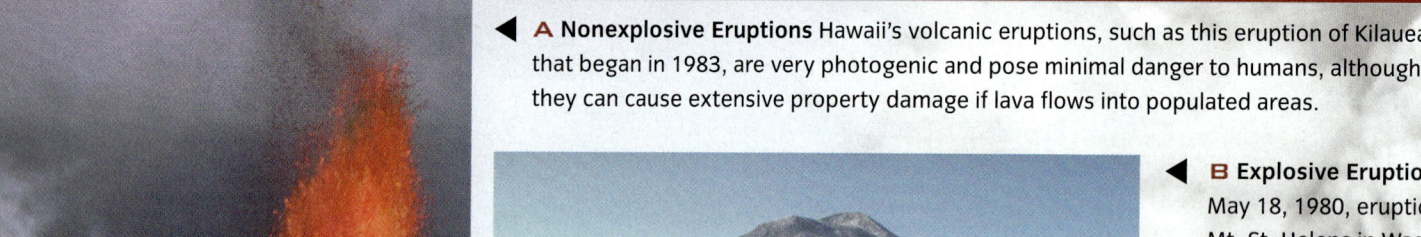

◀ **A Nonexplosive Eruptions** Hawaii's volcanic eruptions, such as this eruption of Kilauea that began in 1983, are very photogenic and pose minimal danger to humans, although they can cause extensive property damage if lava flows into populated areas.

◀ **B Explosive Eruptions** The May 18, 1980, eruption of Mt. St. Helens in Washington was much more violent than the relatively harmless eruptions of Kilauea in Hawaii. The entire top of Mt. St. Helens was destroyed in a cataclysmic explosion, as seen in these *before* (**B**) and *after* (**C**) pictures.

C Here you can see the large crater where the top of the mountain used to be. At least 57 people were killed in the eruption. ▶

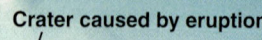

Crater caused by eruption

ERUPTIONS, LANDFORMS, AND MATERIALS

Named for Vulcan, the Roman god of fire, volcanoes are all dangerous at some level. However, as Figure 9.1 illustrates, there is a range of eruption styles. We can distinguish among several types of eruptions on the basis of their explosiveness and the materials they produce.

Types of eruptions
The differences between volcanoes are reflected in the kinds of terrain they build. **FIGURE 9.2** on page 274 shows several types of volcanoes, arranged in order from least to most explosive. The eruptive style of volcanoes can change from year to year, month to month, or even from one hour to the next; there is no such thing as a completely safe volcano or a completely predictable one.

Hawaiian eruptions consist of low-**viscosity** lava that flows easily from a volcanic vent. These flows gradually build up to form broad, flat-topped volcanoes with very gently sloping sides, called **shield volcanoes** (**FIGURE 9.2A**). They can grow to enormous size, and resemble a warrior's shield lying flat. Mauna Kea and Mauna Loa, on the big island of Hawaii, rise more than 10 km from their bases (6 km below sea level) to their peaks, 4 km above sea level. That makes them the tallest mountains on Earth, measured from base to peak.

Sometimes the same kind of low-viscosity lava that forms shield volcanoes rises to the surface through long fissures rather than central craters. Fissure flows can produce vast, flat lava plains called *flood basalts* or *basalt plateaus* (**FIGURE 9.2B**). Shield volcanoes like Mauna Loa often display some fissure activity as well as central crater flows.

Strombolian eruptions (**FIGURE 9.2C**) are more dramatic than Hawaiian eruptions. The lava is more viscous than the lava of shield volcanoes, and the eruption consists of fountains of lava and showers of lava fragments reaching hundreds of feet in the air. This type of eruption creates cones of loose volcanic rock called *spatter cones* or *cinder cones*.

Vulcanian eruptions are explosive kinds of eruptions (**FIGURE 9.2D**). Incandescent fragments of lava are blasted from the crater, and sometimes **pyroclastic flows** of hot cinders and ash sweep down the mountainside like an avalanche. These travel much faster than flowing lava, and they are the most dangerous consequence of a volcanic eruption.

The most violent, the most dangerous, and also the most famous eruptions are *Plinian*, named after Pliny the Elder, a Roman scholar who died during the eruption of Mt. Vesuvius in 79 A.D., and his nephew, Pliny the Younger, who lived and wrote of the eruption (**FIGURE 9.2E**). These eruptions produce ash columns that reach into the stratosphere (20 km or more) and also create pyroclastic flows. Both Vulcanian and Plinian eruptions tend to build steep-sided volcanoes called **stratovolcanoes** (**FIGURE 9.3** on page 275).

What causes this diversity of eruptions? The answer lies mostly in the properties of the magma that provides the source for the volcano. Two factors are important: the viscosity of the magma and the amount of gas dissolved in it. Viscosity is determined by temperature and composition. The higher the temperature, the lower the viscosity and the more readily magma flows. The more silica-rich a magma, the more viscous it is. If gas is present in the magma, it must escape somehow. If the magma has a low viscosity and is fairly runny, the dissolved gas will escape relatively easily. The lava may bubble and fountain dramatically as the gas escapes, especially at the beginning of an eruption, but the volcano will not explode. However, if the magma is relatively thick and viscous, it is harder for gas bubbles to form and escape. When the gas finally does escape, it usually vents explosively and shatters the magma into hot, glassy fragments.

---

viscosity The degree to which a substance resists flow; a less viscous liquid is runny and flows rapidly, whereas a more viscous liquid is thick and flows slowly.

shield volcano A broad volcano with gently sloping sides, built of successive flows of low-viscosity lava, generally of basaltic composition.

pyroclastic flow A stream of hot volcanic fragments (tephra) that are buoyed by heat and volcanic gases and flow very rapidly.

stratovolcano A volcano composed of solidified lava flows interlayered with pyroclastic material, generally of andesitic or rhyolitic composition. Such volcanoes have steep sides that curve upward.

Visualizing
Volcanoes and eruptions FIGURE 9.2

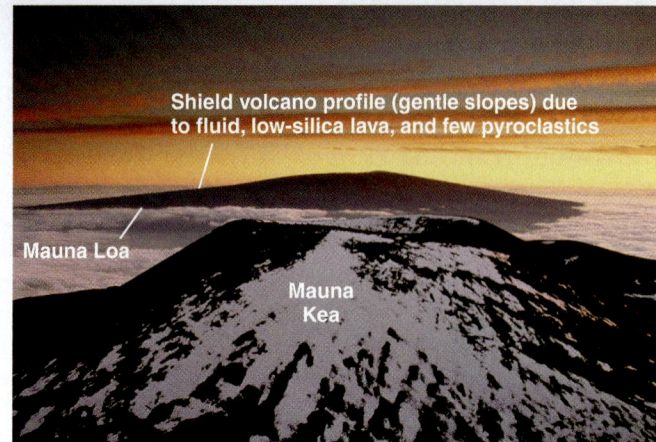

A Mauna Kea and Mauna Loa in Hawaii are shield volcanoes formed by successive eruptions of lava.

B These flood basalts are on the Snake River Plains, Idaho. They were erupted from a fracture in the crust.

C In 2002, Mt. Etna in Sicily experienced a Strombolian eruption.

D The 1993 Vulcanian eruption of Mt. Mayon in the Philippines included pyroclastic flows.

E The Plinian eruption of Mt. St. Helens in May 1980 produced an ash column and released destructive pyroclastic flows and debris avalanches down the steeply sloping sides of the volcano.

Process Diagram

How a stratovolcano is formed FIGURE 9.3

The classic volcano profile is that of a stratovolcano, which builds up over time from pyroclastic flows released in explosive eruptions, as well as from less frequent lava flows. This process creates steep-sided volcanoes like Mt. Fuji (inset), a national symbol of Japan. Fuji is a sleeping beauty now, but its steep upward-curving profile is a sign of a violent past.

1 Magma is very viscous, preventing the escape of gas bubbles. The trapped gas and upward movement of magma create increasing pressure inside the volcano. Thick deposits of pyroclastic material, like these, are often a sign of a past violent explosion—and a sign of possible future eruptions.

2 Gas continues to build and magma rises until the pressure causes an explosion. Small bits of lava and rock (called tephra) are ejected in all directions.

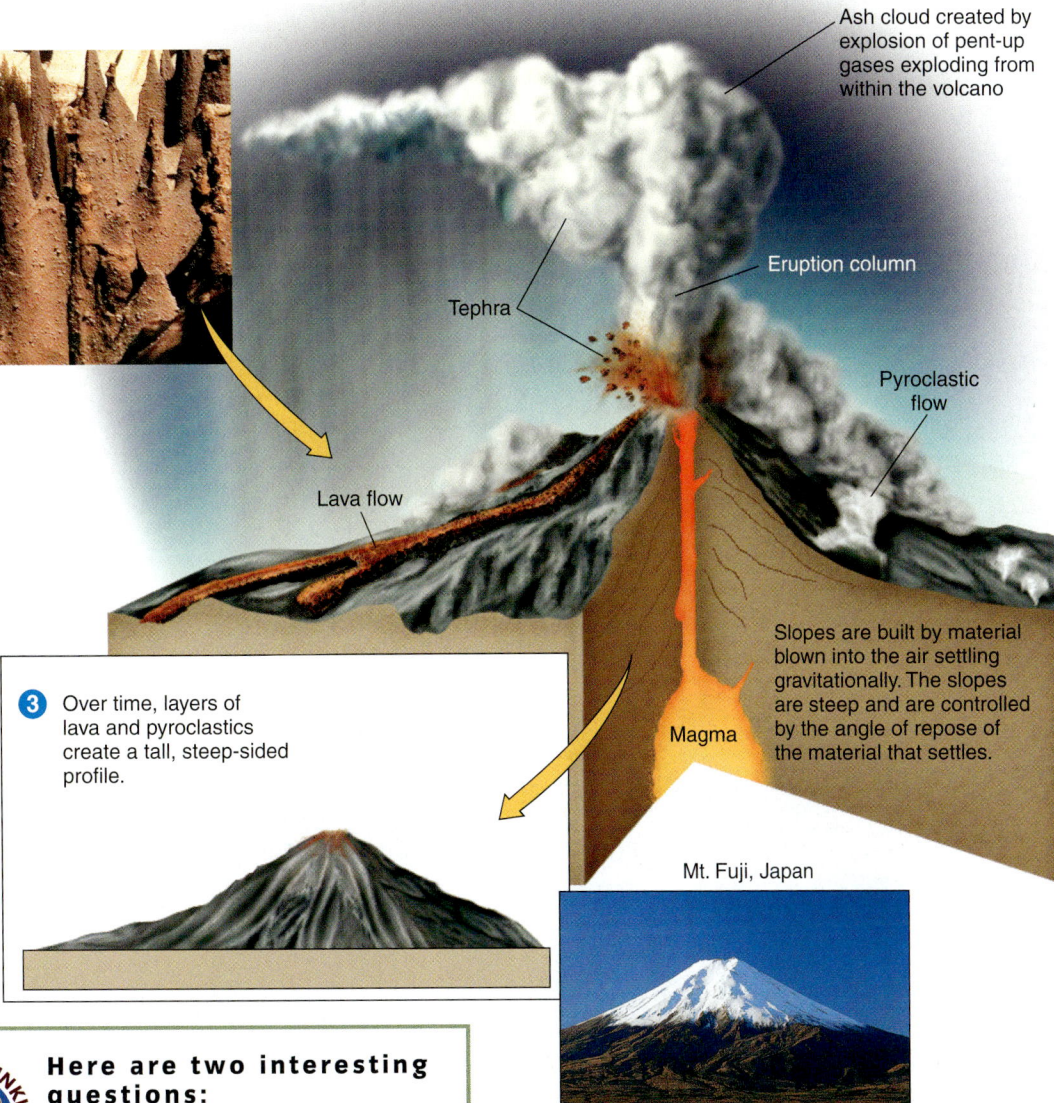

3 Over time, layers of lava and pyroclastics create a tall, steep-sided profile.

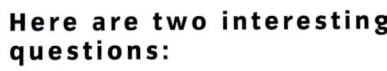

CRITICAL THINKING

Here are two interesting questions:
- At what point does a hot ash cloud cease to rise and flatten out?
- Which of the three eruption products, lava flows, pyroclastic flows, and ash clouds, are likely to cause the greatest economic damage?

Volcanoes and Volcanic Hazards

As we discussed in Chapter 3, a fragment of rock or magma ejected during a volcanic eruption is called a *pyroclast* (from the Greek words meaning "fire broken"). Collectively, all the ejecta from a volcano are known as *tephra*, and they can range from car-sized rocks to ultrafine *volcanic ash* whose individual particles can be seen only under a microscope (Figure 3.5, page 71).

Loose pyroclasts are often welded together during the eruption, or cemented together afterwards, forming pyroclastic rocks. These rocks are called *agglomerates* when the tephra particles are large and *tuff* when the particles are small. Spatter cones and tephra cones are made of exactly this kind of rock. On the other hand, stratovolcanoes have a somewhat more complicated structure, with alternating layers of pyroclastic material and solidified lava flows. Their height stems from the layers of hardened lava, which act as a cement that holds the pyroclasts together.

Other volcanic features

Near the summit of most volcanoes is a *crater*, the funnel-shaped depression from which gas, tephra, and lava are ejected. Some volcanoes have a much larger depression known as a *caldera*, a roughly circular, steep-walled basin that may be several kilometers in diameter. Calderas form when the chamber of magma underlying a volcano partially empties due to eruption, and the unsupported roof of the chamber collapses under its own weight. Crater Lake in Oregon (see *What an Earth Scientist Sees*) occupies a caldera 8 km in diameter that formed after immense eruptions about 6,600 years ago. Tephra deposits from those eruptions can still be seen in Crater Lake National Park and over a vast area of the northwestern United States and southwestern Canada.

Volcanoes do not necessarily become inactive after a major eruption. If magma begins to enter the chamber again, it may lift the floor of the caldera or crater and form a *resurgent dome*. The caldera of Mt. St. Helens contains a dome that has been growing since the eruption of 1980, indicating that the volcano is still active and may erupt again in the future.

What an Earth Scientist Sees

Crater Lake

A Beautiful Crater Lake, Oregon, the deepest lake in the United States, is all that remains of a once-lofty stratovolcano that scientists have named Mt. Mazama. Wizard Island, a small tephra cone in the middle of the lake, was formed by resurgent activity after the collapse that created the caldera.

B Tephra erupted during the Mt. Mazama eruption. Notice the mixed particle sizes, which is characteristic of tephra deposits immediately adjacent to the volcano. The columnar structures are a result of weathering.

CRITICAL THINKING

Here are two interesting questions:
- It is apparent in Figure B that the tephra is layered. How might the layering have formed?
- Why has weathering of the tephra formed spires?

A The geyser from which all other geysers take their name is called The Great Geysir, and it is located in Iceland.

B Sulfurous fumes pouring from a line of fumaroles associated with an eruption in Hawaii in 1965. The yellow color is due to sulfur condensing from the fumes.

Geysers and fumaroles FIGURE 9.4

When volcanism finally ceases, the magma chamber still contains hot (though not necessarily molten) rock for hundreds of thousands of years. When groundwater comes into contact with this hot rock, it heats up and may create a *thermal spring*. Many such springs have been turned into famous health spas. Some thermal springs have a natural system of plumbing that allows intermittent eruptions of water and steam. These are called *geysers*, a name that comes from the Icelandic word *geysir*, meaning "to gush" (FIGURE 9.4A). Finally, some volcanic vents emit only gas—usually water vapor that's sometimes mixed with foul-smelling sulfur compounds. These features are known as *fumaroles* (FIGURE 9.4B).

Where are volcanoes located? Refer back to Figure 7.9 (pages 212–213), Our tectonic planet. The locations of all volcanic eruptions known to have occurred over the past 10,000 years are plotted, together with Earth's major plate tectonic boundaries. Note that a large number of the volcanoes are clustered around the margin of the Pacific Ocean and occur in groupings parallel to convergent plate boundaries. Now look at FIGURE 9.5, in which the different kinds of volcanoes are plotted. Note that dangerous stratovolcanoes are clustered along the convergent plate boundaries that surround the Pacific Ocean rim, forming the so-called *Rim of Fire*. Convergent plate boundaries are subduction zones, and rock melting caused by subducted

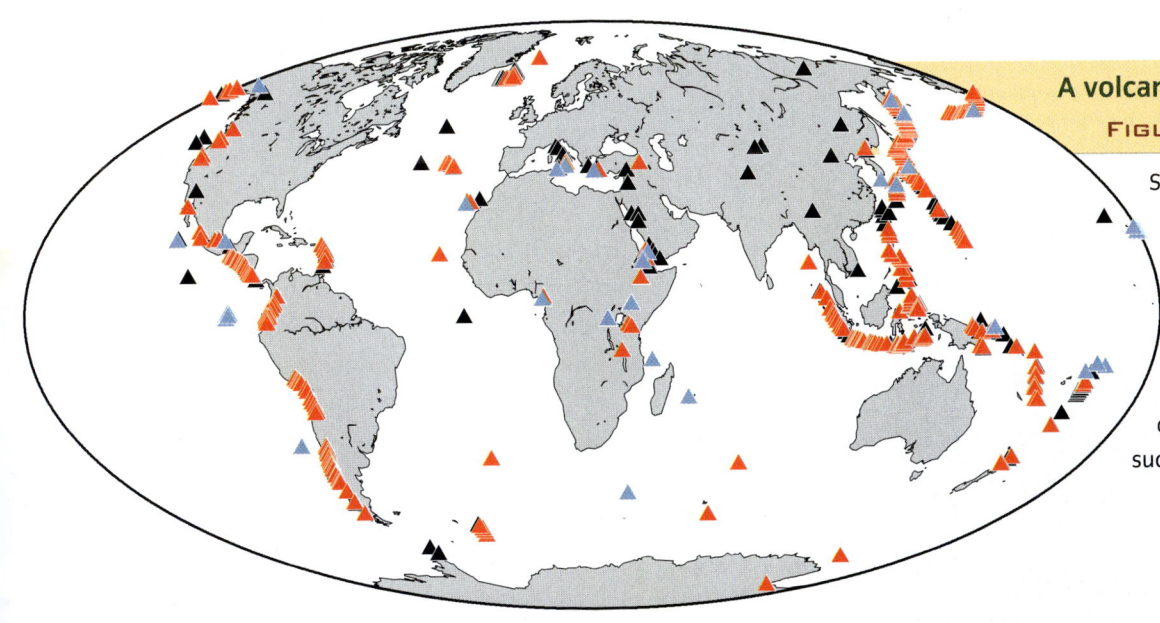

A volcanically active planet
FIGURE 9.5

Shown on this map are locations of volcanoes that have erupted in historic times and are considered to be still active. Red triangles are stratovolcanoes, blue triangles are shield volcanoes, and black triangles are other kinds of eruptive centers, such as fissures and cinder cones.

oceanic crust creates viscous, silica-rich magma that is erupted—most of it violently—and forms stratovolcanoes. Volcanoes related to divergent plate boundaries, as in midocean ridges, the Red Sea, and the African Rift Valley, commonly erupt basaltic lava, as do the less common volcanoes associated with transform faults, as in Central Asia. Giant shield volcanoes, such as those on Hawaii, Iceland, and Tahiti, are thought to be related to plume-generated hot spots, though an alternate hypothesis is that some may be caused by magma rising up from the mantle through giant fractures.

HAZARDS AND PREDICTION

Like other natural hazards, such as earthquakes, a volcanic eruption has *primary effects,* which are directly caused by the eruption itself; *secondary effects,* which are indirectly triggered by the eruption; and *tertiary effects,* which are long-lasting or even permanent changes brought about by the eruption.

Primary effects
People are usually able to outrun lava flows, even the least viscous and most rapidly flowing ones, so lava flows typically cause more property damage than injuries. In Hawaii, where Kilauea has erupted almost continuously for more than three decades, homes, cars, roads, and forests have been burned or buried by lava, but not a single life has been lost (**FIGURE 9.6**). It is sometimes possible to control a lava flow, at least partially, with retaining walls or a water spray, but nothing can be done to stop a volcanic eruption from occurring.

The greatest threats to human life during volcanic eruptions do not come from lava but from pyroclastic

Lava flow FIGURE 9.6
This house in Kalapana, Hawaii, is about to succumb to the slow but unstoppable advance of a lava flow in June 1989. The grass of the lawn burns on contact with the molten rock.

Victim of Mt. Vesuvius FIGURE 9.7
This is not an actual body but a plaster cast of a citizen of Pompeii, Italy, killed during the eruption of Mt. Vesuvius in 79 A.D. Death was caused by poisonous gases; then the body was encased by pyroclastic material. The body decayed, but a mold of its shape remained in the tephra. The cast was made when the natural mold was discovered during modern archaeological excavations.

flows and volcanic gases. Unlike slowly moving lava, pyroclastic flows move extremely rapidly and can easily outrace a running (or even a driving) human. The most destructive pyroclastic flow in modern times (in terms of lives lost) occurred on the island of Martinique in 1902 during a Plinian eruption, when an avalanche of searing ash descended Mt. Pelée at a speed of more than 160 km/hr and killed 29,000 people (there were two survivors). In 79 A.D., during the eruption of Mt. Vesuvius, the Italian towns of Pompeii and Herculaneum were buried under hot pyroclastic material, entombing the bodies and buildings in a natural time capsule (**FIGURE 9.7**). However, most of these people were dead already, due to another hazard of volcanic eruptions: poisonous gases. More recently, at least 1700 people and 3000 cattle lost their lives when poisonous gas erupted from a volcano at Lake Nyos in Cameroon (see the *Case Study*). In 1783, the Laki eruption in Iceland released so much acidic gas that nearly a third of the people and half the domestic animals in the country perished.

Secondary effects
Secondary effects include fires (which are often caused by lava flows) and flooding (which may happen if a river channel is blocked or a crater lake bursts). The famous eruption of the volcanic island at Krakatau, Indonesia, in 1883 claimed most of its victims due to a tsunami, the same kind of ocean wave that can also be caused by earthquakes, and in this case flooded coastal towns on nearby Indonesian islands. Volcanoes can also produce volcanic tremors, a type of seismic activity that helps scientists predict eruptions but rarely poses a threat itself.

CASE STUDY

Lakes of Death in Cameroon

Two small volcanic lakes in a remote part of Cameroon, a country in central Africa (**A**), made international news in the mid-1980s when they emitted lethal and invisible clouds of carbon dioxide from deep beneath their surface. The first gas discharge, which occurred at Lake Monoun (not shown) in 1984, asphyxiated 37 people. The second, which occurred at Lake Nyos in 1986, released a highly concentrated cloud of carbon dioxide that killed more than 1700 people. Both occurred at night during the rainy season, both involved volcanic crater lakes, and both are likely to happen again if technological intervention is not successful.

After the incidents, scientists discovered that water in the lakes was stratified, and the bottom layers had huge reservoirs of carbon dioxide dissolved in them. Some minor event—a landslide, or perhaps nothing more than winds at the surface—disturbed the layers, and the stratified column of water turned over. Like a newly opened soda bottle, reduction of pressure on the gas-rich waters allowed the carbon dioxide to bubble to

Global Locator

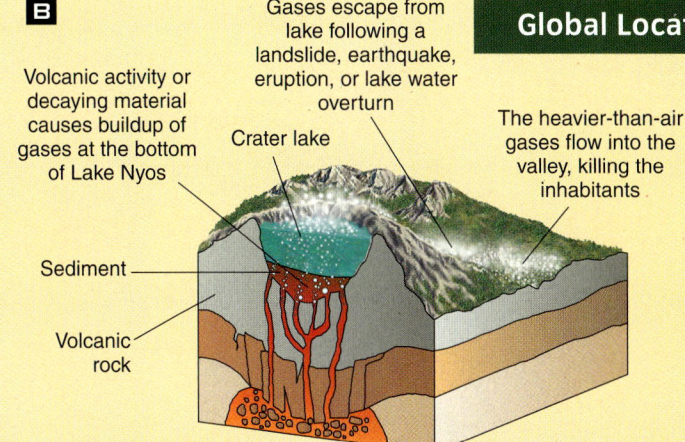

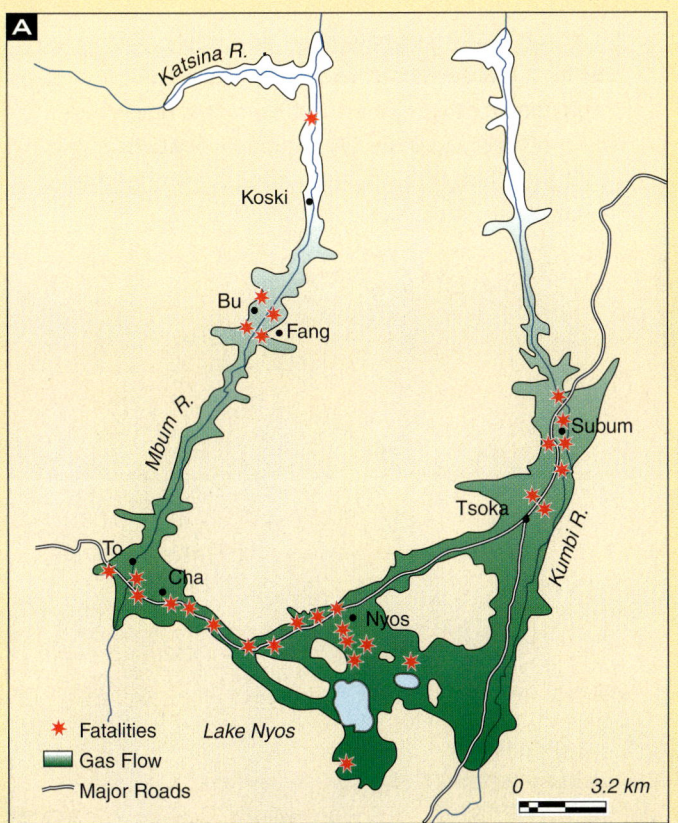

the surface. In the case of Lake Nyos, approximately 100 million cubic meters of carbon dioxide were released in just two hours (**B**). Because carbon dioxide is heavier than air, it flowed down the mountainside in a ground-hugging layer that displaced the oxygen that both cattle (**C**) and people needed to breathe.

The supply of gas at the lake bottom is constantly being replenished from the magma chamber. Scientists estimated that Lake Monoun could experience another violent degassing event within 10 years, and Lake Nyos is at risk within 20 years. It may be possible to siphon off the excess gas by installing a subsurface network of pipes. A prototype system was tested successfully at Lake Monoun in 1992. However, Lake Nyos poses a more complicated problem because of its greater depth and its fragile stratification. Engineers must be very careful not to initiate the very situation that they are trying to prevent.

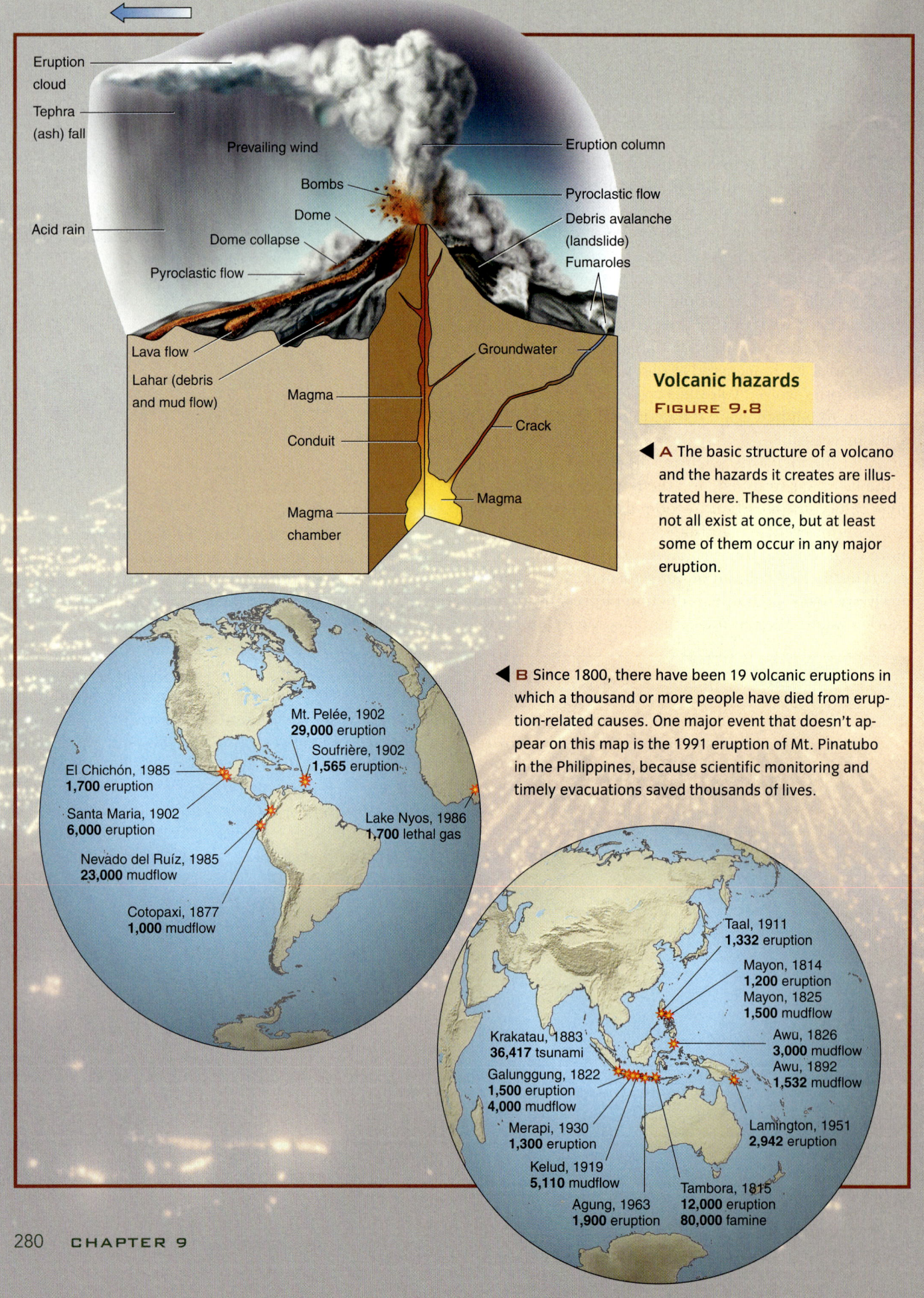

Volcanic hazards
FIGURE 9.8

A The basic structure of a volcano and the hazards it creates are illustrated here. These conditions need not all exist at once, but at least some of them occur in any major eruption.

B Since 1800, there have been 19 volcanic eruptions in which a thousand or more people have died from eruption-related causes. One major event that doesn't appear on this map is the 1991 eruption of Mt. Pinatubo in the Philippines, because scientific monitoring and timely evacuations saved thousands of lives.

Volcanoes and climate FIGURE 9.9

◀ **A** The fissure eruption of Laki, a volcano in Iceland, lasted from 1783 to 1784 and was the largest flow of lava in recorded history. Primary and secondary effects killed a third of Iceland's population, and the effects of the eruption were felt widely around the Northern Hemisphere.

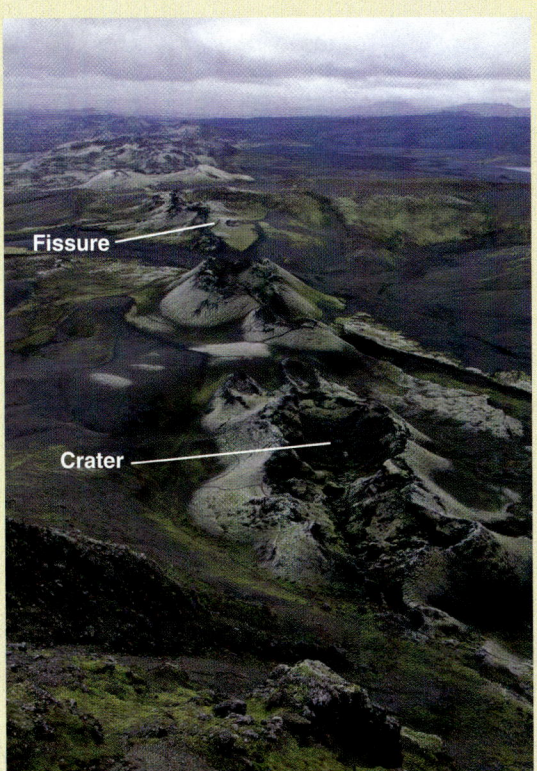

▼ **B** In the winter after Laki's eruption, the average temperature in the Northern Hemisphere was about 1°C below normal. In the eastern United States, the decrease was closer to 2.5°C. At the same time, ice cores from Greenland record a dramatic spike in acidity, due to acid precipitation.

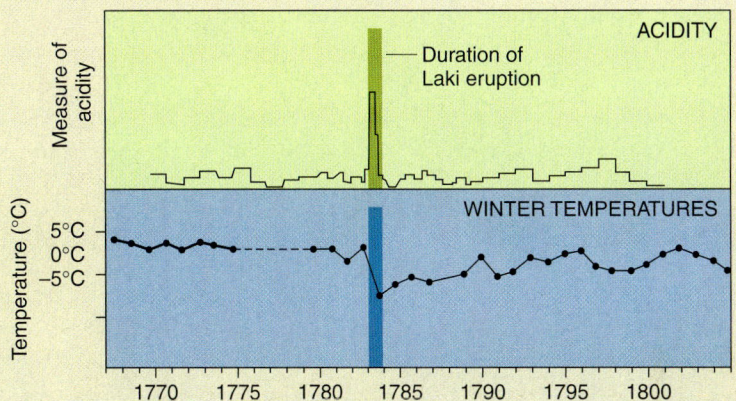

Mudslides have often been a major cause of volcano-related deaths. When volcanic ash mixes with snow at the volcano's summit (or vice versa, when rain falls on recently deposited volcanic ash), it can start a deadly mudflow called a *lahar*. The account at the beginning of the chapter describes a lahar from the eruption of Mt. Pinatubo. Lahars can occur months after the eruption. A related phenomenon is a volcanic *debris avalanche,* in which many different types of material, such as mud, pyroclastic material, and downed trees, are mixed together. A devastating debris avalanche caused much of the damage from the 1980 eruption of Mt. St. Helens.

FIGURE 9.8 summarizes the various primary and secondary hazards from an eruption and shows where the most deadly eruptions of the past two centuries occurred. Note that in many cases the secondary effects, such as the tsunami associated with Krakatau, are responsible for the greatest losses of life.

Tertiary effects Volcanic activity can change a landscape. Eruptions can block river channels and divert the flow of water. They can dramatically alter a mountain's appearance, as in the 1980 eruption of Mt. St. Helens (Figure 9.1). They can form new land, such as the black sand beaches of Hawaii, which are made of dark pyroclastic fragments, or the volcanic island of Surtsey, which emerged from the ocean near Iceland in 1963 and is composed of both lava flows and pyroclastic cones.

Volcanoes can also affect the climate on a regional and global scale (see **FIGURE 9.9**). Major eruptions can cause toxic and acidic rain, spectacular sunsets, or extended periods of darkness. Sulfur dioxide, a common gaseous emission of volcanoes, forms small droplets or *aerosols*. If they get into the stratosphere, these aerosols spread around the world, absorb sunlight, and cool Earth's surface. An example from recent times is the 1815 eruption of Tambora in Indonesia, which caused three days of near darkness as far away as Australia. The following year was so cool in Europe and North America that it was called "the year without a summer." Farther back in time the eruption of flood basalts, such as the Deccan Traps in India and the Siberian Traps in Russia, may have caused or contributed to several of the mass extinctions that divide geologic periods.

Beneficial effects Not all the effects of volcanoes are negative, and it is no accident that people choose to live near active volcanoes. Periodic volcanic eruptions renew the mineral content of soils and replenish their fertility—some of the most fertile agricultural lands in the world are adjacent to active volcanoes (see **FIGURE 9.10** on pages 282–283). Volcanism also provides

Living with danger FIGURE 9.10

Farming villages speckle the slopes of Mt. Merapi, an active volcano on the island of Java, Indonesia. Merapi has a long history of dangerous, often fatal, eruptions, but the fertility of its soils lures farmers to its hazardous slopes. Although volcanic soils cover just 1% of Earth's land surface, they support 10% of the world's population.

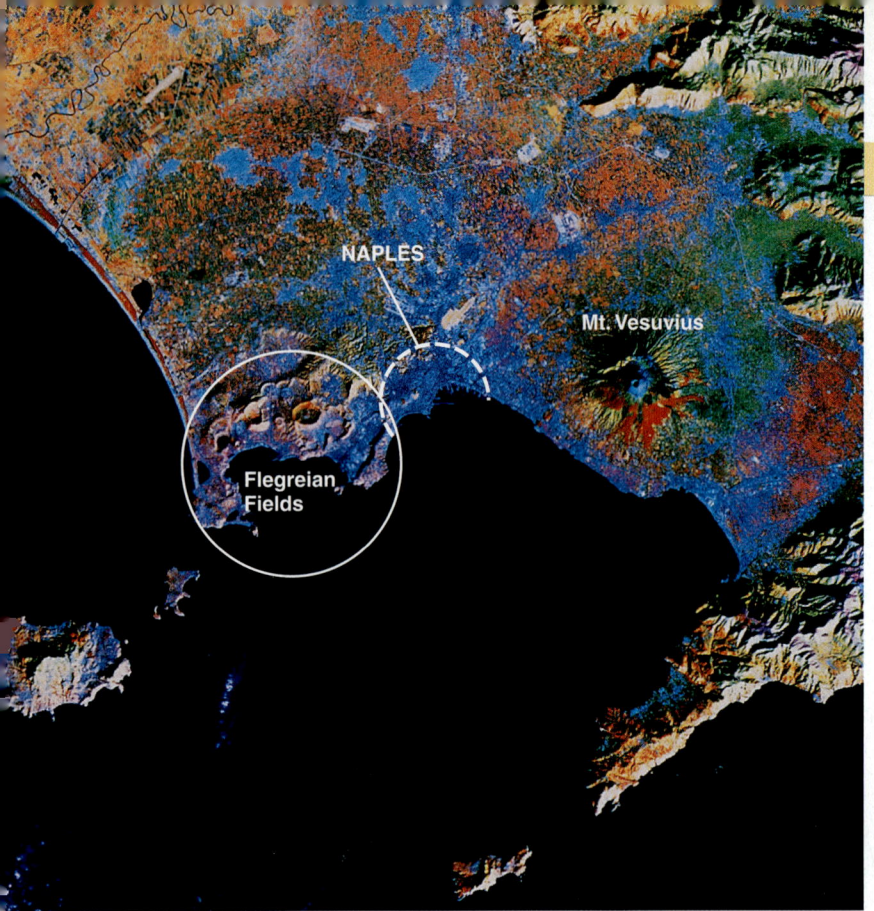

Satellite monitoring FIGURE 9.11

This false-color satellite image shows the area around Mt. Vesuvius (center right) and the Bay of Naples, Italy, a densely populated region. Recent lava flows show up bright red in this image, which records infrared radiation (i.e., heat). Older lavas and volcanic ash show up as shades of yellow and orange. The dark blue and purple region at the head of the bay is the city of Naples. West of Naples lies a cluster of smaller volcanoes called the Flegreian Fields. By comparing successive satellite images, an Earth scientist can detect changes in ground temperature.

geothermal energy and some types of mineral deposits. One rare kind of volcanism brings up diamond-bearing magma from deep in the mantle. All natural gem-quality diamonds on Earth reach the surface by volcanism.

Predicting eruptions It isn't possible to stop volcanic eruptions, but it is sometimes possible to predict them. The first step in prediction is to identify a volcano as active, dormant, or extinct. An *active* volcano has erupted within recent history; a *dormant* one has not erupted in recent history. A volcano is called *extinct* when it shows no signs of activity and is deeply eroded. Mt. Pinatubo in the Philippines had been dormant for about 500 years prior to its awakening in 1991.

Another important step in prediction is identifying the past eruptive style of a volcano. For example, Mt. Pinatubo is surrounded by thick deposits of pyroclastic material, a sign that the volcano erupted violently in the past. Subduction zone volcanoes such as Mt. St. Helens are more likely to erupt explosively than shield volcanoes and fissure eruptions. The type of rock that has solidified from past eruptions, either silica-rich or silica-poor, also indicates a volcano's style of eruption.

When a volcano shows warning signs of increasing activity, scientists monitor it more closely. Some of these warning signs include changes in the shape or elevation of the ground, such as bulging, swelling, or the formation of a dome. The presence of these features suggests that the underground reservoir of magma is growing. The release of gases can also be a warning sign, as can changes in the temperature of crater lakes, well water, or hot springs. A sudden increase in local seismic activity is also a warning sign. Earth scientists monitor volcanic activity by using tiltmeters to detect bulging, satellite images, and devices that identify gas emissions or changes in the temperature of the ground (**FIGURE 9.11**).

CONCEPT CHECK STOP

Which kinds of eruptions pose the greatest risk to humans?

Which kinds of eruptions pose comparatively low risk?

Why do pyroclastic flows travel faster and kill more people than do lava flows?

What is the difference between a crater and a caldera?

Why do explosive eruptions occasionally affect global climate?

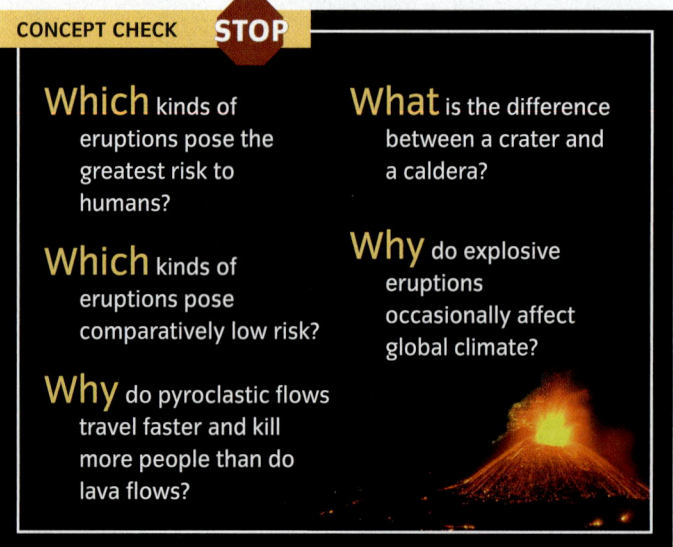

284 CHAPTER 9 Volcanism and Other Igneous Processes

Why and How Rocks Melt

> **LEARNING OBJECTIVES**
>
> **Describe** how temperature, pressure, and water content affect a rock's melting point.
>
> **Define** fractional melting.
>
> **Identify** three properties that distinguish one lava or magma from another.

Underneath every active volcano lies a reservoir of magma, called a *magma chamber*. Understanding volcanism includes understanding how rocks melt to become magma. Fortunately, rocks can be melted artificially as well as naturally (**FIGURE 9.12**). We can thus learn about the behavior of molten rock from laboratory experiments.

At Earth's surface, common types of rock, such as granite and shale, begin to liquefy when heated to a temperature between about 800°C and 1000°C. However, rock (unlike ice, for example) typically consists of many different minerals, each with its own characteristic melting temperature. Thus, we cannot talk about a single melting point for a rock; rather there is a temperature range across which melting occurs. Melting may start at 800°C, but complete melting is commonly attained by about 1200°C. Two other factors beside temperature also strongly affect melting: pressure and the presence of water in the rock.

HEAT AND PRESSURE INSIDE EARTH

If you descend into a mine, it becomes apparent that the farther down you go, the hotter it gets. The rate at which temperature increases with depth, called the **geothermal gradient**, is quite different underneath continental surfaces than it is under the seafloor. Continental crust is thick, and the temperature underneath it increases more gently, starting at about 20°C per km and increasing more rapidly at depth, for an average rate of about 7.0°C per kilometer, reaching 1000°C at a depth of 150 km, which is the base of the lithosphere. Underneath the ocean floor, the rate of increase

> **geothermal gradient** The rate at which temperature increases with depth below Earth's surface.

Molten rock: artificial vs. natural FIGURE 9.12

A In a steel mill, workers heat metal ores to the melting point in order to separate the metal from the surrounding rock.

B An Earth scientist in a protective suit measures the temperature of lava erupting from Mauna Loa, Hawaii. Bright orange, yellow, and white lava is hotter, whereas dull red, brown, and black colors indicate cooler lava.

Geothermal gradient FIGURE 9.13

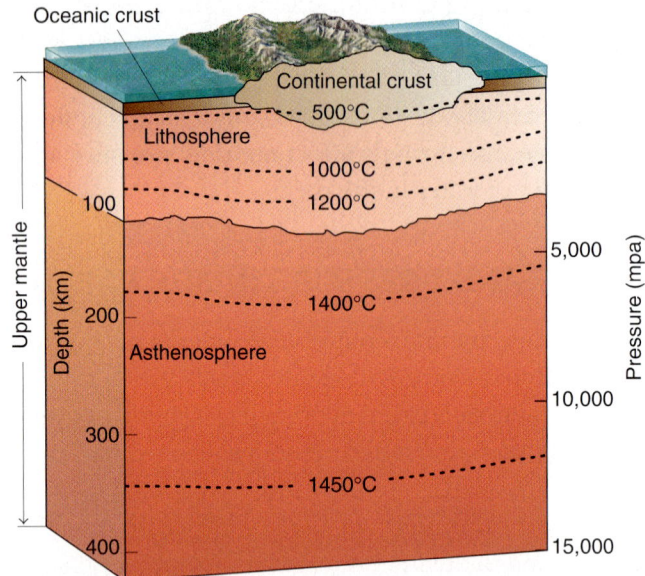

A Temperature increases with depth. The dashed lines are *isotherms*, lines of equal temperature. Notice how the lines "sag" underneath the continental crust, because the rate of increase of temperature is slower there.

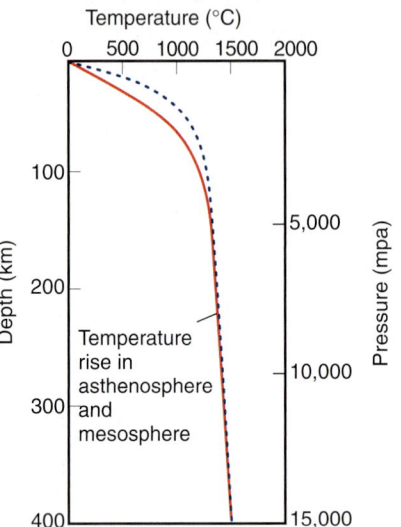

B This graph represents the same information as shown in **A**. Earth's surface is at the top, so depth (and pressure) increases as you move down. The dashed curve shows the geothermal gradient under oceans, and the solid curve shows the gradient under continental crust. Note that the two curves merge (and the isotherms become level) below 200 km.

---

is about twice as rapid. As with the continental crust, the temperature increases more rapidly with depth, for an average rate of 13°C/km, reaching 1000°C at a comparatively shallow depth of 80 km (**FIGURE 9.13**). Below the asthenosphere–lithosphere boundary, heat is transferred by convection and the geothermal gradient becomes more gradual (0.5°C/km), and the temperature difference between suboceanic and subcontinental rock disappears.

As you can see in Figure 9.13, the temperature in the upper mantle is higher than the temperature at which most rocks melt at Earth's surface. Yet the upper mantle is mostly solid. How is this possible?

The answer is that the pressure also rises very dramatically with increasing depth, and increasing pressure causes rock to resist melting (**FIGURE 9.14A**). For example, albite, a common rock-forming mineral (a feldspar), melts at 1104°C at the surface. At a depth of 100 km, the pressure is 35,000 times greater than it is at sea level. At that pressure, the melting temperature of albite rises to 1440°C, which still slightly exceeds the normal temperature at that depth. Thus, albite remains solid when it is beneath the surface.

The presence of water (or water vapor) in a rock dramatically reduces its melting temperature (**FIGURE 9.14B**). By analogy, as anyone who lives in a cold climate knows, salt can melt the ice on an icy road because a mixture of salt and ice has a lower melting temperature than pure ice. Similarly, a mineral-and-water mixture has a lower melting temperature than the dry mineral alone because water plays the same role as salt in a salt-ice mixture.

The effect of water on the melting of a rock becomes particularly important in subduction zones, where water is carried down into the mantle by oceanic crust, as described in Chapter 7.

Effects of temperature and pressure on melting FIGURE 9.14

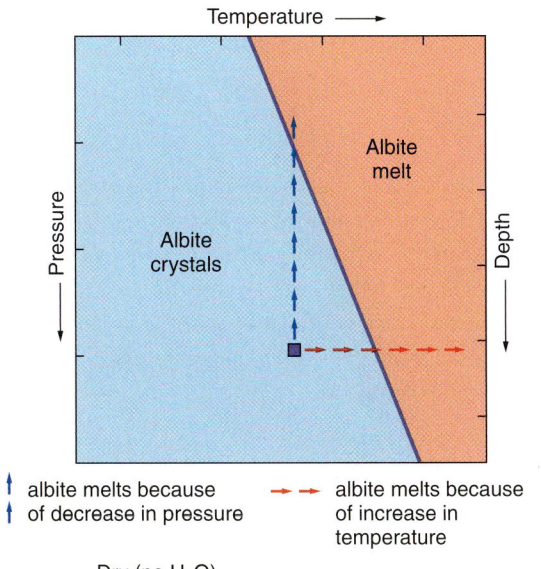

A The melting temperature of a dry mineral (albite, in this case) increases at high pressures. A mineral at depth (shown by the small square) can melt in two different ways: either by an increase in temperature (red arrows) or by a decrease in pressure (blue arrows). The latter effect is called *decompression melting*, and it is an important reason why much magma stays molten all the way to the surface.

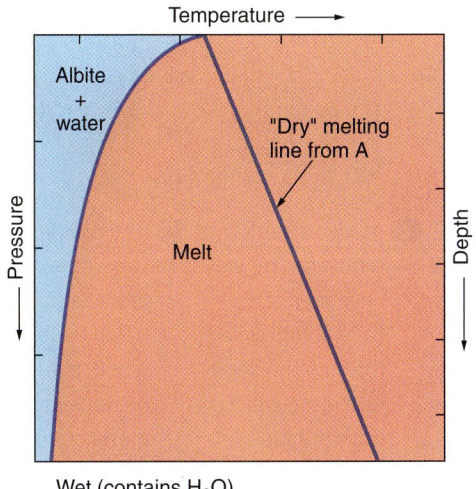

B The melting temperature of a mineral in the presence of water typically decreases as pressure increases. This is exactly the opposite of what happens to dry minerals. Magmas containing dissolved water typically solidify before they reach the surface.

FRACTIONAL MELTING

Because rocks are composed of many different minerals, they melt over a range of temperatures. This means that the boundary between solid and melt is not crisp, but blurry, as in FIGURE 9.15 on page 288. When the temperature rises enough for part of the materials in a rock to melt and part to remain solid, it becomes a **fractional melt**. Only if the temperature continues to increase, or the pressure to decrease, will the rock melt completely. **Fractionation**, an important process that can lead to the development of a diversity of rock types, is caused by fractional melting.

> **fractional melt** A mixture of molten and solid rock.
>
> **fractionation** Separation of melted materials from the remaining solid matter during the course of melting.

MAGMA AND LAVA

As mentioned earlier, molten rock is called *magma*. When magma reaches the surface, it is called *lava*. A lot of magma never reaches the surface, but instead remains underground, trapped in a magma chamber, until it crystallizes and hardens to igneous rock. We cannot study magma underground in its natural setting, but we can study lava and we can experiment with synthetic magma. From our direct observations of lava, we know that magmas differ in *composition, temperature,* and *viscosity*.

Composition Most magma is dominated by silicon, aluminum, iron, calcium, magnesium, sodium, potassium, hydrogen, and oxygen—the most abundant elements in the mantle and crust. Oxygen combines with the others to form oxides, such as SiO_2, Al_2O_3, CaO, and H_2O. Silica, or SiO_2, usually accounts for 45% to 75% of the magma, by weight. In addition, a small amount of dissolved gas (between 0.2% and 3% of the magma, by weight) is usually present, primarily water vapor (H_2O) and carbon dioxide (CO_2). Despite their low abundance, these gases strongly influence the properties of magma. The proportion of silica (SiO_2) also has a strong effect on the magma's appearance and

Why and How Rocks Melt

Process Diagram

Effect of pressure and temperature on rocks FIGURE 9.15

VIEW THIS IN ACTION in your WileyPLUS course

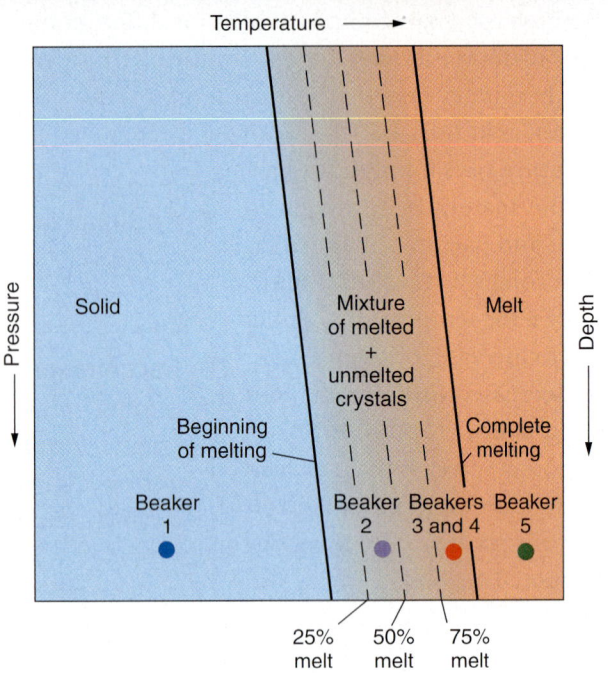

◀ Because almost all rocks contain a mixture of materials, they do not melt all at once, at a single temperature; instead, there is a range of temperatures and pressures in which they contain a mixture of melted and unmelted crystals. The dots on this diagram show the stages in the melting process, which are illustrated further below.

▼ The process of fractional melting can lead to another effect called *fractionation* (illustrated for convenience in a laboratory beaker rather than buried in Earth's mantle).

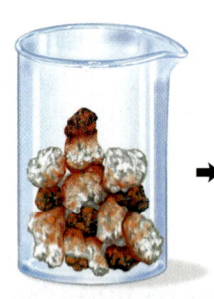
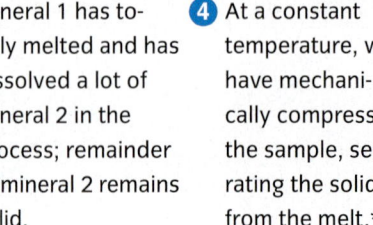

① The first beaker shows a mixture of two minerals. At a low temperature, both are solid.

② As the temperature increases, mineral 1 (the dark mineral) begins to melt and as it does so, it dissolves some of mineral 2.

③ Mineral 1 has totally melted and has dissolved a lot of mineral 2 in the process; remainder of mineral 2 remains solid.

④ At a constant temperature, we have mechanically compressed the sample, separating the solid from the melt.*

⑤ If temperature were to continue to increase, the material in the beaker would eventually become completely melted.

*In the lithosphere, this mechanical separation can occur as a result of the tectonic motion of plates. If the melt now cooled again, we would have two separate deposits, one of mineral 1 plus a little of mineral 2, crystallizing from the melt, and one of pure mineral 2, the solid remnant in Beaker 4.

Here's an interesting question:
- How could the mixture in Beaker 2 become the same as the mixture in Beakers 3 and 4 without any change in temperature?

288 CHAPTER 9 Volcanism and Other Igneous Processes

Viscosity of lava   FIGURE 9.16

A Lavas at high temperature and of low silica content tend to flow more freely. This stream of low-viscosity lava was erupted in Hawaii in 1983. The temperature of the lava was about 1100°C.

B Two basaltic lava flows with identical compositions but different viscosities are visible. Each has a Hawaiian name for the texture. The scientist is standing on the smooth, ropy surface of an old, low-viscosity flow of *pahoehoe* (pronounced pa-hoy-hoy), similar to the flow in **A**. He is sampling a viscous, slow-moving flow of chunky *aa* (pronounced ah-ah). The temperature of the aa flow was 1000°C.

properties. High-silica magmas and lavas tend to have high viscosities.

Temperature   We know from direct measurements at erupting volcanoes that lavas vary in temperature from about 800°C to 1200°C. From laboratory experiments with synthetic magma, scientists know that magma temperatures in the mantle must rise as high as 1400°C. They also know that magmas with high H_2O content tend to melt at lower temperatures. High-temperature magmas and lavas tend to have low viscosities.

Viscosity   All magma is liquid and has the ability to flow, but magmas differ to a marked extent in how readily they flow. This is certainly true for lavas. As shown in **FIGURE 9.16**, some lava is very fluid, almost like a stream of water. But other lava creeps along slowly and inexorably, like molasses. In particular, lavas with high silica content tend to flow slowly because of the tendency of silica molecules to polymerize, or form long chains (see Chapter 2). Thick, slow-moving lavas have high viscosity, and these lavas have the greatest tendency to erupt explosively.

CONCEPT CHECK STOP

Why is it not possible to assign a single melting temperature to a rock?

How does water affect the melting temperature of a rock?

When and how can a rock melt without increasing in temperature?

How does fractional melting separate different minerals from one another?

Why and How Rocks Melt   289

Cooling and Crystallization of Magma

LEARNING OBJECTIVES

Explain how different igneous rocks may form from the same magma through magmatic differentiation by fractional crystallization.

Identify the two processes that operate in Bowen's reaction series.

Demonstrate how a rhyolitic magma can develop from a basaltic magma through the combination of a continuous and a discontinuous reaction series.

R efer back to Table 3.1, in which the six most common igneous rocks are listed. These rocks, three volcanic and three plutonic, are formed by **crystallization** of the three most common kinds of magma: *rhyolitic*, *andesitic*, and *basaltic*. In addition to the six most common types, hundreds of other igneous rock occur on Earth. Most are rare, but the fact that they occur suggests an important point: A magma of a given composition can crystallize into many different kinds of igneous rock. This is called **magmatic differentiation**.

FRACTIONAL CRYSTALLIZATION

How does magmatic differentiation occur? In other words, how can different rocks form from a single magma? A particularly important mechanism is the process of **fractional crystallization**. Just as *fractional melting* influences the composition of magma, so can the opposite effect, fractional crystallization, determine the composition of an igneous rock. Let's examine this process more closely.

Recall how fractionation occurs through fractional melting (see Figure 9.16). A newly forming melt is separated from a solid residue with a different composition; the result is magma and a rock with different compositions.

Crystallization in a cooling magma occurs over the same range of temperatures as melting, and the last minerals to melt are the first to crystallize. As shown in **FIGURE 9.17**, the newly formed crystals can become

crystallization The process whereby mineral grains form and grow in a cooling magma (or lava).

magmatic differentiation The formation of many different kinds of igneous rock from a single magma.

fractional crystallization Separation of crystals from liquids during crystallization.

Separating crystals from melts FIGURE 9.17

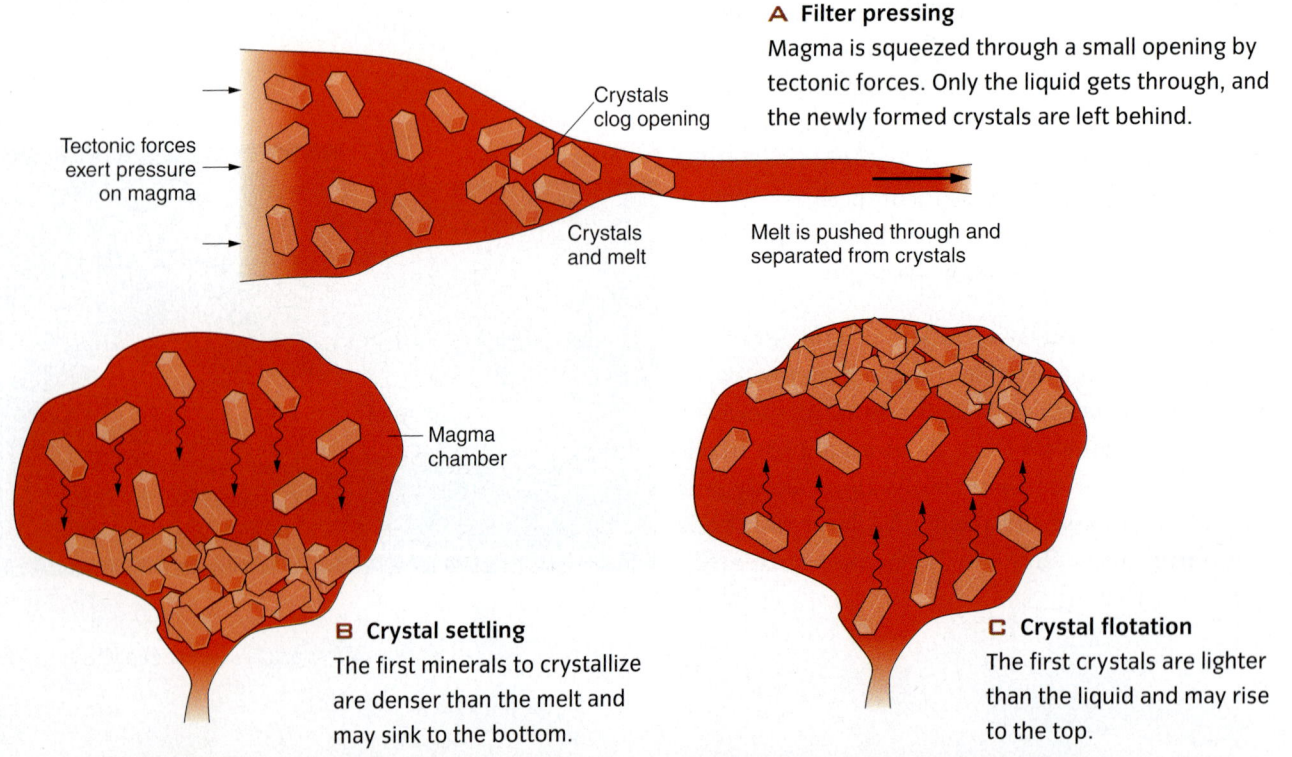

A Filter pressing
Magma is squeezed through a small opening by tectonic forces. Only the liquid gets through, and the newly formed crystals are left behind.

B Crystal settling
The first minerals to crystallize are denser than the melt and may sink to the bottom.

C Crystal flotation
The first crystals are lighter than the liquid and may rise to the top.

A This zoned feldspar grain, in andesite, is viewed through a microscope in polarized light in order to reveal the growth zoning. The core of the grain is calcium-rich feldspar; the rim is sodium-rich feldspar. Zones in between are intermediate in composition.

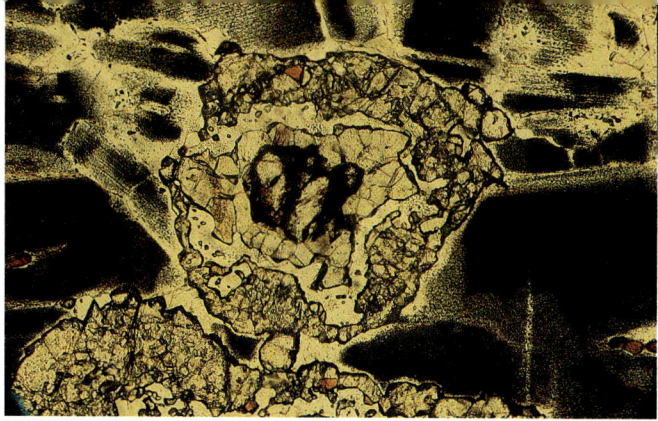

B The center of this mantled grain is olivine (O), surrounded by pyroxene (P), and on the outside, by amphibole (A). Olivine was the first mineral to crystallize; a mantle of pyroxene formed when olivine reacted with the residual magma. Still further reaction produced the mantle of amphibole.

Proof that compositions can vary as a result of crystallization FIGURE 9.18

separated from the remaining magma in several different ways. The result is a rock and a magma with different compositions, both different from the original magma. When you combine the different magma compositions, rates of cooling, and modes of fractional crystallization, you can see why there are so many different types of igneous rocks.

There are two more processes of fractional crystallization; they are *crystal zonation* and *mantling*. Both processes involve the growth of a protective rim around a newly forming crystal, sheltering the original core of the crystal from further contact with the magma. Crystal zonation happens in minerals that change composition through atomic substitution (Chapter 2) without change of crystal structure (**FIGURE 9.18A**). Mantling happens when an early-formed mineral reacts with the remaining magma and develops a mantle of a different mineral (**FIGURE 9.18B**). The person who first realized the importance of zonation and mantling to magmatic differentiation was the Canadian scientist N. L. Bowen.

BOWEN'S REACTION SERIES

Because magma of basaltic composition is by far the most common magma, Bowen hypothesized that all other magmas may be derived from basaltic magma by magmatic differentiation through fractional crystallization. The sequence of magma compositions and rock types that form during such a process is known as **Bowen's reaction series**.

> **Bowen's reaction series** The order in which minerals crystallize from, and subsequently react with, a cooling magma.

Continuous reaction series Bowen carried out a series of experiments to test his hypothesis. He knew that the first feldspar to crystallize from basaltic magma is calcium-rich (a plagioclase feldspar called *anorthite*). He also knew that feldspar crystallizing from rhyolitic magma is sodium-rich (a plagioclase feldspar called *albite*). Bowen's experiments confirmed that the composition of the first feldspar to crystallize from basaltic magma is calcium-rich. But the feldspar's composition changed and became richer in sodium as crystallization proceeded. Bowen referred to this change in composition as a *continuous reaction series*, by which he meant that the composition of a mineral in a crystallizing magma changed continuously (reacting chemically with the magma), even though its crystal structure remained unchanged.

For a feldspar crystal to change composition while maintaining its crystal structure, ions must *diffuse* into and out of the crystal. Diffusion is a slow process. Because cooling and crystallization rates are much faster than diffusion rates, a balance with diffusion is rarely attained. The result is zoned crystals of feldspar (Figure 9.18A). The early-formed calcium feldspar is surrounded by later-formed sodium feldspar. Here is the important point. Because a balance between crystallization and diffusion is not maintained, the feldspar is

Cooling and Crystallization of Magma

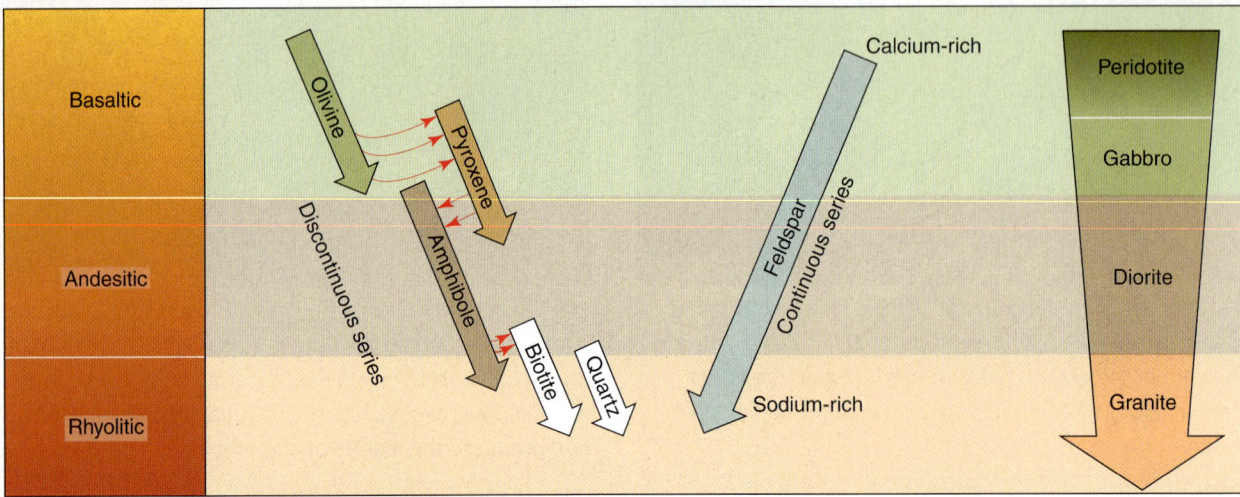

Bowen's reaction series FIGURE 9.19

The earliest minerals to crystallize from magma of basaltic composition are calcium-rich feldspar and olivine. As cooling and crystallization proceed, the calcium-rich feldspar reacts with the residual melt and continually changes composition, becoming richer in sodium. Meanwhile, the early-crystallized olivine re-

acts to form pyroxene. Pyroxene reacts in turn to form amphibole, and amphibole reacts to form biotite (mica). The composition of the remaining melt becomes increasingly silica-rich, and eventually the final small fraction of melt has the composition of a rhyolitic magma.

richer in calcium than it should be and the residual magma is richer in sodium than it should be: *If a sodium-rich melt were somehow separated from zoned, early-formed calcium feldspar crystals, the result would be a sodium-rich magma and a calcium-rich rock.*

Discontinuous reaction series

Bowen also identified compositional changes involving other minerals besides feldspars. One of the earliest minerals to form in basalt magma is olivine. Olivine contains about 40% SiO_2 (silica) by weight, whereas basaltic magma contains about 50% SiO_2. When silica-poor olivine crystallizes, it leaves the remaining magma a little richer in silica. Eventually there is an imbalance between olivine and the remaining magma; the olivine reacts with the magma to form a new, more silica-rich mineral, pyroxene. Figure 9.18B is an example of an olivine changing to pyroxene, which then changes to amphibole. Bowen referred to such chemical reactions, in which early-formed minerals form entirely new minerals through reaction with the melt, as *discontinuous reaction series*. The effect of removing the early-formed olivine from further reaction with the remaining magma through the formation of a mantle of pyroxene is like zoning in

feldspar—*a silica-rich magma and an assemblage of silica-poor minerals.*

Together, the continuous and discontinuous reaction series identified by Bowen are known as Bowen's reaction series (FIGURE 9.19). We now know that Bowen's hypothesis that all magmas are derived from basalt is not correct—fractional crystallization rarely goes to the length shown in Figure 9.19. But a careful examination of almost any igneous rock will reveal evidence that fractional crystallization involving Bowen's reaction series has occurred at some point in its formation. Magmatic differentiation, fractional crystallization, and Bowen's reaction series play crucial roles in the formation of the great variety of igneous rocks we find on Earth.

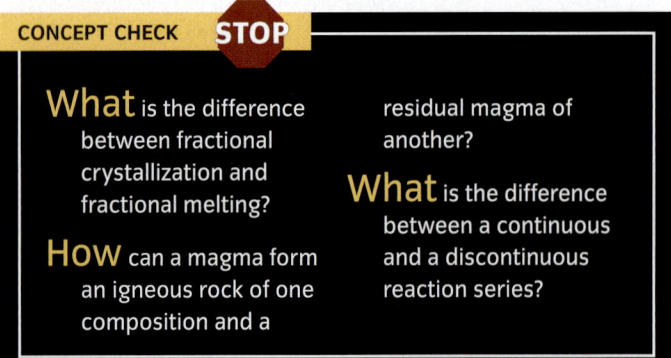

CONCEPT CHECK STOP

What is the difference between fractional crystallization and fractional melting?

How can a magma form an igneous rock of one composition and a residual magma of another?

What is the difference between a continuous and a discontinuous reaction series?

Plutons and Plutonism

LEARNING OBJECTIVES

Describe the most common plutonic formations.

Explain why volcanoes create plutonic rock and plutons.

Although they are formed underground, plutonic rocks give rise to some very dramatic geologic formations, known as **plutons** (FIGURE 9.20). Named after Pluto, the Roman god of the underworld, plutons are always *intrusive* bodies, different from the rock that surrounds them. They always originate as magma underground, but they are exposed as plutons at the surface by erosion.

BATHOLITHS AND STOCKS

Plutons are named according to their shapes and sizes. The largest type of pluton is a **batholith** (from the Greek words meaning "deep rock"). Some batholiths exceed 1000 km in length and 250 km in width (FIGURE 9.21 on page 294).

Where they are visible at the surface due to erosion, the walls of batholiths tend to be nearly vertical. This early observation led scientists to believe that batholiths extend downward to the base of Earth's crust. However, geophysical measurements suggest that this perception is incorrect. Most batholiths seem to be only 20 to 30 km thick.

A smaller version of a batholith, only 10 km or so in its maximum dimension, is called a *stock*. In some cases, as shown in Figure 9.20, a stock may be associated with a batholith that lies underneath it.

Most stocks and batholiths are granitic or between granite and diorite in composition. As mentioned earlier, Bowen's hypothesis for the formation of granite batholiths by fractional crystallization of basaltic magma does not stand up to testing. The magma that

pluton Any body of intrusive igneous rock, regardless of size or shape.

batholith A large, irregularly shaped pluton that cuts across the layering of the rock into which it intrudes.

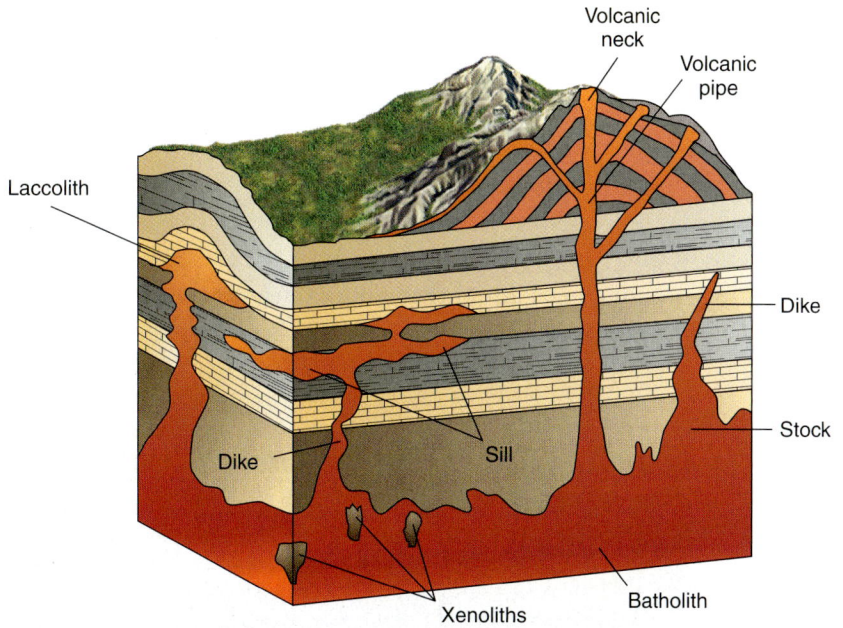

Plutons FIGURE 9.20

This diagram shows the origin of various forms taken by plutons. Note the vertical volcanic necks; sills parallel to the layering in the surrounding rocks; and dikes, which cut across the surrounding rock layers. In every case, the magma intrudes into previously existing rock.

forms batholiths results from extensive fractional melting of the lower continental crust. The heat that causes the melting comes from andesitic magma formed by wet-partial melting of mantle rocks as a result of subduction. Batholiths, such as the Sierra Nevada, were once capped by arcs of andesitic stratovolcanoes. Despite their huge size, the magma bodies that form batholiths migrate upward, squeezing into preexisting fractures and pushing overlying rocks out of their way.

DIKES AND SILLS

Smaller plutons tend to take advantage of fractures in the rock. Two of the most obvious indicators of past igneous activity are **dikes** and **sills** (FIGURE 9.22). A dike forms when magma squeezes into a cross-cutting fracture and then solidifies. If the magma intrudes between two layers and is parallel to them, it forms a sill. Sometimes this intrusion will cause the overlying rock to bulge upward, forming a mushroom-shaped pluton called a *laccolith*. As shown in Figure 9.20, all of these intrusive forms may occur as part of a network of plutonic bodies.

Dikes and sills can be very large. For instance, there is a large and well-known sill-like mass, made of gabbro, in the Palisades, the cliffs that line the Hudson River opposite New York City. The Palisades Intrusive Sheet is about 300 m thick. It formed from multiple charges of magma intruded between layers of sedimentary rock about 200 million years ago. The sheet is visible today because tectonic forces raised that portion of the crust upward, and then the covering sedimentary rocks were largely removed by erosion.

As Figure 9.20 shows, plutonic rocks can also be connected with volcanoes. Beneath every volcano lies a complex network of channels and chambers through which magma reaches the surface. When a volcano becomes extinct, the magma in the channels solidifies into various kinds of plutons. A *volcanic pipe* is the remnant of a channel that originally fed magma to the *volcanic vent*; when exposed by erosion, it is called a *volcanic neck* (Figure 9.22).

Batholiths FIGURE 9.21

Because batholiths are so immense, we cannot show one in a single photo. However, we can illustrate them on a map. The Coast Range batholith of southern Alaska, British Columbia, and Washington dwarfs the largest batholiths of Idaho and California. Many of the individual stocks and batholiths shown on this map are just the exposed tops of much larger intrusive bodies that lie underground.

CONCEPT CHECK STOP

What are some common kinds of plutons?

How do plutons become visible at Earth's surface?

Smaller plutons exposed by erosion FIGURE 9.22

◀ **A** A dike of gabbro cuts across horizontally layered sedimentary rocks in Grand Canyon National Park, Arizona.

▶ **B** The sill is the middle piece of this rock "sandwich," a layer of dark-brown gabbro intruded between layers of sedimentary rock above and below, in Big Bend National Park, Texas.

◀ **C** This volcanic neck called Devil's Tower, in Wyoming, is all that remains of an ancient, eroded volcano. You might remember this location for the role it played in the movie *Close Encounters of the Third Kind*.

Amazing Places: Mt. St. Helens

A Volcanic ash continued to erupt from the crater for nine hours.

C Here you see hummocks left by the largest debris avalanche in recorded history. Immediately after the eruption, this entire landscape was a barren gray. The terminus of the debris avalanche, where the flowing material stopped and hardened, dwarfs a passing truck in this photo.

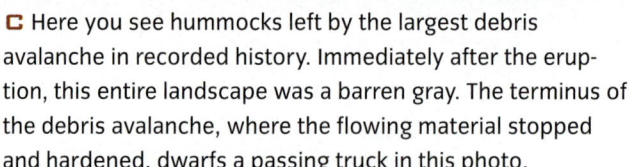

B Weeks after the eruption, not a living thing could be found in the vicinity of the volcano. Here, a forester inspects a forest that was flattened instantaneously by the blast wave, as if the trees were just blades of grass.

Mt. St. Helens is a young and active volcano that has erupted many times over the past 37,000 years. Even so, Earth scientists and nearby residents alike were stunned by the violence of its eruption in 1980. The mountain blew off its top and most of its northern flank. The initial blast rattled windows all over the state of Washington, flattened trees as far as 27 km away, and killed 57 people.

If you visit the Mt. St. Helens National Volcanic Monument, you will see many signs of revival amidst vistas of incredible desolation.

D Today the volcano is quiescent, but you may be able to spot steam rising from the resurgent dome in the center of the crater. This is a reminder that Mt. St. Helens may one day (if it remains active) grow back to its former height and violently explode or erupt again.

SUMMARY

1 Volcanoes and Volcanic Hazards

1. **Volcanoes** eject a wide variety of materials, including **lava** (molten rock, or **magma**, that has reached Earth's surface), gases, and *tephra*.

2. The diversity of volcanic eruption types is mostly due to two factors, the **viscosity** of the magma and the amount of gas present in it. The viscosity of the magma in turn depends on its chemical composition, temperature, and gas content. **Shield volcanoes** and fissure eruptions tend to be quiet and nonexplosive. They gradually build up the volcano through a series of lava flows. **Stratovolcanoes** tend to erupt explosively. They are built up from a series of layers of lava and pyroclastic material.

3. Funnel-shaped *craters* and much larger depressions known as *calderas* are common features at the summits of volcanoes. New magma may push up the floor of a crater or caldera and form a *resurgent dome*. If groundwater in the vicinity of an active magma chamber becomes heated, it may form *thermal springs* or *geysers*.

4. Direct, or *primary*, volcanic hazards of volcanoes are those directly caused by the volcanic eruption. They include **pyroclastic flows**, lava flows, and poisonous gases. *Secondary* hazards are those triggered by the eruption, such as *volcanic tremors* or *lahars*. Some extremely large eruptions produce *tertiary effects*, long-lasting and even permanent changes brought about by the eruption. One example is a worldwide drop in temperature because of *aerosols* in the upper atmosphere. Tertiary effects can also be beneficial, such as the creation of new land.

5. For purposes of prediction, volcanoes are categorized as *active, dormant,* or *extinct*. Volcanologists can often tell an eruption is coming because of changes in seismic activity, gas emissions, ground and water temperature, and slope of a volcano's sides.

2 Why and How Rocks Melt

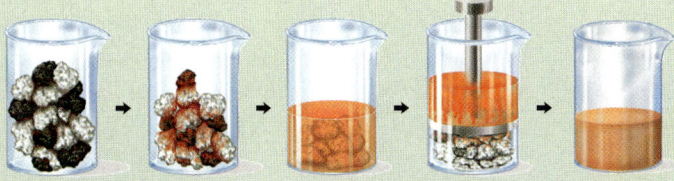

1. Both temperature and pressure increase with depth. The increase of temperature with depth is called the **geothermal gradient**; the rate of increase with depth is less through and under the continental crust than it is through and under the oceanic crust.

2. Minerals can melt in two ways: by an increase in temperature or a decrease in pressure (known as *decompression melting*). The presence of water in a rock typically lowers its melting point.

3. The minerals in a rock melt at different temperatures, so rocks do not have a single melting point; rather, they melt over a range of temperatures. A **fractional melt** is a body of rock in which some materials have melted and others have not. Fractional melting often separates minerals through a process called **fractionation**. When a rock begins to melt, therefore, only a small volume of melt—a partial melt—forms at first. In some circumstances, the melt may become segregated from the remaining solid rock (for example, by *filter pressing*), which may then continue to melt on its own. The separated material will have a different mineral composition from the original rock.

4. Magmas can be distinguished from one another by **composition**, **temperature**, and **viscosity**. Those with high-silica content usually melt at lower temperatures and have higher viscosities.

3 Cooling and Crystallization of Magma

1. **Crystallization**, the process whereby mineral grains form and grow in a cooling magma or lava, influences the final properties of the igneous rock that results. Besides the six common kinds of igneous rock, there are many additional kinds, formed by **magmatic differentiation** of a common type of magma.

2. An important kind of magmatic differentiation happens through **fractional crystallization**, which occurs when mineral grains become separated from the melt from which they are crystallizing. This separation can happen if the crystals sink to the bottom (*crystal settling*) or float to the top of the magma (*crystal flotation*). If the magma flows through an opening that is too constricted to allow the crystals to pass through, separation

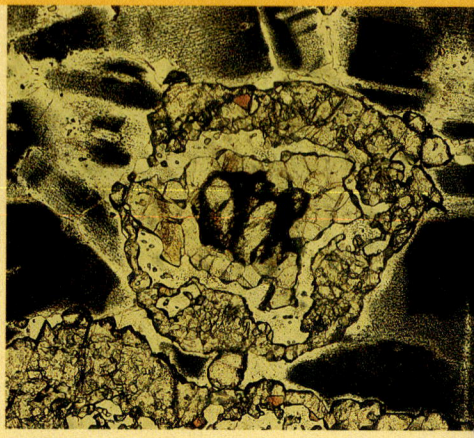

by *filter pressing* may occur. The combination of different magma compositions, rates of cooling, and fractional crystallization leads to the great diversity of igneous rock types on Earth.

3. Fractional crystallization can also occur when early-crystallized minerals react with the remaining melt and form either compositionally zoned (*continuous reaction series*) or mantled (*discontinuous reaction series*) crystals. The combination of continuous and discontinuous reaction series is called **Bowen's reaction series** and can produce small quantities of rhyolitic magma from a starting basaltic composition.

4 Plutons and Plutonism

1. Plutonic rock bodies, or **plutons**, form underground when magma intrudes into preexisting rock layers, either cutting through them or flowing between them.

2. Plutons, which occur in a variety of sizes and shapes, can be exposed at the surface when surrounding rock is stripped away by weathering and erosion. The largest type of pluton is a **batholith**, an irregularly shaped igneous body more than 50 km in diameter, which cuts across the rock it intrudes on. A *stock* is a smaller version of batholith. Two smaller types of plutons, **dikes** and **sills**, are indicators of past magmatic activity.

KEY TERMS

- volcano p. 272
- lava p. 272
- magma p. 272
- viscosity p. 273
- shield volcano p. 273
- pyroclastic flow p. 273
- stratovolcano p. 273
- geothermal gradient p. 285
- fractional melt p. 287
- fractionation p. 287
- crystallization p. 290
- magmatic differentiation p. 290
- fractional crystallization p. 290
- Bowen's reaction series p. 291
- pluton p. 293
- batholith p. 293

CRITICAL AND CREATIVE THINKING QUESTIONS

1. For many years, scientists debated the reasons for the existence of Earth's geothermal gradient. What explanations can you think of for the hotter temperatures toward the center of Earth? Why is the geothermal gradient steeper under the oceans?

2. Volcanic rock is sometimes called *extrusive* and plutonic rock is sometimes called *intrusive*. Why do you think that Earth scientists describe them this way? (You may wish to look up these words in a dictionary.)

3. What factors might prevent magma from reaching Earth's surface?

4. The slopes of active volcanoes tend to be populated. What reasons can you think of for living near a volcano? Do you think the advantages outweigh the disadvantages? Why or why not?

5. Several flood basalt eruptions apparently occurred at roughly the same time as mass extinctions that divide geologic eras or periods. But Earth scientists are not sure yet whether volcanic activity actually causes mass extinctions. What are some arguments for and against this theory?

What is happening in this picture?

Pumice, a volcanic rock, is so light that this woman has no trouble holding a large armful.

- How can a rock be so light?
- What does this rock tell us about the magma from which it came?

SELF-TEST

1. _____ are explosive eruptions characterized by pyroclastic flows and ash plumes that extend into the stratosphere.
   a. Vulcanian
   b. Hawaiian
   c. Plinian
   d. Strombolian

2. _____ eruptions consist of low-viscosity lava that flows easily from a volcanic vent.
   a. Vulcanian
   b. Hawaiian
   c. Plinian
   d. Strombolian

3. These two photographs show Mauna Loa, Mauna Kea, and Mt. Fuji volcanoes. Of these volcanoes, which has the greatest potential for an explosive eruption?
   a. Mauna Loa
   b. Mt. Fuji
   c. Mauna Kea
   d. They all have equal potential for an explosive volcanic eruption.

4. Which of the volcanoes depicted in the photos in Question 3 is being fed by magma with the highest viscosity?
   a. Mauna Loa
   b. Mt. Fuji
   c. Mauna Kea
   d. The magma composition is probably identical for all of these volcanoes.

5. _____ form when the chamber of magma underlying a volcano empties due to eruption and the unsupported roof of the chamber collapses under its own weight.
   a. Fumaroles
   b. Geysers
   c. Calderas
   d. Craters

6. The greatest threats to human life during volcanic eruptions do not come from lava but from _____ and _____.
   a. pyroclastic flows/volcanic gases
   b. pyroclastic flows/ash fall
   c. ash fall/mudflows
   d. ash fall/volcanic gases

7. One way scientists monitor volcanic activity is by studying changes in the shape of volcanic features using _____.
   a. temperature gauges
   b. seismographs
   c. geologic studies of past eruptions
   d. tiltmeters

8. Melting of a rock can occur because of _____ and can be facilitated by the presence of water.
   a. increasing temperature or increasing pressure
   b. increasing temperature or decreasing pressure
   c. decreasing temperature or decreasing pressure
   d. decreasing temperature or increasing pressure

9. The three most common kinds of magma are _____, _____, and _____.
   a. rhyolitic/andesitic/pegmatitic
   b. rhyolitic/andesitic/basaltic
   c. rhyolitic/pyroxenitic/basaltic
   d. andesitic/albitic/basaltic

10. In the process of _____, crystals that have already formed in magma become separated from the remaining melt.
    a. magmatic redistribution
    b. distributed crystallization
    c. fractional crystallization
    d. fractional melting

11. Crystal settling and crystal flotation in a magma is controlled by differences in the _____ of the crystals and magma.
    a. viscosity   c. density
    b. hardness   d. All of the above are correct.

12. A continuous reaction series is one in which a mineral _____.
    a. reacts with other minerals in a mineral assemblage
    b. reacts with magma to form a different mineral
    c. reacts with magma and changes composition but not crystal structure
    d. reacts to changes in temperature and pressure

13. Label this block diagram, depicting various plutonic bodies, using the following terms:

    dike
    volcanic neck
    sill
    stock
    batholith
    xenoliths

    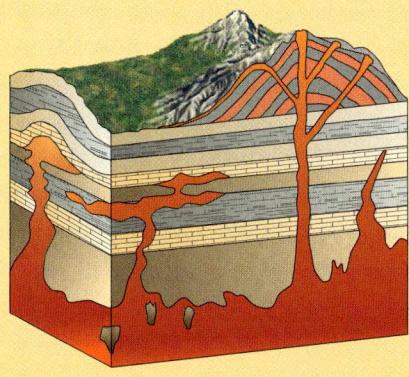

14. How will the plutonic features depicted in the block diagram in Question 13 become exposed at Earth's surface?
    a. Through continued volcanic eruptions.
    b. Through uplift and erosion.
    c. Through melting of the overlying rock.
    d. Only after an explosive volcanic eruption.

15. Batholiths are huge masses of igneous rock that _____.
    a. form the margins of continents
    b. stretch from the surface to the base of the crust
    c. extend for hundreds of miles but are only about 20 miles deep
    d. underlay the midocean ridges

How Old Is Old? The Rock Record and Deep Time

10

Global Locator

What's in a name? Say the word *Jurassic*, and most people probably think of *Jurassic Park*, the blockbuster movie by Steven Spielberg. Next, they might recall the stars of the movie—the terrifying *Tyrannosaurus rex* and the fast and wily velociraptors (for the movie, modeled on close relative *Deinonychus*).

They probably wouldn't think about the Jura Mountains in Switzerland (*right*). These lushly wooded limestone hills contain millions of fossils of extinct sea creatures called *ammonites*, which lived in coiled seashells reminiscent of the modern nautilus (see inset). In the 19th century, Earth scientists gradually realized that these creatures—and the rocks in which they were fossilized—all came from a particular time in Earth's history. They named it the *Jurassic Period*, after the mountain range where they found some of their best and most distinctive samples of ammonites. It is one of the delightful ironies of science that the fossil chosen to define the period was not a mighty dinosaur but a humble sea creature.

(By the way, no tyrannosaurs or velociraptors lived in the Jurassic Period, though many other dinosaurs did. The name "Cretaceous Park" would have been more accurate.)

Fossils and rocks don't come with a date stamped on them. So how do scientists tell when an ancient animal lived or when a particular layer of rocks was deposited? We will describe several methods in this chapter. It has taken scientists two centuries to piece together Earth's chronology, and they are still refining it.

NATIONAL GEOGRAPHIC

CHAPTER OUTLINE

- Relative Age p. 304

- The Geologic Column p. 312

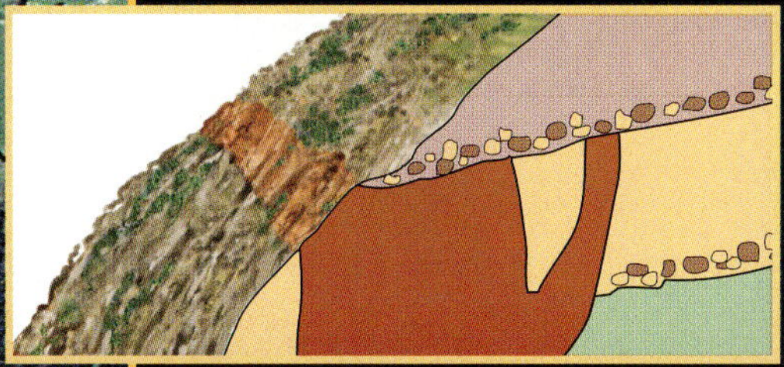

- Numerical Age p. 317

- The Age of Earth p. 325

Relative Age

LEARNING OBJECTIVES

Define relative age and numerical age.

Define stratigraphy and identify the four main principles of stratigraphy.

Explain why there are many gaps in the rock record.

Describe how fossils make it possible for Earth scientists to correlate strata in different places.

Through most of human history, when people thought about the origins of features like mountain ranges and oceans, they tended to think in terms of catastrophic events—mountains being thrust up in a single paroxysm or floods that covered the world. This line of thinking came to be called *catastrophism*. Toward the end of the 18th century, a Scottish scientist, James Hutton, made careful studies of the landscape. He noted that erosion of mountains proceeded very slowly and that features he could see in today's sediments were similar to those he could see in ancient rocks that had once been sediments. This led Hutton to the realization of an important principle of science, which later came to be called **uniformitarianism**. One way to express the principle is "the present is the key to the past."

The principle of uniformitarianism provided the first steps toward understanding Earth's history, and further, because things happen so slowly, to the realization that Earth must be incredibly ancient. The first scientific attempts to determine the numerical extent of Earth's history were made a little over two centuries ago. These early Earth scientists speculated that they might be able to estimate the time needed to erode away a mountain range by measuring the rate at which sediment was transported by streams. Their attempts were all underestimates, but the inescapable conclusion supported Hutton—Earth must be millions of years old because of the great thickness of sedimentary rocks. Hutton was so impressed by the evidence that in 1788 he wrote that for Earth there is "no vestige of a beginning, no prospect of an end."

Hutton made careful observations and used the scientific method. His conclusions were at odds with the belief that catastrophes had shaped the landscape and that life on Earth was recent. Such ideas were more in line with those of James Ussher, Archbishop of Armagh in Ireland. In 1658, Ussher published his book, *Annals of the World*, in which he reported his calculations of the date of Creation. He based his work on the histories of the civilizations of the Middle East and the Mediterranean, and on the Bible, and concluded that the day of Creation was Sunday, October 23, 4004 B.C. Sir John Lightfoot, a contemporary of Ussher and Vice-Chancellor of Cambridge University in England, went a step further; he made similar calculations and concluded that the time of day was 9 A.M.

The scientists who followed Hutton agreed with his conclusion that Earth must be very ancient, but they lacked a precise way to determine exactly how long ago a particular event occurred. The only thing they could do was determine the sequence of past events. They could thus establish the **relative ages** of rock layers or other natural features, which means that they could determine whether a particular layer or feature was older or younger than another layer or feature. Relative ages are derived from three basic principles of **stratigraphy**, generally attributed to Hutton and Steno (see Chapter 2), and to the principle of cross-cutting relationships.

uniformitarianism The concept that the processes governing the Earth system have operated in a similar manner throughout Earth's history and that past events can be explained by phenomena and forces observable today.

relative age The age of a rock layer, fossil, or other natural feature relative to another feature.

stratigraphy The science of rock layers and the processes by which strata are formed.

STRATIGRAPHY

In places where you can find large exposed rock formations, such as the Badlands of South Dakota and the American Southwest, you will often see that the rocks have a banded appearance (see **FIGURE 10.1**).

Visualizing

Strata and the law of horizontality FIGURE 10.1

All of these strata were originally deposited as horizontal layers; the strata in **B** and **C** were disrupted by later tectonic activity.

◀ **A** Horizontal strata in Badlands State Park, South Dakota.

B Tilted strata in Telfer Gold Mine, Great Sandy Desert, Western Australia. The tiny figure of a person walking along a mining road gives an idea of scale.
▼

◀ **C** Folded strata in Hamersley Gorge, Western Australia.

Relative Age   305

What an Earth Scientist Sees

Global Locator

The Principle of Stratigraphic Superposition

The Grand Canyon is considered one of the natural wonders of the world, and so it is not surprising that nearly 5 million people visit it every year. To an Earth scientist, the Grand Canyon is more than just a scenic stop. It is also a beautiful example of the record of sedimentation from millions of years ago.

A panoramic view of the Grand Canyon from the Toroweap Overlook shows a nearly 2-kilometer thickness of horizontal, sedimentary strata lying on top of older strata that were tilted and tectonically deformed before the horizontal strata were deposited. According to the principle of stratigraphic superposition, each stratum is younger than the layer below it (inset). Think of a pile of newspapers as an analogy. If a paper is added to the pile each day, the paper on the bottom will always be the oldest, with the most recent on top. Each stratum in this section of the Grand Canyon is named after its predominant rock type and the location in the Grand Canyon where the best outcroppings occur (such as the Toroweap Overlook). The tilted strata, called the Grand Canyon Supergroup, are now known to be much older than the horizontal ones, but stratigraphy can tell us only the relative ages.

Here's an interesting question:
- Use an average rate of deposition of sediment of 1 cm a year, and estimate the time needed for 2 km of sedimentary rocks to be deposited above the older, tilted rocks. What might the errors be in estimating an age by this method?

These bands are called **strata** (an individual band is a *stratum*), from the Latin word for "layer." The bands are often horizontal, but it is not unusual to see them tilted or bent.

All of the rocks in a typical stratified formation are sedimentary, deposited over the ages by water. This observation leads to the first of four key laws or principles (see Chapter 1) of stratigraphy, the *law of original horizontality*, which states that water-laid sediments are deposited in horizontal strata. You can test this law yourself. Shake up a bottle of muddy water so that all of the particles are suspended. Let the bottle stand and then examine the result—the mud will be deposited at the bottom of the bottle in a horizontal layer. Therefore, whenever we observe water-laid strata that are bent, twisted, or tilted so that they are no longer horizontal, we can infer that some tectonic force must have disturbed the strata after they were deposited (see **FIGURES 10.1B** and **C**).

A second key to stratigraphy, which, like the first, is based on common sense, is the *principle of stratigraphic superposition*. This principle states that in any undisturbed sequence of strata, each stratum is younger than the stratum below it and older than the stratum above it (see *What an Earth Scientist Sees*).

Finally, the third key to stratigraphy is the *principle of lateral continuity* of sedimentary strata. This principle is based on the observation that sediments are deposited in continuous layers. A layer of sediment will extend horizontally as far as it was carried by the water that deposited it. Layers of sediment do not terminate abruptly, though they get thinner and ultimately pinch out altogether at their farthest edges. The same is true of sedimentary rocks; they may thin or pinch out laterally, but they are continuous sheets.

In addition to the three principles of stratigraphy, there is a fourth principle that is important for determining relative ages; it is the *principle of cross-cutting relationships*, which states that a stratum must always be older than any feature that cuts or disrupts it. If a stratum is cut by a fracture, for example, the stratum itself is older than the fracture that cuts across it, as shown in **FIGURE 10.2**. When magma fills a fracture, the result is a vein of igneous rock that cuts across the strata. In this case, too, the sedimentary rocks must be older than the cross-cutting vein. Similarly, a "foreign rock" (called a *xenolith* or an *inclusion*) that is encased within another rock unit must predate the rock that surrounds it.

The principle of cross-cutting relationships FIGURE 10.2

A These fractures are younger than the strata they cut, according to the principle of cross-cutting relationships. The fractures are in a sequence of sandstone strata in Merseyside, United Kingdom.

B At Three Valley Gap in Alberta, Canada, layers of metamorphic rock are sliced by two darker, cross-cutting dikes of igneous rock. An Earth scientist sees four stages: deposition of sediment, formation of sedimentary rocks, metamorphism into gneiss, and intrusion of the dikes. By the principle of cross-cutting relationships, the dikes must be the youngest feature.

Process Diagram

Unconformities: Gaps in the rock record FIGURE 10.3

Any boundary that represents a gap in the sedimentary record is called an unconformity. This diagram illustrates the formation of three common types of unconformities: ❶ nonconformity, ❷ angular unconformity, ❸ disconformity.

STAGE 1 — Layers of sediment deposited on an erosional surface become young sedimentary rocks. Ancient igneous and metamorphic rocks are underlying. (Ocean)

Ocean recedes, exposing new surface rocks

STAGE 2 — Tectonic forces distort strata; erosion carves the surface. New unconformity is in the making. (Tectonic forces)

Ocean returns

STAGE 3 — New sedimentary strata are deposited atop old eroded surface.

Ocean recedes, new surface exposed

STAGE 4 — New erosional surface is carved. Uplift distorts strata.

Ocean returns

STAGE 5 — Cycle continues; new rocks form.

GEOLOGIC TIME: Millions of years ago → Today

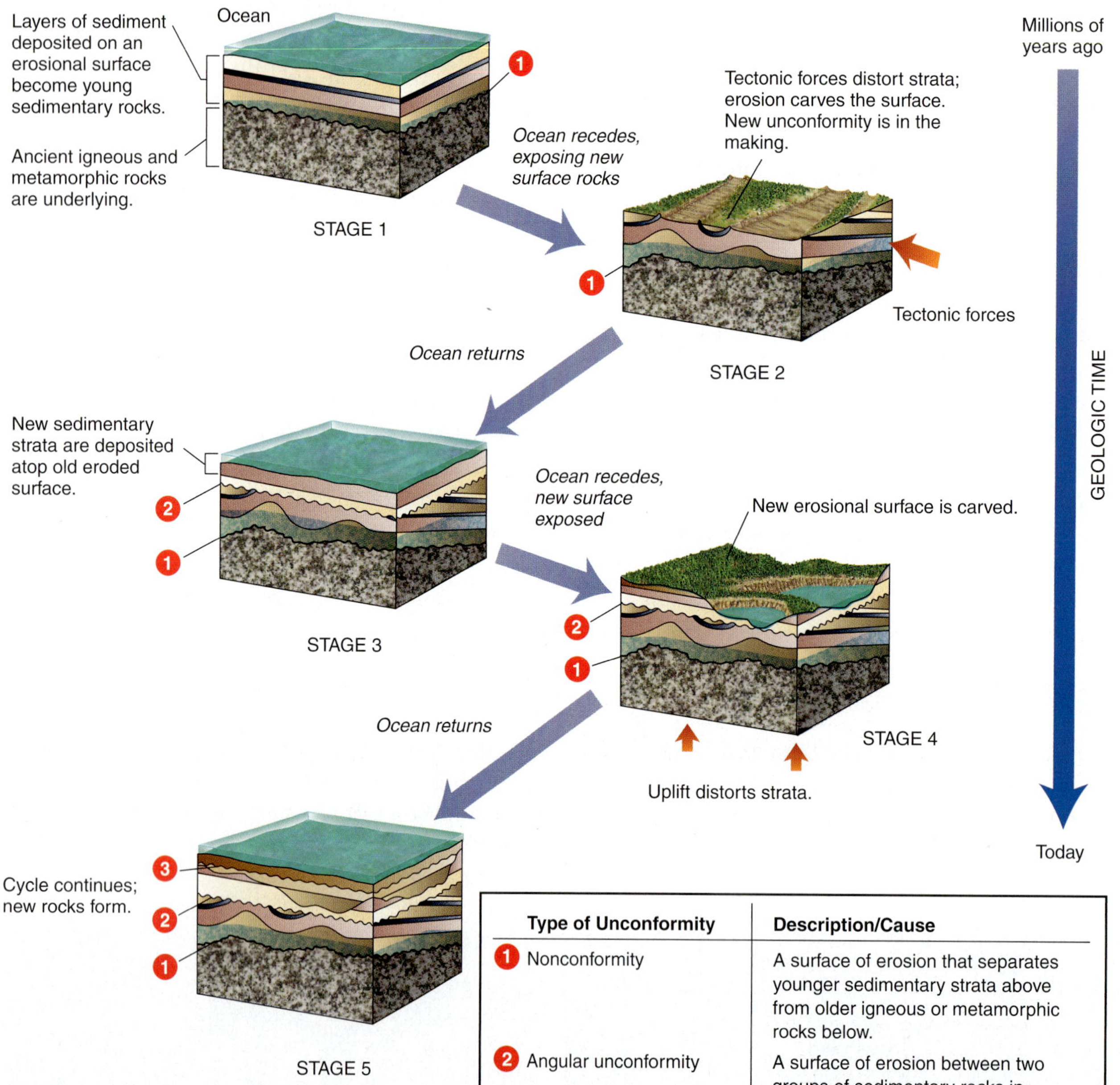

Type of Unconformity	Description/Cause
❶ Nonconformity	A surface of erosion that separates younger sedimentary strata above from older igneous or metamorphic rocks below.
❷ Angular unconformity	A surface of erosion between two groups of sedimentary rocks in which the orientation of older strata, below, are at an angle to younger strata, above.
❸ Disconformity	A surface of erosion in which the orientation of older strata, below, are parallel to younger strata, above.

CRITICAL THINKING — Here are two interesting questions:
- Would a surface between adjacent, parallel layers of sediment be a disconformity if erosion had not occurred?
- What is a common circumstance where such surfaces might be observed?

308 CHAPTER 10 How Old Is Old? The Rock Record and Deep Time

GAPS IN THE RECORD

Nineteenth-century Earth scientists who followed Hutton tried to estimate the **numerical age** of rocks—the exact number of years that have elapsed since a given stratum was deposited. For example, observations might suggest that it would take 1 year for 2 centimeters of sediment to accumulate. If a stratum lies 40 meters (4000 cm) below the ground surface under younger sedimentary strata, then its numerical age could be estimated to be about 2000 years (4000 cm ÷ 2 cm/year). Three assumptions must be true for this kind of approach to work. First, the rate of sedimentation must be constant while the layers are being deposited. Second, the thickness of sediment must have been the same as the thickness of the sedimentary rock that eventually formed from it. Finally, all strata must be *conformable*. This means that each layer was deposited on the one below it without any interruptions—in other words, there are no depositional gaps in the stratigraphic record.

> **numerical age** The age when a rock layer or natural feature was formed, in years before the present.

There are problems with each of these assumptions. Rates of sedimentation vary greatly. Sediments are often compressed while they are turning into rock, resulting in a much thinner stratum than was originally deposited. The conformity assumption also fails; gaps in the stratigraphic record, or **unconformity**, are common and occur for a variety of reasons (see **FIGURE 10.3**). In sum, all of the numerical ages determined by early Earth scientists from the thickness of stratigraphic sequences turned out to be underestimates.

> **unconformity** A substantial gap in a stratigraphic sequence that marks the absence of part of the rock record.

Hutton was particularly impressed by the evidence he observed at Siccar Point, not far from Edinburgh (**FIGURE 10.4**). There he saw ancient sandstone layers, originally horizontal but now standing vertically (**FIGURE 10.4A**) and capped by layers of younger sandstone (**FIGURE 10.4B**). He realized that the boundary between the layers was an ancient *erosion surface*.

The angular unconformity at Siccar Point, Scotland FIGURE 10.4

Here, in 1788, James Hutton first demonstrated that Earth processes of deposition–uplift–erosion repeat cyclically and endlessly. The vertical layers of 450-million-year-old sedimentary rock (A), originally horizontal, were uplifted and tilted to vertical. Exposed to weathering, these rocks eroded to form a new land surface (C), upon which layers of younger sediments (B), were laid down. These younger sediments, now gently sloping, are 370 million years old. The angular unconformity is therefore an 80-million-year gap in the stratigraphic record.

Varieties of fossils FIGURE 10.5

▲ Most fossils begin as hard plant or animal parts, such as the bones of this 23-cm-long Pachypleurosaurus, a marine reptile that lived about 230 million years ago and was preserved in rocks now found in Switzerland.

▲ Other fossils record the impressions of soft tissue on mud that later solidified. The lobe-shaped fins of fish like this *Eusthenopteron,* preserved here in 380-million-year-old rocks found in Quebec, eventually evolved into legs.

◄ These fossils, found in Sweden, record the slithering, sliding, and digging of some marine creatures, identity unknown, about 530 million years ago.

FOSSILS AND CORRELATION

Many strata contain remains of plants and animals that were incorporated into the sediment as it accumulated. **Fossils** usually consist of "hard parts," like shells, bones, or wood, whose forms have been preserved in sedimentary rocks (see FIGURE 10.5). In some cases, the imprints of soft animal tissues, like skin or feathers, or the leaves and flowers of plants, have been preserved. Even the preserved tracks and footprints of animals are considered to be fossils. Many fossils found in strata that are young compared to the age of Earth look similar to plants and animals living today (see FIGURE 10.6). The farther down in the stratigraphic sequence we go, the more likely we are to find fossils of extinct plants or animals, and the less familiar they seem.

Nicolaus Steno, whom we introduced in Chapter 2 in connection with crystals, was also one of the first scientists to conclude that fossils were the remains of ancient life. He published his ideas in a landmark paper in 1669, in which he also stated the principles of stratigraphic superposition and original horizontality. His conclusions were ridiculed at the time, but by the next century, the idea of the plant and animal origin of fossils was widely accepted.

Ancient and modern FIGURE 10.6

A fossilized imprint of a *Lebachia* (A), a conifer that is found in rocks about 250 million years old, strongly resembles a modern Norfolk pine (B).

The study of fossils, called **paleontology**, is hugely important for understanding the history of life on Earth, and we will have much more to say about it in Chapter 11. Aside from their biologic interest, fossils also have great practical value to Earth scientists. About the same time that James Hutton was working in Scotland, a young surveyor named William Smith was laying out canal routes in southern England. As the canals were excavated, Smith noticed that each group of strata contained a specific assemblage of fossils. In time, he could look at a specimen of rock from any sedimentary layer in southern England and name the stratum and its position in the sequence of strata. This skill enabled him to predict what kind of rock the canal excavators would encounter and how long it would take to dig through it. It also earned him the nickname "Strata."

The stratigraphic ordering of fossil assemblages is known as the *principle of faunal and floral succession*. *Fauna* refers to animals, *flora* refers to plants, and *succession* means that new species succeed earlier ones as they evolve. William Smith's practical discovery turned out to be of great scientific importance. Faunal and floral successions can be employed in determining the sequence of strata and therefore are important in fixing the relative ages of strata. But the successions turned out to have an even more important role to play in the development of Earth

> **paleontology** The study of fossils and the record of ancient life on Earth; the use of fossils for the determination of relative ages.

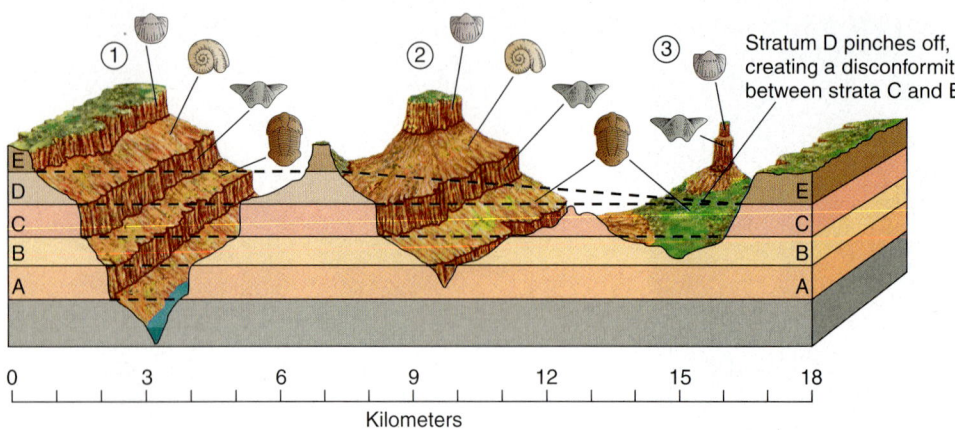

Fossils and correlation
FIGURE 10.7

Earth scientists can use fossils to correlate strata at localities (1, 2, and 3) that are many kilometers apart.

Strata **B, C, D,** and **E** have fossil assemblages that are different from one another but consistent among the three sites. Note that stratum **D** is missing at locality 3, because **E** directly overlies **C**. Why?

Answer: Either **D** was never deposited there or it was deposited and later removed by erosion. Either way, the boundary between **C** and **E** is a disconformity.

science. Scientists soon demonstrated that the faunal succession in northern France is the same as that found by Smith in southern England. By the middle of the 19th century, it had become clear that faunal succession is essentially the same everywhere. Thus Smith's practical observations led to a means of worldwide **correlation**, which is the most important way of filling the gaps in the stratigraphic record (see **FIGURE 10.7**).

■ **correlation**
A method of equating the ages of strata that come from two or more different places.

CONCEPT CHECK STOP

What are the law and three principles on which stratigraphy is based?

How can the law and principles of stratigraphy be used to determine the relative ages of strata?

What is the difference between conformable and unconformable strata?

How can fossils provide information about the relative ages of strata?

The Geologic Column

LEARNING OBJECTIVES

Explain how worldwide observations of strata and the fossils they contain led to a single sequence of relative ages called the *geologic column*.

Distinguish among four units of geologic time: eons, eras, periods, and epochs.

Explain how different eras and periods correspond to different fossil assemblages.

Worldwide stratigraphic correlation (see **FIGURE 10.8**) was one of the greatest successes of 19th-century science. It meant that a gap in the stratigraphic record in one place could be filled using evidence from somewhere else. Through worldwide correlation, 19th-century scientists assembled the **geologic column**, or *stratigraphic time scale*, a composite diagram showing the succession of all known strata, fitted to-

■ **geologic column**
The succession of strata, fitted together in relative chronological order.

gether in chronological order, on the basis of evidence of relative age (see **FIGURE 10.9A** on the following page).

EONS AND ERAS

Even the most cursory inspection of the geologic column reveals how closely our understanding of strata is intertwined with the history of life. The vast majority of Earth's history is divided into three **eons** in which fossils are extremely rare or nonexistent, and in which information about the relative ages of strata is similarly sparse. The earliest eon is that time between Earth's formation and the age of the oldest rocks, so far discovered; because there are no known rocks, there is no accepted name for this earliest time of Earth's history (it is sometimes called the *Hadean*, an allusion to the probable hellish conditions at that time). The most ancient rocks preserved on Earth mark the base of the *Archean* ("ancient") Eon, and the Archean is followed by the *Proterozoic* ("early life") Eon. The Archean is roughly the period when single-celled life developed, and the Proterozoic is when multi-celled, soft-bodied organisms first emerged. We now know that each eon spanned several hundred million years of time, as shown in Figure 10.9A; however, the 19th-century scientists who first worked out the geologic column had no way of determining this information.

At the beginning of the current eon, called the *Phanerozoic* ("visible life"), the fossil record suddenly becomes much more detailed, thanks to the appearance of the first animals with hard shells and skeletons. Thus, paleontologists of the 19th century divided the Phanerozoic Eon into three shorter units called **eras**: the *Paleozoic* ("ancient life"), *Mesozoic* ("middle life"), and *Cenozoic* ("recent life"). These eras were separated by major extinction events, when more than 70% of the species on Earth perished (see **FIGURE 10.9B**). More recently, with the development of radiometric dating methods (discussed later in the chapter), the Archean and Proterozoic Eons have also been divided into eras.

Markers of Earth's history   FIGURE 10.8

These ammonite fossils are encased in limestone and came from the Jura Mountains in Switzerland. Such fossils are typical of those used to correlate strata from place to place. Rocks anywhere in the world that contain the same species of ammonite can be reliably dated to the Jurassic Period (named after the Jura Mountains). Other species of marine invertebrates can be used to correlate rocks from other periods. While dinosaur and other vertebrate fossils may attract more popular attention, marine invertebrate fossils are the ones that have been most helpful in correlation and allowing scientists to reconstruct the history of our planet because they are more common and widely distributed than dinosaur fossils.

A The geologic column in words

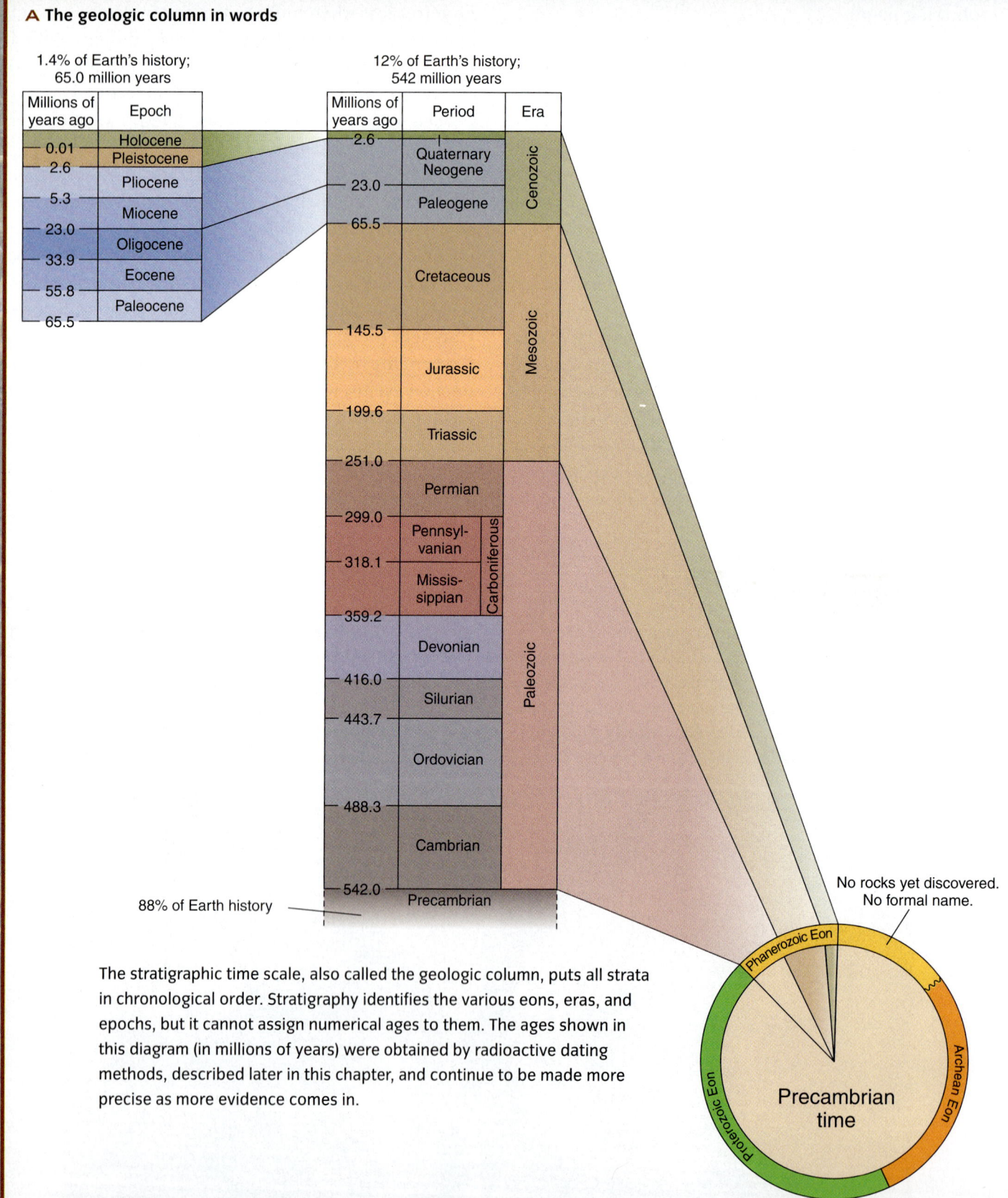

The stratigraphic time scale, also called the geologic column, puts all strata in chronological order. Stratigraphy identifies the various eons, eras, and epochs, but it cannot assign numerical ages to them. The ages shown in this diagram (in millions of years) were obtained by radioactive dating methods, described later in this chapter, and continue to be made more precise as more evidence comes in.

Visualizing

The geologic column FIGURE 10.9

B The geologic column in pictures

Cenozoic Era
In the Cenozoic Era, birds and mammals flourished. In this scene from 55 million years ago in the Paleogene Period, we can see some possible ancestors of primates.

Mesozoic Era
The Mesozoic Era saw the rise of dinosaurs, which were the dominant vertebrates (animals with backbones) on land for many millions of years. If you look closely in the corners, you can see two harbingers of the future that first appeared in the Mesozoic Era: the first magnolia-like flowering plants (*lower left*) and the first shrew-like mammals (*lower right*).

Paleozoic Era
During the Paleozoic Era, the evolution of life progressed from marine invertebrates (animals without backbones) to fish, amphibians, and reptiles. In this scene from 350 million years ago during the Carboniferous Period, a lobe-finned fish like the one in Figure 10.5 comes to life (*foreground*) along with some early amphibians whose fins have evolved into legs, but who still have a fishlike tail.

The Geologic Column

The Cambrian Explosion FIGURE 10.10

Only the sketchiest remains are left of any life-forms from Earth's earliest eons. But in the Cambrian Period there is abundant evidence of a profusion of bizarre and now extinct forms. Trilobites (*below*) ranged from 1 millimeter in size to the largest known specimen, 72 cm.

Anomalocaris was the most fearsome predator of the Cambrian seas, a swimming creature up to 1 m long.

Opabinia, which among other oddities had five eyes, was about 7 cm long and would have been a tasty morsel for *Anomalocaris*.

PERIODS AND EPOCHS

Eras are divided into still shorter units called **periods**. The three eras of the Phanerozoic Eon were the first to be divided into periods, and the divisions were made on the basis of fossils. Some are delineated by massive die-outs, or extinctions, of species. Periods were named in a somewhat haphazard manner, with some named for geographic locations (e.g., *Jurassic*, from the Jura Mountains in Switzerland where this layer was first studied) and others named for the characteristics of their strata (e.g., *Cretaceous*, from the Latin word for chalk).

The earliest period of the Paleozoic Era, the Cambrian Period, is especially noteworthy. Not only is this when animals with hard shells first appeared in the geologic record, but in the opinion of many paleontologists it was a time of unparalleled diversification of life, a phenomenon called the *Cambrian Explosion* (see **FIGURE 10.10**). Rocks that formed during the final period on the Proterozoic Eon, the Ediacaran Period, contain imprints of large, soft-bodied fossils, but all fossils in strata older than those of the Ediacaran, if they exist at all, are generally microscopic. Thus, Earth scientists often lump the enormous 4-billion-year swath of time that precedes the Cambrian Period, and the rocks deposited in them, into one category: **Precambrian**.

Periods, as we now know, lasted for tens of millions of years. Thus, Earth scientists find it helpful to split the periods of the Phanerozoic Eon into still smaller divisions called **epochs**. The names of the epochs of the Paleogene, Neogene, and Quaternary periods are somewhat more familiar than others because human ancestors emerged during these epochs. These recent epochs are not defined by extinction events, but according to the percentage of their fossils that are represented by still-living species. Many plant and animal fossils found in Pliocene strata, for example, have still-living counterparts, but fossils in Eocene strata have few still-living counterparts.

CONCEPT CHECK STOP

What are the major subdivisions of Earth's time scale?

Why does the time scale apply to rocks everywhere on Earth, and not just to those in one locality?

What major biologic event distinguishes the Phanerozoic Eon from the preceding (Precambrian) eons?

Numerical Age

LEARNING OBJECTIVES

Describe how scientists finally arrived at a way of measuring numerical time.

Describe the process of radioactive decay.

Explain how and why radioactive decay can be used to date the time of formation of igneous rocks.

Explain how reversals of magnetic polarity can be used to date both igneous and sedimentary rocks.

The scientists who worked out the geologic column were tantalized by the challenge of numerical time. They wanted to know Earth's age, how fast mountain ranges rise, how long the Paleozoic Era lasted, and most challenging of all, when life first appeared, and how long humans have inhabited Earth.

EARLY ATTEMPTS

Several methods for solving the problem of numerical time were proposed during the 19th century. All of them were unsuccessful, but the reasons for their failure are nonetheless quite illuminating—and they help explain where our currently accepted estimates came from (see **FIGURE 10.1** on the following page). As previously mentioned, an early approach involved estimating rates of sedimentation and multiplying by the thickness of stratigraphic sections. Unfortunately, the resulting estimates for Earth's age varied too widely to be useful, from 3 million to 1.5 billion years.

Early scientists recognized that rivers carry dissolved substances to the sea from the erosion of rocks on land. In 1715, Edmund Halley, for whom Halley's Comet is named, suggested that one could measure the rate at which salts are added to the sea by river input and calculate the time needed to transport all the salts now present in the sea. Halley did not carry out his own suggestion, and it was not until 1899 that John Joly made the necessary measurements and calculations. Joly estimated that the ocean and, therefore, Earth were 90 million years old. Unfortunately, neither Halley nor Joly realized that the ocean, like other parts of the Earth system, is an open system. Salt is removed by reactions between seawater and volcanic rocks on the seafloor, as well as by the evaporation of seawater in isolated basins. The addition and removal of salts have balanced each other for hundreds of millions (even billions) of years.

The 19th century also saw the publication of Charles Darwin's *On the Origin of Species* in 1859. The book intensified the ongoing debate over numerical ages. Darwin understood that the evolution of new species by natural selection must be a very slow process that needed vast amounts of time. The first edition of his book contained a rough estimate of Earth's age, based on the erosion rate of a mountain range, of at least 300 million years.

However, William Thomson (Lord Kelvin), a leading physicist and contemporary of Darwin, emphatically rejected Darwin's estimate. Kelvin used the laws of thermodynamics to calculate how long Earth has been a solid body. Kelvin made two assumptions: Earth was once completely molten and no heat was added to it after it formed. Once it had cooled enough to form a solid outer layer, heat would escape only by conduction (the same way that heat moves through the wall of a coffee cup). Kelvin measured the current-day rate of heat loss and extrapolated backward to figure out the age of Earth's solid crust. He arrived at a figure between 20 and 100 million years. Even Darwin conceded, "Thomson's views on the recent age of the world have been for some time one of my sorest troubles," and removed his age estimate from later editions of *On the Origin of Species*.

Darwin died before scientists vindicated him. Kelvin's mistake was his assumption that no heat

Visualizing

The debate over numerical age FIGURE 10.11

This timeline chronicles some of the major episodes in the long debate over how old Earth is. Note that even the unsuccessful attempts, such as Kelvin's or Joly's, were useful because they pointed out the incorrect assumptions that these scientists and others had made.

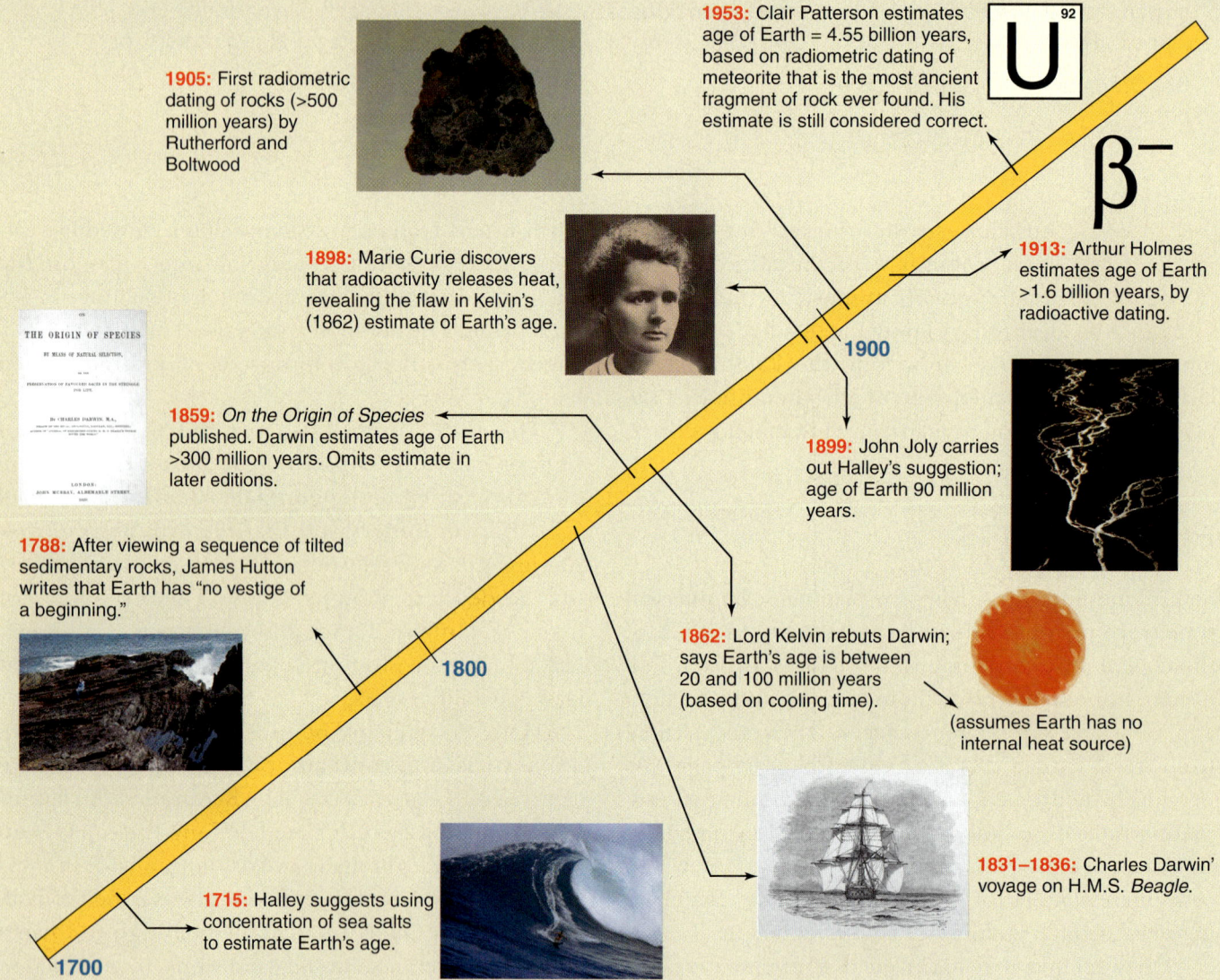

1905: First radiometric dating of rocks (>500 million years) by Rutherford and Boltwood

1953: Clair Patterson estimates age of Earth = 4.55 billion years, based on radiometric dating of meteorite that is the most ancient fragment of rock ever found. His estimate is still considered correct.

1898: Marie Curie discovers that radioactivity releases heat, revealing the flaw in Kelvin's (1862) estimate of Earth's age.

1913: Arthur Holmes estimates age of Earth >1.6 billion years, by radioactive dating.

1859: *On the Origin of Species* published. Darwin estimates age of Earth >300 million years. Omits estimate in later editions.

1899: John Joly carries out Halley's suggestion; age of Earth 90 million years.

1788: After viewing a sequence of tilted sedimentary rocks, James Hutton writes that Earth has "no vestige of a beginning."

1862: Lord Kelvin rebuts Darwin; says Earth's age is between 20 and 100 million years (based on cooling time).

(assumes Earth has no internal heat source)

1831–1836: Charles Darwin' voyage on H.M.S. *Beagle*.

1715: Halley suggests using concentration of sea salts to estimate Earth's age.

---

had been added to Earth's interior since its formation. He did not know about **radioactivity**, a natural physical process that releases heat. Though Earth is a closed system (as we explained in Chapter 1), it does have inputs and outputs of energy, and Kelvin had missed a key input.

What was needed to solve the problem of numerical time was a way to measure events by some process that runs continuously, is not reversible, is not influenced by such factors as chemical reactions and high temperatures, and leaves a continuous record without any gaps. The discovery of radioactivity not only proved that Kelvin's assumptions were wrong, but by fortunate chance, also provided the breakthrough needed to measure numerical time.

> **radioactivity** A process in which an element spontaneously transforms itself into another isotope of the same element, or into a different element.

318 CHAPTER 10 How Old Is Old? The Rock Record and Deep Time

RADIOACTIVITY AND NUMERICAL AGES

To explain how radioactivity allows us to determine the ages of rocks, we need to go back to the fundamentals of chemistry. Chapter 2 stated that most chemical elements have two or more *isotopes* that have the same number of protons per atom but a different number of neutrons per atom. To put it another way, each isotope of an element has the same atomic number but a different mass number.

Most naturally occurring isotopes have stable nuclei. However, a number of them—such as carbon-14 and potassium-40—are unstable. They spontaneously release particles from their nuclei. In the process, they change their mass number, their atomic number, or both (see **FIGURE 10.12**). Any isotope that spontaneously undergoes such nuclear change is said to be *radioactive*, and the process of change is referred to as *radioactive decay*.

Rates of decay

In any radioactive decay system, the number of original radioactive *parent atoms* continuously decreases, while the number of nonradioactive *daughter atoms* produced by the radioactive decay continuously increases. For this reason, many of the radioactive isotopes that were present when Earth was formed have decayed away because they were short-lived. However, radioactive isotopes that decay very slowly are still present.

All decay rates follow the same basic law: The *proportion* of parent atoms that decay during each unit of

Radioactive decay FIGURE 10.12

Because radioactive decay involves the nucleus, it is *not* a chemical process. In fact, it releases far more energy than any chemical reaction, which explains why it is still keeping Earth's interior hot after more than 4 billion years. The decay of radioactive isotopes is not influenced by any chemical process or by any conditions of heat or high pressure within Earth. Thus radioactive decay is a perfect built-in clock, at least for rocks that contain radioactive isotopes. Furthermore, because each radioactive isotope has its own rate of decay, a rock that contains several different radioactive isotopes has numerous built-in clocks that can be checked against each other.

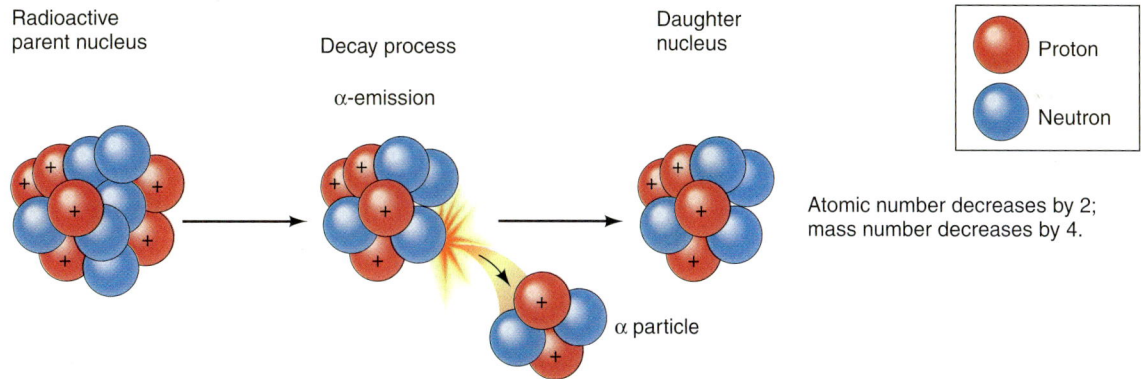

A radioactive nucleus releases an α particle, which consists of two protons and two neutrons.

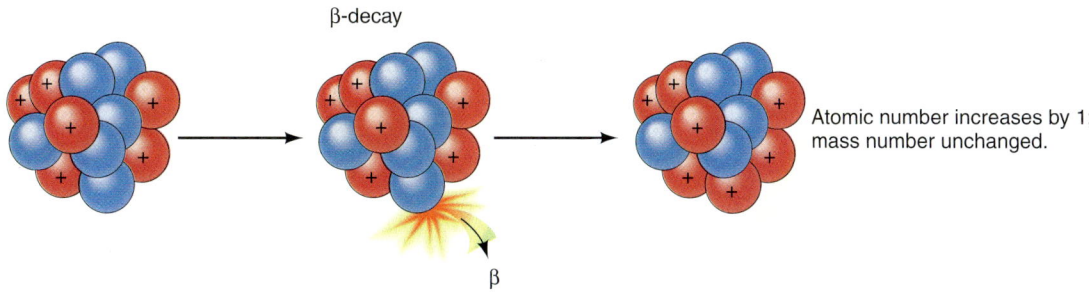

A radioactive nucleus releases a β particle, and one of its neutrons turns into a proton.

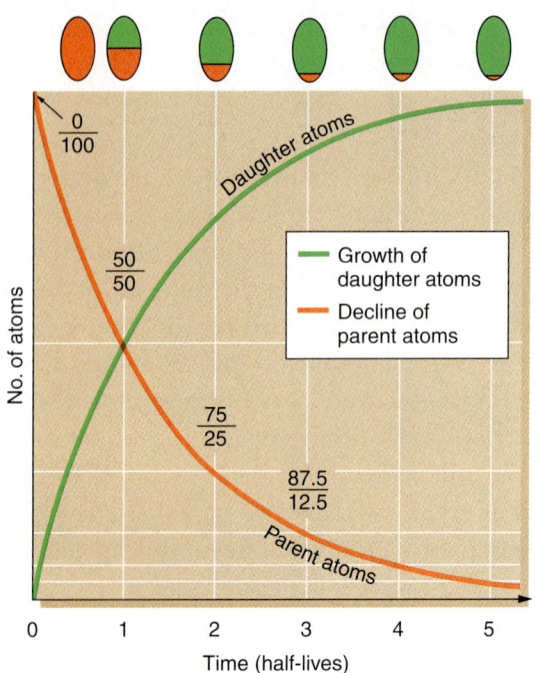

Radioactivity and time FIGURE 10.13

This graph illustrates the basic decay law of radioactivity. Suppose that a given isotope has a half-life of 1 hour. If we started with a sample consisting of 100% radioactive parent atoms, after an hour, only 50% of the parent atoms would remain and an equal number of daughter atoms would have formed. At the end of the second hour, another half of the parent atoms would be gone, so there would be 25% parent atoms and 75% daughter atoms. After the third hour, another half of the parent atoms would have decayed, leaving 12.5% parent atoms and 87.5% daughter atoms. Note that the total number of atoms, parent plus daughter, remains constant. This fact is the key to the use of radioactivity to measure geologic time.

half-life The time needed for half the parent atoms of a radioactive substance to decay into daughter atoms.

time is always the same (see **FIGURE 10.13**). *Proportion* means a fraction, or percentage, not a whole number. The rate of radioactive decay is determined by the **half-life**.

Radiometric dating: not one clock, but many

Following the discovery of radioactivity by Becquerel in 1896, and further work by Marie and Pierre Curie, the first estimates of the ages of rocks using radioactive decay were made in 1905. The long-hoped-for "rock clock" was finally available. The results were, and continue to be, remarkable. **Radiometric dating** has revolutionized the way we think about Earth and its long history.

A radioactive rock clock usually measures the amount of time that has elapsed since the minerals in an igneous rock crystallized. When a new mineral grain forms—for example, a grain of feldspar in cooling lava—all the atoms in the grain become locked into the crystal structure and isolated from the environment outside the grain. In a sense, the atoms in the mineral grain, including any radioactive atoms, are sealed in an atomic time capsule. The trapped radioactive parent atoms decay to daughter atoms at a rate determined by the half-life.

In the simplest case, if no daughter atoms were present in the mineral at the time of formation, we can use Figure 10.13 to work backward and determine how long ago the time capsule was sealed. There can be complications; for example, if the mineral grain trapped some daughter atoms at the time of formation, the process is more difficult. Scientists have developed several ways to estimate the initial contamination of a sample by daughter atoms—most obviously, by looking for daughter atoms in other minerals that do not contain any radioactive atoms. Once that is done, and provided we know the half-life of the radioactive parent, it is a simple matter to calculate how long ago the mineral crystallized. Another complication is exemplified by carbon-14, which decays to nitrogen-14. Nitrogen-14 is abundant in the atmosphere, so all samples are hopelessly contaminated. But carbon-14 is produced in the atmosphere by cosmic rays, and because the atmosphere is rapidly mixed, all living matter takes in the same ratio of carbon-14 to carbon-12. At the moment a plant or animal dies, it stops taking in carbon, and because carbon-12 is a stable isotope, the ratio of 14 to 12 changes at a known rate as carbon-14 decays away. Scientists use different isotopic systems to study rocks, fossils, and biologic materials of different ages and compositions (see **FIGURE 10.14**).

radiometric dating The use of naturally occurring radioactive isotopes to determine the numerical ages of minerals, rocks, and fossils.

Isotopes used in radiometric dating Figure 10.14

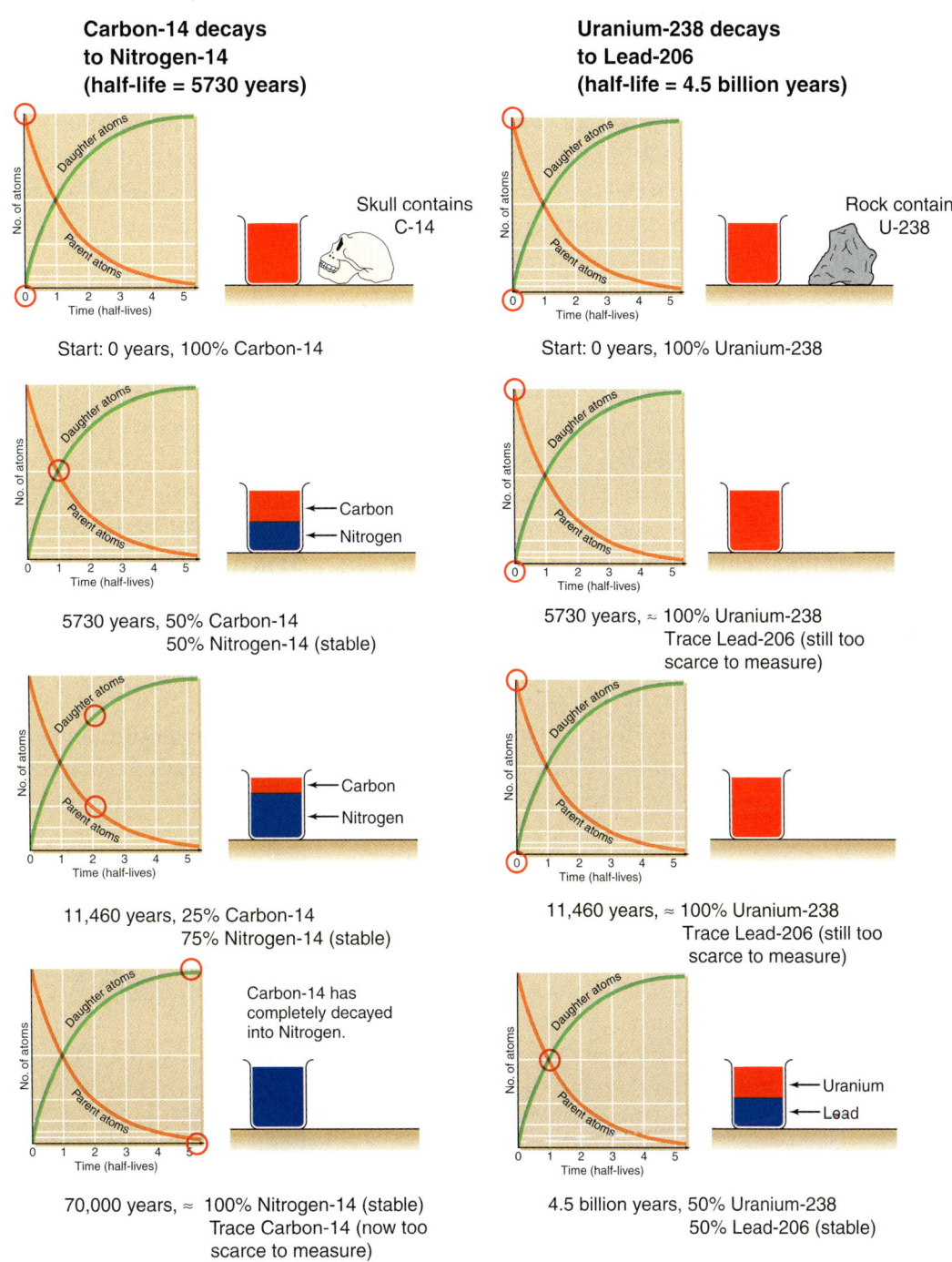

Long-lived isotopes such as uranium-238 are especially useful for dating ancient rocks. Short-lived isotopes such as carbon-14 are useless for dating samples that are more than about 70,000 years old, which include many of the rocks that are of interest to Earth scientists. However, carbon-14 is very useful for dating items younger than 70,000 years of biologic origin, such as human remains and artifacts. Earth scientists have also used it to study the latest ice age, by estimating the ages of wood samples taken from trees killed by the advancing ice sheet.

Radiometric dating has been particularly useful for determining the ages of igneous rocks, because the mineral grains in an igneous rock form at the same time as the rock that contains them. On the other hand, most sedimentary rocks consist largely of mineral grains that were formed long before the strata that contain them were deposited. Radiometric dating will tell how old the grains are but not when the strata were deposited. This makes it difficult to directly date most sedimentary rocks and to infer the numerical age of ancient life-forms fossilized in the sediments.

Numerical time and the geologic column

As scientists worked out the geologic column, they found many locations where layers of solidified lava and volcanic ash are interspersed with sedimentary strata. Through radiometric dating, they could determine the numerical ages of the lavas and volcanic ash layers and thereby bracket the ages of the sedimentary strata (see FIGURE 10.15).

Through a combination of careful examination of stratigraphic relationships, correlations, and radiometric dating, scientists have been able to fill in all the dates in the geologic column as shown in Figure 10.9. The scale is continually being refined, so the numbers given in the figure are considered the best currently available but are subject to change. Further work will make the numbers more precise. It is a tribute to the work of scientists during the 19th century that the geologic column they established by ordering strata according to their relative ages has been fully confirmed by radiometric dating. At the same time, it is humbling to see how one wrong assumption caused Lord Kelvin to be more than 4 billion years off in his estimate of Earth's age.

MAGNETIC POLARITY DATING

Time is so central to the study of Earth that scientists are always seeking new ways to estimate ages. An exciting newer method of dating, developed in the 1960s, involves *paleomagnetism* (see Chapter 7), the study of Earth's past magnetic field (FIGURE 10.16). Both igneous and sedimentary rocks "lock in" information about the magnetic field at their time of formation.

Earth's magnetic field reverses its *polarity* at irregular intervals, but on average, about once every half-million years. This means that the magnetic pole that had been in the Northern Hemisphere moves to a position near Earth's South Pole, and the magnetic pole that had been in the south moves to the north.

Radiometric dating and the geologic column FIGURE 10.15

This idealized drawing shows how radiometric dating can be used to bracket the ages of sedimentary strata. The conduit for the lava flow labeled **A** can be dated directly by radiometric means. The conduit cuts through strata 1, 2, 3, and 4, so it must be younger than all of them. Igneous intrusion **B** can also be dated by radiometric means. It cuts through strata 1, 2, and 3 but not 4, and hence must be older than layer 4. Thus, the sedimentary rocks in layer 4 are between 20 million and 34 million years old.

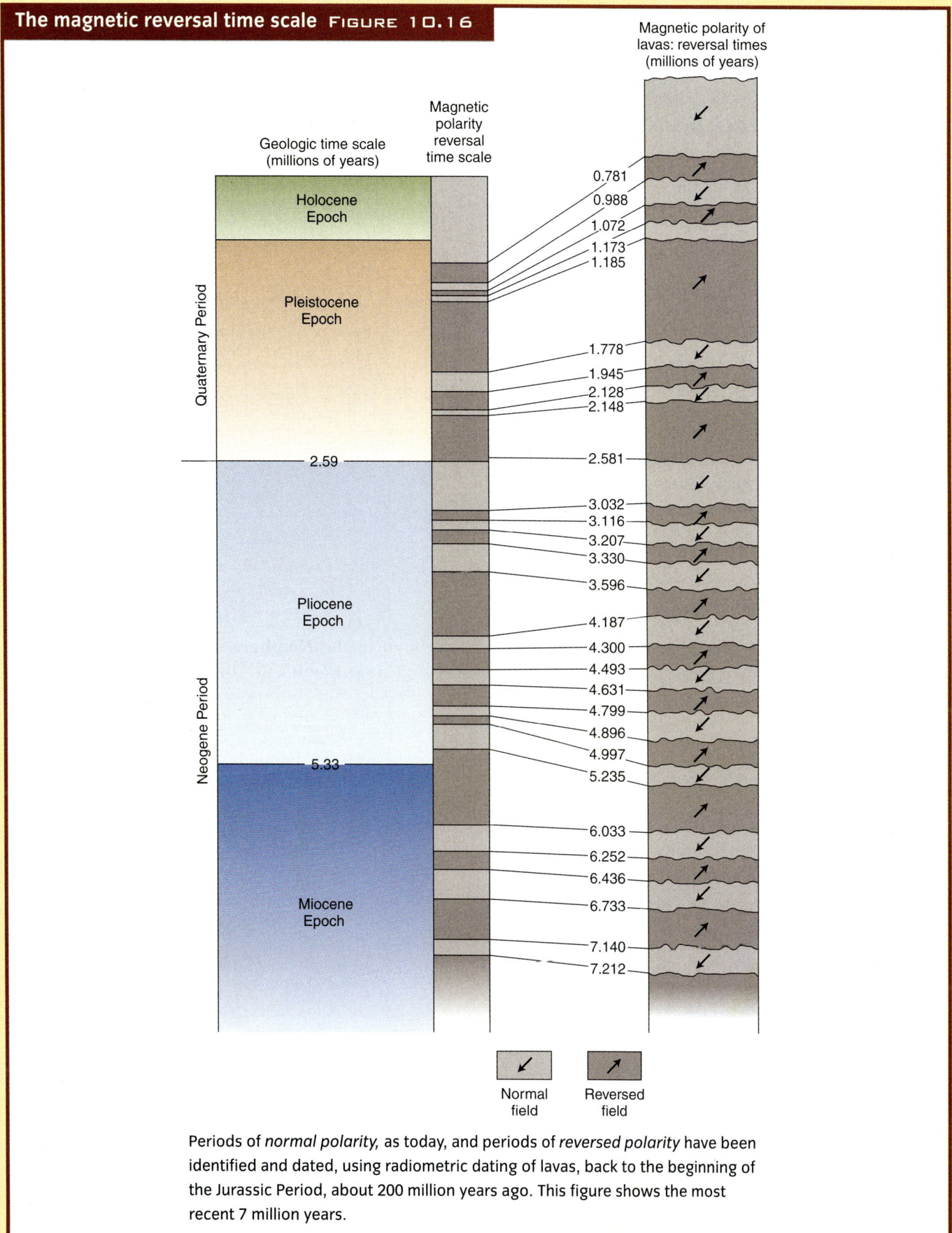

The magnetic reversal time scale FIGURE 10.16

Periods of *normal polarity*, as today, and periods of *reversed polarity* have been identified and dated, using radiometric dating of lavas, back to the beginning of the Jurassic Period, about 200 million years ago. This figure shows the most recent 7 million years.

CASE STUDY

Dating Human Ancestors

The Haddar region of northern Ethiopia has been a fertile site for finding fossils of ancient human ancestors, part of a larger group called *hominids*. Many of the fossils have been found in ancient stream gravels, where they were tumbled and battered by long-ago floodwaters. The fossil-bearing gravels are interlayered with sediments that give good magnetic signals.

The problem is to know where in the magnetic polarity time scale the Haddar sediment falls. Scientists answered this question by using the Potassium–Argon (K–Ar) method to establish the ages of a lava flow that lay below some of the hominid fossils and a layer of volcanic ash that lay above them. The two radiometric dates determined the magnetic reversal ages unambiguously and indicated that early hominids lived in the region until about 3.3 million years ago, during the Pliocene Epoch. Through many studies like this, it is slowly becoming clear that the hominid genus *Homo* (which includes our own species, *Homo sapiens*) evolved a little more than 2 million years ago from an older genus called *Australopithecus*. (Two reconstructed Australopithecus skulls are shown at bottom right.)

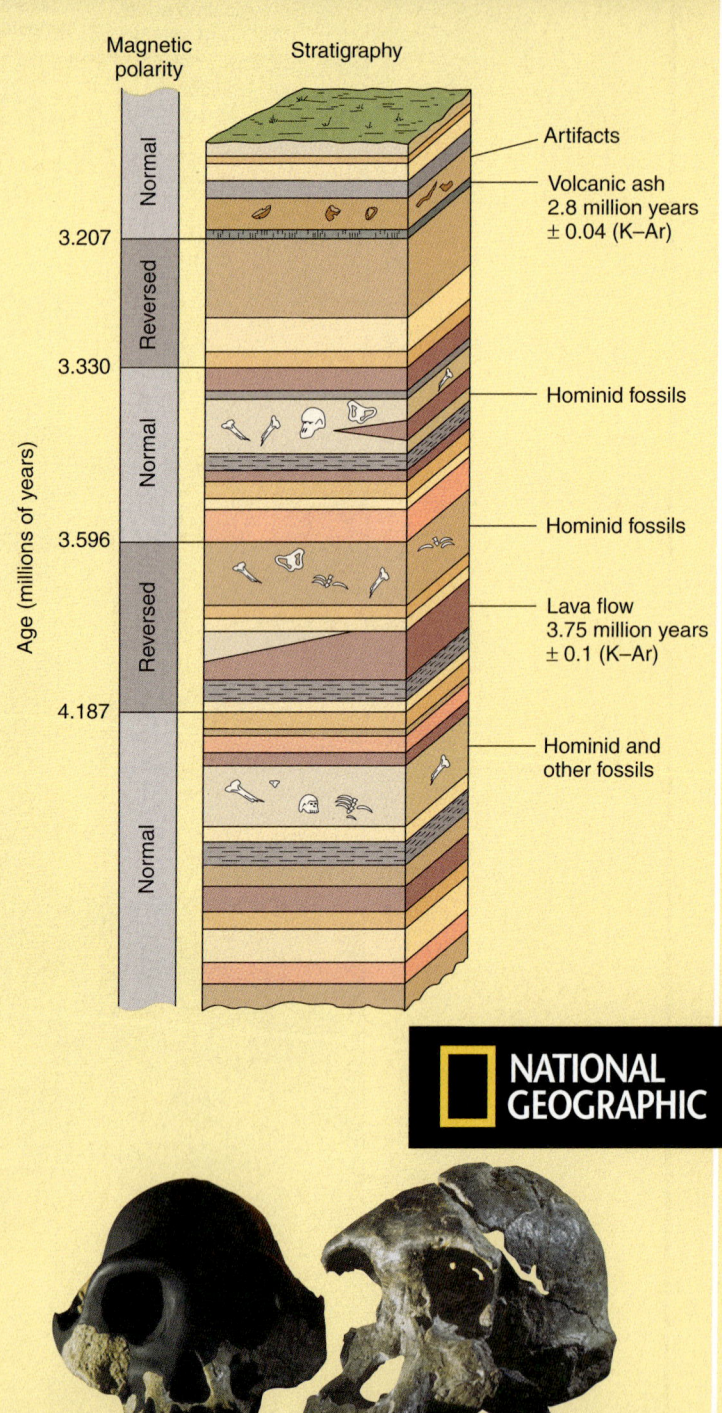

324 CHAPTER 10 How Old Is Old? The Rock Record and Deep Time

Note, however, that Earth's geographic North and South Poles of rotation do not change position. (Earth's magnetic field is explained in additional detail in Chapter 7.) Scientists are still working out details of how or why **magnetic reversals** happen. The two important points are that the reversal happens quickly by Earth's time standards, and that any iron-bearing mineral in a sedimentary or igneous rock retains or "remembers" the magnetic polarity of Earth at the time that the rock was formed—that is, a change in the magnetic field does not affect already-formed minerals. Through a combination of radiometric dating and magnetic polarity measurements, it has been possible to establish a time scale of magnetic polarity reversals dating back to the Jurassic Period (see Figure 10.16). Still earlier reversals are a subject of ongoing research.

> **magnetic reversal** The period of time in which Earth's magnetic polarity reverses itself.

Correlation on the basis of magnetic reversals differs from other stratigraphic correlation methods. One magnetic reversal looks just like any other in the rock record. When evidence of a magnetic reversal is found in a sequence of rocks, the problem lies in knowing which of the many reversals it actually represents. When a continuous record of reversals can be found, starting with the present, it is simply a matter of counting backward. But if not, the technique must be combined with stratigraphic and radioactive dating techniques (see the *Case Study*). Chapter 7 discusses how magnetic polarity studies played a crucial role in the development of plate tectonic theory.

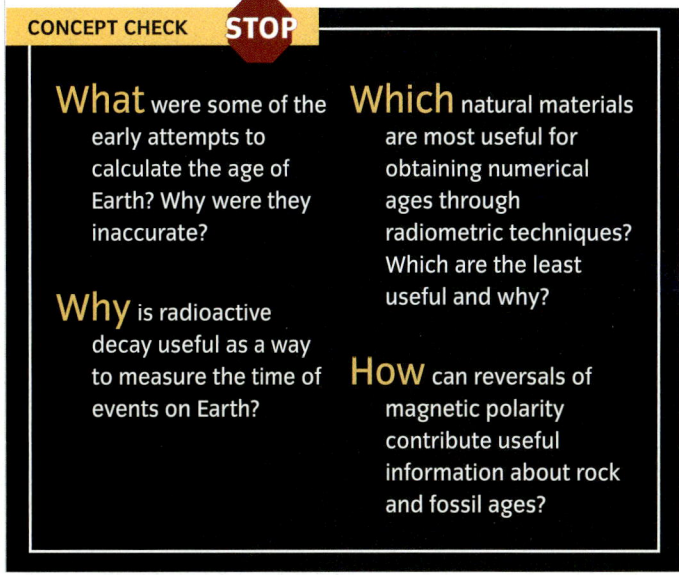

CONCEPT CHECK **STOP**

What were some of the early attempts to calculate the age of Earth? Why were they inaccurate?

Why is radioactive decay useful as a way to measure the time of events on Earth?

Which natural materials are most useful for obtaining numerical ages through radiometric techniques? Which are the least useful and why?

How can reversals of magnetic polarity contribute useful information about rock and fossil ages?

The Age of Earth

LEARNING OBJECTIVES

Explain why the oldest rocks are not necessarily the same age as the planet.

Explain why scientists currently believe Earth is about 4.56 billion years old.

Throughout this book, we mention examples of actual rates of Earth processes. This would not be possible without the numerical dates obtained through radiometric dating and other numerical age methods. In fact, more than any other contribution by scientists, the ability to determine numerical dates has changed the way humans think about the world and the immensity of Earth's history.

Now that we know how to determine the numerical ages of rocks, can we determine Earth's age? It's not as easy as you might think. The continual recycling of Earth's surface by erosion and plate tectonics means that very few, if any, remnants of

Earth's oldest rocks FIGURE 10.17

The Acasta gneiss in northern Canada was formed 4.0 billion years ago. It is the most ancient body of rock so far discovered on Earth.

Earth's original crust remain. Of the many radiometric dates obtained from Precambrian rocks, the oldest is about 4.0 billion years (see **FIGURE 10.17**). Although no rocks older than this have been found, an individual mineral grain from a sedimentary rock in Australia has been dated to 4.4 billion years, so it is conceivable that igneous rocks older than 4.0 billion years may someday be located.

However, because the oldest mineral grain came from a sedimentary rock, there must have been a period of rock formation and erosion before sedimentation, so scientists understand that there is still a gap between the age of the mineral grain and the age of Earth. There is strong evidence that Earth formed at the same time as the Moon, the other planets, and meteorites. Through radiometric dating, it has been possible to determine the ages of meteorites and of "moon dust" brought back by astronauts. The *Apollo* astronauts found rocks and individual grains of Moon dust that are pieces of the Moon's original crust. Such rocks are abundant because the Moon system has been tectonically much less active than the Earth system.

Meteorite ages are especially valuable because some meteorites have remained virtually unaltered since the formation of the solar system. Melting and other types of alteration will reset the radiometric clock. However, some meteorites, such as the Allende Meteorite, which fell to Earth in the Mexican state of Chihuahua on February 8, 1969, belong to a rare category called *carbonaceous chondrites,* which, as far as we can tell, contain unaltered material from the formation of the solar system. It is carbonaceous because it contains tiny amounts of carbon (about 3 parts per 1000). Some of the carbon is in chemical compounds called *amino*

A cosmic interloper FIGURE 10.18

The Allende Meteorite, which fell to Earth in Mexico, is one of the most famous meteorites in history. Note the white spots on the meteorite. Some of these inclusions, which are slightly older than the black carbonaceous material around them, are more than 4.6 billion years old, making them the oldest objects of any kind ever found on Earth.

acids—organic components that are essential for life. The dark, fine-grained part of the meteorite is mostly olivine, with a few flecks of metallic iron and some carbon. The clumps of white material are oxides of calcium and aluminum and are thought to be among the first matter to condense from the gas cloud from which the solar system formed (**FIGURE 10.18**). The white clumps are older than Earth itself.

The ages of many of these objects from the solar system cluster closely around 4.55 billion years. Planetary scientists therefore conclude that Earth, and indeed the Sun's entire planetary system, formed at that time.

Today, more than two centuries after Hutton, it is widely agreed that Earth's age is approximately 4.55 billion years. When will it cease to exist? Astronomers tell us that billions of years in the future, the Sun will become a red giant, at which point it will expand and engulf Earth. However, Hutton is still correct in one sense: Earth's history is profound and evidence here on Earth shows no prospect of an end.

CONCEPT CHECK STOP

What is the oldest age that has been obtained from material found on Earth? Does this match the presumed age of Earth? Why or why not?

How have meteorites and rocks from the Moon helped scientists to determine the age of Earth?

The Age of Earth 327

Amazing Places: The Grand Canyon

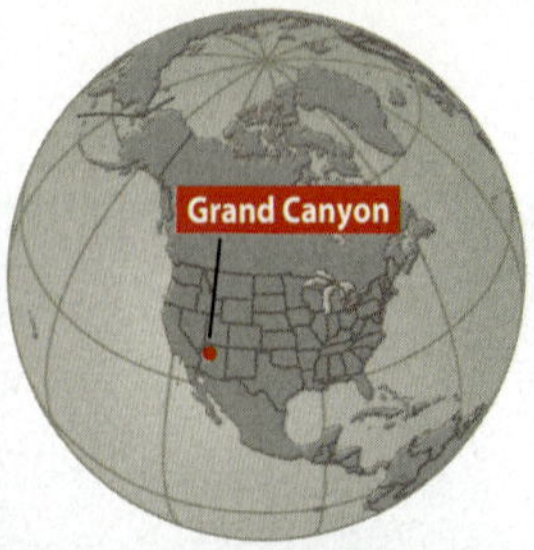

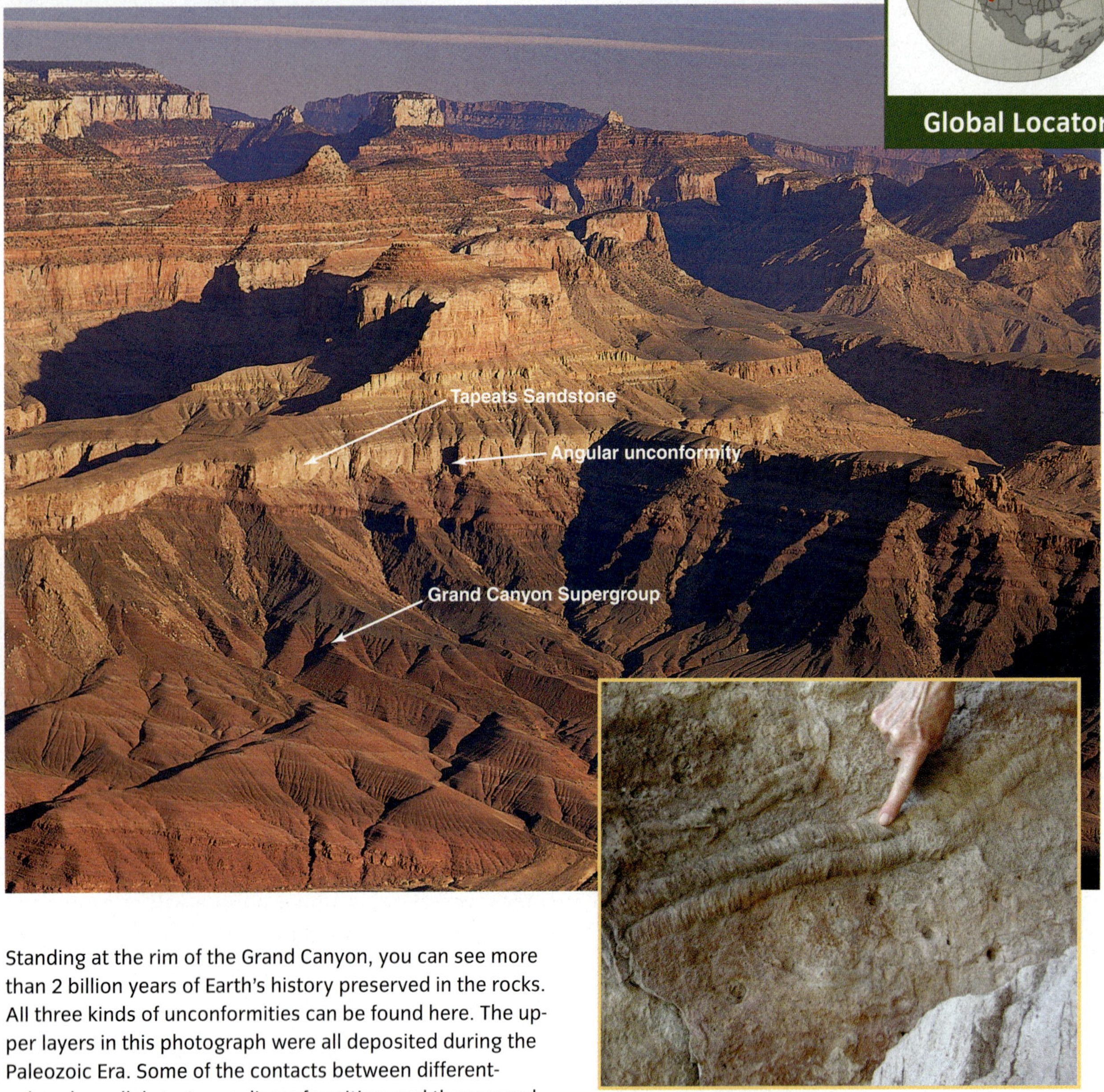

Standing at the rim of the Grand Canyon, you can see more than 2 billion years of Earth's history preserved in the rocks. All three kinds of unconformities can be found here. The upper layers in this photograph were all deposited during the Paleozoic Era. Some of the contacts between different-colored parallel strata are disconformities, and they record the rising and ebbing of seas over this part of the North American continent. Below the Tapeats Sandstone (arrow) you can see an angular unconformity, which also represents a major time gap between Precambrian rocks (deposited about 825 million years ago) and Cambrian rocks (deposited less than 542 million years ago). Finally, there is a nonconformity (not visible in the photo) between the lowermost sedimentary layer of the Grand Canyon Supergroup, the Bass Limestone, and the Vishnu Formation, a foundation of metamorphic and igneous rocks that once lay at the base of an ancient mountain range that eroded away long before the Rocky Mountains came into existence.

As an Earth science student, you owe it to yourself not to stop at the rim, as most tourists do, but to descend to the river level and get a closeup look at 2 billion years of Earth's history. You might be lucky and see trilobite tracks, like these, in the Tapeats Sandstone. The tracks were made when trilobites (like the one shown in Figure 10.10) extended their legs sideways, pulled in mud, then, under the safety of their hard shells, picked over the mud for food.

SUMMARY

1 Relative Age

1. Earth scientists study the chronologic sequence of natural events, that is, their **relative age**. Relative age is derived from **stratigraphy**, the study of rock layers and how those layers are formed.

2. There are four basic principles used to determine relative ages. **Strata**, or sedimentary rock layers, are horizontal when they are deposited as water-laid sediment (*law of original horizontality*). Strata accumulate in sequence, from the oldest on the bottom to the youngest on the top (*principle of stratigraphic superposition*). A rock stratum is always older than any feature, such as a fracture, that cuts across it (*principle of cross-cutting relationships*). Finally, rock strata extend outward horizontally; they may thin or pinch out at their farthest edges, but they generally do not terminate abruptly unless cut by a younger rock unit (*principle of lateral continuity*).

3. **Numerical age**, the exact number of years of a natural feature or event, is more difficult to determine. One difficulty that arises is that the sequence of strata in any particular location is not necessarily continuous in time. An **unconformity** is a break or gap in the normal stratigraphic sequence. It usually marks a period during which sedimentation ceased and erosion removed some of the previously laid strata. The three common types of unconformities are nonconformities, angular unconformties, and disconformities.

4. **Correlation** of strata is the establishment of the time equivalence of strata in different places. **Fossil** assemblages, usually consisting of hard shells, bones, and plant material, have been the primary key to correlation of strata across long distances. The study of fossils and the record of ancient life on Earth is called **paleontology**. The principle of *faunal and floral successions* (animals and plants, respectively) is the stratigraphic ordering of fossil assemblages.

2 The Geologic Column

1. The **geologic column**, a *stratigraphic time scale*, is a composite diagram that shows the succession of all known strata, arranged in chronological order of formation, based on fossils and other age criteria.

2. The geologic column is divided into several different units of time, called **eons**, **eras**, **periods**, and **epochs**. The majority of Earth's history is divided into two eons, in which fossils are very rare or nonexistent. Those eons, each spanning several hundred million years, are the *Archean* and *Proterozoic*. The third and most recent eon, the *Phanerozoic*, is the only eon in which fossils are abundant. Very dramatic changes in fossil assemblages occur between the three eras of the Phanerozoic Eon—the *Paleozoic*, *Mesozoic*, and *Cenozoic*—which were separated by major extinction events. The earliest period of the Paleozoic Era is especially important, as this was a time of unique diversification. This period is known as the *Cambrian Explosion*. Except for the very end of the Proterozoic Eon, rocks formed before the Cambrian Period cannot be differentiated by the fossil record.

3 Numerical Age

1. Determining the age of Earth and of events that have happened on Earth has long been of interest to scientists. Different hypotheses were proposed by Halley, Joly, Darwin, Lord Kelvin, and several other prominent scientists of the later 1800s and early 1900s. Though many of these hypotheses were found to be incorrect, each was an important step in eventually finding a method of numerical dating.

2. **Radioactivity** is the process in which an element spontaneously transforms itself into another *isotope* of the same element, or into a different element, through the release of particles and heat energy. The *radioactive decay* of isotopes of chemical elements provides a basis for radiometric dating, which gives values for the numerical ages (age in years) of rock units and thus values for numerical dates of past events. Because radioactive decay is not influenced by chemical processes or by heat and high pressure in Earth, it is an extremely accurate gauge of numerical age.

3. **Radiometric dating** is based on the principle that in any sample containing a radioactive isotope, half of the atoms of that isotope will change to daughter atoms within a specific length of time, called the **half-life**. (The *proportion* of parent atom decay during a unit of time is always the same.) Radioactive isotopes with a long half-life, such as uranium, are most useful for dating rocks. Carbon-14, which has a much shorter half-life, is most useful for dating organic materials of relatively recent origin (less than 70,000 years).

4. Though radiometric dating is primarily useful for igneous rocks, a complementary technique called *magnetic polarity dating* works for sedimentary rocks, too.

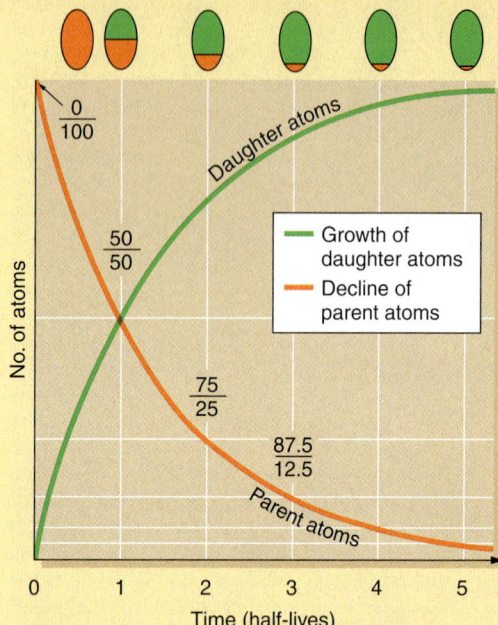

Magnetic polarity dating involves paleomagnetism, the study of reversals in Earth's magnetic field. As yet, **magnetic reversals**, or periods of time in which Earth's magnetic polarity reverses itself, are not fully understood.

4 The Age of Earth

1. Through measures of numerical age, it has become clear that most of Earth's history took place in Precambrian time. The oldest Earth rocks discovered are about 4.0 billion years old.

2. Earth is not a good place to look for the oldest rocks in the solar system. Earth's surface has been subjected to a lot of erosional activity. This has reset some radiometric clocks and destroyed the earliest fragments of the crust. Samples from the Moon and from meteorites indicate that the solar system formed about 4.56 billion years ago, and by inference this is also the age of Earth.

KEY TERMS

- uniformitarianism p. 304
- relative age p. 304
- stratigraphy p. 304
- numerical age p. 309
- unconformity p. 309
- paleontology p. 311
- correlation p. 312
- geologic column p. 312
- radioactivity p. 318
- half-life p. 320
- radiometric dating p. 320
- magnetic reversal p. 325

CRITICAL AND CREATIVE THINKING QUESTIONS

1. Do the principles of stratigraphy apply on the Moon in the same way as they do on Earth? Bear in mind that the processes operating on the Moon have been very different from those on Earth. If you had to determine the relative age of features on the Moon, based entirely on satellite photographs, what would you look for and how might you proceed?

2. Check the area in which you live to see whether there is an excavation—perhaps one associated with a new building or road repair. Visit the excavation and note the various layers, the paving (if the excavation is in a road), and the soil below the surface. Is any bedrock exposed beneath the soil?

3. How old are the rock formations in the area where you live and attend college or university? How can you find out the answer to this question?

4. Choose one of the geologic periods or epochs listed in Figure 10.9A and find out all you can about it: How are rocks from that period identified? What are its most characteristic fossils? Where are the best samples of rocks from your chosen period found?

5. Do some research to determine the ages of the oldest known fossils. What kind of life-forms were they?

What is happening in this picture?

This skier is hauling a sled past a cliff face on Ellesmere Island, Canada.

- Why do you think the rock strata in the background tilt at such a steep angle?
- Why are they wavy instead of straight?

SELF-TEST

1. A _____ is the age of one rock unit or geologic feature compared to another.
   a. numerical age
   b. relative age
   c. radioactivity age
   d. stratigraphic age

2. The principle of cross-cutting relationships says that _____.
   a. waterborne sediments are deposited in nearly horizontal layers
   b. a sediment or sedimentary rock layer is younger than the layers below it and older than the layers that lie above
   c. a rock unit is older than a feature that disrupts it, such as a fault or igneous intrusion
   d. a sediment or sedimentary rock layer is older than the layers below it and younger than the layers that lie above

3. The _____ states that waterborne sediments are deposited in nearly horizontal layers.
   a. law of superposition
   b. principle of faunal succession
   c. law of original horizontality
   d. principle of cross-cutting relationships

4. In a conformable sequence, _____.
   a. each layer must have been deposited on the one below it without any interruptions
   b. there must not be any depositonal gaps in the stratigraphic record
   c. Both a and b are correct.
   d. None of the above is true.

5. An unconformity represents _____.
   a. a gap in the stratigraphic record
   b. a period of erosion or no deposition
   c. Both a and b are correct.
   d. None of the above is true.

6. On this illustration, label each unconformity as one of the following:
   a. nonconformity
   b. angular unconformity
   c. disconformity

7. Fossils found in strata _____.
   a. are the records of ancient life
   b. allow the correlation of strata separated by many miles
   c. have been useful to geologists in creating the geologic column
   d. All of the above statements are correct.

8. The three eras that make up the Phanerozoic Eon are the _____.
   a. Hadean, Archean, and Proterozoic
   b. Paleozoic, Mesozoic, and Cenozoic
   c. Triassic, Jurassic, and Cretaceous
   d. Pliocene, Pleistocene, and Holocene

9. The most distinctive changes in the fossil record occur across the boundaries between _____.
   a. periods
   b. eras
   c. epochs
   d. disconformity

10. The dinosaurs were dominant during _____.
    a. the Cenozoic Era
    b. the Mesozoic Era
    c. the Paleozoic Era
    d. the Precambrian time

11. Label the two decay sequences depicted in this illustration as either alpha emission or beta decay. For each decay sequence, also label the following:
    a. parent nucleus
    b. alpha-particle
    c. daughter nucleus
    d. beta-particle

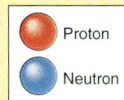

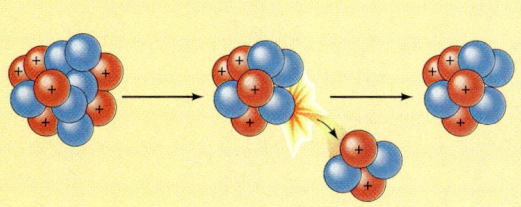

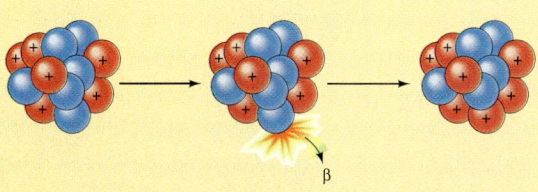

12. Potassium-40 is a naturally occurring radioisotope that decays to Argon-40 and is common in many rocks of the continental crust. The half-life of Potassium-40 is 1.3 billion years. Assuming no contamination, what would be the age of a sample that contained a 1:1 ratio of Potassium-40 to Argon-40?
    a. 1.3 billion years
    b. 650 million years
    c. 2.6 billion years
    d. 325 million years

13. If the sample indicated above showed evidence that it had been heated by contact with a more recent lava flow, what would be the likely error in the determined age?
    a. The sample would appear too young.
    b. The sample would appear too old.
    c. Rocks are a closed system; there would be no error.

14. A gravel deposit containing an important hominid tooth fossil is found in a field location in northern Ethiopia. The gravel deposit has fragments of volcanic rocks dated at 3.75 million years ±0.1 by Potassium–Argon dating (K–Ar) and is known from stratigraphy to be younger than a 2.8-million-year ±0.04 (K–Ar) volcanic ash deposit. Sediments interlayered with the fossil-bearing gravels have good magnetic signals and have a normal polarity. Given the figure in the second column for the region, what is the most likely date for the hominid fossil?
    a. between 4.2 and 3.3 million years
    b. between 3.6 and 3.3 million years
    c. exactly 3.75 million years
    d. younger than 2.8 million years

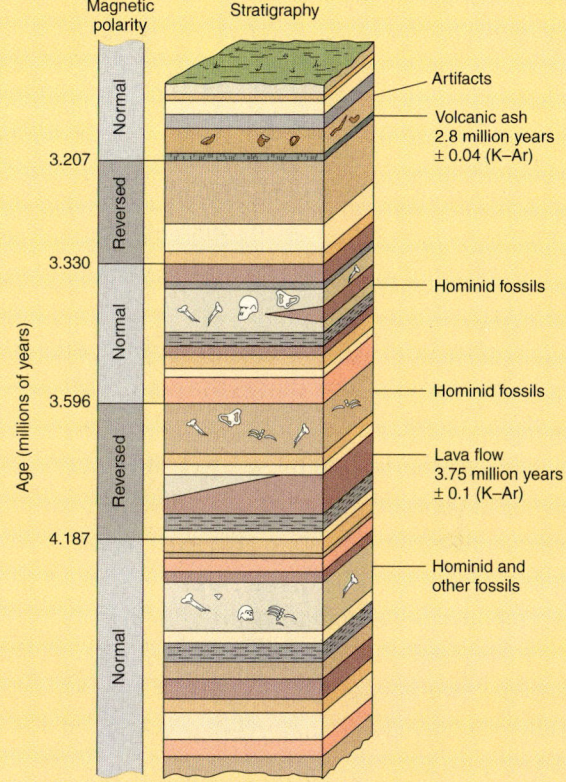

15. Earth is not considered a good place to look for the oldest rocks in the solar system because _____.
    a. contamination from atmospheric tests of nuclear weapons have contaminated the crust of Earth
    b. Earth's magnetic field interferes with the radiometric clocks in most igneous rocks
    c. melting has reset radiometric clocks in the rocks of Earth's crust
    d. the earliest crustal rocks have been destroyed by geologic activity
    e. All of the above are true.
    f. Answers c and d are correct.

A Brief History of Life on Earth   11

The history and diversity of life, from this ferocious *Tyrannosaurus rex* (on exhibit at the American Museum of Natural History) to the humblest microbe, is very much a part of the planet's history. Throughout this book, you will find numerous examples of interactions between the biosphere and other parts of the Earth system. Plants and microorganisms accelerate the mechanical weathering of rocks and formation of soil. The skeletons of plankton sink to the bottom of the sea, where they form sediments that eventually turn into limestone. Land plants form coal, and animals leave fossils that paleontologists use to reconstruct the history of life. Biologic processes, including the transpiration of plants and respiration of animals, regulate the very air that we breathe.

This interaction also works in the opposite direction: The Earth system affects the course of life on this planet. Earth's atmosphere and hydrosphere provided a habitat in which life could develop and prosper. The forms that life takes are governed by the need to survive in a particular environment—the scalding waters around a mid-ocean rift, the arid land of a desert, the humid climate of a tropical rainforest, all of which result from natural processes.

The study of Earth science is thus inseparable from the study of life on Earth. In this chapter, we trace the story of life from its beginnings. We examine in greater detail how life has been affected by its interactions with the atmosphere, hydrosphere, and lithosphere and how, in turn, life has shaped the Earth system.

CHAPTER OUTLINE

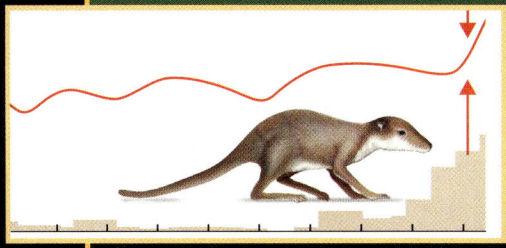

The Ever-Changing Earth p. 336

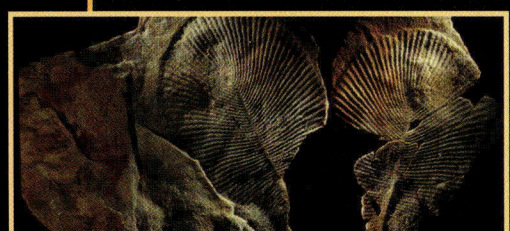

Early Life p. 339

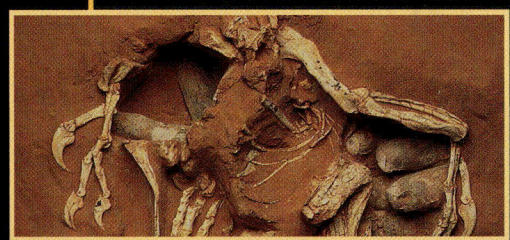

Evolution and the Fossil Record p. 345

Life in the Phanerozoic Eon p. 349

The Ever-Changing Earth

LEARNING OBJECTIVES

Describe how Earth's environment has changed over the last 4 billion years.

Explain how photosynthesis adds oxygen to the atmosphere.

Describe how the oxygen cycle affects life on Earth.

Throughout this book we have seen evidence that Earth is a place of constant change and that negative feedbacks seem to moderate the changes so that Earth continues to be habitable. The most obvious changes, and the ones most readily demonstrated, are the rearrangements of the continents and oceans as a result of plate tectonics. Refer back to Figure 1.11 and review the dramatic continental rearrangements that have occurred over the past 500 years. There is abundant evidence to demonstrate that plate motions have been operating for at least 2 billion years and probably even longer, but the exact locations of continents and oceans during the Precambrian is still uncertain and the focus of much research. In this chapter, we will present evidence for change in Earth's life zone (a summary of the changes is shown in **FIGURE 11.1**).

The changing Earth FIGURE 11.1

From left to right, this diagram illustrates some of the major events in the history of Earth's surface environment during its first 4.6 billion years.

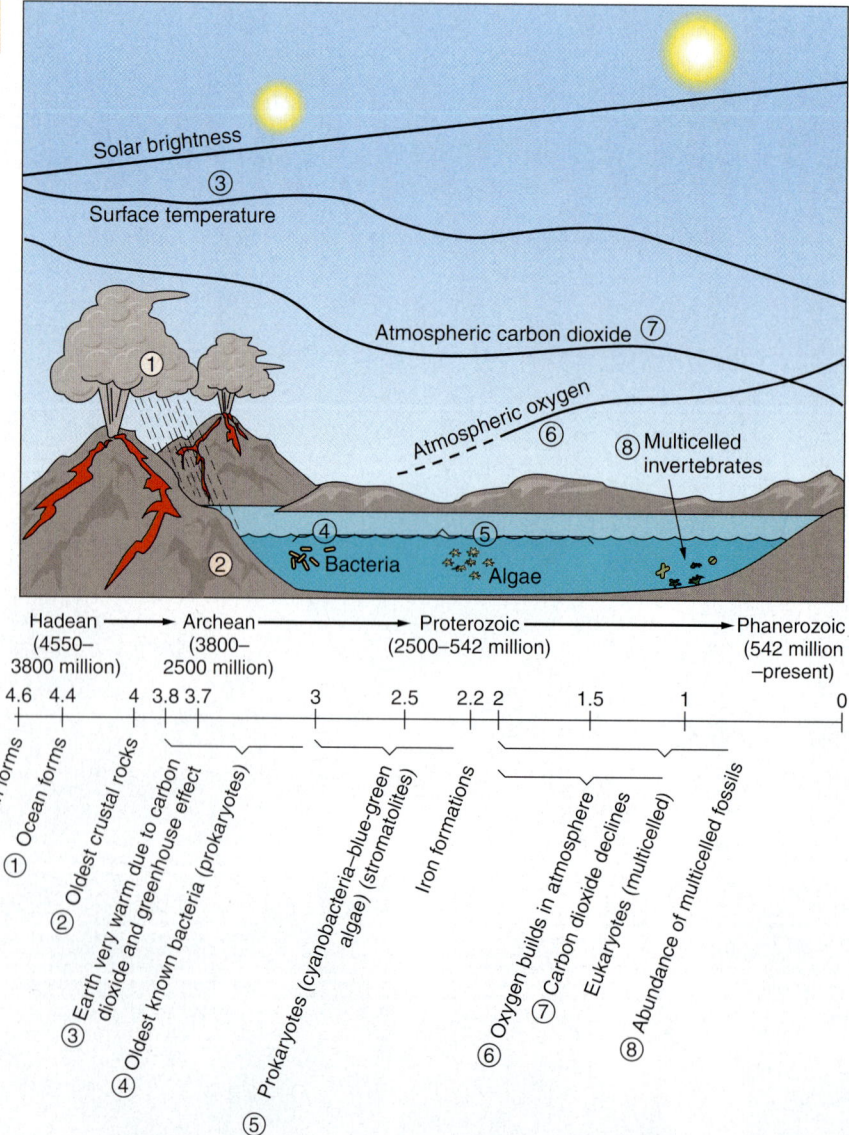

A This pea plant, like all green, leafy plants, produces oxygen by the process of photosynthesis.

Photosynthesis FIGURE 11.2

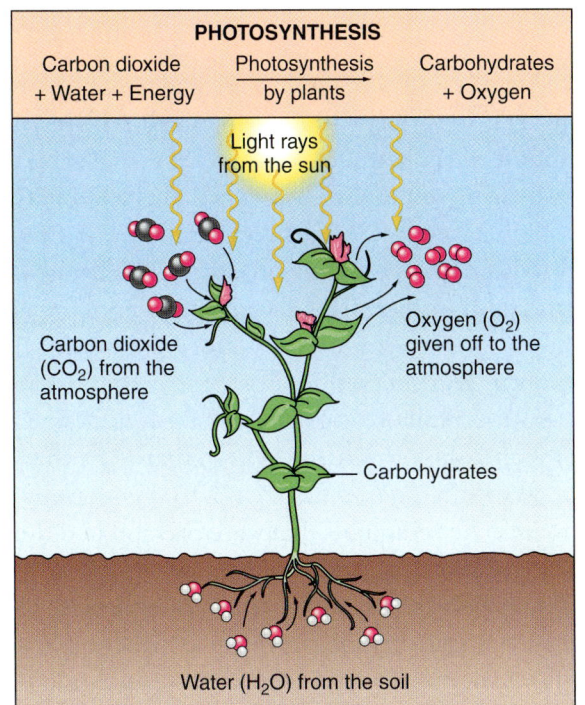

B The photosynthetic reaction combines carbon dioxide (CO_2) and water (H_2O) to make carbohydrates (molecules containing C, H, and O), which the plant needs to grow. The reaction produces oxygen molecules (O_2). The plant releases the oxygen into the atmosphere through pores in its leaves. The reaction does not happen spontaneously, but requires energy from sunlight; that is why it is called *photosynthesis* (*photo-*, meaning "light").

CHANGES IN THE ATMOSPHERE AND HYDROSPHERE

Like other parts of the Earth system, the atmosphere has changed through time (see Figure 11.1). Compared to the Sun, which is representative of the raw materials from which Earth and the other planets of our solar system formed (Chapter 17), Earth contains less of some volatile elements, such as nitrogen, argon, hydrogen, and helium. These elements were lost when the envelope of gases or *primary atmosphere* that surrounded early Earth was stripped away by the solar wind or by meteorite impacts, or both. Little by little, Earth generated a new, *secondary atmosphere,* by volcanic outgassing of volatile materials from its interior.

Volcanic outgassing continues to be the main process by which volatile materials are released from Earth—although it is now going on at a much slower rate. The main chemical constituent of volcanic gases (as much as 97% by volume) is water vapor, with varying amounts of nitrogen, carbon dioxide, and other gases. In fact, the total volume of volcanic gases released over the past 4 billion years or so accounts for the present composition of the atmosphere, with one extremely important exception: oxygen. As you can see in Figure 11.1, Earth had virtually no oxygen in its atmosphere more than 4 billion years ago, but the atmosphere is now approximately 21% oxygen.

Traces of oxygen were probably generated in the early atmosphere by the breakdown of water molecules into oxygen and hydrogen by ultraviolet light (a process called *photodissociation*). Although this is an important process, it doesn't even come close to accounting for the present high levels of oxygen in the atmosphere. Almost all of the free oxygen now in the atmosphere originated through **photosynthesis** (**FIGURE 11.2**).

> **photosynthesis** A chemical reaction whereby plants use light energy to induce carbon dioxide to react with water, producing carbohydrates and oxygen.

The Ever-Changing Earth

Oxygen is a very reactive chemical, so at first most of the free oxygen produced by photosynthesis was combined with iron in ocean water to form iron oxide–bearing minerals. The control exerted by oxygen on iron dissolved in the ocean was discussed in Chapter 3 (see *What an Earth Scientist Sees*, page 76). Evidence of the gradual transition from oxygen-poor to oxygen-rich ocean water is preserved in seafloor sediments. The minerals in seafloor sedimentary rocks that are more than about 2.5 billion years old contain *reduced* (oxygen-poor) iron compounds. In rocks that are less than 1.8 billion years old, *oxidized* (oxygen-rich) compounds predominate. The sediments that were precipitated during the transition contain alternating bands of red (oxidized iron) and black (reduced iron) minerals. These rocks were deposited as chemical sediments and are called *banded iron formations* (Chapter 3). Because ocean water is in constant contact with the atmosphere, and the two systems function together in a state of dynamic equilibrium, the transition from an oxygen-poor to an oxygen-rich atmosphere also must have occurred during this period.

Along with the buildup of molecular oxygen (O_2) came an eventual increase in ozone (O_3) levels in the atmosphere. Because ozone filters out harmful ultraviolet radiation, this made it possible for life to flourish in shallow water and finally on land. This critical stage in the evolution of the atmosphere was reached between 1100 and 542 million years ago. Interestingly, the fossil record shows an explosive diversification of life-forms 542 million years ago (the beginning of the Phanerozoic Eon).

Oxygen has continued to play a key role in the evolution and form of life. Over the last 200 million years, the concentration of oxygen has risen from 10% to as much as 25% of the atmosphere, before settling (probably not permanently) at its current value of 21% (see **FIGURE 11.3**). This increase has benefited humans and all other mammals, because mammals are voracious oxygen consumers. Not only do we require oxygen to fuel our high-energy, warm-blooded metabolism, our unique reproductive system demands even more. An expectant mother's used (venous) blood must still have enough oxygen in it to diffuse through the placenta into her unborn child's bloodstream. It would be very difficult for any mammal species to survive in an atmosphere of only 10% oxygen.

Scientists cannot yet be certain why the atmospheric oxygen levels increased, but they have a hypothesis, and it illustrates the interactions between all parts of the Earth system. First, note that photosynthesis is only one part of the oxygen cycle. The cycle is completed by decomposition, in which organic carbon combines with oxygen and forms carbon dioxide. But if organic matter is buried as sediment before it fully decomposes, its car-

The oxygen content of the atmosphere FIGURE 11.3

Over the last 200 million years, oxygen levels in the atmosphere have increased markedly. The rise of mammals, though aided by the demise of the dinosaurs 65 million years ago, may also have resulted partly from the plentiful oxygen supply.

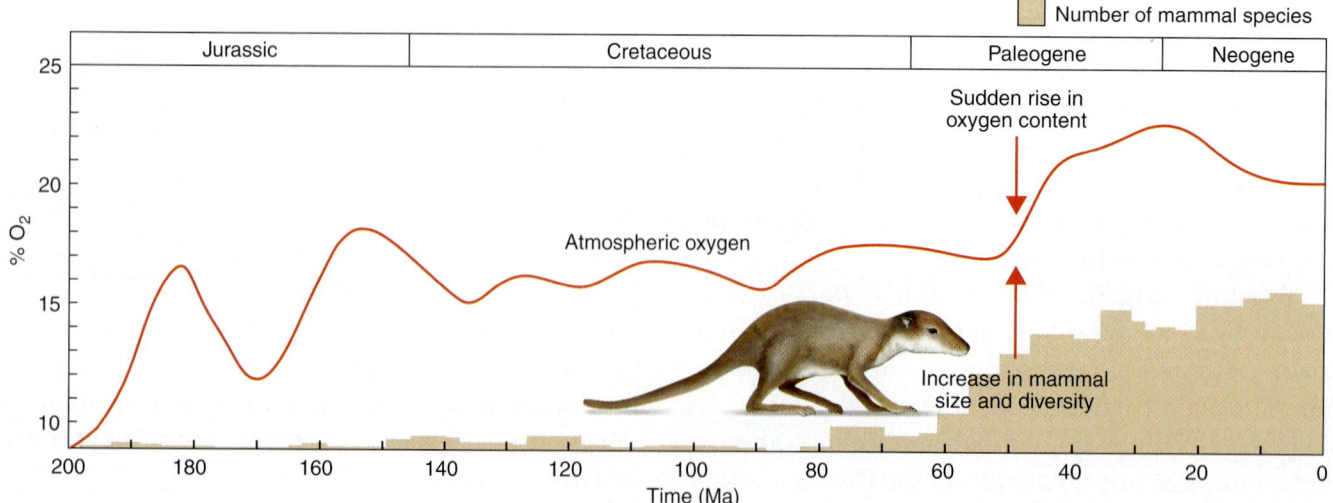

Visualizing
The three steps of life

The first not-so-easy step: Making organic molecules
Figure 11.4

Miller's experiment used a "primordial soup" of methane, ammonia, and hydrogen, in the central flask shown here. Unfortunately, scientists now believe that he used the wrong ingredients: The primitive atmosphere that developed from Earth's outgassing probably consisted mainly of carbon dioxide, water vapor, and nitrogen. No one has come up with a convincing mechanism for producing organic molecules out of these ingredients in conditions that resemble those of early Earth.

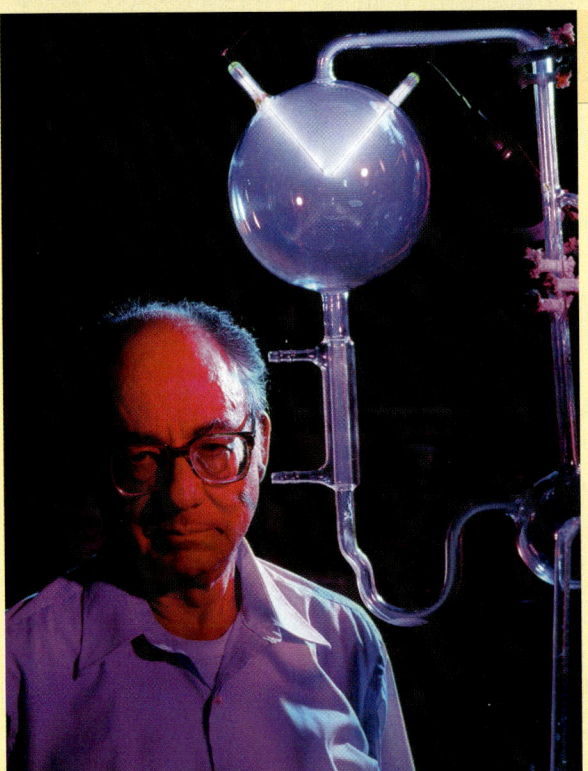

The second not-so-easy step: Replication Figure 11.5

The two strands of the twisted molecule of DNA are held together by organic molecules called *nucleotides*. Nucleotides come in complementary pairs (shown as orange and brown, pink and green). When the two strands are separated, either one can act as a sort of photographic template to recreate the other. This allows for the duplication of genetic information, which is needed for an organism to grow or reproduce.

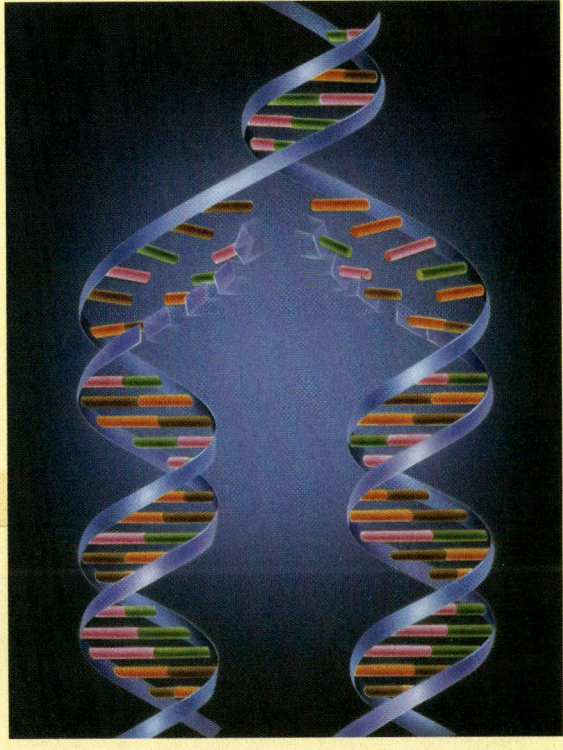

The third not-so-easy step: Metabolism Figure 11.6

Seafloor hydrothermal vents give us an idea of how life might have emerged in the very hot environment of early Earth.

A "Black smokers," such as this vent near the Galápagos Islands, release superheated water and a variety of minerals. Sunlight never penetrates to these depths, so organisms must rely on chemosynthesis rather than photosynthesis to obtain energy. **B** One-celled organisms like these extract energy from the chemicals produced by the vent and form the base of a food chain that includes tubeworms, crabs, and shrimp (**C**).

ARCHEAN AND PROTEROZOIC LIFE

Although, as mentioned earlier in the chapter, some chemical signs of life date as far back as the beginning of the Archean Eon, the most ancient fossils known are the remains of microscopic **prokaryotes**. Prokaryotes are still present on Earth today: All *bacteria* are prokaryotes, and the extremophiles shown in Figure 11.6 are part of a separate **domain** of prokaryotes called *Archaea*. Prokaryotes are unicellular (although they may form colonies, they can survive on their own), and they have rudimentary nuclei. One should not draw the tempting conclusion that bacteria are a "less advanced" life-form than we are. They are well adapted to their own environments, and far more numerous than anything else on Earth. Your own body contains ten times as many bacterial cells as human cells!

For the first two billion years of life, the fossil record is rather sparse, because the only forms of life were microscopic and had no hard parts to preserve. However, we do have one important piece of indirect evidence of ancient life. *Stromatolites* are mound-like structures consisting of many thin layers of calcium carbonate. Similar structures can be found today; they are formed in seawater by the action of photosynthetic bacteria called *cyanobacteria* (see **FIGURE 11.7**). Fossilized stromatolites have been found in rocks up to 3.55 billion years old, providing evidence that photosynthesis and the production of oxygen is an ancient process on Earth.

> **prokaryotes** Single-celled organisms with no distinct nucleus—that is, no membrane separates their DNA from the rest of the cell.

> **domain** The broadest taxonomic category of living organisms; biologists today recognize three domains: bacteria, *Archaea*, and eukaryotes.

An ancient life-form FIGURE 11.7

A These odd-looking bumps in Shark's Bay, Western Australia, are stromatolites, which are formed in warm, shallow seas by photosynthetic bacteria.

B This 2.2-billion-year-old fossilized stromatolite from Michigan shows a pattern of growth rings that is identical to a cross-section of a modern-day stromatolite.

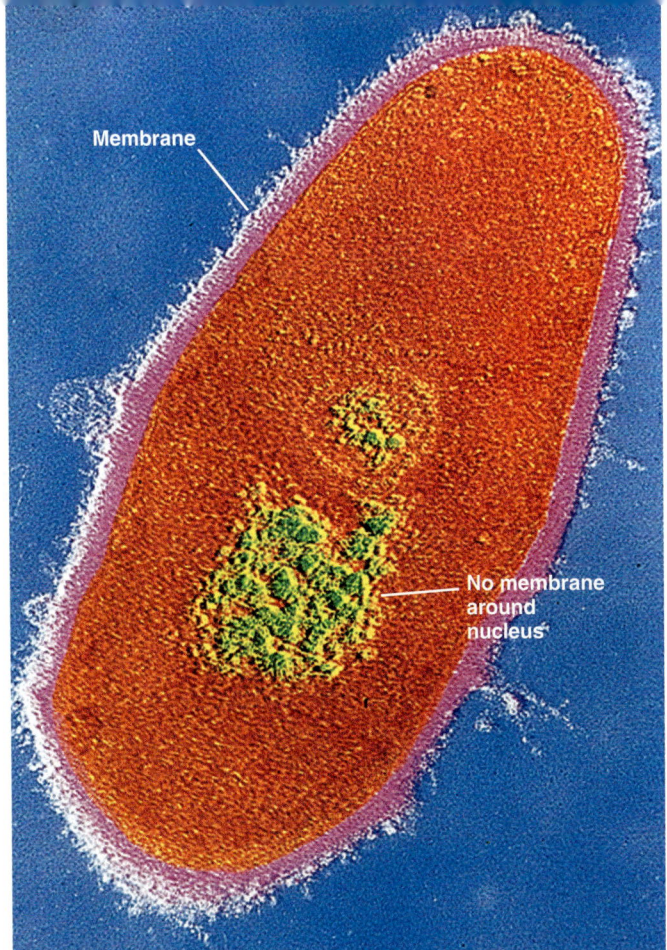

A In prokaryotic cells, such as this bacterium, the nucleus is poorly defined and not contained by a membrane.

B In eukaryotic cells, such as this cell from a plant root, the nucleus and several other structures called *organelles* are clearly defined.

Prokaryotes and eukaryotes FIGURE 11.8

Both cells have been stained to enhance their visibility under a microscope.

At some time in the Proterozoic Eon, a new kind of life-form appeared: **eukaryotes** (from the Greek words meaning "true nucleus"). In addition to a nucleus and membrane, eukaryotes (see **FIGURE 11.8**) contain a number of smaller structures called *organelles*, some of which may originally have been prokaryotic bacteria that were engulfed by the larger eukaryotes and came to live in symbiosis with them.

■ **eukaryotes** Organisms composed of eukaryotic cells—that is, cells that have a well-defined nucleus and organelles.

Eukaryotes appeared at least 1.4 billion years ago, and perhaps as long ago as 2.7 billion years. Like most of the other dates in the early history of life, these dates are uncertain because the first eukaryotes were microscopic in size and lacked hard parts, and therefore are not well preserved as fossils. It is no accident that they emerged after the transition to an oxygenated atmosphere, because they had aerobic metabolisms and thus were better equipped to make use of free oxygen than prokaryotic bacteria had been. The use of oxygen as a fuel had profound ramifications. Aerobic metabolisms are more efficient than anaerobic metabolisms, allowing more energy to be extracted from each molecule of food. Eukaryotic cells grew larger and became more complex than prokaryotic cells. Eukaryotes were also more tolerant of crowding than anaerobic bacteria had been. Although the first eukaryotes were still single-celled organisms, similar to modern slime molds, the stage was set for multicellular life.

Early Life 343

A *Mawsonia spriggi* was probably a floating, disc-shaped animal like a jellyfish, 13 cm in diameter.

B *Dickinsonia costata* was a worm-like creature, 7.5 cm in diameter.

The first multicelled animals FIGURE 11.9

These strangely shaped fossils, specimens of the Ediacara fauna, are the most ancient evidence of multicelled animals.

The earliest fossils of multicellular eukaryotic organisms appear just at the end of the Proterozoic Eon, in rocks about 630 million years old. These fossils, which have now been found in a number of locations, are called the *Ediacara fauna* after the site where they were first discovered, the Ediacara Hills of South Australia. The Ediacara fauna lived in quiet marine bays. They were jelly-like animals with no hard parts (see FIGURE 11.9). These organisms represent a huge jump in complexity from the first unicellular eukaryotes, which appeared at least 800 million years earlier. Scientists still do not know much about what happened during those 800 million years, because fossil evidence is sparse and difficult to interpret.

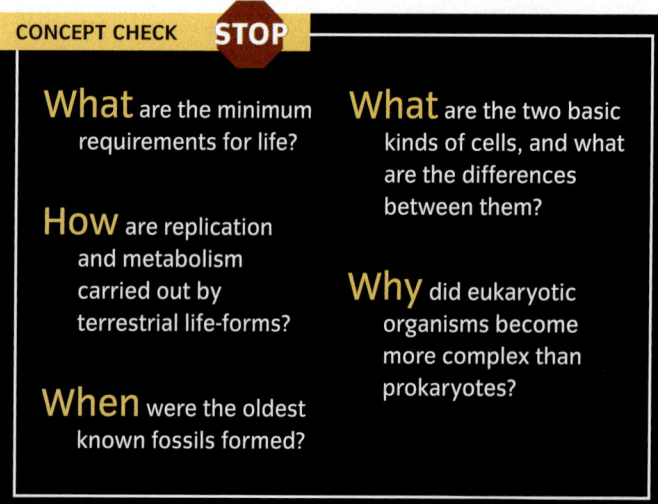

CONCEPT CHECK STOP

What are the minimum requirements for life?

How are replication and metabolism carried out by terrestrial life-forms?

When were the oldest known fossils formed?

What are the two basic kinds of cells, and what are the differences between them?

Why did eukaryotic organisms become more complex than prokaryotes?

344   CHAPTER 11   A Brief History of Life on Earth

Evolution and the Fossil Record

LEARNING OBJECTIVES

Define the theory of natural selection.

Describe how evolution works.

Explain the role of genetics in evolution.

Describe several ways in which fossils form.

At the beginning of the Phanerozoic Eon, about 542 million years ago, the fossil record suddenly shows an explosion in the diversity of species. We have explored some of the probable reasons for this explosion: the oxygenated atmosphere, the protective layer of ozone, and the emergence of eukaryotes and then multicellular organisms. At this point, it seems appropriate to pause and describe *how* new species arise, and how they are preserved as fossils.

EVOLUTION AND NATURAL SELECTION

On December 27, 1831, Charles Darwin set sail from England as an unpaid naturalist and "gentleman's companion" for the captain of the H.M.S. *Beagle* (see **FIGURE 11.10**). Trained as a clergyman, Darwin at the time of his departure believed in the Biblical account of Creation and the fixity of species. By the time he returned in 1836, his views had changed considerably, though he still saw the hand of God at work.

Darwin kept scrupulous notes and made paintings and sketches of the plants, animals, and fossils he saw on the voyage. In 1859, long after his return to England, Darwin published his observations and ideas in a book called *On the Origin of Species by Means of Natural Selection*. He waited so long to publish his findings because he was concerned about the uproar it might—and did—cause. From the time of William "Strata" Smith, and the *principle of faunal and floral succession*, it was widely recognized that life has changed through time. By the time of Darwin, the concept of **evolution** had reached the status of a theory.

In his book, Darwin outlined a hypothesis of **natural selection** to explain how evolution had happened. Darwin proposed that new **species** develop from existing ones by a gradual process of change through inheritable characteristics. All present-day organisms are descendants of different kinds of organisms that existed

evolution The theory that life on Earth has developed gradually, from one or a few simple organisms to more complex organisms.

natural selection The process by which individuals that are well adapted to their environment have a survival advantage, and pass on their favorable characteristics to their offspring.

species A population of genetically and/or morphologically similar individuals that can interbreed and produce fertile offspring.

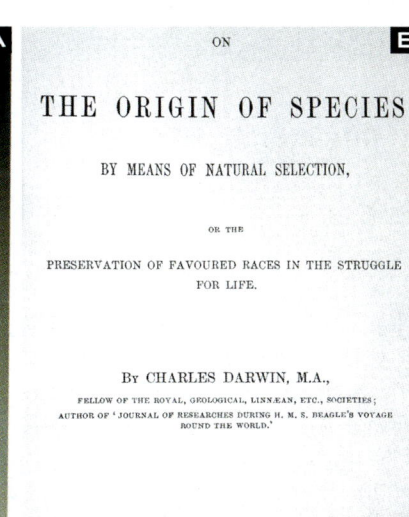

Charles Darwin: Decipherer of evolution's clues FIGURE 11.10

Though he was a retiring and unpretentious man, Charles Darwin (**A**) wrote one of the most controversial and influential books in scientific history, *On the Origin of Species* (**B**). Originally printed in a small edition of 1250 copies, his book has now been through at least 400 printings and translated into at least 29 languages.

Process Diagram

Darwin's finches FIGURE 11.11

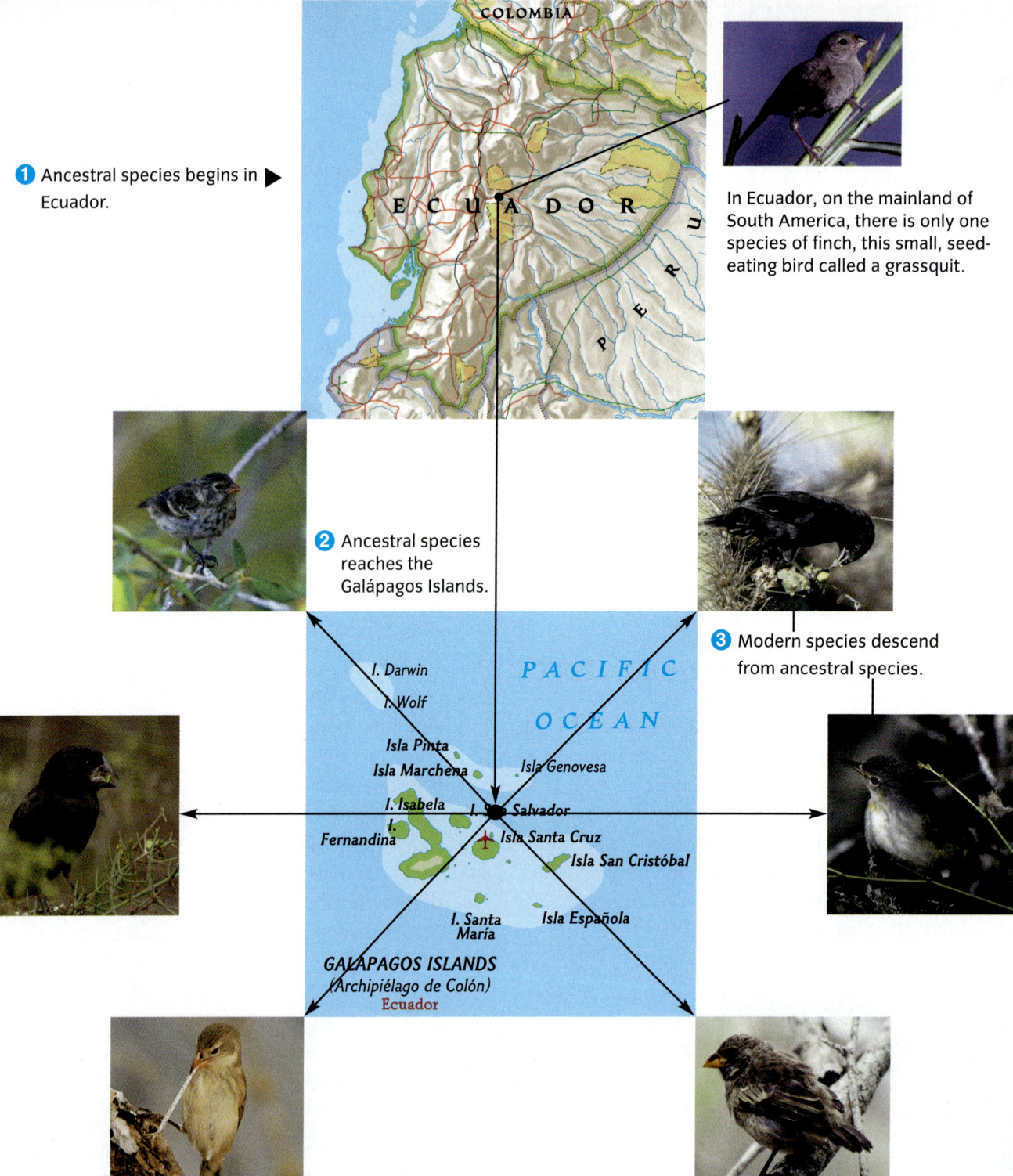

1. Ancestral species begins in Ecuador.

In Ecuador, on the mainland of South America, there is only one species of finch, this small, seed-eating bird called a grassquit.

2. Ancestral species reaches the Galápagos Islands.

3. Modern species descend from ancestral species.

On the Galápagos Islands, Darwin counted 13 species of apparently related birds, but with different beak shapes and different diets. What could account for such a dramatic difference in diversity between the islands and the mainland? Darwin reasoned that long ago, finches from the mainland had colonized the volcanic islands and had subsequently changed as a result of having to adapt to their new environments. Different beak shapes had developed as the birds adapted to different diets. It was a logical idea, but it contradicted the widely believed theological doctrine that no new species had appeared on Earth since the Creation.

CRITICAL THINKING — Here's an interesting question:
- Darwin reasoned that the original grassquits that colonized the islands had become stranded and therefore isolated from the population in Ecuador. How might they have become stranded?

in the past, whose populations slowly changed in response to changing environmental conditions. Darwin was not the first to suggest evolution as an explanation for the variety and distribution of species on Earth, but he was the first to provide such a thorough discussion of the evidence, gathered during his voyage and over the following years (see **FIGURE 11.11**).

Because the word *theory* is often misunderstood (Chapter 1), it is important to emphasize that there is no doubt among legitimate biologists that evolution does occur, has occurred throughout the history of life on this planet, and has repeatedly been shown to occur in laboratory situations. The question is *how* it occurs. The late paleontologist Stephen Jay Gould said that Darwin's two great accomplishments were establishing the *fact* of evolution and proposing the *theory* of natural selection to explain it. There is an analogous situation in physics, where it is a *fact* (established by Galileo) that gravitation exists, but there are different *theories* (proposed by Newton and Einstein) of how it works. Darwin was the Galileo and Newton of biology, rolled into one.

Darwin was motivated to publish *On the Origin of Species* when another young naturalist, Alfred Russell Wallace, independently hit on the same idea of natural selection. According to this theory, any generation of a species will have a broad range of genetic characteristics. Individuals who are better adapted to their surroundings will have more survival success, and more reproductive success. In later generations, their descendants will be more numerous than those of less well-adapted individuals. Over time, the entire population will evolve as natural selection favors the better-adapted individuals (see Figure 11.11).

Natural selection also allows for relatively rapid changes. In modern laboratories, scientists can observe evolution in progress, through the expression of specific genetic traits in organisms that reproduce rapidly, such as fruit flies. In nature, if a useful new trait emerges that provides a competitive advantage for an individual and its descendants—for example, a different beak shape, which might help a finch procure seeds—then after several generations, most individuals will have that trait. *Note that the organism does not develop this trait in response to an environmental need or constraint. The trait occurs by chance, but its occurrence enhances the survival and therefore the reproductive success of the individual.* Since Darwin's time, the science of genetics has greatly strengthened his argument. Darwin did not know how organisms transmit their traits to later generations; now we know that heritable traits are encoded in *genes* on an organism's DNA. The random variation that is essential to Darwin's theory is explained by *mutations* in those genes—accidental substitutions of one nucleotide for another, or deletions or transpositions.

There is still some scientific debate over the relative importance of gradual change versus rapid change in evolution. Darwin favored *gradualism*, whereas Gould, among others, advocated *punctuated equilibrium*, in which species persist for a very long time with few changes and undergo occasional periods of very rapid change. In recent years, scientists have begun to recognize the importance of occasional catastrophic events. The best-documented example (see the end of this chapter) is the meteorite impact that is thought to have wiped out the dinosaurs. These animals died out, in all likelihood, not because they were less well adapted to their environment than other species, but because of a chance event that they had never experienced before and never had an opportunity to adapt to.

HOW FOSSILS FORM

The best evidence for evolution lies in the vast numbers of **fossils** chronicling the succession of species that no longer exist on Earth. Although fossilization can occur in many ways, it is always a relatively rare event. If a dead animal or plant is exposed to air, running water, scavengers, or bacteria, it will decompose or be eaten, and its parts will be scattered or destroyed. Hard parts such as bones, shells, and teeth are less easily destroyed and thus more likely to be preserved than soft or delicate parts such as skin, hair, feathers, or leaves. In any case, for an organism to be preserved as a fossil it must be quickly covered up by a layer of protective material—usually sand or mud, sometimes tree sap, ice, tar, or volcanic ash.

> **fossil** Remains of an organism from a past age, embedded and preserved in Earth's crust.

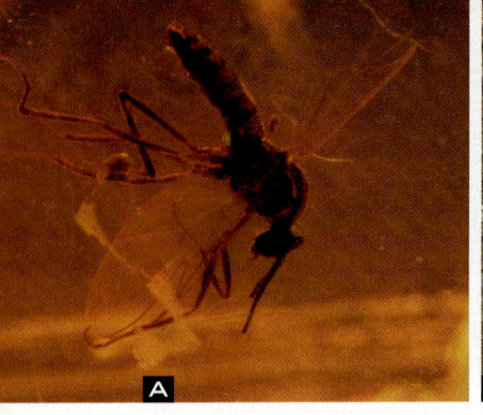

Examples of fossil formation FIGURE 11.12

A Even the delicate legs and wings of this ancient mosquito are preserved in its casing of amber, which is fossilized tree resin. The mosquito is more than 24 million years old.

B Over many millennia, this log in Petrified Forest National Park, Arizona, became mineralized. Although it looks uncannily like fresh wood, preserved even down to the cellular level, it is completely made of stone.

Sometimes a deceased organism is preserved with little or no alteration. For example, insects many millions of years old have been trapped in tree sap, which hardens into amber (see **FIGURE 11.12A**). This seals the specimen off from the elements so completely that parts of its original organic matter can still be recovered. Ice and tar are also excellent preservatives. In dry climates, natural *mummification* can occur, in which the soft parts dry and harden before they have a chance to decompose.

More often, however, fossils reflect the original shape of an organism but do not contain the original materials. Bones and other hard parts are replaced by minerals carried in solution by groundwater (a process called *mineralization*). This is the process that creates *petrified wood* (see **FIGURE 11.12B**). The remains of plants are sometimes preserved by carbonization, which occurs when volatile material in the plant evaporates, leaving behind a thin film of carbon.

Some fossils do not contain any actual parts of the organism. For example, the organism may leave behind an imprint or *mold* in the soft sediment that covered it, as in the Ediacara fauna. Other kinds of indirect evidence of previous animal life include eggshells and **trace fossils** (see **FIGURE 11.13**). Finally, prehistoric animals left behind fecal droppings, which are called *coprolites* when preserved and fossilized. In spite of their unappealing origin, such fossils provide useful clues about the animals' characteristics, habits, and diets.

trace fossil Fossilized evidence of an organism's life processes, such as tracks, footprints, and burrows.

CONCEPT CHECK STOP

Where did Charles Darwin find the evidence that led him to the theory of natural selection?

Why does natural selection rely on random variations in the genetic code?

Why do hard-bodied animals (or hard parts such as teeth) appear more often as fossils than soft-bodied animals or soft parts?

How are soft-bodied animals sometimes preserved?

Not all fossils are bodies FIGURE 11.13

A This two-meter-long fossilized dinosaur (oviraptor) was found curled protectively around a nest containing at least 20 eggs. This is considered to be the first proof that dinosaurs cared for their young.

B Over 65 million years ago, hadrosaurs in what is now Argentina left their tracks in soft, red mud, which turned to rock. The formation of the Andes Mountains tilted it to such an extent that the formerly horizontal mud flat is now a vertical rock wall.

PERIODS MASS EXTINCTIONS ⊘	542 PRECAMBRIAN (millions of years before present)	CAMBRIAN	488
PLANTS, ANIMALS (% of species extinct)	Algae, First multicellular animals	No known land plants... Algae dominate Explosive radiation of marine animals... Hard shells, skeletons... Trilobites dominate	

Life in the Phanerozoic Eon

LEARNING OBJECTIVES

Describe the dramatic change in Earth's biota during the Cambrian Period.

Identify the requirements for a living organism to survive on land.

Describe how plants, amphibians, and reptiles met the above requirements.

Outline the most recent evolutionary steps in the development of humans.

Define mass extinctions and identify two theories of what causes them.

During most of the history of life on Earth—3 billion years of it—the only living organisms were of microscopic size. But about 630 million years ago, with the appearance of larger multicellular animals, the Ediacara fauna, life began to diversify very rapidly into a wide variety of sizes and body types. We will take a whirlwind tour through this last phase of development of life, which continues to the present day. You can follow the major changes by checking the timeline that runs across the top of the pages discussing the Phanerozoic Eon.

The Phanerozoic Eon is named from the Greek *phaneros,* meaning "visible," because it is that part of Earth's history in which evidence of life is abundantly visible. The Phanerozoic Eon is divided into three eras (see Chapter 10): the Paleozoic Era (old life), the Mesozoic Era (middle life), and the Cenozoic Era (recent life).

THE PALEOZOIC ERA

The Paleozoic Era started 542 million years ago with the Cambrian Period. It was a time of incredible diversification of life (see *What an Earth Scientist Sees*). Why was this so? One hypothesis is that sexual reproduction, which developed with the eukaryotes, afforded a more rapid way for new "experimental" forms of life to emerge. Another hypothesis is that there was finally enough oxygen in the atmosphere to support the metabolism of larger organisms. The presence of oxygen led to the

A Sample of Cambrian Life

A Trilobites were one of the first animals to develop a hard covering, presumably to defend against predators. They are ubiquitous among Cambrian fossils. This 34-cm, total-length specimen of Olenellus getzi was found in Lancaster County, PA, and is Lower Cambrian in age.

B This specimen of *Waptia fidensis*, a soft-bodied arthropod, is preserved in the Burgess Shale of British Columbia. This species did not have a hard shell as the trilobites did, and therefore fossils of it are much rarer.

Here's an interesting question:
- Despite their efficient defense against predation, trilobites eventually became extinct. Arthropods, however, are still with us—they are a hugely diverse family. Why have arthropods been so successful?

What an Earth Scientist Sees

development of an ozone layer and this would have shielded Cambrian life-forms from harmful ultraviolet radiation (see Chapter 14). There are other hypotheses, too; one is that predation started and this led to the preservation of creatures that had developed protective hard parts; another is that extreme climactic changes at the end of the Proterozoic Eon led to accelerated evolutionary responses of surviving organisms.

Whatever the reasons, a great many changes began to occur about 542 million years ago, in what has been called the *Cambrian explosion* or *Cambrian radiation*. Compact animals replaced the soft-bodied organisms of Ediacara times. These included trilobites, such as the one in *What an Earth Scientist Sees*, mollusks (clams and sea snails), and echinoderms (sea urchins). All of the new animals were equipped with gills, filters for feeding, efficient guts, and a circulatory system.

The Cambrian Period also saw the development of skeletons, both internal and external. Skeletons gave organisms a selective advantage, protecting them against predators, against drying out, against being injured in turbulent water, and so on. Hard-shelled organisms, such as trilobites, are well-preserved in the fossil record. However, a wide variety of soft-bodied creatures have been found in the Burgess Shale and even richer deposits in Chengjiang, China. From these deposits we now know that every **phylum** in the animal **kingdom** was present in the Cambrian Period, as well as a number of phyla that no longer exist. In this sense, the Cambrian marine environment had the greatest diversity of life-forms in Earth's history.

kingdom The second-broadest taxonomic category. There are six recognized kingdoms, including animals and plants.

FROM SEA TO LAND

The great proliferation of life in the Cambrian explosion was confined to the sea. In order to spread to land, it was essential for living organisms to meet certain requirements. First is some means of *structural support*. Whereas aquatic organisms are buoyed up by water, land-based organisms must contend with gravity. But because water remains critical to all the chemical processes of life, any land-based organism must maintain an *internal aquatic environment*. Living in an environment surrounded by air, rather than water, the organism must develop some way of *exchanging gases* with air instead of water. Finally, all sexually reproducing organisms require a *moist environment for the reproductive system*. The first organisms to overcome these four hurdles were plants.

Plants Evidence suggests that land plants evolved from green algae more than 600 million years ago (see FIGURE 11.14). Eventually, during the Silurian Period, *vascular plants* evolved, which have structural support from stems and limbs (requirement 1) and a set of vessels through which water and dissolved elements are transferred from the roots to the leaves (requirement 2). Gas exchange (requirement 3) occurs by diffusion through adjustable openings in the leaves called *stomata*. When carbon dioxide pressure inside the leaf is high, the stomata open; when it is low, they close. The stomata also close when the plant is short of water, thereby protecting it from drying out.

The earliest life on land? FIGURE 11.14

The first plants to colonize the land may have looked like the green algae shown in this photograph.

DEVONIAN

Mosses, ferns still prevalent... First naked seed plants (gymnosperms)... First trees
Early sharks... Age of fishes... First land vertebrates (amphibians) evolve from lobe-finned fish

The evolution of plant life FIGURE 11.15

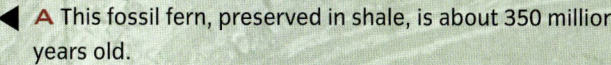

A This fossil fern, preserved in shale, is about 350 million years old.

B Compare the fossil to a modern fern. This photograph shows the spore-producing organs, the dark spots on the underside of the frond.

C Naked-seed plants developed from the seedless plants late in the Devonian Period. These are leaves of modern and fossilized ginkgos.

D Flowers and fruits represented an advantageous adaptation for plants, because they serve as an incentive for insects to do the work of distributing pollen. This 15-million-year-old fossil found in Idaho is shown next to its modern equivalent, the sweet gum fruit.

The means of reproduction in plants (requirement 4) has passed through several evolutionary stages. The earliest plants were seedless (see **FIGURE 11.15**), and some seedless plants still exist, such as mosses and ferns. These can reproduce either sexually or asexually, but for the sexual phase they require a body of water where the male and female reproductive bodies (*spores*) meet and fuse. Thus mosses and seedless plants in

Life in the Phanerozoic Eon   351

general have always depended on a rather moist environment. Seedless plants reached their peak in the Carboniferous Period, and their fossils created the extensive deposits of coal (carbon-rich sedimentary rocks) that gave the period its name.

In the middle Devonian Period, some plants began to evolve their own moist environment to facilitate sexual reproduction. The first plants to do this were **gymnosperms**, which include ginkgos and conifers. The female cell of a gymnosperm is attached to the vascular system and therefore has a supply of moisture. The male cell is carried in a pollen grain with a waxy coating. When the two fuse, a *seed* results. The seed contains moisture and a food supply that sustains the growth of the young plant until it can support itself. With the evolution of seeds, vascular plants were able to spread beyond swampy lowlands to other habitats.

■ **gymnosperm** A naked-seed plant.

Arthropods

Many of the creatures in the Cambrian seas belonged to the phylum *Arthropoda,* named after their jointed legs. Modern arthropods include crabs, spiders, centipedes, and insects—the most diverse phylum on Earth. They were the first creatures to make the transition from sea to land.

With a few exceptions, early arthropods were small and light. They were covered with a hard shell of *chitin* (a fingernail-like material). Thus, they were well adapted to life on land, in regard to the need for structural support and water conservation. The first arthropods on land were probably centipedes and millipedes, in the Silurian Period. By the Mississippian Period, insects were abundant and included dragonflies with a wingspan of up to 60 cm (see **Figure 11.16**). But for all their success as land creatures, the arthropods have very primitive respiratory and vascular systems. For example, insects breathe through tiny tubes that penetrate their outer coating. This mode of respiration severely limits the size of an insect and explains why almost all insects are small.

The "blood" of arthropods is simply body fluid that bathes the internal organs; it does not circulate in blood vessels. The fluid is generally kept in motion by a sluggish "heart" that is little more than a contracting tube. However, primitive does not mean poorly adapted. The arthropods have branched into more than a million species, and they are close to indestructible. Who ever heard of a cockroach having a heart attack?

Ancient insects FIGURE 11.16

A single wing of the giant dragonfly Megatypus schucherti, is 16 cm in length. The largest dragonfly today is only 15 cm in total width. This specimen is Lower Permian in age and is from Dickinson County, KS.

| PERMIAN | 251 | TRIASSIC | 200 |

Pangaea forms... Conifers appear, colonize uplands
First beetles and flies... Mammal-like reptiles (therapsids)   95%

First dinosaurs appear... First true mammals   80%

A When a living coelacanth was caught in the Indian Ocean in 1938, it created a sensation because not only its species but its entire order, the Crossopterygians, had been believed to be extinct. The first fish to haul themselves onto land may have been relatives of the coelacanth. Today's coelacanths, like this one photographed near the Comoro Islands, are exclusively deep-sea creatures.

B Another candidate for the first fish to make the transition to land is the lungfish. Unlike the coelacanth, today's lungfish still survive for short periods on land without water.

Pioneer fish? FIGURE 11.17

Fishes and amphibians

The phylum of greatest interest to most of us, because humans belong to it, is the *chordates*. These are animals that have at least a primitive version of a spinal cord (called a *notochord*). Like all the other animal phyla, chordates can be found in fossils of the Cambrian Period, although they are relatively inconspicuous. The earliest so far discovered is *Haikouichthys*, a jawless fish akin to a hagfish or a lamprey that lived 525 million years ago—only 17 million years after the beginning of the Cambrian explosion.

Jawed fish arrived next. With jaws, fish could lead a predatory lifestyle and grow to much larger sizes; the jawless fish had been limited to filtering food out of the water or dredging it from the sea floor. Among the first large jawed fish were sharks and ray-finned fish. The earliest known intact shark fossil is 409 million years old.

The first fish to venture onto land may have been a member of an obscure order called *Crossopterygii*, or lobe-finned fish (see **FIGURE 11.17**). These fish had several features that could have enabled them to make the transition to land. Their lobe-like fins contained all the elements of a quadruped limb. They had internal nostrils, characteristic of air-breathing animals. As fish, they had already developed a vascular system that was adequate for life on land.

The first terrestrial chordates, amphibians, have never become wholly independent of aquatic environments, because they have not developed an effective method for conserving water. To this day they retain permeable skins. They have also never really met the reproductive requirement for life on land. In most amphibian species, the female lays her eggs in water and the male fertilizes them there after a courtship ritual. The young (e.g., tadpoles) are fish-like when first hatched. Just like the seedless plants, amphibians have kept one foot (figuratively speaking) on land and one foot in the water. They originated in the Devonian Period.

Life in the Phanerozoic Eon   353

THE MESOZOIC ERA

The Paleozoic Era closed with the extinction of trilobites and a great many other marine creatures. An estimated 96% of all living species disappeared at the end of the Permian Period in the greatest mass extinction in Earth's history. We will return to the question of mass extinctions at the end of this chapter, but the cause of the extinction remains an enigma that has yet to be solved. The era that followed the Paleozoic, the Mesozoic Era, commenced with all the major landmasses on Earth joined together in the supercontinent Pangaea (see Figure 1.11). Then, in the middle of the era, the supercontinent began to split apart in a process that continues today.

Whatever the cause of the great extinction at the end of the Permian, it did not seem to greatly affect plant life on land. Gymnosperms, which had appeared in the Devonian Period, dominated the plant world during the Triassic and Jurassic Periods. Gymnosperms, however, have an important liability. The male cell-carrier, the pollen, is spread through the air. This is extremely inefficient. What chance does a pollen grain in the air have of finding a female cell? Eventually, at the beginning of the Cretaceous period, **angiosperms**—flowering plants—found a more efficient solution. For a small incentive, such as nectar or a share of the pollen, insects deliver pollen directly from one flower to another, or from one part of a plant to another. After pollination, the plant develops a seed in much the same way as gymnosperms. In many cases, birds and other animals help distribute the seeds by eating the plant's seed-bearing fruits and distributing the seeds in their feces. Flowering plants still dominate land plants today.

angiosperm A flowering, or seed-enclosed, plant.

Reptiles, birds, and mammals
The Carboniferous and Permian Periods were times when amphibians were abundant. Most families died out in the end-of-Permian extinction. One branch of the amphibians,

Early birds and mammals FIGURE 11.18

A The skeletons and teeth of *Archaeopteryx* were very similar to those of dinosaurs. However, the very detailed impressions of feathers in this fossil identify *Archaeopteryx* as a bird.

B Discovered in 2002 in China, this shrew-sized *Eomaia scansoria* specimen is the oldest known fossil of a placental mammal (that is, a mammal that gives live birth). It lived 125 million years ago, during the height of the dinosaur age.

however, evolved into reptiles—the first fully terrestrial animals. Reptiles were freed from the water by evolving an egg that contained amniotic fluid for the young to grow in and by developing a watertight skin. These two evolutionary advances enabled them to occupy many terrestrial niches that the amphibians had not been able to exploit because of their need to live near water. The amniotic egg led to an explosion in reptile diversity, much as the evolution of the jaw had done for fishes. Moving out of the Mississippian and Pennsylvanian swamps, some colonized the land, some moved back into the water, and a few took to the air. Not only did reptiles greatly increase in diversity during the Jurassic Period, they also grew to tremendous size. The dinosaurs were the largest land animals that ever lived, possibly ranging up to 100 metric tons in weight and 35 m in length.

Birds first appeared near the end of the Jurassic Period, and they are now considered to be direct descendants of the dinosaurs. An early bird, *Archaeopteryx* (see **FIGURE 11.18A**), would have been classified as a dinosaur if not for the discovery that it had feathers. Even before *Archaeopteryx*, vertebrates had made the transition to the air in the form of pterosaurs—flying reptiles with long wings and tails. Pterosaurs are not considered by most paleontologists to be true dinosaurs, although they were very closely related. The detail of the transition from reptiles to birds remains one of the most highly controversial topics in paleontology today.

Mammals are descended from a class of "mammal-like reptiles" that existed as long ago as the Permian Period. The transition from reptiles to mammals is well understood, though perhaps not quite as well understood as the transitions from fishes to amphibians. It is difficult to pick out a single mammalian adaptation comparable to the jaws of fish, the eggs of reptiles, or the feathers of birds. During the Cretaceous Period, mammals certainly did not outcompete the dinosaurs; they survived by being small and inconspicuous (**FIGURES 11.18B** and **11.18C**).

THE CENOZOIC ERA

Evidence suggests that an accidental catastrophe—a giant meteorite impact—was at least partially responsible for the environmental changes that wiped out the great reptiles and 70% of all other species, too, at the end of the Mesozoic Era. The departure of dinosaurs gave mammals a chance to grow larger and to diversify. In the Cenozoic Era, mammals have also benefited from the atmosphere's high oxygen level, conducive to a fast metabolism. Unlike reptiles, whose brain sizes have not grown (relative to their body size), mammals have continued to evolve toward larger brain size throughout the Cenozoic Era. This may be one key to mammalian success, because it has enabled them to diversify their lifestyles to a much greater extent than reptiles ever could have.

The last frontier for plants—the dry steppes, savannahs, and prairies—was not colonized until the Paleogene Period, when grasses evolved. This process involved the assistance of animals, in particular the great grazing herds that lived on all continents except Antarctica.

C The shape of *Eomaia*'s claws shows that it was a climber that could grasp tree branches. It is now known that small, tree-dwelling mammals began to experiment with gliding flight as early as 130 million years ago-almost as early as the first birds began flying.

65 | TERTIARY | 1.8 | QUATERNARY | Present

Grasses evolve... Mammals increase in diversity and size... Grazing mammals...
High oxygen levels enable large brains, fast metabolism... First hominids... Ice ages
Genus *Homo*... Modern Humans
⊘ Possible mass extinction due to human impact on environment?

Lucy's relatives FIGURE 11.19

A This drawing by Michael Rothman depicts a mother and child of the species *Australopithecus afarensis*. These human-like individuals lived together in small groups, formed lasting bonds with mates, and looked after their children through infancy.

B This 3.3-million-year-old fossil is the skull of an *A. afarensis* baby, who probably looked much like the child in the drawing.

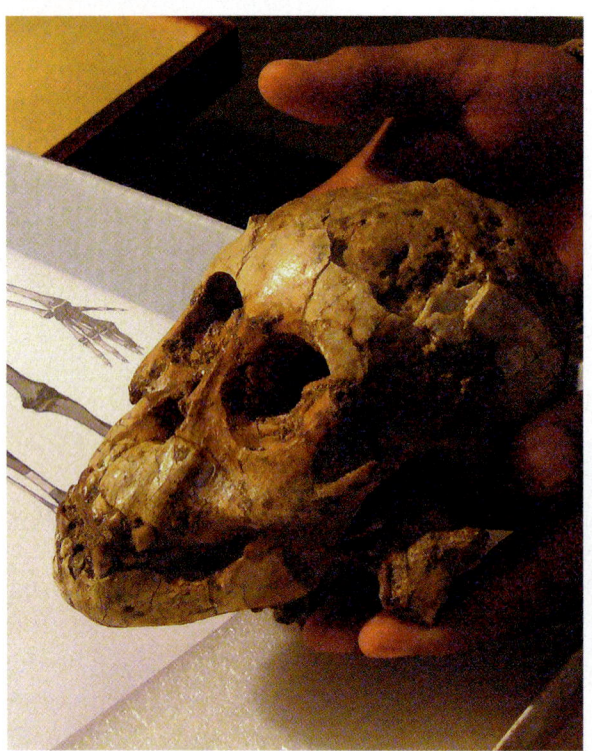

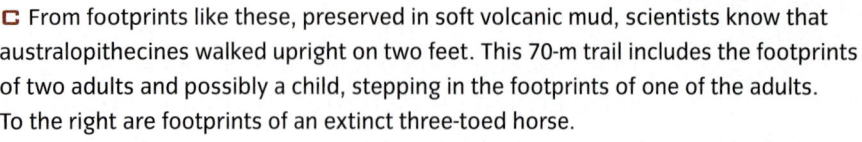

C From footprints like these, preserved in soft volcanic mud, scientists know that australopithecines walked upright on two feet. This 70-m trail includes the footprints of two adults and possibly a child, stepping in the footprints of one of the adults. To the right are footprints of an extinct three-toed horse.

The human family From a human point of view, it is surely true that the most remarkable thing that has happened during the Cenozoic Era is the emergence of our own species. Charles Darwin was often accused of believing that humans are descended from the apes. In fact, the family of humans, *Hominidae,* and the family of apes, *Pongidae,* are both descended from an earlier common ancestor that was neither human nor ape. The emergence of humans is one of the most complex and controversial fields in paleontology, in part because of the paucity of the fossil record and the lack of transitional forms. But it is clear that **hominids**—human-like organisms—are a very recent evolutionary development. The first hominid that was clearly *bipedal* (walked upright) was *Australopithecus*, of which the famous fossil "Lucy" is an example (see **FIGURE 11.19**). These hominids were only about 1.2 m in height, but had a brain capacity larger than chimpanzees. Their fossils range from about 3.9 million to 3.0 million years in age. From the shape of its pelvis and from footprints left in soft volcanic mud (see **FIGURE 11.19C**), we know that *Australopithecus* walked upright, though its skull looked more apelike than human. Lucy's descendents never spread beyond Africa, and they disappeared altogether about 1.1 million years ago.

Homo erectus, probably the first species of our own genus (*Homo*), was more widely traveled than *Australopithecus*. Fossils of *Homo erectus*, dating back about 1.8 million years, have been found in Africa, Europe, China, and Java. Even earlier, a problematic species called *Homo habilis* used stone tools. Because toolmaking is the distinguishing feature of the genus *Homo*, some experts include *habilis* in this genus; others argue that the skull of this species is more like that of the australopithecines.

Homo erectus disappeared 300,000 years ago, and was replaced by *Homo neandertalensis* ("Neandertal man") no later than 230,000 years ago. Unfortunately, the poor fossil record between 400,000 and 100,000 years ago has made it difficult for scientists to determine how this transition occurred. We hypothesize from burial sites that Neandertals might have practiced some form of religion. Because of similarities in teeth and brain size (slightly larger than our own), some experts have argued that Neandertal was part of our own species; however, recent DNA studies suggest that *Homo sapiens* is not a direct descendant of Neandertals. The Neandertals disappeared about 30,000 years ago and were replaced rather suddenly by the biologically modern people, the first indisputable members of our own species, *Homo sapiens*.

Did the *Homo sapiens* evolve from the Neandertals, or were they a distinct species? What happened during the 5000-year period when both kinds of humans were alive, and overlapped geographically in Europe? Did *Homo sapiens* kill the Neandertals? These and many other questions await answers, as paleontologists continue to look for clues in the fossil record.

MASS EXTINCTIONS

At several places in the fossil record of the Phanerozoic Eon, paleontologists have found abrupt and profound changes in the fossil assemblage—not just in one location, but worldwide. These dramatic changes record **mass extinctions**. (In the timeline on the previous eight pages, the boundaries marked in red correspond to mass extinctions.) For Earth scientists, mass extinctions are a convenience, because they delineate so many of the time periods in the geologic column, but they are also a conundrum. What could possibly cause most of the species on Earth to become extinct in a short period of time?

> **mass extinction**
> A catastrophic episode in which a large fraction of living species become extinct within a geologically short time.

The most famous mass extinction occurred 65 million years ago, at the end of the Cretaceous Period. (This is the boundary between the Mesozoic and Cenozoic Eras.) Paleontologists estimate that around 70% of all species died out in the end-of-Cretaceous extinction, including the dinosaurs. Many of the species that survived must also have come perilously close to extinction.

In the last quarter-century, scientific opinion has coalesced around the hypothesis that the end-of-Cretaceous extinction was at least partly caused by environmental changes that resulted from the impact of a large meteorite with Earth (see **FIGURE 11.20** on the following page). When it was proposed by Walter and Luis Alvarez in 1980, the meteorite-impact theory was highly controversial, but many lines of evidence now support it (see **FIGURE 11.21** on page 359). We can be reasonably certain now that the events described in Figure 11.21 did happen, although we cannot be absolutely certain that this event killed all the dinosaurs.

One of the remaining problems with the meteorite impact theory is that the great end-of-Cretaceous extinction was not unique, nor was it even the most dramatic of all extinctions. As mentioned earlier in this chapter, the most devastating mass extinction occurred 245 million years ago, at the end of the Permian Period, when as many as 96% of all species died out. There have been at least 5 and possibly as many as 12 mass extinctions during the last 250 million years: Some of them can be linked to massive meteorite impacts but others cannot. Several other known craters on Earth rival the size of the Chicxulub crater, yet the ages of the craters do not correlate well with the dates of other mass extinctions. On the other hand, some scientists have pointed out that

The day the dinosaurs died? FIGURE 11.20

A About 65 million years ago, a meteorite as large as Mt. Everest struck Earth at a spot just offshore from Mexico's Yucatan Peninsula. It blasted a crater 180 km wide.

B The blast wave itself would have killed animals and plants all over the Western Hemisphere. As broiling-hot debris from the impact rained down, forests ignited, creating continent-wide forest fires.

C Soot from the fire may have remained in the atmosphere for months, or even years, blocking out the sunlight and bringing photosynthesis all over the world to an effective halt.

D A year after the impact, algae and ferns may have begun to grow again, but the forests were still bare.

Visualizing
The effects of a meteorite impact

Evidence of impact FIGURE 11.21

A A thin layer of clay, shown here in a sample from the Raton Basin in Colorado, can be found all over the world in rocks at the end-of-Cretaceous boundary. Here the whitish layer of clay is about 2 cm thick. Above it is a thinner layer containing shocked quartz particles that is enriched in iridium, an element that is rare on Earth but more common in meteorites. This evidence strongly suggests that both layers were formed by a meteorite impact. In addition, the clay layer becomes thicker as one gets closer to the hypothesized impact site in Mexico. Dinosaur fossils can be found beneath the clay layer but not above it.

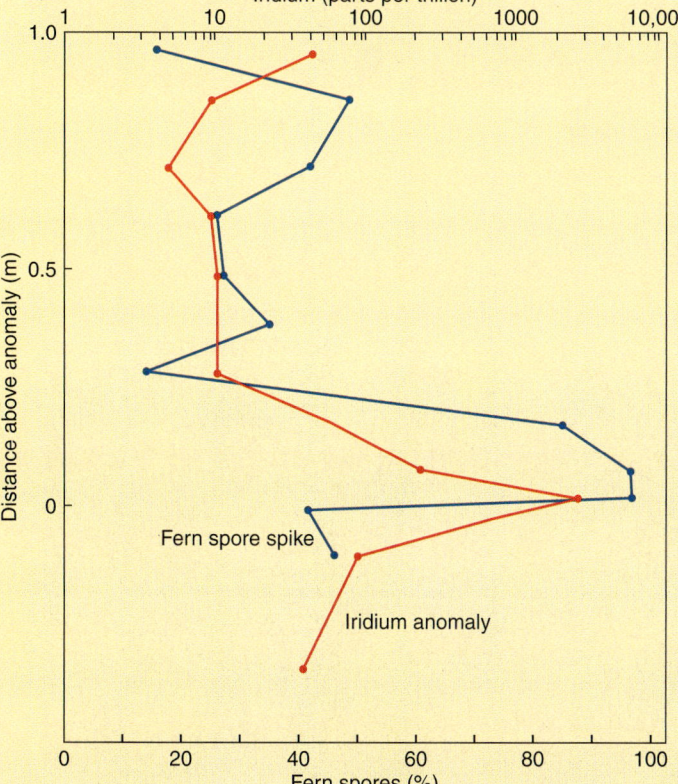

B In layers just above the impact, fern spores are much more prevalent than angiosperm pollen—a sign of a regenerating plant community.

C Scientists believe they have found the site where the meteorite landed. A crater lies underneath Mexico's Yucatan Peninsula. Although the crater is now completely covered by sediments, a satellite radar image shows a faint circular depression of the ground around its edge. The crater is named *Chicxulub* (pronounced tjik'ju'lub) after a Mayan town near the center of the crater. A rough translation of the Mayan name is "tail of the devil."

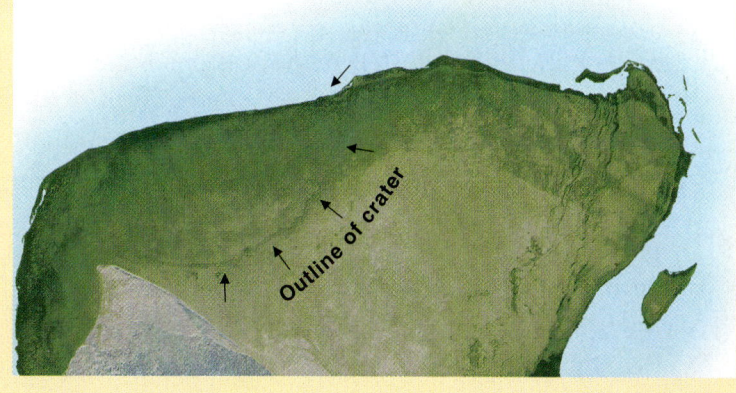

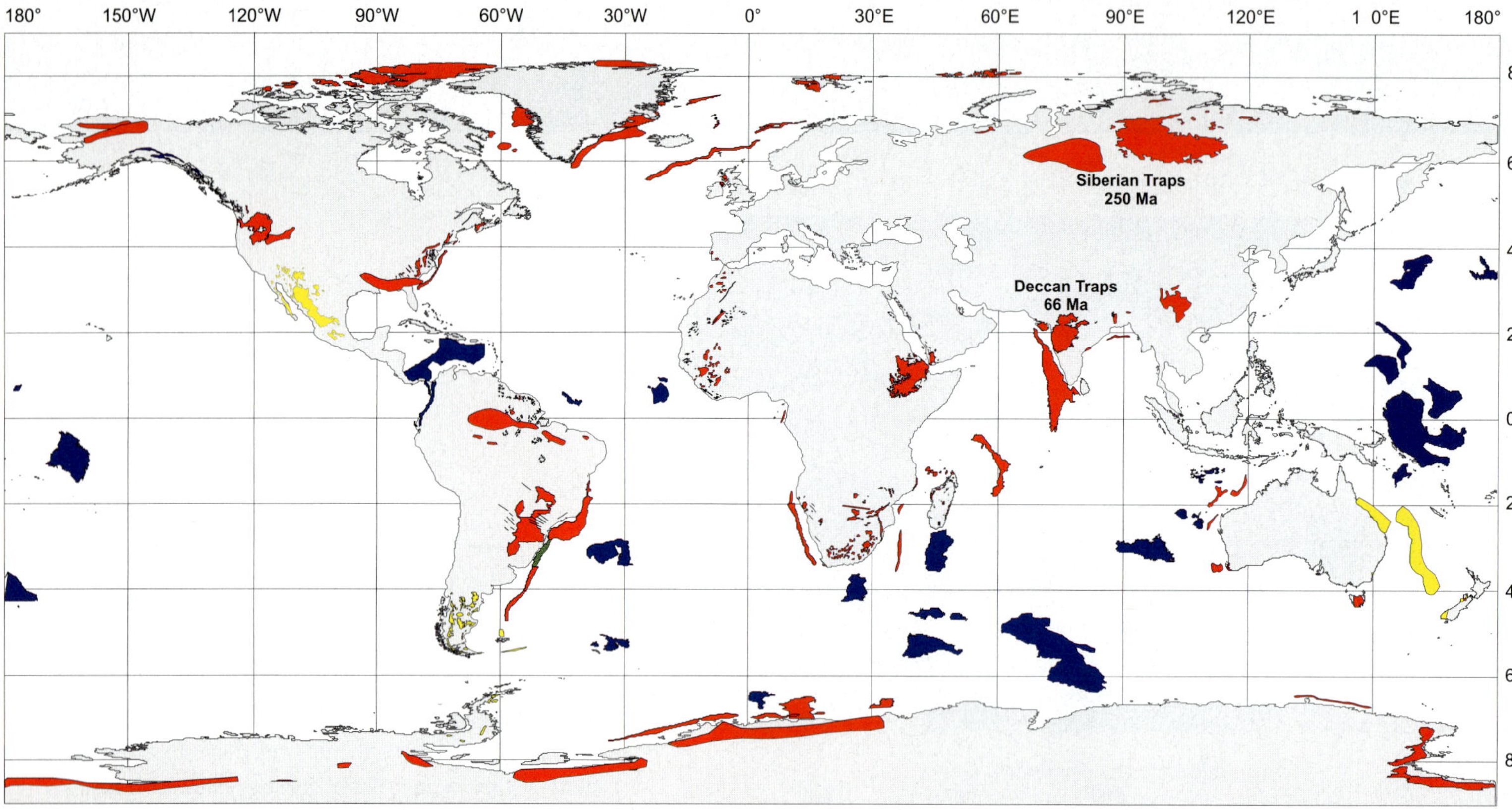

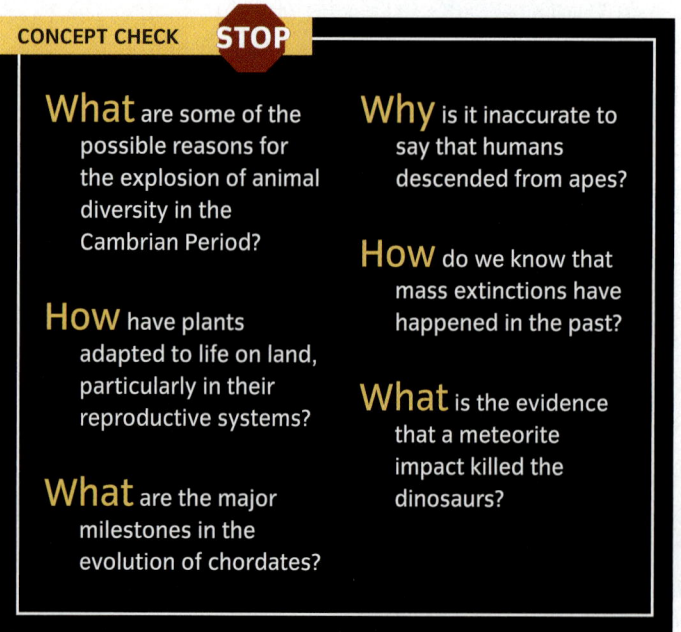

■ Continental flood basalt provinces/Volcanic rifted margins   ■ Rhyolite provinces   ■ Ocean basin flood basalt provinces

Volcanism and mass extinctions FIGURE 11.22

Several of the world's largest flood basalt deposits formed at roughly the same time as mass extinctions. Scientists are still debating whether, and how, an enormous lava flow in one region could cause mass extinctions all over the world. (Ma = millions of years before present.) *Courtesy Dr. Richard Ernst.*

a strong correlation exists between the eruption of flood basalts (Chapter 9) and mass extinctions. The end-of-Cretaceous boundary lies very close in time to the eruption of the Deccan Traps in India, and the Permian extinction occurred around the time of the largest known flood basalts, the Siberian Traps (see FIGURE 11.22).

Perhaps massive outbreaks of volcanism, producing worldwide climate change, are the "normal" agent of mass extinctions, and the end-of-Cretaceous meteorite was merely an interloper that happened to strike Earth at the wrong time. The study of mass extinctions, and the role that impacts may have played in them, continues to be an area of active research. It is especially appropriate today, because the human footprint on planet Earth is so heavy (FIGURE 11.23 on pages 362–363) that the current rate of species extinctions now rivals the rate of extinctions during the previous die-offs.

CONCEPT CHECK STOP

What are some of the possible reasons for the explosion of animal diversity in the Cambrian Period?

How have plants adapted to life on land, particularly in their reproductive systems?

What are the major milestones in the evolution of chordates?

Why is it inaccurate to say that humans descended from apes?

How do we know that mass extinctions have happened in the past?

What is the evidence that a meteorite impact killed the dinosaurs?

360 CHAPTER 11 A Brief History of Life on Earth

3 Evolution and the Fossil Record

1. As a result of **evolution**, all present-day organisms are descendants of different kinds of organisms that existed in the past. The mechanism by which evolution occurs is **natural selection**, in which well-adapted individuals tend to have greater survival and more breeding success, and thus pass along their traits to later generations.

2. New **species** may emerge slowly, as the environment changes gradually over time, or quickly, when a beneficial *mutation* appears or when a population moves to a new habitat. It is not clear which process is more common.

3. Organisms can be preserved as **fossils** in many different ways. The body may be kept essentially intact, or its shape may be preserved while minerals replace its contents. Some organisms leave behind evidence, such as *molds* of footprints or **trace fossils**, without leaving behind any body parts.

4 Life in the Phanerozoic Eon

1. The Cambrian Period, at the beginning of the Phanerozoic Eon, was a time of explosive diversification of life-forms. Several factors may have contributed to the *Cambrian explosion*: the emergence of multicellular life (just before the Cambrian Period), the protective ozone layer, the rise in oxygen levels, and the ability to form shells.

2. All of the extant **phyla** in the animal **kingdom**, as well as several extinct phyla, emerged during the Cambrian Period. All of the Cambrian fauna were aquatic, and some had hard internal or external skeletons. In order to colonize land, plants and animals had to develop a means of *structural support,* an *internal aquatic environment,* and a way of *exchanging gases* with the atmosphere. For sexual reproduction, a *moist environment* was also essential.

3. The earliest land plants were seedless plants, such as ferns. The next major event in the plant kingdom was the emergence of **gymnosperms** or naked-seed plants, which flourished during the age of dinosaurs. **Angiosperms**, or flowering plants, developed last, with a much more efficient reproductive system facilitated by insect pollination.

4. The earliest land animals were arthropods (a phylum that includes insects). Fish first ventured onto land during the Devonian Period, and eventually gave rise to the amphibians. These, however, were limited in their geographic range because they still required water to spawn in.

5. Reptiles evolved a watertight skin and eggs that could be incubated outside of water. This freed them from dependence on watery environments. Both mammals and birds descended from reptiles, by different evolutionary pathways.

(continued)

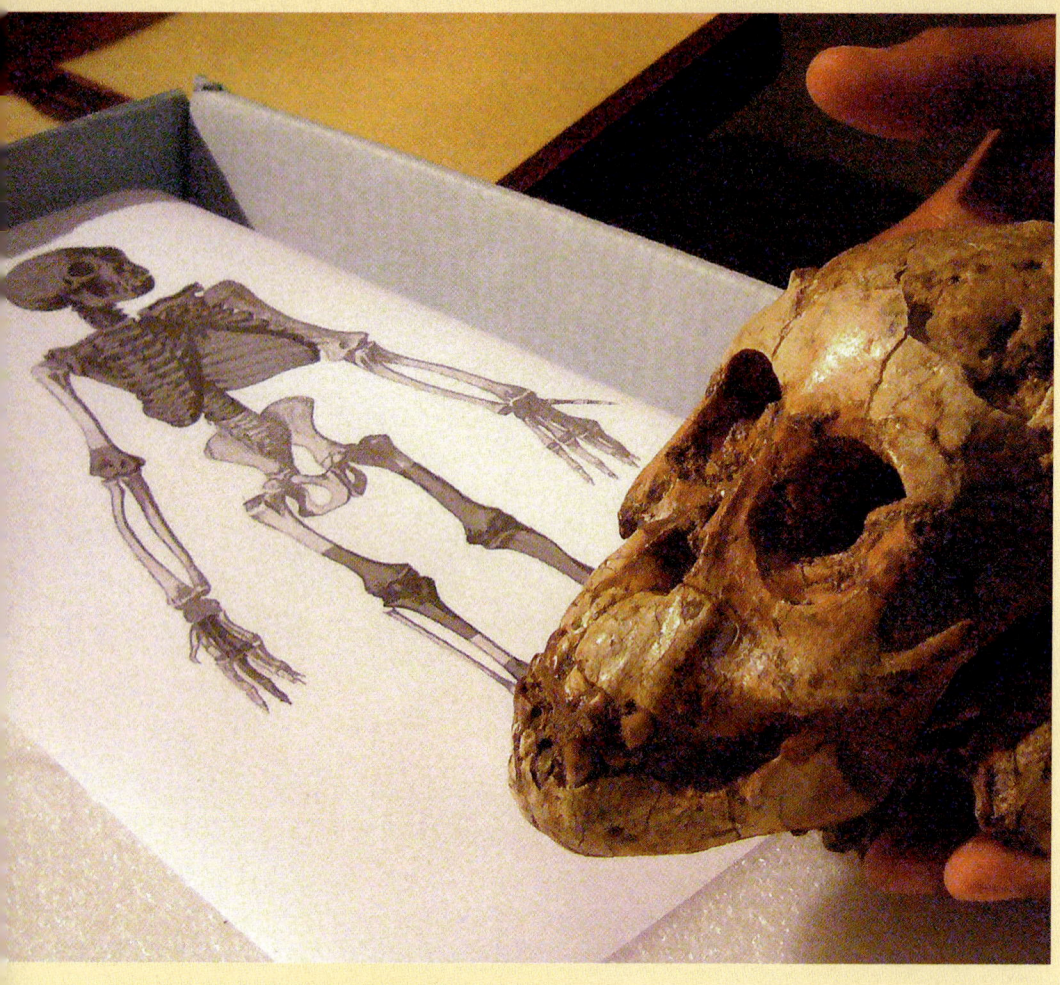

6. Mammals have existed since the Jurassic Period. It was apparently the disappearance of the dinosaurs, along with a rise in atmospheric oxygen, which gave mammals the opportunity to diversify and increase in size. **Hominids** are a very recent family of mammals, which appeared within the last 3.9 million years. Anatomically modern humans appeared only 30,000 years ago. Scientists are not certain whether Neandertals, which died out around the same time, were a distinct species.

7. Several geologic time periods are delineated by **mass extinctions**. The causes of mass extinctions are not thoroughly understood, but most scientists now believe that the end-of-Cretaceous extinction, which killed the dinosaurs, resulted from a meteorite impact. Other mass extinctions may have been caused by volcanism.

KEY TERMS

- photosynthesis p. 337
- cell p. 340
- DNA p. 340
- prokaryotes p. 342
- domain p. 342
- eukaryotes p. 343
- evolution p. 345
- natural selection p. 345
- species p. 345
- fossil p. 347
- trace fossil p. 348
- kingdom p. 350
- gymnosperm p. 352
- angiosperm p. 354
- mass extinction p. 357

CRITICAL AND CREATIVE THINKING QUESTIONS

1. Recall from Chapter 1 (see Table 1.1) that Earth and Venus are so similar in size and overall composition that they are almost "twins." Why did these two planets evolve so differently? Why is Earth's atmosphere rich in oxygen and poor in carbon dioxide, whereas the reverse is true on Venus? What would happen to Earth's oceans if Earth were a little bit closer to the Sun?

2. One classical criticism of the theory of evolution by natural selection was the paucity of transitional species. However, *Archaeopteryx* is such a species. In what ways did it resemble its dinosaur predecessors? In what ways was it like or unlike modern birds?

3. Another common criticism of Darwin's theory was its alleged inconsistency with the Biblical account of Creation. Yet Darwin himself was a Christian, and many scientists since Darwin have had no difficulty in reconciling the evidence for evolution with their religious beliefs. Investigate how they have done so. Do you personally feel that evolution by natural selection conflicts with your religious beliefs?

4. What do you think might have happened to mammals if the end-of-Cretaceous extinction had not wiped out the dinosaurs?

What is happening in these pictures?

In this fossil bird and this fossil frog, found in the Messel Oil Shale near Darmstadt, Germany, the animals' soft tissue, such as feathers and skin, are exceedingly well preserved. (Note the faint imprint of the frog's skin around its bones.)

- Why do you think the soft body parts were preserved in this case, even though they are seldom found in most fossils?

SELF-TEST

1. Which of the following statements best describes changes in Earth's environment over the last 4 billion years?
   a. Solar brightness and oxygen generally increase while surface temperature and atmospheric CO_2 decrease.
   b. Solar brightness and oxygen generally decrease while surface temperature and atmospheric CO_2 increase.
   c. Solar brightness and surface temperature generally increase while oxygen and atmospheric CO_2 generally decrease.
   d. Solar brightness and surface temperature generally decrease while oxygen and atmospheric CO_2 generally increase.

2. Photosynthesis is fundamentally important to life on Earth. On this illustration, label the input and output of the photosynthetic process using the following terms:

   carbon dioxide CO_2 light rays
   water H_2O carbohydrates
   oxygen O_2

3. Although scientists cannot be certain why the atmospheric oxygen levels increased, they hypothesize that atmospheric oxygen has increased over time _____.
   a. as a continued increase in photosynthetic organisms
   b. because of increased rates of seafloor spreading
   c. because organic matter is buried before it decomposes, reducing the formation of carbon dioxide (CO_2)
   d. All of the above statements are correct.

4. The minimum requirements for life are _____.
   a. photosynthesis and a means of replication
   b. mobility and a means of replication
   c. metabolism and a means of replication
   d. mobility and metabolism

5. There are two cells in these photographs. One is from a prokaryotic and the other from a eukaryotic organism. Label structures in each of the cells with the following terms, and then label each cell either *prokaryotic* or *eukaryotic*.

   organelles   cell membrane   nucleus

   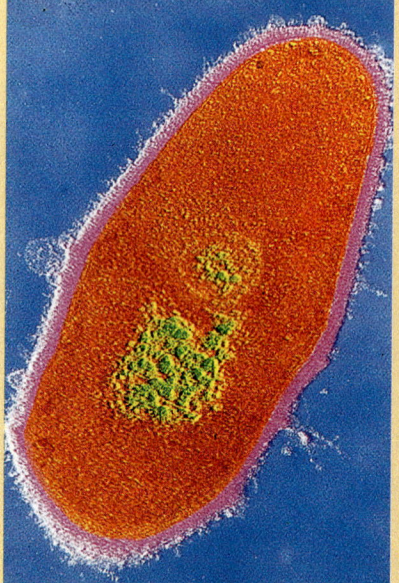 

6. _____ is an example of anaerobic metabolism.
   a. Chemosynthesis   d. Both a and b are correct.
   b. Photosynthesis   e. Both a and c are correct.
   c. Fermentation

7. Natural selection is the process by which _____.
   a. individuals migrate to environments better suited to survival of the species
   b. individuals adapt to their environment over time
   c. well-adapted individuals pass on their survival advantages to their offspring
   d. All of the above are correct.

8. As a mechanism for evolution, natural selection requires the passing on of specific traits from one generation to the next. These specific traits may reflect _____ passed on through an organism's _____.
   a. genetic mutations/metabolism
   b. genetic mutations/DNA
   c. genetic mutations/RNA
   d. metabolism/DNA
   e. metabolism/RNA

9. How are organisms preserved as fossils within the rock record?
   a. through the preservation of the organism in substances such as sap, tar, or ice
   b. through minerals replacing the contents of the dead organism (mineralization)
   c. by the leaving of impressions, molds, or footprints
   d. through mummification in especially dry environments
   e. All of the above are methods through which organisms could be preserved as fossils.

10. Which of the following hypotheses do scientists think best explains the explosion of life-forms represented by the Cambrian fossil record?
    a. Sexual reproduction began in the early Cambrian, leading to a greater diversity of living organisms.
    b. An oxygen-rich atmosphere allowed for the metabolism of larger organisms.
    c. Predation began in the early Cambrian, leading to the development of many more organisms with shells and skeletons that might be better preserved in the fossil record.
    d. Extreme climatic changes at the end of the Proterozoic Eon led to accelerated evolutionary responses in surviving organisms.
    e. All of the above hypotheses are considered equally valid by scientists today.

11. Which of the following is not a requirement for a living organism to survive on land?
    a. structural support
    b. a means of locomotion
    c. an internal aquatic environment
    d. a mechanism of exchanging gases with the air
    e. if sexually reproducing, a moist environment for the reproductive system

12. Vascular plants exchange gas with the air by _____.
    a. diffusion through adjustable openings in the leaves called *stomata*
    b. osmosis involving oxygenated water in the root system
    c. photosynthesis
    d. All of the above are correct.

13. Reptiles were freed from the water by _____.
    a. evolving an egg that contained amniotic fluid for the young to grow in
    b. developing a watertight skin
    c. developing a warm-blooded metabolism
    d. Both a and b are correct.
    e. Both b and c are correct.

14. _____ is probably the first species of our own genus.
    a. *Australopithecus*   c. *Homo neandertalensis*
    b. *Homo erectus*   d. *Homo sapiens*

15. Mass extinctions on Earth may be correlated with _____.
    a. massive eruptions of flood basalts
    b. the impact of massive meteorites
    c. glaciation
    d. Both a and b are correct.
    e. Both b and c are correct.

The Oceans 12

Imagine a dry Earth, devoid of water. Viewed from an orbiting spacecraft, a dry Earth would no longer have a bluish color, the land would lack a vegetation cover, and no clouds would obscure the surface. Earth would appear as desolate as the Moon or Mars.

Circling Earth, we would see several vast, interconnected basins, each floored with oceanic crust and rimmed with continental crust. If these huge basins were slowly filled with water, the scene would be transformed. First, the rising water would fill the deepest parts of the basins, creating a number of shallow seas. These seas would then merge to form an ever-larger ocean that would eventually creep up the slopes of the continental margins. With the ocean basins filled to capacity, more than two-thirds of Earth's surface would be covered by water and Earth, viewed from space, would be a blue planet unique in the solar system.

Under the ocean, beyond the continental slopes, lies the remote world of the deep ocean floor. We now have devices for mapping the ocean floor—with its volcanoes and canyons—and teams of marine scientists have explored it. The romanticist in each of us may regret that beliefs and legends built up through more than 3000 years of human history—monsters, mermaids, strange and threatening sea gods, fabled cities and castles that sank into watery depths—have vanished. But in their place, we are beginning to find a new world rich in unimaginable undersea life.

CHAPTER OUTLINE

The Ocean Basins p. 372

The Composition of Seawater and the Movement of Sediment p. 376

Ocean Water and Its Circulation p. 382

The Ocean Basins

LEARNING OBJECTIVES

Discuss the age and origin of the oceans.

Describe the distribution of water and land over the globe.

Explain how ocean depth is measured.

Identify the important features of ocean basins, including the midoceanic ridge, the axial rift, and the abyssal plains.

Water has been present on Earth's surface since the planet's earliest days. In fact, we do not have any definitive evidence that there was ever a time when Earth lacked water. The oldest rock so far discovered on Earth is gneiss that was once contained in *sedimentary strata*. Since these strata are about 4.0 billion years old and were originally deposited in water, we can be reasonably certain that the ocean formed sometime between 4.56 billion years ago (when Earth was formed) and 4.0 billion years ago. The most ancient mineral grain so far discovered on Earth is a grain of zircon from a sandstone in Western Australia (see Chapter 11). That grain is 4.4 billion years old, and the oxygen isotopes it contains record contact with water, so it is possible that there was an ocean of some sort as much as 4.4 billion years ago. But regardless of when the earliest ocean began to form, just where the water in it came from is still an open question (**FIGURE 12.1**).

The origin of the oceans FIGURE 12.1

Scientists do not yet agree on where Earth's water came from.

A Some of the water was present from the beginning in the materials that formed Earth. In the first few hundred million years after the planet formed, this water was released as steam from volcanoes.

H_2 escapes to space.

O_2, CO_2, H_2, H_2O Vapor, N_2

CO_2, CH_4 remain in atmosphere.

CO_2, NH_2, Methane CH_4, H_2O

H_2O condenses.

Rain forms world oceans.

B More water may have been added to Earth's chemical mix by comets and meteorites coming in from the outer regions of the solar system, after the planet had already formed.

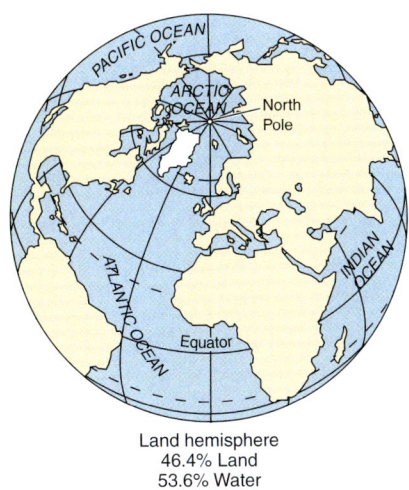
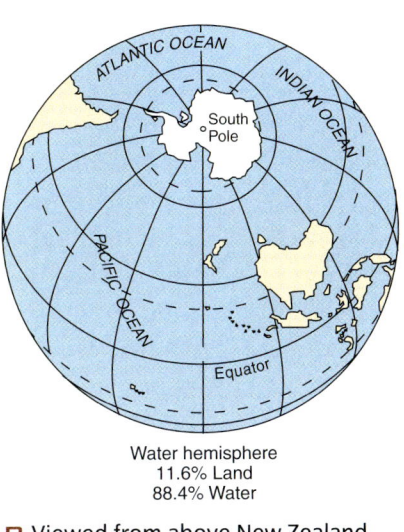

Land and ocean FIGURE 12.2

A Viewed from above Great Britain, over 46% of the Northern Hemisphere is land.

B Viewed from above New Zealand, over 88% of the Southern Hemisphere is water.

OCEAN GEOGRAPHY

Seawater covers 71% of Earth's surface. The land that covers the remaining 29% is unevenly distributed. This uneven distribution is especially striking when we compare two views of the globe: one from a point directly above Great Britain and the other from a point directly above New Zealand (**FIGURE 12.2**). The uneven distribution of land and water plays an important role in determining the paths along which water circulates in the open ocean and its marginal seas. We will look at ocean circulation later in the chapter.

Most of the water on our planet is contained in four huge interconnected **ocean basins**. The Pacific, Atlantic, and Indian oceans are connected with the Southern Ocean, the body of water south of 50°S latitude that completely encircles Antarctica. (The fifth major ocean, the Arctic Ocean, is considered an extension of the North Atlantic.) Collectively, these four bodies of water, together with a number of smaller ones, are often referred to as the *world ocean*.

ocean basins Regions of Earth's crust that are covered by seawater.

The smaller water bodies connected with the Atlantic Ocean include the Mediterranean, Black, North, Baltic, Norwegian, and Caribbean Seas; the Gulf of Mexico; and Baffin and Hudson Bays. The Persian Gulf, Red Sea, and Arabian Sea are part of the Indian Ocean, while among the numerous marginal seas of the Pacific Ocean are the Gulf of California, the Bering Sea, the Sea of Okhotsk, the Sea of Japan, and the East China, South China, Coral, and Tasman Seas. All these seas and gulfs vary considerably in shape and size; some are almost completely surrounded by land, whereas others are only partly enclosed. Each owes its distinctive geography to plate tectonics. This ongoing process, described in Chapter 7, has created many small basins both in and next to the major ocean basins.

DEPTH AND VOLUME OF THE OCEANS

Before the 20th century, we knew little about the depth of the oceans. Sailors determined the depth of water by lowering a weighted hemp line, or a strong wire, until it hit the bottom. This technique was effective and relatively rapid in shallow water, but it could take 8 to 10 hours to recover a weighted wire in water thousands of meters deep. By the close of the 19th century, about 7000 measurements had been made in water more than 2000 m (6500 ft) deep, and fewer than 600 had been made in water more than 9000 m (29,500 ft) deep.

In the 1920s, ship-borne acoustical instruments called *echo sounders* were developed to measure ocean depths. An echo sounder generates a pulse of sound and accurately measures the time it takes for the echo bouncing off the seafloor to return to the instrument. Because we know the speed of sound traveling through water, we can work out the depth of the water beneath the ship.

Topography of the ocean basin By far the longest chain of mountains on Earth—some 64,000 km in length—is hidden from our view. That's because it's underwater, twisting and branching in a complex pattern through the ocean basins. Plate tectonics has given the ocean floor a distinctive topography, and the features of ocean basins are quite different from those of continents. Much of the oceanic crust is less than 60 million years old, and the oldest oceanic crust so far discovered is only 200 million years old. By contrast, the great bulk of the continental crust is over 1 billion years old.

FIGURE 12.3 shows the important relief features of ocean basins. The basin is divided in about half by a *midoceanic ridge* of submarine hills. In the center of the ridge, at its highest point, is a narrow, trench-like feature called the *central rift valley* (also known as the *axial rift*). The location and form of this rift suggest that the crust is being pulled apart along the line of the rift. On either side of the midocean ridge is a broad, deep plain, known as an *abyssal plain*.

Over the past 80 years, the oceans have been crossed many thousands of times by ships carrying echo sounders. As a result, we have been able to map the topography of the seafloor and the depth of the water lying directly above it in considerable detail for all but the most remote parts of the ocean basins. The greatest ocean depth yet measured (11,035 m; 36,205 ft) lies in the Mariana Trench, near the island of Guam in the western Pacific. This is more than 2 km (6500 ft) farther below sea level than Mt. Everest rises above sea level. Based on recent satellite measurements, the average depth of the sea is about 4500 m (14,760 ft), compared to an average height of the land of only 750 m (2460 ft). The present volume of seawater is about 1.35 billion cubic kilometers (324 million mi^3); more than half this volume resides in the Pacific Ocean. We say *present* volume, because the amount of water in the ocean fluctuates over thousands of years, mainly because of the growth and melting of continental glaciers (see Chapter 6).

Figure 12.3 shows a symmetrical ocean-floor model that fits the North Atlantic, South Atlantic, Indian, and Arctic Ocean Basins well. These oceans have *passive continental margins* that have not been subjected to strong tectonic and volcanic activity during the last 50 million years. This relative inactivity is due to the fact that the continental and oceanic lithospheres that join at a passive continental margin are part of the same lithospheric plate and move together, away from the axial rift.

But unlike the symmetrical ocean-floor model of the North Atlantic, the margins of the Pacific Ocean Basin have deep offshore oceanic trenches (see *What an Earth Scientist Sees*). We call these trenched ocean-basin edges *active continental margins*. Here, oceanic crust is being bent downward and is sinking under continental crust, creating trenches and inducing volcanic activity.

Ocean basins
FIGURE 12.3

This schematic block diagram shows the main features of ocean basins. It applies particularly well to the North and South Atlantic Oceans.

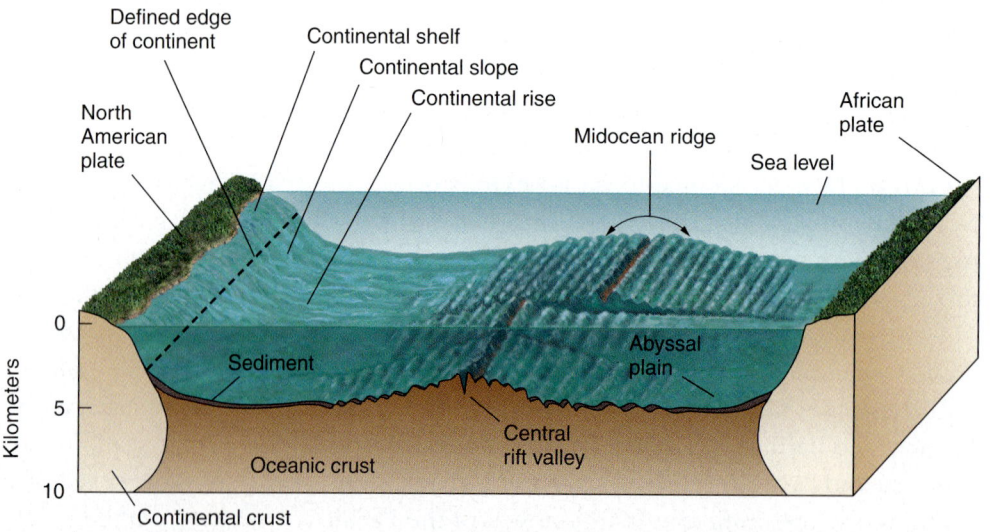

What an Earth Scientist Sees

Undersea Topography

These images of the ocean floor were constructed from radar data that measured the surface height of the ocean very precisely. Deeper regions are shown in tones of purple, blue, and green, while shallower regions are shown in tones of yellow and reddish brown. In **A**, you can see a prominent ring of trenches in the western Pacific Basin. An Earth scientist would recognize that these are *subduction* trenches, created, as shown in **B**, when oceanic crust is being bent downward and forced under continental crust. In **C**, you can see the mid-Atlantic ridge, which is the result of seafloor spreading (**D**). Data were acquired by the U.S. Navy Geosat satellite altimeter.

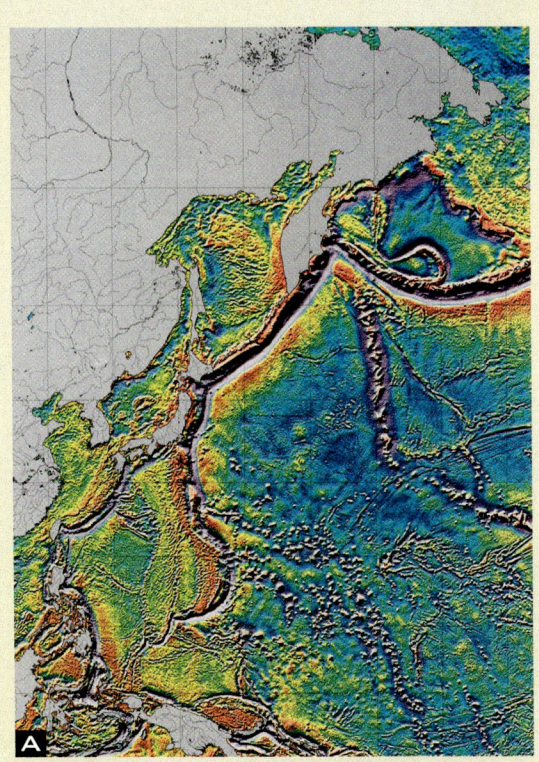

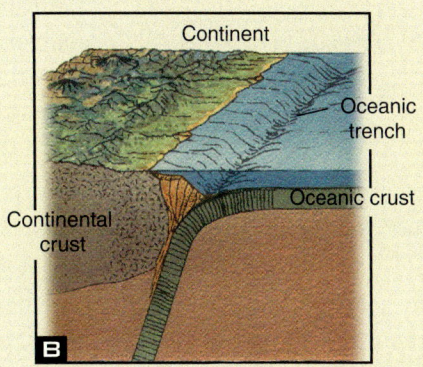

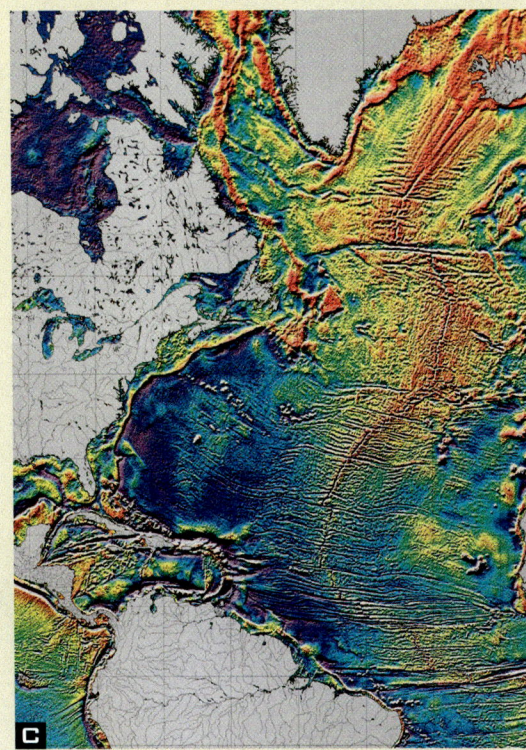

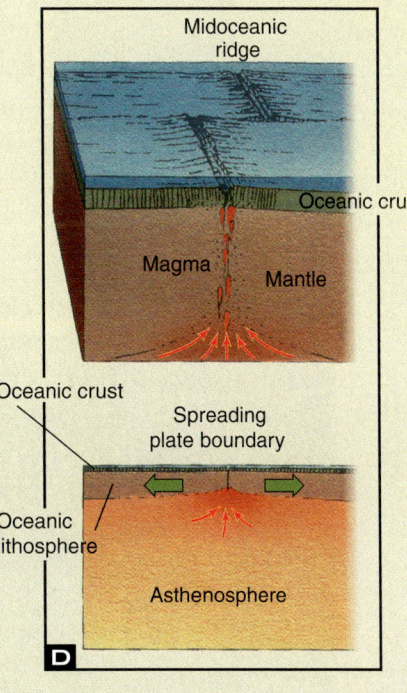

Here's an interesting question:

- In **Figure A**, there is a prominent volcanic ridge—it is the underwater continuation of the Hawaiian Islands chain. What might explain the sharp bend in the ridge?

CONCEPT CHECK STOP

When did the oceans form?

Where did the water come from?

How do we measure ocean depth?

What are the important features of ocean basins?

The Composition of Seawater and the Movement of Sediment

LEARNING OBJECTIVES

Define salinity and **explain** how seawater maintains its salinity.

Discuss the movement of sediment through the ocean.

Define *turbidity current* and *turbidite*.

Identify the principal biotic zones and deep-sea sediments.

If you have ever swum in the ocean, you'll know that seawater contains a lot of salt, making it not only unpalatable but also dangerous for human consumption. In fact, if you drank only salt water you would become dehydrated because your body would try to rid itself of the excessive amounts of ions, such as sodium and potassium, contained in salt water.

The **salinity**, or salt content, of seawater ranges from 3.3 to 3.7%. This may sound quite small, but it is about 70 times the salinity of tap water. Not surprisingly, when seawater is evaporated, more than three-quarters of the dissolved matter is sodium chloride—that is, table salt. Seawater contains most of the other natural elements as well, but many of them are present in such low concentrations that only extremely sensitive analytical instruments can detect them.

salinity A measure of the salt content of a solution.

The elements dissolved in seawater come from several sources. Chemical weathering of rock releases soluble materials such as salts of sodium, potassium, and sulfur. Having been dissolved out of the rock, the soluble compounds become part of the dissolved load in river water flowing to the sea. Volcanic eruptions, both on land and beneath the sea, also add soluble compounds to water via volcanic gases and hot springs. The *black smoker* vents found at spreading centers add dissolved minerals from the hot rocks in the midocean rift (**FIGURE 12.4**). Two other processes, evaporation of surface water and freezing of seawater, tend to make seawater saltier because they remove fresh water while leaving the salts behind.

Why, then, doesn't the sea become saltier over time? The answer is that it constantly receives fresh water from *precipitation* and *river flow* (**FIGURE 12.5**). Also, aquatic plants and animals withdraw some elements, such as silicon, calcium, and phosphorus, to build their shells or skeletons. Other elements precipitate out in mineral form and settle to the seafloor, and chemical reactions between seawater and hot volcanic rocks remove some salts from solution. All of these processes balance each other, keeping the composition of seawater essentially unchanged. This is the problem that John Joly ran into when he tried to use the salinity of seawater to estimate the age of the ocean (Chapter 10).

Black smoker vents FIGURE 12.4

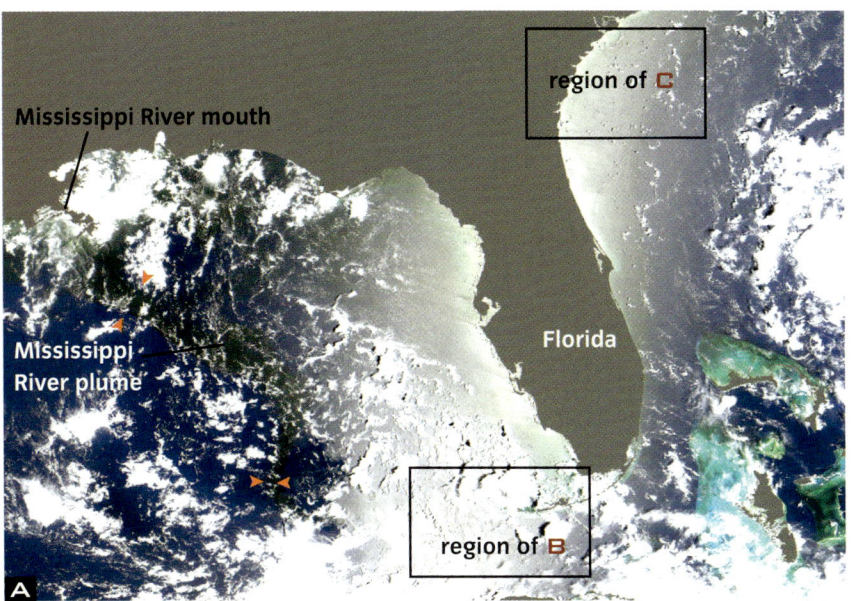
July 30, 2004

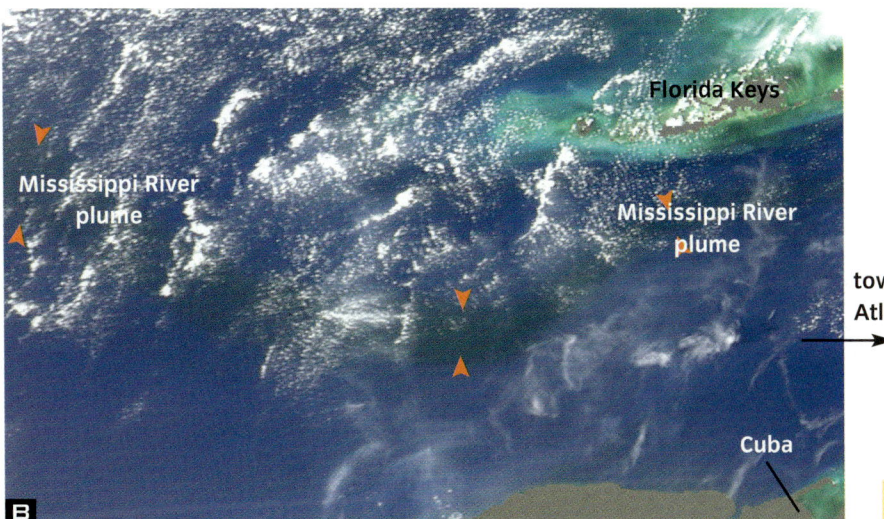
August 4, 2004

August 9, 2004

A river runs through it
FIGURE 12.5

These satellite photos, taken in 2004, show a "river" (or, as Earth scientists call it, a *plume*) of less salty water running from the mouth of the Mississippi River through (A) the Gulf of Mexico and (B) the Florida Straits and into (C) the Atlantic Ocean, before it finally mixes sufficiently with ocean water that it cannot be seen any more. The river water appears brown in the photos. Shipboard measurements showed that the plume was 50 km wide and 10 to 20 m deep. It had lower salinity, higher temperature, and double the chlorophyll concentration of the other water in the Florida Straits. Scientists estimate that 23% of the Mississippi River's discharge flows directly to the Atlantic Ocean in this manner.

TURBIDITY CURRENTS

In 1929, an earthquake occurred in the Grand Banks region off the coast of Newfoundland, triggering a submarine landslide. This generated an underwater current, laden with sediment, and within the next 13 hours, the passing current had snapped a series of transatlantic telephone cables. Scientists knew the positions of the breaks in the cables and the times at which they broke, so they could work out that the current traveled at a speed of about 75 km/h.

To understand what happened, let's look at how seawater transports sediment in more detail. In Figure 12.3, we can see that the shallowly submerged *continental shelves* pass abruptly into *continental slopes*. These slopes descend to depths of several kilometers. Sediment deposited on the edge of the shelf is poised for further transport down the slope and onto the adjacent continental rise.

Marine scientists have found thick bodies of coarse sediment lying at the foot of the continental slope at depths as great as 5 km (**FIGURE 12.6A**). At first, it was difficult to explain where these coarse accumulations came from, but eventually scientists demonstrated that the sediments could be deposited by **turbidity currents**. These are gravity-driven currents that contain dilute mixtures of sediment and water with a density greater than that of the surrounding water. Similar currents have been reproduced in the laboratory, using dense mixtures of water, silt, and clay in water-filled tanks. They have also been discovered moving across the floors of lakes and reservoirs (**FIGURE 12.6B**). In the oceans, turbidity currents are set off by earthquakes, as in the Grand Banks example, or by landslides, and major coastal storms. Off the mouths of rivers, they can be set in motion by large floods.

> **turbidity current**
> A gravity-driven current consisting of a dilute mixture of sediment and water with a density greater than that of the surrounding water.

Deep-sea turbidites FIGURE 12.6

A These deep-sea turbidite beds have been tilted, uplifted, and exposed in a wave-eroded beach along the coast of the Olympic Peninsula in Washington.

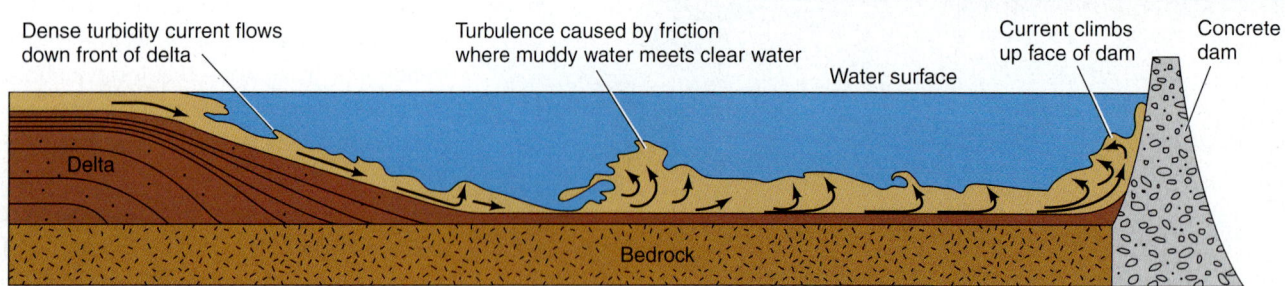

▲ B A turbidity current generated by a surge of sediment-laden water enters the quiet water of a reservoir behind a dam. Moving rapidly down the face of a delta, the current passes along the lake floor and climbs up the face of the dam before subsiding. As the sediment settles to the bottom, it forms a graded *turbidite* layer.

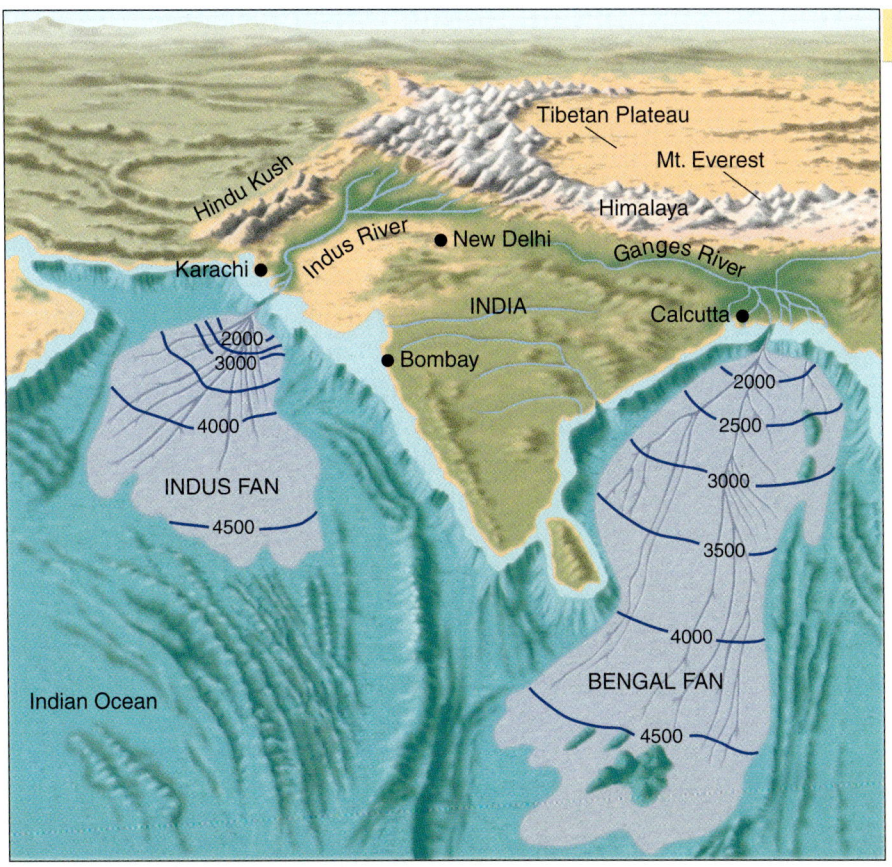

Deep-sea fans FIGURE 12.7

The Indus and Ganges–Brahmaputra river systems have built the vast deep-sea fans on the seafloor adjoining the Indian subcontinent. Most of the sediment in the fans was transported from the high Himalaya to the north as the mountain system was uplifted over tens of millions of years. Contours are water depths in meters.

Turbidity currents are effective geologic agents on continental slopes, where they can reach velocities greater than those of the swiftest streams on land. Some achieve a velocity of more than 90 km/h and transport up to 3 kg/m³ of sediment, spreading it as far as 1000 km from its source.

A turbidity current typically deposits a graded layer of sediment called a **turbidite** (Figure 12.6B). We see these graded layers because the moving current that formed them was continuously, and rapidly, losing energy. As a rapidly flowing turbidity current slows down, it deposits successively finer sediment. Turbidites are deposited relatively rarely at any one site on the continental rise or an adjacent abyssal plain, perhaps only once every few thousand years. In these places, far distant from the source of the sediment, the deposits are thin layers a few millimeters to 30 cm thick. Over millions of years, turbidites can slowly build up to form vast deposits that stretch beyond the continental realm.

Deep-sea fans are a dramatic example of just how far land-derived sediment in the ocean can extend beyond the continental shelves. When shelves are exposed at times of lowered sea level and rivers extend across them nearly to the continental slope, the stage is set for the rapid formation of deep-sea fans. Some large submarine canyons on the continental slopes are aligned with the mouths of major rivers like the Amazon, Congo, Ganges, and Indus. At the base of many such canyons is a huge deep-sea fan, a fan-shaped body of sediment that spreads downward and outward to the deep seafloor (**FIGURE 12.7**). The sediments, which are derived mostly from the land, include fragments of land plants as well as fossils of shallow- and deepwater marine organisms. We also find many graded layers that we can recognize as turbidites.

turbidite A graded layer of sediment that is deposited by a turbidity current.

The Composition of Seawater and the Movement of Sediment 379

BIOTIC ZONES AND DEEP-SEA SEDIMENTS

We can classify plants and animals according to which part of the ocean—that is, which biotic zone—they live in (**Figure 12.8**). Plants and animals living in the uppermost waters of the ocean occupy the *pelagic* zone and are called *pelagic organisms*. This includes animals that swim freely under their own locomotion, such as reptiles, squids, fish, and marine mammals. *Benthic organisms* live on the bottom or in bottom sediments (the *benthic* zone). Floating or drifting (*planktic*) organisms include phytoplankton, which are mainly single-celled plants, and zooplankton, which are tiny animals. Among the most important zooplankton are single-celled foraminifera and radiolaria. Foraminifera have a calcareous shell and, as we will see in Chapter 16, they play an important role in helping Earth scientists identify changes in climate over the years. Radiolarian remains, by contrast, consist of silica.

Because plant life needs enough light energy for *photosynthesis*, it is restricted to the upper 200 m of the ocean (the *photic* zone). Plants also require nutrients, which are available mainly along coasts and shallow continental margins, and as a result plant life is limited over much of the deep ocean.

By analyzing samples from *sediment cores*, Earth scientists can work out the sources of seafloor sediment (**Figure 12.9**). (We will describe sediment cores in detail in Chapter 16.) A large portion of this sediment is produced by biological activity in the surface waters, but some sediment is transported over great distances from continental interiors, especially by winds, and eventually reaches the deep sea (see Figure 1.3).

The deep seafloor is mantled with the skeletal remains of single-celled planktonic and benthic animals and plants, which cover vast areas (look ahead to Figure 16.3A). When more than 30% of the surface sediment is made up of such remains, we call it a *calcareous ooze* or *siliceous ooze*, depending on the chemical compo-

Marine organisms Figure 12.8

A A leopard seal bares teeth in a threat display to protect her kill, Antarctic Peninsula, Antarctica.

B This tube anemone, waving its tentacles, is a benthic organism.

C Antarctic krill (*Euphausia superb*), a small shrimp-like crustacean, is the most important zooplankton in the Antarctic food web, Weddell Sea Antarctica. The yellow color comes from algae in its stomach.

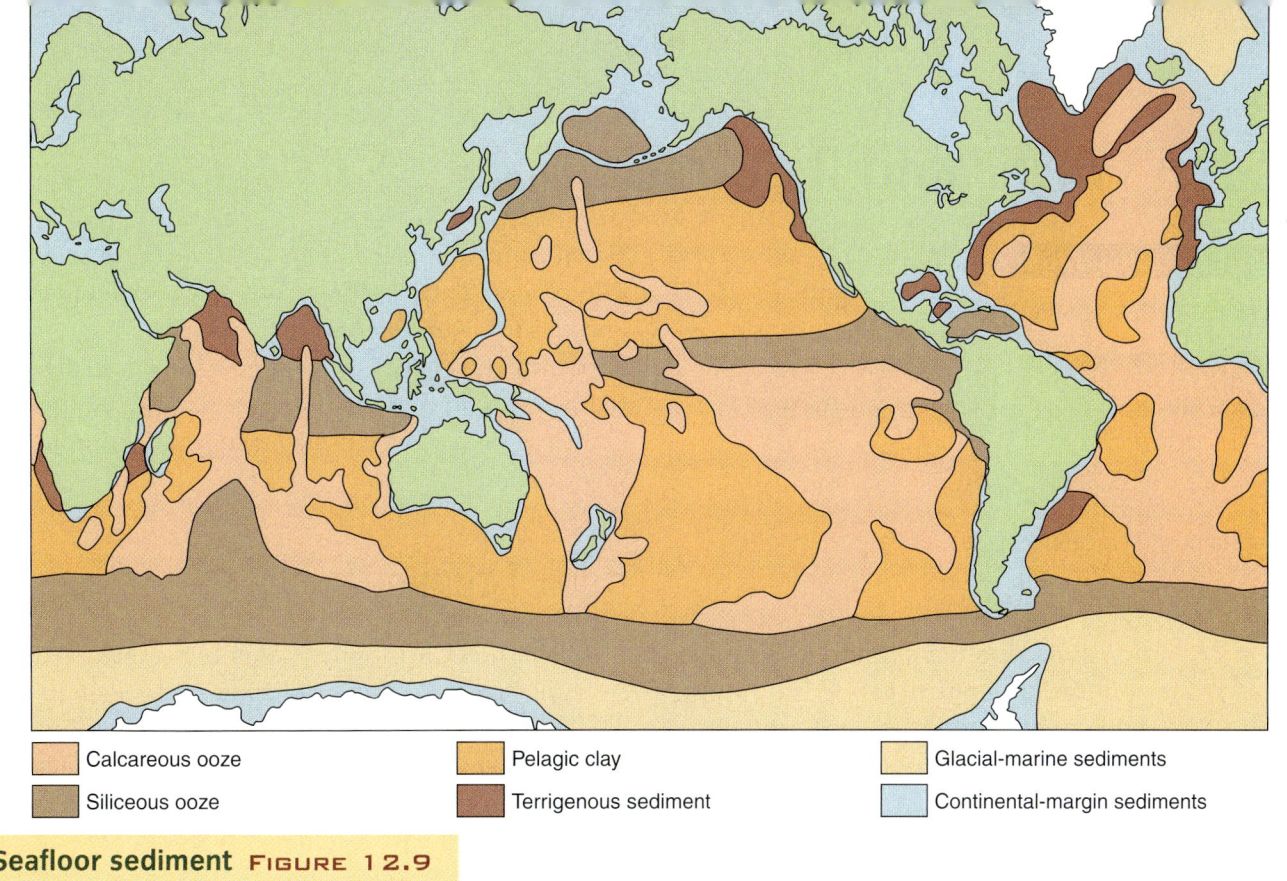

Calcareous ooze	Pelagic clay	Glacial-marine sediments
Siliceous ooze	Terrigenous sediment	Continental-margin sediments

Seafloor sediment FIGURE 12.9

Map of the World Ocean showing the distribution of the principal kinds of sediment on the ocean floor.

sition of its major component. Calcareous ooze covers broad areas of low to middle latitudes, because the warm surface waters in those latitudes favor the growth of carbonate-secreting organisms (Figure 12.9). Their tiny shells settle to the seafloor in vast numbers, but they accumulate at an average rate of only about 1–3 cm per thousand years.

We do not find calcareous ooze in regions where the water is unusually deep, even at low to middle latitudes. That's because the cold, deep ocean water is under high pressure and contains more dissolved carbon dioxide than is contained in shallower waters, making the waters more acidic. For this reason, these deep waters can readily dissolve any carbonate particles that reach their level. In the Pacific Ocean, this level lies at 4000–5000 m, whereas in the Atlantic it is somewhat shallower. This explains why calcareous ooze is missing over large portions of the deep north and south Pacific Ocean and some marginal parts of the Atlantic Ocean.

Siliceous organisms make up the major component of the bottom sediments in broad belts across the equatorial and far northern Pacific Ocean, sectors of the Indian Ocean, and a belt around the Southern Ocean.

Sediment is composed of *lithic* clasts of rock fragments from the continents' mantles—the continental shelves and slopes. These sediments are also found at locations where debris-laden glaciers generate icebergs that raft sediment seaward from glacier margins (see Chapter 6) and where turbidity currents have transferred sediment from the shelves to the deep-sea floor. Finally, far from land in regions of low biological activity, we find very fine-grained reddish or brownish clay, generally called *red clay*. Much of the clay is made up of fine windblown dust. The dust is red because it contains oxidized iron-rich minerals.

CONCEPT CHECK STOP

Where do the materials dissolved in seawater come from?

What is a turbidity current?

What is a turbidite?

What are the biotic zones?

What kinds of organisms are found in each?

The Composition of Seawater and the Movement of Sediment

Ocean Water and Its Circulation

LEARNING OBJECTIVES

Identify the layers of the ocean.

Describe ocean currents and why they occur.

Explain the changes in Pacific Ocean currents that cause El Niño and La Niña.

Discuss the ocean conveyor belt.

Explain how the oceans regulate climate.

The salinity and temperature of seawater combine to control its density: Cold, salty water is dense and will sink, whereas warm, less salty water is less dense and will rise. Looking back at Figure 12.5, we can see this very clearly: The warm and less salty (low-density) water from the Mississippi River remains on top of the water in the Gulf of Mexico for hundreds of kilometers.

Ocean scientists have discovered three major layers, or zones, in the ocean, in each of which the density of the water differs. These are the surface layer, the transitional zone (the *thermocline*), and the deep zone. The differences are caused by changes in both temperature and salinity, with temperature being the major factor (**FIGURE 12.10**). Warm water is less dense than cold water, and saltier water is denser than less salty water. Once established, the pressure differences induce the water to flow.

OCEAN CURRENTS

In 1492, when Christopher Columbus set sail across the Atlantic Ocean in search of China, he took an indirect route. On the outward voyage, he sailed southwest toward the Canary Islands and then west on a course that carried him to the Caribbean Islands, where he first sighted land. In choosing this course, he was taking advantage of favorable winds and surface **ocean currents**, rather than fighting the westerly winds and currents that would have hindered his progress if he had taken a more

ocean current A persistent, dominantly horizontal flow of water in the ocean.

Ocean layers FIGURE 12.10

The three major density zones in the ocean: the surface layer, the transitional zone (*thermocline*), and the deep zone.

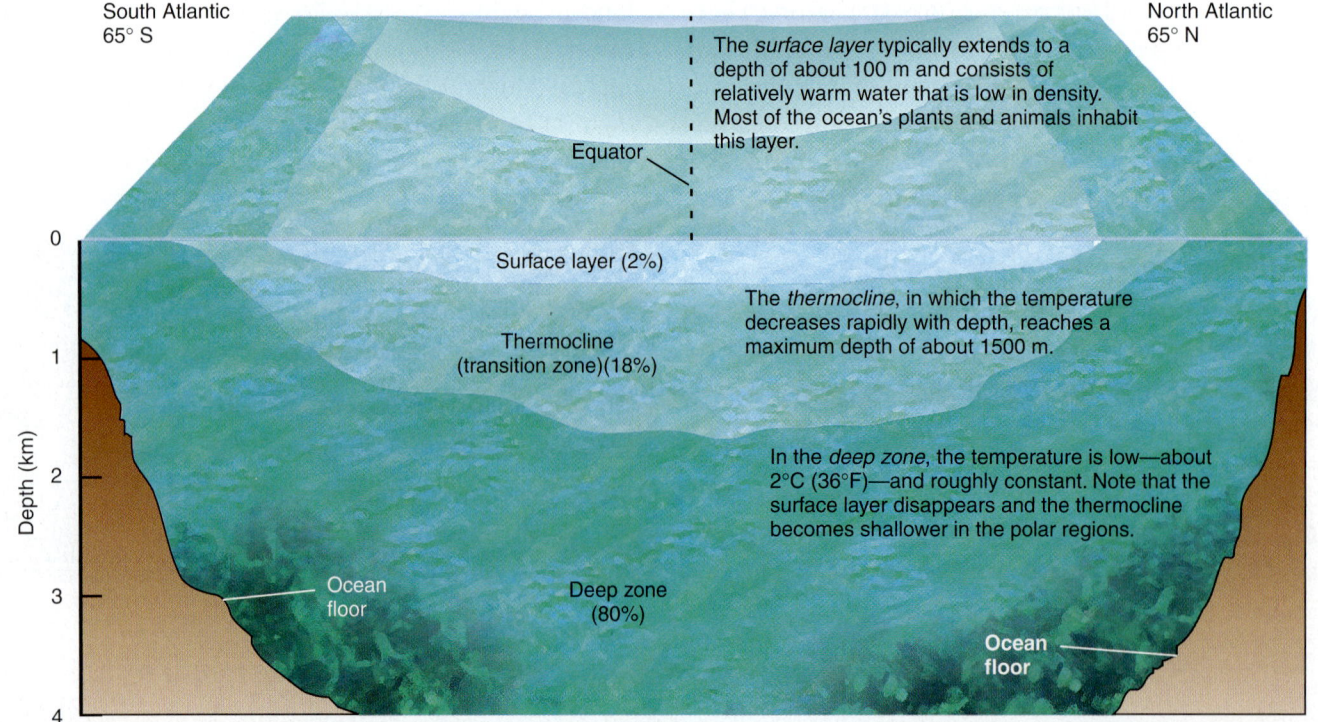

World map of surface ocean currents FIGURE 12.11

Surface drifts and currents of the oceans in January.

Cold western boundary currents
In the Northern Hemisphere, where there are significant continental boundaries in the high latitudes, cold equatorward currents flow along the east sides of continents and meet up with the poleward-flowing western boundary currents in the midlatitudes.

North Atlantic drift
In the northeastern Atlantic Ocean, the west-wind drift forms a relatively warm current, which spreads around the British Isles, into the North Sea, and along the Norwegian coast. Eventually, the current cools and sinks, flowing equatorward along the ocean bottom.

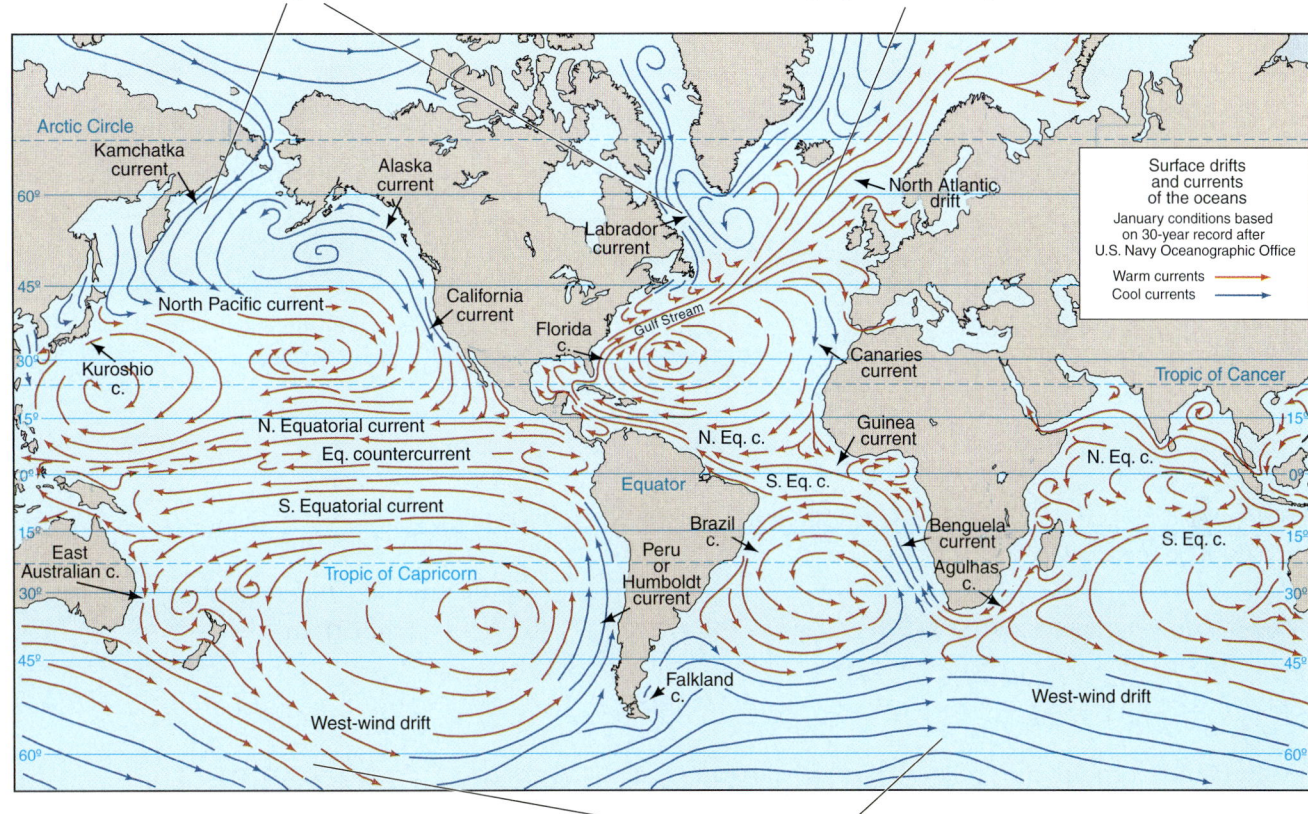

Circumpolar current
In the Southern Hemisphere, the absence of continental boundaries allows the strong west winds around Antarctica to produce a continuously flowing current called the *Antarctic circumpolar current*.

Here's an interesting question:
- In May 1990, a container ship called *Hansa Carrier* encountered a storm in the North Pacific at about 48° north latitude, and lost 21 containers overboard, five of which contained 80,000 Nike shoes. The shoes carried serial numbers that identified them as being from the containers. By the end of 1990, shoes were washing up on Vancouver Island, British Columbia, and Oregon. By the winter of 1991, shoes were appearing on Hawaiian beaches. Which ocean currents were most likely to have transported the shoes?

northerly route. On the return trip, he sailed a more northerly route in order to take advantage of the westerly winds at that latitude. The North Atlantic currents that Columbus exploited are shown in FIGURE 12.11, a world map that illustrates how surface ocean currents are established in different regions.

The force of wind on surface water also creates oceanic circulation. Energy is transferred from the prevailing surface wind to water by the friction of the air blowing over the water. The direction of the water drift is not identical to the direction of the driving wind due to the *Coriolis effect*. We will meet the Coriolis effect in more detail in Chapter 15, where we will learn about its origin due to the rotation of Earth and how it affects circulation patterns in the atmosphere. Here, it is sufficient to note that the Coriolis effect deflects the actual direction of water drift about 45° from the direction of the driving wind.

Ocean Water and Its Circulation 383

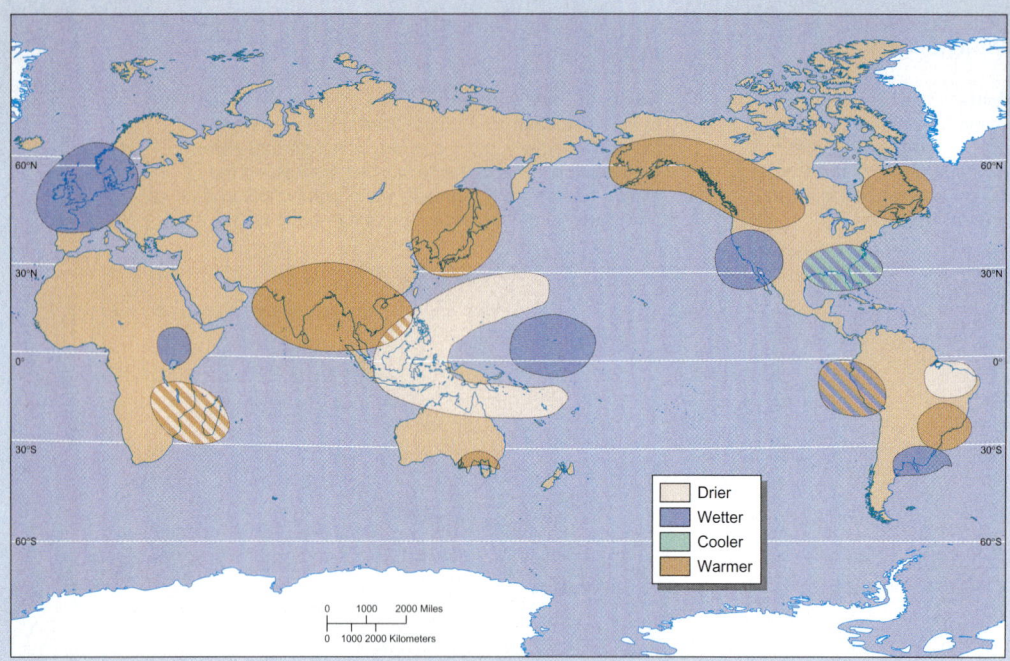

A ENSO events drastically alter climate, even in many areas far from the Pacific Ocean. As a result, some areas are drier, some wetter, some cooler, and some warmer than usual. Typically, northern areas of the contiguous United States are warmer during the winter, whereas southern areas are cooler and wetter.

B ENSO-driven warmer waters off western South America.

El Niño southern oscillation (ENSO) FIGURE 12.12

A dramatic phenomenon related to surface ocean currents is **El Niño**. Under normal circumstances, cool, nutrient-rich water is driven from lower layers to the ocean surface by the wind—a process known as upwelling. During an El Niño event, Pacific Ocean surface currents shift into a characteristic pattern, in which the upwelling along the Peruvian coast ceases and trade winds weaken. This sets up a weak equatorial eastward current, changing precipitation patterns across the globe and bringing floods to some regions and droughts to others. In contrast to El Niño is *La Niña*, in which the normal Peruvian coastal upwelling is enhanced, trade winds strengthen, and cool water is carried far westward in an equatorial plume. FIGURE 12.12 shows sea-surface temperatures observed during the *El Niño southern oscillation* (*ENSO*).

> **El Niño** An episodic cessation of the normal upwelling of cold water off the coast of Peru.

THE OCEAN CONVEYOR BELT

The surface layer of the ocean is only about 50 to 100 m deep. Friction between the movement of wind above the ocean surface and the water generates ocean currents in this layer. The pattern of water flow in the deep ocean is quite different and is best viewed in three dimensions, as shown in FIGURE 12.13. Deep currents are powered by changes in temperature and salinity in surface waters. This process, termed the **thermohaline circulation**, is driven by variations in ocean temperature (*thermo*) and salinity (*haline*). Once again, the directions of the resulting currents are controlled by the Coriolis effect.

> **thermohaline circulation** The rising and sinking of water driven by contrasts in water density created by differences in temperature and salinity; this circulation is also known as the *ocean conveyor belt*.

Visualizing

Ocean circulation and sea-surface temperature FIGURE 12.14

SEA-SURFACE TEMPERATURE

Because the heat-holding capacity of water is so high, sea-surface temperature has an important effect on both weather and climate. Slow changes in ocean–atmosphere systems, such as El Niño, the North Atlantic Oscillation, and the Pacific Decadal Oscillation, can be felt in climate cycles over many years.

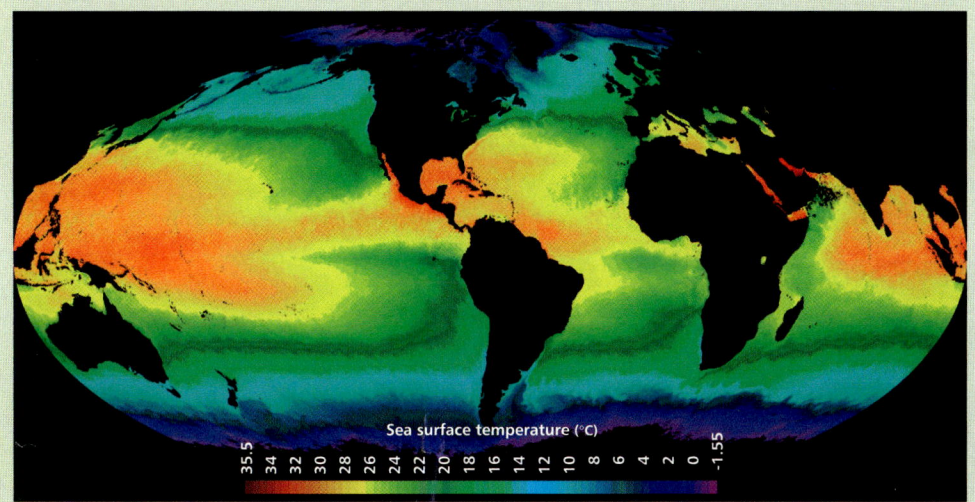

◀ **Sea-surface temperature**
While ocean temperatures are generally warm near the Equator and cold at high latitudes, winds and currents distort this simple picture. Thus, the western equatorial Pacific is warmer than the eastern half, and warm equatorial currents move poleward in the form of major ocean currents such as the Gulf Stream.

▼ **El Niño–La Niña**
El Niño is the appearance of warm water in the eastern equatorial Pacific (left) near Christmas time. El Niño is often followed by La Niña, characterized by cold ocean temperatures across the equatorial Pacific. These local ocean properties are connected to global changes in atmospheric patterns.

▼ **North Atlantic Oscillation**
The weather of northern Europe has dramatic swings every 5–10 years in connection with the North Atlantic Oscillation. When the air pressure over Portugal becomes larger than that over Iceland, the westerly wind in the Atlantic becomes stronger and brings warm and wet marine air to northern Europe for a mild winter. The opposite brings a colder and drier winter. The extreme warm or cool phases can last for as long as two decades. Ocean temperatures suggest a reversal to cool Pacific Decadal Oscillation conditions in 1998.

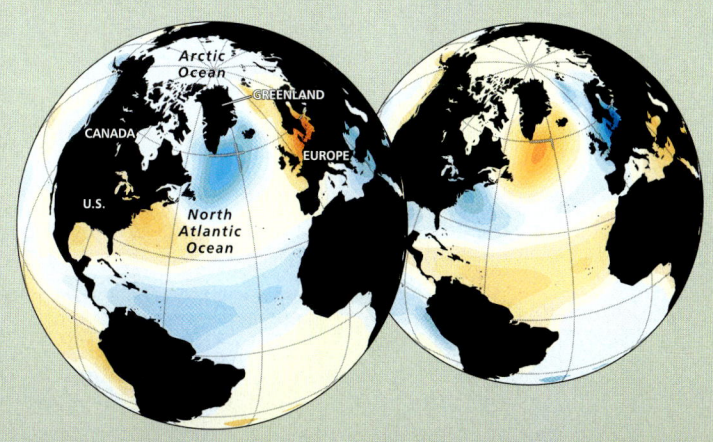

Ocean Water and Its Circulation 387

heat and maintains the temperature of the water around it. Both of these processes tend to reduce the amount of variation in ocean temperatures (**FIGURE 12.15**). The ocean in turn moderates the climate of coastal regions on land. Along the Pacific coast of Washington and British Columbia, for example, winter air temperatures seldom drop to freezing, whereas inland temperatures plunge to –30°C or lower.

The thermohaline portion of the ocean's global conveyor belt is one of the most important natural influences on climate. Because it redistributes heat from the tropics to the poles, any weakening of the circulation would be expected to increase the temperature contrast between the two. The cause and effect may work the other way, too: During the last ice age, the polar ice caps may have cut off the "conveyor belt," which would have made the glaciation longer-lasting and more severe. In Chapter 16, we will look at the possibility of rapid climate change triggered by a shutdown of the thermohaline circulation.

The global air temperature distribution FIGURE 12.15

These maps show the average daily air temperature in (**A**) January and (**B**) July over both land and sea. Notice the much broader range of temperatures over land in both maps. Also, note that there is a pronounced seasonal change in land temperatures, whereas in the world's oceans there is little difference between summer and winter (especially in the tropics).

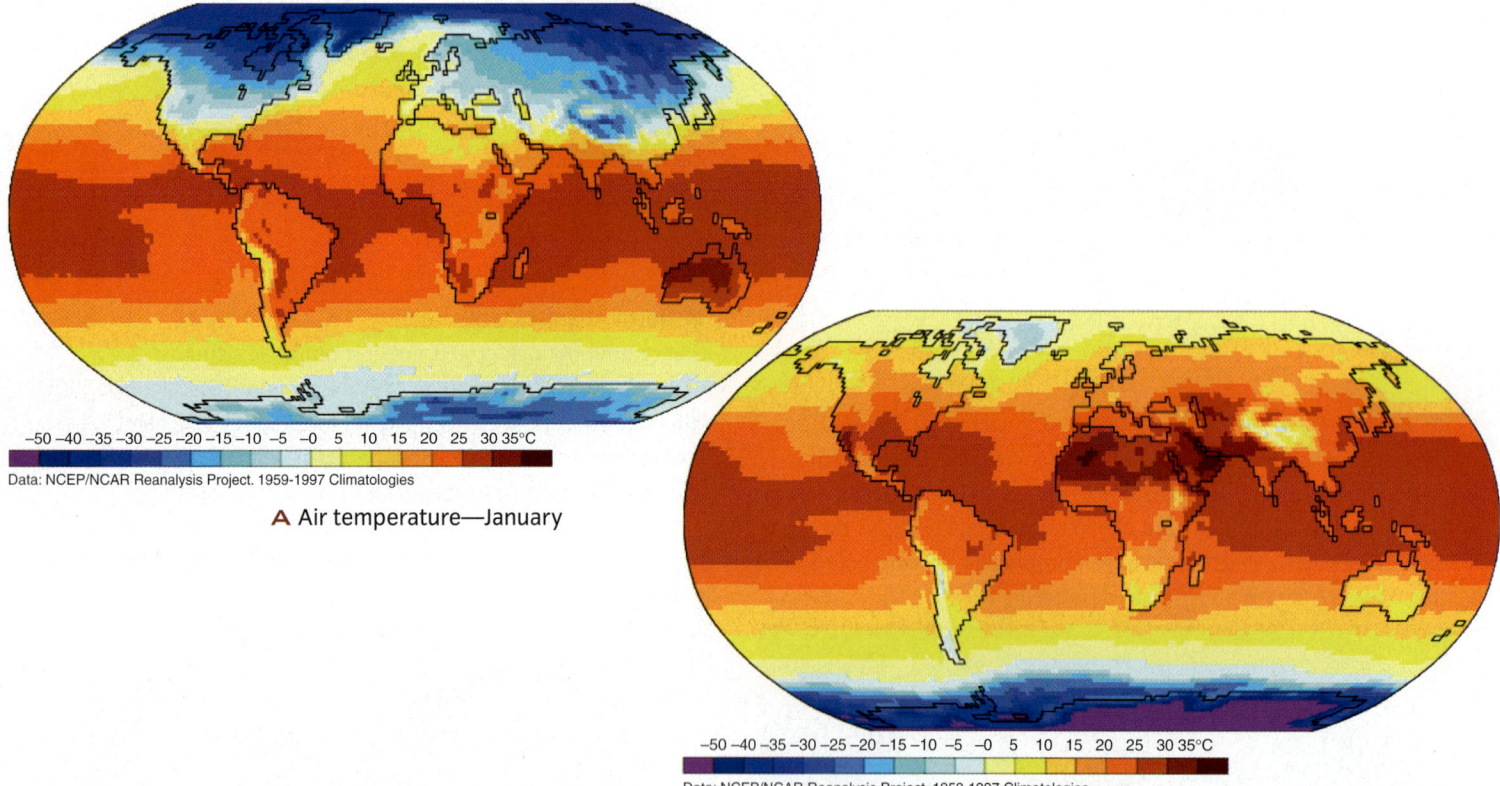

–50 –40 –35 –30 –25 –20 –15 –10 –5 –0 5 10 15 20 25 30 35°C
Data: NCEP/NCAR Reanalysis Project. 1959-1997 Climatologies

A Air temperature—January

–50 –40 –35 –30 –25 –20 –15 –10 –5 –0 5 10 15 20 25 30 35°C
Data: NCEP/NCAR Reanalysis Project. 1959-1997 Climatologies

B Air temperature—July *Data: NCEP/NCAR Reanalysis Project; 1959–1997, Climatologies.*

CONCEPT CHECK **STOP**

What are the three layers of the ocean?

What drives surface ocean currents?

How are pressure differences created in ocean water?

How do El Niño and La Niña conditions affect Peruvian coastal upwelling and trade winds?

What is the ocean conveyor belt?

Amazing Places: Monterey Bay, California

Just offshore from the Monterey Peninsula, California, the Monterey Submarine Canyon bisects Monterey Bay, plunging to about 1.8 km, making it comparable in depth to the Grand Canyon (**A**). The near-shore presence of a deepwater canyon provides a coldwater upwelling—produced by offshore winds—that is rich in nutrients that can support unusually abundant food for seabirds, whales, dolphins, and sea otters (**B**). The floor of Monterey Bay is host to a number of other weird creatures. This three-inch long nudibranch (**C**) is called the sea shawl. The nudibranch is a carnivorous organism, with some species even eating other nudibranchs. A rosy rockfish swims by a basket starfish (**D**).

Global Locator

Ocean Water and Its Circulation

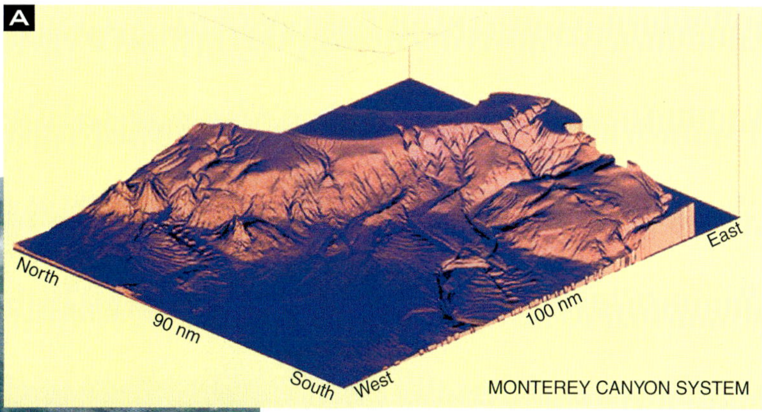

SUMMARY

1 The Ocean Basins

1. There has been an ocean on Earth for at least 4 billion years.

2. Most of the ocean water is contained in four huge, interconnected **basins**—the Pacific, Atlantic, Indian, and Southern Oceans.

3. The water in the ocean may have condensed from steam produced by primordial volcanic eruptions, been delivered to the planet's surface via cometary impacts, or both.

4. Ocean basins are marked by a *mid-oceanic ridge* with a *central axial rift* and by *abyssal plains*.

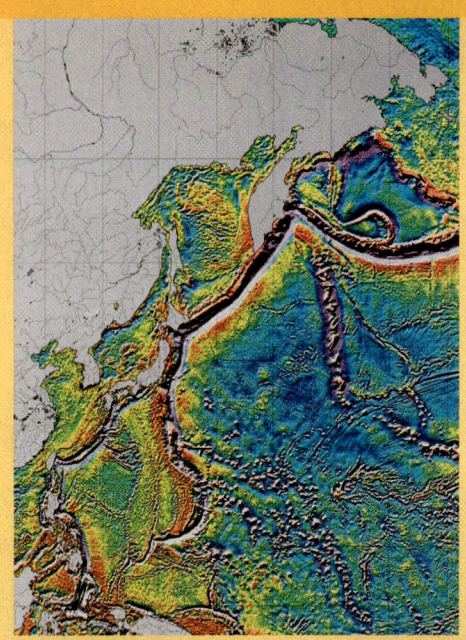

2 The Composition of Seawater and the Movement of Sediment

1. Seawater ranges in **salinity** from 3.3 to 3.7%. Freezing and evaporation make the water saltier, whereas rain, snow, and *river flow* make it less salty.

2. **Turbidity currents** have built thick deposits of sediment at the base of the *continental slope* and on the adjacent abyssal plain.

3. The chief kinds of sediment on the deep seafloor are brownish or reddish clay (blown in from continents by the global wind patterns), calcareous ooze, and siliceous ooze. The distribution of oozes is related to surface-water temperature and water depth.

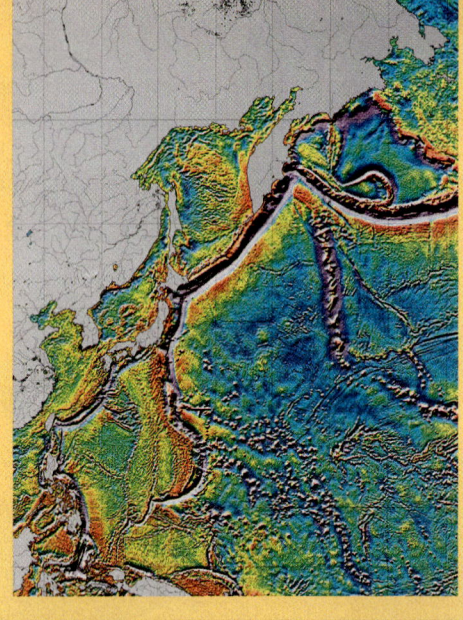

3 Ocean Water and Its Circulation

1. Ocean water forms layers based on density, which is controlled by temperature and salinity. These layers are the surface layer, the thermocline, and the deep layer.

2. Surface **ocean currents** are set in motion by the prevailing winds. **El Niño** is an example.

3. The **thermohaline circulation** involves deep currents that are powered by changes in temperature and salinity in surface waters, which change the water's density and thereby cause it to rise or sink. It is also known as the *ocean conveyor belt*. The thermohaline circulation helps regulate climate.

KEY TERMS

- ocean basins, p. 373
- salinity, p. 376
- turbidity current, p. 378
- turbidite, p. 379
- ocean current, p. 382
- El Niño, p. 384
- thermohaline circulation, p. 384

CRITICAL AND CREATIVE THINKING QUESTIONS

1. What kind of evidence might be sought to establish the existence of an ocean early in Earth's history?

2. Refer back to Figure 1.11 and look at the distribution of land and oceans 200 million years ago. How might surface ocean currents have flowed in the world ocean at the time? What would the climate have been in the center of the large landmass (Pangaea)?

3. Some scientists have proposed controlling climate change by increasing the photosynthetic productivity of microscopic marine algae in the oceans. How could marine algae influence the climate? How could their productivity be increased? Do you think that this is a realistic way to counter climate change?

4. Earth scientists have proposed that the eruptions of the Indonesian volcanoes Krakatoa in 1883 and Tambora in 1815 had lasting effects on the behavior of the oceans, leading to a slight drop in sea level. How could a volcanic eruption have led to this effect?

5. During the early Tertiary Period, North and South America were not connected. How might that have changed the global pattern of ocean circulation? Do some research and suggest some ways of testing your hypothesis.

What is happening in this picture?

This striking image, acquired by a NASA satellite, shows land and sea surface temperatures for a week in April. The eastern coast of North America is on the left and the Atlantic Ocean is on the right. The color scale ranges from deep red (hot) to violet (cold). The Gulf Stream, which carries warm water northward, stands out as a tongue of red and yellow extending from the Caribbean along the coasts of Florida and the Carolinas.

- Can you pick out the tip of the Labrador current, which carries cold water southward?

- Looking at the land, note the purple colors of the northern Great Lakes. What might this coloration indicate?

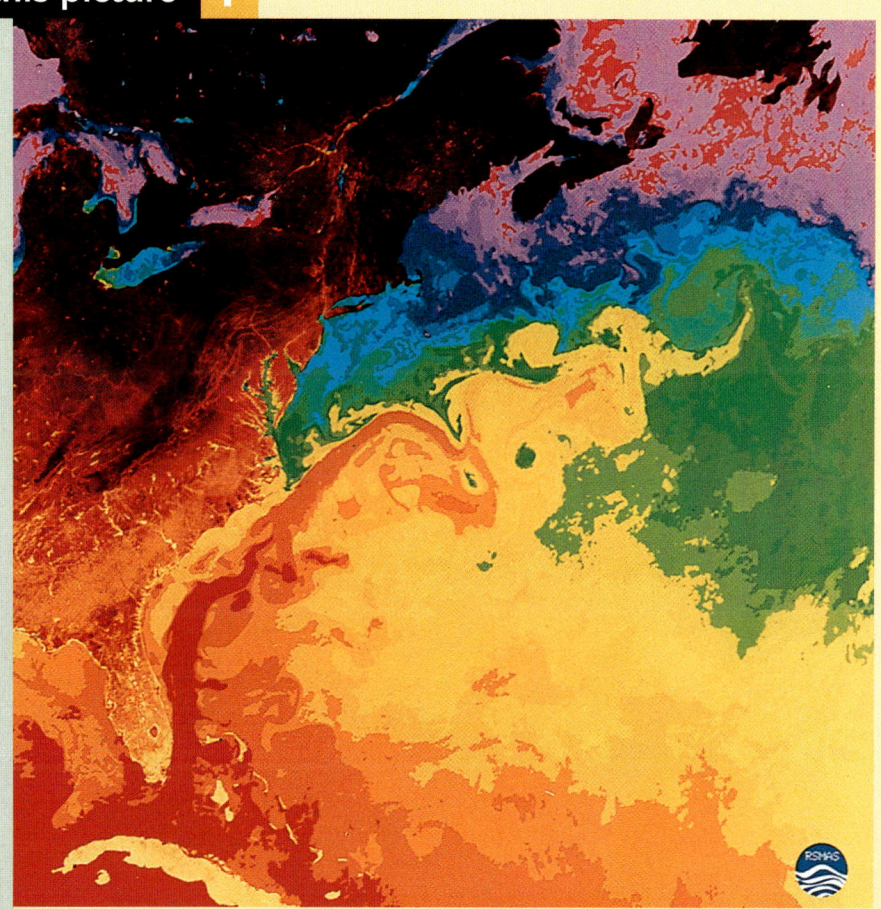

SELF-TEST

1. The ocean formed between _____ billion and _____ billion years ago.
   a. 3.0/3.56
   b. 3.56/4.0
   c. 4.0/4.56
   d. 4.56/5.0

2. Today, scientists measure the depth of oceans with _____.
   a. acoustical instruments
   b. weighted wires
   c. laser imaging
   d. optical instruments

3. Much of the oceanic crust is less than _____ years old, while the bulk of the continental crust is over _____ years old.
   a. 50 million/5 billion
   b. 100 million/4 billion
   c. 40 million//1 billion
   d. 60 million/1 billion

4. The diagram shows a _____ plate boundary.
   a. collision
   b. spreading
   c. subduction
   d. transform

5. Some continental margins are _____ and accumulate thick deposits of continental sediments, while other continental margins are _____ and have trenches marking the location at which ocean crust is sliding beneath continental crust.
   a. passive/active
   b. passive/tectonic
   c. active/passive
   d. submerging/emerging

6. The salinity of seawater is roughly _____ times the salinity of tapwater.
   a. 0.7   b. 7   c. 70   d. 700

7. Which of the following processes increases the salinity of seawater?
   a. freezing
   b. rain, snow, and river runoff
   c. evaporation
   d. Both a and c are correct.

8. Which of the following processes decreases the salinity of seawater?
   a. freezing
   b. rain, snow, and river runoff
   c. evaporation
   d. Both a and c are correct.

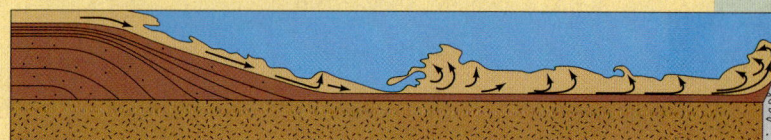

9. Turbidity currents (above) are driven by _____.
   a. prevailing winds
   b. gravity
   c. the Coriolis effect
   d. temperature differences

10. The uppermost waters contain _____ organisms, while _____ organisms occupy the bottom sediments.
    a. benthic/planktic
    b. pelagic/planktic
    c. planktic/benthic
    d. pelagic/benthic

11. Forminifera shells form part of the _____.
    a. calcareous ooze
    b. siliceous ooze
    c. red clay
    d. lithic sediment

12. Ocean water in the _____ extends to a depth of 100 m, is relatively warm, and moves in broad, wind-driven currents.
    a. deep zone
    b. surface layer
    c. thermocline
    d. transition zone

13. Which of the following factors does not affect surface ocean currents?
    a. prevailing winds
    b. gravity
    c. the Coriolis effect
    d. salinity differences
    e. temperature differences

14. During an El Niño event, the normal Pacific upwelling along the Peruvian coast _____ and trade winds _____. This sets up a weak equatorial _____ current that brings floods to some regions and droughts to others.
    a. ceases/weaken/eastward
    b. increases/strengthen/westward
    c. ceases/strengthen/eastward
    d. increases/weaken/westward

15. During the last ice age, the polar ice caps may have _____ the "ocean conveyor belt," which would have made the glaciation _____.
    a. enhanced/longer lasting
    b. cut off/longer lasting
    c. enhanced/shorter lasting
    d. cut off/shorter lasting

Where Ocean Meets Land

13

Living on the coast can be idyllic, but as residents of Galveston, Texas, have discovered, it can also be deadly.

Built on a barrier island that protects Galveston Bay, the busiest harbor in Texas, Galveston was the largest city in Texas in 1900. It had weathered many storms without trouble, and its 42,000 citizens confidently faced the future.

On September 7, 1900, Cuban observers used primitive telephone connections to send word of a hurricane heading into the Gulf of Mexico. The U.S. Weather Bureau predicted a different path, one that followed the Atlantic coast. By the afternoon of the 9th, it was clear that the Cubans were correct—a hurricane was loose in the Gulf. Warning flags were raised in Galveston, but few heeded them. The Great Storm struck that evening.

The Great Storm was a category 4 hurricane, and Galveston took a head-on hit. Wind-driven waves battered the shore, and then a hurricane-driven storm surge submerged the town under 7 ft of water. The town was destroyed, and at least 8,000 lost their lives. As seen in the photograph, it took rescuers many days to search for bodies in the worst natural disaster in U.S. history.

Galveston was rebuilt, and seawalls were installed to lessen storm damage. Then on September 14, 2008, Hurricane Ike visited, and Galveston suffered again. Ike was a category 2 hurricane, but it moved so slowly, the battering lasted longer. Galveston was better prepared this time, but damage was severe (see inset). Fortunately, most people heeded evacuation warnings, and only 51 deaths were reported for the entire Texas coast.

NATIONAL GEOGRAPHIC

CHAPTER OUTLINE

- Changes in Sea Level p. 396
- Waves p. 400
- Shorelines and Coastal Landforms p. 405
- Humans versus the Sea p. 410

Changes in Sea Level

LEARNING OBJECTIVES

Explain why worldwide sea levels change.

Describe the possible long-term consequences of sea-level rise brought on by global warming.

Discuss what causes tides.

If you visit almost any coastline on two occasions a year apart, you will see changes. Sometimes the changes are small, but often they are substantial. Large sand dunes may have shifted. Sand may have built up behind barriers or been eroded away. Steep sections of coastline may have collapsed. New channels may also have broken through from the sea to lagoons on the landward side.

Water plays a powerful role in shaping coastlines. Before we turn to the landforms created by water's action, let's look at the processes themselves. These

The Bering land bridge FIGURE 13.1

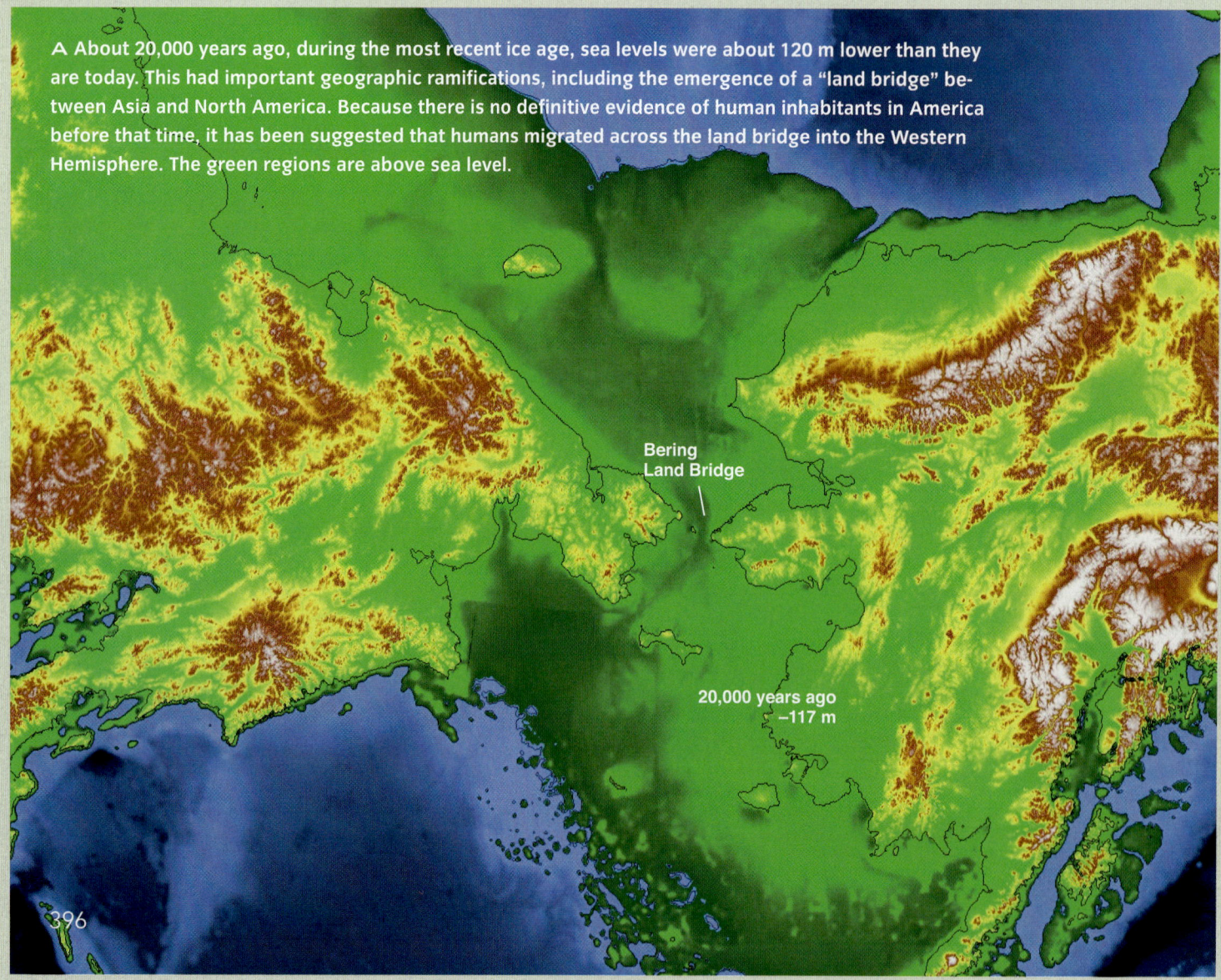

A About 20,000 years ago, during the most recent ice age, sea levels were about 120 m lower than they are today. This had important geographic ramifications, including the emergence of a "land bridge" between Asia and North America. Because there is no definitive evidence of human inhabitants in America before that time, it has been suggested that humans migrated across the land bridge into the Western Hemisphere. The green regions are above sea level.

processes operate on very different time scales. On a very long time scale, over many thousands of years, sea level can rise or fall by hundreds of meters. On a shorter time scale—twice a day—tides change the water level by several meters in certain places. On the shortest time scale, ocean waves constantly stir up sediment along the coast, and especially during storm season, waves batter the coastline and can produce dramatic effects.

GLOBAL CHANGES IN VOLUME

One of the most important factors determining the long-term position of any given shoreline is the volume of water in the ocean. This in turn is strongly tied to changes in the global climate. When the climate warms, water stored on continents in glaciers and ice caps melts and returns to the sea, and the upper layers of the ocean expand as they grow warmer. Both effects cause a worldwide rise in sea level. Conversely, during cold climatic periods, sea level drops as the ocean cools and contracts, glaciers and ice caps expand, and water is withdrawn from the ocean and stored on land in the form of ice. At present, the warming of the global climate is believed to be responsible for a worldwide rise in sea level of approximately 2.4 mm/yr. Although such a trend is nearly imperceptible on the scale of a human lifetime, over longer periods it can account for major changes in the position of a shoreline (**FIGURE 13.1**).

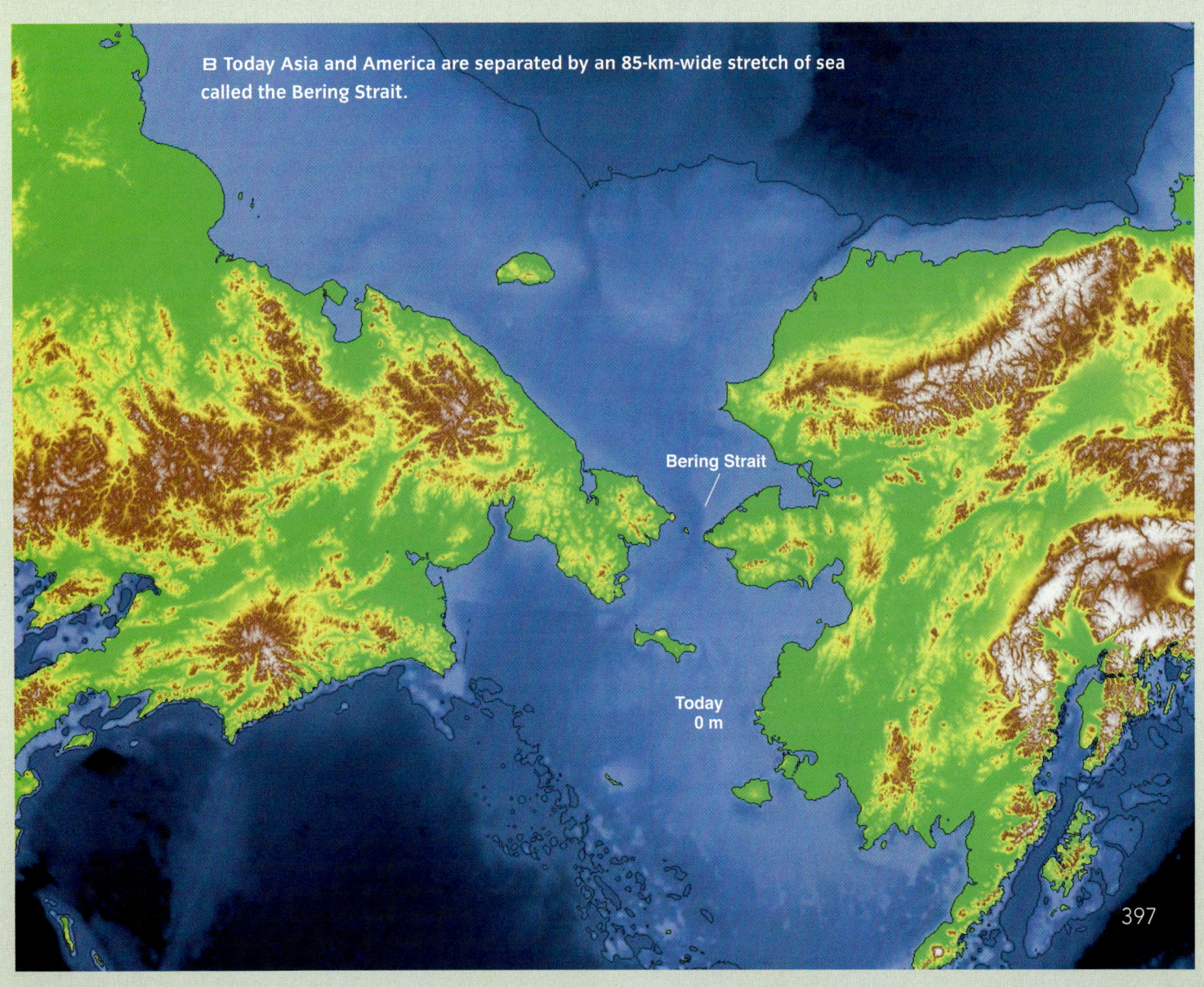

B Today Asia and America are separated by an 85-km-wide stretch of sea called the Bering Strait.

Bering Strait

Today
0 m

Rising sea levels FIGURE 13.2

A **New inlet** Storm waves from Hurricane Isabel breached the North Carolina beach to create a new inlet, as shown in this aerial photo from September 2003. Notice the widespread destruction of the shoreline in the foreground, with streaks of sand carried far inland by wind and wave action. As sea level rises, ocean waves will attack beaches with increasing frequency and ferocity.

B **Coastal erosion** Wave action has undermined the bluff beneath these two buildings, depositing them on the beach below.

According to current estimates, global warming will cause the overall sea level to rise by about 20 to 60 cm (8 to 24 in.) between now and 2100. Rising sea levels will have a number of effects. Most coastal erosion occurs in severe storms. Global warming will increase extreme weather conditions, and a higher sea level rise will push beaches, salt marshes, and estuaries landward (**FIGURE 13.2**). More land will subside, or sink to lower levels, and recent estimates suggest that as much as 22% of the world's coastal wetlands could be lost by the 2080s. These changes would have a major effect on commercially important fish and shellfish populations.

TIDES

A much more obvious change in sea level is a result of **tides**. The gravitational attraction of the Moon exerts a pull on the ocean water on the side of Earth nearest the Moon, making it bulge outward. There is also a bulge away from the Moon in the ocean water on the opposite side of Earth, caused by an *inertial force* (the same force that pulls the string of a yo-yo tight if you swing it around your finger). The Sun also affects the tides, but because it is so much farther away than the Moon, its tide-producing force is only about half as strong.

To visualize how tides work, consider the tidal bulges oriented with their maximum height lying along a line running through the center of Earth and the center of the Moon, as shown in **FIGURE 13.3**. Whereas Earth rotates around its axis, the tidal bulges remain stationary opposite the Moon. Thus, any given coastline will move eastward through both tidal bulges each day. Every time a landmass encounters a tidal bulge, the water level along the coast rises. As Earth rotates, the coast passes through the highest point of the tidal bulge (high tide) and the water level begins to fall. Along most coastlines we see two high tides and two low tides. The effect of the Sun on tides is not as large as the effect of the Moon. The Sun's pull can enhance or lessen the Moon's effect, depending on the relative positions of the Moon, Sun, and Earth.

> **tide** A regular, daily cycle of rising and falling sea level that results from the gravitational attraction between the Moon, the Sun, and Earth.

What causes the tides?
FIGURE 13.3

The Moon's gravitational attraction and the inertia of the rotating Earth–Moon system combine to produce tides. In this greatly exaggerated sketch, the two forces stretch Earth into an oblong shape, with bulges directed toward and away from the Moon. The bulges remain essentially stationary while Earth rotates through them, creating two high tides and two low tides per day.

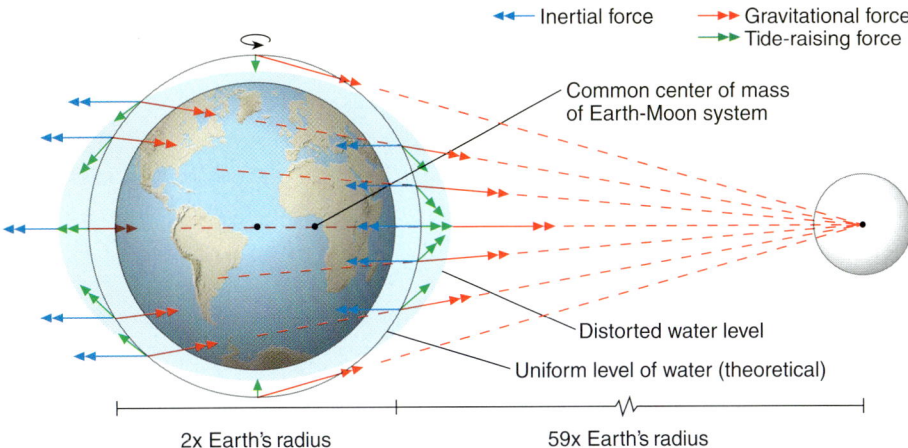

In the open sea, tides are small. However, the shape of a coastline can greatly influence the height of the tidal *run-up*, the highest elevation reached by the incoming water. Narrow openings into bays, rivers and estuaries can amplify normal tidal fluctuations. At the Bay of Fundy in Nova Scotia, for example, the *tidal range* (the difference between high and low tide) is up to 16 m (**FIGURE 13.4**). The bay is very long and narrow, a configuration that causes the incoming tide to form a steep-fronted wall of water called a *tidal bore* that can move faster than a person can run. Other places have almost no tides, or even once-a-day instead of twice-a-day tides, all depending on local conditions.

In most places, though, the tides act too slowly to have much impact on a shoreline. Perhaps their most important effect is global in scale. Remember that a tidal bulge is fixed opposite the Moon, but Earth rotates on its axis. Each time the edge of a continent meets a tidal bulge there is a minuscule slowing of Earth's rotation due to friction of the tide. The effect is slow and gradual, but the tides are slowing Earth's rotation, causing the length of the day to increase. In the early Phanerozoic Eon, Earth rotated faster, and the day was about 22 hours long; as a result there were nearly 400 days in a year.

> **CONCEPT CHECK   STOP**
>
> **What** effects cause changes in sea level? On what time scales do they act?
>
> **What** are the possible consequences of rising sea levels due to global warming?
>
> **Why** do tides form?

The Bay of Fundy FIGURE 13.4

The tidal range in the Bay of Fundy in eastern Canada is one of the greatest in the world.

A These fishing boats are grounded at low tide. At high tide, the water will rise to the level where the color on the posts in the photograph changes.

B The tides are so high because the frequency of the tides closely matches the natural period of oscillation of water in the long, thin bay. It is like sloshing the water in a bathtub—if you move back and forth at just the right frequency, the water will overflow.

Waves

LEARNING OBJECTIVES

Describe surf and breaking waves.

Explain how waves and littoral drift transport sediment and shape beaches.

Discuss tidal currents.

Ocean waves receive energy from winds. The size of a wave depends on how fast, how far, and how long the wind blows across the water's surface. A gentle breeze blowing across a bay may ripple the water or form low waves less than a meter high. In contrast, storm waves whipped up by intense winds over several days may tower over ships unfortunate enough to be caught in them (**FIGURE 13.5**). The waves generated by a storm can travel great distances. For example, surfers in California keep an eye on the weather report for storms in the tropical Pacific because they know that a good storm creates a "swell" that will arrive at their shores several days later.

Rogue waves generated by storms FIGURE 13.5

Aug 20, 1996
Overhead view of ocean surface (satellite radar)

◀ **A** In a storm, the highest waves can wash over the deck of a ship.

▼ **B** For centuries, sailors have told tales about *rogue waves* as high as a 10-story building, which are big enough to capsize a ship. For years, scientists did not believe such stories. However, evidence such as this radar image, taken from a satellite in 1996, has removed all doubt. In the center of the white bar, you can see a dark trough that goes down more than 10 m below sea level, next to a bright peak that reaches more than 15 m above sea level. A ship caught in the trough would have seen a wall of water more than 25 m high (inset). Note that the rogue wave is quite localized; all around it are waves that are less than 10 m high from trough to crest.

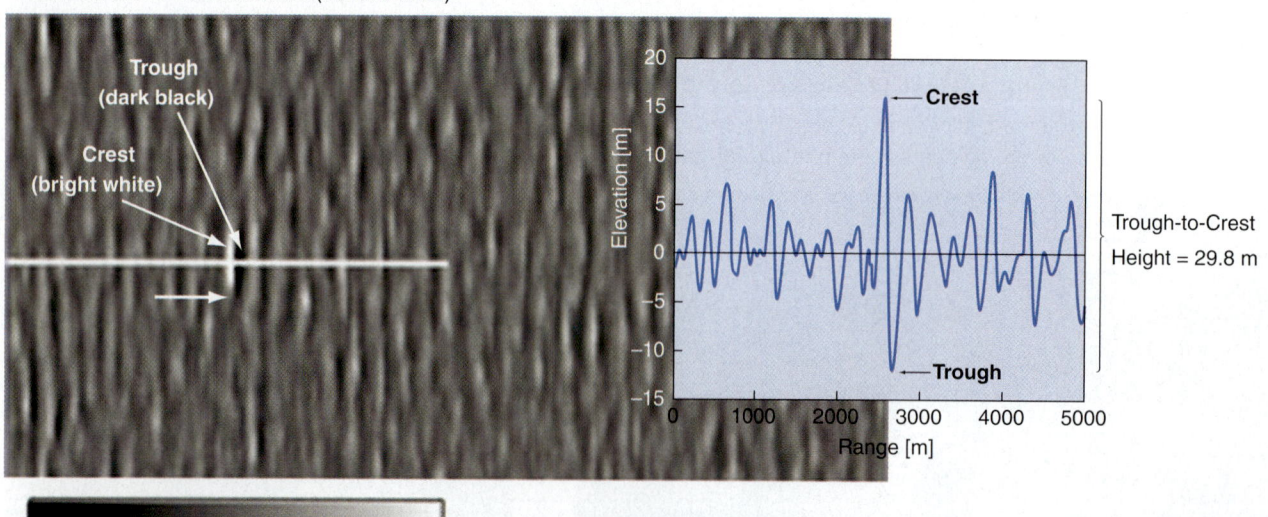

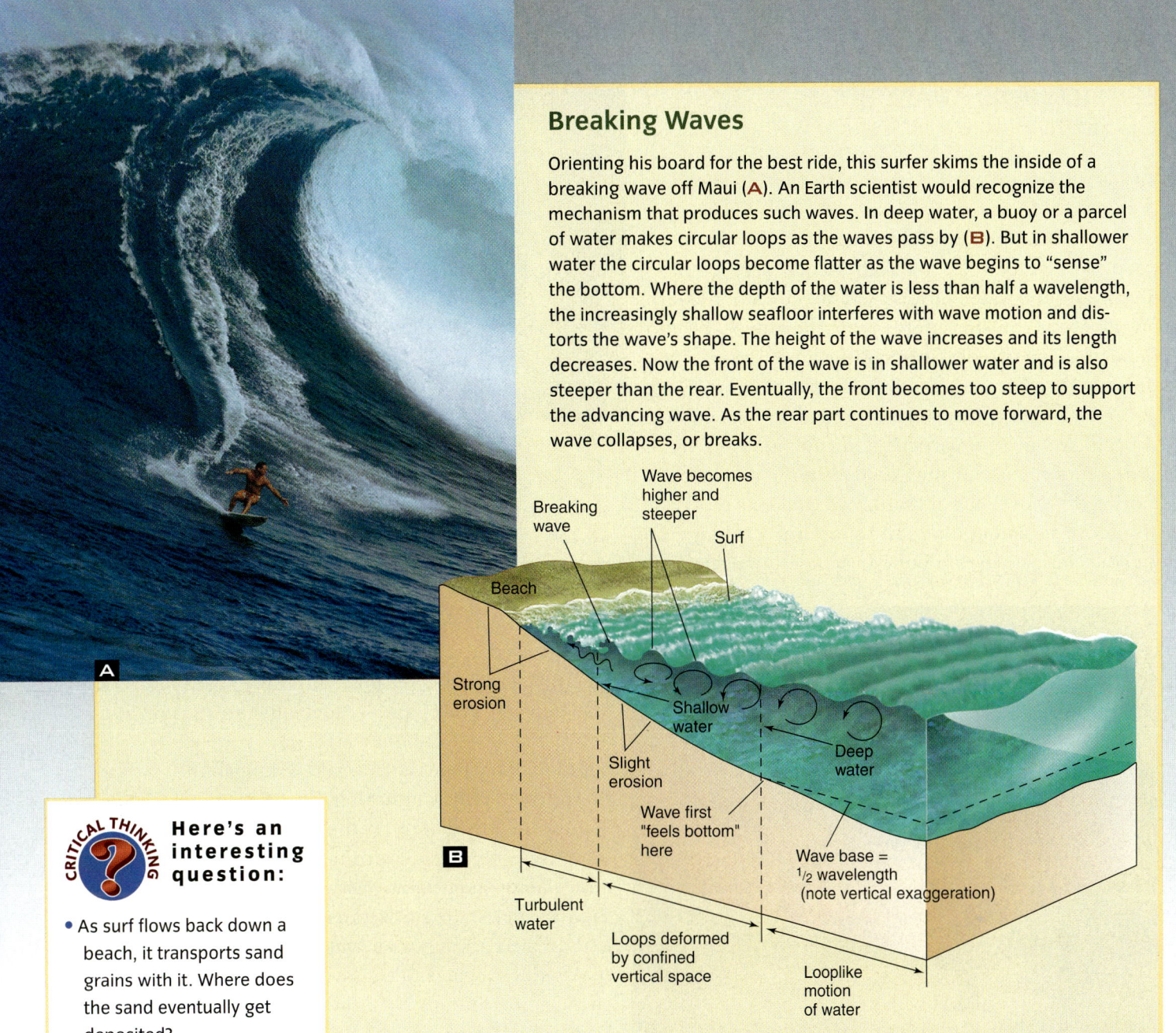

What an Earth Scientist Sees

Breaking Waves

Orienting his board for the best ride, this surfer skims the inside of a breaking wave off Maui (**A**). An Earth scientist would recognize the mechanism that produces such waves. In deep water, a buoy or a parcel of water makes circular loops as the waves pass by (**B**). But in shallower water the circular loops become flatter as the wave begins to "sense" the bottom. Where the depth of the water is less than half a wavelength, the increasingly shallow seafloor interferes with wave motion and distorts the wave's shape. The height of the wave increases and its length decreases. Now the front of the wave is in shallower water and is also steeper than the rear. Eventually, the front becomes too steep to support the advancing wave. As the rear part continues to move forward, the wave collapses, or breaks.

CRITICAL THINKING — Here's an interesting question:

- As surf flows back down a beach, it transports sand grains with it. Where does the sand eventually get deposited?

WAVE ACTION ALONG COASTLINES

As a wave approaches the shore, it undergoes a rapid transformation—one that is often exploited by surfers. When a wave breaks, as described in *What an Earth Scientist Sees*, the motion of the water instantly becomes turbulent, like that of a swift river. **Surf** is found between the line of breakers and the shore, forming an area known as the *surf zone*. Each wave finally dashes against

surf The "broken" turbulent water found between a line of breakers and the shore.

the rocks or rushes up a sloping beach until its energy is expended; then the water flows back toward the open sea. Water that has piled up against the shore returns seaward in an irregular and complex way, partly as a broad sheet along the bottom and partly in localized narrow channels known as *rip currents*. These can sweep unwary swimmers out to sea. However, experienced swimmers know that the rip current is not wide. By swimming sideways to the current, rather than against it, they can get out of trouble.

Waves

EROSION AND TRANSPORT OF SEDIMENT BY WAVES

Surf is a powerful erosive force because it retains most of the original energy of the waves that created it (**Figure 13.6**). When winds blow over broad expanses of water, they generate waves. Both friction between moving air and the water surface and direct wind pressure on the waves transfer energy from the atmosphere to the water. The size of waves and the direction in which they travel are determined by winds far offshore. Most of the wave energy is eventually used up in the constant churning of mineral particles and water as waves break at the shore. This churning erodes shoreline materials, moving the shoreline landward. Waves and currents can also move sediment along the shoreline for long distances. This activity can build beaches outward as well as forming *barrier islands* just offshore. (Barrier islands are discussed in the next section.)

Storm waves Figure 13.6

High surf can rapidly reshape a beach, flattening the beach slope and moving beach sediment just offshore. These large waves at Humbug Mountain State Park, Oregon, were most likely generated by an offshore storm.

Wave erosion takes place not only at sea level but also below sea level and—especially during storms—above sea level. In the surf zone, rock particles are worn down, becoming smoother, rounder, and smaller. At the same time, through continuous rubbing and grinding with these particles, the surf scours and deepens the bottom. Onshore, the surf acts like a saw cutting horizontally into the land. Its energy is eventually consumed in turbulence, in friction at the bottom, and in the movement of sediment thrown up from the bottom. Two processes, one on land and one underwater, transport the sediment: the **longshore current** and **beach drift**. **Figure 13.7** shows a Brazilian beach that has been carved by these processes. **Figure 13.8** describes these two effects, which together are known as **littoral drift**.

Beach carved by littoral drift Figure 13.7

Surf swashes obliquely onto a Brazilian beach and forms a series of arc-shapes in the sand. A grain of sand moves along a zigzag path as successive waves wash up on the shore at an angle and retreat back downslope in a direction nearly perpendicular to the shoreline.

> **longshore current** A current within the surf zone that flows parallel to the coast.
>
> **beach drift** The movement of particles along a beach as they are driven up and down the beach slope by wave action.
>
> **littoral drift** Transport of sediment parallel to the shoreline by the combined action of beach drift and longshore current transport.

Littoral drift FIGURE 13.8

Breaking waves create a cycle of *swash*, as they surge forward, and *backwash*, as the water retreats after having spent its energy. This process transports sand and shapes beaches, in a process known as littoral drift.

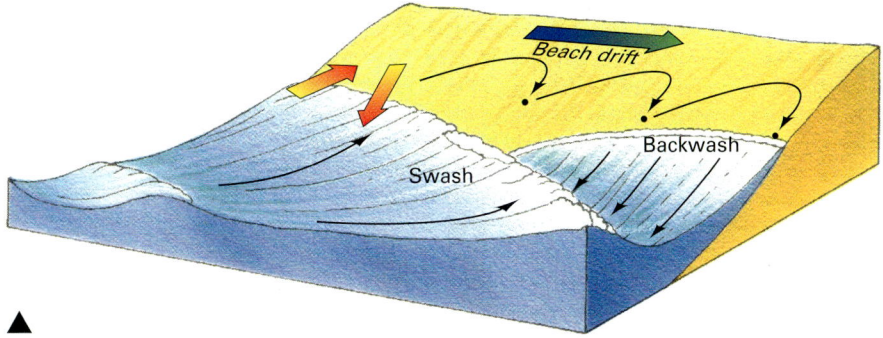

▲
Beach drift
Swash tends to approach the shore at an oblique angle, carrying its burden of sand with it. But the backwash flows back along the most direct downhill direction. So, sand particles come to rest at a position to one side of the starting place. This movement is repeated many times, so individual rock particles are transported long distances along the shore.

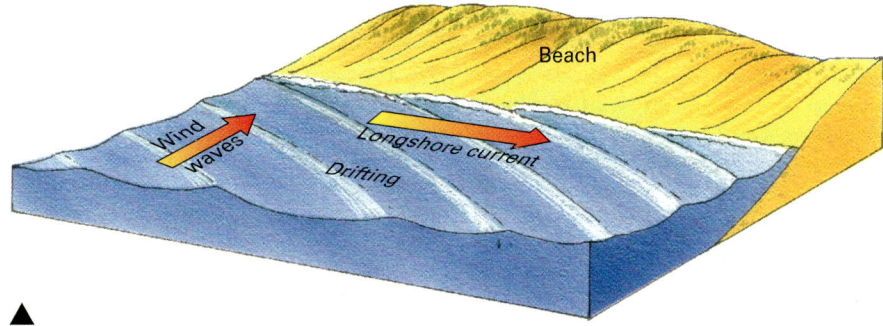

▲
Longshore drift
When waves approach a shoreline at an angle to the beach, a current is set up parallel to the shore in a direction away from the wind. This is known as a longshore current. When wave and wind conditions are favorable, this current is capable of carrying sand along the sea bottom. This is *longshore drift*.

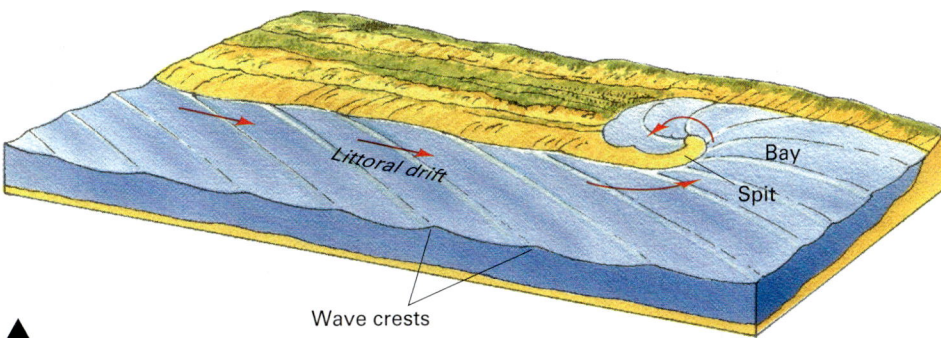

▲
Littoral drift
Beach drift and longshore drift, acting together, move particles in the same direction for a given set of onshore winds. The total process is called *littoral drift*. ("Littoral" means "pertaining to a coast or shore.") Where the shoreline is straight or broadly curved for many kilometers at a stretch, littoral drift moves the sand along the beach in one direction for a given set of prevailing winds. If there is a bay, the sand is carried out into open water as a long finger, or sandspit. As the sandspit grows, it forms a barrier, called a bar, across the mouth of the bay.

CRITICAL THINKING
Here's an interesting question:
- If you interfere with littoral drift by building a jetty or breakwater out from the shore, what would happen to the beach downdrift from the structure?

Salt marsh FIGURE 13.9

Saltwater marshes are abundant on the southeastern coastal plain of the United States. This marsh, photographed at high water, is near Brunswick, Georgia. At low tide, receding water will reveal the muddy bottom of the broad channel.

The processes of erosion and deposition along a coast are anything but steady. A single winter storm can cause more erosion of cliffs or beaches than a full year's worth of ordinary surf. The balance between erosion and deposition can change with the seasons. Along parts of the Pacific coast of North America, for example, winter storm surf tends to carry away fine sediment and the beach becomes narrow and steep. In calm summer weather, fine sediment drifts in and the beach takes on a gentler profile.

TIDAL CURRENTS

We saw earlier that the shape of a coastline can heavily influence tidal run-up, as described in Figure 13.4 for the Bay of Fundy. Tides can also play a role in shaping coastlines, provided that the tides are large. In bays and estuaries, the changing tide sets up *tidal currents*. When the tide begins to fall, an *ebb current* sets in. This flow ceases at about the time when the tide is at its lowest point. As the tide begins to rise, a landward current, the *flood current*, begins to flow.

Ebb and flood currents perform several important functions along a shoreline. First, the currents that flow in and out of bays through narrow inlets are very swift, scouring the inlet. This scouring keeps the inlet open despite the wave-driven shore processes that try to close it with sand.

Second, tidal currents hold large amounts of fine silt and clay suspended in the water. This fine sediment comes from streams that enter the bays or from mud that has been scoured up from the bottom by storm waves. The sediment, which contains large amounts of organic matter, settles to the floors of the bays and estuaries, where it builds up in layers and gradually fills the bay.

In time, tidal sediments fill bays and produce *mud flats*, barren expanses of silt and clay. These are exposed at low tide but covered at high tide. Eventually salt-tolerant plants start growing on the mud flat. The plant stems trap more sediment and the flat builds up, becoming a *salt marsh* (**FIGURE 13.9**). A thick layer of peat eventually forms at the surface of the marsh.

CONCEPT CHECK STOP

When do waves expend most of their energy?

What is littoral drift?

Why is littoral drift important?

How are tidal currents related to tide levels?

Shorelines and Coastal Landforms

LEARNING OBJECTIVES

Describe rocky coasts and pocket beaches.

Explain the formation of raised shorelines and marine terraces and the features associated with wave-cut cliffs.

Discuss beaches, barrier islands, lagoons, and deltas.

Describe coral reef coastlines.

The constant interplay between erosional and depositional forces along coastlines creates a wide variety of shorelines and coastal landforms. Their forms depend on the processes at work, on the composition of the coastal rocks, and how long the processes have been operating. Changes in sea level can also influence the development of coastal features. Many coastal land formations show clear evidence of different sea levels at different times in the past (**FIGURE 13.10**).

Despite the variability of coasts and shorelines, three basic types are most common: the *rocky (cliffed) coast*, the *lowland beach* and *barrier island coast*, and the *coral reef*. Each has its own particular set of erosional and depositional landforms.

ROCKY COASTS

The most common type of coast, comprising about 80% of ocean coasts worldwide, is a rocky or cliffed coast. Seen in profile, the usual elements of a cliffed coast are a vertical **wave-cut cliff** and a horizontal *wave-cut bench* at its base, both products of erosion. As the upper part of the cliff is undermined, it collapses and the resulting debris is redistributed by waves. An undercut cliff that has not yet collapsed may have a well-developed notch at its base, carved out by striking waves. The bench may be covered by sand or the

wave-cut cliff A coastal cliff cut by wave action at the base of a rocky coast.

Wave-cut beaches FIGURE 13.10

The coast of New Zealand at Tongue Point has two terraces, or *benches*. These are both former seafloor that was elevated above sea level by tectonic uplift on two different occasions. In this photograph you can see two other common features of rocky coasts: *headlands*, which jut out into the sea, and *pocket beaches*.

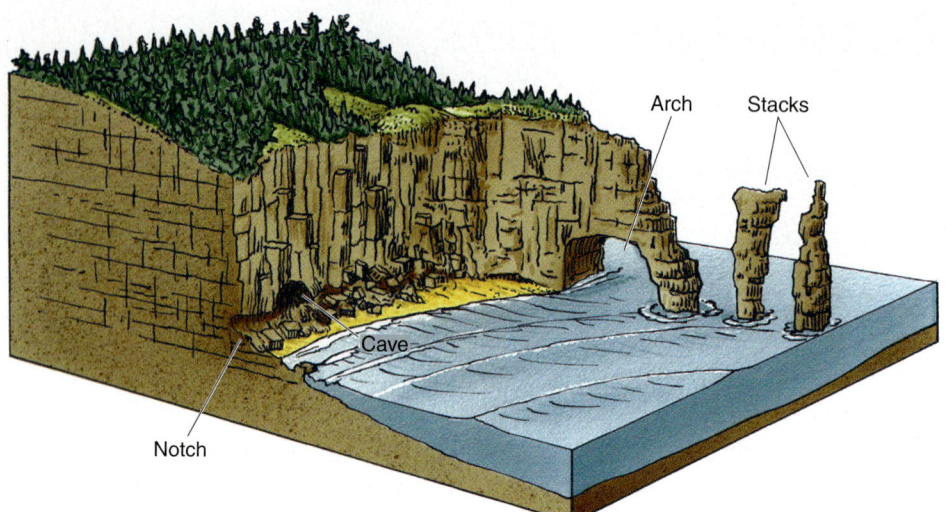

Sea caves, sea stacks, and arches FIGURE 13.11

Wave action undercuts the coastal cliff, creating landforms such as sea stacks and arches.

bedrock may be exposed, especially at low tide. FIGURE 13.11 shows some of the landforms that can be created from a cliff by wave action.

If the coast has been uplifted by tectonic activity, a wave-cut bench and its sediment cover can be lifted out of the water and become a *marine terrace*. In some locations, such as Tongue Point in New Zealand, shown in Figure 13.10, you can see two or more terraces ascending out of the ocean like a giant staircase.

BEACHES AND BARRIER ISLANDS

Beaches are a striking feature of many coasts. Most people think of a beach as the sand surface above the water along a shore. Actually, a **beach** also includes sediment in the surf zone, which is underwater and therefore continuously in motion. At low tide, when a large part of a beach is exposed, onshore winds may blow beach sand inland to form belts of coastal dunes.

A landform commonly associated with beaches is the **barrier island** (FIGURE 13.12). Barrier islands are found along most lowland coasts. The Atlantic and Gulf coasts of the United States consist mainly of a series of barrier islands ranging from 15 to 30 km in length and 1.5 to 5 km in width, located 3 to 30 km offshore. Coney Island, New York, and the long chain of islands known as the Outer Banks of North Carolina are examples. Sand dunes are typically the highest points on a barrier island.

Because barrier islands are topographically low, they are very susceptible to flooding. During a major storm, surf washes across the low places and erodes them, cutting *tidal inlets* that may remain open permanently. At such times, fine sediment is washed between the barrier island and the mainland. Because of this deposition and erosion of sediment, the length and shape of barrier islands are always changing. Unfortunately, the ever-changing nature of barrier island coasts conflicts with people's desire to erect permanent buildings on them. Property owners often protect their properties with artificial seawalls. Although these alleviate problems locally, they hasten erosion or deposition processes elsewhere on the coast.

Barrier island beaches typically exhibit depositional landforms such as *spits* (elongated ridges of sand or gravel that project from land into the open water of an embayment along the coast), *tombolos* (spitlike ridges of sand and gravel that join an island to the mainland), and *bay barriers*, which may completely close off the mouth of a small bay. A well-known example of a large, complex spit is Cape Cod, Massachusetts.

> **beach** Wave-washed sediment along a coast.
>
> **barrier island** A long, narrow, sandy island lying offshore and parallel to a lowland coast.

Visualizing
Barrier islands, spits, and lagoons FIGURE 13.12

A This aerial view of the Outer Banks of North Carolina shows a series of barrier islands. The action of wind and waves constantly pushes the sediment on the island toward the mainland (on the left in this photo).

B Salishan Spit in Oregon, shown here, is attached to the mainland at one end while the other end terminates at a tidal inlet.

Tidal inlet

C Spits and barrier islands often create a sheltered lagoon on their landward side. This lagoon is located on Cayo Costa Island in Florida.

Shorelines and Coastal Landforms   407

The elongated bay lying inshore from a barrier island or other low, enclosing strip of land (such as a coral reef) is called a *lagoon* (Figure 13.12). Lagoons are commonly fed by *estuaries*, the wide, fan-shaped mouths of rivers in the tidal zone where fresh and saltwater meet. Lagoons and estuaries are important habitats for a wide variety of plants, birds, and animals. They also play an important role in the protection of mainland shorelines because they serve as buffers against storm waves. Human activities can adversely affect these sensitive environments, as we will see later in the chapter.

DELTA COASTS

When a stream or river flows into a body of standing water, the current is rapidly slowed. As the current pushes out into the standing water, it deposits clay, silt, and sand, forming a **delta** (FIGURE 13.13). The river channel divides and subdivides into lesser channels called *distributaries*. The coarser sand and silt particles settle out first, while the fine clays continue out farthest and eventually come to rest in fairly deep water. When the fine clay particles in fresh water come into contact with saltwater, they clot together into larger particles that settle to the seafloor.

Deltas can grow rapidly, at rates ranging from 3 m (about 10 ft) per year for the Nile Delta to 60 m (about 200 ft) per year for the Mississippi Delta. Some cities and towns that were at river mouths several hundred years ago are today several kilometers inland.

delta Sediment deposit built by a stream entering a body of standing water.

The delta coast FIGURE 13.13

The Mississippi Delta has long branching fingers that grow far out into the Gulf of Mexico at the end of distributaries. This Landsat image shows the great quantity of suspended clay and fine silt being discharged by the river into the Gulf—about 1 million metric tons per day. (See also Figure 5.7.)

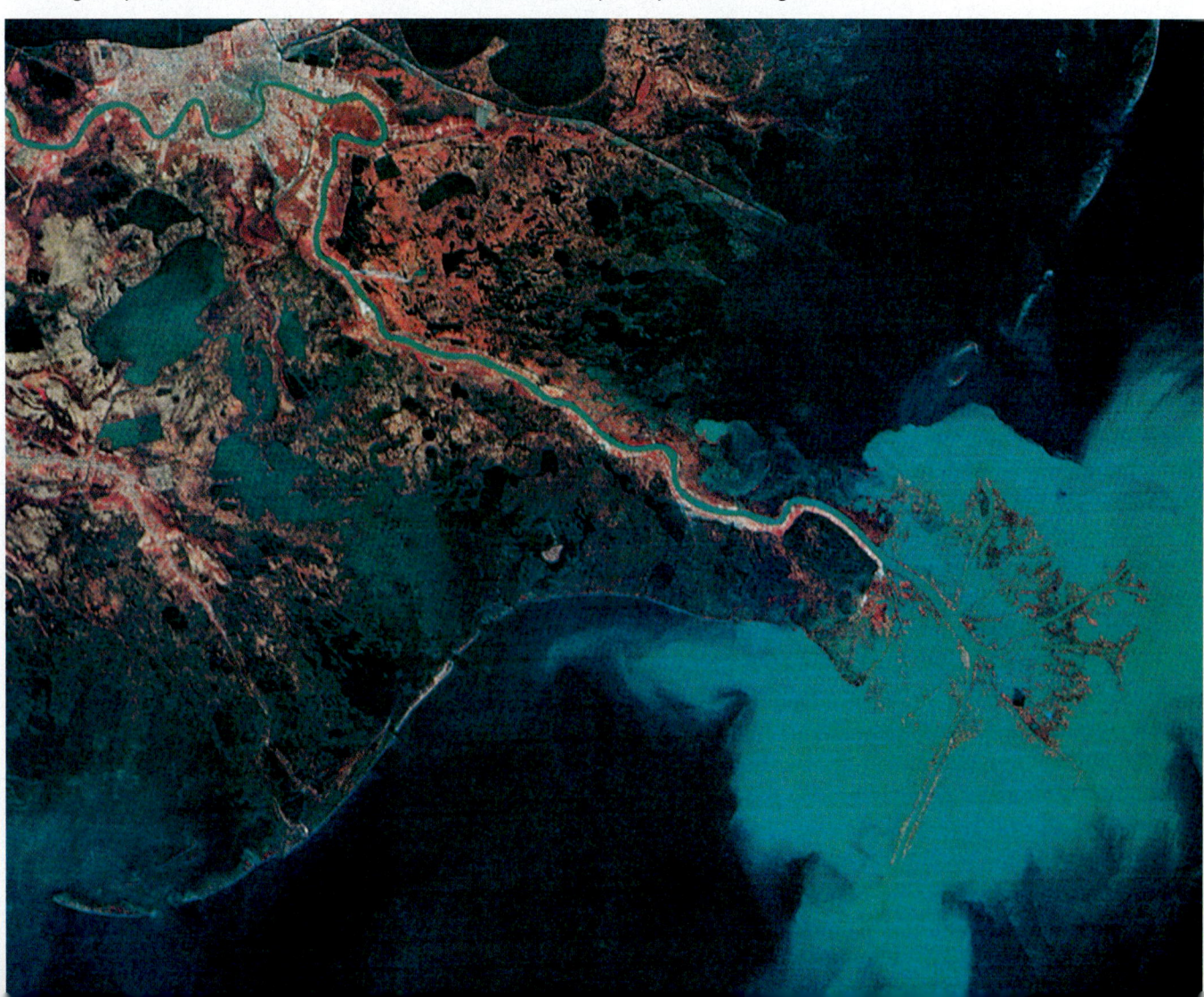

Coral reefs host bountiful life FIGURE 13.14

Anthias fish swim through a multicolored coral reef off the coast of Fiji, an island in the Pacific. Reefs are a vital component of many ocean ecosystems because they provide shelter and nutrients for many kinds of fish.

CORAL REEFS

Many of the world's tropical coastlines are made up of limestone **reefs** built by vast colonies of organisms, principally corals, which secrete calcium carbonate (the main chemical constituent of limestone) as their skeletal material (FIGURE 13.14). Coral reefs are built up very slowly over thousands of years. Each of the tiny coral animals, called *polyps*, deposits a protective layer of calcium carbonate; over time the layers build up, forming a complex reef structure.

■ **reef** A hard structure on a shallow ocean floor, usually but not always built by coral.

Fringing reefs form coastlines that closely border the adjacent land. *Barrier reefs* are separated from the land by a lagoon, as in the case of the Great Barrier Reef off Queensland, Australia. Reefs are highly productive ecosystems that support a diversity of marine life-forms. They also perform an important role in recycling nutrients in shallow coastal environments. They provide physical barriers that dissipate the force of waves, protecting the ports, lagoons, and beaches that lie behind them, and are an important aesthetic and economic resource.

Corals require shallow, clear water in which the temperature remains above 18°C but does not exceed 30°C. Reefs therefore are formed only at or close to sea level and are characteristic of warm, low latitudes. They also favor places where deep sea upwelling provides abundant nutrients, salinity is normal—that is, the water is neither too salty nor too fresh—and there is no sediment deposited by large streams. Because of their very specific requirements, coral reefs are highly susceptible to damage from human activities as well as from natural causes such as tropical storms.

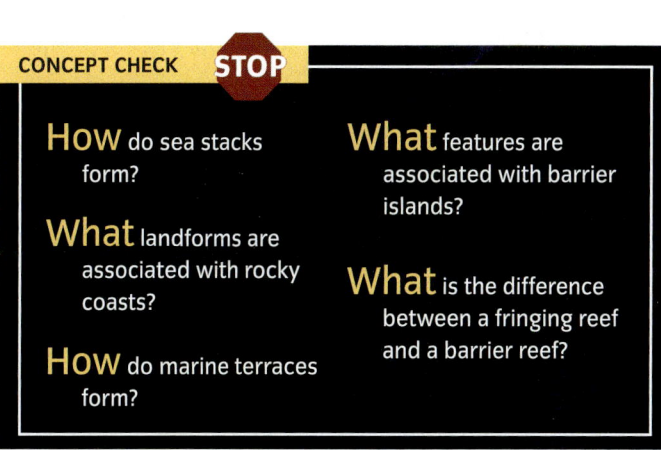

CONCEPT CHECK STOP

How do sea stacks form?

What landforms are associated with rocky coasts?

How do marine terraces form?

What features are associated with barrier islands?

What is the difference between a fringing reef and a barrier reef?

Shorelines and Coastal Landforms

Humans versus the Sea

LEARNING OBJECTIVES

Describe the major hazards that threaten inhabitants of coastal areas.

Explain some of the measures employed to protect shorelines from erosion.

Discuss the unwanted impacts of human interference on shorelines.

The majority of the world's population lives within 100 km of a coastline. Thirty of the 50 states in the United States have coastlines on the Atlantic or Pacific oceans, the Gulf of Mexico, or the Great Lakes. These 30 states contain about 85% of the nation's population, and about half of these people live in the coastal zones. However, the concentration of large numbers of people in coastal areas means that the coastal environment must absorb the impacts of a wide range of human activities. It also means that human vulnerability to natural hazards is particularly high in coastal zones. Clearly, large numbers of people and a great deal of property are at risk, and severe storms and other coastal hazards can cause major damage and loss of life, as discussed at the beginning of the chapter.

COASTAL HAZARDS

People who choose to live along a coast are often exposed to natural hazards that can prove devastating. The dynamic nature of the coastal environment can become especially obvious during storms. Exceptionally strong storms erode cliffs and beaches at a faster rate than usual. During a single storm in 1944, some cliffs on Cape Cod retreated landward by up to 5 m—more than 50 times the rate of retreat seen in a normal year.

Although these rapid bursts of erosion are rare, they play an important role in the natural evolution of a coast and can also have a significant impact on coastal inhabitants. Atlantic hurricanes that reach the eastern coast of the United States can be extremely devastating. The largest to strike the coast in the last decade have caused property damage running to tens of billions of dollars.

Tsunamis As we saw in the account in Chapter 8, a **tsunami** can be a huge threat to coastal populations. Tsunamis are caused by a strong earthquake or other brief, large-scale disturbance of the ocean floor, such as a landslide or volcanic eruption. The sudden movement of the seafloor near an earthquake source generates a train of water waves (see the *Case Study* in Chapter 8, *The Sumatra-Andaman Tsunami of 2004*). These waves travel over the ocean in ever-widening circles. They can travel at great speeds (as much as 950 km/h) and have long wavelengths (up to 200 km).

> **tsunami** A train of sea waves traveling over the ocean surface, triggered by an earthquake or other disturbance of the seafloor.

While tsunami waves are crossing the deep water of the open ocean, their height is relatively low, but as they move into shallow water they rapidly pile up to heights of up to 30 m, creating a towering wall of water that can hit the coast with great force. When you consider that each cubic meter of water weighs about 1000 kg (65 lb/ft^3), you can appreciate the enormous power of the water surge. Ocean waters rush landward and surge far inland, destroying coastal structures and killing inhabitants. After some minutes the waters retreat, causing further devastation. Several surging sea waves can follow one another.

In deep ocean waters, a tsunami's motion is normally too gentle to be noticed, making tsunamis hard to detect in open water. Without a system in place to monitor undersea earthquakes, the tsunami's arrival at the shore seems sudden and without warning, and hence can cause numerous fatalities and considerable damage. In the case of the Sumatra-Andaman tsunami, which originated in the Java Trench, west of Sumatra in the Indian Ocean (**FIGURE 13.15**), no warning network was in place to alert governments and citizens of

The Indian Ocean tsunami of 2004 FIGURE 13.15

A Before the tsunami, Banda Aceh, Indonesia, seen from space on June 23, 2004, is a city of small buildings, parks, and trees.

B Two days after the tsunami, the devastation caused by the earthquake and tsunami is complete. All that remains of most structures are their concrete floors. Trees are uprooted or stripped of leaves and branches.

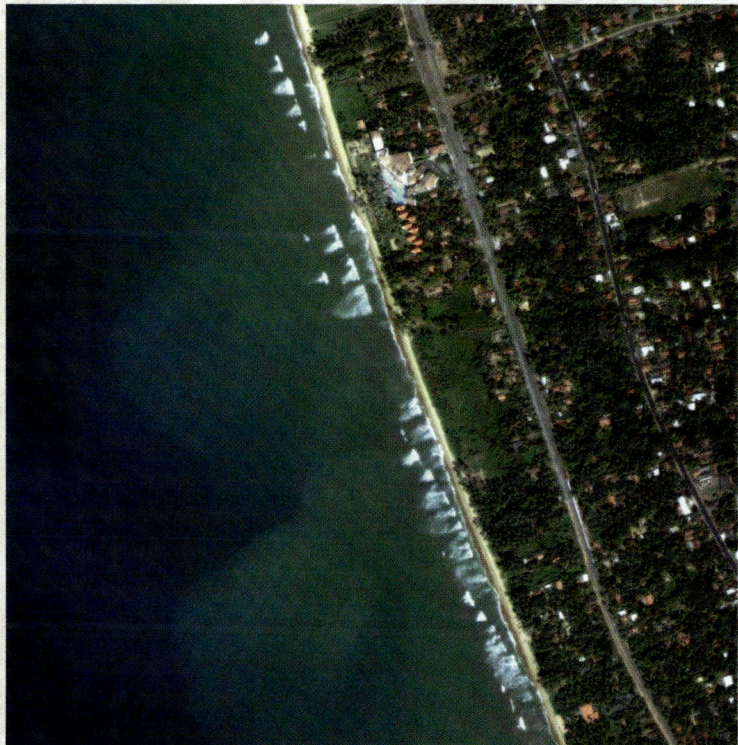

C This satellite image shows Kalutra Beach, Sri Lanka, as seen before the tsunami of 2004.

D An image of Kalutra Beach during the tsunami. Brown floodwaters still cover the land as the wave retreats, drawing streams of floodwater back to the ocean in surging currents.

Humans versus the Sea

Expensive follies FIGURE 13.16

A Houses dangle on the edge of a cliff in Pacifica, California, after winter storms caused a section of the cliff to collapse. Note the rubble on the beach and the scar on the cliff where the debris slide occurred.

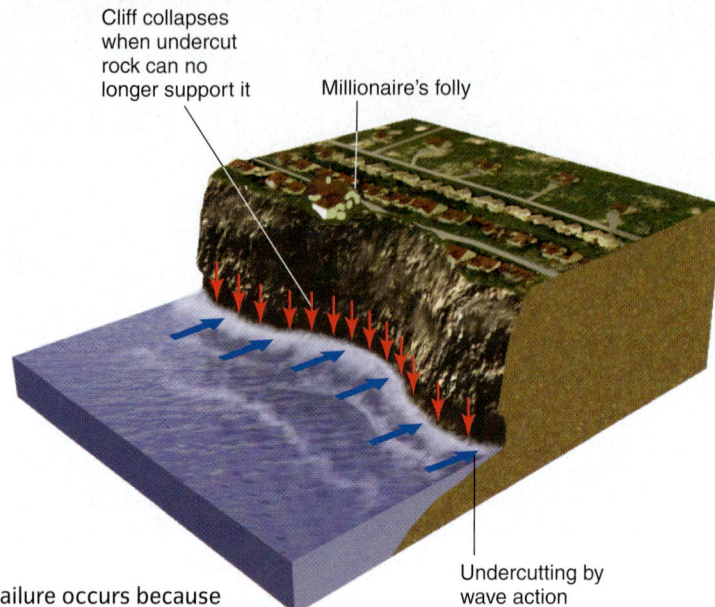

B Typically this kind of slope failure occurs because waves have undercut the base of the cliff. Unfortunately, this scenario will continue to repeat itself as long as people choose home sites based on the desire for "million-dollar views" rather than on sound evaluation of the erosional processes.

the impending disaster. After the catastrophe, plans were made to start building such a network. With luck, the next great earthquake in the Java Trench will find the world better prepared for its aftermath—another giant, deadly tsunami.

In July 2006, the Java Trench rumbled again. The resulting 3 m (10 ft) tsunami was fortunately localized to a 180-km (110-mi) stretch of the south coast of Java. The earthquake was detected by the Pacific Tsunami Warning Center in Hawaii, which issued a tsunami warning bulletin 24 minutes before the wave came ashore. But no local warning system was in place. About 1000 people were killed or reported missing in the aftermath of the tsunami and earthquake.

Landslides Rocky-cliffed coasts can look deceptively stable to people seeking a dramatic site for their home. They see a rocky cliff as a sign of permanence and stability, whereas in fact quite the reverse is the case. Shorelines with cliffs are susceptible to frequent landslides and rock falls as erosion eats away at the base of the cliff. Roads, buildings, and other structures built too close to such cliffs can be damaged or destroyed when sliding occurs (**FIGURE 13.16**).

Landslides on cliffed shorelines sometimes give rise to giant waves that are even more destructive than the slides themselves. When Earth scientists unexpectedly found coral-bearing gravel at altitudes of up to 326 m on the Hawaiian island of Lanai, they suspected that a giant wave had deposited the coral fragments high above sea level. This wave is believed to have been generated by a large submarine landslide off the western coast of the island. Based on dating of the corals, the landslide occurred about 105,000 years ago. If such an event happened today in the islands' densely populated coastal zones, it would wreak great havoc and loss of life.

Very large waves have also been produced by massive coastal landslides at Lituya Bay, which lies along the Fairweather Fault on the southern coast of Alaska (**FIGURE 13.17**).

The Lituya Bay tsunami
FIGURE 13.17

In 1958, a magnitude 8.3 earthquake caused a coastal cliff to collapse into an arm of Lituya Bay, on the southern coast of Alaska. The massive landslide created a wave that rose more than 500 m against the opposite wall of the fjord. The wave swept away vegetation, producing a sharp trimline in the forest, and moved rapidly down the bay toward the sea. A boat anchored in the bay was caught up by the onrushing wave, carried over the spit at the bay mouth, and transported out into the open ocean.

PROTECTION AGAINST SHORELINE EROSION

Despite the dangers, much oceanfront land is considered prime real estate. People who build houses and towns along seacoasts are not eager to lose their investment because of the actions of an unruly ocean, so they take a number of steps to protect their land and property. Unfortunately, there is always a consequence to any action taken.

It's not easy to protect a strip of shoreline that consists of easily eroded rock or sediment. One strategy is to cover the cliff with tightly packed boulders so large that they can withstand the onslaught of storm waves. A cliff can also be defended by a strong seawall built parallel to the shore on foundations deep enough to prevent undermining by surf during storms (FIGURE 13.18). Both structures offer cliffs some protection against ordinary storms, but they are expensive and may not be effective against the largest storms or hurricanes.

A seawall protects a California cliff
FIGURE 13.18

Disturbing the shoreline FIGURE 13.19

Human actions can shift the balance between erosion and deposition. In Ocean City, Maryland (right), the construction of a jetty has caused sand to accumulate and make the beach wider. Just south of Ocean City, Assateague Island in Virginia (left) has been deprived of the sand that the longshore current would have deposited on its shore. As a result, the beach on Assateague is retreating inland.

Beaches also require protection from erosion. Because of their great popularity, it's easier to justify the great expense of maintaining beaches in densely populated regions than the expense of protecting sparsely populated headlands. But beaches present a special sort of problem. As we have seen, beaches represent a delicate balance between the forces of erosion and deposition. Because of beach drift, what happens on one part of a beach affects other parts in the downdrift direction. For example, if we build a seawall, dock, or any other structure at the updrift end of a beach, we reduce the amount of sand available for beach drift (**FIGURE 13.19**). The amount of sand that the water can carry is related to its speed, and so if the longshore current becomes underloaded it will become more erosive and will erode the beach. Small beaches have been completely destroyed by this process in only a few years.

We use breakwaters and groins to protect beaches and boat anchorages from incoming waves. *Breakwaters* are offshore barriers, but they too can upset the natural balance of the adjacent beach, leading to changes in the shoreline (**FIGURE 13.20A**). Building *groins* at short intervals along the beach can keep erosion in check to some extent. A groin is a low wall built out into the water at a right angle to the shoreline (**FIGURE 13.20B**). A groin acts as a check on the rate of beach drift because it traps sand carried to it along the shore. But erosion tends to occur on the downdrift side of a

Shoreline protection   FIGURE 13.20

A Breakwaters Breakwaters constructed along the shore at Tel Aviv in Israel protect the beach from the onslaught of waves, but they have turned a straight coastline into a scalloped one. Wave action around the barriers has added sediment to the beach behind each breakwater, producing a scalloped coastline.

B Groins Severe storms during the winter of 1993 carved out an inlet in this barrier beach on the south shore of Long Island, New York. A system of groins has trapped sand, protecting the far stretch of beach. The beach in the foreground, without groins, has receded well inland of the houses that were once located on its edge.

groin, where the beach sand is not being replenished. The net effect, once again, is to protect one part of a beach at the expense of another part.

Another way to protect an eroding beach is to bring in sand and pile it on the beach at the updrift end. Surf then erodes the pile and drifts the new sand down the length of the beach. As you can imagine, constantly feeding a beach with sand in this way can be expensive.

EFFECTS OF HUMAN INTERFERENCE

Many beaches around the world are deteriorating because of human interference. In Southern California, for example, most of the sand on beaches is supplied not by erosion of wave-cut cliffs but by alluvium carried to the sea at times of flood. But inhabitants have built dams across the streams to protect buildings and other struc-

CASE STUDY

The Black Sea Coast

A dramatic example of human interference can be seen along the Russian coast of the Black Sea (**A**). Ninety percent of the sand and pebbles that form the natural beaches there used to be supplied by rivers as they entered the sea. During the 1940s and 1950s, three things happened: Large resort developments were built at the beaches (**B**); large breakwaters were constructed so that two major harbors could be extended into the sea; and dams were built across some rivers inland from the coast (**C**).

All this construction upset the balance among the supply of sediment to the coast, longshore currents and beach drift, and the deposition of sediment on beaches. By 1960, the combined area of all beaches along the coast had decreased by half. Then beachfront buildings began to sag or collapse as the surf ate away at their foundations. Ironically, many of the resort buildings had been constructed from a concrete aggregate made from large volumes of sand and gravel that had been removed from the beaches. The same concrete had also been used to build the dams that cut off the supply of sediment to the coast.

tures in stream valleys that are vulnerable to floods. Of course, these dams also trap the sand and gravel carried by the streams, preventing the sediment from reaching the sea. Halting the through-flow of sand has upset the natural balance in the longshore transport of sediment, leading some beaches to become significantly eroded. The *Case Study* describes the unwanted effects of efforts to protect the Russian coast of the Black Sea.

CONCEPT CHECK STOP

What causes tsunamis?

Why are tsunamis so dangerous?

What other hazards threaten coastal populations and property?

How do inhabitants of coastal areas attempt to protect shorelines?

What unwanted side effects can these measures have?

Humans versus the Sea 417

Amazing Places: The Florida Keys Reef

The world's third-longest coral barrier reef, and the only living coral reef in the continental United States, runs parallel to the Florida Keys. It stretches for 320 km and in most places is just a short boat ride offshore. Here, you can see Earth science in action. Unlike other animals, whose skeletons become stone only after thousands of years of burial, the skeletons of corals are limestone from day one. Over time, they gradually add to the landmass of the continent.

In the Florida Keys Reef you can find a variety of coral, such as fire coral, mustard hill coral (**A**, *foreground*), and brain coral (**A**, *background*). But the coral's health is declining. In **B** you can see how part of Carysfort Reef progressed from a healthy state in 1975 (*left*) to a sick one in 1985 (*middle*); by 1995, it was almost completely dead (*right*). By 2000, Carysfort Reef had lost 90% of its original coral cover.

What caused this decline? Coral polyps can be killed by natural causes, such as hurricanes, but the reef usually recovers from such episodes over the long term. It's much harder for the coral to recover from injuries due to human activities, such as scarring caused by boats that run aground on the reef. In recent years, a pervasive problem has been coral bleaching (**C**). When the water is too warm, coral polyps expel the symbiotic algae that ordinarily give them their bright colors. The bleached coral is not necessarily dead, but without its algae, it will die soon. The white parts of the staghorn coral in this photo are bleached, whereas the brown parts are still healthy (but probably in danger).

Long-term global warming could devastate coral life. Earth scientists are also worried about the impact of rising carbon dioxide levels in the atmosphere. As more CO_2 dissolves in seawater, it becomes harder for corals to build their skeleton structure.

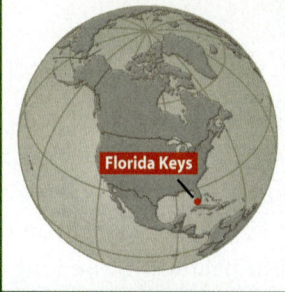

Global Locator

SUMMARY

1 Changes in Sea Level

1. The ocean's level varies over a wide range of time scales and for various reasons. Worldwide changes in sea level due to the melting or growth of polar ice sheets take place over hundreds to thousands of years. Local changes due to **tides** occur twice a day along most coasts.

2. Global warming is predicted to increase extreme weather conditions and cause a rise in sea level. As a result, more land will subside and beaches, salt marshes, and estuaries will be pushed landward.

3. **Tides** result from the gravitational attraction of the Moon and Sun. They are produced as Earth's rotating surface passes through tidal bulges on opposite sides of the planet.

2 Waves

1. When a wave moves onto the shore, its motion is distorted as the shallow bottom interferes with the circular motion of water in the wave. It eventually breaks, creating turbulent **surf**.

2. **Beaches**, usually formed of sand, are shaped by waves. Wave action produces **littoral drift**, accomplished by **longshore currents** (in the *surf zone*) and **beach drift** (on land). These processes move sediment parallel to the beach.

3. Tidal currents in bays and estuaries redistribute fine sediments. The *flood current* flows landward, while the *ebb current* sets in when the tide begins to fall.

4. Tidal sediments produce *mud flats* and *salt marshes*.

3 Shorelines and Coastal Landforms

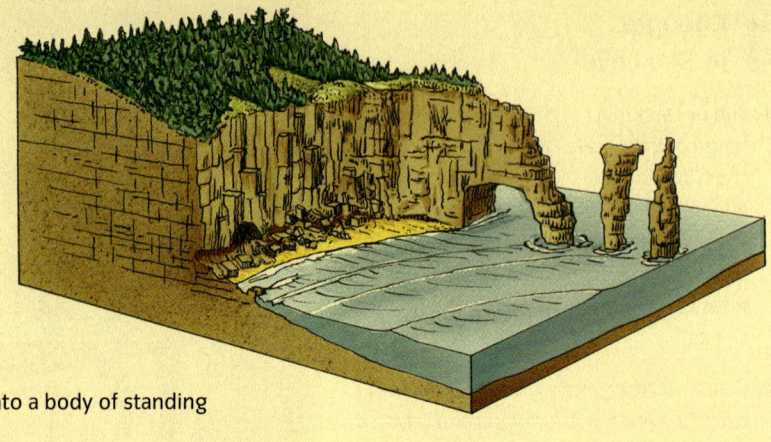

1. Shorelines and coastal landforms are shaped by a combination of erosive and depositional processes.

2. Shorelines are highly variable, but three basic types are most common: rocky coasts and **wave-cut cliffs**, lowland **beaches** and **barrier islands**, and **coral reefs**. Along some coasts, rapid uplift has occurred, creating raised shorelines and marine terraces. Sand dunes are typically the highest points on a barrier island.

3. **Delta** coasts are produced when a stream or river flows into a body of standing water, rapidly slowing and depositing sediment.

4. Storms and artificial structures often accelerate the process of coastal erosion.

4 Humans versus the Sea

1. Rare but powerful storms, **tsunamis**, and large landslides are significant threats to people and structures in coastal areas.

2. Cliffs can be protected to some extent by seawalls or boulders. Beach erosion is a serious problem along many inhabited coasts, but beaches can be temporarily protected by breakwaters, groins, or artificial nourishment.

3. Human interference can have a detrimental effect on beaches in the long term. Dams, breakwaters, and other beach defenses disturb the natural flow of sediment, disrupting natural beach replenishment.

KEY TERMS

- **tide** p. 398
- **surf** p. 401
- **longshore current** p. 402
- **beach drift** p. 402
- **littoral drift** p. 402
- **wave-cut cliff** p. 405
- **beach** p. 406
- **barrier island** p. 406
- **delta** p. 408
- **reef** p. 409
- **tsunami** p. 410

CRITICAL AND CREATIVE THINKING QUESTIONS

1. When the height between high and low tide is large, as in the Bay of Fundy, engineers sometimes consider using the tide as a way to generate electricity. What might some of the negative consequences be from the installation of tidal power plants?

2. If you live or go to school near a coastline, identify the kind of coastline that is near. How will the rise of a meter in sea level affect your area?

3. Scientists who study climate predict that as the climate gets warmer, severe storms and particularly hurricanes, will become stronger and more frequent. Which regions of North America are most vulnerable and likely to be affected by the changes?

4. When building a house, it is important to assess potential hazards. What advice would you offer to someone planning to build a shoreline home in the following places: (a) southern Georgia; (b) Oregon, south of the Columbia River; (c) the Yucatan Peninsula, Mexico; (d) the south side of Nantucket Island, Massachusetts.

5. Tsunamis have been recorded in the Atlantic Ocean, but they are much less common than in the Pacific and Indian Oceans. Why is this so?

What is happening in this picture?

This photo shows the aftermath of Hurricane Dennis, in 1999, on the barrier beach of North Carolina's Outer Banks. A battered house, suspended on stilts, sits in the surf on a sandy Atlantic beach. Its windows and doors are sealed with plywood, and waves break underneath it. A pile of wooden debris is heaped up nearby.

- What might have happened here?

SELF-TEST

1. Which of the following statements about global sea level is true?
   a. Global sea level is constant over both human and extended time periods.
   b. Global sea level can vary by tens to hundreds of meters over extended periods.
   c. Global sea level varies by only a few meters over extended periods.
   d. Large variations in global sea level can be an effect of climate change.
   e. Both (b) and (d) are correct.

2. This illustration depicts the forces involved in the generation of tides on Earth. Label the illustration with the following terms:
   a. distorted water level
   b. common center of mass of Earth–Moon system
   c. inertial force
   d. uniform level of water (theoretical)
   e. gravitational force
   f. tide-raising force

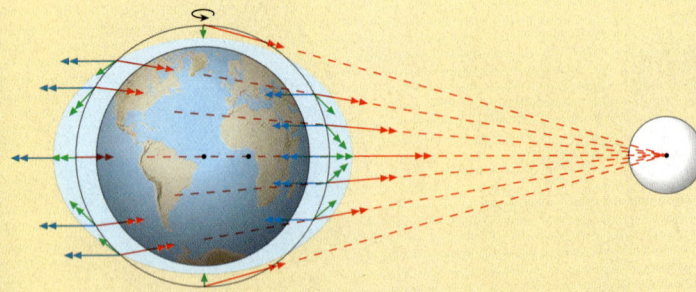

3. The most important agent shaping coastal landforms is the action of _____.
   a. storms   c. salinization
   b. streams   d. waves

4. A _____ is a localized narrow channel of water that can pull unwary swimmers out to sea.
   a. rip current   c. breaking current
   b. surf zone   d. tidal run-up

5. When a wave reaches the shore it breaks, expending most of its energy. Label the following features or regions on the diagram: (a) breaking wave, (b) surf, (c) turbulent water region, (d) the region where the wave first "feels bottom."

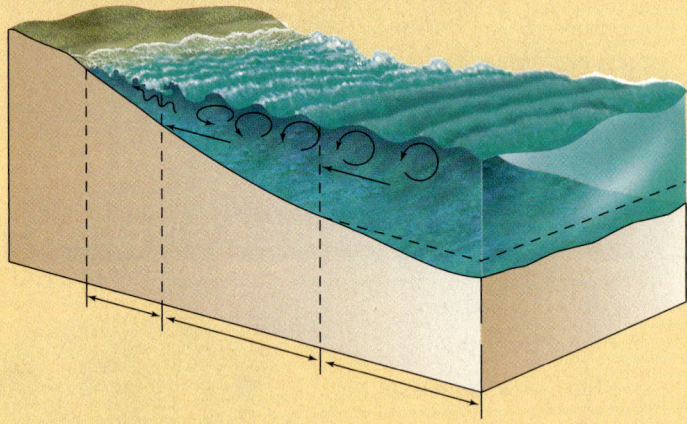

6. Littoral drift, the process through which sediment is transported along a beach, includes _____ and _____.
   a. beach drift/ebb tide
   b. ebb tide/longshore drift
   c. flood tide/ebb tide
   d. longshore current/beach drift

7. Tidal currents are made up of two opposing currents called _____ currents.
   a. longshore and littoral   c. longshore and flood
   b. ebb and flood   d. ebb and littoral

8. The most common type of coast is the _____.
   a. barrier island coast   c. lowland beach
   b. coral-reef coast   d. rocky coast

9. This diagram shows some of the landforms created by wave action against a marine cliff. Label the following features on this diagram: (a) arch, (b) cave, (c) notch, (d) stack.

10. Broad expanses of isolated shallow water called _____ are common features immediately adjacent to barrier-island coastlines.
    a. salt marshes   c. lagoons
    b. marine terraces   d. tidal inlets

11. A _____ is a ridge of sand or gravel that joins a barrier island to the mainland.
    a. spit   c. bay barrier
    b. tombolo   d. lagoon

12. Coral-reef coasts are built up of _____ secreted by organisms as their skeletal material.
    a. calcium carbide   c. calcium chloride
    b. calcium carbonate   d. calcium chlorate

13. Which of the following events could not trigger a tsunami?
    a. hurricane   c. landslide
    b. earthquake   d. volcanic eruption

14. _____ is often used to protect beaches, as shown in this photo of a barrier beach on the south shore of Long Island, New York.
    a. Artificial nourishment   c. A system of groins
    b. A system of breakwaters   d. A seawall

15. Which of the following factors did not contribute to the deterioration of the Russian coast of the Black Sea during the 1940s and 1950s?
    a. construction of beach resorts
    b. construction of breakwaters
    c. construction of dams
    d. construction of groins

Self-Test   423

The Atmosphere: Composition, Structure, and Clouds 14

Around 65 million years ago, the last of the dinosaurs died out. Today, we hypothesize that the extinction was probably triggered by a great meteorite impact, but it's less well known that dinosaur numbers were declining long before the meteorite struck. Why was this? One hypothesis is that the dinosaurs were suffocated by subtle changes in our planet's atmosphere (here viewed from above by astronauts).

Like us, the dinosaurs required oxygen to breathe. But the oxygen content of the atmosphere has varied through Earth's long history. By exploring several lines of evidence, especially by analyzing samples of air trapped in amber (a fossil tree resin), scientists discovered that 100 million years ago, when the dinosaurs were in their prime, the oxygen content of the atmosphere was 40% higher than it is today. Thanks to an oxygen-rich atmosphere, the dinosaurs needed only relatively small lungs to breathe.

However, near the end of the dinosaurs' reign, the oxygen level was declining. It is hypothesized that dinosaurs, with large bodies and small lungs, were unable to adjust to the decline.

Humans are less vulnerable to such changes—we can survive with oxygen levels ranging from 40% above to 44% below those in today's atmosphere. If we could turn the clock back, we could breathe the same air as the dinosaurs. However, we shouldn't be too complacent. The balance of gases in our atmosphere also protects us from harmful solar rays and plays a crucial role in warming our planet and transporting water around the globe. We must be careful not to upset this balance.

CHAPTER OUTLINE

- **Earth's Atmosphere** p. 426

- **Moisture in the Atmosphere** p. 431

- **The Global Energy System** p. 434

- **Formation of Clouds** p. 438

- **Precipitation** p. 442

Earth's Atmosphere

LEARNING OBJECTIVES

Describe the main chemical constituents of Earth's atmosphere.

Identify the four layers of the atmosphere.

Discuss how temperature varies through these layers.

Explain the vital role of ozone to life on Earth and the impact of human activities on the ozone layer.

Earth is surrounded by *air*—a mixture of various gases that extends above the surface to a height of many kilometers. This envelope of air makes up Earth's *atmosphere* (**FIGURE 14.1**). It is held in place by Earth's gravity. Almost all the atmosphere (97%) lies within 30 km (19 mi) of Earth's surface. The upper limit of the atmosphere is approximately 10,000 km (about 6000 mi) above the surface—a distance nearly as great as Earth's diameter.

Air is an invisible, normally odorless mixture of gases and suspended particles. The concentrations of two components of air—aerosols and water vapor—can vary greatly. *Aerosols* are liquid droplets or solid particles that are so small that they remain suspended in the air. They are swept into the air from dry desert plains, lakebeds, and beaches, or released by exploding volcanoes, among other sources. We will see later in the chapter that aerosols are important for cloud formation.

Water vapor is always present in the air, and its concentration can vary widely. For this reason, the relative amounts of the remaining gases are generally reported as if the air were entirely lacking in water vapor and aerosols. When these two components are ignored, three gases—nitrogen (78%), oxygen (21%), and argon (0.93%)—make up 99.96% of dry air by volume (**FIGURE 14.1B**). The remaining gases (carbon dioxide, 0.035%; neon, 0.0018%; and six others) are present in very small quantities. However, some of these minor gases are profoundly important for life on Earth. As we will see, they absorb certain wavelengths of sunlight. They act both as a warming blanket and as a shield against deadly ultraviolet radiation.

Air and sky FIGURE 14.1

A Air is the gaseous envelope that surrounds our planet, making up the atmosphere.

B Air contains two substances whose concentration varies from place to place and day to day: water vapor and aerosols. Water vapor can condense into clouds and liquid water droplets. The rest of the atmosphere consists primarily of nitrogen and oxygen, with a smattering of other gases.

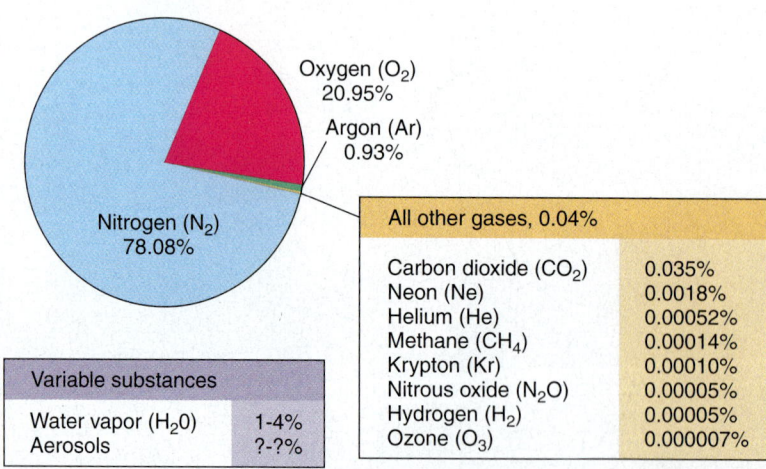

426 **CHAPTER 14** The Atmosphere: Composition, Structure, and Clouds

Oxygen and carbon dioxide levels are maintained by complex chemical cycles; both gases are constantly being either removed from the atmosphere or replenished by plants, animals, and chemical processes such as weathering. If anything affects the rates of replenishment or removal, the makeup of the atmosphere will change.

THE THERMAL STRUCTURE OF THE ATMOSPHERE

Most incoming solar radiation passes through the atmosphere and is absorbed by Earth's surface. The atmosphere is then warmed by heat from the surface (see *What an Earth Scientist Sees*). So it makes sense that, in

What an Earth Scientist Sees

The Bora Bora Sunset

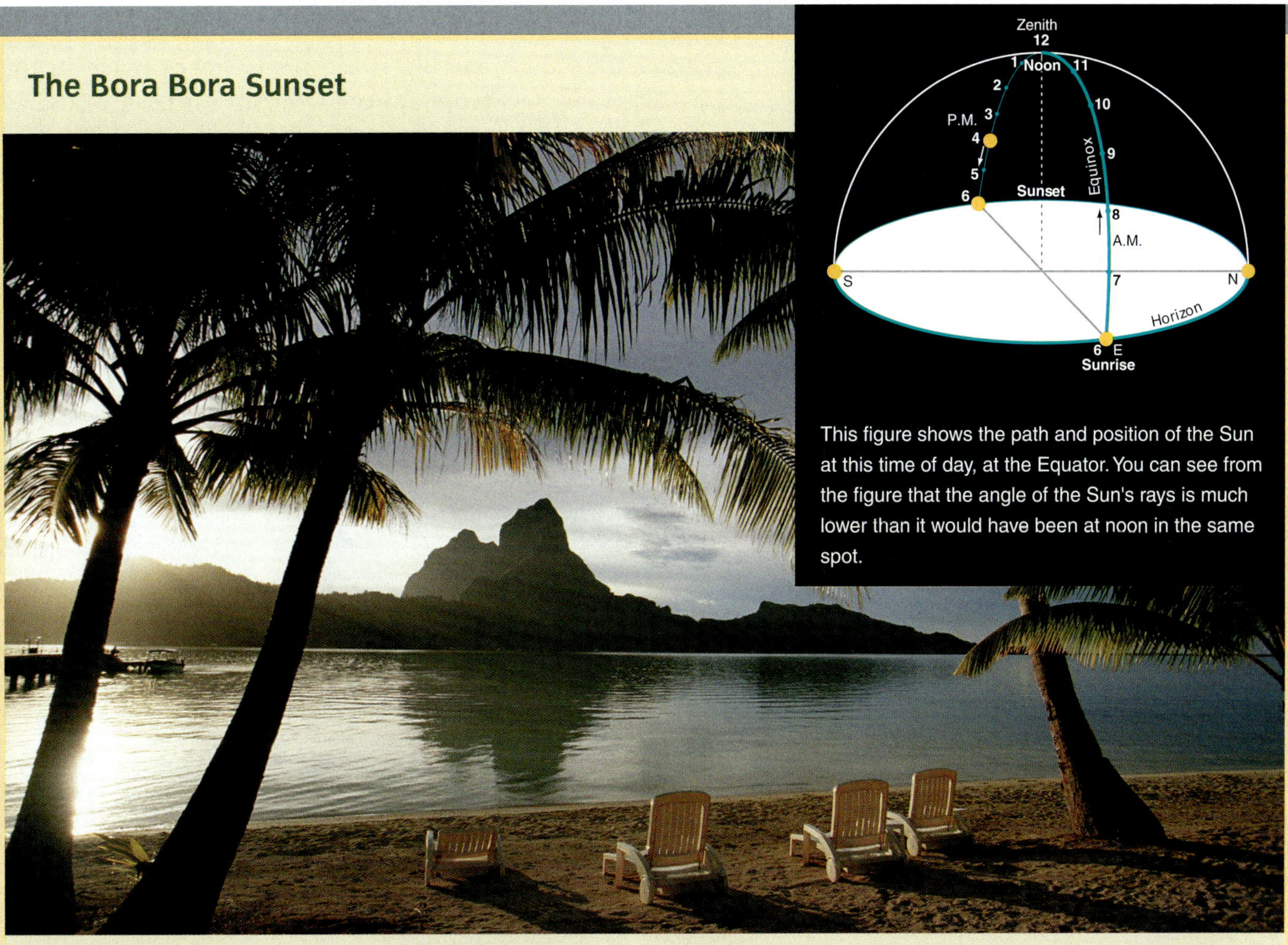

This figure shows the path and position of the Sun at this time of day, at the Equator. You can see from the figure that the angle of the Sun's rays is much lower than it would have been at noon in the same spot.

Imagine yourself sitting in a beach chair on Bora Bora, watching the sunset. Will you be warm or cool?

The beach at sunset will be quite different than it was at noon, a few hours earlier. The long shadows and the position of the glow in the sky show that the Sun is not far from the horizon. As a result, incoming solar radiation will be less than at noon, when the Sun is near the top of the sky.

The Sun will also seem weaker because at such a low angle its direct rays pass through a larger portion of the atmosphere. As a result, more of the solar beam will be absorbed or scattered than would be the case at noon.

Seated in your chair, directly facing the Sun, however, you will be doubly warmed—first by the direct rays of the Sun on your body, and second by the sunlight reflected off the still water.

With this analysis, an Earth scientist could safely conclude that the temperature will be . . . just about perfect!

Here's an interesting question:
- If you sat on the beach in Bora Bora at sunrise, would you be warmer, cooler, or the same as at sunset?

Earth's Atmosphere 427

general, air that is farther from Earth's surface is cooler (**FIGURE 14.2**). We call the decrease in air temperature with increasing altitude the **lapse rate**.

We measure the atmospheric temperature drop in degrees C per 1000 m (or degrees F per 1000 ft). The temperature decreases at an average rate of 6.4°C/1000 m (3.5°F/1000 ft). This average value is known as the *environmental temperature lapse rate*. Looking at the graph in Figure 14.2, we see that when the air near the surface is just below a pleasant 20°C (68°F), the air at an altitude of 12 km (40,000 ft) will be a bone-chilling −55°C (−67°F). Keep in mind that the environmental temperature lapse rate is an average value and that on any given day the observed lapse rate might be quite different.

Scientists have discovered that the atmosphere is composed of four layers with distinct temperature profiles, each separated from the others by boundaries called *pauses*, which are indicated in Figure 14.2.

The troposphere
All human activity, barring space travel, takes place in the **troposphere**—the lowest layer of the atmosphere. The troposphere stretches out to about 12 km (7 mi) above the surface. Everyday weather phenomena, such as clouds and storms, happen mainly in the troposphere. Here temperature decreases with increasing elevation. The tro-

> **lapse rate** The rate at which air temperature decreases with increasing altitude.
>
> **troposphere** The lowest layer of the atmosphere, in which temperature falls steadily with increasing altitude.

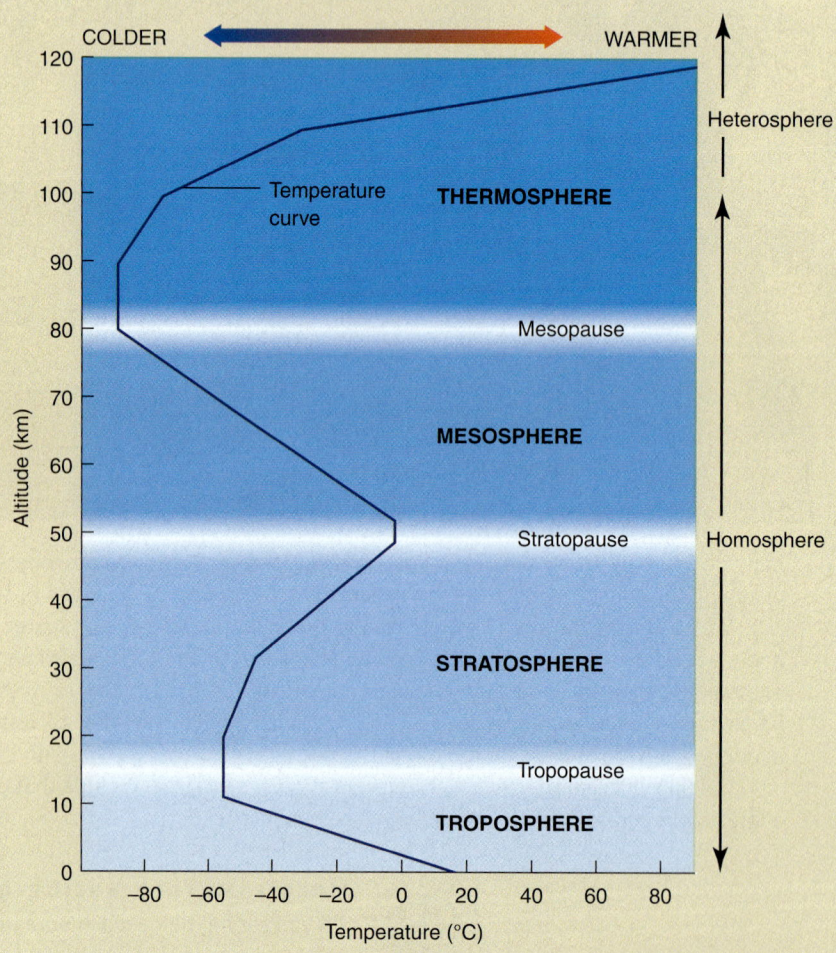

Atmospheric layers FIGURE 14.2

Temperature varies with altitude in the atmosphere. In the lowest level, called the *troposphere*, the temperature drops rapidly with increasing altitude. In the next layer, the *stratosphere*, the reverse is true. Two more reversals occur in the *mesosphere* and *thermosphere*. The air in the thermosphere is so tenuous that this zone is defined by rocket scientists (but not by Earth scientists) as part of "space." NASA awards a space-flight badge to anybody who flies above an altitude of 80 km.

428 CHAPTER 14 The Atmosphere: Composition, Structure, and Clouds

posphere is thickest in the equatorial and tropical regions, where it stretches from sea level to about 16 km (10 mi). It thins toward the poles, where it is only about 6 km (4 mi) thick.

The troposphere contains significant amounts of water vapor. When the water vapor content of the atmosphere is high, vapor can condense into water droplets, forming low clouds and fog, or the vapor can be deposited as ice crystals, forming high clouds. Rain, snow, hail, or sleet—collectively termed *precipitation*—are produced when condensation or deposition happens rapidly. Places where water vapor content is high throughout the year have moist climates. In desert regions, water vapor content is low, so there is little precipitation.

When water vapor absorbs and reradiates heat emitted by Earth's surface, it helps create the *greenhouse effect*—a natural phenomenon that we will discuss in more detail later in the chapter. The greenhouse effect is responsible for warming Earth to temperatures that allow life to exist.

Air in the troposphere contains countless tiny particles, or aerosols. These are important because water vapor can condense on them to form tiny droplets. When these droplets grow large and occur in high concentration, they are visible as clouds or fog. Aerosol particles scatter sunlight, brightening the whole sky while slightly reducing the intensity of the solar beam.

Between 12 and 15 km (7 and 9 mi) above Earth's surface, the temperature stops decreasing and the troposphere gives way to the stratosphere; the transition between the troposphere and the stratosphere is called the *tropopause*. The tropopause varies in height with latitude, and to some extent with season, so the troposphere is not uniformly thick at any given location.

stratosphere The layer of atmosphere above the troposphere; here temperature increases slowly with altitude.

The stratosphere and the upper layers The **stratosphere** lies above the tropopause and extends up to roughly 50 km (about 30 mi) above Earth's surface. It is the home of strong, persistent winds that blow from west to east. Air doesn't really mix between the troposphere and stratosphere, so the stratosphere normally holds very little water vapor or dust. Air in the stratosphere becomes slightly warmer as altitude increases.

The stratosphere contains the ozone layer, which shields earthly life from intense, harmful ultraviolet energy. Ozone molecules warm the stratosphere as they absorb solar energy, causing temperature to increase with altitude.

Temperatures stop increasing at the stratopause. Above the stratopause, we find the *mesosphere*, shown in Figure 14.2. Here, temperature *decreases* with increased elevation. This layer ends at the mesopause, the level at which temperature stops decreasing. The layer above it is the *thermosphere*. Here, temperature again increases with altitude, but because the density of air in this layer is very low, the air holds little heat.

The gas composition of the atmosphere is uniform for about the first 100 km of altitude, which includes the troposphere, the stratosphere, the mesosphere, and the lower portion of the thermosphere. We call this region the *homosphere*. Above 100 km, gas molecules tend to be sorted into layers by molecular weight and electric charge. This region is called the *heterosphere* (Figure 14.2).

Ozone in the upper atmosphere

Another small but important constituent of the atmosphere is **ozone**—a form of oxygen in which three oxygen atoms are bonded together (O_3). Ozone is found mostly in the upper part of the atmosphere, in the *stratosphere*, about 15 to 55 km (9 to 31 mi) above the surface.

ozone A form of oxygen with a molecule consisting of three atoms of oxygen (O_3).

Ozone is most concentrated in a layer that begins at an altitude of about 15 km (about 9 mi) and extends to about 55 km (about 34 mi) above the surface. It is produced in gaseous chemical reactions that require energy in the form of ultraviolet radiation. The reactions are quite complicated, but the net effect is that ozone (O_3), molecular oxygen (O_2), and atomic oxygen (O) are constantly formed, destroyed, and reformed in the ozone layer, absorbing ultraviolet radiation with each transformation.

A shield against radiation FIGURE 14.3

The ultraviolet radiation coming from the Sun can be harmful or lethal; generally speaking, the shorter the wavelength, the more harmful the radiation. Fortunately, the atmosphere protects us from almost all of these rays because they are absorbed by three kinds of oxygen—O, O_2, and O_3 (ozone).

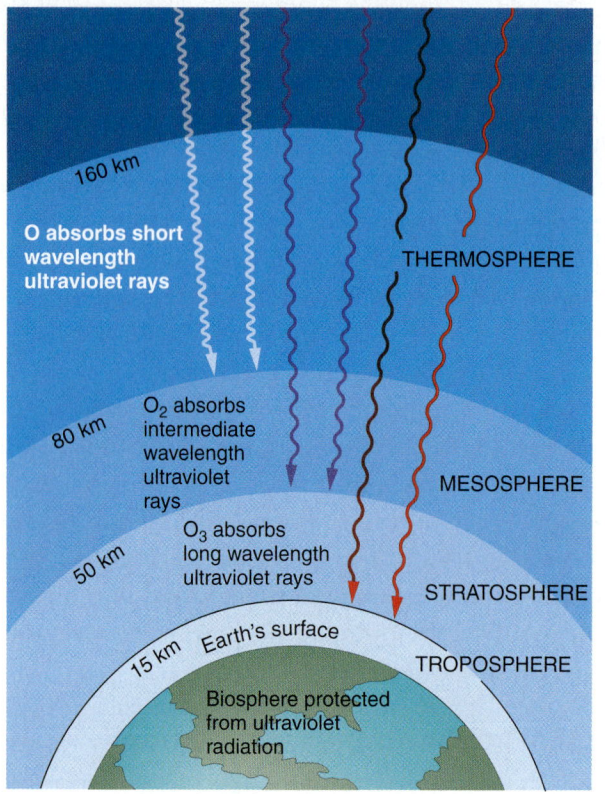

Because the ozone layer absorbs ultraviolet light from the Sun, it protects Earth's surface from this damaging form of radiation (FIGURE 14.3). The presence of the ozone layer thus is essential for the survival of life. If the full intensity of solar ultraviolet radiation ever hit Earth's surface, it would destroy bacteria and severely damage animal tissue.

Certain gases, such as chlorofluorocarbons (CFCs), substantially reduce ozone concentrations. CFCs are synthetic industrial chemical compounds containing chlorine, fluorine, and carbon atoms. Although they were banned in aerosol sprays in the United States in 1976, they are still used as cooling fluids in some refrigeration systems. When appliances containing CFCs leak or are discarded, the CFCs are released into the air.

CFC molecules move up to the ozone layer, where they decompose to form chlorine oxide (ClO). Chlorine oxide attacks ozone, converting it into ordinary oxygen. With less ozone, there is less absorption of ultraviolet radiation.

In the mid-1980s, scientists discovered a "hole" in the ozone layer over Antarctica (FIGURE 14.4). In recent years, the ozone layer has thinned during early spring in the Southern Hemisphere, reaching its minimum thickness during September or October. Typically, the ozone hole slowly shrinks and ultimately disappears in early December.

The ozone hole FIGURE 14.4

In 1985, scientists discovered that a previously unnoticed gap (blue) in the ozone layer was forming over Antarctica every winter. (A smaller hole also formed over the Arctic Ocean.) For several years, the hole grew larger and the depletion at its center grew more severe (purple). The Antarctic ozone hole of 2006 was the largest on record, covering about 29.5 million sq km (about 11.4 million sq mi). Ozone concentration is measured in Dobson units, and October 8, 2006, saw its lowest value—85 units. However, in recent years it has apparently begun to stabilize.

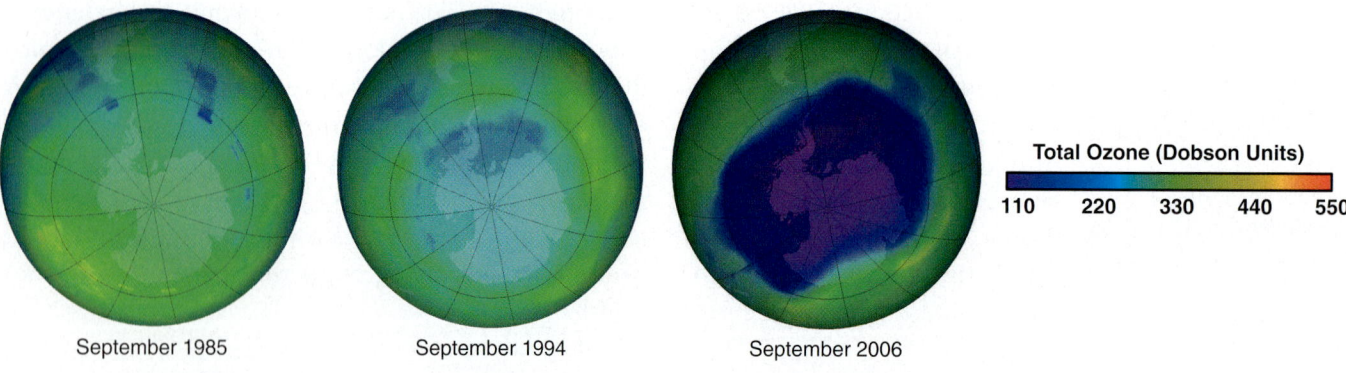

CHAPTER 14 The Atmosphere: Composition, Structure, and Clouds

Since 1978, surface-level ultraviolet radiation has been increasing. Over most of North America, the increase has been about 4% per decade. Crop yields and some forms of aquatic life may suffer as a result, as may humans. Today we are all aware of the dangers of harmful ultraviolet rays to our skin and the importance of using sunscreen when going outdoors.

In 1987, in response to the global threat of ozone depletion, 23 nations signed a treaty to cut global CFC usage by 50% by 1999. The treaty was effective. By late 1999, scientists confirmed that stratospheric chlorine concentrations had peaked in 1997 and had since begun to decline. Although the ozone layer is not expected to be completely restored until the middle of the twenty-first century, this is a welcome observation.

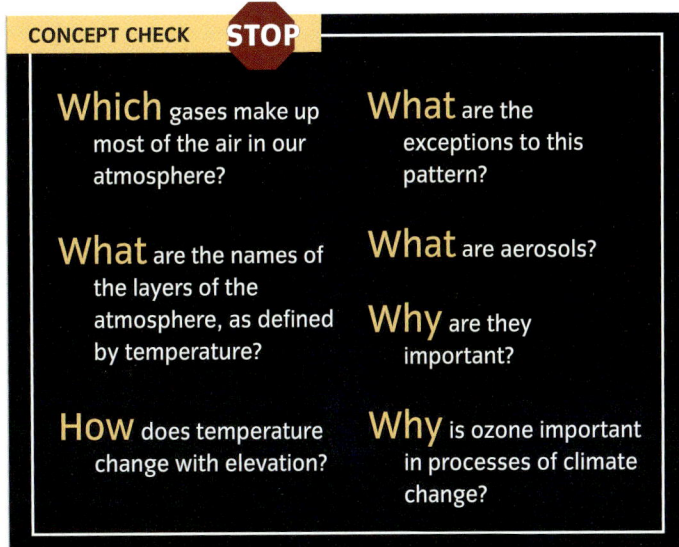

CONCEPT CHECK STOP

Which gases make up most of the air in our atmosphere?

What are the names of the layers of the atmosphere, as defined by temperature?

How does temperature change with elevation?

What are the exceptions to this pattern?

What are aerosols?

Why are they important?

Why is ozone important in processes of climate change?

Moisture in the Atmosphere

LEARNING OBJECTIVES

Discuss the three states of H_2O.

Define humidity.

Describe the difference between specific humidity and relative humidity.

Explain the importance of the dew-point temperature.

The chemical compound H_2O is so familiar that we sometimes forget what an unusual substance it is. In truth, H_2O is the most remarkable substance around—the Earth system works the way it does because H_2O has the properties it does.

All compounds can form solids, liquids, and gases under different temperature and pressure conditions, but H_2O is the only naturally occurring compound that can exist as a solid (ice), a liquid (water), and a gas (water vapor) at Earth's surface. Let's explore what happens when H_2O changes from one state to another.

CHANGES OF STATE

Whenever matter changes from one state to another, energy is either absorbed or released (**FIGURE 14.5**). In going from a more ordered state (a solid) to a less ordered one (a liquid) or to a fully disordered one (a gas), energy is absorbed. Heat is released when

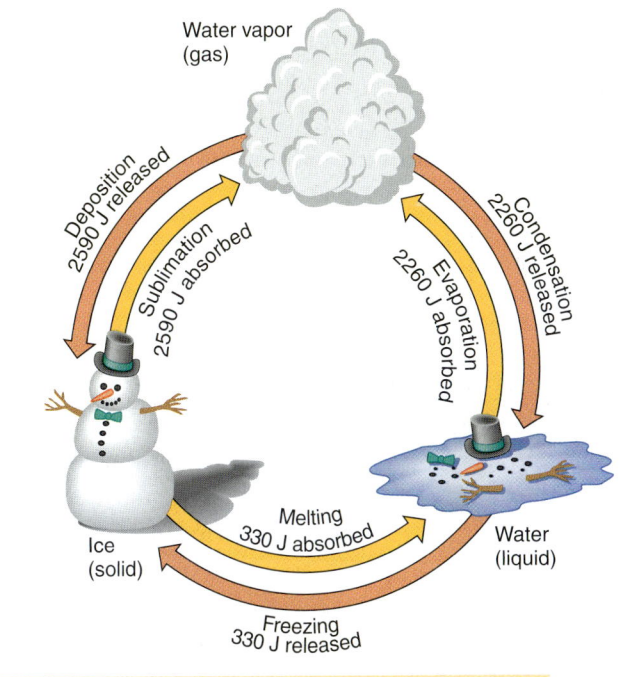

The three states of water FIGURE 14.5

Heat is added to, or released from, H_2O as it changes its state.

the change is from a less ordered state to a more ordered one. The amount of heat released or absorbed per gram during a change of state is known as **latent heat** (from the Latin word *latens*, meaning "hidden"; hence, "hidden heat").

latent heat Heat that is absorbed and stored in a gas or liquid during the processes of evaporation or condensation, melting or freezing, sublimation or deposition.

One familiar phenomenon involving a change of state is *evaporation*. We measure energy in units called joules (J). The 2260 J of energy needed to evaporate a gram of water has to come from somewhere. The reason you feel cool after splashing yourself with cold water on a hot day is that some of the heat needed for evaporation is absorbed from your skin and your body temperature drops as a result. Similarly, the latent heat of *condensation* (less ordered water vapor condensing to more ordered liquid water) is 2260 J/g, while the latent heat of freezing (again from less ordered to more ordered) and melting (more ordered solid ice melting to less ordered fluid water) is 330 J/g.

The six changes of state shown in Figure 14.5 (the six arrows) all play a role in weather, but evaporation and condensation are far more important than the others because they give rise to clouds, fog, and rain, and because they are the means by which huge amounts of heat are moved from equatorial regions toward the poles. To understand these phenomena, we need to look at humidity.

HUMIDITY

Blistering summer heat waves can be deadly, with elderly and ill people at the greatest risk. But even healthy young people need to be careful, especially in hot, humid weather. High **humidity** slows the evaporation of sweat from the body, reducing its cooling effect. Humidity varies widely from place to place and from time to time. In the cold, dry air of arctic regions in winter, the humidity is almost zero, whereas it can reach up to as much as 4 or 5% of a given volume of air in the warm, wet regions near the Equator.

humidity The amount of water vapor in the air.

Water vapor gets into the air through evaporation, as fast-moving liquid molecules of H_2O manage to escape from the sea, lakes, or streams and pass into the atmosphere above. Because molecules in a gas move randomly in all directions, some of the H_2O molecules in the air will also move back into the water (that is, they will condense). When the rate at which molecules are evaporating equals the rate at which they are condensing, the air is said to be *saturated*. This is the point at which the air holds the maximum concentration of H_2O molecules for a specified temperature.

An important principle concerning humidity states that the maximum quantity of moisture that can be held at any time in the air is dependent on air temperature (**FIGURE 14.6**).

Humidity and temperature FIGURE 14.6

The maximum specific humidity of a mass of air increases sharply with rising temperature. As a parcel of air cools along line A–B, it reaches saturation at the intersection, B, and that temperature is the *dew point*.

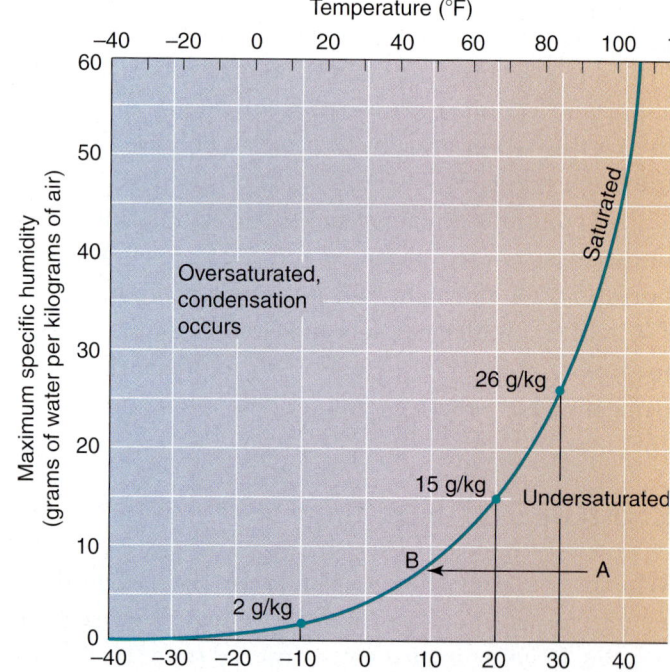

Warm air can hold more water vapor than cold air—a lot more. Air at room temperature (20°C, 68°F) can hold about three times as much water vapor as freezing air (0°C, 32°F).

Relative humidity When radio or television weather forecasters speak of humidity, they are referring to *relative humidity*. This measure compares the amount of water vapor present in the air to the maximum amount it can carry at that temperature. It is expressed as a percentage. So, for example, if the air currently carries half the moisture it could carry at the present temperature, the relative humidity is 50%. When the relative humidity reaches 100%, the air is saturated.

The relative humidity of the atmosphere can change in one of two ways. First, the atmosphere can directly gain or lose water vapor. For example, additional water vapor can enter the air from an exposed water surface or from wet soil. This is a slow process because the water vapor molecules have to diffuse upward from the surface into the air layer above.

The second way is through a change of temperature. When temperature falls, relative humidity rises, even if no water vapor is added. This occurs because the capacity of air to carry water vapor depends on temperature. When the air is cooled, this capacity is reduced. The existing amount of water vapor will then represent a higher percentage of total capacity.

Specific humidity The actual quantity of water vapor carried by a parcel of air is known as its specific humidity and is expressed as grams of water vapor per kilogram of air (g/kg).

Climatologists often use the term *specific humidity* to describe the moisture in a large mass of air. Specific humidity is largest at the equatorial zones and falls off rapidly toward the poles, as we can see in **FIGURE 14.7**. Extremely cold, dry air over arctic regions in winter may have a specific humidity as low as 0.2 g/kg, while the extremely warm, moist air of equatorial regions often holds as much as 18 g/kg. More solar energy is available at lower latitudes to evaporate water in

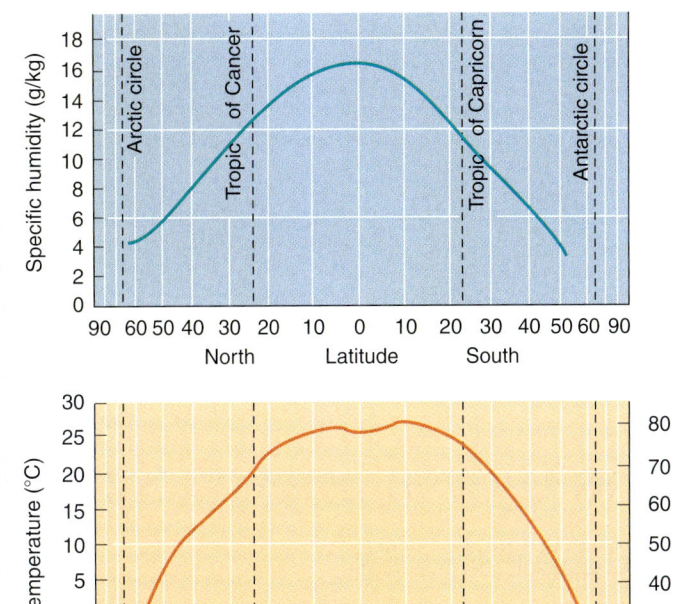

Global specific humidity and temperature
FIGURE 14.7

Pole-to-pole profiles of specific humidity (above) and temperature (below) show similar trends because the ability of air to carry water vapor (measured by specific humidity) is limited by temperature.

oceans or on moist land surfaces, so specific humidity values tend to be higher at low latitudes than at high latitudes.

We can see that the global profile of mean (average) surface air temperature (also shown in Figure 14.7) has a similar shape to that of the specific humidity profile. This is because air temperature and maximum specific humidity vary together.

Specific humidity is also the Earth scientist's yardstick for a basic natural resource—water. It is a measure of the quantity of water in the atmosphere that can be extracted as precipitation. Cold, moist air can supply only a small quantity of rain or snow, but warm, moist air is capable of supplying large amounts.

Moisture in the Atmosphere 433

Another way of describing the water vapor content of air is in terms of its *dew-point temperature*. If air is slowly chilled, it will eventually reach saturation. The temperature at which this occurs is referred to as the dew-point temperature. At this temperature, the air carries the maximum possible amount of water vapor. If the air is cooled further, condensation will begin and dew will start to form. Moist air has a higher dew-point temperature than dry air, as shown in Figure 14.6.

> **CONCEPT CHECK STOP**
>
> **What** are the three states of H_2O?
>
> **What** are the six processes through which H_2O changes its state? In which processes is latent heat taken in? When is it released?
>
> **What** is the difference between specific humidity and relative humidity?
>
> **What** is the dew-point temperature?

The Global Energy System

LEARNING OBJECTIVES

Describe the fate of solar radiation as it passes through the atmosphere.

Define albedo.

Define counterradiation and explain how it leads to the greenhouse effect.

Human activity has changed the planet's surface cover by adding carbon dioxide to the atmosphere faster than it is being extracted by natural processes. Have we irrevocably shifted the balance of energy flows? Is Earth absorbing more solar energy and becoming warmer? Or is it absorbing less and becoming cooler? If we want to understand the impact of humans on the Earth-atmosphere system, we need to examine the energy balance in detail.

SOLAR ENERGY LOSSES IN THE ATMOSPHERE

Let's examine the flow of solar energy, or **insolation**, through the atmosphere on its way to the surface. **FIGURE 14.8** gives typical values for losses of incoming shortwave radiation in the solar beam as it penetrates the atmosphere. Gamma rays and X-rays from the Sun are almost completely absorbed by the thin outer layers of the atmosphere, while much of the ultraviolet radiation is also absorbed, particularly by ozone.

> **insolation** The flow of incoming solar radiation intercepted by an exposed surface, assuming a uniformly spherical Earth with no atmosphere.

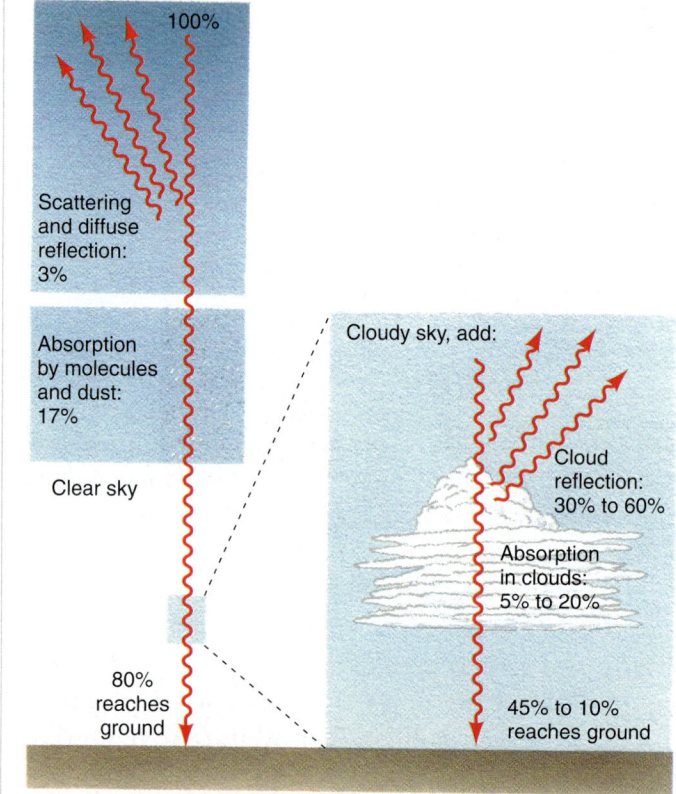

The fate of incoming solar radiation
FIGURE 14.8

Losses of incoming solar energy are much lower with clear skies (left) than with cloud cover (right).

434 CHAPTER 14 The Atmosphere: Composition, Structure, and Clouds

As the radiation moves deeper into the denser layers of the atmosphere, it can be scattered by gas molecules, dust, or other particles in the air, which may deflect it in any direction. Apart from this change in direction, the radiation is unchanged. Some scattered radiation flows down to Earth's surface, while some flows upward. This upward flow escaping back into space amounts to about 3% of incoming solar radiation.

What about absorption? As we saw earlier, molecules and particles can absorb radiation as it passes through the atmosphere. Carbon dioxide and water are the biggest radiation absorbers, and because the water vapor content of air can vary greatly, absorption also varies from one global environment to another. About 17% of incoming solar radiation is absorbed, raising the temperature of the atmosphere. After taking into account absorption and scattering, about 80% of incoming solar radiation reaches the ground.

Clouds can greatly increase the amount of incoming solar radiation that is reflected back to space. Reflection from the bright white surfaces of thick low clouds deflects about 30 to 60% of incoming radiation back into space. Clouds also absorb as much as 20% of the radiation.

ALBEDO

The proportion of incoming radiation that is scattered upward by a surface is called its *albedo* and is measured on a scale of 0 to 1 (**FIGURE 14.9**). The energy absorbed by a surface warms the air immediately above it by means of conduction and convection, so surface temperatures are warmer over low-albedo surfaces than over high-albedo surfaces. Fields, forests, and bare ground have albedos ranging from 0.03 to 0.25.

Certain orbiting satellites carry instruments that can measure shortwave and longwave radiation at the top of the atmosphere, helping us estimate Earth's average albedo. The Earth-atmosphere system reflects slightly less than one-third of the solar radiation it receives back into space.

Albedo contrasts FIGURE 14.9

A Bright snow A layer of new, fresh snow reflects most of the sunlight it receives and has an albedo of 0.45 to 0.85. Only a small portion is absorbed.

B Blacktop road Asphalt paving, with an albedo of 0.03, reflects little light, so it appears dark or black. It absorbs nearly all of the solar radiation it receives.

COUNTERRADIATION AND THE GREENHOUSE EFFECT

In addition to being warmed by shortwave radiation from the Sun, Earth's surface is significantly heated by the longwave radiation emitted by the atmosphere and absorbed by the ground. Let's look at this phenomenon in more detail.

FIGURE 14.10 shows the energy flows between the surface, atmosphere, and space. On the left, we can see the flow of shortwave radiation from the Sun to the surface. Some of this radiation is reflected back to space, but much is absorbed, warming the surface.

Meanwhile, Earth's surface emits longwave radiation upward. Some of this radiation escapes directly into space, while the remainder is absorbed by the atmosphere. The atmosphere also emits longwave radiation in all directions. Some radiates upward into space, while the remainder radiates downward toward Earth's surface. We call this downward flow *counterradiation*. It replaces some of the heat emitted by the surface.

Counterradiation depends strongly on the presence of carbon dioxide, water vapor, and methane in the atmosphere. Much of the longwave radiation emitted upward from Earth's surface is absorbed by these gases. This absorbed energy raises the temperature of the atmosphere, causing it to emit more counterradiation. So, the lower atmosphere, with its longwave-absorbing gases, acts like a blanket that traps heat underneath it. Because liquid water is also a strong absorber of longwave radiation, cloud layers, which are composed of tiny water droplets, are even more important than carbon dioxide and water vapor in producing a blanketing effect. This mechanism is known as the **greenhouse effect** because it is similar to the principle used in greenhouses to trap solar heat, as shown in FIGURE 14.11.

greenhouse effect The accumulation of heat in the lower atmosphere through the absorption of longwave radiation from Earth's surface.

Since energy must be conserved, the flows of energy between the Sun and Earth's atmosphere and surface must balance over the long term. We can keep track of the energy coming into and out of the system by thinking of a global energy budget. The budget takes into account all the important energy flows discussed so far, and helps us understand how global change might affect Earth's climate.

How might human activities influence the greenhouse effect? Suppose, for example, that clearing forests for agriculture and turning agricultural lands into urban and suburban areas decreases surface albedo. In that case, more energy would be absorbed by the ground, raising its temperature. That in turn would increase the flow of surface longwave radiation to the atmosphere, which would be absorbed and would then

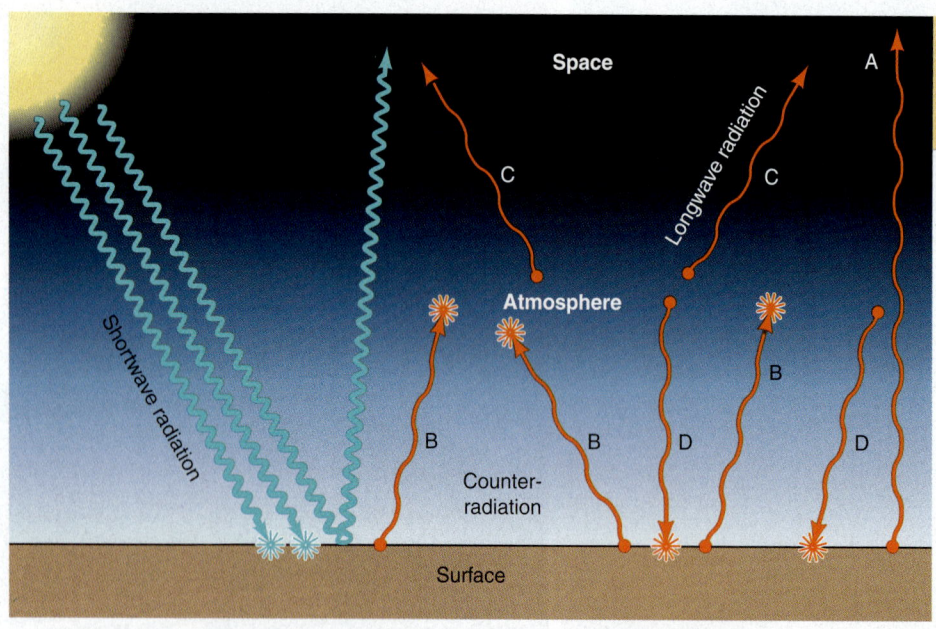

Counterradiation and the greenhouse effect
FIGURE 14.10

Shortwave radiation passes through the atmosphere and is absorbed at the surface. This warms the surface, which emits longwave radiation. Some of this flow passes directly to space (A), but most is absorbed by the atmosphere (B). In turn, the atmosphere radiates longwave energy back to the surface as counterradiation (D) and also into space (C). The counterradiation is the greenhouse effect.

CHAPTER 14 The Atmosphere: Composition, Structure, and Clouds

Process Diagram

How a greenhouse works FIGURE 14.11

Earth's atmosphere acts like the glass in a greenhouse, admitting solar shortwave radiation but blocking the passage of outgoing longwave radiation.

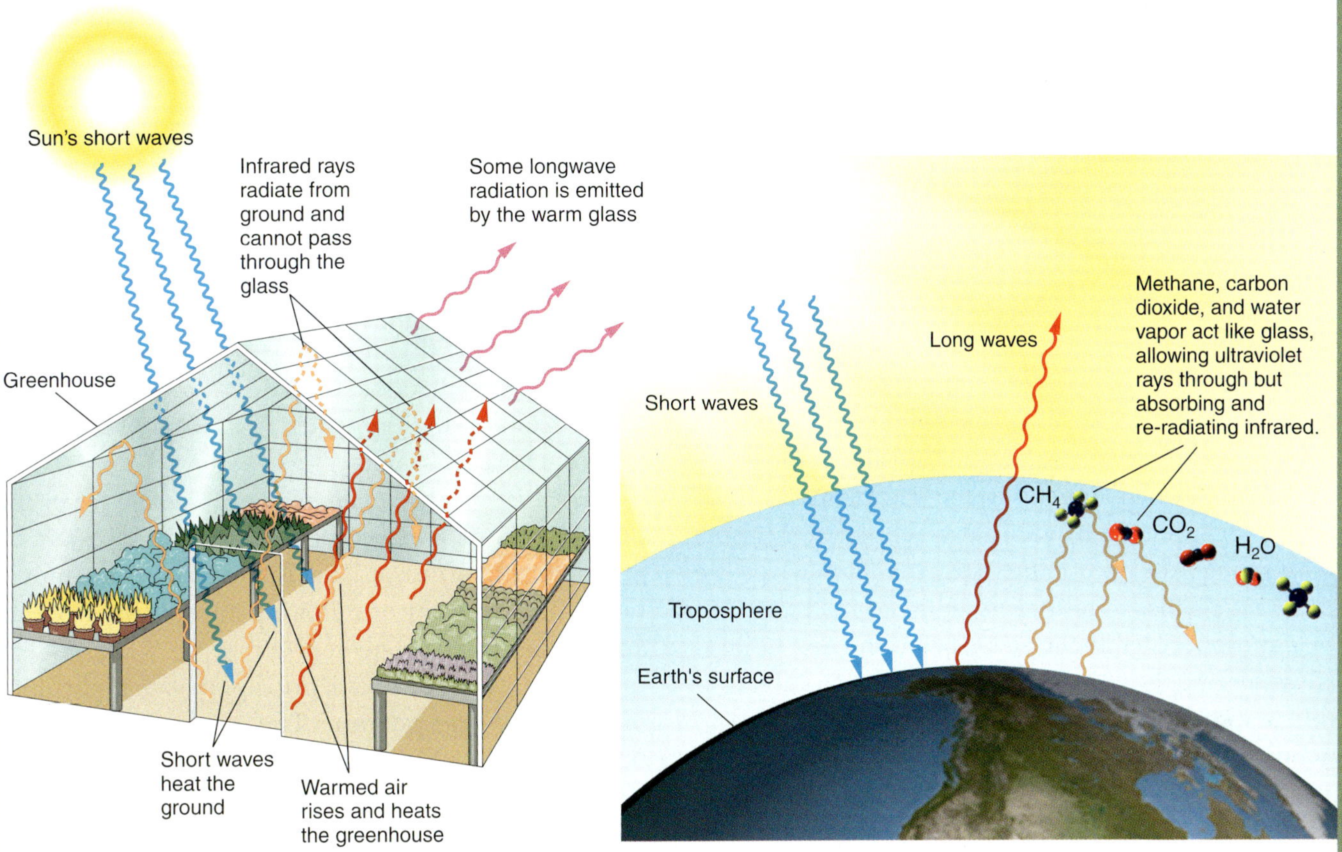

Sun's short waves

Infrared rays radiate from ground and cannot pass through the glass

Some longwave radiation is emitted by the warm glass

Greenhouse

Short waves heat the ground

Warmed air rises and heats the greenhouse

Short waves

Long waves

Methane, carbon dioxide, and water vapor act like glass, allowing ultraviolet rays through but absorbing and re-radiating infrared.

CH_4 CO_2 H_2O

Troposphere

Earth's surface

CRITICAL THINKING — Here's an interesting question:
- Is Earth's greenhouse effect strongest over the land or the ocean?

---

boost counterradiation. The total effect would probably be to amplify warming through the greenhouse effect.

What if air pollution causes low, thick clouds to form? Since low clouds increase shortwave reflection back into space, Earth's surface and atmosphere will cool. What about increasing condensation trails from jet aircraft? These could create high, thin clouds, which absorb more longwave energy than they reflect shortwave energy. This would make the atmosphere warmer, boosting counterradiation and increasing the greenhouse effect. As these examples make clear, the energy flow linkages among the Sun, Earth's surface, the atmosphere, and space are critical components of our climate system.

CONCEPT CHECK STOP

Describe two processes through which solar energy is lost in the atmosphere.

What is albedo? Give one example of an object with a high albedo and one of an object with a low albedo.

Why, on a hot day, does ground in the country feel much cooler than an urban sidewalk?

What is counterradiation?

How does it lead to the greenhouse effect?

Formation of Clouds

> **LEARNING OBJECTIVES**
>
> **Describe** the adiabatic principle.
>
> **Explain** the role of the adiabatic process in the formation of clouds.
>
> **Discuss** how condensation nuclei help clouds form.
>
> **Describe** how clouds are classified.

What makes the water vapor held in the air turn into liquid or solid particles that can fall to Earth? The answer is that the air is naturally cooled. Remember that when air cools to the dew point, it is saturated with water. Think about extracting water from a moist sponge. To release the water, you have to squeeze the sponge—that is, reduce its ability to hold water. In the atmosphere, chilling the air below the dew point is like squeezing the sponge—it reduces the air's ability to hold water, forcing some water vapor molecules to change their state and form water droplets or ice crystals.

One mechanism for chilling air is nighttime cooling. The ground surface can become quite cold on a clear night as it loses longwave radiation. But this is not enough to form precipitation. Precipitation forms only when a substantial mass of air experiences a steady drop in temperature below the dew point. This happens when air is lifted to a higher level in the atmosphere, as we will see in more detail later in the chapter.

THE ADIABATIC PRINCIPLE

If you have ever pumped up a bicycle tire using a hand pump, you might have noticed that the pump gets hot. If so, you have observed the **adiabatic principle**. This is an important law that states that when a gas expands, its temperature drops. And conversely, when a gas is compressed, its temperature increases. So as you pump vigorously, compressing the air, the bicycle pump gets hot. In the same way, when a small jet of air escapes from a high-pressure hose, the hose feels cool.

> **adiabatic principle** A principle of science that states that a gas cools as it expands and warms as it is compressed, provided that no heat flows into or out of the gas during the process.

Scientists use the term *adiabatic process* to refer to a heating or cooling process that occurs solely as a result of a change in pressure. That is, the change in temperature is not caused by heat flowing into or away from the air, but only by a change in pressure.

How does the adiabatic principle relate to the uplift of air and to precipitation? The missing link is simply that atmospheric pressure decreases as altitude increases. So, as a parcel of air is uplifted, the atmospheric pressure on the parcel becomes lower and it expands and cools. As a parcel of air descends, atmospheric pressure becomes higher, and the air is compressed and warmed (**FIGURE 14.12**).

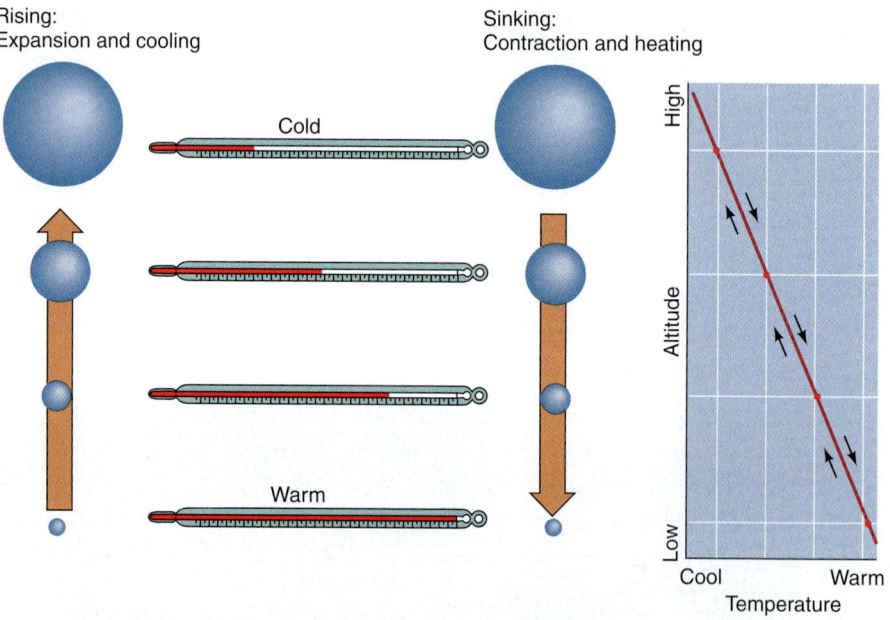

Adiabatic cooling and heating
FIGURE 14.12

When air is forced to rise, it expands and its temperature decreases. When air is forced to descend, its temperature increases.

438 CHAPTER 14 The Atmosphere: Composition, Structure, and Clouds

Cloud formation and the adiabatic process FIGURE 14.13

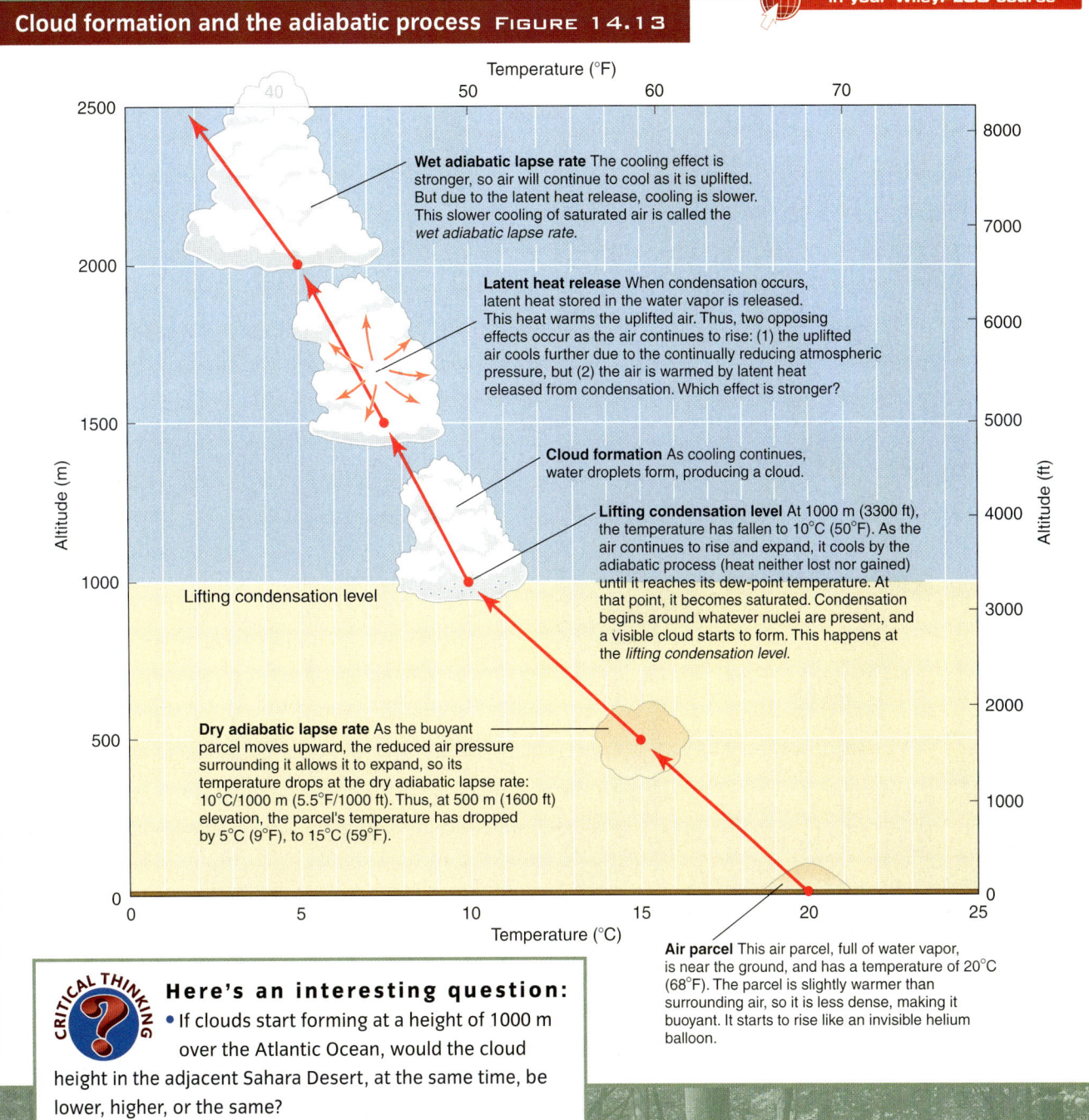

Wet adiabatic lapse rate The cooling effect is stronger, so air will continue to cool as it is uplifted. But due to the latent heat release, cooling is slower. This slower cooling of saturated air is called the *wet adiabatic lapse rate*.

Latent heat release When condensation occurs, latent heat stored in the water vapor is released. This heat warms the uplifted air. Thus, two opposing effects occur as the air continues to rise: (1) the uplifted air cools further due to the continually reducing atmospheric pressure, but (2) the air is warmed by latent heat released from condensation. Which effect is stronger?

Cloud formation As cooling continues, water droplets form, producing a cloud.

Lifting condensation level At 1000 m (3300 ft), the temperature has fallen to 10°C (50°F). As the air continues to rise and expand, it cools by the adiabatic process (heat neither lost nor gained) until it reaches its dew-point temperature. At that point, it becomes saturated. Condensation begins around whatever nuclei are present, and a visible cloud starts to form. This happens at the *lifting condensation level*.

Dry adiabatic lapse rate As the buoyant parcel moves upward, the reduced air pressure surrounding it allows it to expand, so its temperature drops at the dry adiabatic lapse rate: 10°C/1000 m (5.5°F/1000 ft). Thus, at 500 m (1600 ft) elevation, the parcel's temperature has dropped by 5°C (9°F), to 15°C (59°F).

Air parcel This air parcel, full of water vapor, is near the ground, and has a temperature of 20°C (68°F). The parcel is slightly warmer than surrounding air, so it is less dense, making it buoyant. It starts to rise like an invisible helium balloon.

CRITICAL THINKING — Here's an interesting question:
- If clouds start forming at a height of 1000 m over the Atlantic Ocean, would the cloud height in the adjacent Sahara Desert, at the same time, be lower, higher, or the same?

We describe this behavior using the *dry adiabatic lapse rate* for a rising air parcel that has not reached saturation. This rate has a value of about 10°C per 1000 m (5.5°F per 1000 ft) of vertical rise. That is, if a parcel of air is raised 1 km, its temperature will drop by 10°C. Or, in English units, if it is raised 1000 ft, its temperature will drop by 5.5°F. This is the "dry" rate because there's no condensation.

The adiabatic process is responsible for the formation of clouds. As we see in FIGURE 14.13, once a parcel of air reaches saturation, it cools at a different rate, known as the *wet adiabatic lapse rate*. Unlike the dry adiabatic lapse rate, which remains constant, the wet adiabatic lapse rate is variable because it depends on the temperature and pressure of the air, moisture content, and release of latent heat. Nevertheless, for most situations we can use a value of 5°C/1000 m (2.7°F/1000 ft). In Figure 14.13, the wet adiabatic lapse rate is shown as a slightly curving line to indicate that its value increases with altitude.

Formation of Clouds

Ocean aerosols FIGURE 14.14

Breaking or spilling waves in the open ocean, shown here from the deck of a ship, are an important source of aerosols.

CLOUDS

Images of Earth taken from space show that about half of our planet is blanketed in cloud. Clouds are made up of water droplets, ice particles, or a mixture of both, suspended in air. These particles are between 20 and 50 μm (0.0008 and 0.002 in.) in diameter. Cloud particles grow around a tiny center of solid matter. Such a speck of matter is called a *condensation nucleus* and typically has a diameter of between 0.1 and 1 μm (0.000004 and 0.00004 in.).

The surface of the sea is an important source of aerosols that act as condensation nuclei (**FIGURE 14.14**). Droplets of spray from the crests of the waves are carried upward by turbulent air. When these droplets evaporate, they leave behind a tiny residue of salt suspended in the air. This aerosol strongly attracts water molecules, beginning the process of cloud formation. Nuclei are also thrown into the atmosphere as dust in polluted air over cities, aiding condensation and the formation of clouds and fog.

Anyone who has looked up at the sky knows that clouds come in many shapes and sizes. They range from the small, white puffy clouds often seen in summer to the gray layers that produce rain. Meteorologists classify clouds into four families, arranged by their height in the sky. They are: high, middle, and low clouds, and clouds with vertical development. They also group clouds into types according to their shape, as described in **FIGURE 14.15**.

We can also group clouds into two major classes: stratiform, or layered clouds, and cumuliform, or globular clouds. Stratiform clouds are blanket-like and cover large areas. A common type is stratus, which covers the entire sky. Dense, thick stratiform clouds can produce large amounts of rain or snow.

Cumuliform clouds are globular masses of cloud that are associated with parcels of rising air. The most common cloud of this type is the cumulus cloud. Cumulus clouds can be dense and tall clouds and can produce thunderstorms. This form of cloud is the cumulonimbus. (*Nimbus* is a Latin word meaning "rain cloud" or "storm.")

CONCEPT CHECK STOP

What is the adiabatic principle?

How is the adiabatic process involved in the formation of clouds?

What is the difference between the dry and wet adiabatic lapse rates?

What are condensation nuclei? How do they help clouds form?

How are clouds classified? Name four cloud families, two broad cloud forms, and three specific cloud types.

440 CHAPTER 14 The Atmosphere: Composition, Structure, and Clouds

Visualizing
Clouds FIGURE 14.15

A Cloud families and types Clouds are grouped into families on the basis of height. Individual cloud types are named according to their form.

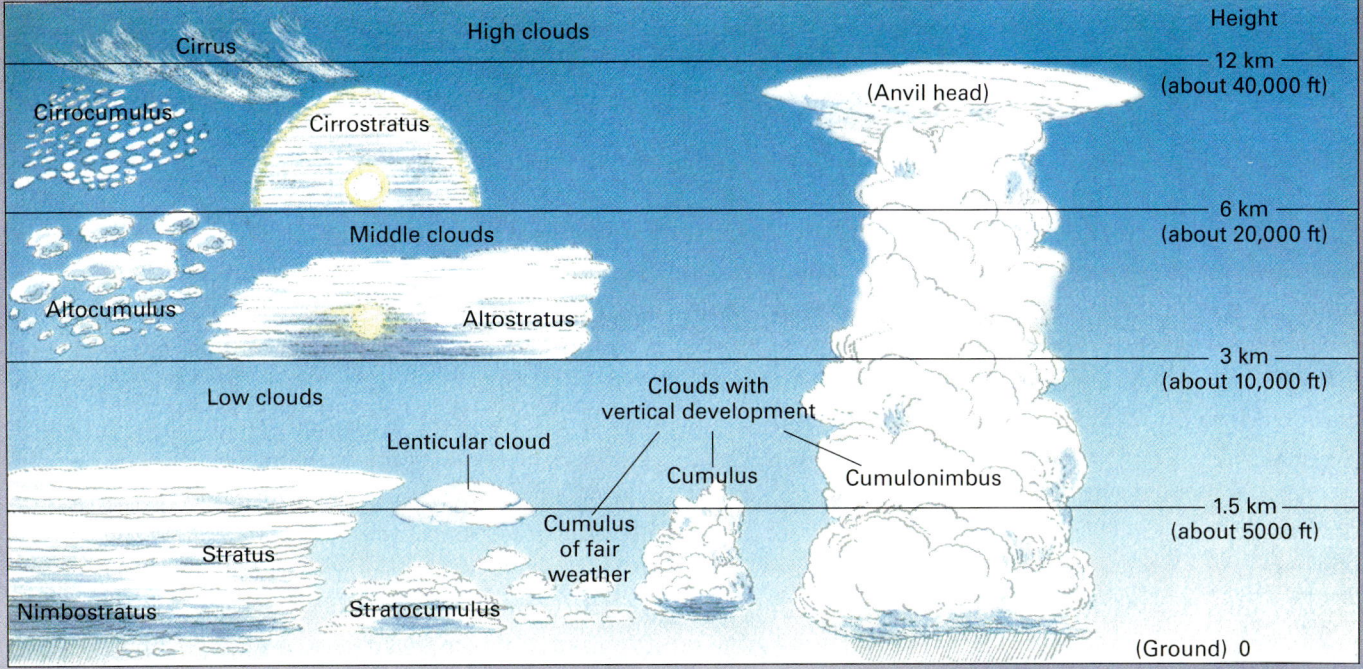

B Cirrus High, thin, wispy clouds drawn out into streaks are cirrus clouds. They are composed of ice crystals and form when moisture is present high in the air.

C Lenticular cloud A lenticular, or lens-shaped, cloud forms as moist air flows up and over a mountain peak or range.

D Cumulus Puffy, fair-weather cumulus clouds fill the sky above a prairie.

E Altocumulus High cumulus clouds, in a pattern sometimes called a *mackerel sky*, as photographed near sunset in Boston.

Precipitation

> **LEARNING OBJECTIVES**
>
> **Describe** the causes of precipitation.
>
> **Name** the four types of precipitation processes and describe the differences among them.
>
> **Explain** the conditions that cause thunderstorms.

Depending on the circumstances, precipitation can be a welcome relief, a minor annoyance, or a life-threatening hazard (**FIGURE 14.16**). Whether it is a source of relief or a hazard, precipitation provides the fresh water that is essential for most forms of terrestrial life. In warm clouds, raindrops form by condensation and grow by collision. **FIGURE 14.17** explains this process.

Snow is formed within cool clouds. Usually we think of water freezing at 0°C (32°F), but within clouds, liquid water droplets can be *supercooled*—that is, they can exist as water at temperatures as low as −12°C (10°F). Cool clouds are a mixture of ice crystals and supercooled water droplets. The ice crystals take up water vapor and grow by deposition. At the same time, the supercooled water droplets lose water vapor through evaporation and shrink. When an ice crystal collides with a droplet of supercooled water, it freezes the droplet. The ice crystals then coalesce to form snow particles, which can become heavy enough to fall from the cloud. Snowflakes formed entirely by deposition can have intricate crystal structures, but most particles of snow have undergone collisions and coalesced with one another, losing their shape and becoming simple lumps of ice.

By the time this precipitation reaches the ground, its form may have changed. When raindrops fall

Precipitation FIGURE 14.16

Monsoon rains drench refugees crossing a flooded field in Bangladesh.

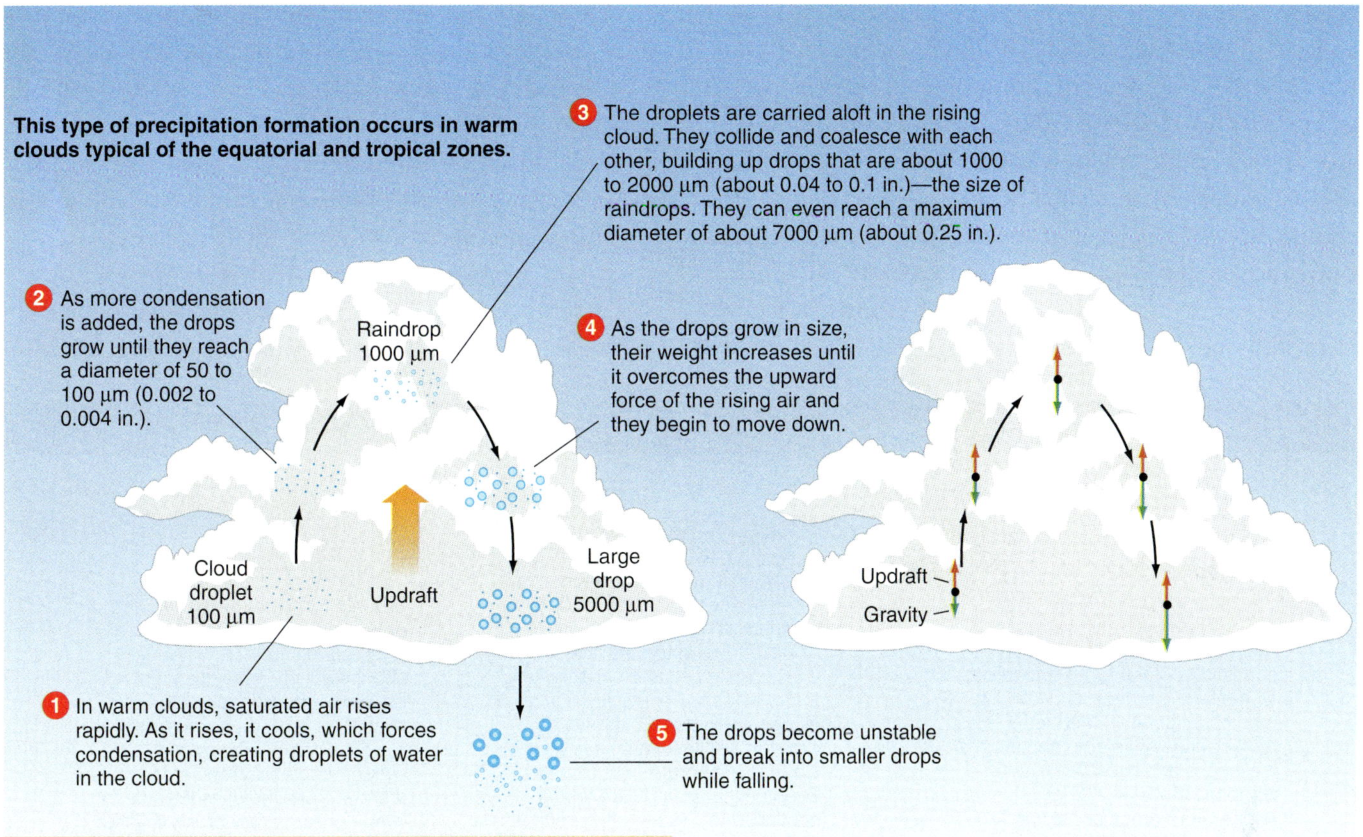

Rain formation in warm clouds FIGURE 14.17

This type of precipitation formation occurs in warm clouds typical of the equatorial and tropical zones.

1 In warm clouds, saturated air rises rapidly. As it rises, it cools, which forces condensation, creating droplets of water in the cloud.

2 As more condensation is added, the drops grow until they reach a diameter of 50 to 100 μm (0.002 to 0.004 in.).

3 The droplets are carried aloft in the rising cloud. They collide and coalesce with each other, building up drops that are about 1000 to 2000 μm (about 0.04 to 0.1 in.)—the size of raindrops. They can even reach a maximum diameter of about 7000 μm (about 0.25 in.).

4 As the drops grow in size, their weight increases until it overcomes the upward force of the rising air and they begin to move down.

5 The drops become unstable and break into smaller drops while falling.

Cloud droplet 100 μm — Updraft — Raindrop 1000 μm — Large drop 5000 μm — Updraft — Gravity

---

through a layer of cold air, they freeze into pellets or grains of ice. When snow falls through a layer of warm air, it melts and arrives as rain. Perhaps you have experienced an ice storm (**FIGURE 14.18**). Ice storms occur when the ground is frozen and the lowest air layer is also below freezing. Rain falling through the cold air layer is chilled and freezes onto ground surfaces, forming a clear, slippery glaze that makes roads and sidewalks extremely hazardous. Ice storms cause great damage, especially to telephone and power lines and to tree limbs pulled down by the weight of the ice.

Ice storm FIGURE 14.18

When rain falls into a surface layer of air whose temperature is below freezing, clear ice coats the ground. The weight of the ice brings down power lines and tree limbs, creating hazardous driving conditions, as in this 1988 photo taken in Watertown, New York.

Hail is another common type of precipitation. It consists of pea- to grapefruit-sized lumps of ice—that is, lumps with a diameter of 5 mm (0.2 in.) or larger. We will discuss how hail is produced later in the chapter.

We talk about precipitation in terms of the depth of the total amount that falls during a certain time. For example, we speak of centimeters or inches of rainfall per hour or per day. A centimeter (inch) of rainfall would cover the ground to a depth of 1 cm (1 in.) if the water did not run off or sink into the soil.

Snowfall is measured by melting a sample column of snow and reducing it to an equivalent in rainfall. In this way, we can combine rainfall and snowfall into a single record of precipitation. Ordinarily, a 10-cm (or 10-in.) layer of snow is assumed to be equivalent to 1 cm (or 1 in.) of rainfall, but this ratio may range from 30 to 1 in very loose snow to 2 to 1 in old, partly melted snow.

So far, we have seen how air that is moving upward will be chilled by the adiabatic process, eventually leading to precipitation. But a key piece of the precipitation

Orographic precipitation FIGURE 14.19

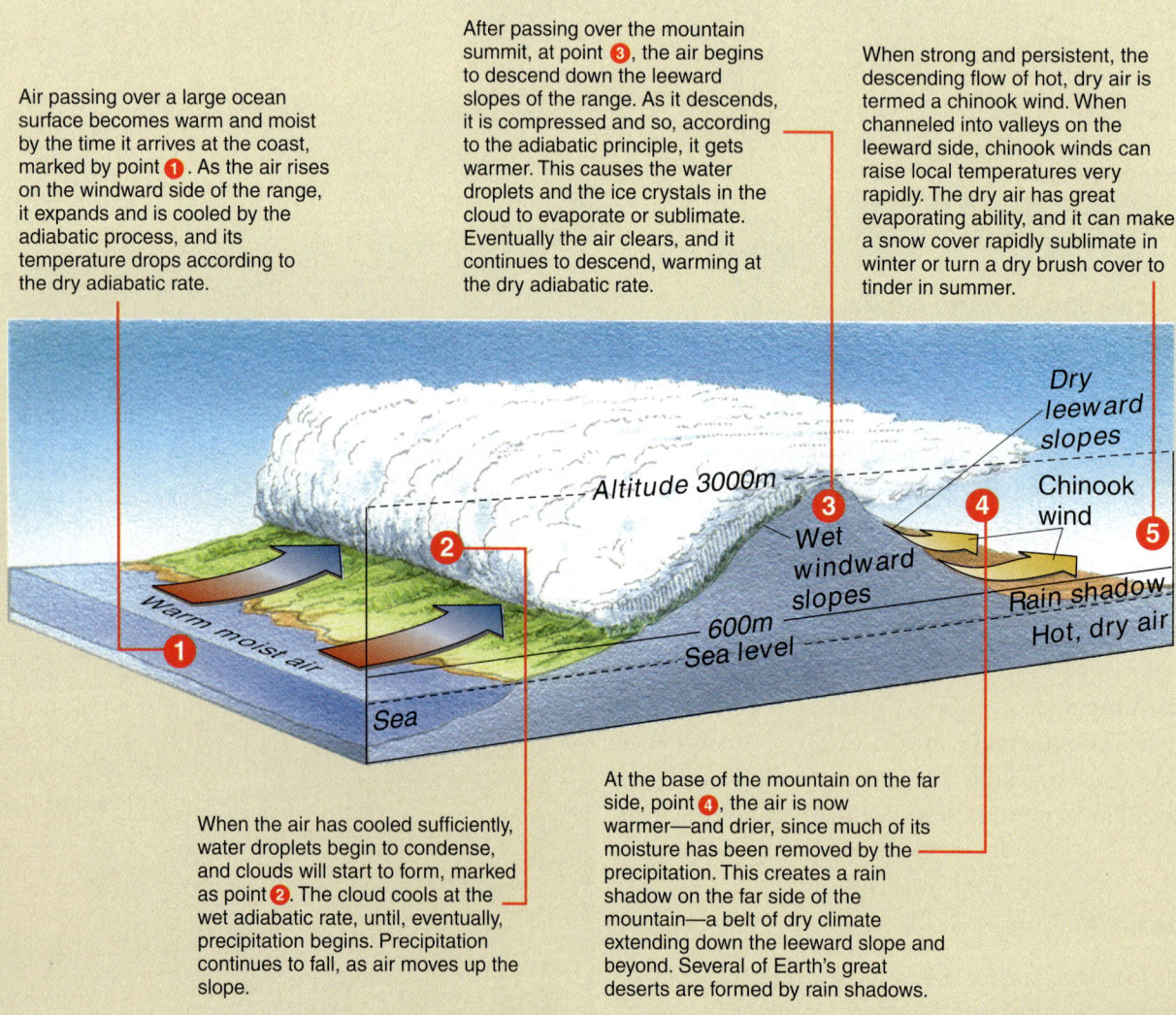

Air passing over a large ocean surface becomes warm and moist by the time it arrives at the coast, marked by point ❶. As the air rises on the windward side of the range, it expands and is cooled by the adiabatic process, and its temperature drops according to the dry adiabatic rate.

After passing over the mountain summit, at point ❸, the air begins to descend down the leeward slopes of the range. As it descends, it is compressed and so, according to the adiabatic principle, it gets warmer. This causes the water droplets and the ice crystals in the cloud to evaporate or sublimate. Eventually the air clears, and it continues to descend, warming at the dry adiabatic rate.

When strong and persistent, the descending flow of hot, dry air is termed a chinook wind. When channeled into valleys on the leeward side, chinook winds can raise local temperatures very rapidly. The dry air has great evaporating ability, and it can make a snow cover rapidly sublimate in winter or turn a dry brush cover to tinder in summer.

When the air has cooled sufficiently, water droplets begin to condense, and clouds will start to form, marked as point ❷. The cloud cools at the wet adiabatic rate, until, eventually, precipitation begins. Precipitation continues to fall, as air moves up the slope.

At the base of the mountain on the far side, point ❹, the air is now warmer—and drier, since much of its moisture has been removed by the precipitation. This creates a rain shadow on the far side of the mountain—a belt of dry climate extending down the leeward slope and beyond. Several of Earth's great deserts are formed by rain shadows.

puzzle is missing: What causes air to move upward in the first place?

Air can move upward in four ways. Here, we will discuss the first two in detail: *orographic precipitation* and *convectional precipitation*. A third way for air to be forced upward is through the movement of air masses. This type of process usually occurs during cyclones (Chapter 16), so we call it *cyclonic precipitation*. The fourth way is by *convergence*, in which air currents coming from different directions converge and air "piles up" and is forced upward.

OROGRAPHIC PRECIPITATION

Orographic precipitation occurs when a current of moist air is forced to move upward (**FIGURE 14.19**). The term *orographic* means "related to mountains." Think of what happens to a mass of air moving up and over a mountain

> ■ **orographic precipitation**
> Precipitation that is induced when moist air is forced over a mountain barrier.

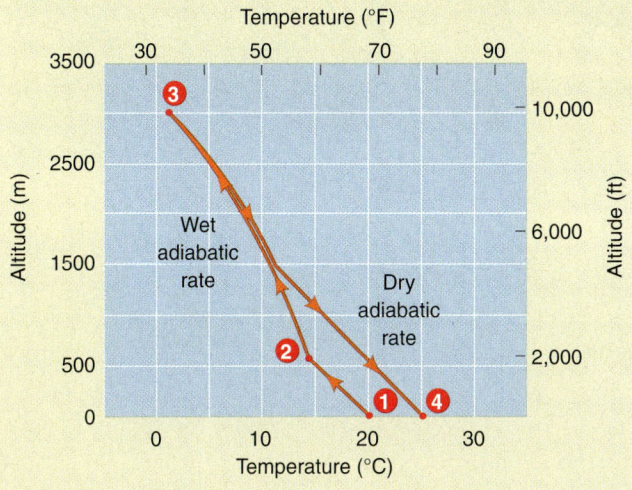

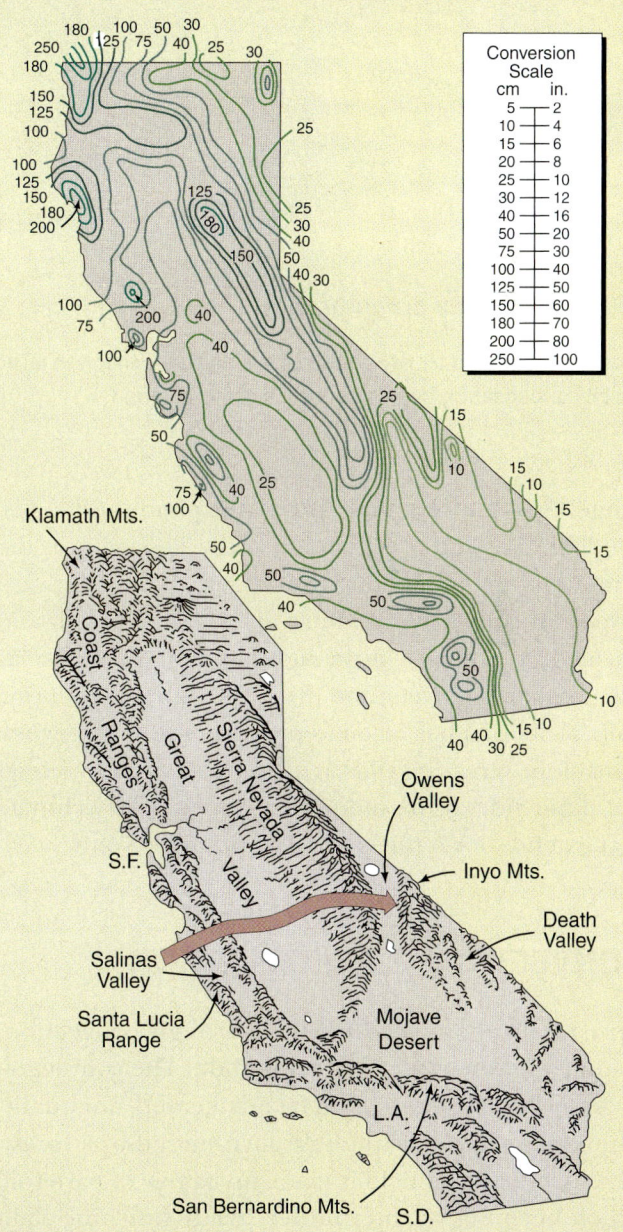

California mountain ranges have a strong effect on precipitation because of the prevailing flow of moist oceanic air from west to east. The upper diagram shows lines of equal precipitation. We can see that centers of high precipitation coincide with the western slopes of mountain ranges, including the coast ranges and Sierra Nevada. The desert regions lie to the east, in their rain shadows.

Precipitation 445

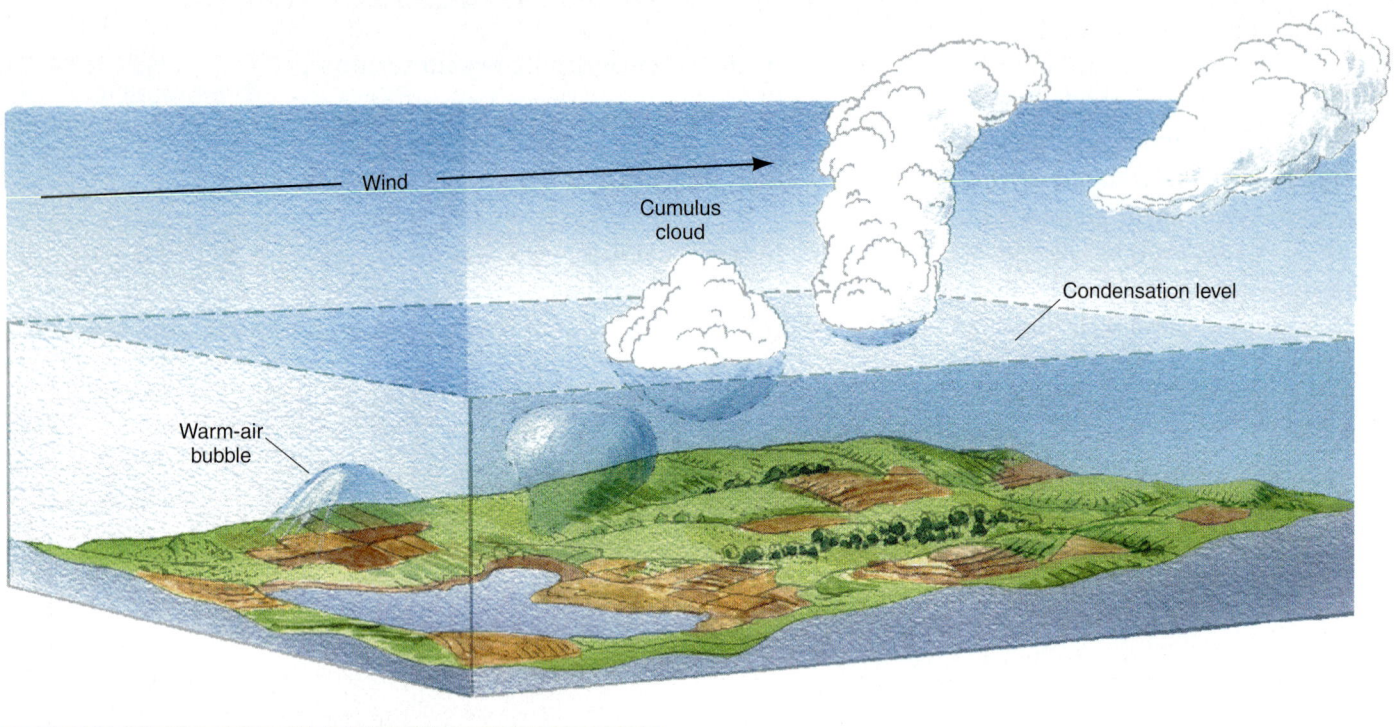

Formation of a cumulus cloud FIGURE 14.20

A bubble of heated air rises above the *lifting condensation level*, the level at which condensation begins, to form a cumulus cloud.

range. As the moist air is lifted, it expands and cools; condensation and precipitation occur. Passing over the summits of the mountains, the air descends the leeward slopes of the range, where it is compressed and warmed. At the base of the mountain range on the far side, the air is warmer and drier. The moisture it originally contained has fallen as precipitation on the windward slope, creating what is known as a *rain shadow* on the other side of the mountain—a belt of dry climate that extends down the leeward slope and beyond.

CONVECTIONAL PRECIPITATION

Air can also be forced upward through convection, leading to **convectional precipitation**. The convection process begins when a surface is heated unequally. Think of an agricultural field surrounded by a forest. The surface of the field is made up largely of bare soil, which becomes warmer under steady sunshine than does vegetation in the adjacent forest. This means that as the day progresses, the air above the field will grow warmer than the air above the forest.

The density of air depends on its temperature—warm air is less dense than cooler air. Because it is less dense, warm air rises above cold air. This process is termed *convection*, and it is the principle on which hot-air balloons operate. In the case of our field, a bubble of air will form over the field, rise, and break free from the surface. FIGURE 14.20 shows this process.

As the bubble of air rises, it cools adiabatically, so its temperature decreases as it rises. We also know that the temperature of the surrounding air will decrease with altitude. Nevertheless, as long as

convectional precipitation Precipitation that is induced when warm, moist air is heated at the ground surface, rises, cools, and condenses to form water droplets, raindrops, and, eventually, rainfall.

the bubble is still warmer than the surrounding air, it will be less dense, so it will continue to rise.

If the bubble remains warmer than the surrounding air and uplift continues, adiabatic cooling chills the bubble below the dew point, and condensation sets in. The rising air column becomes a puffy cumulus cloud. In Figure 14.20, the flat base of the cloud marks the *lifting condensation level*—the level at which condensation begins. The bulging "cauliflower" top of the cloud is the top of the rising column of warm air pushing into higher levels of the atmosphere. Once aloft, the small cumulus cloud will normally encounter winds that mix it into the local air. After drifting some distance downwind, the cloud evaporates.

THUNDERSTORMS AND UNSTABLE AIR

Thunderstorms are awe-inspiring events. Intense rain and violent winds, lightning, and thunder renew our respect for nature's power (**FIGURE 14.21**). But what causes thunderstorms? The answer is that when convection continues strongly, air can become unstable, creating dense cumulonimbus clouds or thunderstorms.

> **thunderstorms**
> Intense local storms associated with a tall, dense cumulonimbus cloud containing very strong updrafts.

Two environmental conditions encourage the development of thunderstorms: (1) air that is very warm and moist, and (2) an *environmental temperature lapse rate* in which temperature decreases more rapidly with altitude than it does for either the dry or the wet adiabatic lapse rates. Air with these characteristics is referred to as unstable.

The key to convectional precipitation in unstable air is latent heat. As we saw at the beginning of the chapter, when water vapor condenses into droplets or ice particles, it releases latent heat. This heat keeps a rising parcel of air warmer than the surrounding air, fueling the convection process and driving the parcel ever higher. When the parcel reaches a high altitude, most of its water will have condensed. As adiabatic cooling continues, less latent heat will be released and the uplift will weaken. Eventually uplift stops, since the energy source, latent heat, is gone. The parcel dissipates into the surrounding air.

Hot air masses in the central and southeastern United States are often unstable. Summer weather patterns sweep warm, humid air from the Gulf of Mexico over the continent. Over a period of days, the intense insolation strongly heats the layer of air near the ground, producing a steep environmental temperature lapse rate. Thus, both of the conditions that create unstable air are present, making thunderstorms very common in these regions during the summer.

Thunderstorms FIGURE 14.21

Lightning strikes an isolated farmhouse near Inverness, Scotland.

ANATOMY OF A THUNDERSTORM

A single thunderstorm typically consists of several individual convection cells. A single convection cell is shown in **FIGURE 14.22A**. A succession of bubble-like air parcels rise within the cell. As a result of the adiabatic process, these bubbles are intensely cooled, producing precipitation. This precipitation can take the form of water if the clouds are at the lower levels, mixed water and snow at intermediate levels, and snow at high levels, where cloud temperatures are coldest.

As the air parcels reach high levels, which may be 6 to 12 km (about 20,000 to 40,000 ft) above the surface or even higher, the rate of rising slows. At such high altitudes, the winds are typically strong, dragging the cloud top downwind and giving the thunderstorm cloud its distinctive shape—resembling an old-fashioned blacksmith's anvil (**FIGURE 14.22B**).

Ice particles falling from the cloud top act as nuclei for freezing and deposition at lower levels. Large ice crystals form and begin to sink rapidly. As they melt, they coalesce into large, falling droplets. As these raindrops rapidly fall adjacent to the rising air bubbles, they pull the air downward, feeding a downdraft within the convection cell. This downdraft of cool air emerges from the cloud base laden with precipitation. It approaches the surface and spreads out in all directions. As part of the downdraft moves forward, warm, moist surface air can rise and enter the updraft portion of the storm. This effect helps perpetuate the storm. The downdraft also creates strong local winds.

Thunderstorm FIGURE 14.22

A Anatomy of a thunderstorm cell Successive bubbles of moist condensing air push upward in the cell. Their upward movement creates a corresponding downdraft, expelling rain, hail, and cool air from the storm as it moves forward.

B Anvil top Large thunderstorms often show a top plume of cloud swept downwind by strong winds at high altitudes. This storm, over Lake Tanganyika in Gombe Stream National Park, Tanzania, is moving from right to left.

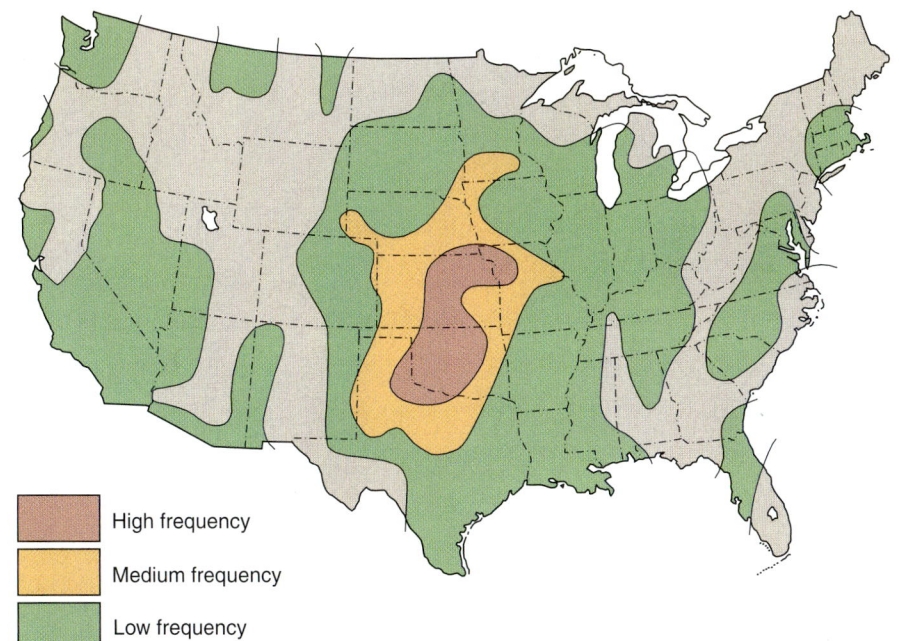

Frequency of severe hailstorms FIGURE 14.23

As shown in this map of the 48 contiguous United States, severe hailstorms are most frequent in the midwestern plains states of Oklahoma and Kansas. A severe hailstorm is defined as a local convective storm producing hailstones equal to or greater than 1.9 cm (0.75 in.) in diameter.

- High frequency
- Medium frequency
- Low frequency

We are familiar with the lightning, thunder, heavy rains, and powerful wind gusts—sometimes violent enough to topple trees and tear the roofs off buildings—that accompany thunderstorms. Lightning is one of the most deadly natural phenomena. In an average year, more people are killed around the world by lightning than by tornadoes or hurricanes. Each year an estimated 400 people survive lightning strikes in the United States; about 200 others do not survive. Between 1940 and 1991, lightning killed 8316 people in the United States.

Approximately 90% of all lightning flashes occur between clouds, but the remaining 10% shoot from clouds to Earth's surface. In fact, Earth is struck by lightning about 100 times every second. During a lightning strike, electricity flows between the sky and the ground, traveling at about one-third of the speed of light and packing a punch of 60,000 to 100,000 amps—thousands of times as much as a household electrical circuit.

A lightning bolt is actually an arc of atmospheric gas that is heated to a temperature as high as 27,000°C (50,000°F)—hotter than the surface of the Sun—as the current flows through it. The hot gas emits light as a flash and expands explosively to create the familiar crack of thunder. We will discuss how lightning forms in Chapter 15.

In addition, thunderstorms can produce hail. Hailstones are formed when layers of ice build up on ice pellets suspended in the strong updrafts of the thunderstorm. They can reach diameters of 3 to 5 cm (1.2 to 2.0 in.) or larger. When they become too heavy for the updraft to support, they fall to Earth.

Crop destruction caused by hailstorms can add up to losses of several hundred million dollars a year. Damage to wheat and corn crops is particularly severe in the Great Plains, running through Nebraska, Kansas, Missouri, Oklahoma, and northern Texas (FIGURE 14.23).

CONCEPT CHECK STOP

What are the four types of precipitation processes?

How does hail form?

How does orographic precipitation differ from convectional precipitation?

What is the lifting condensation level?

What conditions lead to thunderstorms?

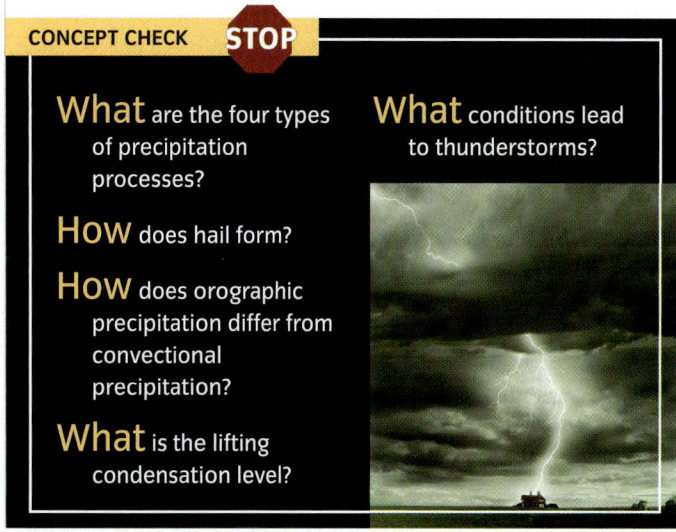

Precipitation

Amazing Places: Hole Punch Clouds in Mobile, Alabama

Hole punch clouds are rare. So under what conditions do they arise? The cloud blanket shown in **A** is made up of *supercooled* clouds, which can contain liquid water droplets at temperatures as low as −12°C (10°F). When aircraft pass through these clouds, tiny particles from their exhaust disturb these supercooled droplets, and they instantly freeze, dropping out of the cloud and leaving behind a gaping hole.

The ice crystals do not reach the ground, however. Instead, they sublimate, turning directly into water vapor.

Global Locator

A The inhabitants of Mobile, Alabama were treated to a strange and unusual cloud formation in the sky on January 12, 2004.

B Images taken by the Moderate Resolution Imaging Spectroradiometer (MODIS) satellite show that many such cloud patterns, dubbed *hole punch* clouds, can occur over a wide area.

C An aircraft contrail can be seen near this hole punch cloud.

450 CHAPTER 14 The Atmosphere: Composition, Structure, and Clouds

SUMMARY

1 Earth's Atmosphere

1. The composition of air, excluding water vapor and aerosols, is 99.96% by volume nitrogen, oxygen, and argon.

2. There are four distinct layers in the atmosphere, each with a distinct temperature profile. From the bottom up, they are the **troposphere**, the **stratosphere**, the *mesosphere*, and the *thermosphere*.

3. Most weather-related phenomena occur in the troposphere. The troposphere also contains most of the gases that play a role in producing Earth's **greenhouse effect**. The **ozone** layer, which protects life on Earth by absorbing harmful incoming ultraviolet radiation, is a concentration of O_3 in the stratosphere.

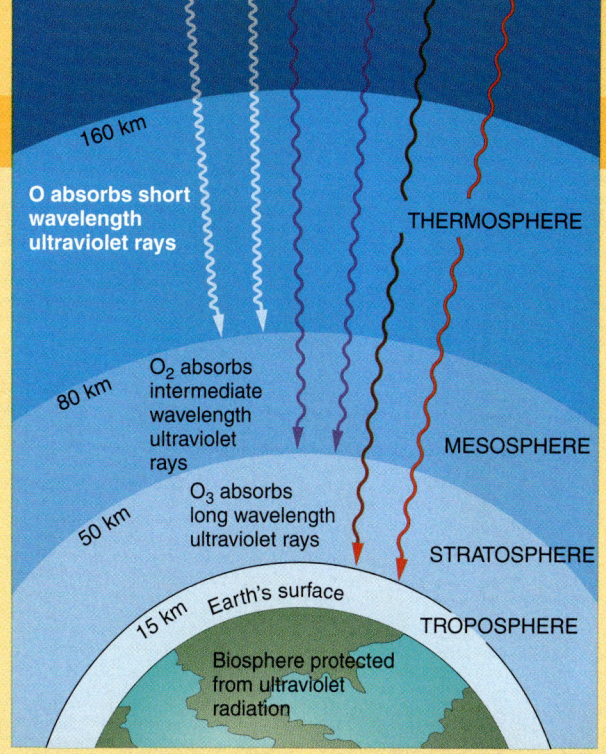

2 Moisture in the Atmosphere

1. H_2O can change its state by evaporating, condensing, melting, or freezing, and through *sublimation* and *deposition*. Each transition is accompanied by the absorption or release of **latent heat**.

2. Warm air can carry much more water vapor than can cold air. *Specific humidity* measures the mass of water vapor in a mass of air. *Relative humidity* measures water vapor in the air as the percentage of the maximum amount of water vapor that can be carried at a given air temperature.

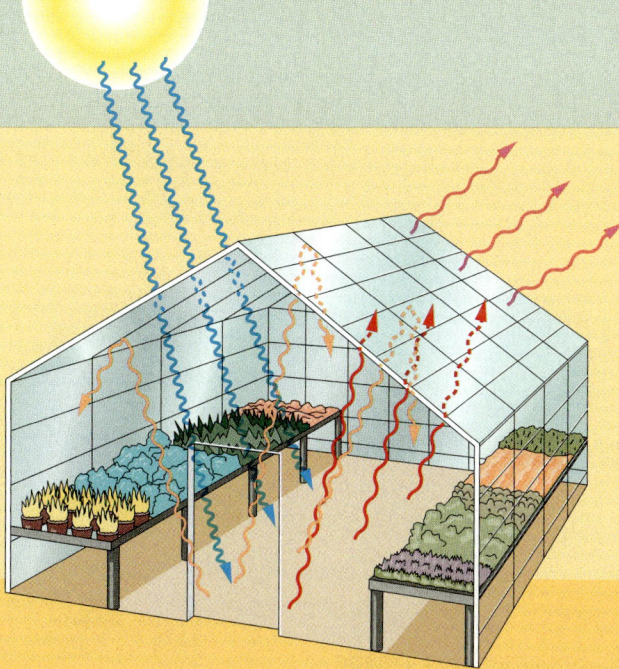

3 The Global Energy System

1. Incoming solar radiation is partly absorbed or scattered by molecules in the atmosphere.

2. The *albedo* of a surface is the proportion of solar radiation that it reflects.

3. The atmosphere absorbs longwave energy from Earth and counterradiates some of it back to Earth, creating the **greenhouse effect**.

4 Formation of Clouds

1. The **adiabatic principle** states that when a gas is compressed, it warms, and when a gas expands, it cools.

2. When a parcel of air moves upward in the atmosphere, it encounters lower pressure, so it expands and cools.

3. The *dry adiabatic lapse rate* describes the rate of cooling with altitude. When condensation or deposition is occurring, the rate of cooling is described as the *wet adiabatic lapse rate*.

4. Clouds are composed of droplets of water or crystals of ice that form on *condensation nuclei*. They occur in layers, as stratiform clouds, or in globular masses, as cumuliform clouds.

5 Precipitation

1. Precipitation from clouds occurs as rain, hail, snow, or sleet.

2. There are four types of precipitation processes: orographic, convectional, cyclonic, and convergent.

3. In **orographic precipitation**, air moves up and over a mountain barrier. As it moves up, it is cooled adiabatically and rain forms. As it descends the far side of the mountain, it is warmed, producing a rain shadow effect.

4. When a surface is heated unequally, a parcel of air can become warmer and less dense than the surrounding air. Because it is less dense, it rises. As it moves upward, it cools, and condensation with precipitation may occur. This is known as **convectional precipitation**.

5. If the air is unstable, **thunderstorms** can form, generating hail and lightning.

KEY TERMS

- **lapse rate** p. 428
- **troposphere** p. 428
- **stratosphere** p. 429
- **ozone** p. 429
- **latent heat** p. 432
- **humidity** p. 432
- **insolation** p. 434
- **greenhouse effect** p. 436
- **adiabatic principle** p. 438
- **orographic precipitation** p. 445
- **convectional precipitation** p. 446
- **thunderstorms** p. 447

CRITICAL AND CREATIVE THINKING QUESTIONS

1. The amount of dangerous ultraviolet radiation that reaches the land surface is a function of latitude, altitude, and clear skies. Where in the world are people most likely to have to take measures to prevent exposure to ultraviolet radiation? If you were to visit such a place, what precautions should you take?

2. The terms *greenhouse effect* and *global warming* are often used interchangeably. Is that a correct thing to do? How would you describe the possible connection between the two if you were to give an explanation to some students?

3. Within the state or region in which you live, is precipitation principally orographic or principally convectional? Under what conditions could convectional precipitation exist in an orographic situation?

4. Some Earth scientists who work in the tundra regions beyond the tree line are concerned that organic matter in the permafrost region will start to decay and release methane if the global temperature rises. Some of the organic matter will also oxidize, and release carbon dioxide. Which would produce the most potent greenhouse effect, a doubling of the methane or a doubling of the carbon dioxide content of the atmosphere?

5. The term *carbon footprint* is used to describe the impact a person has with respect to the increase of carbon dioxide in the atmosphere as a result of daily activities. Work out your personal carbon footprint over the past month (you may need to do some research to complete the calculation). What measures are being taken to help individuals and organizations in your area reduce their carbon footprints?

What is happening in this picture ?

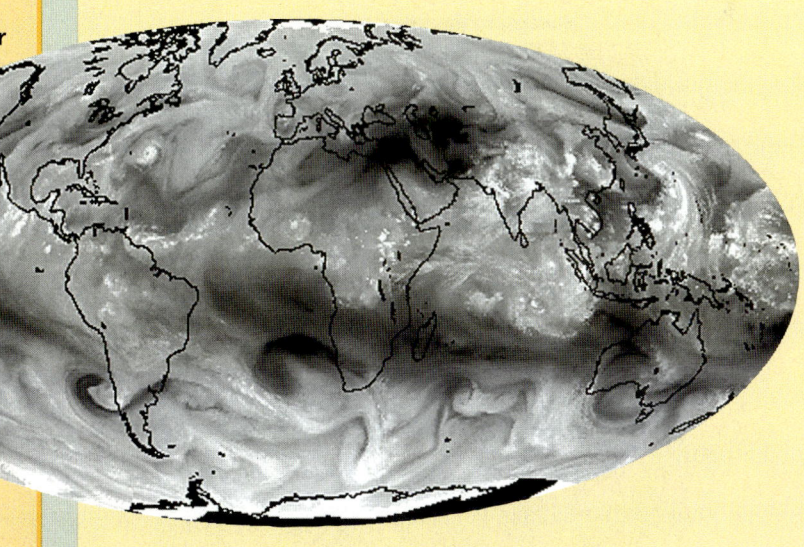

The Geostationary Operational Environmental Satellite (GOES) system is an important tool for weather forecasting. The satellites can show the movement of clouds over Earth's surface or, as in this case, track global water vapor patterns. This is the water vapor image from April 11, 2000. The brightest areas show regions of active precipitation. Dark areas show regions with low water vapor content.

- Are the areas of precipitation and low water vapor located where you would expect?

- Cyclones are marked by dry air spiraling inward with moist air. Can you spot any cyclonic systems building up?

SELF-TEST

1. The diagram shows the proportions of gases in the atmosphere. Label the three main gases that form the bulk of the atmosphere.

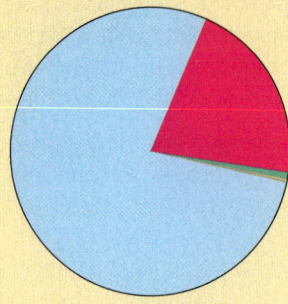

2. Gases are well mixed in the _____ region of the atmosphere.
   a. homosphere
   b. mesosphere
   c. heterosphere
   d. thermosphere

3. The chemical formula _____ represents ozone.
   a. O_2 b. O^2 c. O_3 d. O_4

4. Air temperature changes with altitude in the atmosphere. On the graph, label (a) the tropopause, (b) the troposphere, and (c) the stratosphere.

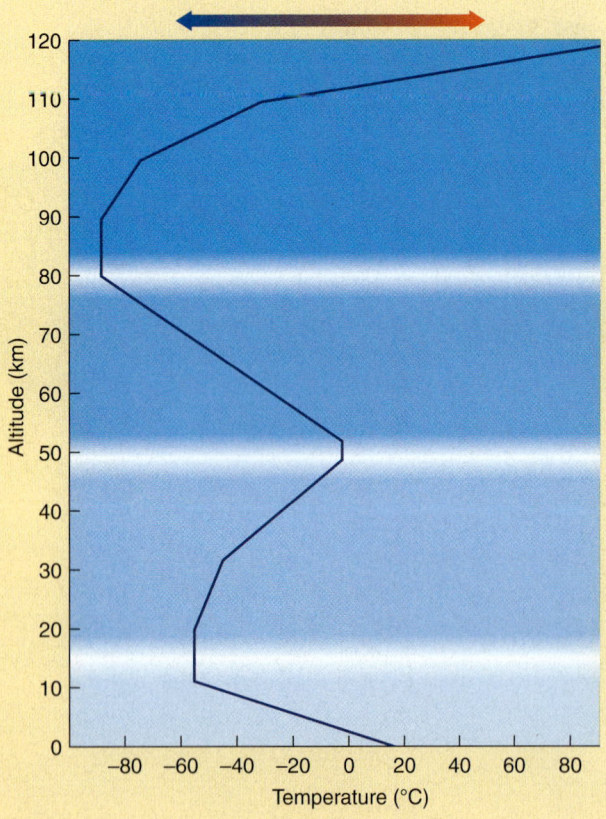

5. The _____ of the air represents the actual quantity of water vapor held by the air.
   a. relative humidity
   b. saturation level
   c. specific humidity
   d. absolute saturation level

6. An _____ temperature change occurs solely as a result of air expansion or compression. A rising air mass, as shown here, expands and _____, while a sinking air mass contracts and _____.
   a. isobaric/cools/heats
   b. adiabatic/cools/heats
   c. isobaric/heats/cools
   d. adiabatic/heats/cools

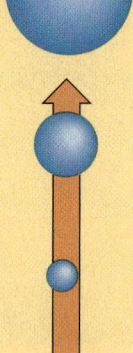

7. The proportion of shortwave radiant energy scattered upward by a surface is termed its _____.
   a. outflow
   b. albedo
   c. output
   d. reflection

8. While Earth's surface can radiate only longwave radiation _____, the atmosphere radiates longwave radiation _____.
   a. upward, downward
   b. upward, in all directions
   c. downward, upward
   d. upward, at right angles to the upward flow

9. _____ from the atmosphere helps warm Earth's surface through a process known as the _____.
   a. Outbound longwave radiation/ozone effect
   b. Insolation/greenhouse effect
   c. Counterradiation/greenhouse effect
   d. Counterradiation/ozone effect

10. Since rising air cools less rapidly when condensation is occurring as a result of the release of latent heat, the _____ has a lesser absolute value than the _____.
    a. dry adiabatic lapse rate/wet adiabatic lapse rate
    b. dry adiabatic lapse rate/environmental adiabatic lapse rate
    c. environmental adiabatic lapse rate/wet adiabatic lapse rate
    d. wet adiabatic lapse rate/dry adiabatic lapse rate

11. The photo shows a _____ cloud, which forms at a _____ level in the sky.
    a. stratocumulus/low   c. altocumulus/middle
    b. nimbostratus/low   d. cirrus/high

12. _____ is the type of precipitation that forms as rain freezes during its descent through the atmosphere.
    a. Freezing rain   c. Hail
    b. Snow   d. Sleet

13. _____ precipitation is a result of air being lifted over a highland area.
    a. Convective   c. Convergence
    b. Orographic   d. Frontal

14. _____ lifting of air is a result of heating.
    a. Convective   c. Frontal
    b. Orographic   d. Adiabatic

15. The two conditions that promote thunderstorm development are _____.
    a. warm, moist air and a decreasing lapse rate
    b. cool, dry air colliding with a colder air mass
    c. warm, moist air and an environmental lapse rate with an absolute value greater than the wet and dry rates
    d. cold, wet air, and an environmental lapse rate with an absolute value greater than the wet and dry rates

Global Circulation and Weather Systems 15

In the last weeks of 1997, extreme weather throughout the world killed about 2100 people and left property damage worth $33 billion in its wake. It began innocently enough with a seesawing in atmospheric pressure in the western Pacific. But that change set in motion a chain of events that wreaked havoc around the world. Forest fires raged in Sumatra, Malaysia, and Borneo, while large portions of Australia and the East Indies were plunged into drought. Torrential rains drenched Peruvian and Ecuadorian coastal ranges, and ice storms left 4 million people in Quebec and the northeastern United States stranded without power. This was the work of El Niño.

El Niño occurs every three to eight years. Pictured here are fires that swept across the vast Indonesian island of Borneo during 2006 and 2007. *El Niño* was given its Spanish name, which means "the little boy" or "Christ Child," by Peruvian fishermen whose anchovy harvests plummeted around Christmastime. Under normal conditions, strong winds blow westward, "piling up" very warm ocean water in the western Pacific. Along the South American coast, this water is replaced by cool bottom water that is filled with nutrients for feeding marine life. But a chain of El Niño events kills these winds, and without the pressure of these winds to hold them back, warmer waters, lacking in nutrients, surge eastward, lowering the anchovy population.

We don't know exactly what causes El Niño, but it is a striking example of our planet's dynamic patterns of wind and water circulation.

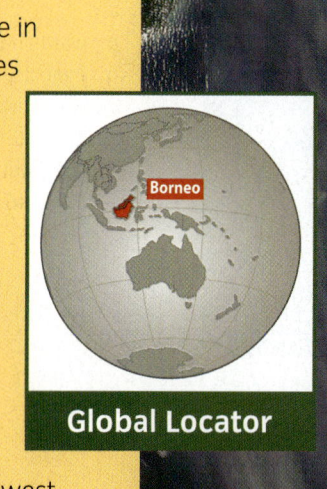

Global Locator

CHAPTER OUTLINE

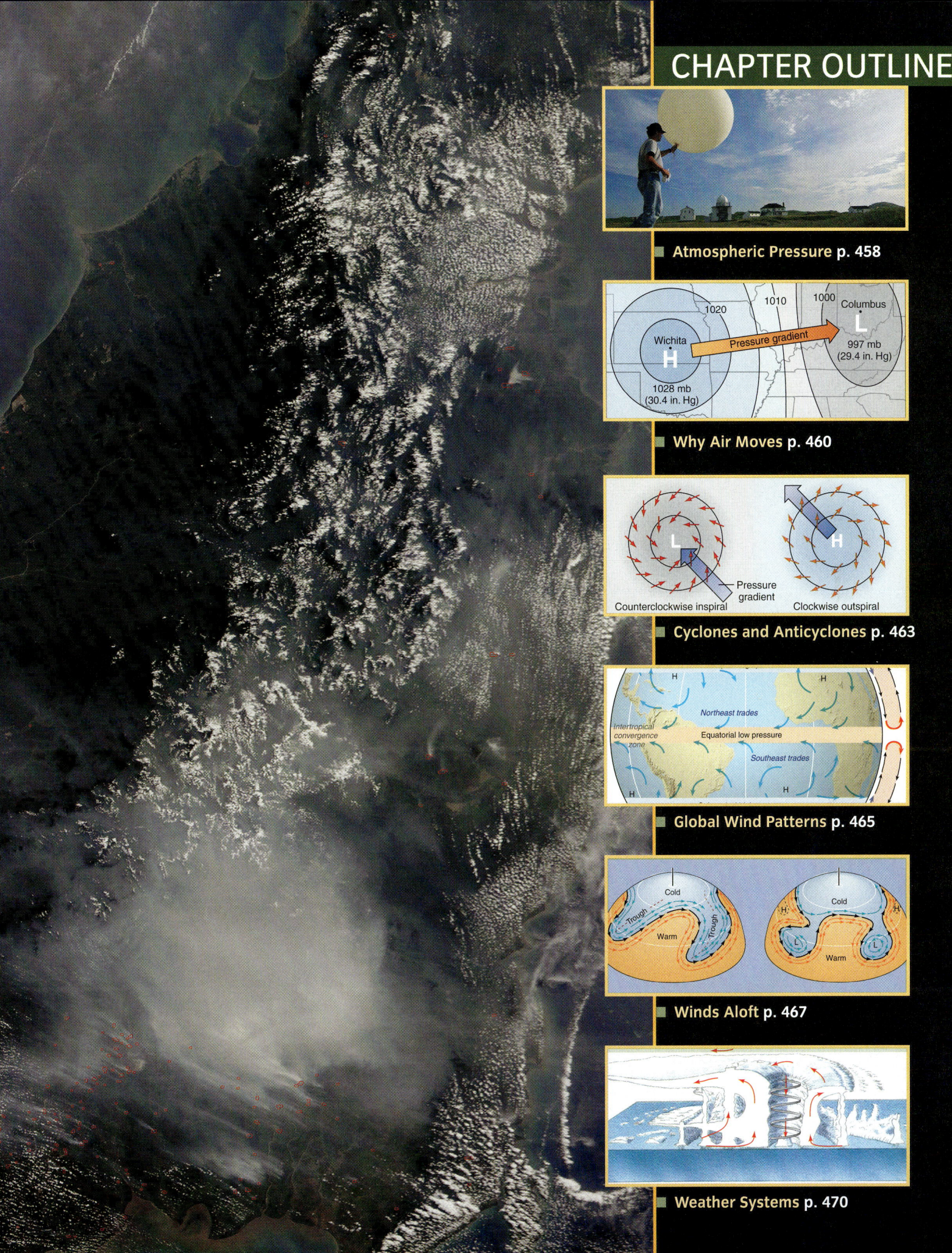

- **Atmospheric Pressure** p. 458
- **Why Air Moves** p. 460
- **Cyclones and Anticyclones** p. 463
- **Global Wind Patterns** p. 465
- **Winds Aloft** p. 467
- **Weather Systems** p. 470

Atmospheric Pressure

LEARNING OBJECTIVES

Define atmospheric pressure.

Explain how a barometer works.

Discuss how air pressure varies with altitude.

We live at the bottom of a vast ocean of air—Earth's *atmosphere* (**FIGURE 15.1**). Like the water in the ocean, the air in the atmosphere is constantly pressing on Earth's surface and on objects on the surface.

The atmosphere exerts pressure because *gravity* pulls the gas molecules in the air toward Earth. Gravity is a force of attraction that acts between all masses—in this case, between gas molecules and Earth's vast bulk.

Atmospheric pressure is produced by the weight of a column of air above a specified area of Earth's surface.

When TV weather forecasters talk about "highs" and "lows," they are referring to air pressure that is higher or lower than average. At sea level, about 1 kilogram of air presses down on each square centimeter of surface (1 kg/cm^2)—about 15 pounds on each square inch of surface (15 lb/in.^{2}).

The basic metric unit of pressure is the *pascal* (Pa). Pressure is measured in terms of the force (measured in units called *newtons* [N]) bearing down on a certain area, and 1 Pa = 1 N per m^2. At sea level, the average pressure of air is 101,320 Pa. Many atmospheric pressure measurements are reported in *bars* and *millibars* (mb) (1 bar = 1000 mb = 10,000 Pa). In this book, we will use the millibar as the metric unit of atmospheric pressure. Standard sea-level atmospheric pressure is 1013.2 mb.

atmospheric pressure Pressure exerted by the atmosphere at Earth's surface because of the force of gravity acting on the overlying column of air.

An ocean of air FIGURE 15.1

Earth's terrestrial inhabitants live at the bottom of an ocean of air. This buoyant weather balloon, known as a *radiosonde*, is on Sable Island, Nova Scotia. It carries instruments that measure temperature and pressure at high atmospheric levels and radios the data to scientists on the ground.

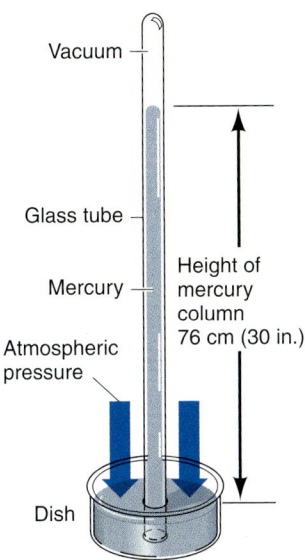

The mercury barometer FIGURE 15.2

Atmospheric pressure pushes the mercury upward into the tube, balancing the pressure exerted by the weight of the mercury column. As atmospheric pressure changes, the level of mercury in the tube rises and falls.

You probably know that a **barometer** measures atmospheric pressure. But do you know how it works? It's based on the same principle as drinking soda through a straw. When using a straw, you create a partial vacuum in your mouth by lowering your jaw and moving your tongue. The pressure of the atmosphere then forces soda up through the straw. A similar process occurs in a barometer (**FIGURE 15.2**).

> **barometer** An instrument that measures atmospheric pressure.

AIR PRESSURE AND ALTITUDE

If you have felt your ears "pop" during an elevator ride in a tall building or on an airplane that is climbing or descending, you've experienced a change in air pressure related to altitude.

In 1658, a young French scientist, Blaise Pascal (1623–1662), carried out an important experiment to prove that air pressure decreases with altitude. He arranged for rock climbers to ascend a prominent volcanic rock known as Puy-de-Dôme, measuring the air pressure at several places during the ascent. Despite the inconvenience of carrying a meter-long glass tube and a flask of mercury up the steep slope of the Puy, the climbers performed the task successfully, and their measurements demonstrated that air pressure decreases with altitude (**FIGURE 15.3**).

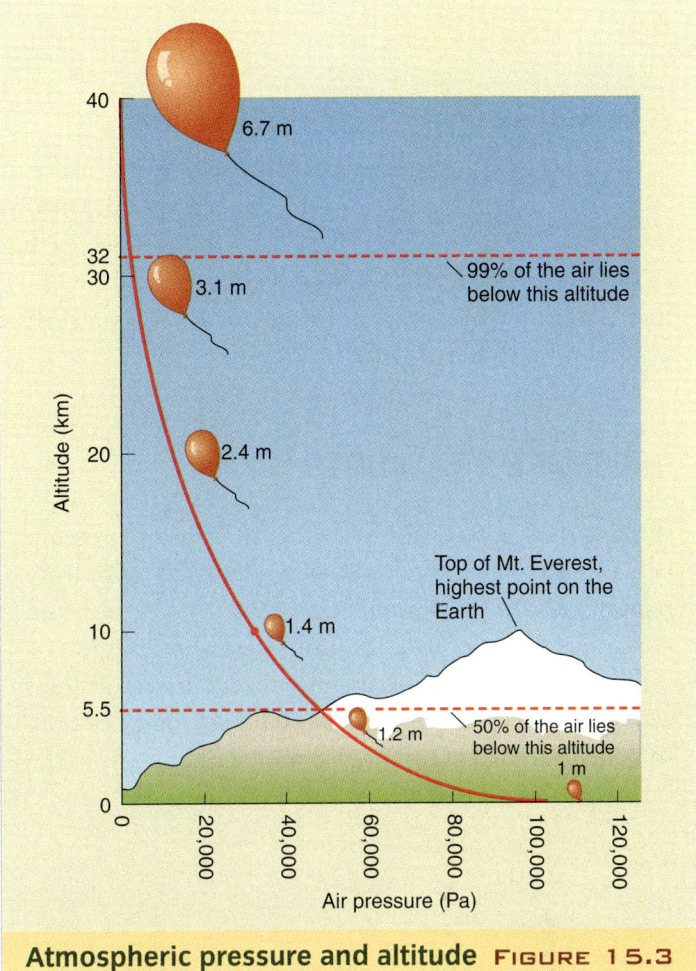

Atmospheric pressure and altitude FIGURE 15.3

Air pressure decreases steadily with altitude. If a helium balloon 1 m in diameter is released at sea level, it expands as it floats upward because of the decrease in pressure. If the balloon did not burst, it would be 6.7 m in diameter at a height of 40 km.

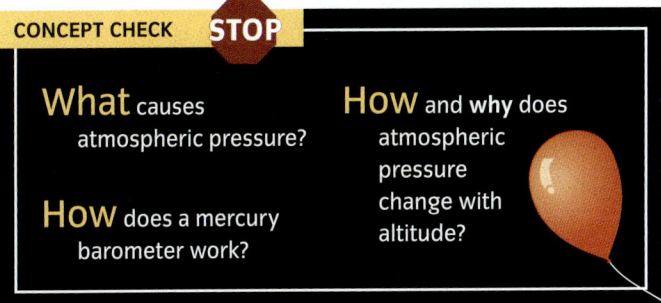

CONCEPT CHECK STOP

What causes atmospheric pressure?

How does a mercury barometer work?

How and **why** does atmospheric pressure change with altitude?

Atmospheric Pressure 459

Why Air Moves

LEARNING OBJECTIVES

Describe how pressure gradients drive wind.

Discuss local wind systems.

Explain convection loops.

ind is defined as air moving horizontally over Earth's surface. Air movements can also be vertical, but these are referred to by other terms, such as *updrafts* or *downdrafts*. Wind direction is identified by the direction from which the wind comes—a westerly wind blows *from* west *to* east, for example.

PRESSURE GRADIENTS

Wind is caused by differences in atmospheric pressure between one place and another. Air tends to move from regions of high pressure to regions of low pressure, and continues to do so until the pressure in both regions is uniform. On a weather map, lines that connect locations with equal pressure are called **isobars**. A change of pressure, or **pressure gradient**, occurs at a right angle to the isobars (**FIGURE 15.4**).

Pressure gradients develop because of unequal heating in the atmosphere. We can see how this occurs by examining the development of **convection loops**, as shown in **FIGURE 15.5**.

LOCAL WINDS

If you're from Southern California, you're probably familiar with the *Santa Ana*, a fierce, searing wind that often drives raging wildfires into foothill communities (**FIGURE 15.6**). In October 2007, fires driven by Santa Ana

> **isobars** Lines on a map drawn through all points having the same atmospheric pressure.
>
> **pressure gradient** A change of atmospheric pressure, measured along a line at right angles to the isobars.
>
> **convection loop** A circuit of moving fluid, such as air or water, created by unequal heating of the fluid.

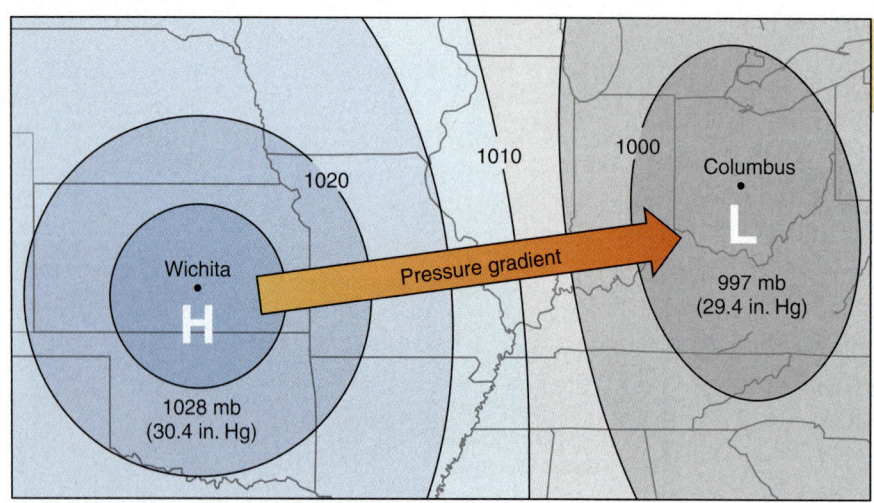

Isobars and a pressure gradient FIGURE 15.4

This figure shows a pressure gradient. Because atmospheric pressure is higher at Wichita than at Columbus, the pressure gradient will push air toward Columbus, producing wind. A greater pressure difference between the two locations would produce a greater force and a stronger wind.

460 CHAPTER 15 Global Circulation and Weather Systems

Convection loops FIGURE 15.5

Uniform atmosphere (heated equally) Imagine a uniform atmosphere above a ground surface. The isobaric levels (or "surfaces") are parallel with the ground surface. (*Isobar* = equal pressure)

Uneven heating Imagine now that the underlying ground surface, an island, is warmed by the Sun, with cool ocean water surrounding it. The warm surface air rises and mixes in with the air above, warming the column of air above the island. Since the warmer air occupies a large volume, the isobaric levels are pushed upward.

Pressure gradient The result is that a pressure gradient is created and air at higher pressure (above the island) flows toward lower pressure (above the ocean).

Surface pressure Because air is moving away from the island and over the ocean surfaces, the surface pressure changes. There is less air above the island, so the ground pressure there drops. Since more air is now over the ocean surfaces, the pressure there rises.

Convection loops The new pressure gradient at the surface moves air from the ocean surfaces toward the island, while air moving in the opposite direction at upper levels completes the two loops.

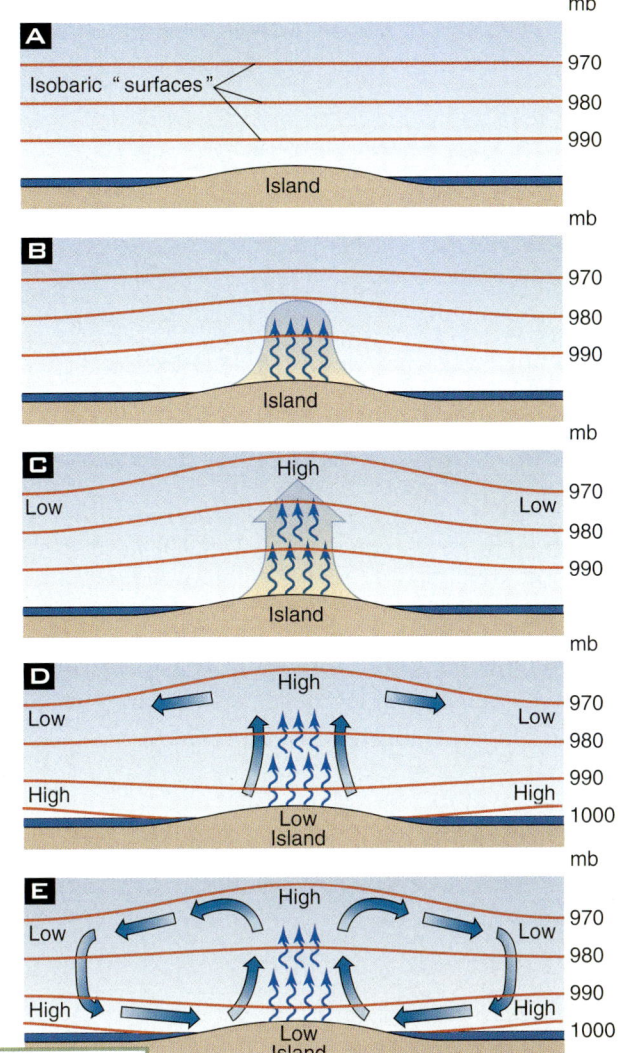

Process Diagram

CRITICAL THINKING — Here's an interesting question:
- Suppose the island was in the Arctic and covered by ice; would the convection loop be the same or different?

winds destroyed more than 1500 homes and killed at least nine people in a huge crescent of wildfire that extended over 500,000 acres from Santa Barbara County to the Mexican border. The Santa Ana is an example of a *local wind* system—one that is generated by local effects. Other local winds include the *Chinook*, a warm, dry wind that occurs in North America where the Canadian Prairies and Great Plains meet various mountain ranges.

California fires FIGURE 15.6

When months of rainless weather team up with a hot, dry Santa Ana wind, the conditions are right for devastating brushfires that can set residential areas aflame.

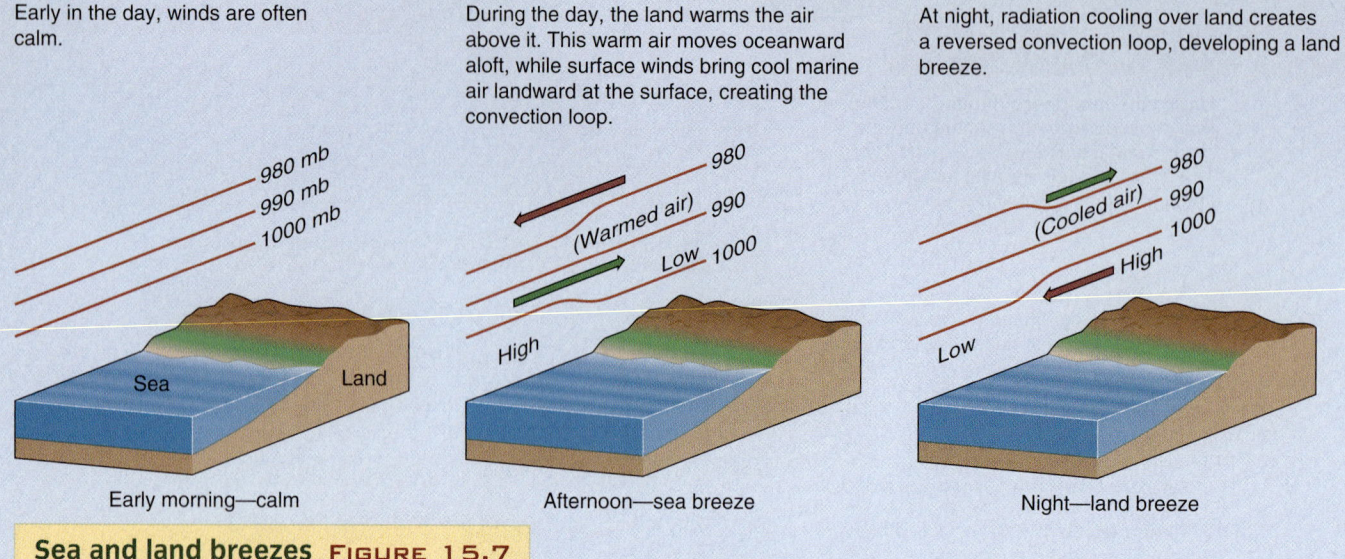

Early in the day, winds are often calm.

During the day, the land warms the air above it. This warm air moves oceanward aloft, while surface winds bring cool marine air landward at the surface, creating the convection loop.

At night, radiation cooling over land creates a reversed convection loop, developing a land breeze.

Early morning—calm

Afternoon—sea breeze

Night—land breeze

Sea and land breezes FIGURE 15.7

Since land surfaces heat and cool more rapidly than does water, temperature contrasts often develop along the coastline.

Local winds, such as the Santa Ana and the Chinook, are generated when the air is unevenly heated, setting up convection loops. FIGURE 15.7 provides a simple example of how such convection loops form and create *sea* and *land breezes*. A similar situation creates *mountain* and *valley winds* (FIGURE 15.8).

Valley and mountain breezes FIGURE 15.8

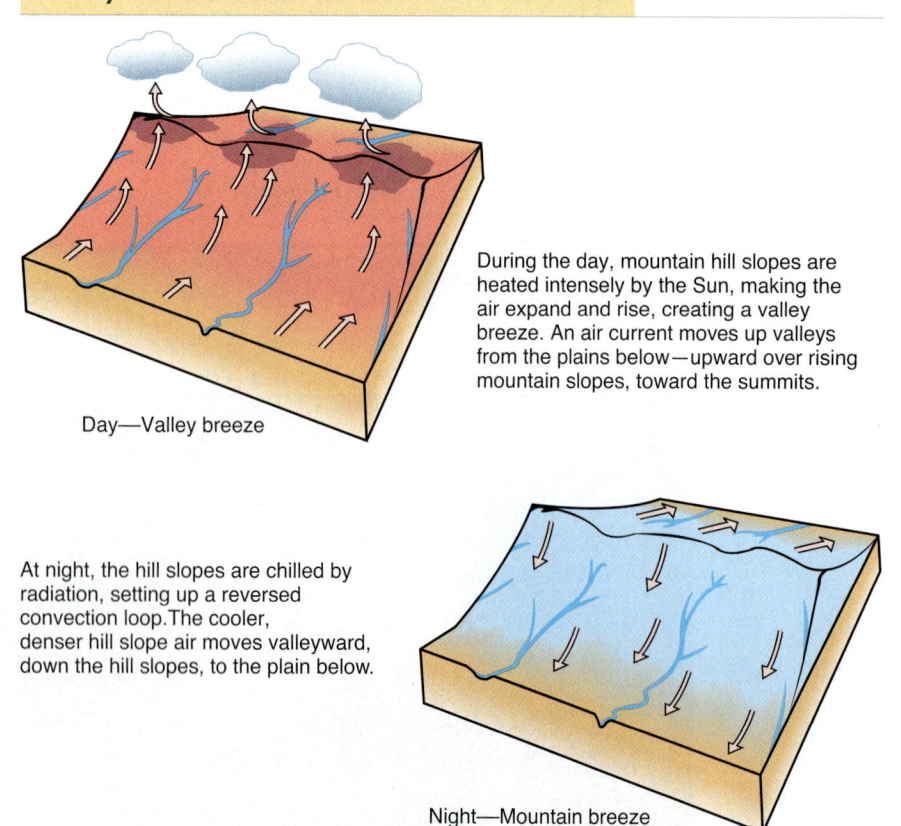

Day—Valley breeze

During the day, mountain hill slopes are heated intensely by the Sun, making the air expand and rise, creating a valley breeze. An air current moves up valleys from the plains below—upward over rising mountain slopes, toward the summits.

At night, the hill slopes are chilled by radiation, setting up a reversed convection loop. The cooler, denser hill slope air moves valleyward, down the hill slopes, to the plain below.

Night—Mountain breeze

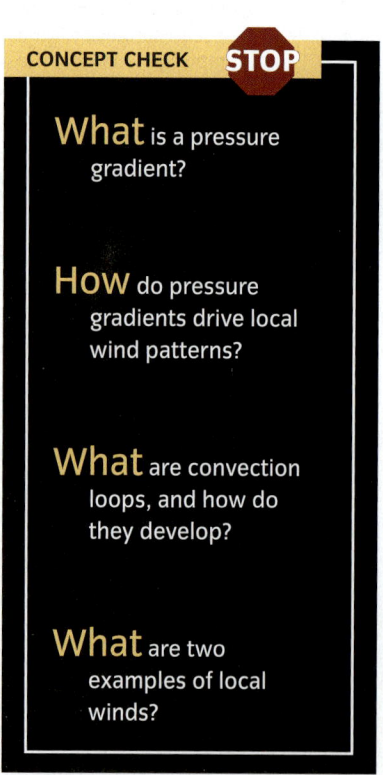

CONCEPT CHECK STOP

What is a pressure gradient?

How do pressure gradients drive local wind patterns?

What are convection loops, and how do they develop?

What are two examples of local winds?

Cyclones and Anticyclones

LEARNING OBJECTIVES

Explain the Coriolis effect.

Describe how cyclones and anticyclones develop.

Discuss the weather patterns associated with cyclones and anticyclones.

We've seen that a pressure gradient forces air to move from a high-pressure region to a low-pressure one. For sea and land breezes, which are local in nature, the wind travels in approximately the same direction as the pressure gradient. But on global scales the direction of air motion is more complicated. The difference is caused by the Earth's rotation, through what is known as the **Coriolis effect** (FIGURE 15.9).

Coriolis effect The effect of Earth's rotation, which acts like a force to deflect a moving object on the surface to the right in the Northern Hemisphere and to the left in the Southern Hemisphere.

The Coriolis effect was identified in 1835 by the French scientist Gustave-Gaspard de Coriolis. Because of the Coriolis effect, an object in the Northern Hemisphere moves as if a force were pulling it to the right (FIGURE 15.9B). In the Southern Hemisphere, objects move as if they were being pulled to the left. This apparent deflection doesn't depend on the direction of motion—it occurs whether the object is moving toward the north, south, east, or west.

The Coriolis effect FIGURE 15.9

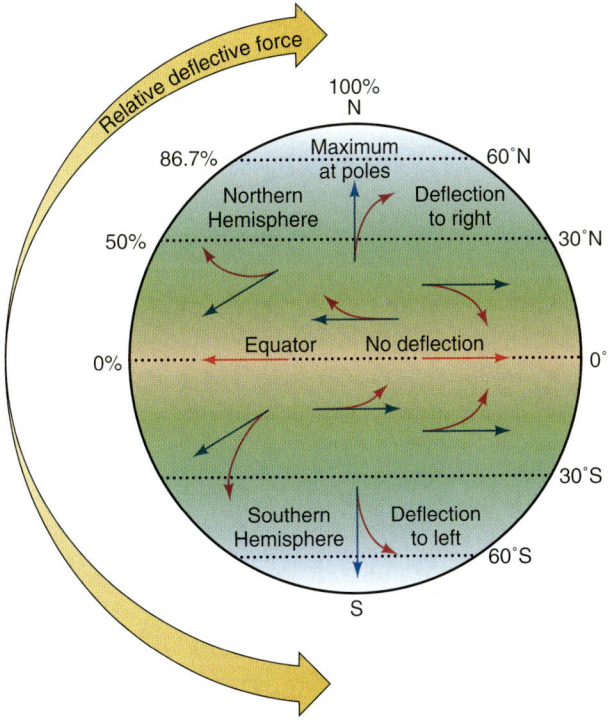

A Imagine that a rocket is launched from the North Pole toward New York City, aimed along the 74° W longitude meridian. As it travels, Earth rotates from west to east beneath its flight path. If you were standing on the rotating Earth below the launch point, you would see the rocket's trajectory curve to the right, away from New York and toward Chicago—despite the fact that the rocket has been flying in a straight line from the viewpoint of space. This apparent deflection is the Coriolis effect.

B The Coriolis effect appears to deflect winds and ocean currents to the right in the Northern Hemisphere and to the left in the Southern Hemisphere. Blue arrows show the direction of initial motion, and red arrows show the direction of motion that is apparent to an observer on Earth. The Coriolis effect is strongest near the poles and decreases to zero at the Equator.

Cyclones and anticyclones FIGURE 15.10

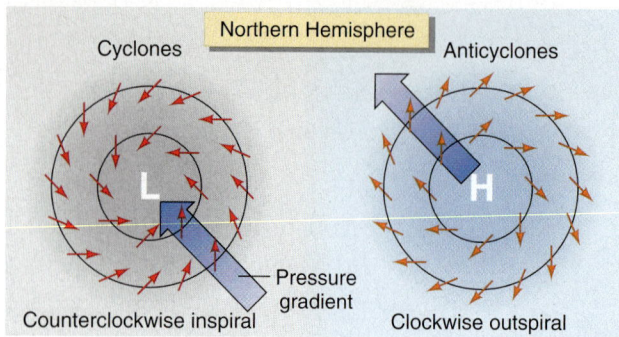

A In a cyclone, low pressure is at the center, so the pressure gradient is straight inward. In an anticyclone, high pressure is at the center, so the gradient is straight outward. But because of the rightward Coriolis force and friction with the surface, the surface air moves at an angle across the gradient, creating a counterclockwise inward-spiraling motion and a clockwise outward-spiraling motion.

B In the Southern Hemisphere, the cyclonic spiral will be clockwise because the Coriolis force acts to the left. For anticyclones, the situation is reversed.

Earth scientists usually analyze air masses or ocean currents from the viewpoint of an observer on Earth, not from the viewpoint of space. So, as a shortcut, we treat the Coriolis effect as a sideward-turning force that always acts at right angles to the direction of motion. The strength of this Coriolis "force" increases with speed of motion but decreases with latitude.

CYCLONES AND ANTICYCLONES

Weather forecasters often refer to low- and high-pressure *centers* on weather maps. To understand what these centers are, look at **FIGURE 15.10**. You can think of them as marking vast whirls of air in spiraling motion. In low-pressure centers, known as **cyclones**, air spirals inward and upward. This inward-spiraling motion is called *convergence*. In high-pressure centers, known as **anticyclones**, air spirals downward and outward. This outward-spiraling motion is called *divergence*.

Low-pressure centers (cyclones) are often associated with cloudy or rainy weather, whereas high-pressure centers (anticyclones) are often associated with fair weather. Why? As we saw in Chapter 14, when air is forced upward, it is cooled according to the *adiabatic principle*, allowing condensation and precipitation to begin. So cloudy and rainy weather often accompanies the inward and upward air motion of cyclones. In contrast, in anticyclones the air sinks and spirals outward. When air descends, it is warmed by the adiabatic process, so condensation can't occur. That is why anticyclones are often associated with fair weather.

Cyclones and anticyclones can cover an area of a thousand kilometers (about 600 mi) or more. A fair-weather system—an anticyclone—may stretch from the Rockies to the Appalachians. Cyclones and anticyclones can remain more or less stationary, or they can move, sometimes rapidly, thus creating weather disturbances.

- **cyclone** A center of low atmospheric pressure.
- **anticyclone** A center of high atmospheric pressure.

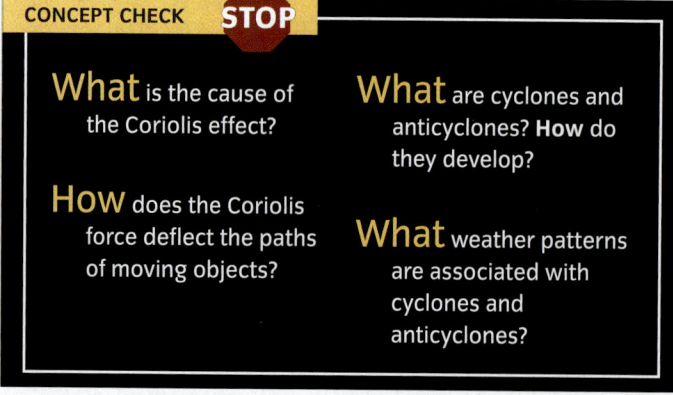

CONCEPT CHECK STOP

What is the cause of the Coriolis effect?

How does the Coriolis force deflect the paths of moving objects?

What are cyclones and anticyclones? **How** do they develop?

What weather patterns are associated with cyclones and anticyclones?

Global Wind Patterns

LEARNING OBJECTIVES

Describe how Earth–Sun relations influence the development of convection cells.

Explain the role of the Coriolis effect in global wind patterns.

Define the intertropical convergence zone and the trade winds.

Two things energize the atmosphere: the Sun's heat and Earth's rotation. Because Earth is a sphere, the Sun does not warm every place on Earth equally. Only at places where the Sun is directly overhead is the maximum amount of heat received per unit of surface area. That's because at those locations the Sun's rays strike Earth perpendicularly. At all other locations the surface is at an angle to the incoming rays, so they receive less heat per unit of surface area (**FIGURE 15.11**).

As we will see in more detail in Chapter 17, the tilt of Earth's axis also helps determine which points on the surface receive the most heat at different times of the year. For now, it is important to understand that at the spring and fall equinoxes the maximum solar energy (about 1366 watts per square meter) falls on the Equator. In December, because of the 23.5° tilt of Earth's axis of rotation to the plane of its passage around the Sun, solar input is most intense at 23.5° S, the Tropic of Capricorn. In June, the incoming solar energy is most intense at 23.5° N, the Tropic of Cancer (see Figure 17.4 on page 520).

Winds and ocean currents are the natural processes by which the Earth system redistributes the heat more evenly. Currents move heat from the Equator—where the input of solar heat is greatest—toward the poles, where it is least. Unequal heating of Earth's surface causes convection loops in the atmosphere, as we saw in Figure 15.5. Heated air near the Equator expands, becomes lighter, and rises. Near the top of the *troposphere*

Solar intensity and latitude FIGURE 15.11

The angle at which the Sun's rays strike Earth's surface varies from one geographic location to another, owing to Earth's spherical shape and its inclination on its axis.

A Sunlight (represented by the flashlight) that shines vertically near the Equator is concentrated on Earth's surface.

B–C Toward the poles, light hits the surface more and more obliquely, spreading the same amount of radiation over larger and larger areas.

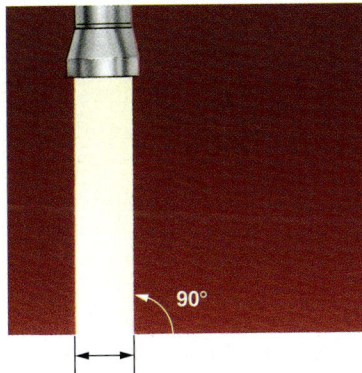

1 unit of surface area

One unit of light is concentrated over 1 unit of surface area.

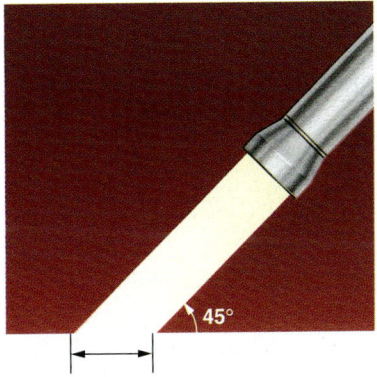

1.4 units of surface area

One unit of light is dispersed over 1.4 units of surface area.

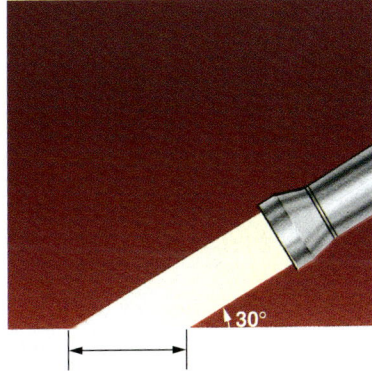

2 units of surface area

One unit of light is dispersed over 2 units of surface area.

Global atmospheric circulation
FIGURE 15.12

Huge convection cells transfer heat from equatorial regions, where the input of solar energy is greatest, toward the poles, where the solar input is least. Because Earth is rotating, the flow of air toward the poles and the return flow toward the Equator are constantly deflected sideways, creating the circulating air masses you may have seen on weather maps. The convection cells are permanent features of Earth's atmosphere and therefore have a great influence both on day-to-day weather and on long-term climate.

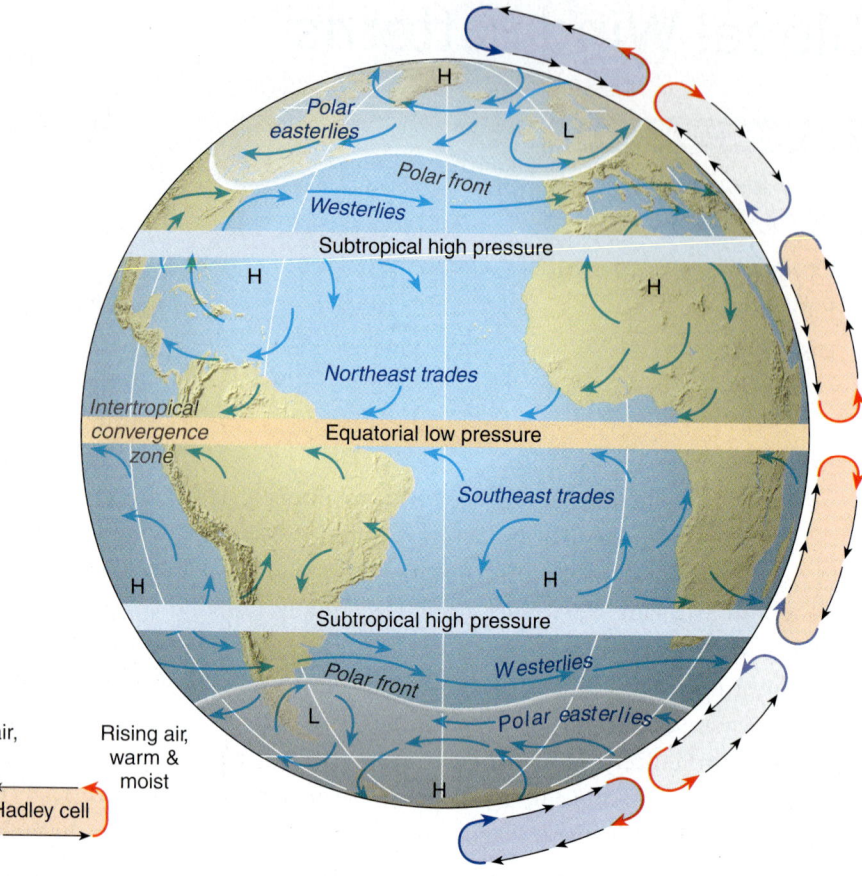

it spreads outward toward the poles. As the upper air travels northward and southward toward the poles, it gradually cools, becomes heavier, and sinks. Upon reaching the surface, this cool air flows back toward the Equator, warms up, and rises again, thereby completing a convective cycle. In reality, it is not quite so simple: Global atmospheric circulation actually organizes itself into three *convection cells* that interlock like gears (**FIGURE 15.12**). We met these same convection cells in Chapter 6 (see Figure 6.1), because they play a major role in the locations of deserts.

WIND SYSTEMS

The Coriolis effect further complicates the picture, breaking up the flow of convective air between the Equator and the poles into belts, as shown in Figure 15.12. For example, a large belt or cell of circulating air lies between the Equator (0°) and about 30° latitude in both the Northern and Southern Hemispheres. Warm air rises near the Equator, creating a low-pressure zone called the **intertropical convergence zone**. The air rises to the top of the troposphere and begins to flow toward the poles, but it veers off course as a result of the Coriolis effect. By the time it reaches a latitude of 30°, the high-altitude air mass has cooled and started to sink. The descending air flows back across Earth's surface toward the Equator. As it flows, the land and sea warm the air so that it eventually becomes warm enough to rise again.

intertropical convergence zone (ITCZ) A zone of convergence of air masses along the equatorial trough.

The low-latitude convection cells (from 0° to 30° N and S) created by this circulation pattern are called *Hadley cells.* The prevailing surface winds in Hadley cells are deflected by the Coriolis effect so that they blow northeasterly in the Northern Hemisphere (that is,

466 **CHAPTER 15** Global Circulation and Weather Systems

they flow from the northeast toward the southwest), whereas in the Southern Hemisphere they are southeasterly. These winds are called **trade winds** because their consistent direction and flow carried trading ships across the tropical oceans at a time when winds were the chief source of power.

trade winds Surface winds that blow from about 30° north and south latitude toward the intertropical convergence zone.

A second set of convecting air cells, called *polar cells*, lies over the polar regions. In a polar cell, frigid air flows across the surface away from the pole and toward the Equator, slowly warming as it moves. When the polar air has reached about latitude 60° north or south, it has warmed enough to rise convectively high into the troposphere and flow back toward the pole, where it cools and descends again, thereby completing the convection cell. Because of the Coriolis effect, the cold air that flows away from the poles is deflected to the right, giving rise to a wind system called the *polar easterlies*.

Between the Hadley cells and the polar cells lies a third, less well-defined set of convection cells. Between about latitude 30° and latitude 60° north and south, we find midlatitude cells, called *Ferrel cells*, in which air flows toward the north and therefore is deflected to the right by the Coriolis effect, creating a wind that blows from the west. That is why weather systems in the continental United States (and southern Canada) travel from west to east. If you take some time to study Figure 15.12, these circulation patterns will become clearer.

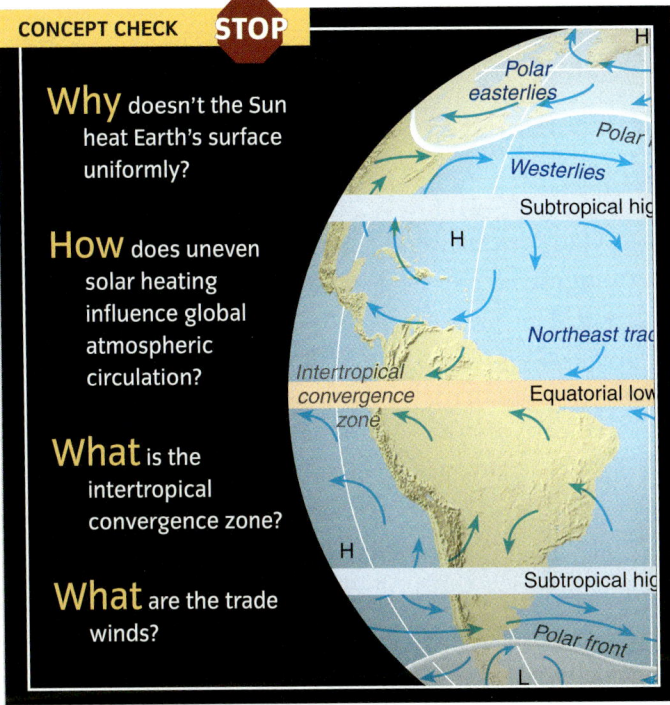

CONCEPT CHECK STOP

Why doesn't the Sun heat Earth's surface uniformly?

How does uneven solar heating influence global atmospheric circulation?

What is the intertropical convergence zone?

What are the trade winds?

Winds Aloft

LEARNING OBJECTIVES

Explain how pressure gradients develop at upper atmospheric levels.

Discuss the geostrophic wind.

Show how Rossby waves develop and grow.

Define jet streams.

We've looked at air flows at or near Earth's surface, including both local and global wind patterns. But how does air move at the higher levels of the troposphere? As with air near the surface, winds at upper levels of the atmosphere move in response to pressure gradients and are influenced by the Coriolis effect.

A simple physical principle states that pressure decreases less rapidly with height in warmer air than in colder air. Also recall that the solar energy reaching Earth is greatest near the Equator and least near the poles, resulting in a temperature gradient from the Equator to the poles. This gives rise to a pressure gradient; because the atmosphere is warmer near the Equator than the poles, a pressure-gradient force pushes air toward the poles.

Winds Aloft 467

THE GEOSTROPHIC WIND

How does a pressure-gradient force pushing toward the poles produce wind, and what will the wind's direction be? Any wind motion is subject to the Coriolis force, which turns it to the right in the Northern Hemisphere and to the left in the Southern Hemisphere. So poleward air motion is toward the east, creating westerly winds in both hemispheres.

Unlike air moving close to the surface, an upper air parcel moves without encountering friction because it is so far from the source of friction—the surface. So there are only two forces operating on the air parcel: the pressure-gradient force and the Coriolis force.

Imagine a parcel of air, as shown in **FIGURE 15.13**. The air parcel begins to move poleward in response to the pressure-gradient force. As it accelerates, the Coriolis force pulls it increasingly toward the right. As its velocity increases, the parcel turns increasingly rightward until the Coriolis force just balances the gradient force. At that point, the sum of forces on the parcel is zero, so its speed and direction remain constant. We call this type of air flow the **geostrophic wind**. It occurs at upper levels in the atmosphere, and we can see from the diagram that it flows parallel to the isobars.

> **geostrophic wind**
> Wind at high levels above Earth's surface blowing parallel with a system of straight, parallel isobars.

GLOBAL CIRCULATION AT UPPER LEVELS

Figure 15.12 sketches the general air-flow pattern at higher levels in the troposphere. It has four major features: weak equatorial easterlies, tropical high-pressure belts, upper-air westerlies, and a polar low. We've seen that the general temperature gradient from the tropics to the poles creates a pressure-gradient force that generates westerly winds in the upper atmosphere. These *upper-air westerlies* blow in a complete circuit about the Earth, from about 25° latitude almost to the poles. So the overall picture of upper-air wind patterns is quite simple—a band of weak easterly winds in the equatorial

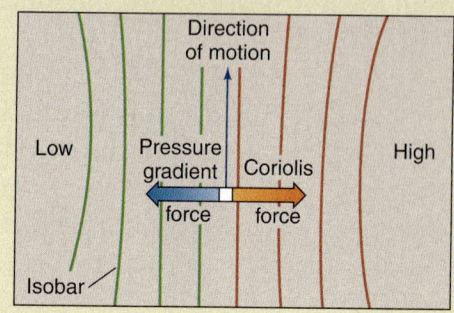

A At upper levels in the atmosphere, a parcel of air is subjected to a pressure-gradient force and a Coriolis force.

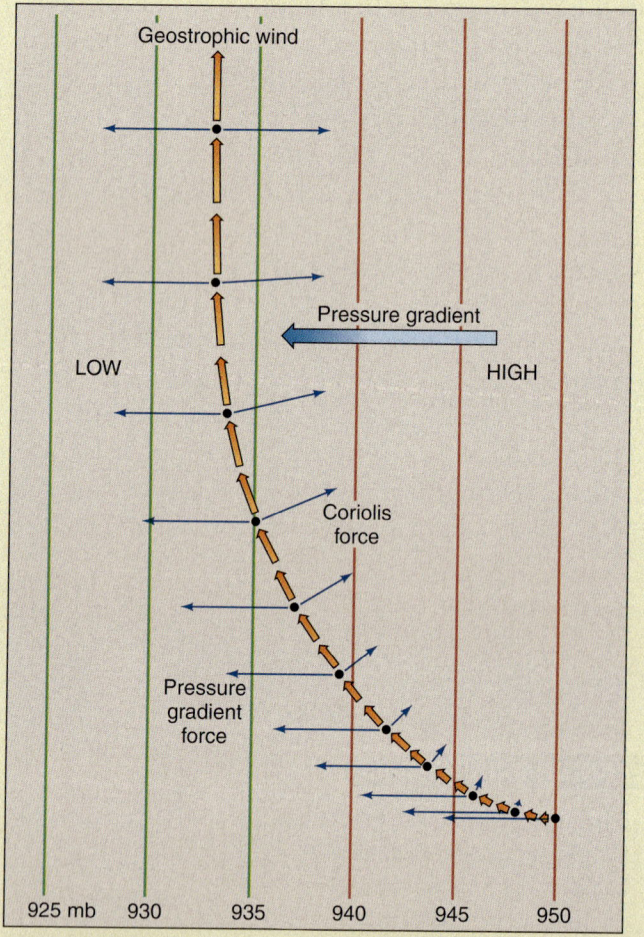

B The parcel of air moves in response to a pressure gradient. At the same time, it is turned progressively sideways until the pressure-gradient force and the Coriolis force balance, producing the geostrophic wind.

The geostrophic wind FIGURE 15.13

zone, belts of high pressure near the Tropics of Cancer and Capricorn, and westerly winds, with some variation in direction, spiraling around polar lows.

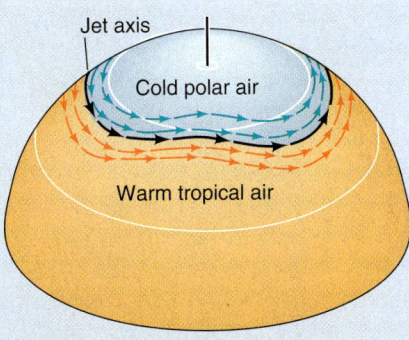

A The flow of air along the front begins to undulate.

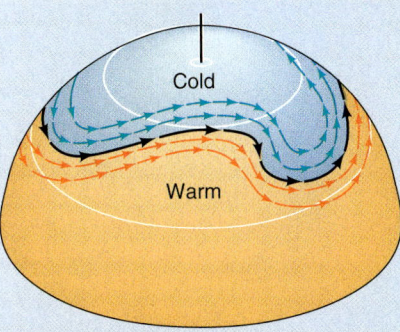

B As the undulation becomes stronger, Rossby waves form. Warm air pushes toward the pole, while cold air is brought to the south.

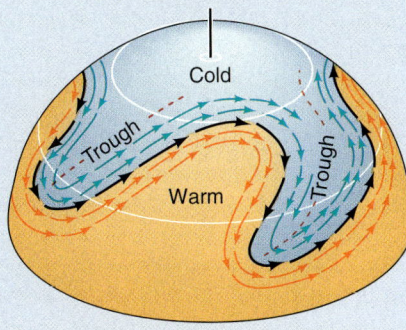

C The waves become stronger, bringing a tongue of cold air to the south as warm air is carried northward.

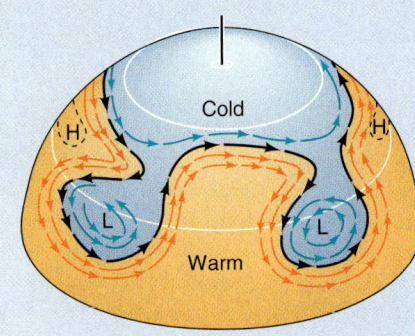

D The tongue is pinched off, leaving a pool of cold air far south of its original location. The waves form cyclones of cold air that may last for up to a week.

Rossby waves FIGURE 15.14

Rossby waves form at the boundary between cold polar air and warm tropical air.

ROSSBY WAVES, JET STREAMS, AND THE POLAR FRONT

The upper-air westerlies undulate back and forth in giant meanders, which are called **Rossby waves**. These waves arise in the zone where cold polar air meets warm tropical air, called the *polar front*. FIGURE 15.14 describes their formation cycle. Rossby waves are the reason that weather in the midlatitudes is often so variable: They cause pools of warm, moist air and cold, dry air to alternately invade midlatitude land masses.

Jet streams are wind streams that reach great speeds in narrow zones at a high altitude (see *What an Earth Scientist Sees* on page 470). They occur in areas where atmospheric pressure gradients are strong. Along a jet stream, the air moves in pulses along broadly curving tracks. The greatest wind speeds occur in the center of the jet stream, with velocities decreasing farther away from it.

The type of jet stream that is closest to the poles is located along the polar front. It is called the *polar-front jet stream* (or simply the *polar jet*) and is generally located between 35° and 65° latitude in both hemispheres. It follows the edges of Rossby waves at the boundary between cold polar air and warm subtropical air. The polar jet is typically found at altitudes of 10 to 12 km (about 30,000 to 40,000 ft), and wind speeds within it range from 75 to as much as 125 m/s (about 170 to 280 mi/hr).

> **Rossby waves** Horizontal undulations in the flow path of the upper-air westerlies; also known as *upper-air waves*.
>
> **jet streams** High-speed air flows in narrow bands within the upper-air westerlies and along certain other global latitude zones at high levels.

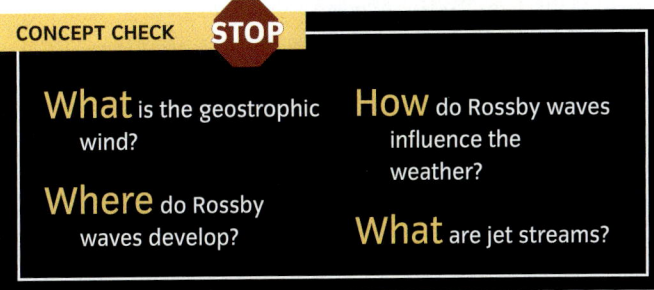

CONCEPT CHECK STOP

What is the geostrophic wind?

Where do Rossby waves develop?

How do Rossby waves influence the weather?

What are jet streams?

What an Earth Scientist Sees

Jet Stream Clouds

In this photo taken from space, the astronauts aimed their camera to take in the Nile River Valley and the Red Sea. At the left you can see the tip of the Sinai Peninsula. They also captured a beautiful band of cirrus clouds at about 25° north latitude. An Earth scientist would identify this as a band of jet stream clouds that occur on the side of the jet toward the Equator. The jet stream is moving from west to east (from the viewer's right to left) at an altitude of about 12 km (40,000 ft).

CRITICAL THINKING

Here's an interesting question:
- To the left of the jet stream clouds there are many small, fluffy clouds. What kind of clouds are they? Are they higher or lower than the jet stream clouds?

Weather Systems

LEARNING OBJECTIVES

Define air mass and **explain** how air masses are classified.

Describe cold, warm, and occluded fronts.

Discuss thunderstorms, wave cyclones, tornadoes, and hurricanes.

Weather systems are often associated with the motion of **air masses**—large bodies of air with fairly uniform temperature and moisture characteristics. An air mass can be several thousand kilometers or miles across and can extend upward to the top of the troposphere. Air masses can be searing hot, icy cold, or any temperature in between, and they can have wildly varying moisture content. They pick up their temperature and moisture characteristics in *source regions*, areas where the air moves slowly or is stationary.

> **air mass** An extensive body of air in which temperature and moisture characteristics are fairly uniform over a large area.

470 CHAPTER 15 Global Circulation and Weather Systems

We classify air masses by latitude and by source region. Latitude is important because it determines the surface temperature and the environmental temperature lapse rate of the air mass. The moisture content is usually determined by whether the air mass originated over a continent or over an ocean. There are seven important types of air masses: maritime equatorial (mE), maritime tropical (mT), continental tropical (cT), continental polar (cP), maritime polar (mP), continental arctic (cA), and continental antarctic (cAA).

FIGURE 15.15 shows the air masses that form near North America and their source regions. These air masses have a strong influence on the continent's weather.

North American air mass source regions and trajectories FIGURE 15.15

Air masses acquire temperature and moisture characteristics in their source regions and then move across the continent.

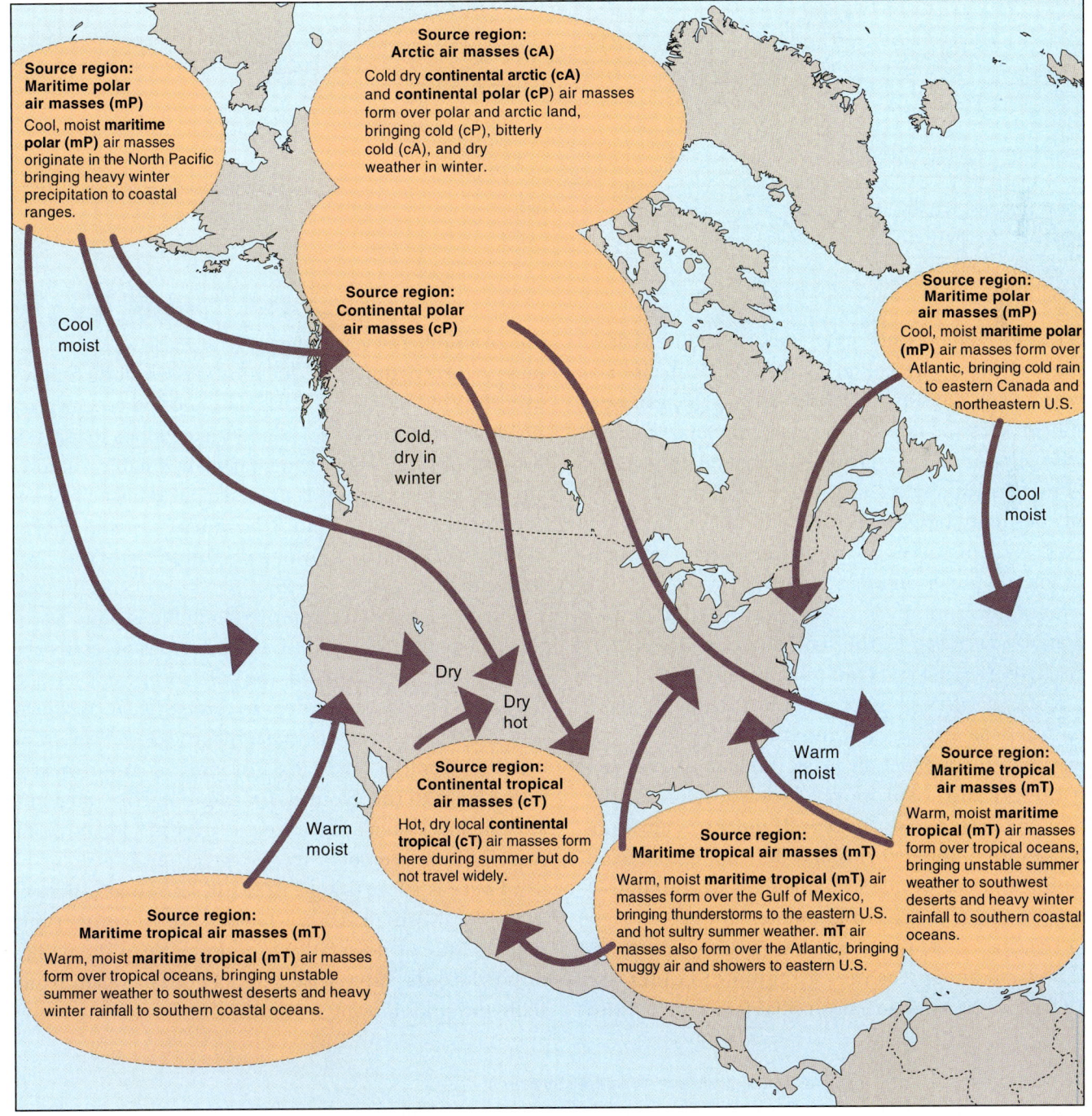

A cold front FIGURE 15.16

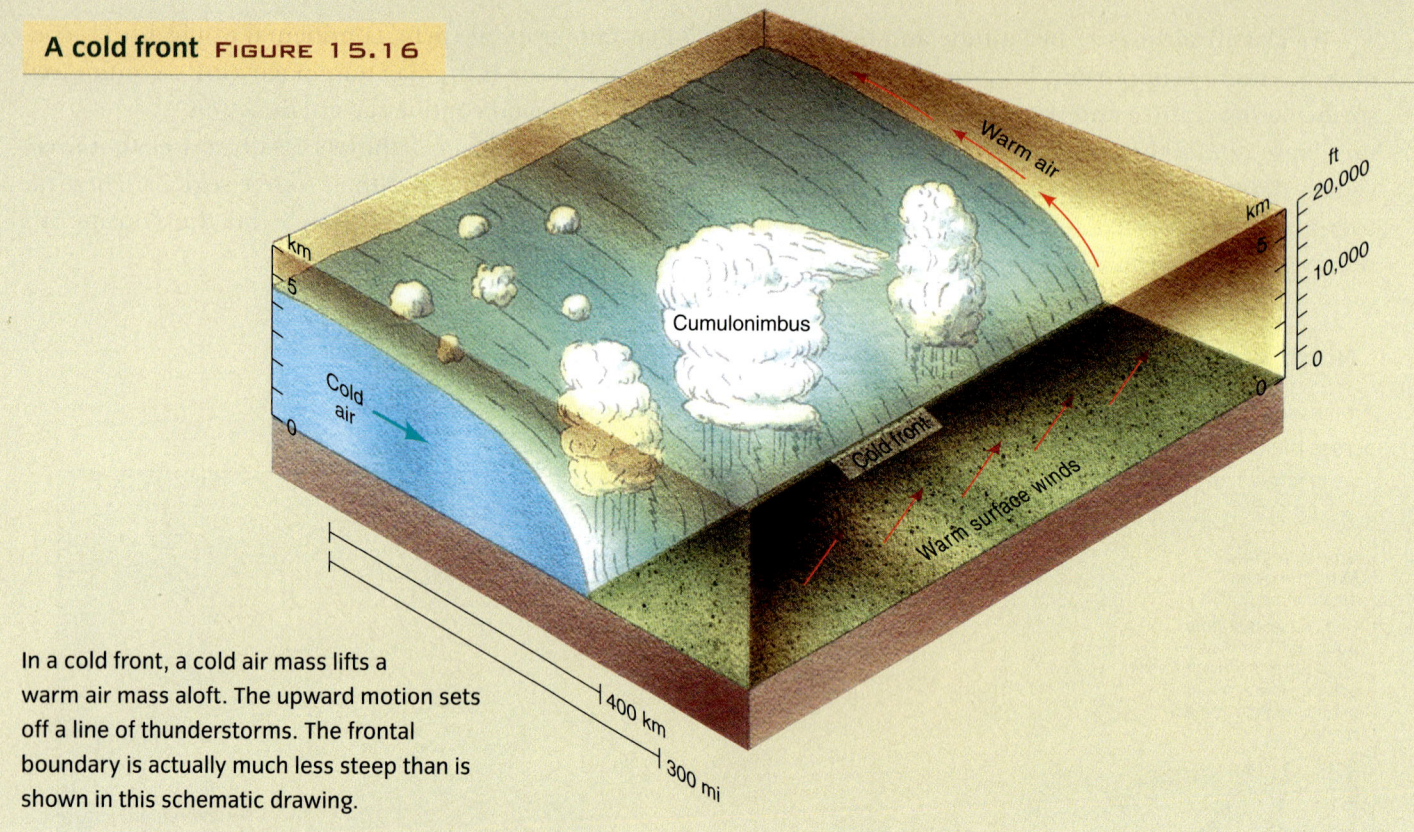

In a cold front, a cold air mass lifts a warm air mass aloft. The upward motion sets off a line of thunderstorms. The frontal boundary is actually much less steep than is shown in this schematic drawing.

Pressure gradients and upper-level wind patterns drive air masses from one region to another. When an air mass moves to a new area, its properties will change because it is influenced by the new surface environment. For example, the air mass may lose heat or take up water vapor.

There is usually a sharply defined boundary, or **front**, between a given air mass and neighboring air masses. A front serves as the leading edge of an air mass, like a bumper on a car.

> **front** The surface of contact between two dissimilar air masses.

FIGURE 15.16 shows the structure of a *cold front*. A cold air mass invades a zone occupied by a warm air mass. Because the colder air mass is denser than the warmer air mass, it remains in contact with the ground. As it moves forward, it forces the warmer air mass to rise above it. If the warm air is unstable, severe thunderstorms may develop. A cold front often forms a line of massive *cumulus*—or globular—clouds that may be tens of kilometers long.

FIGURE 15.17 diagrams a *warm front,* in which warm air moves into a region of colder air. Here again, the cold air mass remains in contact with the ground because it is denser. As before, the warm air mass is forced aloft, but this time it rises along a ramp over the cold air below. This rising motion creates *stratus* clouds—large, dense, blanket-like clouds that often produce precipitation. If the warm air is stable, the precipitation will be steady. If the warm air is unstable, convection cells may develop, producing *cumulonimbus* clouds with heavy showers or thunderstorms (not shown in the figure).

Cold fronts normally move along the ground faster than warm fronts do. So when both types of fronts are in the same neighborhood, the cold front can overtake the warm front. The result is an *occluded front,* as shown in FIGURE 15.18. (*Occluded* means "closed" or "shut off.") The colder air of the fast-moving cold front remains next to the ground, forcing both the warm air and the less cold air ahead to rise over it. The warm air mass is lifted completely free of the ground.

There is a fourth type of front, known as a *stationary front*, in which two air masses are in contact but there is little or no relative motion between them. Stationary fronts often arise when a cold or warm front stalls and stops moving forward.

472 **CHAPTER 15** Global Circulation and Weather Systems

A warm front FIGURE 15.17

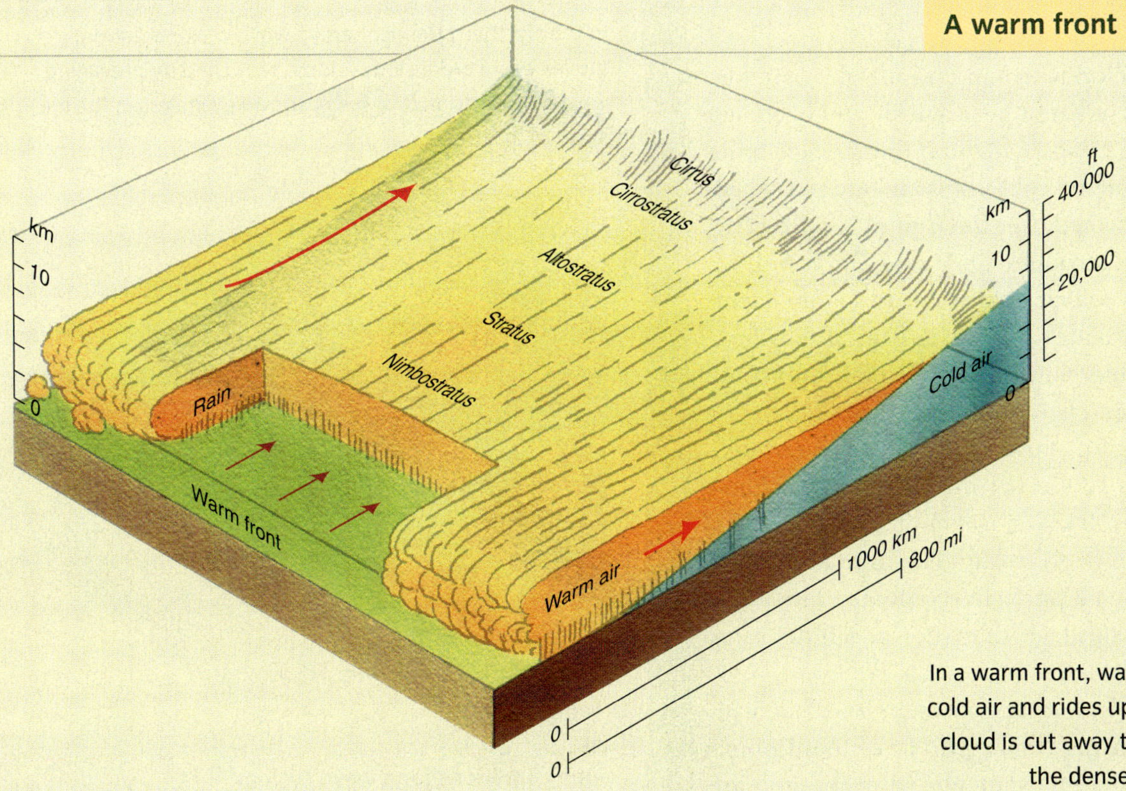

In a warm front, warm air advances toward cold air and rides up and over it. A notch of cloud is cut away to show rain falling from the dense layer of stratus clouds.

The violent disturbances of the atmosphere that mark severe weather and storms can have many causes, but most occur along cold fronts. We will look in detail at three kinds of severe weather—thunderstorms, tornadoes, and hurricanes—that can cause large amounts of damage and even loss of life.

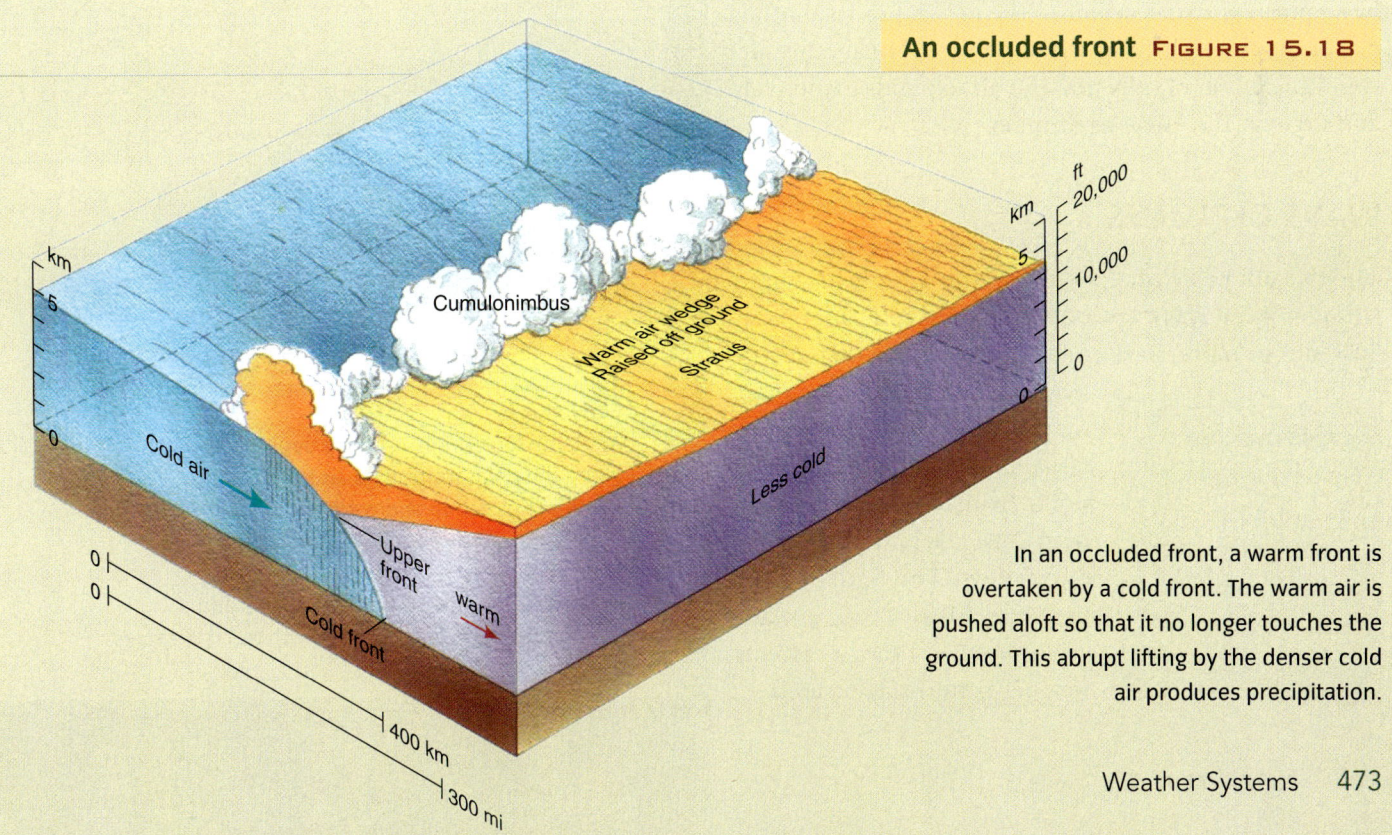

An occluded front FIGURE 15.18

In an occluded front, a warm front is overtaken by a cold front. The warm air is pushed aloft so that it no longer touches the ground. This abrupt lifting by the denser cold air produces precipitation.

Weather Systems 473

THUNDERSTORMS

Thunderstorms develop when an updraft of warm, humid air releases a lot of latent heat quickly and becomes unstable, as we saw in Chapter 14. Most thunderstorms in North America form along cold fronts and are associated with mT air masses formed over the Gulf of Mexico. The released heat causes stronger updrafts, which pull in more warm, moist air, which in turn releases more latent heat. The process escalates and the updraft intensifies, creating cumulonimbus clouds, heavy rainfall, and commonly hail, thunder, and lightning (see Figure 14.22).

Towering masses of dark cumulonimbus clouds, can reach as high as 18 km (11 mi). Winds in a thunderhead can exceed 100 km/h (60 mi/h), and updrafts in a thunderhead can be so strong that large hailstones can form when tiny ice particles coalesce. The hailstones are held aloft until a sudden downdraft deposits them on the ground.

Lightning and *thunder* are caused by electrical charges that form during the growth of a cumulonimbus cloud. The turbulent movement of precipitation inside the cloud causes particles in the upper part to become positively charged and particles in the lower part to become negatively charged. We don't know exactly how the charge builds up, but it can reach hundreds of millions of volts. The charges can be released by a lightning strike, either to the ground or to another cloud. As the lightning strike passes, it heats the surrounding air so rapidly that the air expands explosively and we hear the effect as thunder.

WAVE CYCLONES

The dominant weather system in the middle and high latitudes is the **wave cyclone**. Spanning 1000 km (about 6000 mi) or more, wave cyclones are the *lows* that meteorologists show on weather maps. They form when two large anticyclones meet at the polar front, as shown in **FIGURE 15.19**, and typically last 3 to 6 days. The figure shows the life history of a wave cyclone and how it affects weather patterns as it travels.

wave cyclone
Traveling cyclone of the midlatitudes involving interaction of cold and warm air masses along sharply defined fronts.

Life history of a wave cyclone "low" FIGURE 15.19

In the Northern Hemisphere, a wave cyclone normally moves eastward as it develops, propelled by prevailing westerly winds aloft. ①–④ show key stages in the life cycle of a cyclone.

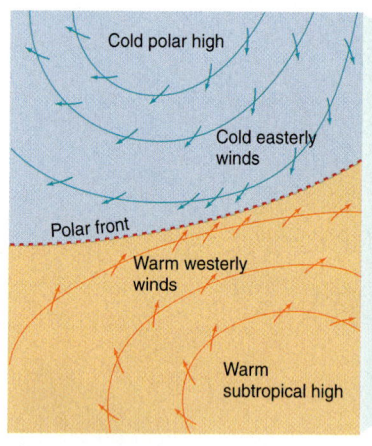

▲ **Ripe conditions for a wave cyclone**
Along the polar front, cold, dry polar air, flowing from the northeast, meets warm, humid subtropical air flowing from the southwest.

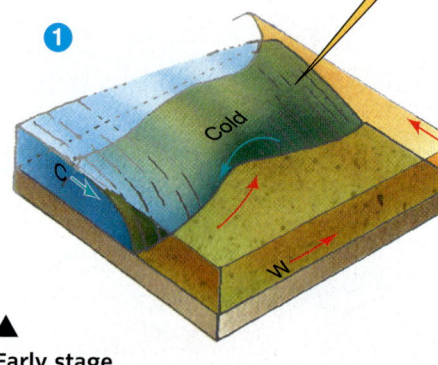

▲ **Early stage**
An undulation or disturbance begins at a point along the polar front. Cold air is turned in a southerly direction and warm air in a northerly direction, so that each advances on the other. This creates two fronts—a cold front indicated by a line with blue triangles and a warm front indicated by a line with red half-circles.

Process Diagram

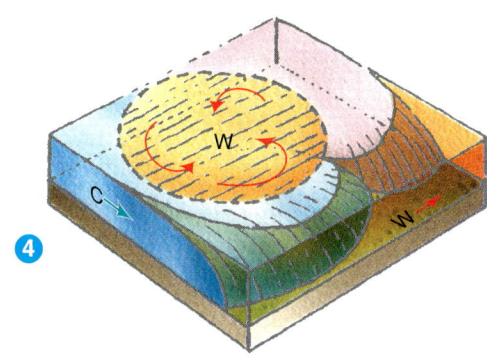

Dissolving stage
Eventually, the polar front is reestablished, but a pool of warm, moist air remains aloft. As its moisture content reduces, precipitation dies out, and the clouds gradually dissolve. Soon, another wave cyclone will form along the polar front and move across the continent.

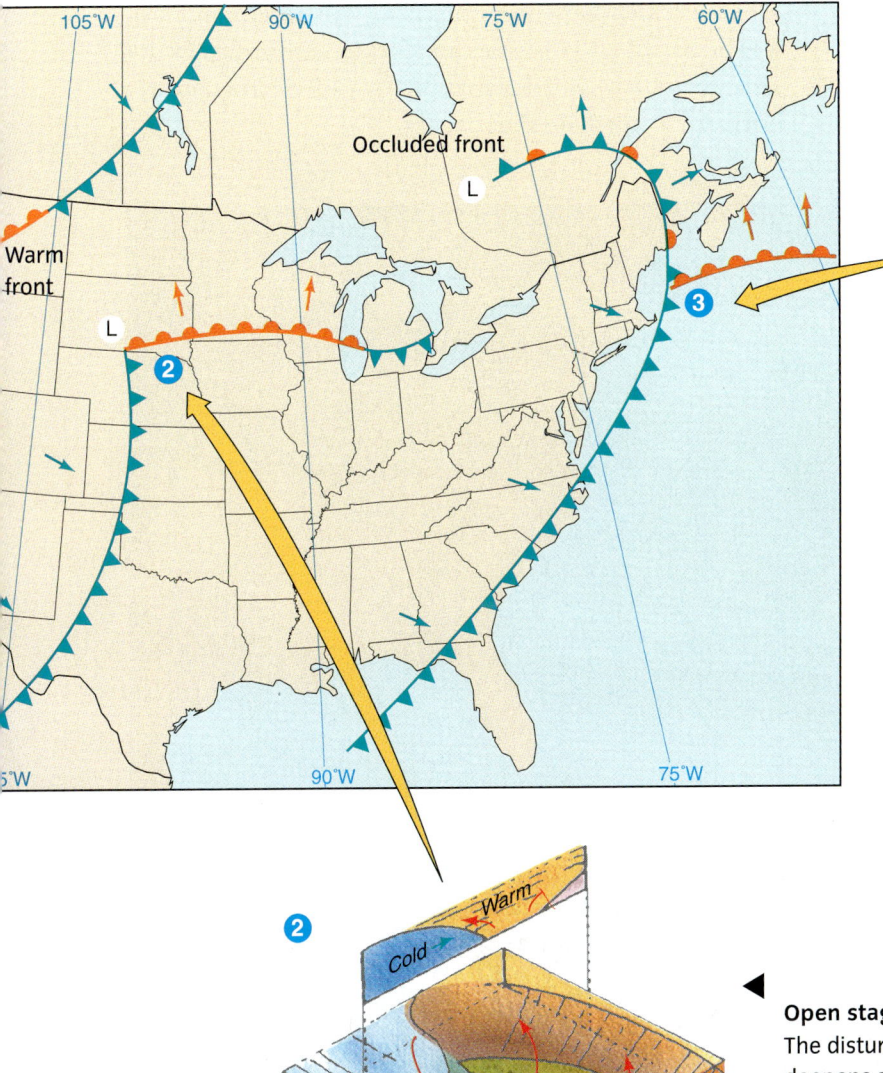

Occluded stage
The faster-moving cold front overtakes the warm front, lifting the warm, moist air mass at the center completely off the ground. Because the warm air is shut off from the ground, this is called an occluded front and is indicated by a line with alternating triangles and half-circles. Precipitation continues to occur as warm air is lifted ahead of and behind the occluded front.

Open stage
The disturbance along the cold and warm fronts deepens and intensifies. Cold air actively pushes southeastward along the cold front, and warm air actively moves northward along the warm front. Precipitation zones along the two fronts are now strongly developed. The precipitation zone along the warm front is wider than the zone along the cold front.

CRITICAL THINKING

Here's an interesting question:
- Imagine you are a weather forecaster. Using the map above, what are your weather forecasts for the next three days in Seattle, Washington; Chicago, Illinois; and Philadelphia, Pennsylvania?

TORNADOES

Tornadoes are violent windstorms produced by a spiraling column of air that extends downward from a cumulonimbus cloud (**Figure 15.20**). An average of 780 tornadoes occur each year in the United States. They can occur in all states, at any time of year, but there is a distinctly intense period of tornado activity from April to August, with a peak in May.

Because tornadoes are so violent, it is very difficult to study them, and we still have not established many details of their formation. The severe thunderstorms that spawn tornadoes form along cold fronts, and the most violent cold fronts are those associated with cP air from the Canadian Arctic and mT air from the Gulf of Mexico. Because the two air masses are most likely to meet in the midcontinent states, that is where most tornadoes occur.

Wind speeds in a tornado run as high as 100 m/s (about 225 mi/hr), exceeding the known speeds of winds in any other storm. Only the strongest buildings constructed of concrete and steel can withstand these violent winds. The National Weather Service (NWS) maintains a tornado forecasting and warning system. Whenever weather conditions favor the development of tornadoes, the NWS issues an alert to authorities in the area at risk and activates systems for observing and reporting any tornado that occurs.

Tornado Figure 15.20

A tornado crossing the plains of North Dakota. At its lower end, the tornado's funnel may be 100 to 450 m (about 300 to 1500 ft) in diameter. The funnel appears dark because of condensing moisture, dust, and debris swept up by the wind. As the tornado moves across the countryside, the funnel writhes and twists.

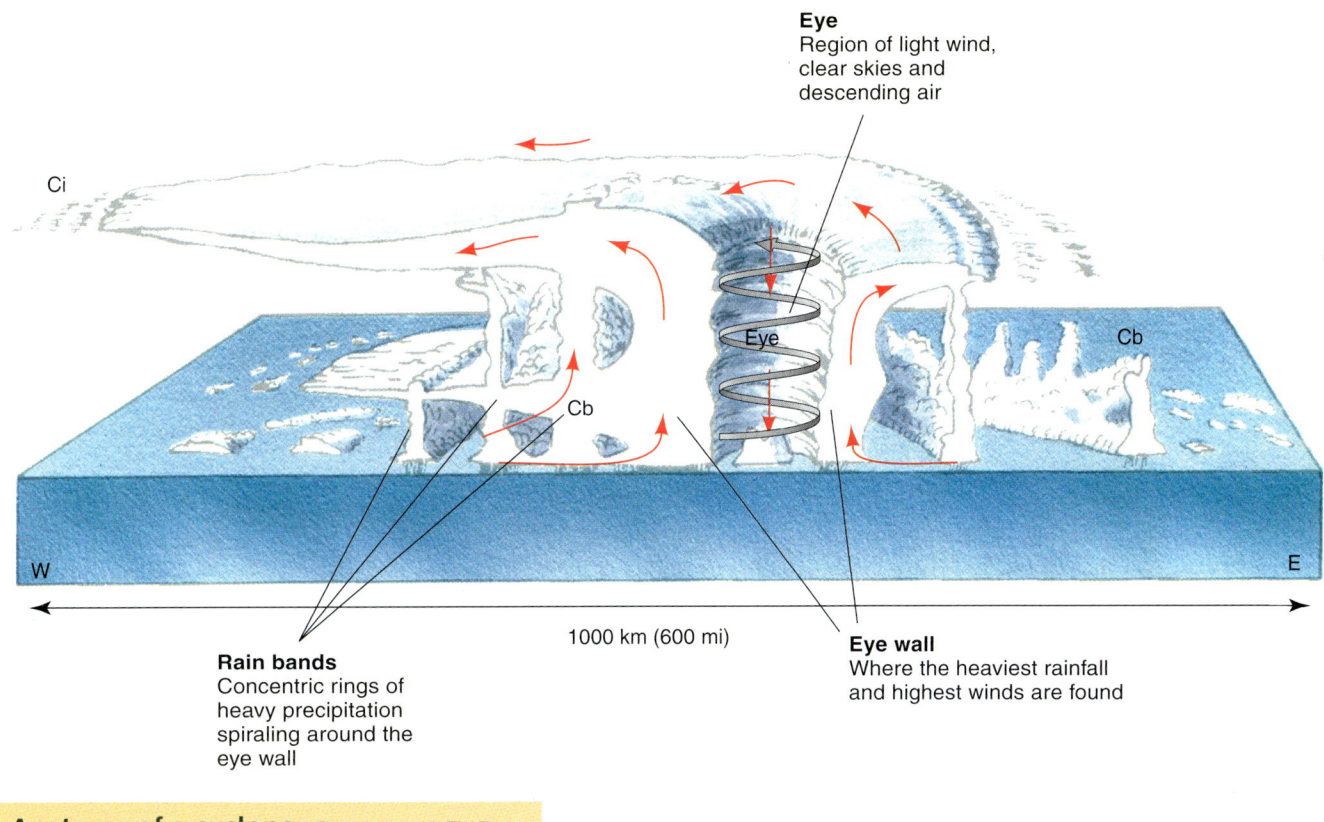

Anatomy of a cyclone FIGURE 15.21

In this schematic diagram, cumulonimbus (Cb) clouds in concentric rings rise through dense stratiform clouds. Cirrus clouds (Ci) spread out ahead of the storm. The width of the diagram represents about 100 km (about 600 mi).

TROPICAL CYCLONES

The **tropical cyclone** is the most powerful and destructive type of cyclonic storm. It is known as a *hurricane* in the Western Hemisphere, a *typhoon* in the western Pacific off the coast of Asia, and a *cyclone* in the Indian Ocean. Because they develop from cyclones, these storms can be spawned only in latitudes where the Coriolis effect is strong enough for cyclonic circulation to develop—that is, in regions higher than about latitude 5°.

A characteristic feature of a well-developed tropical cyclone is its central *eye* (**FIGURE 15.21**). The eye is a cloud-free vortex produced by the intense spiraling of the storm. In the eye, air descends from high altitudes. As we saw in Chapter 14, air that is forced to descend warms. As the eye passes over the surface, calm prevails and the sky clears. It may take about half an hour for the eye to pass, after which the storm strikes with renewed ferocity but with winds traveling in the opposite direction. Wind speeds are highest along the cloud wall of the eye.

Hurricanes start as cyclones over warm ocean water whose temperature is at least 26.5°C (80°F). The water evaporates and subsequently condenses in the hurricane, releasing latent heat. Hurricanes die over land or over bodies of cold water because they no longer have a warm-water energy source to sustain them.

> **tropical cyclone** An intense traveling cyclone occurring in tropical and subtropical latitudes, accompanied by high winds and heavy rainfall.

Hurricanes, by definition, have maximum wind speeds in excess of 119 km/h (74 mi/h). Besides wind damage, two other effects of hurricanes can be devastating. The first is a *storm surge*, a local, exceptional flood of ocean water (**FIGURE 15.22**). The center of a hurricane is a region of very low air pressure—below 920 mb in the greatest hurricanes—and a drop in air pressure raises local sea level. In the eye of a great hurricane, sea level may be 6 m or more (20 ft) above normal, and when hurricane-force winds drive such high seas onshore, extensive flooding may occur. As discussed in Chapter 13, a sudden storm surge generated by a severe hurricane flooded Galveston, Texas, in 1900, and drowned about 8000 people—the largest death toll in a natural disaster yet experienced within the United States.

A second effect associated with hurricanes is torrential rain. Although this rainfall is a valuable water resource, it can also cause fresh-water flooding, raising rivers and streams out of their banks. On steep slopes, soil saturation and high winds can topple trees and produce disastrous earthflows and landslides.

The intensity of tropical cyclones is ranked from category 1 (weak) to 5 (devastating) on what is known as the *Saffir-Simpson scale* (**FIGURE 15.23**) after the scientists who developed it. For convenience, tropical cyclones are given names as they are being tracked by weather forecasters. Male and female names are alternated in an alphabetical sequence that begins anew each year. Different sets of names are used in different regions, such as the western Atlantic, western Pacific, or Australian regions. Names may be reused, but the names of storms that cause significant damage or destruction, such as Andrew or Katrina, are retired from further use.

Tropical cyclone activity varies from year to year and from decade to decade. In 1995, a new phase of Atlantic hurricane activity began. Since 1995, sea surface temperatures in the northern Atlantic from August through October have averaged about 0.5°C (about 1°F) warmer than they did during 1970–1994, providing more latent heat to fuel cyclones. The result has been a period of hurricane activity unequaled in the historical record. By 2005, the average number of named storms had increased to 13 per year, compared to 8.6 a year during 1970–1994. The number of hurricanes had increased from 5 to 7.7 per year, with 3.6 major hurricanes compared to 1.5 in prior years. In 2005 alone, there were 27 named storms, with a record 4 storms reaching category 5. The most recent climatological studies suggest that this enhanced hurricane activity is linked to slowly changing patterns of ocean currents and will persist for another 15 to 20 years.

Stranded vessel FIGURE 15.22

Hurricane Andrew's storm surge and high winds stranded this large motor vessel far from its berth in Homestead, Florida (1992).

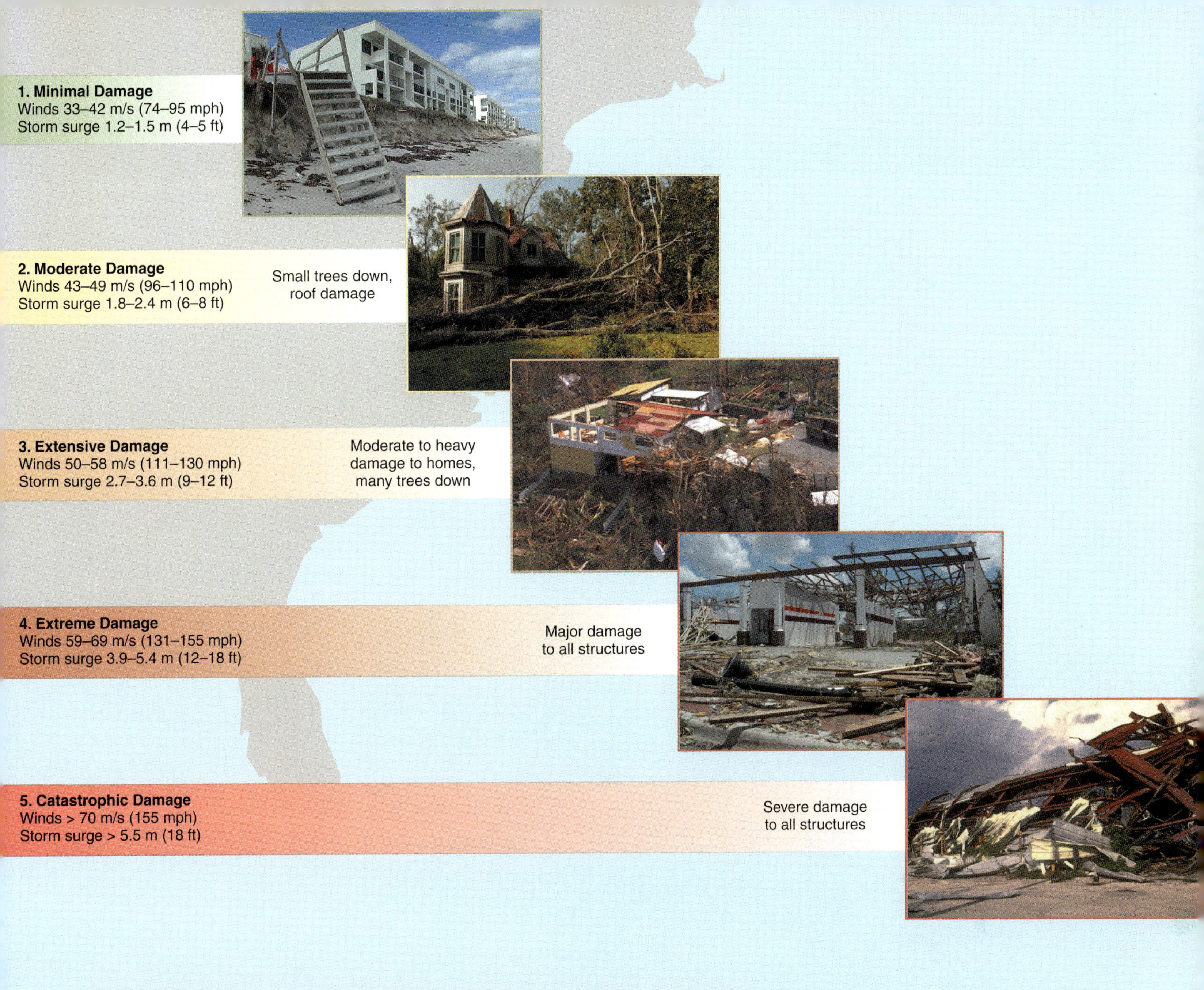

1. Minimal Damage
Winds 33–42 m/s (74–95 mph)
Storm surge 1.2–1.5 m (4–5 ft)

2. Moderate Damage
Winds 43–49 m/s (96–110 mph)
Storm surge 1.8–2.4 m (6–8 ft)

Small trees down, roof damage

3. Extensive Damage
Winds 50–58 m/s (111–130 mph)
Storm surge 2.7–3.6 m (9–12 ft)

Moderate to heavy damage to homes, many trees down

4. Extreme Damage
Winds 59–69 m/s (131–155 mph)
Storm surge 3.9–5.4 m (12–18 ft)

Major damage to all structures

5. Catastrophic Damage
Winds > 70 m/s (155 mph)
Storm surge > 5.5 m (18 ft)

Severe damage to all structures

Saffir-Simpson scale of tropical cyclone intensity FIGURE 15.23

Recent scientific evidence suggests that global warming will increase the intensity of tropical cyclones and hurricanes. Tropical cyclones generate strong winds that typically churn up ocean waters, mixing cooler deeper water with the warmer surface waters. This cold water tends to weaken storms; however, if deeper waters become too warm due to global warming, this natural breaking mechanism will be disrupted.

FIGURE 15.24 on the next page compares wave cyclones with tropical cyclones.

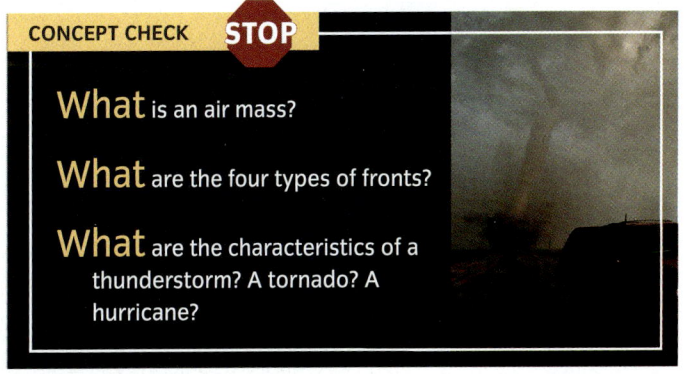

CONCEPT CHECK STOP

What is an air mass?

What are the four types of fronts?

What are the characteristics of a thunderstorm? A tornado? A hurricane?

Weather Systems 479

WAVE CYCLONES

Wave cyclones are common features of midlatitude weather. They can be strong storms, bringing rain, snow, and winds. Strong wave cyclones off the eastern North American coast are known as *nor'easters*.

◄ **The Perfect Storm**
This intense wave cyclone is the first phase of the "Perfect Storm," imaged here on October 30, 1991, by the NOAA GOES-7 weather satellite. Yellow and red areas indicate intense convection where the storm swallowed the remnants of Hurricane Grace. Later it drifted southward, picking up energy from waters of the warm Gulf Stream, to become a tropical cyclone—the "unnamed hurricane" of 1991.

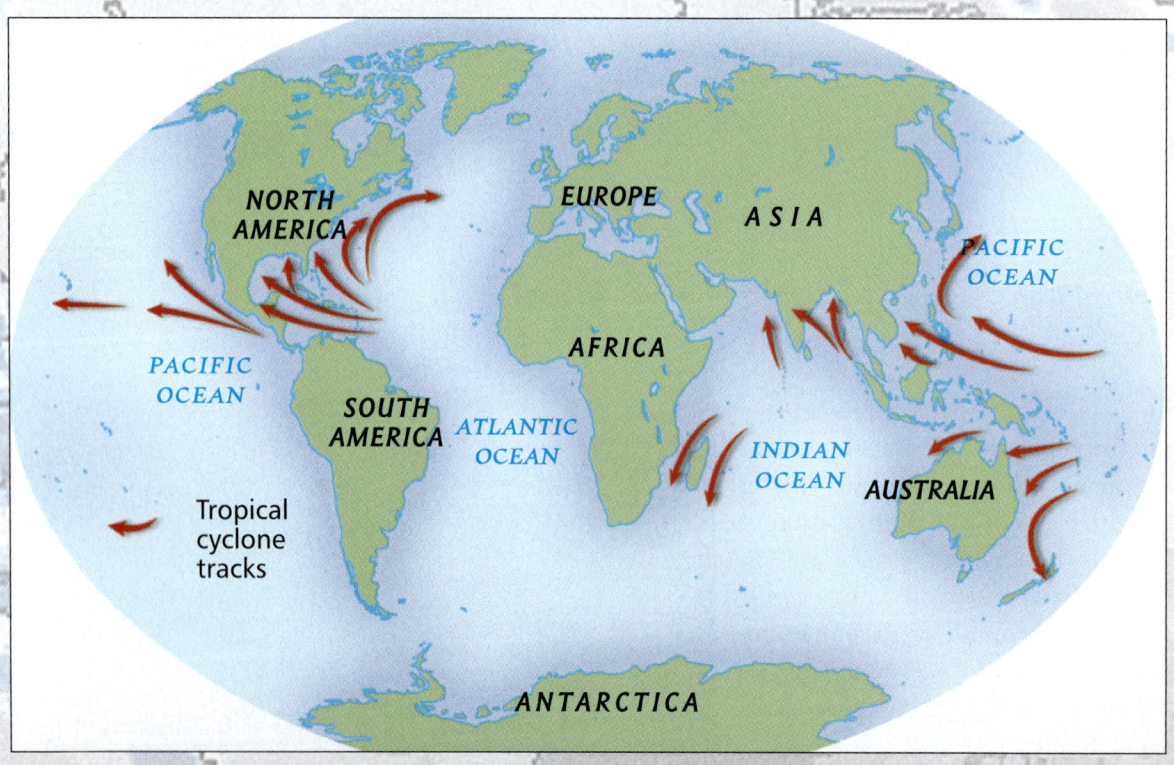

◄ **Cyclone paths**
Tropical cyclones form over warm, tropical oceans and are typically carried westward by easterly trade winds. Many eventually turn poleward and eastward.

Visualizing

Wave cyclones and tropical cyclones FIGURE 15.24

TROPICAL CYCLONES

Tropical cyclones form over warm tropical oceans. They can be very intense storms that devastate islands and coasts. In the Atlantic, they are known as *hurricanes*. In the Pacific, they are *typhoons*, and they are known simply as *cyclones* in the Indian Ocean.

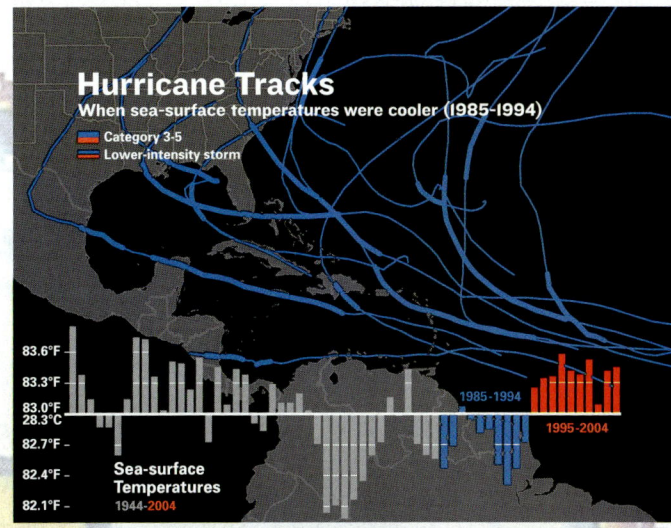

▶ **Hurricane tracks**
Along the southeastern coast of the U.S., many tropical cyclones approach from the east then turn and head back out to sea. These figures show hurricane tracks for 1985–1994 (blue, *upper map*) and 1995–2004 (red, *lower map*). The thicker the line of the track, the more intense the hurricane.

The bar chart in the upper map shows sea-surface temperatures for both periods. Comparing the upper and lower maps, you can see that the number and intensity of storms increases when sea surface temperatures increase.

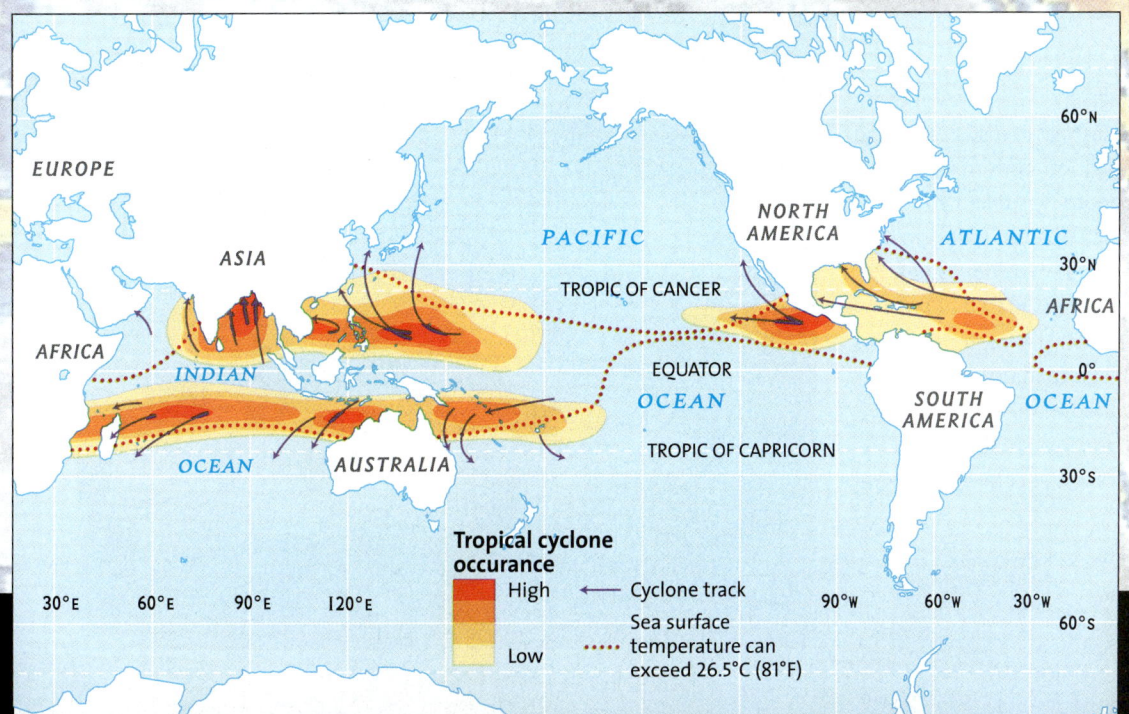

◀ **Oceans and Cyclones**
Tropical cyclones are most likely to occur in areas of greatest heating. Dotted lines show where the sea surface temperature can be greater than 26.5°C (81°F). Cyclones last until they move over cooler waters or hit land. When a cyclone encounters warmer waters, as Hurricane Katrina did in 2005 in the Gulf of Mexico, it picks up energy and intensifies.

Amazing Places: New Orleans Before and After Katrina

In 2005, Hurricane Katrina laid waste to the city of New Orleans and much of the Louisiana and Mississippi Gulf coasts. Originating southeast of the Bahamas, the hurricane first crossed the South Florida Peninsula as a category 1 storm and then moved into the Gulf of Mexico, where it intensified to a category 5 storm. After the storm weakened somewhat as it approached the Gulf Coast, its eye came ashore at Grand Isle, Louisiana, with sustained winds of 56 m/s (125 mi/hr) early on August 29.

A New Orleans, shown here before Katrina struck, is particularly vulnerable to hurricane flooding because it was built largely on the floodplain of the Mississippi River. Most of its land area has slowly sunk below sea level as underlying river sediments have compacted over time. Levees protect the city from Mississippi River floods, as well as from ocean waters along the saline Lakes Borgne (on the east) and Pontchartrain (on the north).

B Katrina wrought a degree of devastation that was unparalleled in the history of disasters in the United States. Many neighborhoods in New Orleans lay in ruins. This photo was taken about two weeks after the storm. The red object in the foreground is a barge. The devastation is nearly complete. Total losses were estimated at more than $125 billion, and the official death toll exceeded 1800. Adding insult to injury, much of New Orleans was reflooded three weeks later by Hurricane Rita, a category 3 storm that made landfall on September 24 at the Louisiana–Texas border.

SUMMARY

1 Atmospheric Pressure

1. The term **atmospheric pressure** describes the weight of air pressing on a unit of surface area. Atmospheric pressure is measured using a barometer.

2. Atmospheric pressure decreases rapidly as altitude increases.

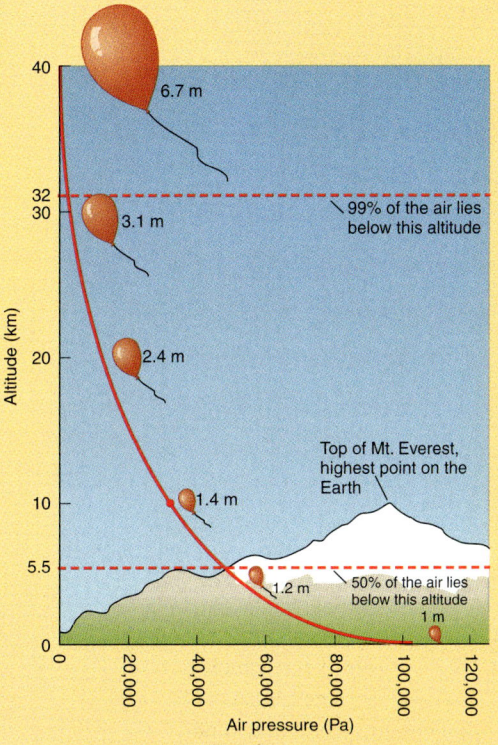

2 Why Air Moves

1. Air motion is produced by **pressure gradients**.

2. Pressure gradients form when air is heated unevenly, creating **convection loops**.

3. Sea and land breezes are examples of convection loops formed from unequal heating and cooling of land and water surfaces. Mountain and valley winds, the Santa Ana, and Chinooks are other examples of local winds.

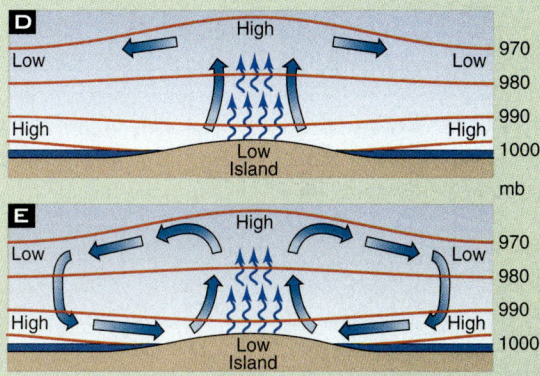

3 Cyclones and Anticyclones

1. Earth's rotation creates the **Coriolis effect**.

2. The *Coriolis force* deflects wind motion, making air spiral around **cyclones** (centers of low pressure and convergence) and **anticyclones** (centers of high pressure and divergence).

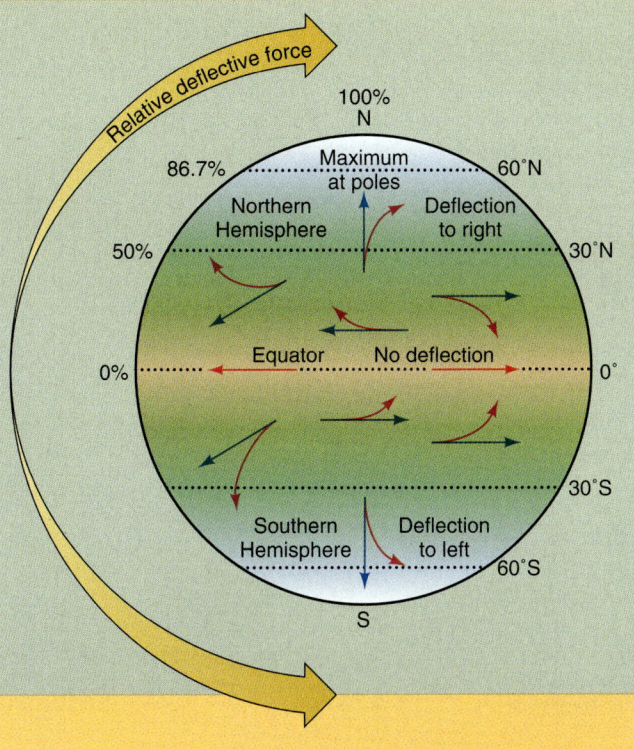

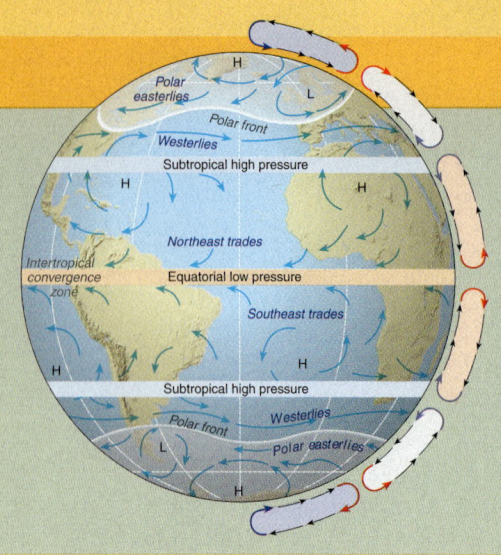

4 Global Wind Patterns

1. Equatorial and tropical regions are heated more intensely than the higher latitudes, setting up convection loops.

2. These loops drive the **trade winds** and the convergence and lifting of air at the **intertropical convergence zone (ITCZ)**.

5 Winds Aloft

1. Winds in the atmosphere are dominated by a global pressure-gradient force between the tropics and the pole in each hemisphere.

2. The global pressure-gradient force and the Coriolis force generate strong westerly **geostrophic winds** in the upper air.

3. **Rossby waves** develop in the upper-air westerlies, bringing cold polar air toward the Equator and warmer air toward the poles. The polar-front and subtropical **jet streams** are concentrated westerly wind streams with high wind speeds.

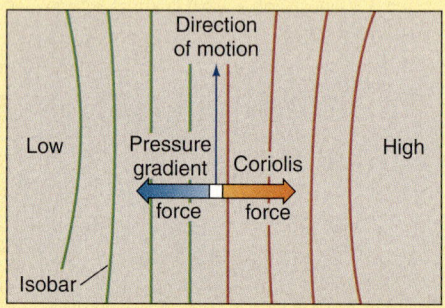

6 Weather Systems

1. **Air masses** are distinguished by their latitudinal location and their source regions. **Fronts** are the boundaries between air masses. They include cold and warm fronts, where cold or warm air masses are advancing. In the occluded front, a cold front overtakes a warm front, pushing a pool of warm, moist air above the surface.

2. Thunderstorms can form if air is unstable, creating hail and lightning. *Tornadoes* are very small, intense cyclones that occur as a part of thunderstorm activity.

3. **Wave cyclones** are produced when anticyclones meet at the polar front. They form the dominant weather system in the middle and high latitudes.

4. **Tropical cyclones**, also known as *hurricanes* or *typhoons*, develop over very warm tropical oceans and can intensify to become vast inward-spiraling systems of very high winds with very low central pressures.

KEY TERMS

- atmospheric pressure p. 458
- barometer p. 459
- isobars p. 460
- pressure gradient p. 460
- convection loop p. 460
- Coriolis effect p. 463
- cyclone p. 464
- anticyclone p. 464
- intertropical convergence zone p. 466
- trade winds p. 467
- geostrophic wind p. 468
- Rossby waves p. 469
- jet streams p. 469
- air mass p. 470
- front p. 472
- wave cyclone p. 474
- tropical cyclone p. 477

CRITICAL AND CREATIVE THINKING QUESTIONS

1. Earth and Venus are almost identical in size and mass, but the atmospheric pressure on the surface of Venus is 90 times greater than air pressure at sea level on Earth. Why?

2. Consult the weather map from your daily newspaper, or download one from a weather station on TV. (Do not look at the weather forecast.) Predict the weather for the week ahead in the place you live or go to school.

3. If air and ocean surface temperatures increase as a result of global climate change, scientists predict that the severity and frequency of hurricanes in the North Atlantic will increase. What reasoning leads them to make such a prediction?

4. Some of the strongest and most sustained winds on record have been recorded at research stations located on the shore at the foot of the Antarctic Ice Cap. What special conditions explain this phenomenon?

5. Imagine you are a commercial airline pilot and you are scheduled to fly nonstop from Vancouver, British Columbia, to Sydney, Australia. What upper-atmosphere wind conditions would you expect to encounter as you travel? What jet streams will you encounter? Will the jet streams slow or speed your aircraft on its way?

What is happening in this picture?

The photograph shows a line of cumulus clouds advancing from left to right. The clouds were formed when warm, moist air was pushed aloft.

- Why is the warm air forced upward?

- The clouds mark an advancing front. What type of front might this be?

SELF-TEST

1. The boiling point of water lowers as one goes higher in elevation because _____.
   a. water is less dense at higher elevations
   b. air is denser at higher elevations
   c. air pressure is less at higher elevations
   d. upward water pressure is much greater

2. In the diagram shown, the air will move from the _____ pressure center at _____ to the _____ pressure center at _____.
   a. high/Columbus/low/Wichita
   b. low/Columbus/high/Wichita
   c. high/Wichita/low/Columbus
   d. low/Wichita/high/Columbus

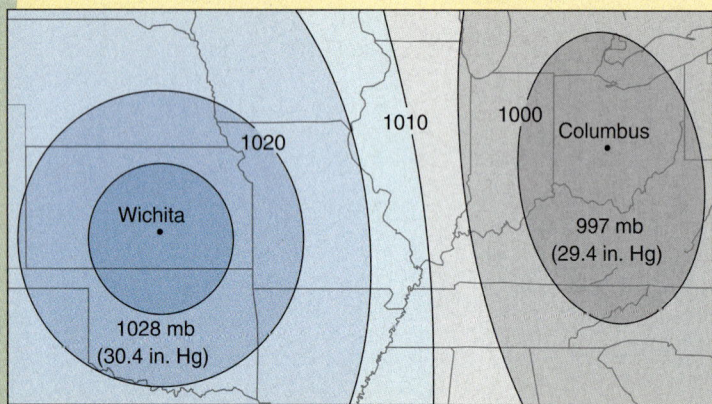

3. A land breeze generally occurs _____.
   a. at night, when the land cools below the surface temperature of the sea
   b. when strong winds blow in from the sea over the land
   c. only during certain restricted seasons
   d. during the day, when the land heats above the surface temperature of the sea

4. A parcel of air at the surface is subjected to three forces and the balance among the pressure gradient, Coriolis, and _____ forces determines the direction of motion of the parcel of air.
   a. gravitational
   b. centrifugal
   c. frictional
   d. divergent

5. The Coriolis effect is _____.
   a. a result of Earth's rotation from east to west and causes objects to curve to the right in the Northern Hemisphere
   b. a result of Earth's rotation from west to east and causes objects to curve to the right in the Northern Hemisphere
   c. a result of Earth's rotation from west to east and causes objects to curve to the left in the Northern Hemisphere
   d. a result of Earth's rotation from east to west and causes objects to curve to the left in the Northern Hemisphere

6. The diagram shows the motion of air in cyclones and anticyclones. Identify which figures represent: (a) a Northern Hemisphere anticyclone, (b) a Southern Hemisphere anticyclone, (c) a Northern Hemisphere cyclone, and (d) a Southern Hemisphere cyclone.

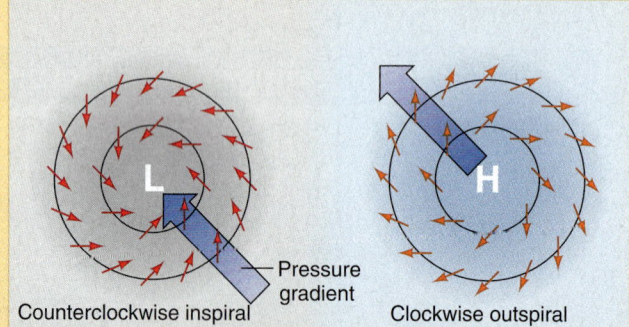

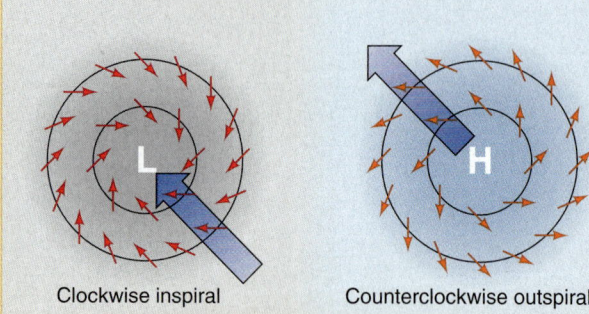

7. Cloudy and rainy weather is often associated with the inward and upward convergence of air within _____.
   a. anticyclones
   b. warm fronts
   c. cold fronts
   d. cyclones

8. Label the following features on this illustration of global atmospheric circulation:

   intertropical convergence zone
   polar front
   northeast trade winds
   southeast trade winds
   subtropical high pressure
   polar easterlies
   westerlies

9. At upper levels in the atmosphere, as a parcel of air moves in response to a pressure gradient, it is turned progressively sideways until the gradient and Coriolis forces balance, thus producing the _____.
   a. geostrophic wind
   b. upper-air westerlies
   c. tropospheric wind
   d. equatorial easterlies

10. Jet streams are _____.
    a. narrow zones at a high altitude in which wind streams reach speeds greater than the speed of sound
    b. narrow zones at a high altitude in which wind streams sometimes reach speeds of over 150 miles per hour
    c. rivers of wind that exist only along the Equator and travel at fairly high velocities
    d. rivers of wind that circulate around the poles

11. What type of front is shown here?
    a. cold
    b. warm
    c. occluded
    d. stationary

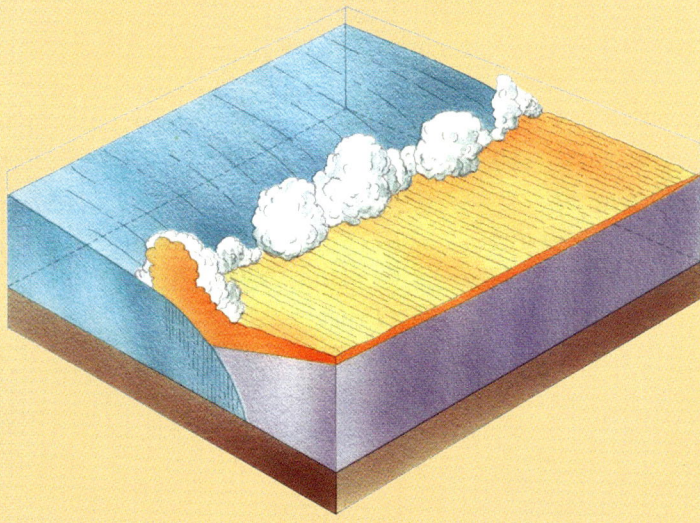

12. A(n) _____ is a center of high pressure and is generally responsible for fair weather.
    a. anticyclone
    b. cyclone
    c. trade wind
    d. midlatitude storm front

13. A _____ is a small but intense cyclonic vortex with very high wind speeds.
    a. hurricane
    b. typhoon
    c. tornado
    d. cyclone

14. Hurricanes and typhoons generally develop within the _____ latitudinal zones.
    a. 15° to 30° north and south
    b. 10° to 20° north and south
    c. 8° to 15° north and south
    d. 30° to 45° north and south

15. A _____ is a sudden rise of water level caused by a hurricane.
    a. storm surge
    b. tsunami
    c. flood
    d. tidal flood

Self-Test 487

Global Climates Past and Present 16

In the late summer of 1991, two German trekkers made a gruesome discovery high in the Tyrolean Alps. The mummified body of a prehistoric man was seen protruding from slowly melting ice near the margin of the Similaun Glacier at 3200 m (10,500 ft) altitude.

Scientific studies have determined that Ötzi the Iceman (see inset), as he has been dubbed, was about 45 years of age when he died. Radiocarbon dating of his skin and bones reveal that he died over 5000 years ago, making him a member of the Late Neolithic and Bronze age population of south-central Europe.

Ötzi's discovery provided unprecedented riches to help archaeologists reconstruct what everyday Neolithic life was like. Along with his corpse were a fur robe, a woven grass cape, leather shoes, a flint dagger, a copper ax, a wooden bow, and 14 arrows.

The discovery was also a boost for climate scientists, confirming the picture of the region they had pieced together from other evidence. At the time that Ötzi lived, the climate was getting cooler and glaciers, such as the now-retreating Bertrab Glacier in South Georgia pictured here, were growing larger. When he died near the margin of the Similaun glacier, his remains became entombed in an Alpine deep-freeze that kept him perfectly preserved for more than 5 millennia. The following years saw a succession of cool periods, the last of which is known as the Little Ice Age. However, none of the intervening mild periods was warm enough to release Ötzi, until the recent warming exposed him to view.

NATIONAL GEOGRAPHIC

CHAPTER OUTLINE

Global Climate Change in the Past p. 490

Global Climates Today p. 498

Present-Day Climate Change p. 505

Global Climate Change in the Past

LEARNING OBJECTIVES

Describe the trends in Earth's climate over the last few million years and the last few thousand years.

Explain how Earth scientists learn about past climatic conditions.

Identify several known or suspected causes of climate change.

Discuss rapid climate change and the Younger Dryas event.

ou have probably read about global warming in newspapers and magazines. Because it is a politically and emotionally fraught issue, as well as a scientifically complex one, it is important to emphasize the difference between what we *know* is happening and what we *think* is happening—a distinction that is often lost in the speculative reports one sees on tele-

Past climate change FIGURE 16.1

This graph shows an estimate of global temperatures, based on deep-ocean sediments, over **A** the last 60 million years, **B** the last 10 million years, and **C** the last 150,000 years.

A Global ocean temperatures, 60 million years ago to present
At the beginning of the Cenozoic Era, Earth's surface was largely free of ice. Sea levels were higher (note that southeastern North America was underwater), and seawater could circulate freely.

B Global ocean temperatures, 10 million years ago to present
As plate motions moved the major landmasses near their present locations, temperatures fell and glaciers appeared at the poles. In the last 800,000 years (the blue band in the temperature graph), the climate has fluctuated eight times between ice ages and warm interglacial periods.

C Global ocean temperatures, 150,000 years ago to present
At the peak of the last ice age 18,000 years ago, glaciers blanketed most of North America. Earth is now warmer than it has been at any time in the last 100,000 years, and roughly at the same temperature as it was in the last interglacial period 120,000 years ago.

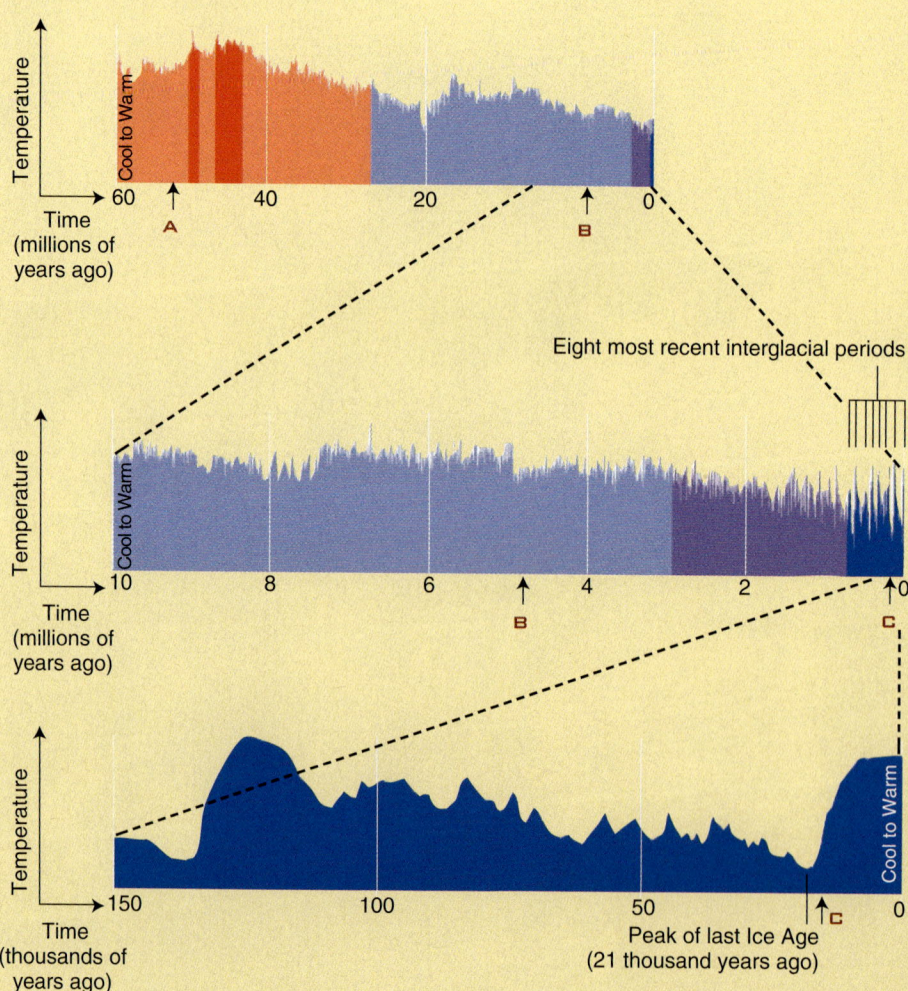

490 CHAPTER 16 Global Climates Past and Present

climate The annual cycle of prevailing weather conditions at a given place based on statistics recorded over a long period.

vision, reads in the press, or hears in the doomsday accounts in fictional movies.

Earth's climate system is complex, comprising multiple interacting parts and subsystems in a state of dynamic equilibrium. **Climate** can be influenced by a wide variety of processes, both natural and human. All these processes, working together, cause climate to vary in a cyclical fashion and on a number of different time scales. Before we can understand the human role in climate change, present or future, we must try to understand how natural Earth processes determine climate and control climatic variations. To do this, we must look back in time and study how climate has changed over the course of Earth's history.

WHAT WE KNOW

In FIGURE 16.1, you can see the trend in worldwide temperatures over the last 60 million years, roughly since the beginning of the Cenozoic Era. This figure highlights three very important points. First, climate change is the norm, not the exception, on our planet. No matter how much global warming occurs in the next century, it is unlikely to show up as more than a barely perceptible blip on this graph. Second, the overall trend in temperature in the Cenozoic Era has

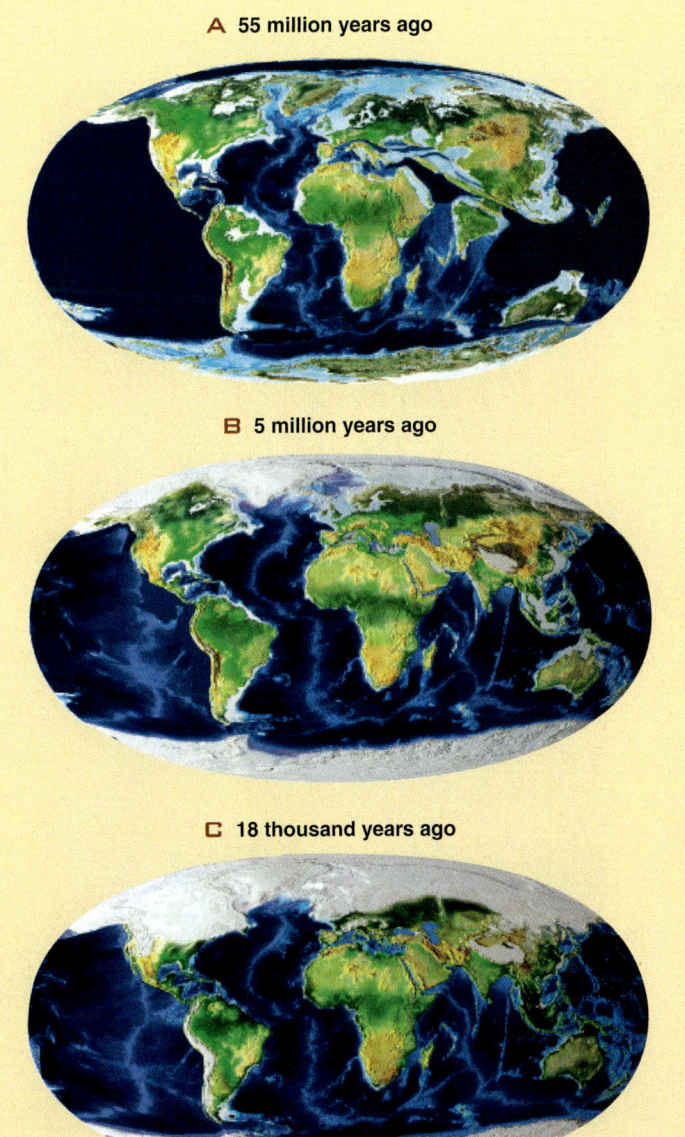

A 55 million years ago

B 5 million years ago

C 18 thousand years ago

Polar
Tundra
Cold temperate
Warm temperate
Arid
Tropical everwet

Global Climate Change in the Past

Visualizing

Reconstructing temperature records FIGURE 16.2

Temperature variation over the last thousand years

Direct measurements of the temperature in the Northern Hemisphere from the middle of the 19th century onward are shown by the black curve. The other curves are based on indirect reconstructions. Indirect techniques include examining tree rings (green), studying the length of glaciers (dark blue), and analyzing ice core data (light blue). The multi-proxy curves (yellow, red, and purple) were reconstructed using a combination of different methods. All curves show that temperatures during the last few decades of the 20th century were higher than during any comparable period in the last thousand years.

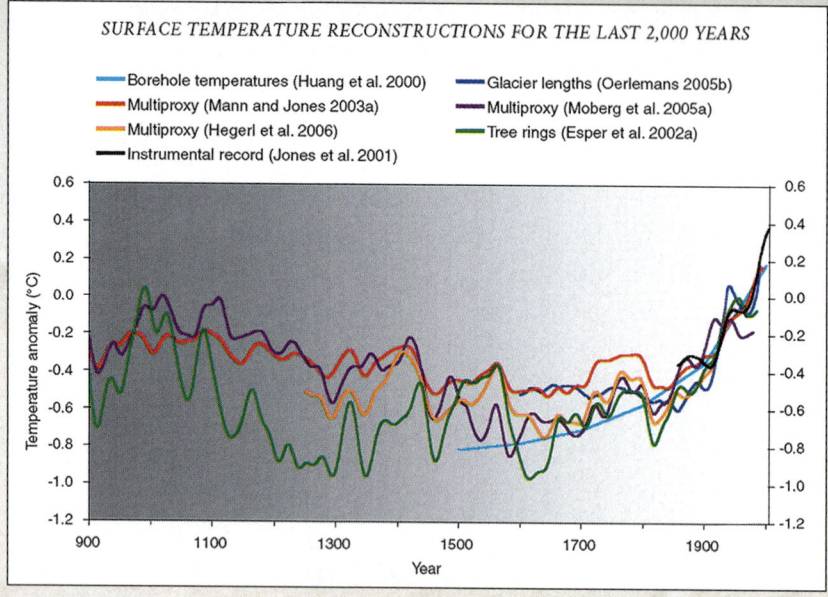

Ice cores
Chemical analysis of ice cores taken from glaciers and ice caps can provide a record of temperature at the time the ice was formed.

Tree rings
The width of annual tree rings varies with the growing conditions and in some locations can indicate temperature.

Coral growth
The constant growth of some corals produces structures similar to the annual rings of trees. Sampling the chemical composition of the coral can provide a temperature record. Black lines drawn on the lower section mark years, and red and blue lines mark quarters.

been downward. Viewed in a longer-term context, the current global warming episode will do nothing to change that. Third, in the last 800,000 years the planet has alternated between ice ages, or **glaciations**, and warm *interglacial periods*, each cycle lasting about 100,000 years. In all, more than 20 ice ages have occurred over the past 2 million years. We are currently near the peak of a warm interglacial cycle. In 50,000 years or so, Earth will probably experience another ice age—and the current global warming trend is unlikely to delay or hasten it. At most, it may create a sort of "superinterglacial spike."

> **glaciation** The covering of large land areas by glaciers or ice sheets.

HOW WE KNOW IT: THE TEMPERATURE RECORD

How do scientists know about Earth's climate and surface temperature from 20,000 years ago, or even 1 million or 100 million years ago? Measured records of air temperatures date back only to the middle of the 19th century. If we want to know about temperatures at earlier times, we need to use indirect methods, such as tree-ring, coral, and ice-core analysis (**FIGURE 16.2**).

Deep-sea sediments provide some of the best indirect evidence we have of past climatic changes (**FIGURE 16.3**). The temperature graph in Figure 16.1 was derived from the isotopes of oxygen in deep-sea fossils. Another important piece of evidence is the extent of glacial deposits such as terminal moraines, described in Chapter 6. Layered sediments also provide information about climate. For example, paleontologists can use plant and animal fossils to infer past climates. Fossilized pollen spores in old bogs and lake-bottom sediments have been particularly useful for reconstructing past climatic changes on a fine scale. Sedimentologists and stratigraphers study ancient soil horizons, called *paleosols*, which represent former land surfaces and provide information about climate and weather at the time they formed. The minerals present in paleosols allow geochemists to determine the chemical composition of the ambient air and water.

As you can see, the evidence of past temperature fluctuations and glaciations comes from multiple sources. No single source is definitive.

Deep-sea microfossils FIGURE 16.3

A Seafloor sediments obtained by drilling contain fossils of tiny sea creatures, foraminifera, that once lived in surface waters. These microfossils speak volumes about the chemistry and temperature of the ocean. Foraminifera have calcium carbonate shells that contain oxygen.

B The ratio of oxygen-16 (the lighter isotope of oxygen) to oxygen-18 is a measure of the temperature of the seawater in which the microscopic creatures lived.

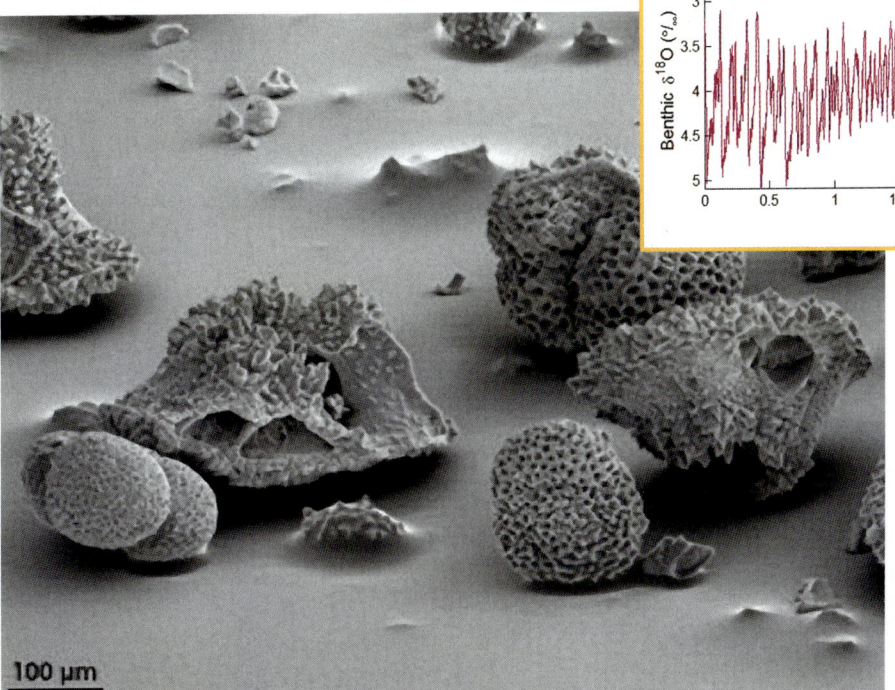

What an Earth Scientist Sees

The Gubbio Sediments

Rock near the Italian town of Gubbio reveals secrets about changes in Earth's orbit over millions of years. The surrounding hills expose thousands of thin layers of limestone, and a sedimentary rock that is rich in clay and organic carbon, called *marl* (A). An Earth scientist looking at the light and dark stripes would realize that for tens of millions of years the seafloor chemistry cycled between an oxygen-rich and an oxygen-poor environment. These sediments accumulated from a slow rain of dead marine organisms. When dissolved oxygen was present, scavenger organisms picked the skeletal material clean, leaving behind calcium carbonate to form limestone. Without oxygen, organic carbon was preserved, leaving a dark marl.

Earth scientists have found that these variations match the signature of Milankovitch cycles (B). These alter the pattern of solar heating by just a few percentage points, comparable to the recent heating due to greenhouse gases. The rock record shows that when pushed, Earth's climate can jump.

A

B Three kinds of orbital change affect Earth's climate. When these three factors are added together, they greatly affect how much sunlight reaches Earth, and where, at any given time.

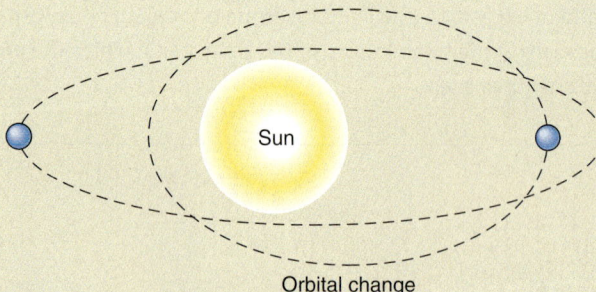

Orbital change
Earth's orbit becomes more and less elongated over a period of 100,000 years.

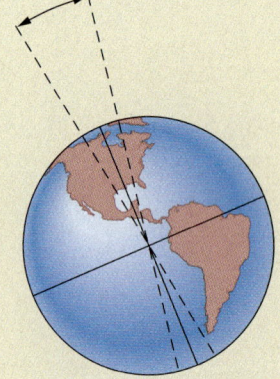

Axial tilt change
Earth's axis changes its tilt over a period of 41,000 years.

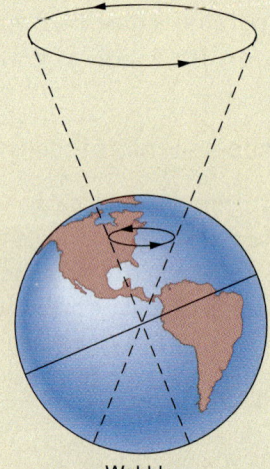

Wobble
Earth's axis wobbles in a circle once every 26,000 years.

Here's an interesting question:

- It was a scientist studying the cycling of the sedimentary layers who first noticed a thin layer that differed from all the rest. Research showed it to be a layer of debris from a meteorite impact. This was the first clue that led to the hypothesis of dinosaur extinction due to a giant impact. How likely is it that such a thin layer—no more than a few centimeters—would actually be noticed?

CAUSES OF CLIMATE CHANGE

It is clear that climate has changed dramatically in the past and will continue to do so in the future. However, the reasons for climate change are somewhat murkier, and this makes it hard for us to predict *how* the climate will change and at what rate.

Several mechanisms cause natural climate changes. Some of these are geographic changes. For example, during the Cenozoic Era, tectonic activity created the Isthmus of Panama, which joins North and South America. This land bridge severed the connection between the Atlantic and Pacific Oceans, altered oceanic and atmospheric circulation, and thus had a major impact on global climate. Other factors include changes in solar activity and volcanic eruptions. As solar output increases, global temperatures rise. Volcanic gases released into the atmosphere can condense into tiny liquid droplets, known as *aerosols*, that can have a cooling effect; aerosols from other sources, such as forest fires or power plants that release carbon particles, can sometimes have a warming effect.

Astronomical factors are also believed to affect Earth's climate. Small changes in the *eccentricity* (departure from circularity) of Earth's orbit, the *tilt of* the planet's axis of rotation, and the *precession* (wobbling) of the axis affect how much solar radiation reaches Earth's surface and at what times of the year. These variations, called **Milankovitch cycles**, correspond reasonably well to the periods of past glaciations (see *What an Earth Scientist Sees*). The combined effect of tilt and precession is roughly correlated with 20,000- to 40,000-year interglacial cycles, and variations in eccentricity may contribute to cycles that last 100,000 years. Earth's orbit is discussed in more detail in Chapter 17.

> **Milankovitch cycles** Climate cycles that occur over tens to hundreds of thousands of years because of changes in Earth's orbit and tilt.

All these influences on climate are tempered by the *greenhouse effect*, which we discussed in detail in Chapter 14. Water vapor, carbon dioxide, and methane in the atmosphere store heat that radiates from Earth's surface and radiate it back downward. Without this natural greenhouse warming by the atmosphere, the average temperature on the surface would be much cooler.

RAPID CLIMATE CHANGE: THE YOUNGER DRYAS EVENT

At the end of the last glaciations, about 11,000 to 10,000 years ago, the climate in the North Atlantic and adjacent lands experienced a rapid and remarkable change (**FIGURE 16.4**). For 2000 years, the

Evidence for the Younger Dryas event FIGURE 16.4

A Under full-glacial conditions, plants that are currently limited to polar and high-altitude regions could move into forests in northwestern Europe. Among these plants is *Dryas octopetala*, shown here. A large amount of *Dryas* pollen was found in organic deposits dating to the cold period, now known as the Younger Dryas event.

B Measurements of oxygen isotopes in sediments taken from a Swiss lake and an ice core from the Greenland Ice Sheet show that both the onset and the end of the Younger Dryas event were rapid. You can view the curves in the graph as marking changes in temperature. At the end of the event, the climate over Greenland warmed by about 7°C in only 40 years—a rate that exceeds that predicted by climate models for the coming century.

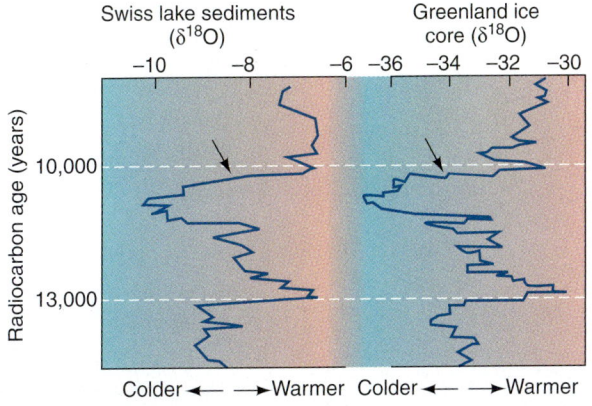

climate had been warming, causing ice sheets in North America and Europe to retreat and allowing plants and animals to reoccupy deglaciated land. Then, very abruptly, the climate cooled. Water temperatures in North America fell, and the ice sheets stopped retreating and began advancing again. This cold episode is known as the **Younger Dryas** event.

Younger Dryas A cold period that occurred between about 11,000 and 10,000 years ago, during the generally mild epoch.

What caused such a dramatic and rapid change in climate? The effects of Younger Dryas cooling are most pronounced around the North Atlantic, so it makes sense to look to that region for clues. It turns out that the solution lies in the interactions between the melting ice sheets and ocean circulation patterns, as described in FIGURE 16.5.

Process Diagram

Causes of the Younger Dryas event FIGURE 16.5

Ocean circulation in the North Atlantic plays an important role in controlling climate, as we saw in Chapter 12, where we met the thermohaline circulation system, also known as the *ocean conveyor belt*.

1 Meltwater lakes As the ice sheet over eastern North America retreated, vast meltwater lakes holding icy water were created. At the same time, the ocean conveyor belt was at work in the North Atlantic. Wind-driven warm surface currents, such as the Gulf Stream, headed toward the poles, cooling and eventually sinking at high latitudes, having transferred heat energy around the globe.

2 Drainage As the ice shrank further, it uncovered a natural drainageway between the meltwater lakes and the North Atlantic. The meltwater flooded rapidly into the ocean, forming a freshwater lid over the denser salty seawater. The cold surface meltwater in turn reduced the salinity of the water and the rate of evaporation from the ocean surface, shutting down the normal pattern of ocean circulation, known as the *thermohaline conveyor system*.

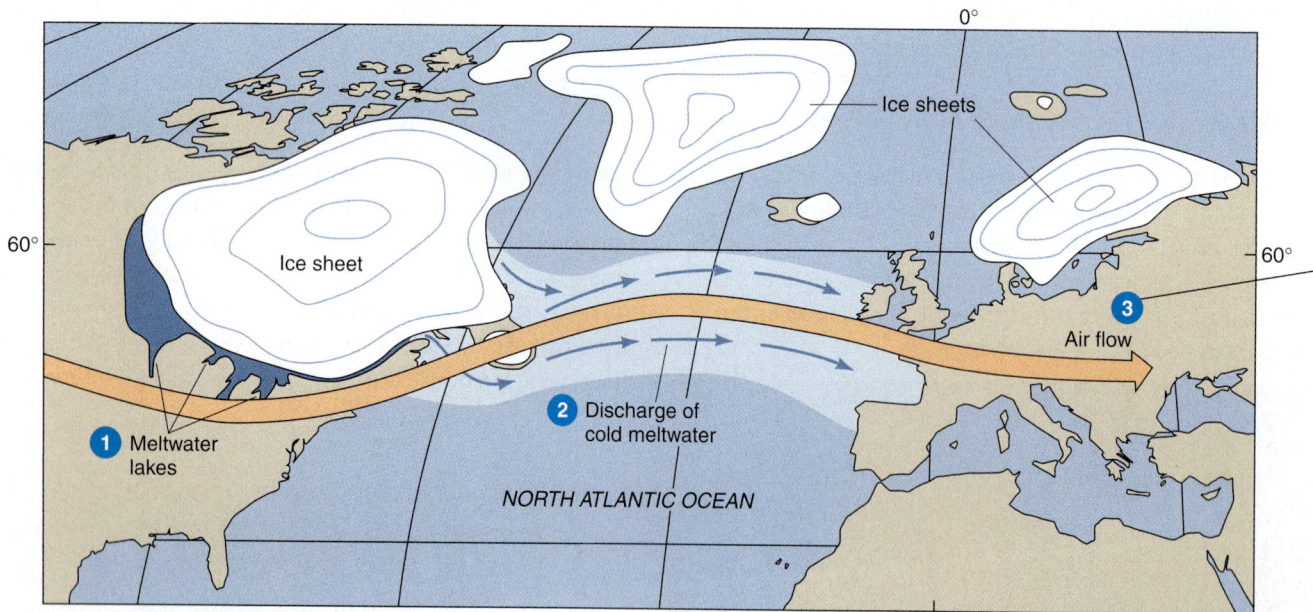

496 CHAPTER 16 Global Climates Past and Present

CONCEPT CHECK STOP

What are glaciations? How long do they last?

How does the current global climate change trend compare with previous climate fluctuations?

How do Earth scientists construct global temperature records stretching back many centuries?

What was the Younger Dryas event?

What are Milankovitch cycles? What combination of changes in the Milankovitch cycles would cause the highest and lowest summer temperatures at the poles?

Today, this system is active and the movement of warm water plays an important role in heating polar regions. During the Younger Dryas event, this system shut down, allowing the frigid ice-age conditions to return.

4 End of Younger Dryas As the meltwater lakes drained, the meltwater flow slowed and the thermohaline circulation gradually started up again. A warmer climate returned to the North Atlantic region, leading to a rapid termination of the ice age.

3 Younger Dryas Without the thermohaline circulation system, air passing over the cold North Atlantic brought colder conditions to northwestern Europe that led to the growth of glaciers and a major change in vegetation.

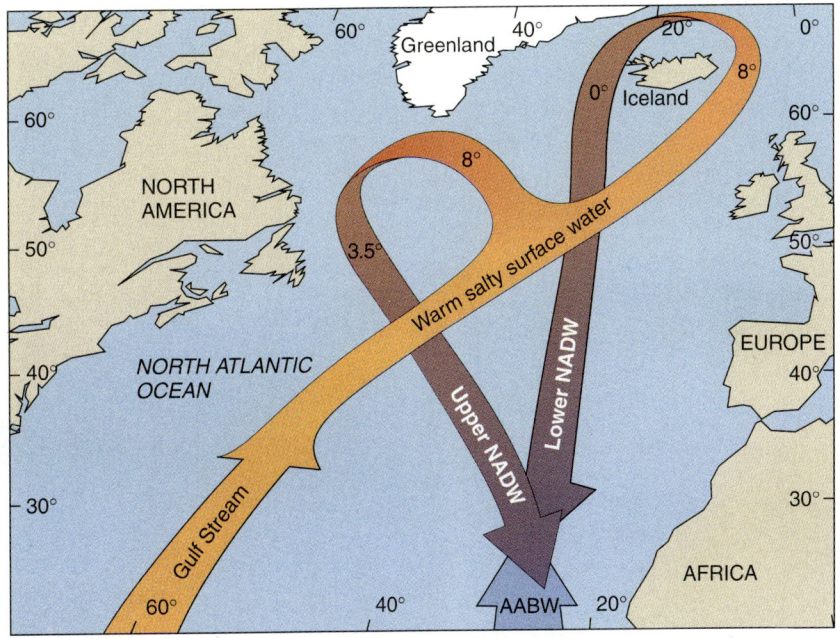

NADW: North Atlantic Deep Water
AABW: Antarctic Bottom Water

CRITICAL THINKING

Here's an interesting question:
- The cause of the Younger Dryas presented here is still a hypothesis. What kind of evidence could be sought to test the hypothesis that giant lakes of glacial meltwater suddenly drained into the North Atlantic?

Global Climate Change in the Past

Global Climates Today

LEARNING OBJECTIVES

Explain the factors that affect temperature around the globe.

Discuss how the movement of masses of air can affect precipitation patterns.

Describe low-latitude, midlatitude, and high-latitude climates.

The primary driving force for the weather is solar energy, as we saw in Chapter 14. Therefore, to understand how climate varies around the globe, we must understand how the flow of solar energy, or *insolation*, received by Earth differs from one region to another. **FIGURE 16.6** describes the three most important influences on climate and temperature: latitude, coastal or continental location, and moisture.

To understand climates, we need to think about the movement of air around the globe. *Air masses* are defined by their source region, as we saw in Chapter 15. The latitude of the source region determines the temperature of the air mass, and whether it originated over land or sea determines its moisture content. The *frontal zones*—the regions where air masses meet—can create cyclonic precipitation, as we saw in Chapter 14. Considering these factors, we can split the globe into three

Influences on temperature FIGURE 16.6

A Latitude
The annual cycle of insolation varies with latitude. In turn, the annual cycle of temperature at any place depends on its latitude. Near the Equator, temperatures are warmer and the annual range is low. Toward the poles, temperatures are colder and the annual range is greater. Ellesmere Island, Nunavut, Canada.

B Coastal–continental location
Ocean-surface temperatures vary less with the seasons than do temperatures of land surfaces. So regions near the coast show a smaller annual variation in temperature, while the variation is larger inland. Aerial view of Cornwall, England.

C Moisture
Air temperature has an important effect on precipitation. Warm air can hold more moisture than cold air. This means that colder regions generally have lower precipitation than do warmer regions. In addition, precipitation will tend to be greater during the warmer months of the temperature cycle. Mt. Des Voeux, Tavenui Island, Fiji.

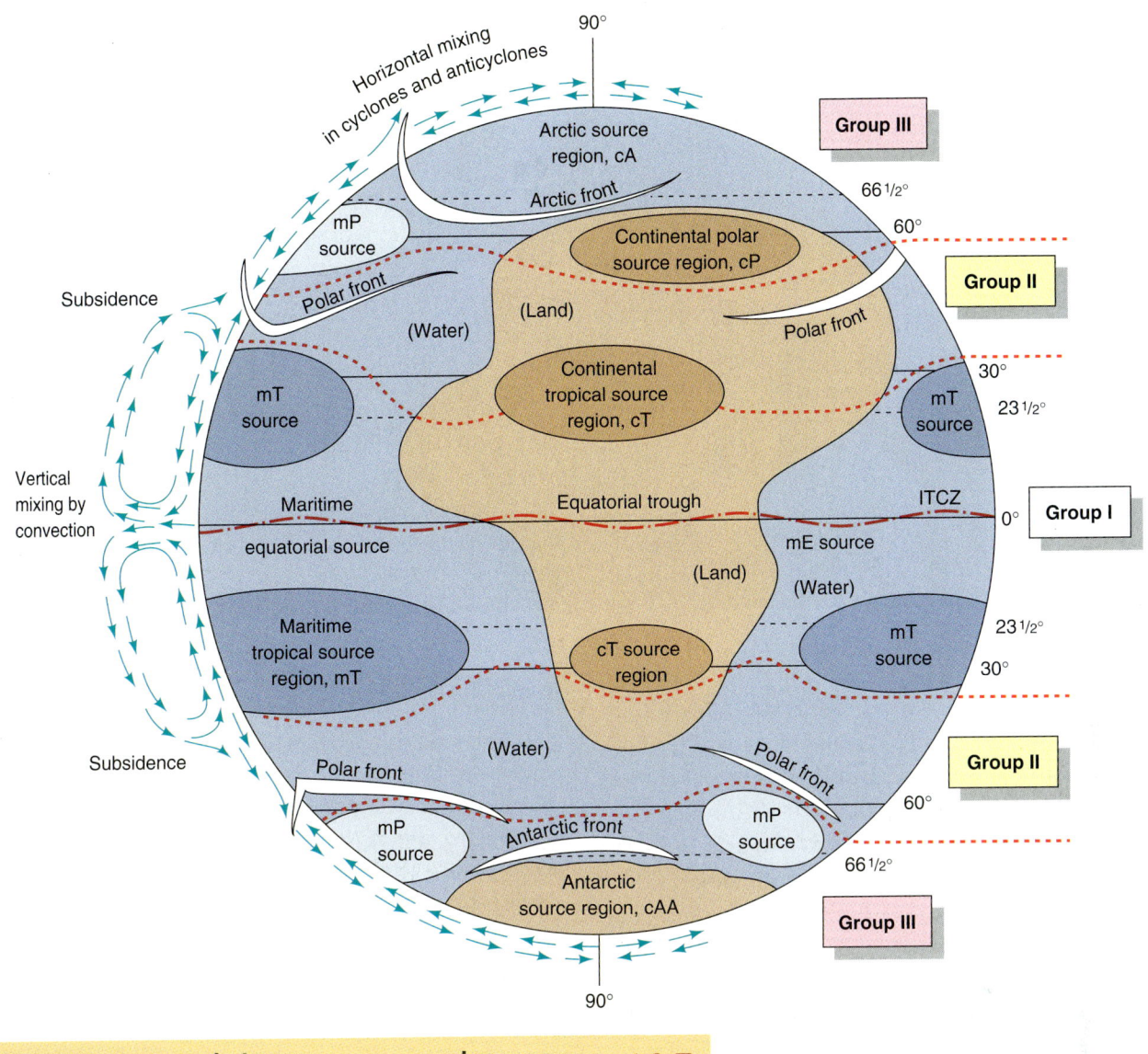

Climate groups and air mass source regions FIGURE 16.7

Using the map of air mass source regions, we can identify five global bands associated with three major climate groups. Each group has a set of distinctive climates with unique characteristics that are explained by the movements of air masses and frontal zones.

broad climate bands: low-latitude (Group I), midlatitude (Group II), and high-latitude (Group III), as shown in FIGURE 16.7.

LOW-LATITUDE CLIMATES

The low-latitude climates are typically warm and can range from extremely wet to severely dry. They lie for the most part between the Tropics of Cancer and Capricorn, occupying all of the equatorial zone (10° N to 10° S), most of the tropical zone (10°–15° N and S), and part of the subtropical zone. These regions include the equatorial trough of the *intertropical convergence zone (ITCZ)*, the belt of tropical *easterlies* (northeast and southeast *trades*), and large portions of the *oceanic subtropical high-pressure belt*, described in detail in Chapter 15.

Global Climates Today 499

Iquitos, Peru FIGURE 16.8

Frequent rainfall is a way of life on the Amazon River near Iquitos, Peru. Iquitos is located in the upper Amazon lowland, close to the Equator. Temperatures differ very little from month to month, and there is plenty of rainfall throughout the year.

Near the Equator, between 10° N and 10° S, the climate is controlled by warm, moist air masses that formed over equatorial and tropical oceans. This creates heavy rainfall, as seen in the Amazon lowland of South America, the Congo Basin of Africa, and the East Indies, from Sumatra to New Guinea (**FIGURE 16.8**).

Between 5° and 25° N and S, we find climates that are warm throughout the year but have strong seasonal patterns of rainfall because of the movement of the ITCZ and the *trade winds* (the winds that blow toward the Equator) and the reversing Asian wind pattern known as the *monsoon winds* (see Figure 15.17). Trade-wind coasts—the coasts that are exposed to the trade winds—are found along the eastern sides of Central and South America, the Caribbean Islands, Madagascar (Malagasy), Southeast Asia, the Philippines, and northeast Australia. Monsoon winds blow from southwest to northeast, so it is the western coasts of landmasses that are exposed to this moist air flow. Western India, Myanmar (formerly Burma), and Bangladesh suffer from very heavy monsoon rains.

A combination of factors gives rise to another, very special environment—the **low-latitude rainforest** (**FIGURE 16.9**).

> **low-latitude rainforest** Dense evergreen forest of low-latitude equatorial and tropical zones with abundant rainfall.

Low-latitude rainforest FIGURE 16.9

In the rainforest, temperatures are uniformly warm through the year and rainfall is high. Streams flow abundantly throughout most of the year, and the river banks are lined with dense forest vegetation. The river shown here is a tributary of the Amazon, near Manaus, Brazil.

Global Locator

Savannah FIGURE 16.10

Most plants in the savannah are rain-green vegetation—staying dormant during the dry period, then bursting into leaf and bloom with the rains. Here, coarse grasses occupy the open space between the rough-barked and thorny trees. There may also be large expanses of grassland. In the dry season, the grasses turn to straw and many of the tree species shed their leaves to cope with the drought.

As we move farther toward the poles, the seasonal cycles of rainfall and temperature become stronger. Between 5° and 20° N and S in Africa and the Americas, and between 10° and 30° N in Asia, the climate consists of a very dry season with cool temperatures giving way to a very hot period before the heavy rains begin. This situation produces a *savanna environment* with sparse vegetation (**FIGURE 16.10**). When the rains fail, there can be devastating famine.

When we reach the Sahara–Saudi Arabia–Iran–Thar desert belt of North Africa and southern Asia, we meet some of the driest regions on Earth. Another large desert region is the desert of central Australia. The west coast of South America, including portions of Ecuador, Peru, and Chile, also has a dry climate, but temperatures there are moderated by a cool marine air layer that blankets the coast.

MIDLATITUDE CLIMATES

The *midlatitude climates* are more variable than the low-latitude climates thanks to the interaction between conflicting masses of air. Tongues of warmer, moist maritime tropical (mT) air masses enter the midlatitude zone from the subtropical zone, where they meet and interact with cold tongues of maritime polar (mP) and continental polar (cP) air from above the poles, creating midlatitude cyclones (centers of low atmospheric pressure, described in Chapter 15) along the *polar-front* zone. This interaction creates a range of climates—from those with strong wet and dry seasons to those with uniform precipitation. Temperature cycles are also quite varied in this region.

Unlike the low-latitude climates, which are about equally distributed between the Northern and Southern Hemispheres, nearly all of the midlatitude climate area is found in the Northern Hemisphere. In the Southern Hemisphere, the land area poleward of the 40th parallel is so small that climates there are dominated by the great southern ocean.

If we travel northward to about 34° N in North America, we arrive at the interior Mojave Desert of southeastern California. Here we encounter environmental features that are significantly different from those of the low-latitude deserts of tropical Africa, Arabia, and northern Australia. Although the summer heat is strong—comparable to conditions in the Sahara

Global Climates Today

The Mojave Desert FIGURE 16.11

The odd-looking plants here are Joshua trees, which are abundant in the Mojave Desert. Most areas of this desert have fewer plants.

Desert—there is a cool winter season that does not occur in the tropical deserts. Here, cyclonic precipitation occurs in most months, including the cool months, and desert plants and animals have adapted to the dry environment (**FIGURE 16.11**).

Along the eastern sides of continents between latitude 20° and 35° N and S, the climate is moist, with plenty of rainfall in the summer and occasional tropical cyclones. We find such climates in South America, including parts of Uruguay, Brazil, and Argentina.

The *Mediterranean climate*, found between latitude 30° and 45° N and S, is unique because it has a wet but mild winter and a warm to hot, dry summer (**FIGURE 16.12**). In the Southern Hemisphere, it occurs along the coast of Chile, in the Cape Town region of South Africa, and along the southern and western coasts of Australia. In North America, it is found in central and Southern California. In Europe, this climate type surrounds the Mediterranean Sea, giving this type of climate its name.

Mediterranean climate
FIGURE 16.12

As we can see in this photo of Monterey harbor, the west coasts of regions with a Mediterranean climate often support extensive fisheries. Monterey, California, is located at latitude 36° and has a very weak annual temperature cycle because of its closeness to the Pacific Ocean. The summer is very dry, and it is frequently foggy.

Yakutsk, Siberia
FIGURE 16.13

These warmly dressed pedestrians are making their way down a snowy street in winter in Yakutsk, a Siberian city at latitude 62° N. Extremely cold winters are a characteristic of high-latitude climates, with the temperature in Yakutsk reaching about −42°C (−44°F) in January.

Where coasts are mountainous, the orographic effect, described in Chapter 14, causes large amounts of precipitation. Precipitation is plentiful in all months, especially in winter, while the marine influence keeps winter temperatures milder than those prevailing at inland locations at equivalent latitudes. In North America, such climates are found from the western coast of Oregon to northern British Columbia.

Another variation in climate can be found in the interior regions of North America and Eurasia that lie near mountain ranges on the west or south. The ranges effectively block the eastward flow of maritime air, so air from the polar continent dominates the climate in winter. Summers are warm to hot, but winters are cold to very cold. A similar climate stretches from the southern republics of the former Soviet Union to the Gobi Desert and northern China. The latitude range of this climate is 35° to 55° N.

The central and eastern parts of North America and Eurasia in the midlatitudes, meanwhile, lie in the polar-front zone—the battleground of polar and tropical air masses. The temperature contrast from season to season is strong, day-to-day weather is highly variable, and there is ample precipitation throughout the year.

HIGH-LATITUDE CLIMATES

As we move to higher latitudes, the climate becomes colder and progressively dryer. The high-latitude climates coincide with the belt of prevailing westerly winds that circles each pole. In the Northern Hemisphere, this circulation sweeps maritime polar (mP) air masses, formed over the northern oceans, into conflict with continental polar (cP) and continental arctic (cA) air masses on the continents. This interaction creates cyclones along the *arctic-front* zone. Cyclones that are formed along fronts are called *wave cyclones* (see Chapter 15).

Between about 50° and 70° N latitude, we find a climate with long, bitterly cold winters and short, cool summers (**FIGURE 16.13**). The annual temperature range is greater than that associated with any other climate, with the greatest range seen in Siberia, in Russia. Precipitation increases substantially in summer, but total annual precipitation is low. Although the climate in much of the forest region is moist, large areas of western Canada and Siberia have low annual precipitation and are therefore cold and dry.

Global Climates Today

Upernavik, Greenland
FIGURE 16.14

A chunk of glacial ice floats past the village of Upernavik, situated on the tundra-covered slopes of a small island in Baffin Bay. Upernavik, located on the west coast of Greenland (latitude 73° N), has short mild periods with above-freezing temperatures—equivalent to a summer season in lower latitudes. The long winter is very cold. In July, the sea ice melts and the ocean water warms, raising the moisture content of the local air mass and increasing precipitation.

The winters become longer and more severe as we move up to latitudes approaching 75° N and S, where we find the *tundra*. However, nearby ocean water moderates winter temperatures, so they don't fall to the extreme lows found in the continental interior (**FIGURE 16.14**). The tundra climate extends to latitudes higher than 80° N for the northern coast of Greenland. There is a very short mild season, but many climatologists do not recognize this as a true summer. This climate surrounds the Arctic Ocean and extends across the island region of northern Canada. It includes the Alaskan north slope, the Hudson Bay region, and the Greenland coast in North America. In Eurasia, it occupies the northernmost fringe of the Scandinavian Peninsula and the Siberian coast.

Because of the cold temperatures experienced in the tundra and northern boreal forest climate zones, the ground in these regions is typically frozen to a great depth. This perennially frozen ground, or *permafrost*, prevails throughout the tundra region. Normally, the top layer of the ground, 0.6 to 4 m (2 to 13 ft) thick, thaws each year during the mild season.

Approaching the poles, we find the vast, high ice sheets of Greenland and Antarctica and sea ice in the Arctic Ocean. Mean annual temperatures here are much lower than those associated with any other climate. In Antarctica and Greenland, the high surface altitude of the ice sheets intensifies the cold and there are frequent strong cyclones with blizzard winds. There is very little precipitation, and most of what there is takes the form of snow. But the snow accumulates because of the continuous cold. **FIGURE 16.15** shows an ice sheet in Antarctica, the coldest place on Earth.

CONCEPT CHECK STOP

What are the three key factors that influence temperature around the globe?

What are air masses? Why are they important for understanding climate?

What are the three climate groups?

Ice sheet FIGURE 16.15

Snow and ice accumulate at higher elevations in the Dry Valleys region of Victoria Land, Antarctica. Most of Antarctica is completely covered by ice sheets of the polar ice cap, and temperatures in the interior of Antarctica are far lower than at any other place on Earth. A low of −88.3°C (−127°F) was recorded in 1958 at Vostok, about 1300 km (about 800 mi) from the South Pole at an altitude of about 3500 m (11,500 ft).

Present-Day Climate Change

LEARNING OBJECTIVES

Examine the evidence for anthropogenic climate change.

Discuss the effects of climate change that we have already seen.

Describe the possible impact of global warming in the future.

As we have seen, climatic fluctuations are a natural part of the functioning of the atmosphere and climate system. However, within the past two centuries a new player has been added to the system: an industrialized human society that is now capable of significantly affecting climate worldwide.

WHAT WE KNOW

The first clear evidence of a human-induced, or **anthropogenic**, global effect on the atmosphere was the steady rise in carbon dioxide levels observed over the past 50 years at Mauna Loa Observatory in Hawaii (**FIGURE 16.16**). Not only are carbon dioxide levels rising, but from evidence in the ice cores, we know that they are rising much more rapidly than at any time in the past 100,000 years. Something altogether new is happening, and it is difficult to escape the conclusion that the new ingredient is human consumption of fossil fuels. Humans pump roughly 8 billion tons of carbon, most of it as carbon dioxide, into the atmosphere per year—which is actually more than enough to explain the observed increase.

Carbon dioxide slows the escape of heat, so both logic and past climate records suggest that an increase in carbon dioxide content of the atmosphere should be accompanied by an increase in average global temperature. The actual evidence is becoming more conclusive year by year. Worldwide temperature records show an increase of about 0.6°C in the past century,

anthropogenic Produced by human activities.

Anthropogenic climate change FIGURE 16.16

A The measurements shown here, made by Charles David Keeling, record the steady rise of carbon dioxide levels in the atmosphere over a 47-year period (1958–2005). The wobbles reflect the fact that growing plants absorb more carbon in the summer and spring and then decay, releasing carbon in the fall and winter. The Keeling curve is the world's longest continuous record of atmospheric carbon dioxide and was the first to confirm the rise of atmospheric carbon dioxide released from the burning of fossil fuels.

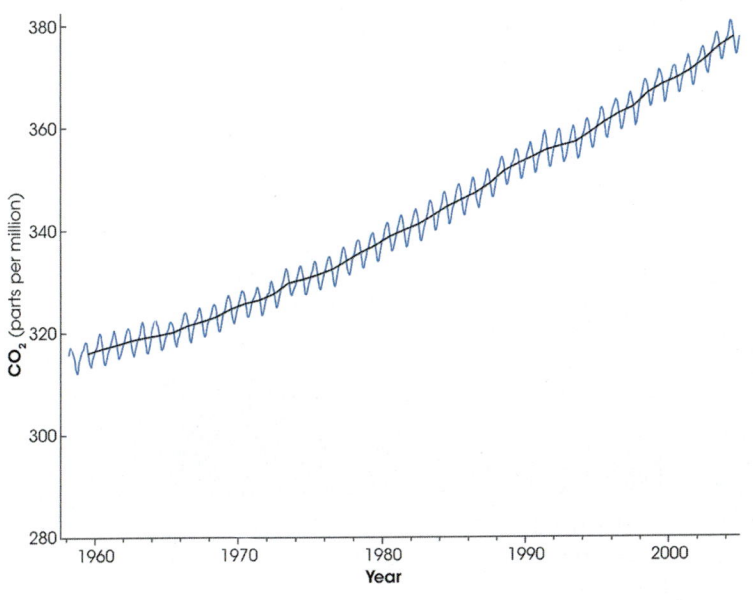

B The data were collected at the Mauna Loa Observatory, perched 3397 m above sea level on the northern slopes of Earth's largest volcano. The observatory is located away from dense vegetation or human population that might skew the measurements.

and the five warmest years on record (through 2005) were 1998, 2002, 2003, 2004, and 2005. However, even if there is an upward trend, it is not necessarily all due to human activities. Natural climatic fluctuations of this size and duration have certainly happened in the past; we can see the evidence in the growth rings of trees and coral. Such variations may be caused by changes in cloud cover, solar brightness, ocean circulation, or other, unknown factors.

Whatever its cause (or causes), the temperature increase has been greatest in the Arctic, and it is beginning to have real and observable effects. These include the retreat of glaciers, the calving of large icebergs from ice shelves, and the shrinking of some animal habitats and expansion of others (**Figure 16.17**).

WHAT WE THINK

In recent years, climatologists have developed general circulation models that attempt to link processes in the atmosphere, the hydrosphere, and the biosphere. However, because we still do not fully understand many of the linkages in the climate system, it is difficult to include them in a model. For instance, computer models do not yet adequately portray the dynamics of ocean circulation or cloud formation, two of the most important elements of the climate system. Despite their limitations, these models have been successful in simulating the general character of present-day climates and have greatly improved weather forecasting. This success encourages us to use them to obtain a general picture of future climate change.

Observed effects of climate change Figure 16.17

A The pack ice in the Beaufort Sea, off the North Slope of Alaska, now breaks up weeks earlier in the spring than it once did.

B On the West Antarctic Peninsula, the rise in temperatures has meant an influx of Gentoo penguins, which prefer warmer subarctic temperatures, and a sharp decline in the numbers of Adélie penguins.

C The tiny golden frog, which makes its home in Costa Rica's misty forests, is in danger of extinction as its habitat is becoming drier.

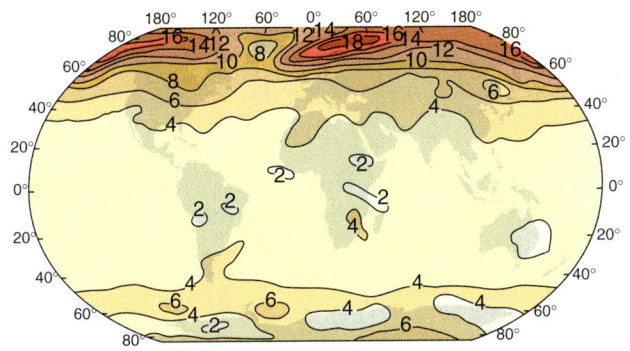

A Temperatures in winter would increase by 10°C or more through most of the Arctic. (Lines labeled "4" show where the projected temperature increase would be 4°C, "6" shows where the increase would be 6°C, and so on.)

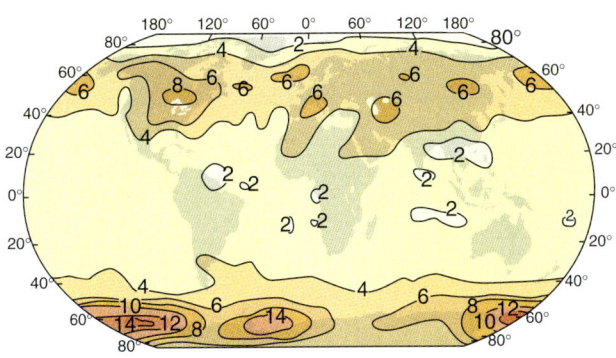

B Temperatures in June, July, and August would increase by 6°C or more in most of Antarctica.

Temperature rise if carbon dioxide doubles FIGURE 16.18

A computer model forecasts the changes in surface air temperature at different times of the year (in °C) that would result from a doubling of the amount of carbon dioxide in the atmosphere.

General circulation models allow us to incorporate different assumptions about the anthropogenic factor—that is, the burning of fossil fuels. The models differ in various details, but they all predict that if greenhouse gas levels stay where they are today—an optimistic assumption—we can expect global temperatures to rise by 0.5° to 1.5°C during the next 50 years, roughly consistent with the increase in temperature in the past 50 years.

However, if fossil-fuel consumption continues to grow at an increasing rate, atmospheric carbon dioxide levels are projected to double by 2100, if not sooner. If this proves to be the case, global climate models predict that average global temperatures will rise by between 1.5° and 4.5°C (FIGURE 16.18).

The indirect effects of such a warming are more difficult to quantify, but they include the rising of sea levels due to the melting of glacial ice; increased intensity of severe hurricanes and typhoons, which feed off warm seawater; and significant changes in animal and plant populations due to death or migration. Warming will also cause more methane to be released into the atmosphere from ocean deposits. Since methane is an even more potent greenhouse gas than carbon dioxide, this will further amplify warming. This phenomenon is an example of a feedback loop, in which rising temperatures change the environment in ways that then encourage more warming. Earth scientists also report that as rising temperatures dry out forests, decaying vegetation releases even more carbon dioxide into the atmosphere. Beyond a certain point, known as the *tipping point*, there may be no way to stop these changes.

The effects of warming will not be uniform all over the globe but will continue to be much greater in polar regions. FIGURE 16.19, on the following page, looks at how various regions of the world will be affected by global warming.

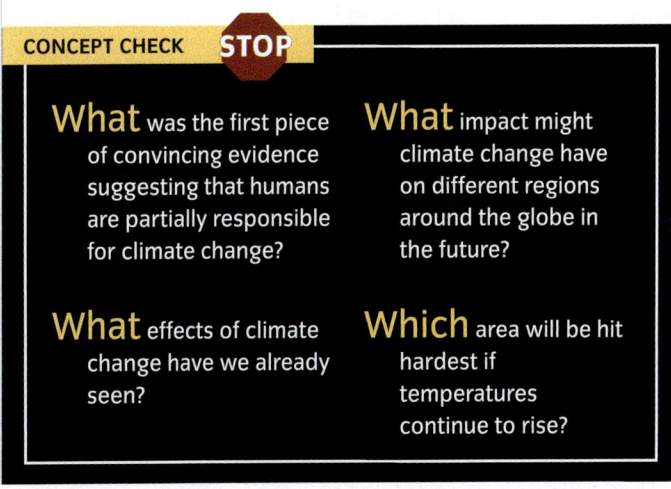

CONCEPT CHECK STOP

What was the first piece of convincing evidence suggesting that humans are partially responsible for climate change?

What effects of climate change have we already seen?

What impact might climate change have on different regions around the globe in the future?

Which area will be hit hardest if temperatures continue to rise?

Present-Day Climate Change

▼ Europe

A century ago, the Rhône Glacier, source of the Rhone River, terminated at this spot in Valais, Switzerland. Now the terminus is barely visible in the distance. As temperatures rise in the next century, ice and permafrost at higher European latitudes will melt, the tundra will become drier, and forests will migrate north. A changing climate will dry out southern Europe and boost agricultural production in central and northern areas. The region's key vulnerabilities will be water and land resources, semi-natural ecosystems and forests, and agriculture and fisheries.

▲ North America

Although individual storms or stormy seasons cannot be blamed on global warming, unusually warm water temperatures in the Gulf of Mexico in the fall of 2005 apparently contributed to Hurricane Katrina's becoming a monster storm. Climatologists estimate that floods, droughts, storms, and landslides will increase in frequency, severity, and duration in North America. There will be more very hot days in summer and fewer cold snaps in winter. Higher seas and bigger storms will lead to increased coastal erosion. The Great Plains in the United States and the Canadian Prairies may face increased drought, but overall North American food production will rise.

◄ Latin America

The Amazonian rainforests will dry out, encouraging wildfires that will threaten the continent's rich biological diversity. *El Niño* events, described in detail in Chapter 12, could become more frequent in a warmer world. Andean glaciers will retreat farther, and Mexico's droughts will become more frequent.

Visualizing

Future impacts of climate change FIGURE 16.19

Global warming due to climate change will have repercussions around the world.

◀ **Asia**
Crop production in northern and midlatitude Asia will improve, thanks to the warmer weather. However, South and Southeast Asia's many developing countries will suffer from intolerably high temperatures that will reduce food production and from a drop in rainfall and water supply. In arid and semiarid Asia, higher temperatures and more evaporation will reduce rice yields dramatically.

◀ **Small Island States**
The tens of thousands of small islands scattered across the world's oceans are particularly vulnerable to climate change. Many rise only 1 or 2 m above sea level. Male, the capital of the Maldives, could find itself underwater if sea levels rise. The Maldives, an island nation in the Indian Ocean, are built on coral reefs that grow over time but cannot possibly grow fast enough to keep up with a rapid change in sea level.

◀
Australia and New Zealand
Crops in arid Australia are already growing near their maximum heat tolerance, so warmer weather will affect food production in these regions. New Zealand, which is cooler and wetter, could initially benefit from warmer temperatures.

◀ **Africa**
Over the next century, East Africa could receive more rain, but southern Africa is forecast to become much drier. Floods and storms will increase in frequency, and food and water shortages will be a growing problem.

Amazing Places: Barrow, Alaska

Global Locator

The people of Barrow, Alaska, live in one of the most extreme environments on our planet (**A**). Situated 539 km north of the Arctic Circle, this is the northernmost settlement on the North American mainland. In winter, the Arctic Ocean freezes right up to the coast, and when the sun sets in mid-November, it does not rise again until January.

The Inupiat inhabitants are acutely vulnerable to climate change. They travel out on the sea ice in spring and summer to hunt whales, seals, and walrus for sustenance (**B**). However, as weather conditions warm, the amount of sea ice is decreasing, and hunting is becoming increasingly precarious. The way that families store food is also affected, as higher temperatures make it harder to store food in natural freezers below the ground.

The poles are warming much faster than the rest of the planet. Average temperatures in Barrow have increased by around 2.5°C in the past 30 years. Earth scientists long ago predicted that the most visible impacts from a globally warmer climate would strike high latitudes first. They predicted that air and sea temperatures would rise, that the snow would melt earlier and the ice would freeze later, and that there would be an increase in storm intensity. Now all of those impacts have been documented in Alaska, which stands as an early warning system for the rest of the planet.

SUMMARY

1 Global Climate Change in the Past

1. **Climate** change is nothing new. Over the past 2 million years, Earth has experienced repeated **glaciations** and warm interglacial periods. We are now in an interglacial period. The last glaciation, or ice age, peaked about 18,000 years ago. The current trend of global warming may create a more pronounced interglacial period but is unlikely significantly to delay the next ice age, which can be expected in 50,000 years or so.

2. There are direct records of air temperature dating to the middle of the 19th century. To reconstruct past climates beyond that, Earth scientists analyze fossils, sediments, tree rings, corals, and ice cores. The composition of the shells of microscopic fossils, found in ocean sediments, reflects changes in the proportions of warm water and cold water. Fossilized plants also reveal changes in world temperatures. Ice cores preserve trapped bubbles of "fossil air" that help scientists determine the concentration of greenhouse gases in the atmosphere in the past.

3. Climate change can come about naturally for many reasons, making it difficult to predict exactly how the climate will change. Tectonic activity can shift land masses, leading to changes in ocean and atmosphere circulation. Volcanic activity can release aerosols into the air, which affect the amount of heat absorbed and emitted in the atmosphere. Changes in solar output and astronomical cycles—called **Milankovitch cycles**—also influence how much solar energy reaches Earth. Greenhouse gases have always affected the climate.

4. Climate can also change rapidly. During the **Younger-Dryas** event, which occurred between 10,000 and 11,000 years ago, the climate in the North Atlantic abruptly cooled. This happened because melting glaciars shutdown an important ocean circulation system, the thermohaline circulation.

2 Global Climates Today

1. Climate depends on latitude, location, and moisture. Latitude determines the annual pattern of insolation; near the Equator, temperatures are warmer and toward the poles, temperatures are colder. Location—continental or maritime—can enhance or moderate the annual insolation cycle because ocean-surface temperatures vary less with the seasons than land-surface temperatures. Air temperature also has an important effect on precipitation because warm air can hold more moisture than cold air.

2. Global precipitation patterns are determined largely by **air masses** and their movements. The latitude at which the air mass originates determines its temperature. Its original coastal or continental location determines its moisture content. Cyclonic precipitation is created at the *frontal zones* where air masses meet.

3. The globe can be split into three broad climate bands, arranged by latitude. Low-latitude climates (Group I) are typically warm and can range from extremely wet to severely dry. **Low-latitude rainforests** are found in this climate group. Midlatitude climates (Group II) are more variable because conflicting air masses tend to meet and interact in these regions. High-latitude climates (Group III) become colder and dryer approaching the poles.

3 Present-Day Climate Change

1. The possibility of **anthropogenic** climate change due to the burning of fossil fuels has arisen only in the past century. Ice core analysis suggests that carbon dioxide levels are rising much more rapidly than at any time in the past 100,000 years. Such a rise can be accounted for by the 8 billion tons of carbon that humans pump into the atmosphere each year. Many scientists believe that global warming will continue through the 21st century and that its extent will depend on human actions to limit the production of greenhouse gases.

2. So far, the effects of the warming are most pronounced in the polar regions, where glaciers are retreating and ice caps are melting.

3. If fossil-fuel consumption continues to grow at an increasing rate, atmospheric carbon dioxide levels are projected to double by 2100 and average global temperatures are predicted to rise by between 1.5° and 4.5°C. Worldwide effects in the future will vary according to location but may include rising sea levels, increases in the severity of weather systems, and changes in animal habitats that will cause some species to die and force others to migrate to new territory.

KEY TERMS

- climate p. 491
- glaciation p. 493
- Milankovitch cycles p. 495
- Younger Dryas p. 496
- low-latitude rainforest p. 500
- anthropogenic p. 505

CRITICAL AND CREATIVE THINKING QUESTIONS

1. If you were to set out to make a model that might be used to predict global climates 50 years into the future, what kind of climate data would you put into your model? Once you have developed your model, how might you test it?

2. Suppose Earth's axis was perpendicular to the plane of Earth's orbit around the Sun. How would the climate at the place you live differ from the climate today? How would it change if Earth's orbit became much more elliptical?

3. Is there any evidence in the landforms around the area in which you live or go to school to indicate that the area was glaciated during the most recent ice age? How thick was the ice at maximum glaciation?

4. There is evidence in the ice cores that there have been very rapid increases and decreases in temperatures in the past. Suppose that climate modelers predict that 50 years from now there will be a time of rapid temperature rise of as much as 5°C over 10 years. Choose one country from each of the three climate groups discussed in the chapter and suggest what changes they should expect.

5. Find out whether your city, state, province, or country has set goals for the reduction of carbon dioxide emissions in order to limit their potential contribution to global warming. What steps have been taken to meet these goals?

What is happening in this picture?

The photo shows bubbles of air locked in an ice core extracted from the rapidly retreating Quelccaya ice cap in Peru. As ice in a glacier recrystallizes, tiny bubbles of air become permanently trapped in the ice. When the ice is melted under controlled conditions in the laboratory, this fossil air can be analyzed.

- What measurements would you ask scientists to make?
- What information could they gain from them?

SELF-TEST

1. Study of Earth's climate record indicates that temperatures have _____ over the past two million years.
   a. fluctuated greatly
   b. remained constant
   c. been slowly decreasing
   d. been slowly increasing

2. From the standpoint of global climate, Earth is now _____.
   a. at the beginning of an interglacial period
   b. at the end of a glaciation
   c. at the peak of an interglacial period
   d. 20,000 years into a superinterglacial period

3. Earth scientists use a number of techniques to study paleoclimate, including data collected from _____.
   a. fossil pollens from peat bogs and lakes
   b. stable oxygen isotope ratios
   c. ice cores
   d. All of the above are techniques used to study paleoclimate.

4. These microfossils help scientists study climate change because they record changes in the concentration of _____ in the ocean, which can be linked to changes in temperature.
   a. calcium carbonate
   b. oxygen
   c. carbon dioxide
   d. hydrogen

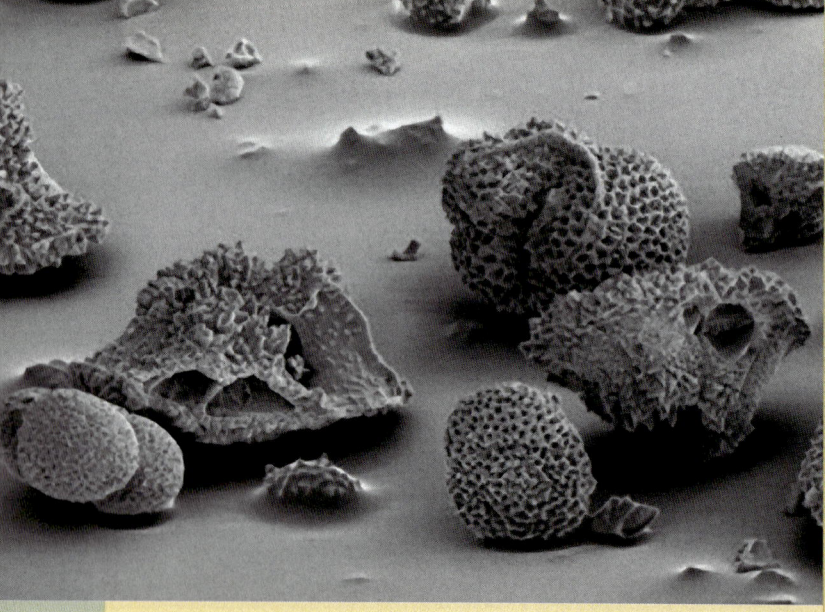

5. Earth science studies have revealed a number of natural agents that can act together to cause climate change. Which of the following are considered possible causes of climate change?
   a. changes in the eccentricity of Earth's orbit
   b. plate tectonic activity
   c. interruptions of ocean circulation
   d. the tilt and precession of Earth's rotational axis
   e. All of the above are considered possible natural causes of climate change.

6. The layers in this rock provide evidence of _____.
   a. the Younger-Dryas event
   b. superinterglacial spikes
   c. Milankovitch cycles
   d. tipping points

7. Approximately when did the Younger Dryas event occur?
   a. 5000 years ago
   b. 10,000 years ago
   c. 50,000 years ago
   d. 100,000 years ago

8. Which of these is not a key factor in controlling climate?
   a. coastal–continental location
   b. plate tectonic activity
   c. latitude
   d. longitude

9. Name the three climate groups and sketch their rough boundaries on the diagram.

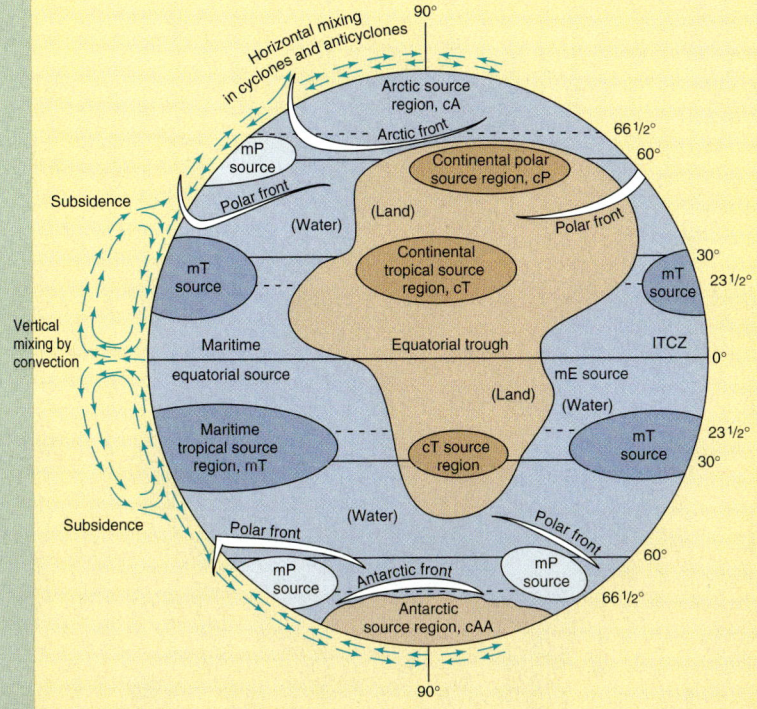

10. In which climate group would you expect to find these trees?
    a. Group I
    b. Group II
    c. Group III

11. In which climate group would you find the Mediterranean climate and the tundra (respectively)?
    a. Group I and Group II
    b. Group I and Group III
    c. Group II and Group III
    d. Group III and Group I

12. Anthropogenic carbon dioxide in the atmosphere results from the continued burning of fossil fuels. How far back does the recent record of atmospheric carbon dioxide measurements extend?
    a. 100 years   c. 50 years
    b. 75 years   d. 10 years

13. Atmospheric carbon dioxide is projected to double by _____, if not sooner, leading to a predicted global temperature rise of between _____° and _____°C.
    a. 2050/3.5/7.5   c. 2100/1.5/4.5
    b. 2070/2.5/5.0   d. 2200/2.5/4.5

14. What effects would continued global warming have over the next century?
    a. retreat of glaciers
    b. calving of large icebergs from ice shelves
    c. shrinking of some animal habitats
    d. expansion of some animal habitats
    e. All of the above are possible effects of global warming.

15. Food production in South and Southeast Asia is predicted to _____ as rainfall _____, affecting rice yields. In North and mid-latitude Asia, food production will _____.
    a. increase/increases/increase
    b. increase/decreases/decrease
    c. decrease/increases/decrease
    d. decrease/decreases/increase

Earth's Place in Space

17

In 2006, our solar system received a makeover: Pluto was demoted from its position as the ninth planet and reclassified as a dwarf planet.

Pluto's story begins with a case of mistaken identity. In the 1840s, Urbain Le Verrier predicted the position of the then-undiscovered Neptune, after analyzing unexpected deviations in the orbit of Uranus that could be explained by the gravitational tug of a nearby massive object. Soon after, Neptune was indeed discovered. When astronomers noticed that its orbit seemed to be similarly disrupted, they began to hunt for another new planet, and on February 18, 1930, Pluto was discovered by Clyde Tombaugh. As it turned out, Pluto's tiny mass was too small to significantly disrupt Neptune, and it was later found that, in fact, no missing planet is needed to explain Neptune's orbit. However, Pluto, pictured in this Hubble Space Telescope image surrounded by three moons, took its place as the ninth planet.

Nevertheless, Pluto never quite fit in. Lying farthest from the Sun, it is little larger than the Moon and is dwarfed by its neighbors, Jupiter, Saturn, Uranus, and Neptune. Its fate was sealed with the discovery in 2005 of Eris, an object slightly bigger than Pluto that also orbits the Sun. Rather than officially inducting Eris as the tenth planet—and risk opening the planetary club to many other such small bodies—the International Astronomical Union created a new definition of *planet*. This included the criterion that the body must have cleared comparable-sized objects from the neighborhood of its orbit, and both Pluto and Eris failed this condition.

CHAPTER OUTLINE

■ Astronomy and the Scientific Revolution p. 518

■ The Solar System p. 524

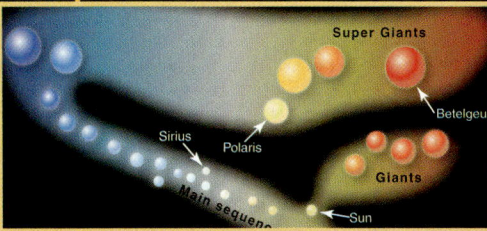

■ Stars and Stellar Evolution p. 538

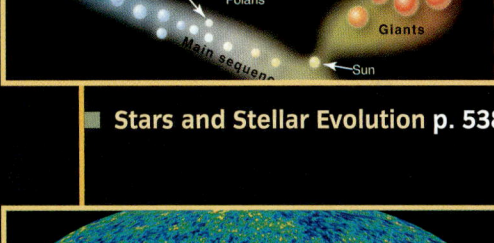

■ The Universe and How It Came to Be p. 540

Astronomy and the Scientific Revolution

LEARNING OBJECTIVES

Discuss evidence supporting the heliocentric universe.

Describe retrograde motion.

Explain why Earth has seasons.

Describe planetary orbits in terms of Kepler's laws and Newton's discovery about gravity.

For much of human history, people believed that everything in the universe revolved around Earth—for good reason. Observing the Sun rising and setting, while Earth seemingly stood still, they reasoned that the universe was **geocentric**, with the Sun revolving around a stationary Earth. Today we are taught that Earth revolves around the Sun. But just how do we know that? The search for evidence that we live in a **heliocentric** universe was a major factor in the rise of modern science.

geocentric A model of the universe in which a stationary Earth is at the center and all other celestial bodies revolve around it.

heliocentric A model of the universe in which a stationary Sun is at the center and the planets revolve around it.

IDEAS FROM ANTIQUITY

Ancient Greek civilization, which flourished for 800 years from about 650 B.C. to A.D. 150, spawned many famous philosophers. One of the most influential of these, Aristotle (384–322 B.C.), favored the notion of a geocentric universe (**FIGURE 17.1**). However, a few others, most notably Aristarchus (312–230 B.C.), realized that the daily rising and setting of the Sun and the apparent movement of the star sphere across the sky could be explained just as well if the stars were fixed and Earth rotated on its axis once every day. Using two of the branches of mathematics discovered by the Greeks, geometry and trigonometry, Aristarchus determined that the Sun was huge compared to a relatively small Earth and tiny Moon. The idea of a huge Sun revolving around a small Earth did not make sense to him. In addition, he noticed that a heliocentric universe helped to explain the origin of the seasons, which would arise naturally if the axis of rotation was tilted and Earth revolved around the Sun. But

The celestial spheres FIGURE 17.1

Aristotle pictured the Sun, the Moon, the five visible planets, and the stars as being suspended on concentric, hollow spheres that rotate about an imaginary axis extending outward from the two poles of Earth. Beyond the star sphere lay the realm of the gods.

Aristarchus could not win others over to his heliocentric suggestion, and by the 16th century a geocentric universe had come to be accepted as a divine fact—to be protected by the Catholic Church.

COPERNICUS'S CHALLENGE

The hardest challenge for the concept of a geocentric universe was explaining the motions of the planets. To the naked eye, the five visible planets—Mercury, Venus, Mars, Jupiter, and Saturn—look like stars, except that they seem to wander in relation to the fixed stars. Indeed, the very name *planet* comes from *planetai*, an ancient Greek word meaning "wanderers." The paths followed by these wanderers are odd ones. The planets move a bit farther east each evening, but periodically they slow down and briefly reverse direction before resuming their eastward motion. The temporary reversal of direction is known as *retrograde motion*.

During the 1490s, Nicolaus Copernicus (1473–1543), a student at the University of Bologna, Italy, realized that a heliocentric system could easily explain retrograde motion. The appearance of retrograde motion would result from differences between the time it takes Earth to orbit the Sun and the time it takes any other planet to orbit the Sun, as shown in **Figure 17.2**. Moreover, Copernicus suggested that because Mars has a larger retrograde motion than does Jupiter or Saturn, it must be the closest of the three planets to Earth, while Saturn, with the smallest retrograde motion, must be the most distant. This proposal could be tested by suitable astronomical measurements.

Copernicus also offered two other major hypotheses. First, he suggested that the positions of the planets at any given time in the future could be accurately predicted by assuming that they move in circular orbits around the Sun. Second, he revived the old suggestion of Aristarchus that Earth spins on its axis and the daily

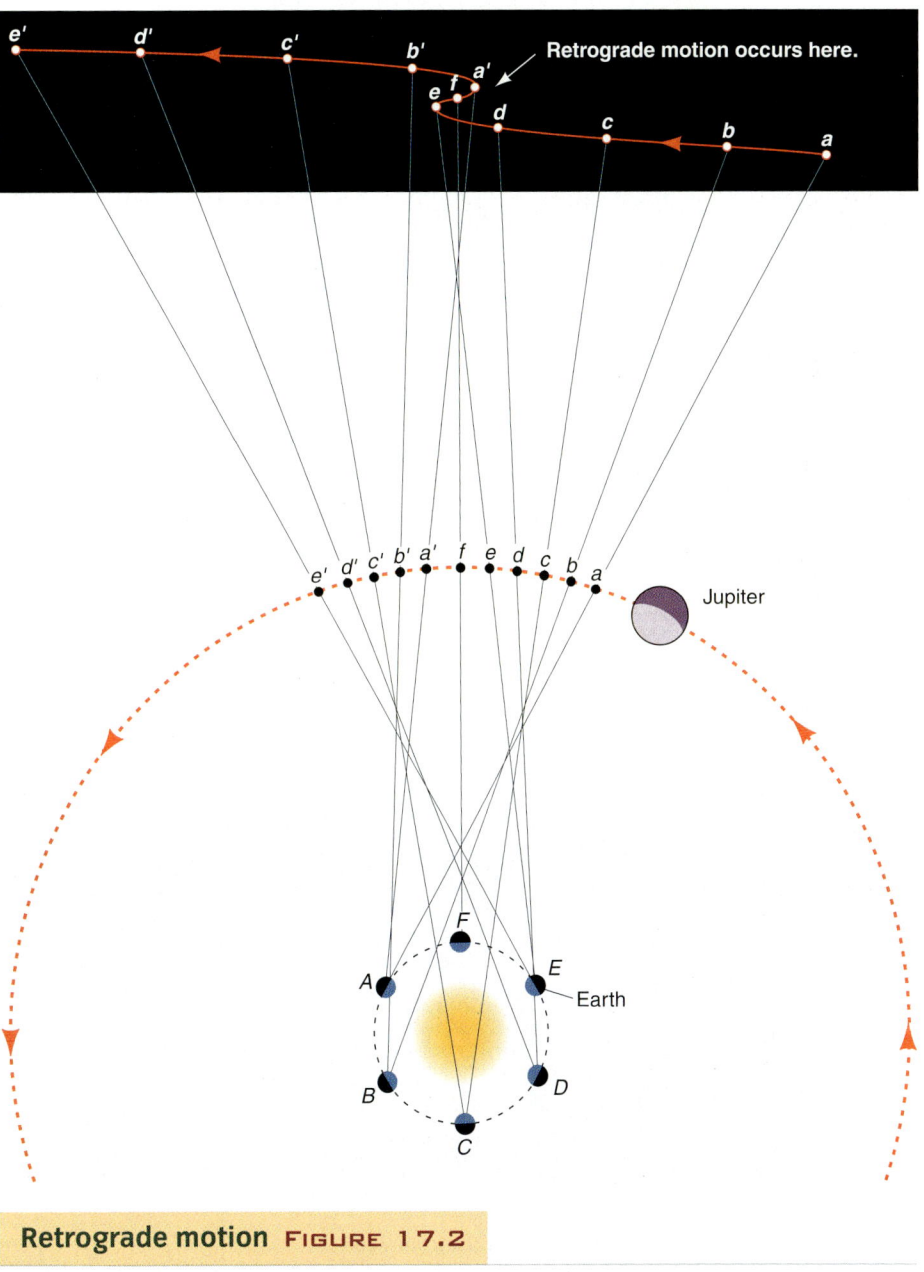

Retrograde motion Figure 17.2

As Jupiter moves from point *a* to point *e′* in its orbit, Earth moves counterclockwise from point *A*, completely around the Sun and back to *A*, and then on to point *E*. When one watches Jupiter from Earth, the planet seems to wander in relation to the fixed stars.

rising and setting of the Sun is a result of that rotation. It took until 1851 to demonstrate that Earth really does spin on its axis; this was accomplished using Foucault's pendulum, as described in FIGURE 17.3. By extension, Copernicus could explain the seasonality of Earth's climate (FIGURE 17.4).

Copernicus's view of the universe directly challenged the Roman Catholic Church, which had built the concept of a geocentric universe into Church doctrine. Looking back, we can see that Copernicus sowed the seeds that finally separated science from religion,

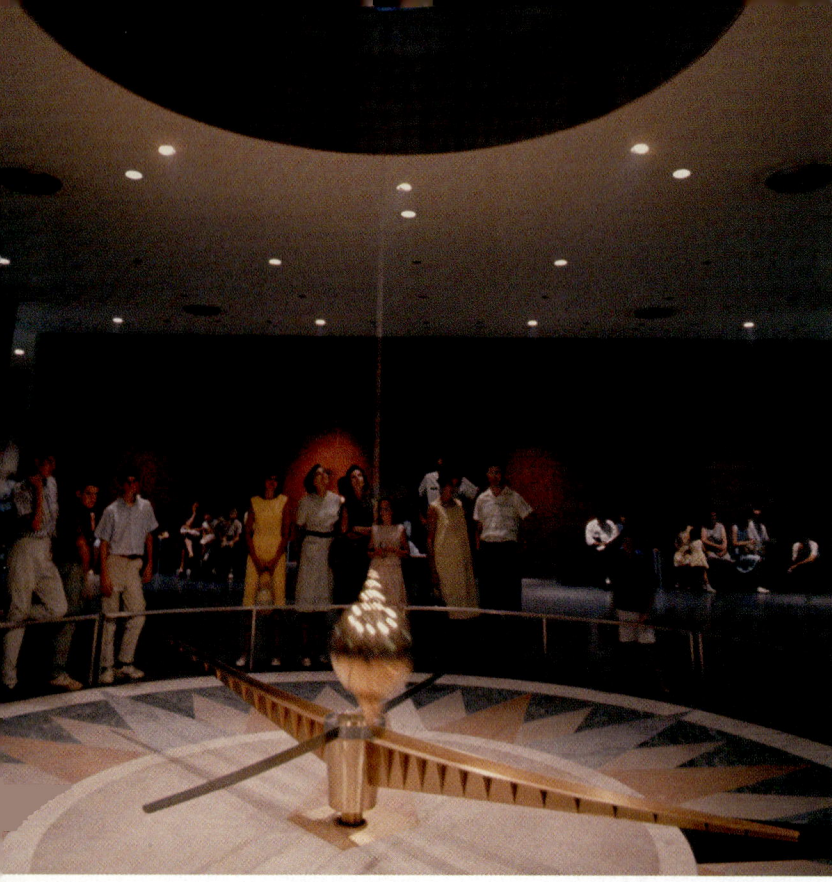

Foucault's pendulum FIGURE 17.3

Léon Foucault provided the first demonstration of the rotation of Earth in 1851 using a huge pendulum, with a 28 kilogram bob and a 67 meter wire, hung from the dome of the Panthéon in Paris. The pendulum was free to swing in any vertical plane. Because Earth rotated under the pendulum, the direction along which the pendulum swung appeared to shift, rotating clockwise 11° per hour and making a full circle in 32 hours.

Process Diagram

The seasons: Earth–Sun relations through the year FIGURE 17.4

VIEW THIS IN ACTION in your WileyPLUS course

The four seasons occur because Earth's axis is tilted by 23 1/2°.

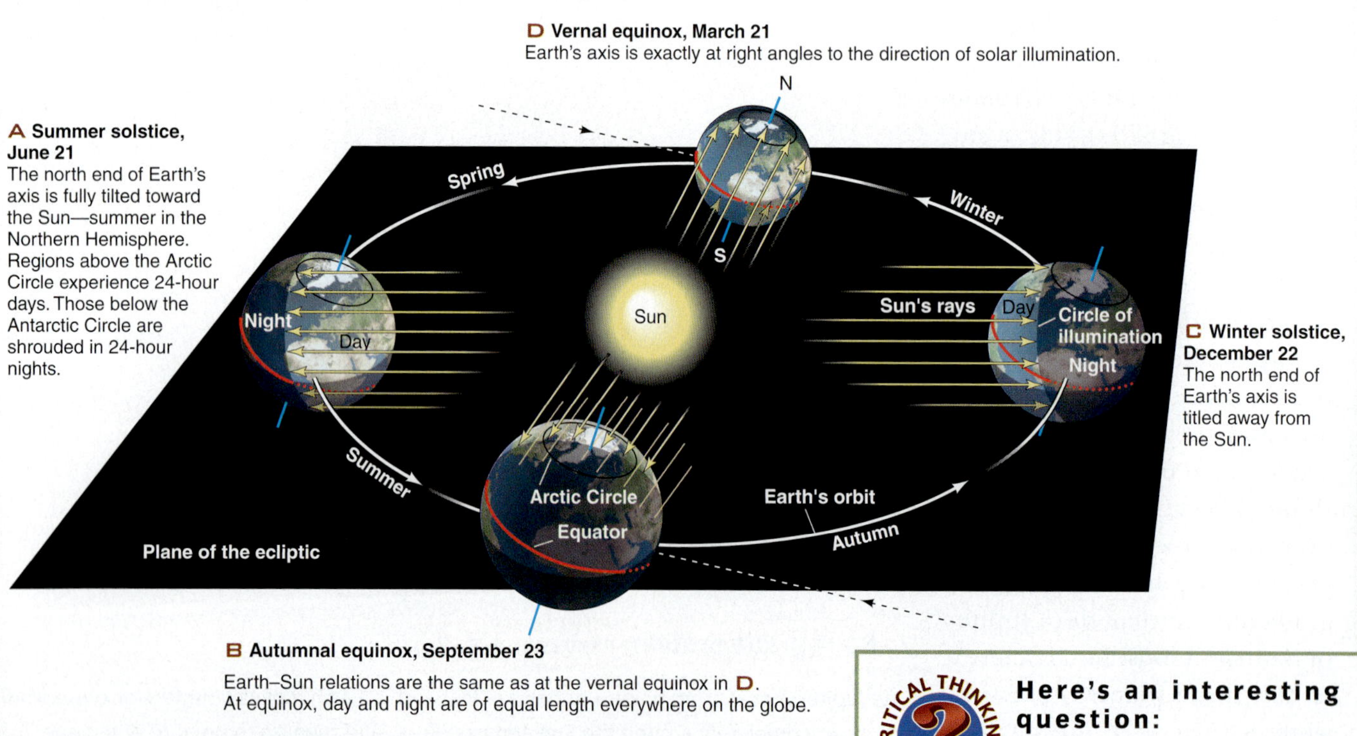

D Vernal equinox, March 21
Earth's axis is exactly at right angles to the direction of solar illumination.
N
S
A Summer solstice, June 21
The north end of Earth's axis is fully tilted toward the Sun—summer in the Northern Hemisphere. Regions above the Arctic Circle experience 24-hour days. Those below the Antarctic Circle are shrouded in 24-hour nights.
Spring
Winter
Night
Day
Sun
Sun's rays
Day
Circle of illumination
Night
C Winter solstice, December 22
The north end of Earth's axis is titled away from the Sun.
Summer
Arctic Circle
Equator
Earth's orbit
Autumn
Plane of the ecliptic
B Autumnal equinox, September 23
Earth–Sun relations are the same as at the vernal equinox in D.
At equinox, day and night are of equal length everywhere on the globe.

CRITICAL THINKING

Here's an interesting question:

- How would the seasons differ in the place where you live if Earth's axis were perpendicular to the plane of its orbit around the Sun?

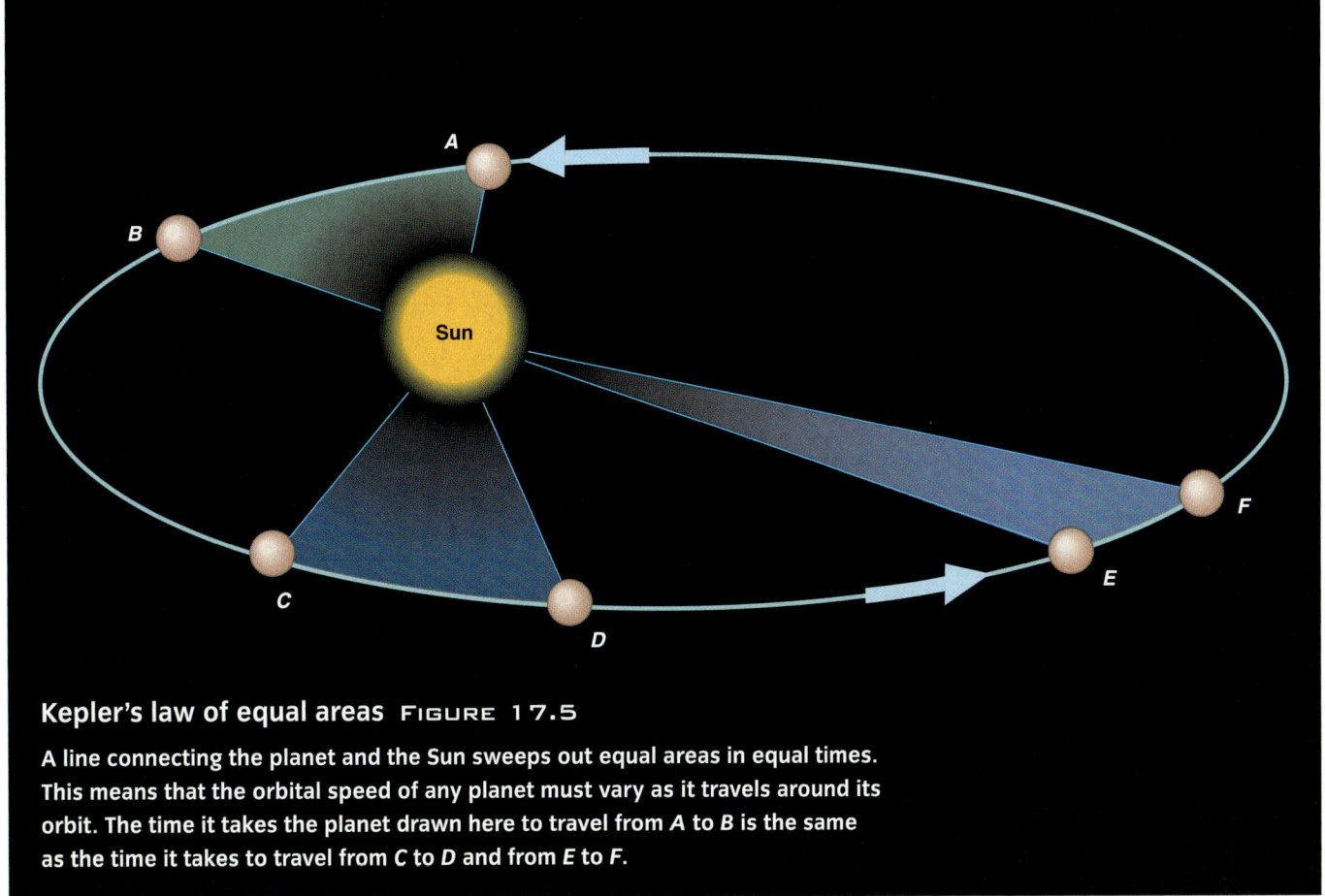

Kepler's law of equal areas FIGURE 17.5

A line connecting the planet and the Sun sweeps out equal areas in equal times. This means that the orbital speed of any planet must vary as it travels around its orbit. The time it takes the planet drawn here to travel from A to B is the same as the time it takes to travel from C to D and from E to F.

spawning the scientific revolution that shaped the society in which we live today.

KEPLER AND THE NEW ASTRONOMY

Although Copernicus was correct in proposing that we live in a heliocentric universe, he was wrong about the shape of the planets' orbits. But it took years of painstaking naked-eye measurements of planetary positions to reveal his mistake. In 1572, Tycho Brahe (1546–1601) built the first modern astronomical observatory on the Danish island of Hven. Tycho was skeptical of the heliocentric hypothesis, and to prove Copernicus wrong he began collecting the most accurate planetary data up to that time.

In 1597, Tycho hired a young German mathematician, Johannes Kepler (1571–1630), to carry out astronomical calculations. Kepler, unlike Tycho, thought Copernicus might be right; he also gave a great deal of thought to a problem Copernicus had not treated: What is the nature of the force that keeps the planets moving around the Sun? Why do they revolve in orbits instead of moving in straight lines out into space?

Because the planets closest to the Sun move faster than those far away, Kepler suggested that a mysterious force must reside in the Sun and have a greater effect on closer objects. Today we know that the force is gravity, but in Kepler's day gravity was an unknown concept. Kepler suggested that the force might be magnetism.

Try as he might, Kepler could not make planetary positions calculated from circular orbits agree with Tycho's measurements. Eventually he tried calculating the position of Mars based on an elliptical orbit, with far better success.

Kepler discovered three laws that describe planetary motion:

1. *The law of ellipses.* The orbit of each planet is an ellipse with the Sun at one focus.

2. *The law of equal areas.* A line drawn from a planet to the Sun sweeps out equal areas in equal amounts of time (FIGURE 17.5). Because of this, the orbital speeds of the planets are not uniform. A planet moves rapidly when close to the Sun and slowly when far away from the Sun.

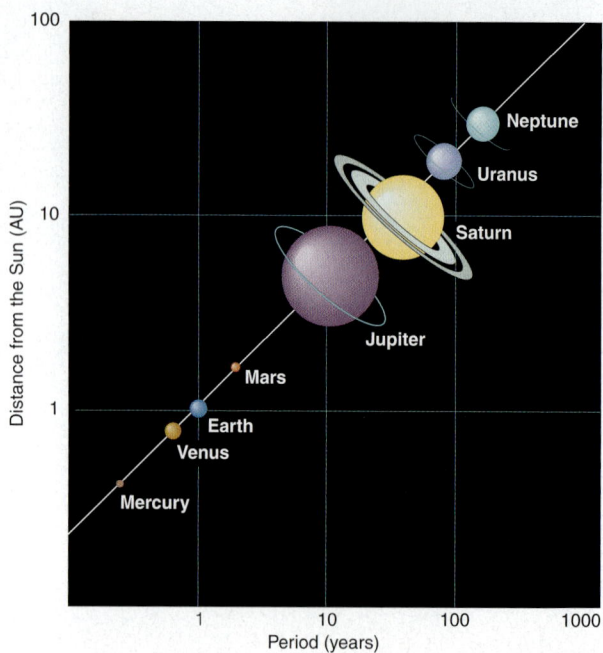

Kepler's law of orbital harmony FIGURE 17.6

Kepler noted that the distance of the planet from the Sun is related to the period of the planet's orbit around the Sun. The unit of distance is the *astronomical unit*, the average distance of Earth from the Sun. You can see that there is a gap between Mars and Jupiter. This is where the asteroid belt lies. The asteroid belt contains small, rocky fragments that did not cluster together to form planets.

3. *The law of orbital harmony.* For any planet, the square of the orbital period in years is proportional to the cube of the planet's average distance from the Sun (FIGURE 17.6). The period is the time a planet takes to make one complete revolution around the Sun. For example, the period of Earth is 365.24219 days.

GALILEO AND NEWTON

Galileo Galilei (1564–1642) was an extraordinary man who made a great many scientific discoveries. In 1609, he constructed a small telescope that magnified objects by 30 times. With this device, he helped put to rest the notion of a geocentric universe. Among other discoveries, Galileo saw four moons orbiting Jupiter, proving that Earth is not the center of all orbital motion. He also found that Venus has phases, just like the Moon, and also changes greatly in apparent size—a fact that can be explained only if Venus and Earth are in orbit around the Sun (FIGURE 17.7).

Galileo also made major contributions to our understanding of moving bodies and gravity. He reasoned that bodies move when they are acted on by forces, and once a body is moving, it will stop or change direction only if another force is applied to it. He also concluded that gravity pulls all falling bodies with the same accel-

Phases of Venus

FIGURE 17.7

When Earth and Venus are on the same side of the Sun, Venus is seen as a crescent. When Earth and Venus are on opposite sides of the Sun, Venus is seen as a full disk, but it is only one-seventh the diameter of the crescent because it is so far away. In a geocentric system, with Venus in orbit around Earth, the apparent size of Venus should change very little.

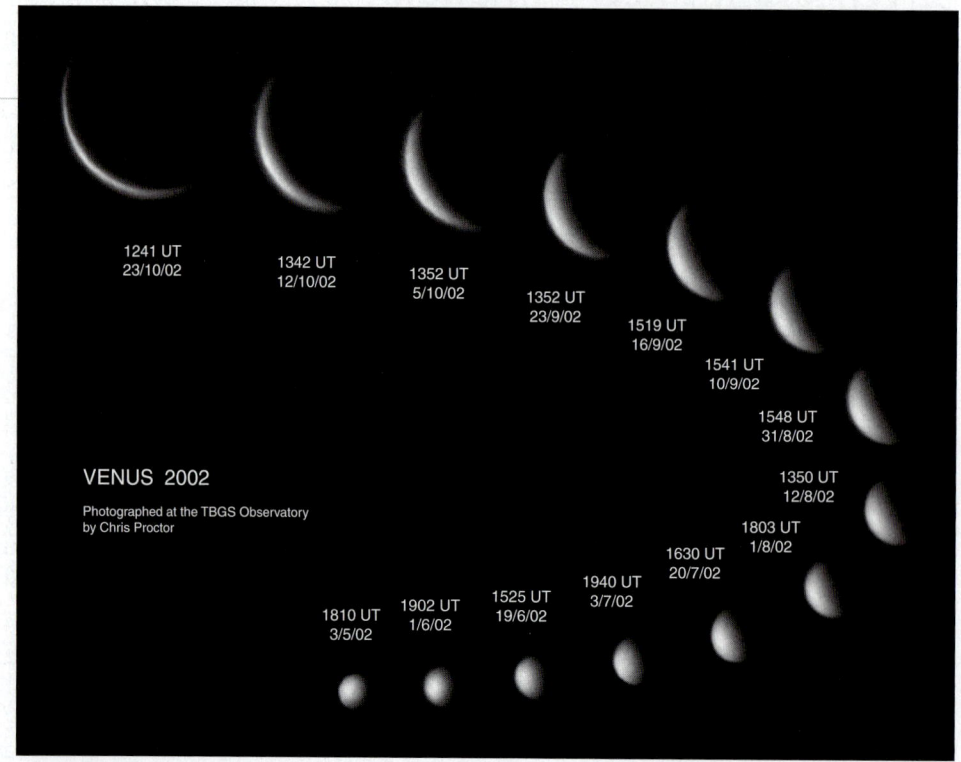

Earth's pull on the Moon   FIGURE 17.8

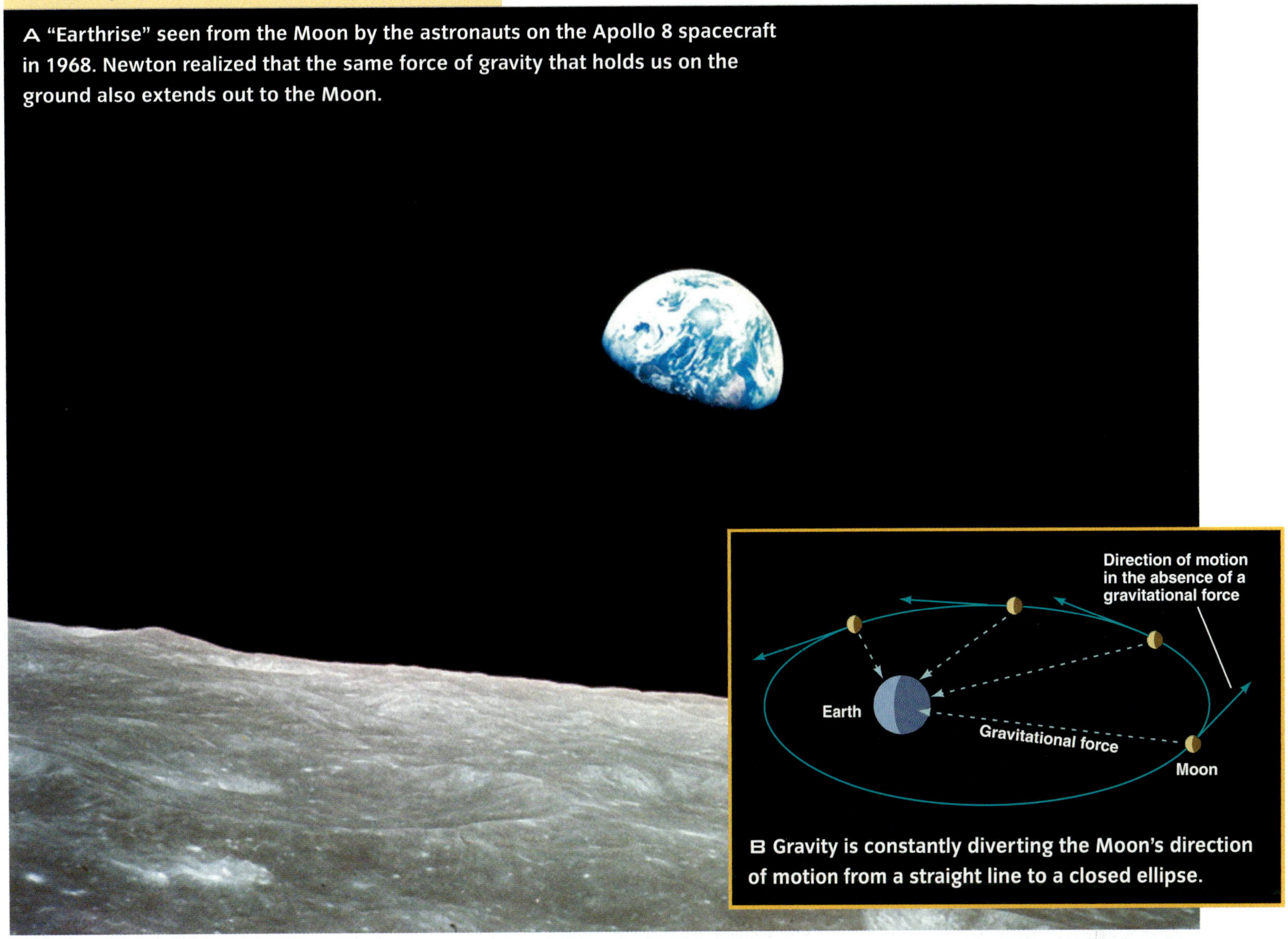

A "Earthrise" seen from the Moon by the astronauts on the Apollo 8 spacecraft in 1968. Newton realized that the same force of gravity that holds us on the ground also extends out to the Moon.

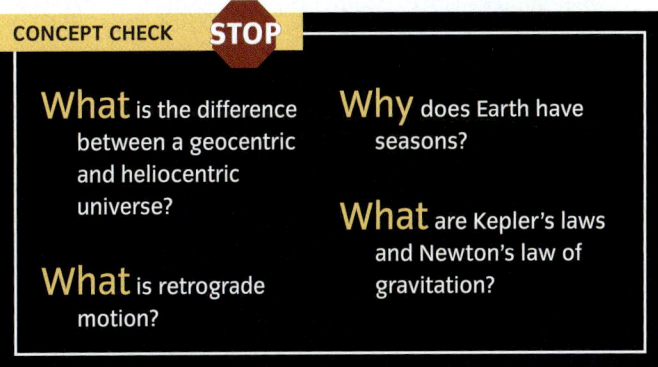

B Gravity is constantly diverting the Moon's direction of motion from a straight line to a closed ellipse.

eration, regardless of their mass. His insight provided one of the key steps to explaining the law of gravity and the motions of planets. The person who pulled all the pieces together, however, was Isaac Newton (1642–1727).

According to legend, Newton started thinking about gravity when he saw an apple fall from a tree. He reasoned that if the force of gravity acts on an apple, that force must also extend "to the orb of the Moon." The Moon revolves around Earth instead of moving through space in a straight line because the force of Earth's gravity exerts a small pull on it (**FIGURE 17.8**). The force that Kepler had misidentified as magnetism was actually gravity. Newton had discovered that gravity acts between all bodies.

With his discovery, Newton had finally united the heavens, which once were believed to belong to the gods, with Earth, showing that the same forces are at play in both realms.

CONCEPT CHECK STOP

What is the difference between a geocentric and heliocentric universe?

What is retrograde motion?

Why does Earth have seasons?

What are Kepler's laws and Newton's law of gravitation?

Astronomy and the Scientific Revolution

The Solar System

LEARNING OBJECTIVES

Explain the nebula hypothesis and the origin of the solar system.

Describe the structure of the Sun and **explain** how it produces energy.

Identify the characteristics of the terrestrial and Jovian planets.

Describe meteorites, asteroids, comets, and the role of impacts in the formation of the Moon.

Discuss other solar systems and the possibility of life on other planets.

Just how did the solar system that we call home form? Any theory for the origin of the solar system must be able to explain its two most striking characteristics. In Chapter 1, we met the first of these features: The planets can be separated into two distinct groups based on their density and closeness to the Sun (see Figure 1.9). The terrestrial planets—Mercury, Venus, Earth, and Mars—lie closest to the Sun and are small, dense, and rocky. The asteroids, which orbit the Sun in the asteroid belt, as we saw in Figure 17.6, are also rocky, dense bodies, but they are too small to be called planets. The Jovian planets lie farther from the Sun than Mars and are much larger than the terrestrial planets, yet much less dense.

If you were to view the solar system from a spaceship, you would notice a second remarkable feature: Almost everything—planets, moons, Sun—revolves and rotates in the same direction. All the planets revolve around the Sun and all the moons revolve around their respective planets in approximately the same plane. If your vantage point is high above the North Pole, the

Process Diagram

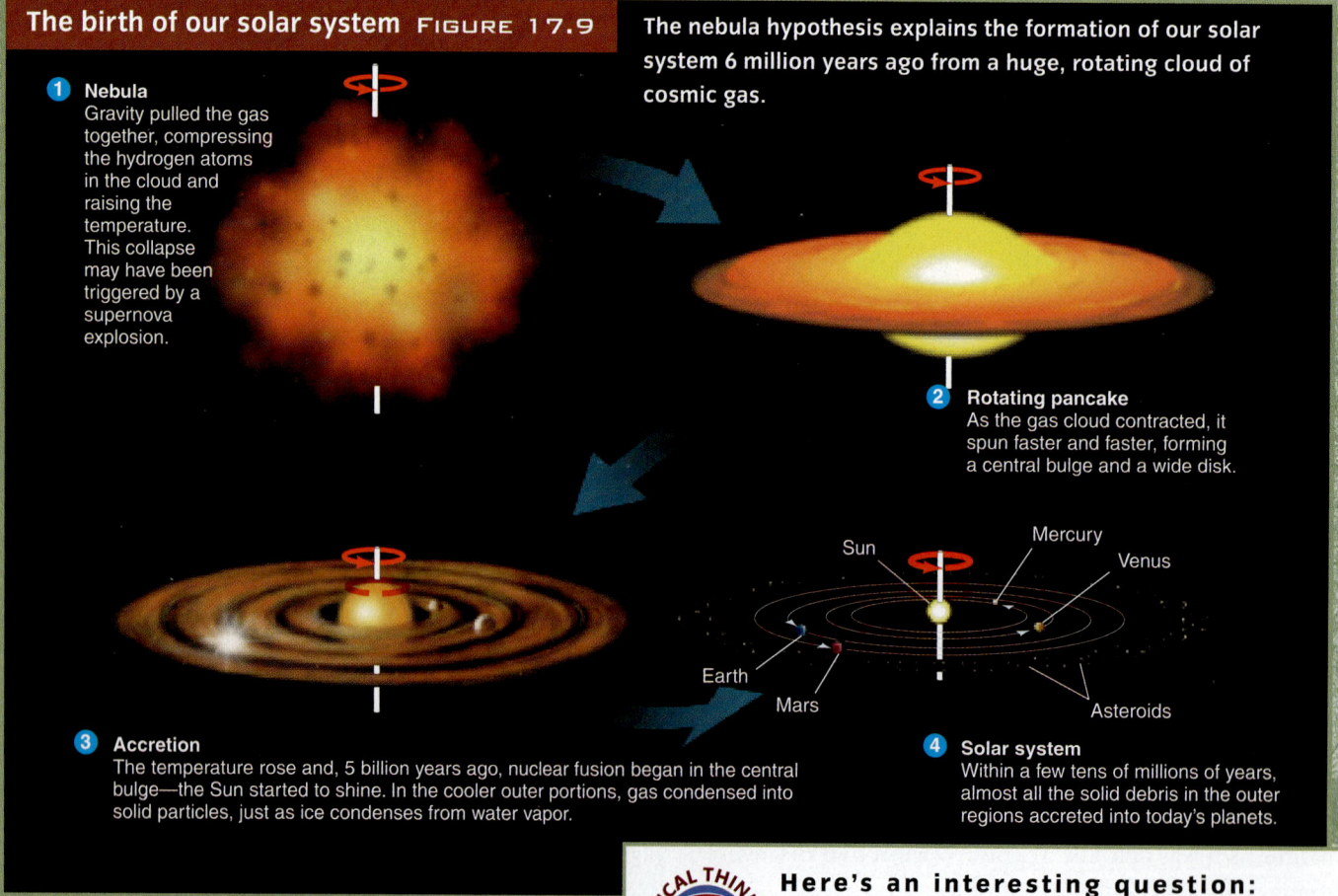

The birth of our solar system FIGURE 17.9

The nebula hypothesis explains the formation of our solar system 6 million years ago from a huge, rotating cloud of cosmic gas.

1 Nebula
Gravity pulled the gas together, compressing the hydrogen atoms in the cloud and raising the temperature. This collapse may have been triggered by a supernova explosion.

2 Rotating pancake
As the gas cloud contracted, it spun faster and faster, forming a central bulge and a wide disk.

3 Accretion
The temperature rose and, 5 billion years ago, nuclear fusion began in the central bulge—the Sun started to shine. In the cooler outer portions, gas condensed into solid particles, just as ice condenses from water vapor.

4 Solar system
Within a few tens of millions of years, almost all the solid debris in the outer regions accreted into today's planets.

CRITICAL THINKING

Here's an interesting question:
- Why don't the following negate the nebular hypothesis: Venus rotates slowly in the opposite direction to the other planets, Earth's axis is tilted to the plane of its orbit, and Uranus's axis is near the plane of its orbit?

What an Earth Scientist Sees

The Crab Nebula

On July 4, A.D. 1054, Chinese and Arab astronomers recorded the sudden appearance of a bright light within the constellation Taurus. The light was so intense that for a few years it could be seen even during the day. We now know that they were observing a supernova explosion—the death throes of a star. The remnants of that supernova, known as the Crab Nebula, are shown in this image from the Hubble Space Telescope (HST).

Earth scientists recognize that such supernovae are vital for the formation of planets such as Earth. Stars are the cauldrons in which the heavy chemical elements—carbon, oxygen, silicon, iron, and so on—were cooked up from the hydrogen and helium that were produced after the big bang.

In order for rocky planets to form, the heavy elements made inside old stars must somehow get into a cosmic gas cloud. We now believe that supernovae do this job, scattering the heavy elements into space. The HST image of the Crab Nebula shows gaseous filaments streaming out from the center of the explosion at up to half the speed of light. The principal element in many of these filaments is identified by its color: Hydrogen is shown in orange; nitrogen in red; sulfur in pink; oxygen in green.

 **Here's an interesting question:**
- Could the first stars formed after the big bang have had planets in orbit around them?

---

direction of rotation will be counterclockwise. Moreover, the Sun and most of the planets rotate around their axes in the same counterclockwise direction. Venus is an exception, rotating slowly in a clockwise direction, although we do not know why. Both the extraordinarily consistent motions of the planets and moons and the grouping of the planets into two classes—terrestrial and Jovian—must be explained.

To date, the best model for the origin of the solar system is the **nebula hypothesis**, which proposes that the Sun and planets formed from a huge, rotating cloud of cosmic gas (**FIGURE 17.9**). This gas cloud had to contain not only the light elements hydrogen and helium, found in the Sun, but also heavier elements such as carbon, oxygen, silicon, and iron, which make up the bulk of the rocky planets. As we discuss later in the chapter, the light elements were created soon after the big bang. Heavier elements are forged only within stars, and scientists believe that they must have been released into the gas cloud when an old star exploded as a supernova (see *What an Earth Scientist Sees*).

The nebula hypothesis successfully explains why the inner planets are rocky and the outer planets contain higher proportions of gases and ice. According to the hypothesis, the temperature was higher in the portion

nebula hypothesis A model that explains the formation of the solar system from a large cloud of gas and dust floating in space 4.56 billion years ago.

The Solar System 525

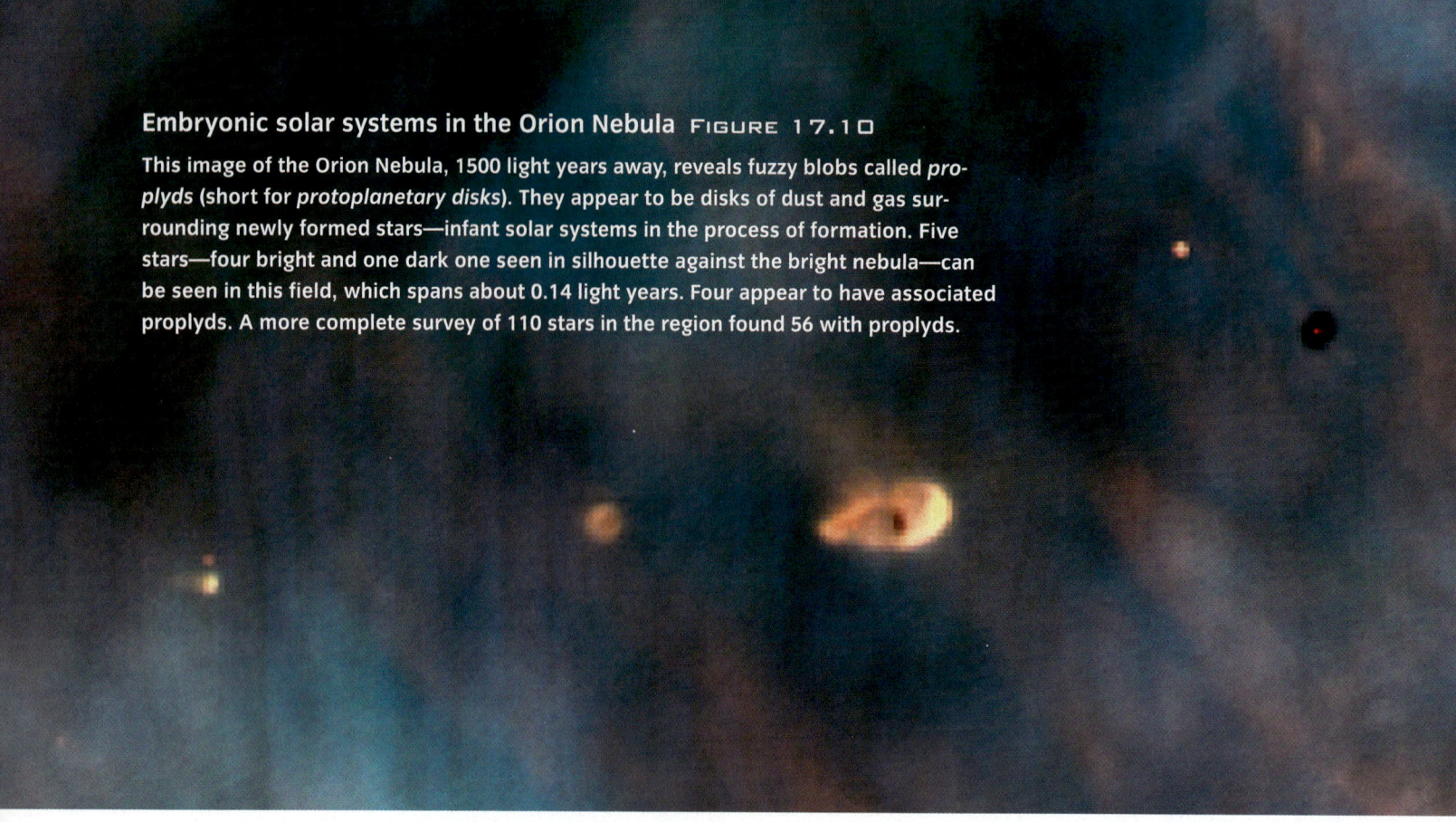

Embryonic solar systems in the Orion Nebula FIGURE 17.10

This image of the Orion Nebula, 1500 light years away, reveals fuzzy blobs called *proplyds* (short for *protoplanetary disks*). They appear to be disks of dust and gas surrounding newly formed stars—infant solar systems in the process of formation. Five stars—four bright and one dark one seen in silhouette against the bright nebula—can be seen in this field, which spans about 0.14 light years. Four appear to have associated proplyds. A more complete survey of 110 stars in the region found 56 with proplyds.

of the nebula that was closest to the young Sun, which was to become the innermost part of the solar system. Only elements with a high melting point, such as iron, aluminum, and silicon—the chemicals that make up rock—would have been able to condense from the regions of the cloud nearest the Sun. Meanwhile, a strong solar wind stripped much of the lighter gases, such as helium and hydrogen, from the inner planets. But because this solar wind was not strong enough to disrupt the outer planets, they could grow into gas giants. The moons and smaller bodies of the outer solar system are made up of water ice, methane ice, and other kinds of ice because volatile compounds, such as water vapor and methane, could condense at lower temperatures farther from the Sun. FIGURE 17.10 is an image of embryonic solar systems in the Orion Nebula.

THE SUN

Within our vast universe, our Sun ranks as just an "average" star, in terms of size and age (FIGURE 17.11 on pages 528–529). But this fiery ball of constantly churning gas is the largest thing in our solar system—containing more than 98% of the solar system's mass (FIGURE 17.11A). It has a radius of 690,000 km—more than a hundred times that of Earth—and a mass of about 2×10^{30} kg. It is the powerhouse that controls Earth's climate, provides the ultimate energy source behind the food we eat and the fossil fuels we burn, and drives Earth's wind and wave patterns, as we saw in earlier chapters.

FIGURE 17.11B shows the structure of the Sun, divided into six main parts: the solar core, the radiation and convection zones, and the photosphere, the chromosphere, and the corona. Let's look at the Sun's structure and method of energy production in more detail.

The solar interior At the Sun's core, which makes up about 10% of the star's volume, the temperature reaches 15 million °C (27 million °F), and the pressure is 340 billion times Earth's air pressure at sea level—conditions intense enough to cause the nuclear reactions that give the Sun its immense power. The most common material in the universe is hydrogen, and hydrogen atoms form the bulk of most stars, including the Sun. The Sun's energy is created when the nuclei of four hydrogen atoms (*protons*) combine to form the nucleus of a helium atom, in a process called the **proton–proton chain**. The final helium atom actually has a slightly smaller mass than would be expected by simply adding together the masses of four hydrogen atoms. This "lost" matter has been converted to energy during the process, according to Einstein's famous formula, $E = mc^2$, in which E is equal to the energy generated, m is equal to the mass that has been converted, and c is the speed of light (299,800,000 m/s).

> **proton–proton chain** A chain of nuclear reactions in which hydrogen nuclei (or protons) fuse together to form helium nuclei, releasing energy.

The Sun consumes about 600 million tons of hydrogen every second. Although only a tiny fraction (0.7%) of this hydrogen is converted to energy, a huge amount of energy is generated overall because the speed of light is so large. Scientists have calculated that the Sun could continue burning for another 75 billion years before it runs out of fuel. However, as we will see later in the chapter, the Sun is expected to die long before it consumes all its hydrogen. Our best estimate for its lifetime is about 11 billion years. Our Sun is already over 4.5 billion years old, and so it is now in middle age.

Energy generated in the core takes over a hundred thousand years to reach the Sun's swirling surface. Atoms in the zone surrounding the core absorb the generated energy, store it for a while, and then later emit that energy as new radiation. In this manner, the energy that is produced in the core is passed from atom to atom, and so this region is known as the *radiation zone* (Figure 17.11B).

The energy-transfer mechanism changes in the upper 200,000 km of the Sun and energy is transferred by *convection*. We met convection in Chapter 14. In the solar *convection zone*, hot gas rises toward the surface, because it is less dense than the cooler surrounding gas.

The Sun's outer layers So far, we have discussed the solar interior, which is hidden from our view. We can directly observe only the Sun's thin outer layer, which is about 150 km deep. The part of this surface layer that emits the most light is the Sun's *photosphere* (Figure 17.11B). Here, the temperature is about 5500°C (about 10,000°F). Due to the convection process described above, regions of hot gas rise up, creating bright *granules*—typically the size of Texas—on the surface (**FIGURE 17.11C**). When this gas spreads out and cools, it sinks once more. Because the Sun is a ball of gas, there are no sharp boundaries between its layers. The photosphere gives way to a thin layer of gas called the *chromosphere*, which is a few thousand kilometers thick. Usually, we cannot see the chromosphere because it is washed out by the bright light of the photosphere. But during a total solar eclipse, it can be seen to have a reddish tint (**FIGURE 17.11D**). The most common features seen in the chromosphere are *spicules*, flamelike structures that jut out to about 10,000 km, like blades of grass. Spectacular solar *prominences*, made up of huge tongues of chromospheric gas, can also erupt from the surface, stretching out to hundreds of thousands of kilometers.

The outermost portion of the solar atmosphere forms an envelope of gas, reaching millions of kilometers out from the Sun, called the *corona* (**FIGURE 17.11E**). It produces a glow almost half as bright as the full Moon.

The Sun constantly emits a stream of particles that are fast enough to overcome the gravitational pull of the Sun and escape from the corona into interstellar space as part of the *solar wind*. We see the effects of this wind on Earth, as auroras, which are created when the solar wind interacts with the outer reaches of Earth's atmosphere (**FIGURE 17.11F**). The most explosive ejection of particles occurs in *solar flares*, which last for about an hour and heat the solar gas to tens of millions of degrees Celsius. The flares accelerate particles to almost the speed of light. When these particles reach Earth, they ionize atoms in the gases of the atmosphere, which in turn disrupts radio communication.

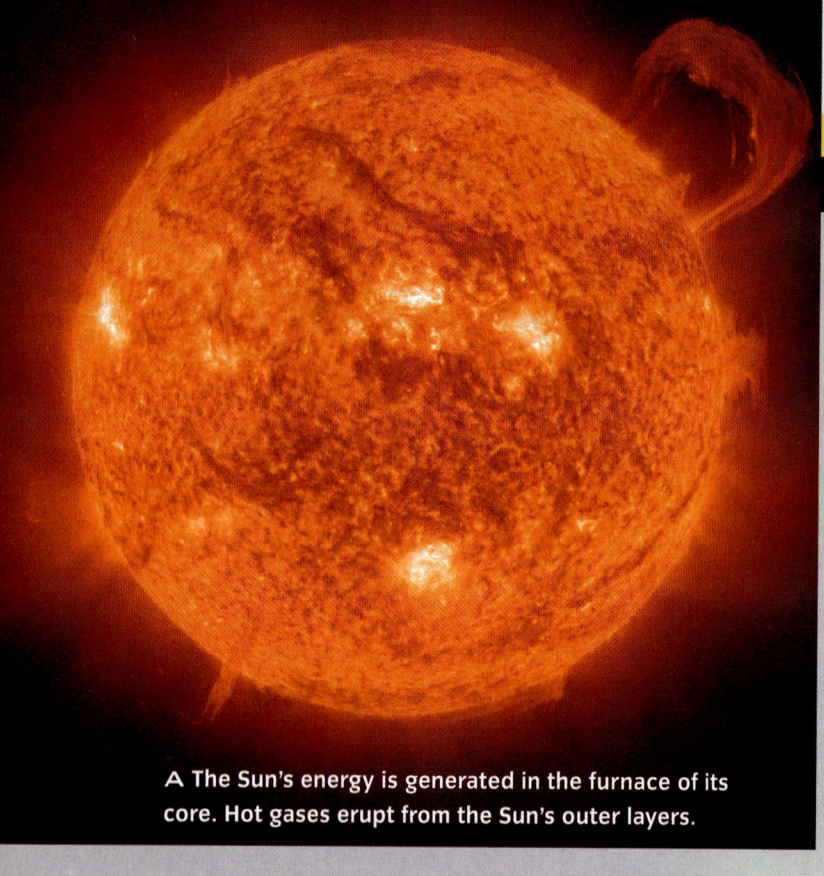

A The Sun's energy is generated in the furnace of its core. Hot gases erupt from the Sun's outer layers.

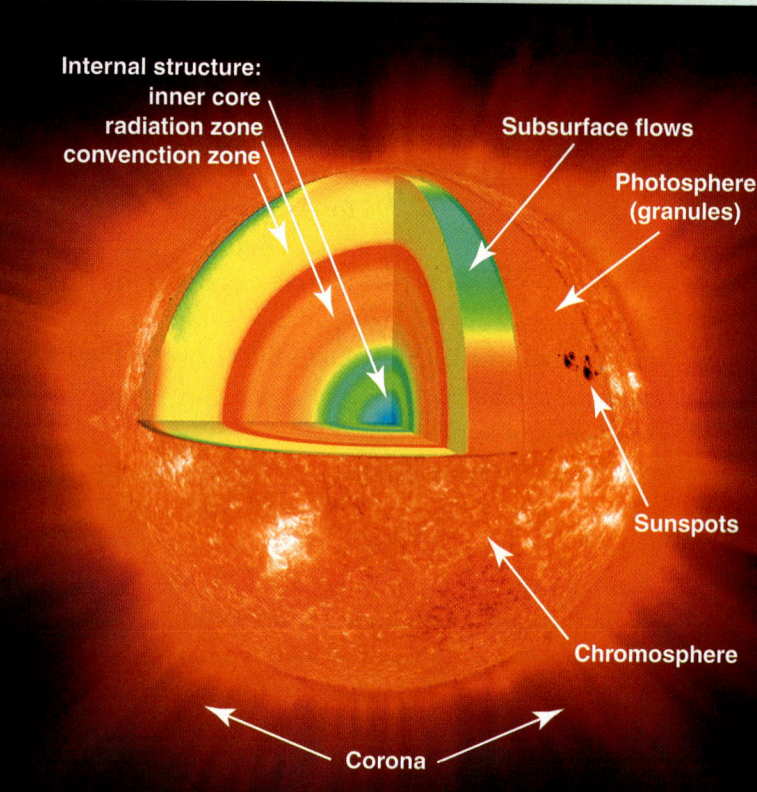

B The solar interior comprises the core and the radiation and convection zones. The photosphere, the chromosphere, and corona form the Sun's outer layers. Sunspots, granules, and *prominences* are features seen on the Sun's active surface.

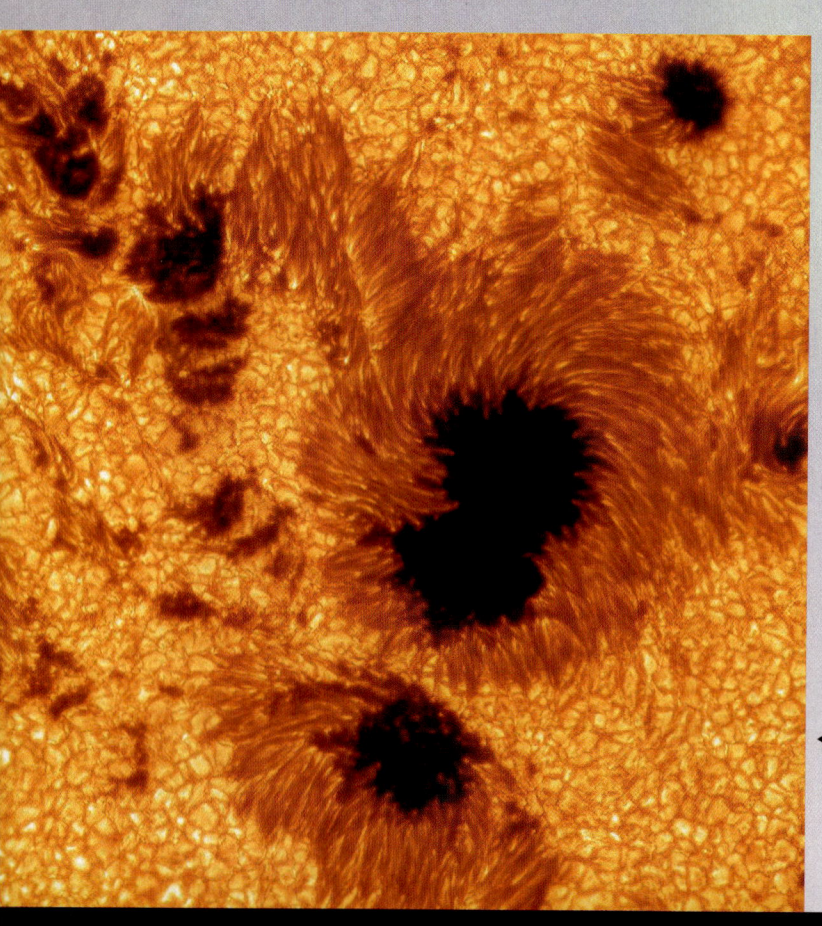

C The photosphere is a bubbling layer of gas. Hot granules of gas rise to the surface, due to convection, and then sink again as they cool. The dark patches in the image are regions of slightly cooler gas, known as *sunspots*.

Visualizing
The Sun FIGURE 17.11

The Sun's outermost atmospheric layers—**D**, the *chromosphere* and **E**, the *corona*—can be seen only during a total solar eclipse, when light from the brilliant photosphere is blocked.

D Chromosphere

E Corona

F Auroras are produced when charged particles from the solar wind collide with particles in Earth's atmosphere.

THE PLANETS

As we saw earlier, the planets in our solar system can be divided into two groups: the innermost terrestrial planets, and the outer Jovian planets. Earth scientists have built up a hypothesis of planetary accretion that can explain the differences between the two groups. The *planetary accretion hypothesis*, a supplement to the nebula hypothesis we met earlier, states that planets assembled themselves from rocky, metallic, and icy debris 4.56 billion years ago, shortly after the Sun itself was formed.

The terrestrial planets: Mercury, Venus, Earth, and Mars

Earth scientists hypothesize that every terrestrial or rocky planet probably started out hot enough to melt either partially or completely, and it was during this period that its interior separated into layers with different chemical compositions, a process called *differentiation*. As we saw in Chapter 1, Earth differentiated into three layers: a relatively thin, low-density, rocky *crust*; a rocky, intermediate-density *mantle*; and a metallic, high-density *core*. Similar layers are present in Mercury, Venus, Mars, and our Moon, although they differ in size and composition. The most remarkable of these bodies is the innermost planet Mercury. Barely larger than the Moon, it has a core that makes up 42% of its volume and an estimated 80% of its mass (**FIGURE 17.12**). All the terrestrial planets have undergone intense impact cratering, but while the signs of this process are well hidden on Venus and Earth, they are particularly apparent on Mercury.

We do not know whether any of the terrestrial planets besides Earth have molten or partially molten cores. The molten outer core and the relatively rapid rotation of Earth give rise to Earth's strong magnetic field. Magnetic fields do exist on the other terrestrial planets, but they are much weaker than Earth's.

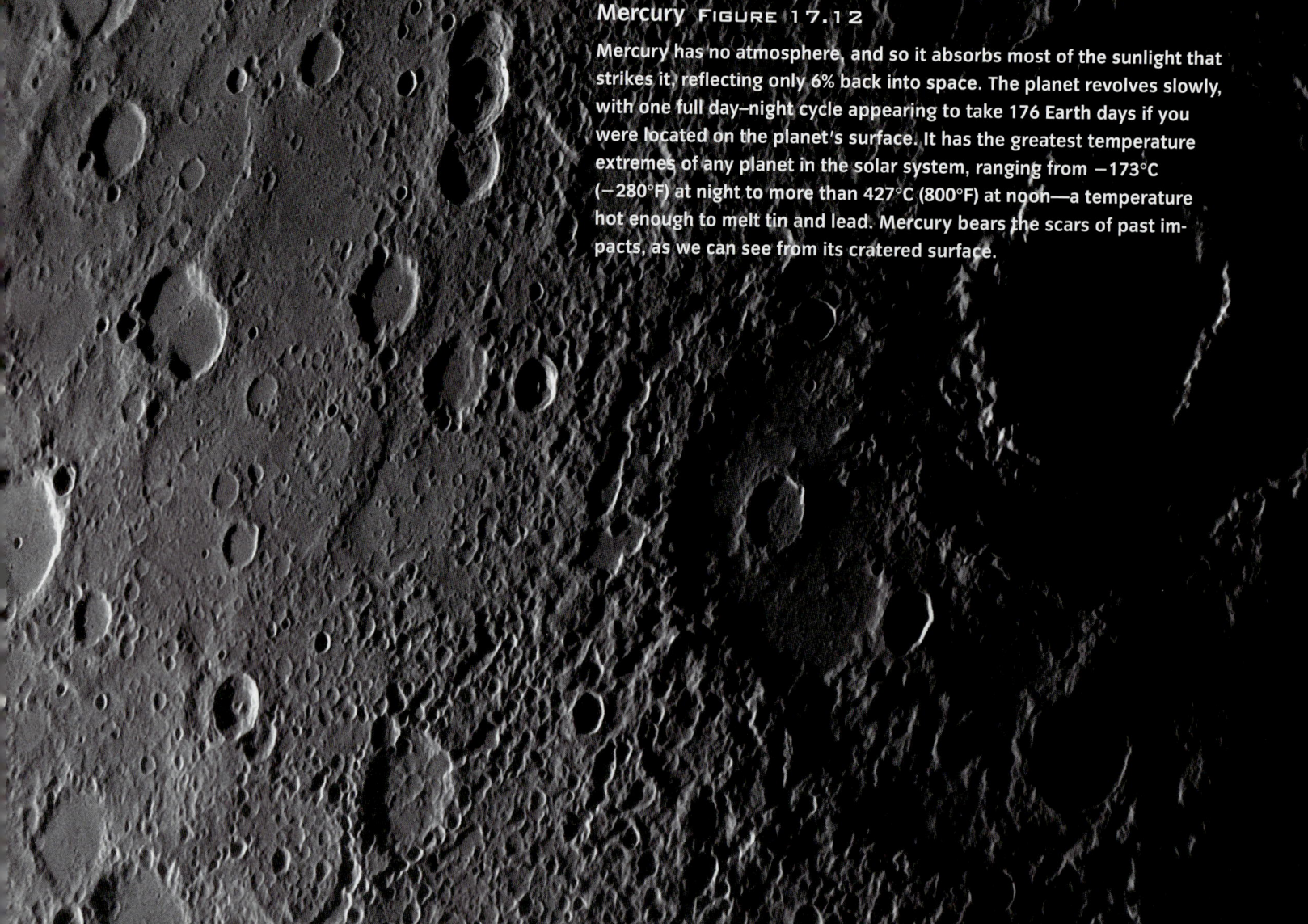

Mercury FIGURE 17.12

Mercury has no atmosphere, and so it absorbs most of the sunlight that strikes it, reflecting only 6% back into space. The planet revolves slowly, with one full day–night cycle appearing to take 176 Earth days if you were located on the planet's surface. It has the greatest temperature extremes of any planet in the solar system, ranging from −173°C (−280°F) at night to more than 427°C (800°F) at noon—a temperature hot enough to melt tin and lead. Mercury bears the scars of past impacts, as we can see from its cratered surface.

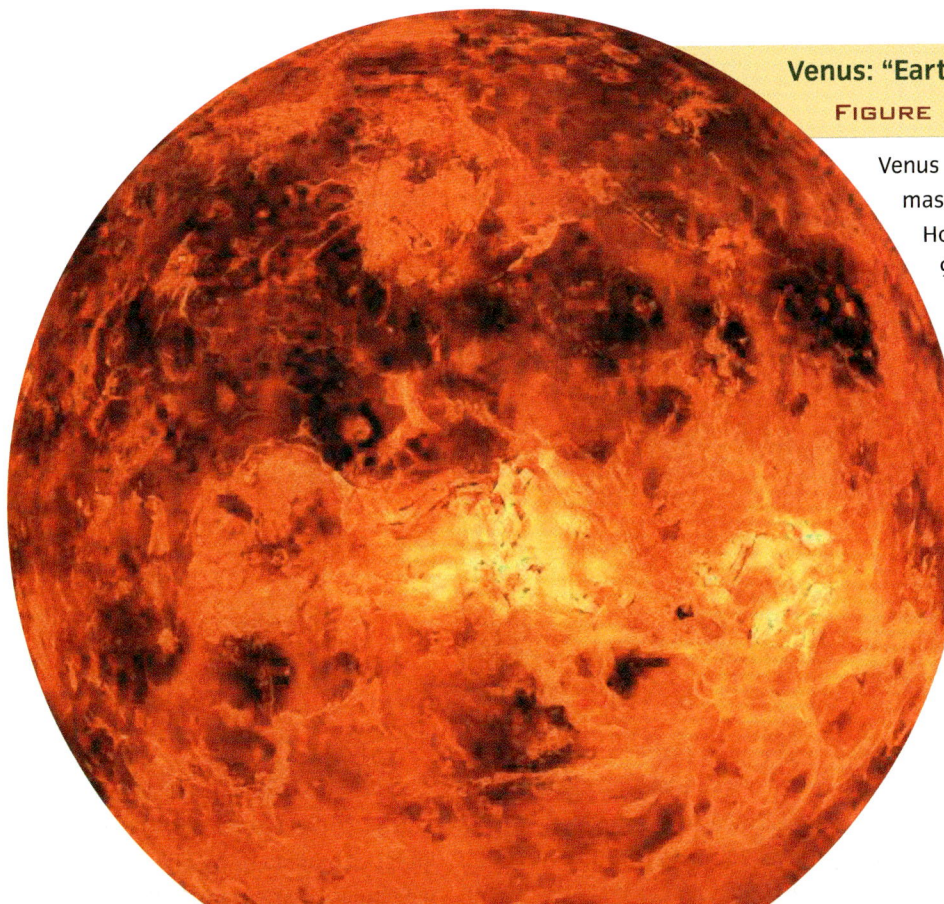

Venus: "Earth's twin"

FIGURE 17.13

Venus is similar to Earth in size, mass, and distance from the Sun. However, its atmosphere is 97% carbon dioxide, and its surface reaches 475°C (900°F), making it inhospitable to life. The Magellan mission used radar to see through the dense clouds cloaking the planet and map the surface. The bright band running across the planet is a highland area of mountains and canyons. Many of the dark areas represent lava flows.

There are other important similarities among the terrestrial planets. All of them have experienced volcanic activity, Venus being a particularly good example (FIGURE 17.13). This means that they either have, or have had, an internal heat source, as we saw in Chapter 8. The volcanism is dominated by the formation of basalt, which forms most of the surface of Venus, Mars, and the dark-colored "seas" of Earth's Moon.

It is more difficult to determine whether each terrestrial planet has *plate tectonic* activity (described in Chapter 7). Simple observation of planets and their moons reveals that rocks on the surface of each planet fracture and deform as they do on Earth, which suggests that the planets have *lithospheres*. What little evidence we have suggests that *asthenospheres* and lithospheres probably are present in each terrestrial planet but that the asthenosphere of Earth is unusually near the surface and the lithosphere unusually thin. In fact, it is very likely that Earth is such a dynamic planet *because* its lithosphere is thin. The other terrestrial planets seem to have much thicker lithospheres and to be much less dynamic than Earth.

Finally, all terrestrial planets have lost their primordial atmospheres of hydrogen and helium. Earth, Mars, and Venus have new atmospheres that leaked from their interiors via volcanoes and were trapped by the planets' gravity. Mercury and the Moon are too small to have held onto the gases given off by volcanoes, so they lack atmospheres.

Given that the terrestrial planets are similar in composition and overall structure, why is life not present on Venus, or on Mars (highlighted in the *Amazing Places* feature at the end of the chapter)? The answer probably lies in two factors: (1) Earth's size and the fact that its interior is still hot leads to volcanism and the continual addition of new nutrients to the surface in the form of lava; (2) the distance of Earth from the Sun is such that H_2O can exist as water, ice, and water vapor. Not only do we need liquid water to live, but the fact that H_2O can exist as ice and water vapor plays a vital role in determining the character of our atmosphere and in creating habitable climates around the globe, as we saw in Chapter 14.

Jupiter Figure 17.14

None of the gases of Jupiter's atmosphere have been able to escape its gravity, and storms are common. The wind systems create Jupiter's distinctive bright and dark bands. They are driven by heat generated in Jupiter's interior, in contrast to Earth's wind systems, which are driven by energy from the Sun.

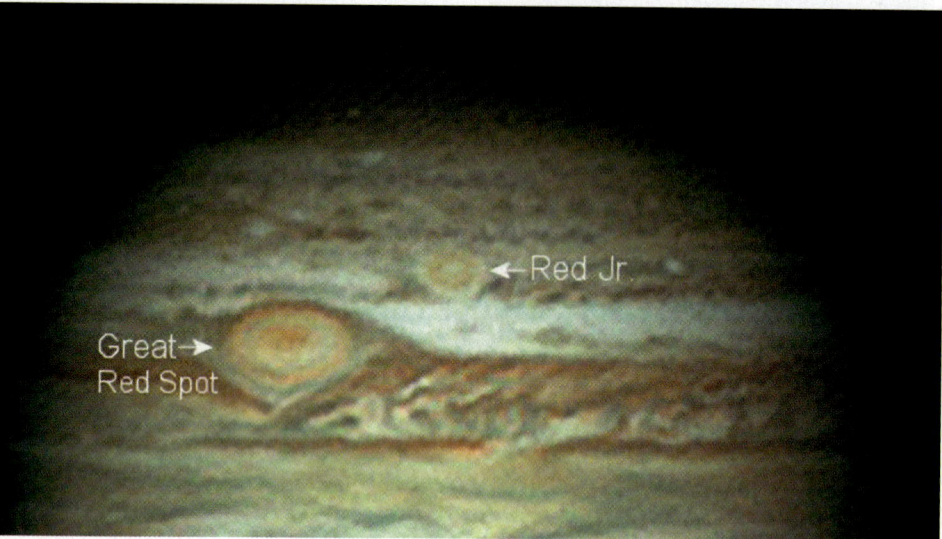

A The famed Great Red Spot on Jupiter, which is twice as wide as Earth, marks a storm that may have been raging for over 300 years. In 2006, an amateur astronomer noted a new storm, dubbed "Red Junior."

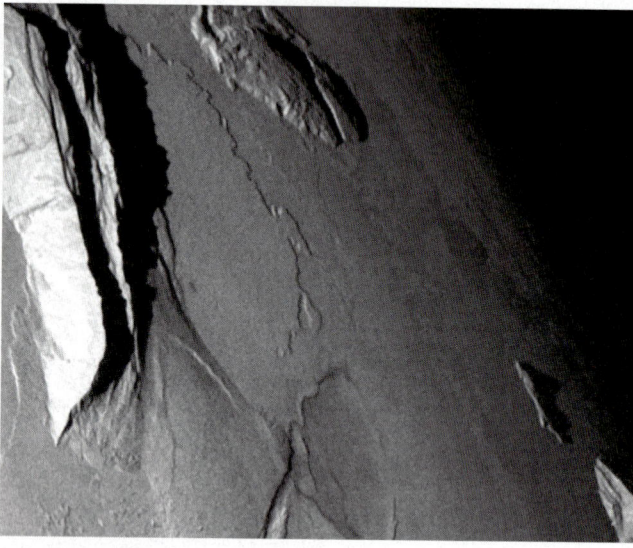

B NASA's Galileo spacecraft captured this dramatic image of mountains on one of Jupiter's moons, Io, in February 2000. Mongibello Mons, the jagged ridge at the left of the image, rises 23,000 ft above the plains of Io, higher than any mountain in North America.

The Jovian planets We cannot see anything that lies below the thick blankets of atmosphere—rich in hydrogen and helium—that cover the Jovian planets. But we can work out their internal structure based on remote-sensing measurements of the planets and their moons. Their cores are inferred to be rocky, surrounded by a thick layer of ice. Jupiter is a gas giant (**Figure 17.14A**) that is so large it has a mass that is two and a half times that of all the other planets in the solar system combined. If the planet had been 10 times larger, it would probably have developed into a small star—as described in the next section. Indeed, Jupiter lies at the center of its own satellite system, with more than 60 moons discovered so far (**Figure 17.14B**).

What an Earth Scientist Sees describes some recent observations of the next Jovian planet in line from the Sun, Saturn. The planet's famous ring system was first discovered by Galileo, with his primitive telescope. It was recently discovered that Jupiter, Neptune, and Uranus also have ring systems. The origin of these rings is still under debate. Saturn is similar in composition to Jupiter and also has a turbulent atmosphere.

Because pressures inside the Jovian planets must be enormous, we think that deep in their interiors hydrogen may be so tightly squeezed that it is condensed to a liquid. Still deeper inside Jupiter and Saturn, pressures equivalent to 30 million times the pressure at the surface of Earth are reached. Under such conditions, electrons and protons become less closely linked and hydrogen becomes metallic. In Jupiter, it is possible that pressures may even reach levels high enough for solid metallic hydrogen to form a sheath around the ice core.

What an Earth Scientist Sees

Saturn's Rings

The Cassini probe took this spectacular view of Saturn in 2006 (A). The unique view of Saturn's rings was made possible because Saturn passed directly between the probe and the Sun, leaving the planet in silhouette. Viewing the rings through a visual and infrared mapping spectrometer reveals that the grains range from very small, like powdery snow on Earth, to larger grains, like more granular snow, related to their distance from the planet (B). Cassini also confirmed that the grains are frozen water molecules, which supports the notion that Saturn may have a rocky core surrounded by a layer of ice.

The probe also glimpsed a tiny blue orb peeking out between Saturn's rings—Earth (C). Our world has an atmosphere that scatters blue light, and as a result it is the only planet that appears blue when viewed from deep space. Scientists looking for habitable planets orbiting neighboring suns will be encouraged if they spot a similar blue dot. But colors are only clues. A planet with an atmosphere might not appear blue. Mars has a thin red atmosphere. The blue light in the Martian skies is soaked up by iron-containing molecules on the surface that radiate red light.

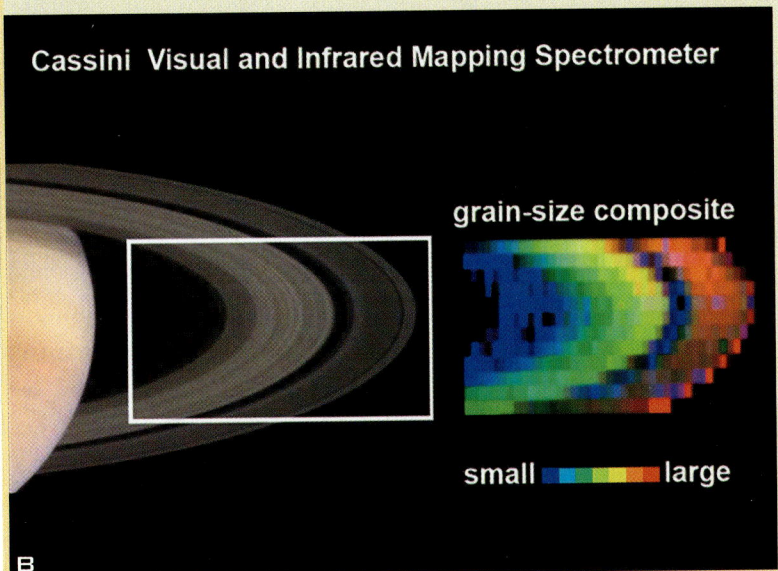

CRITICAL THINKING

Here's an interesting question:
- What clues, in addition to color of the atmosphere, would you seek if you were a scientist looking for planets potentially habitable for humans?

The Solar System 533

A Uranus was the first planet discovered with the aid of a telescope, in 1781, and takes 84 years to complete one orbit. The planet's atmosphere is made up largely of hydrogen and helium, with a small amount of methane—which gives its blue-green color—and traces of water and ammonia.

B False colors emphasize bands of "smog" above Uranus's south pole.

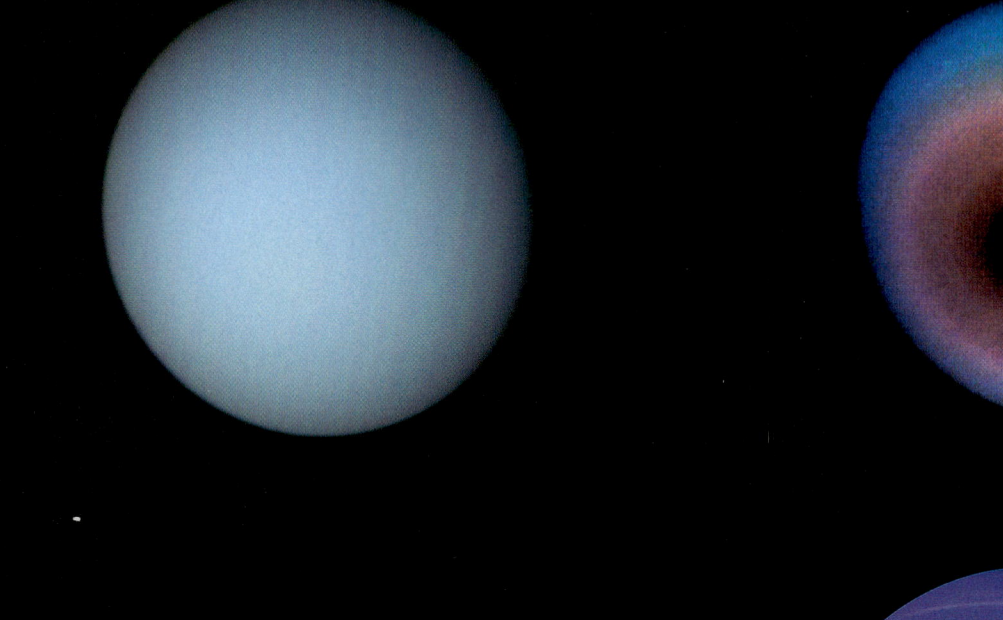

C Neptune is almost identical in size to Uranus, with only 1% difference in diameter. It is one of the windiest planets, with winds exceeding 1000 km/h (600 mph). This image from the Voyager spacecraft shows white, cirrus-like clouds, which are probably made of frozen methane.

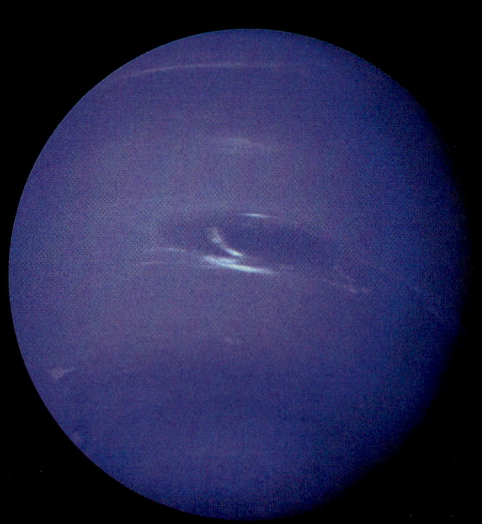

Uranus and Neptune FIGURE 17.15

Neptune and Uranus are thought to be similar in composition to Jupiter and Saturn, although neither is large enough for pressures to reach the levels required for the formation of metallic hydrogen (FIGURE 17.15).

The dwarf planets

As we saw in the chapter opener, Pluto—once classified as the ninth planet in the solar system—was reclassified by the International Astronomical Union (IAU) as a "dwarf planet" in 2006. Like normal planets, dwarf planets orbit the Sun and have a large enough mass to pull their matter into a roughly spherical shape. However, unlike planets, they have not been able to clear their orbits of other objects of a similar size. There are currently two other objects that are classified as dwarf planets, Eris and Ceres. Ceres lies in the *asteroid belt*, which we will discuss in the next section.

Asteroids, meteorites, and the formation of the Moon The planetary accretion hypothesis also accounts for the existence of meteoroids and asteroids—small solar system bodies that are not massive enough to have pulled themselves into a round shape. Today's *meteoroids* are the debris that was never swept up to form a planet. Sometimes pieces of this debris happen to fall to Earth as **meteorites**. As such, meteorites are fascinating relics of the early days of the solar system (**FIGURE 17.16A**).

Asteroids are rocky space objects that orbit the Sun, and can be as large as several hundred kilometers across. Most are concentrated in a belt between Mars and Jupiter, and take between three and six years to orbit the Sun. A few large asteroids pass close to Earth and the Moon, and probably caused the most recent impact craters on our planet.

It was only after the Moon landings in 1969–1972 that scientists began to grasp the great importance of violent collisions in the solar system's history. Every crater on the Moon, as far as we know, was formed by a meteoroid impact. (The astronauts looked for volcanic craters but found none.) Scientists recognize impact craters on Earth, too, although they are harder to find because erosion and other processes cover them up or erase them (**FIGURE 17.16B**).

> **meteorite** A fragment of extraterrestrial material that falls to Earth.

Meteorites—messengers from space FIGURE 17.16

A This boy is examining a large iron meteorite in the American Museum of Natural History in New York City. Information gained from the study of such meteorites helps scientists interpret the distribution of chemicals in our solar system.

B Manicouagan Crater in Quebec was created about 210 million years ago by the impact of a much larger meteorite. The original crater, now marked by a ring lake, was 100 km in diameter.

The Solar System

Most planetary scientists now conclude that Earth collided with a planetary body roughly the same size as Mars about 4.5 billion years ago. That impact tilted Earth's axis of rotation at an angle of 23.5 degrees to the plane of its orbit around the Sun. As we saw earlier, this tilt explains why we have seasons. The impact must also have melted most of Earth's surface due to the tremendous amount of energy it released. (At the hyperspeeds typical of cosmic impacts, every ton of an impactor strikes Earth with an energy equivalent to 100 tons of dynamite.) The collision completely destroyed the other planet and blasted so much debris into orbit that for a while Earth had rings much denser than Saturn's. Eventually the rings of debris coalesced to form the Moon (**Figure 17.17**). This hypothesis explains the existence of a magma ocean early in lunar history (shown by rocks retrieved from the Moon). It also explains our Moon's relatively large size in contrast to other moons, which are many times smaller than their parent bodies, and accounts for certain chemical discrepancies and similarities between Earth and its Moon.

Such giant collisions were the inevitable final stage of planetary accretion, when most of the debris had been swept up and only larger objects remained. Signs of giant impacts abound. One such impact probably caused Uranus's axis to tip over on its side. Pluto's moon, Charon, shown in the photo at the beginning of the chapter, was probably created by an impact, because it is unusually large compared to Pluto. Perhaps Venus experienced a giant impact, too. Although it lacks a moon, it is the only planet that rotates clockwise on its axis, an effect that could have been produced by a large glancing blow that essentially turned the planet upside down.

Comets *Comets* have been dubbed "dirty snowballs" because they are balls of frozen gases (water, ammonia, methane, carbon dioxide, and carbon monoxide) and small pieces of rocky and metallic materials. They measure just a few kilometers or tens of kilometers across. The orbits of most comets are extremely elongated, taking hundreds of thousands of years to travel around the Sun and carrying them far beyond Pluto. However, a few comets, such as Halley's comet, have short-periods of less than 200 years, and regularly enter the vicinity of the planets. As comets approach the inner solar system, radiation from the Sun vaporizes their frozen gases, producing their distinctive glowing head, or *coma*. As they approach the Sun, some also develop a tail that can stream behind them for millions of kilometers (**Figure 17.18**).

Formation of the Moon Figure 17.17

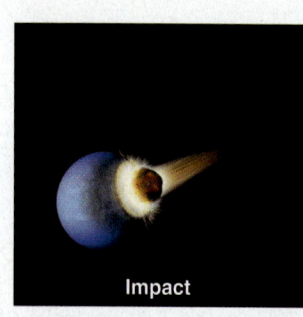

Impact

Some 4.5 billion years ago, the still-forming Earth runs into another growing planet, which scientists have dubbed Theia.

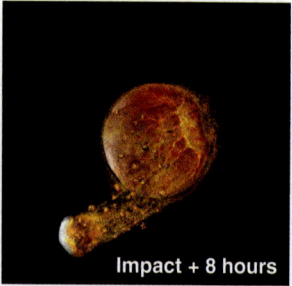

Impact + 8 hours

Theia is destroyed, and its remnants—along with a good chunk of Earth's mantle—are blasted into orbit around Earth. The off-center impact knocks Earth's axis of rotation askew.

Impact + 24 hours

The debris spreads itself into a ring and begins to clump together.

Impact + 1 year

The largest clump starts to attract other fragments and is well on its way toward becoming the Moon.

Comet Hale-Bopp FIGURE 17.18

Short-period comets are thought to have originated in a belt that lies just beyond Neptune, known as the *Kuiper Belt*. Long-period comets are believed to have originated in a region known as the Oort cloud, on the outer edge of the solar system.

OTHER SOLAR SYSTEMS

For centuries, human beings have pondered the question of whether life might exist on other planets. By the early 1990s, technology had advanced sufficiently for astronomers to pick out signs of planetary systems orbiting other stars. Such planets, known as extrasolar planets or **exoplanets**, are extremely difficult to spot. Unlike stars, they do not shine, so they are not easily visible. In most cases, astronomers cannot see exoplanets directly but can infer that they are orbiting stars because their gravitational pull causes their parent star to wobble in a characteristic way.

■ **exoplanet** Short for *extrasolar planet*, a planet outside our solar system.

So far, more than 250 exoplanets have been found. Most exoplanets are giant planets and probably resemble Jupiter. In 2007, astronomers announced the discovery of Gliese 581c, a small exoplanet orbiting the star Gliese, about 20 light years away from Earth, near the constellation Libra. Gliese 581c is the best current candidate for a terrestrial exoplanet. It lies in its star's habitable zone, which means that its distance from the star is such that its temperature might be in the right range potentially to harbor liquid water. But after its discovery it was shown that the atmosphere of Gliese 581c might be undergoing a runaway greenhouse effect, so the planet might not be habitable.

The European Space Agency's COROT telescope, launched in 2006, is currently searching for rocky exoplanets. NASA's Kepler Mission, scheduled to launch in 2009, will tell us whether planets like Earth are common or rare in our galaxy. Thus, the hunt for other Earths continues.

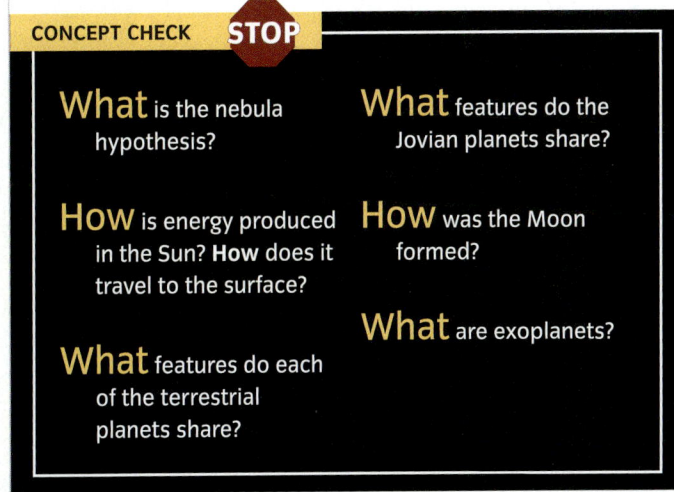

CONCEPT CHECK STOP

What is the nebula hypothesis?

How is energy produced in the Sun? **How** does it travel to the surface?

What features do each of the terrestrial planets share?

What features do the Jovian planets share?

How was the Moon formed?

What are exoplanets?

The Solar System

Stars and Stellar Evolution

LEARNING OBJECTIVES

Describe how stars are characterized.

Explain how the Hertzsprung-Russell diagram helps to categorize stars.

Discuss the possible fates of stars.

Define red giant, white dwarf, supernova, neutron star, pulsar, and black hole.

Our Sun is just one of a host of stars in the universe. All stars share certain characteristics, but they do not all share the same pattern of evolution. Let's examine how we classify stars and what their features tell us about their ultimate fate.

Since stars are so far away, astronomers must infer their characteristics indirectly. They can calculate the distance to nearby stars using simple geometry, by measuring how the angle from Earth to the star changes when Earth moves around its orbit. Astronomers need to know how far away stars are, in order to relate their *apparent brightness*, that is, how bright they appear when viewed from Earth, to the star's actual energy output or *luminosity*. Astronomers must also calculate a star's surface temperature, which they infer from the star's color—blue stars are hotter than red stars. A star's luminosity and surface temperature allow astronomers to rank stars on a Hertzsprung-Russell diagram.

THE HERTZSPRUNG-RUSSELL DIAGRAM

Early in the twentieth century, Ejnar Hertzsprung and Henry Russell independently discovered a way to order the stars. They plotted the stars on a graph, known as a **Hertzsprung-Russell (H-R) diagram**, according to their luminosity and surface temperature and found certain important groupings (**FIGURE 17.19**). Most stars, including our Sun, lie along a single band. Such stars are all in their hydrogen-burning phase and are called *main-sequence stars*.

> **Hertzsprung-Russell (H-R) diagram** A plot of stars according to their luminosity and surface temperature.

In the upper-right corner of the H-R diagram are a group of stars that emit a lot of energy and hence have large luminosities, but are surprisingly cool. Since temperature is related to the energy emitted over a certain

Hertzsprung-Russell diagram
FIGURE 17.19

When stars are plotted on a graph according to their luminosity and surface temperature, most—including our Sun—lie on a band stretching from the upper left to the lower right. This band contains *main-sequence stars*. There are two other prominent groupings: the giants and the white dwarfs.

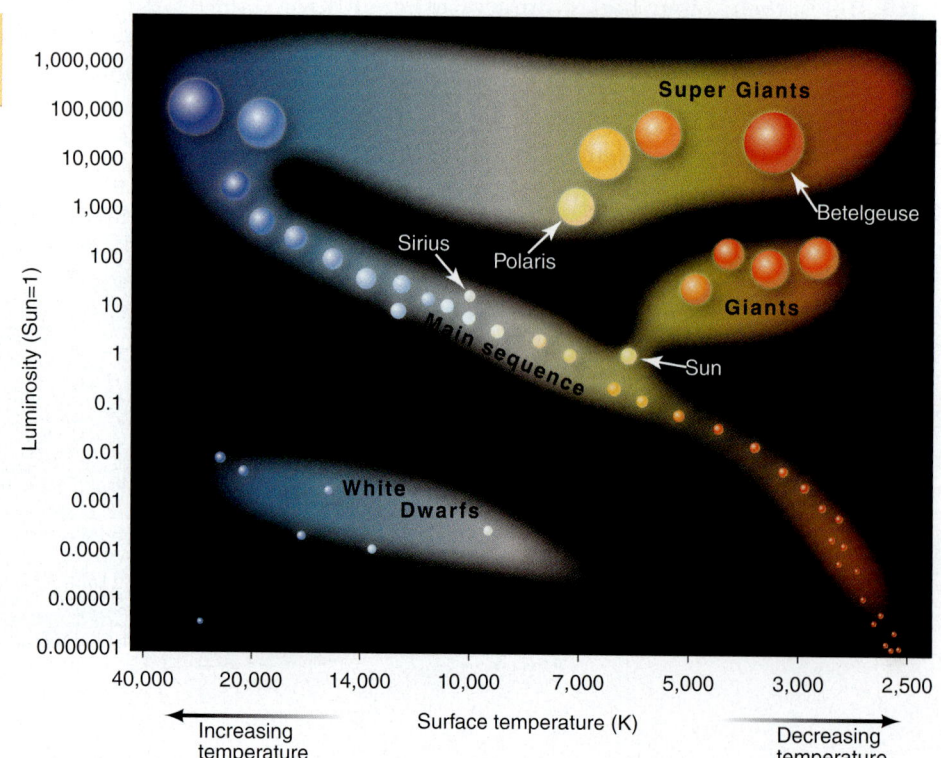

area, these stars must be very large—many times the size of the Sun—to account for their low temperature. They are known as giants, and *red giants* have a reddish color.

At the other end of the H-R diagram lie *white dwarfs*. These emit low amounts of energy, but have very high surface temperatures, indicating that they must be small—typically the size of Earth. Both red giants and white dwarfs play an important role in the life cycle of stars.

STELLAR EVOLUTION: BIRTH TO DEATH

In Figure 17.9, we saw how our solar system is thought to have formed from a nebula. All stars are believed to begin life in this way, when a nebula collapses to form an immense ball of hydrogen and helium. Once the central mass reaches a critical size, the temperature and pressure will be high enough for nuclear fusion reactions to begin—and the star will be born. How that star dies, however, depends on its initial mass.

Stars much less massive than the Sun

Stars that contain about 10% of the Sun's mass are called *brown dwarfs*. They burn hydrogen extremely slowly and shine faintly compared with the Sun. Because their nuclear reactions proceed so slowly, they can burn for hundreds of billions of years, with little change in their size, temperature, and energy output.

Stars with a mass similar to the Sun

Stars like our Sun will have a much more eventful life. Any star is involved in a constant battle against the force of gravity, which is trying to cause the star to collapse inward. When it begins to burn hydrogen in its core, the star's temperature and central pressure rise, countering the inward pull of gravity. However, as the star uses up the hydrogen in its core, it produces less energy to counter gravity, and the star will begin to contract once more. As the star contracts, the temperature surrounding the core rises, triggering the burning of hydrogen in the shell around the hydrogen-depleted core. The temperature within the core will also increase to a level high enough for helium to begin undergoing nuclear fusion reactions, a process called *helium burning*.

The increased energy from nuclear fusion causes the star to balloon out, and it leaves the main sequence, becoming a red giant. When the Sun becomes a red giant, for instance, its radius will swell out past the current orbit of Venus.

Helium will continue to burn within the core, creating carbon. The star will continue to pour out enormous amounts of energy and its carbon core will shrink, until its constituent particles cannot be squeezed together more tightly. At this point, the star will be about the size of Earth, but hundreds of thousands of times more massive—it will be a white dwarf.

Very large stars

Stars that are much more massive than the Sun end their lives in immense explosions. In these stars, the pull of gravity is so large that after helium burning has created carbon in the star's core, nuclear fusion reactions still continue, converting the carbon to heavier elements, such as oxygen, magnesium, and silicon. The largest stars produce elements as heavy as iron through a chain of nuclear reactions.

When the core becomes iron-rich, the star undergoes a catastrophic collapse. Protons and electrons within the iron nuclei recombine to form neutrons, and as these neutrons are squeezed tightly, they exert a counterpressure against gravity. As the outer gaseous envelope tries to contract, its constituent particles collide with the rebounding neutrons, creating intense shockwaves, and the outer part of the star explodes. This is a **supernova**.

> **supernova** A stellar explosion marking the death of a massive star.

Neutron stars and black holes

In the aftermath of a supernova, the ejected material forms a cloud around the dead core of the former star, but then dissipates into interstellar space. The remaining core is formed of neutrons, and is typically about 20 km across—the size of a city. The object is called a *neutron star*.

Neutron stars can have very strong magnetic fields and can rotate at a prodigious rate. Some neutron stars in our galaxy spin at 1000 times a second, in contrast to the Sun, which spins once every 26 days. The combination of the large magnetic field with the star's fast rotation creates radio waves from the neutron star's magnetic poles, like a radio beacon. If Earth lies in the path of these beams, the radio source would appear to blink on and off as the neutron star's radio beam rotated past Earth. In the 1970s, just such an object, dubbed a *pulsar*, emitting short pulses of radio waves was discovered in the Crab Nebula (the nebula is shown in *What an Earth Scientist Sees* on page 525). All pulsars are neutron stars, but not all neutron stars are observed as pulsars.

Stars that are about 10 times as massive as the Sun will become neutron stars upon death. For even more massive stars, perhaps 50 times the mass of our Sun, the neutron star can crunch down further, forming a *black hole*—an object so dense that even light cannot escape its gravitational pull. We cannot directly see black holes, but we can see their effects on objects around them. Huge amounts of energy are released when material falls into a black hole, in the form of X-rays and other high-energy emissions, and these can be seen by space-based telescopes.

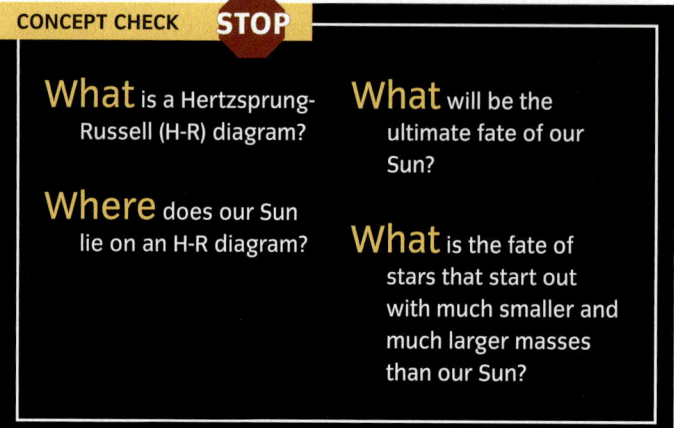

CONCEPT CHECK STOP

What is a Hertzsprung-Russell (H-R) diagram?

Where does our Sun lie on an H-R diagram?

What will be the ultimate fate of our Sun?

What is the fate of stars that start out with much smaller and much larger masses than our Sun?

The Universe and How It Came to Be

LEARNING OBJECTIVES

Explain the big bang scenario and discuss evidence supporting it.

Describe the galaxies, quasars, and black holes found in the universe today.

Discuss dark matter and dark energy.

Define what the cosmic microwave background is and explain why it is useful.

In the centuries after Galileo turned his primitive telescope to the skies and revolutionized our view of Earth's place in the universe, telescopes gradually became more advanced and revealed more surprises about our galaxy and the nature of the universe itself. By the 20th century, we were forced to accept that our galaxy, the Milky Way, is only one of many. And more detailed observations of neighboring galaxies overturned our notion of the origin of the universe.

EDWIN HUBBLE AND THE DISCOVERY OF GALAXIES

If we wanted to give the full address of our solar system, we could say that it lies in the Orion arm of the Milky Way (**FIGURE 17.20**). Within this arm are our Sun; eight planets; a vast number of small, rocky bodies called *asteroids*; millions of comets; innumerable small fragments of rock and dust called *meteoroids*; and numerous moons.

Looking up at night, we can see the hazy band of the Milky Way arching across the sky. A modest telescope reveals that this band is actually composed of millions of stars. We can also see other distant, cloudlike objects called *nebulae* (**FIGURE 17.21**). Until the early 20th century, we did not have telescopes that were powerful enough to resolve whether nebulae were distant collections of stars or nearby dust clouds.

The issue was resolved by American astronomer Edwin Hubble (1889–1953), but the answer in turn raised more puzzles. In 1919, Hubble went to work at the Mount Wilson Observatory in California, where he had access to a newly built telescope that could magnify images to an unprecedented level. With it, Hubble could

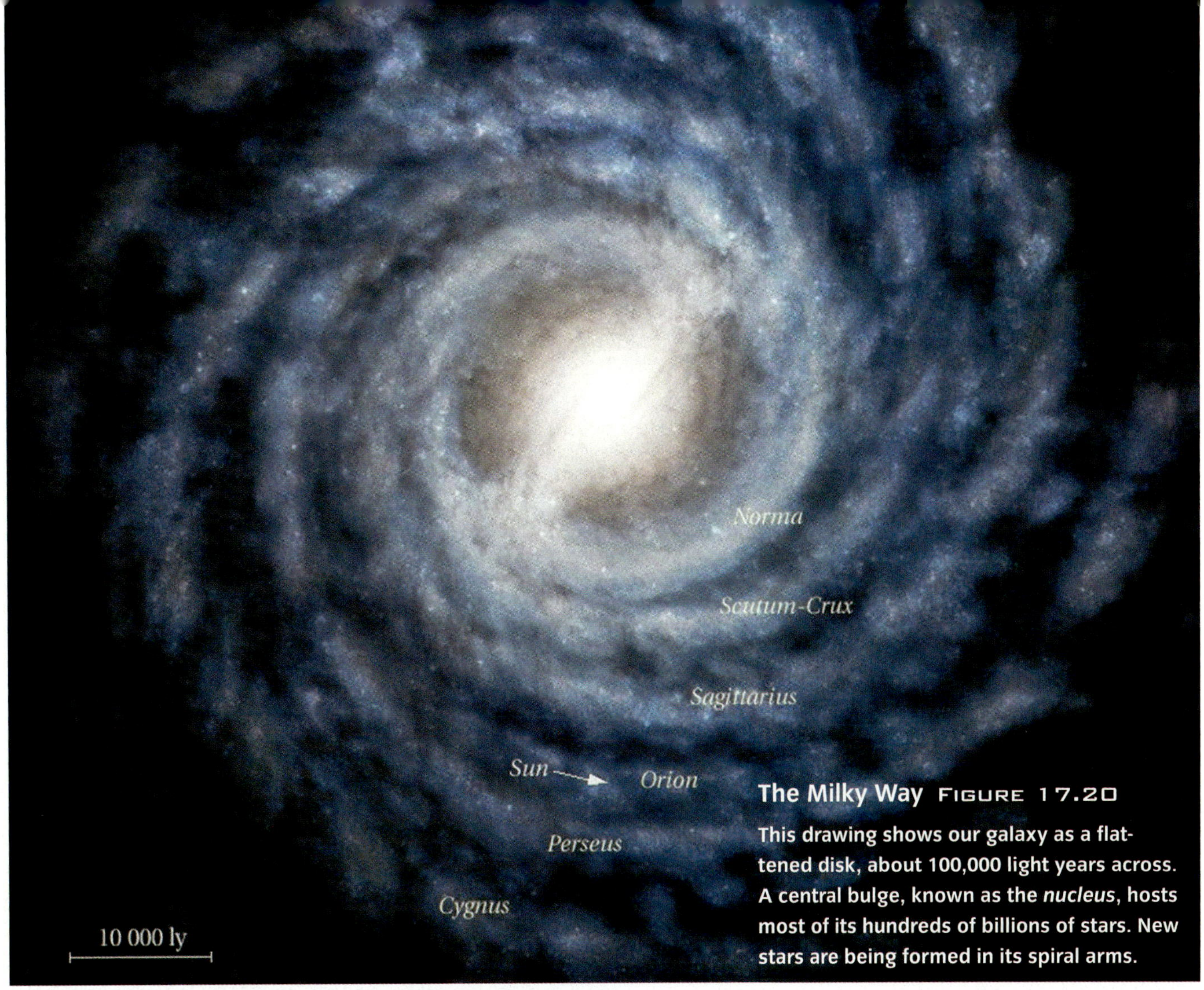

The Milky Way FIGURE 17.20

This drawing shows our galaxy as a flattened disk, about 100,000 light years across. A central bulge, known as the *nucleus*, hosts most of its hundreds of billions of stars. New stars are being formed in its spiral arms.

The Cat's Eye Nebula FIGURE 17.21

This nebula lies 3000 light years from Earth. This image from the Hubble Space Telescope reveals some of the most complex structure seen in any nebula. Astronomers believe that the bright central object may be a binary star—that is, two stars in orbit around each other.

locate individual stars within some nebulae. Some of these stars were of a special type, known as *Cepheid variables*, that astronomers can use to work out the distance to the star. Cepheids brighten and dim on a regular cycle and the rate of this variation in brightness can be related to their distance from Earth. Hubble calculated that the nearest Cepheid lay in the Andromeda nebula, some 2 million light years away. With this observation, he established that stars exist outside our galaxy and that the universe must contain other galaxies. Our galaxy, the Milky Way, is just one of these.

Hubble's law **FIGURE 17.22**

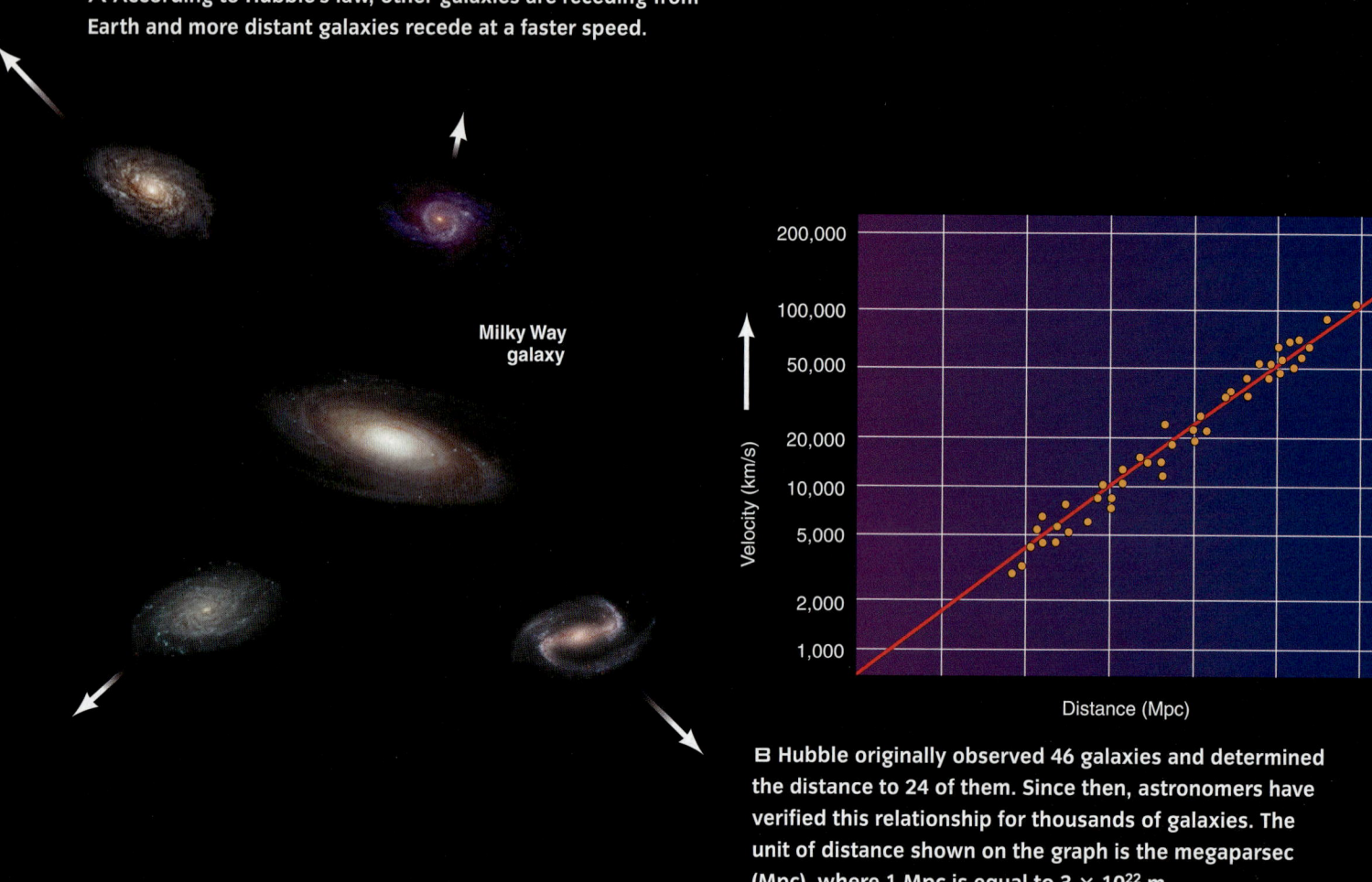

A According to Hubble's law, other galaxies are receding from Earth and more distant galaxies recede at a faster speed.

B Hubble originally observed 46 galaxies and determined the distance to 24 of them. Since then, astronomers have verified this relationship for thousands of galaxies. The unit of distance shown on the graph is the megaparsec (Mpc), where 1 Mpc is equal to 3×10^{22} m.

Hubble not only was able to work out the distance to other galaxies but also could tell how fast they were moving by looking at the light emitted by chemical elements in the galaxies. Hubble's calculations were based on a phenomenon known as the *Doppler effect*. If you have heard a police car siren change pitch as the car speeds past you, you are familiar with this effect. Because the source of the sound waves, the siren, is moving relative to you, the pitch or frequency of the sound waves appears to change. Similarly, if distant galaxies are moving toward Earth or away from it, the frequency of light waves emitted by various chemical elements in those galaxies will appear to shift in a distinct way that depends on their speed.

Hubble found that no matter which galaxy he looked at, the light waves were shifted to longer wavelengths, indicating that all other galaxies are moving away from Earth. He also calculated that the farther away a galaxy was, the faster it was receding from Earth (**FIGURE 17.22**).

It may seem like a modest discovery, but **Hubble's law** laid the groundwork for the theory of the origin of the universe known as the **big bang theory**.

Hubble's law The observed relationship that the velocity of recession of a galaxy is proportional to its distance.

542 CHAPTER 17 *Earth's Place in Space*

THE BIG BANG

The relationship between galaxy distance and speed had a startling implication: If nearby galaxies are moving away from us and faraway galaxies are moving even faster, the universe must be expanding, almost like a balloon that is being inflated (**Figure 17.23**). Running that picture backward in time suggests that our universe started out as a very tiny object. This is the **big bang theory**—the universe began at a specific time in the past as an infinitely hot, dense concentration of energy and matter, and has been expanding ever since.

■ **big bang theory**
The theory that the universe began to expand after an explosion of concentrated matter and energy.

■ **cosmic microwave background (CMB) radiation**
Radiation left over from the big bang, detected in the microwave portion of the spectrum with a temperature of only 2.7 K.

Hubble's observation of receding galaxies provides support for the big bang theory but is not conclusive evidence for it. More substantial evidence for the theory came in the form of the **cosmic microwave background (CMB) radiation**, the faint afterglow of radiation that physicists at Princeton University had predicted would be left behind by the big bang. The distribution of visible matter in galaxies today can, in theory, be traced back to the ripples in the CMB, 380,000 years after the big bang (see the *Case Study* on the next page).

Another important piece of evidence for the big bang theory is the amount of the three lightest chemical elements—hydrogen, helium, and lithium—in the universe. (As we learned earlier in the chapter, heavier elements are produced inside stars.) For these light elements to form, subatomic matter had to be densely packed together in the early universe so that particles could collide, producing *fusion* reactions. The temperature conditions also had to be just right—high enough for the fusion reactions to happen, but not so hot that the newly formed elements would be broken apart again in further collisions. In the big bang model, the density of matter drops as the universe expands, leaving only a brief period for the light elements to form. Scientists have been able to predict the abundance of light elements that should have been produced under these conditions, and their predictions fit very well with actual observations.

The inflating balloon analogy of the expanding universe Figure 17.23

Imagine that you live on the surface of a balloon. Evenly spaced points are dotted over the balloon, representing different galaxies. As the universe expands—represented by the balloon inflating—every other galaxy appears to move away from you. The farther another galaxy is from you, the faster it recedes. Notice that it does not matter which galaxy point you live on, because every galaxy is moving away from every other galaxy.

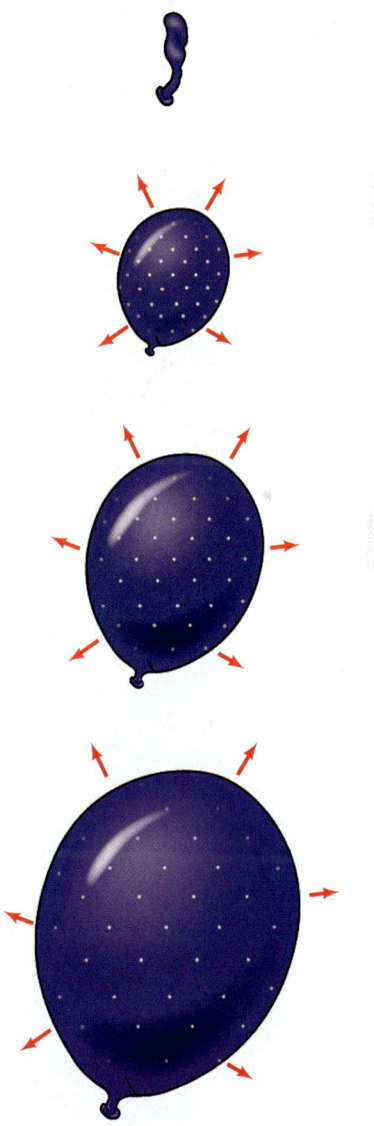

The Universe and How It Came to Be

CASE STUDY

The Afterglow of the Big Bang

In 1965, Arno Penzias and Robert W. Wilson at Bell Laboratories in New Jersey made an accidental discovery that helped to rewrite the way we think of the universe (FIGURE A). They were scanning the skies for background radio signals that might interfere with radio broadcasts. Regardless of the direction in which they pointed their receiver, they picked up a faint microwave hiss.

A year earlier, physicists at Princeton University had predicted that just such a signal should have been left behind by the big bang. This is because every object in the universe with a temperature above absolute zero emits some sort of radiation. Just as a lump of hot coal may burn white hot and change color to yellow, orange, and then red as it cools, the big bang left behind electromagnetic radiation that we can still pick up in the microwave range—the cosmic microwave background (CMB)—with a temperature of 2.7 K.

The CMB is a snapshot of the temperature of the universe just 380,000 years after the big bang. In 2001, NASA launched the Wilkinson Microwave Anisotropy Probe (WMAP) satellite to measure slight temperature differences in the CMB. The light and dark patches on the map correspond to relatively cool and hot patches in the CMB (FIGURE B).

The patterns in the WMAP data allowed astronomers to calculate their best estimate for the age of the universe so far, 13.7 billion years, and construct a timeline for the formation of stars, galaxies, and planets (FIGURE C).

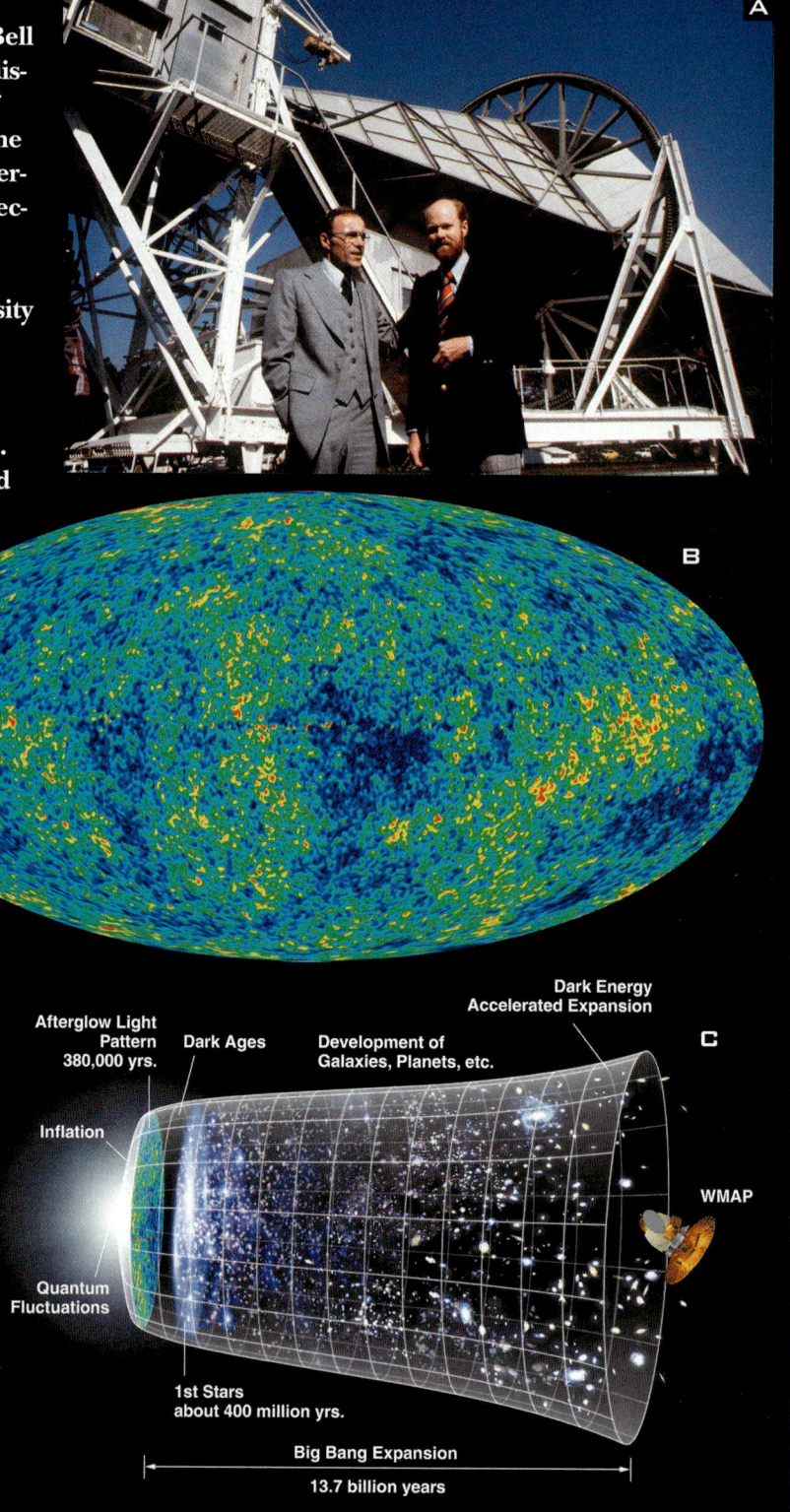

544 CHAPTER 17 Earth's Place in Space

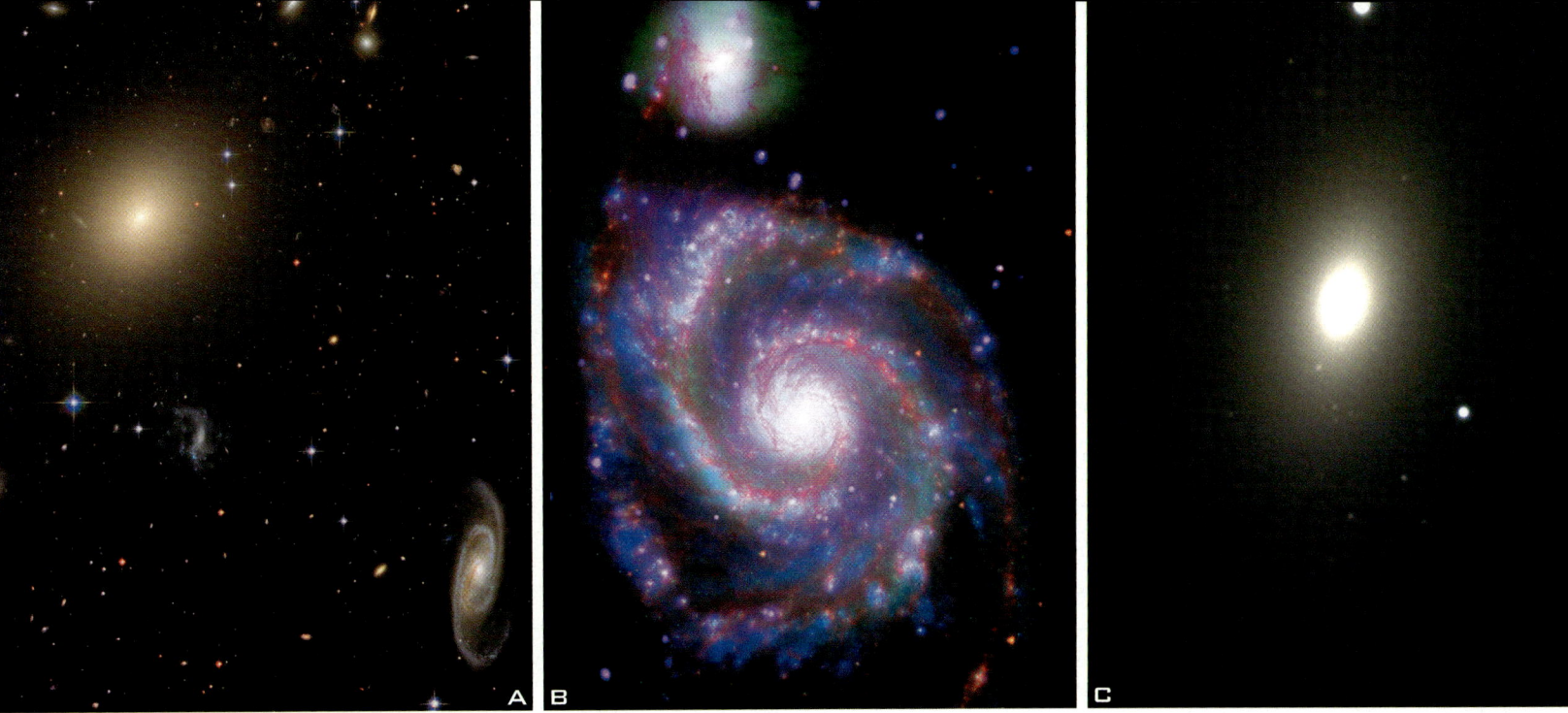

Galaxies FIGURE 17.24

A Taking long-exposure photographs over several hours, the Hubble Space Telescope reveals hundreds of galaxies in what would otherwise appear to be "empty" sky. These deep-field images provide evidence that the universe holds tens of billions of galaxies. Most are spiral galaxies, such as the Whirlpool Galaxy, shown in **B**. Another common type is the elliptical galaxy, such as this galaxy from the Virgo cluster, shown in **C**.

THE UNIVERSE TODAY

We now know that the universe contains countless galaxies (**FIGURE 17.24**). These galaxies fall into two main types: *spirals* and *elliptical galaxies*. About 75% of the brighter galaxies in the sky are spirals. Another 20% are elliptical galaxies, resembling cosmic footballs. The universe is also littered with irregular and dwarf galaxies, which could be the most common types of galaxies in the universe, but because they are faint, they are hard to detect. The Milky Way is an example of a spiral galaxy, as shown in Figure 17.20, with new stars forming in its spiral arms.

Star formation is usually an orderly process, but things are different at the heart of a small, but violent, minority of galaxies. About 10,000 *active galaxies* house *quasars* (quasi-stellar radio sources) at their cores. Quasars are relatively compact objects that spew vast amounts of energy into space. Astronomers do not yet understand the source of this energy, but many think that it could be due to a supermassive black hole lurking inside a quasar. As mentioned earlier, although we cannot see black holes directly, astronomers expect that huge amounts of energy would be released when matter falls into a black hole.

CONCEPT CHECK STOP

What is Hubble's law?

How does Hubble's law lead to the big bang scenario for the origin of the universe?

What are the two main types of galaxies?

What is the cosmic microwave background? What does it tell us about the universe?

Amazing Places: Mars

Humans have always been fascinated by Mars. For the past 40 years, NASA spacecraft have been increasing our understanding the red planet's climate and features.

A NASA's Hubble Space Telescope took this closeup of the red planet in August 2003, when it was just 34,648,840 mi (55,760,220 km) away, just 11 hours before the planet made its closest approach to Earth in 60,000 years. The planet's northern hemisphere is home to volcanoes that may have been active about 1 billion years ago. These volcanoes resurfaced the landscape, perhaps filling in many impact craters. In contrast, the southern hemisphere is pockmarked with ancient-impact craters.

B In recent years, the Mars rovers, Spirit and Opportunity, and the Phoenix lander have been exploring the planet and sending home unique postcards showing its surface in unprecedented detail. In May 2008, Phoenix returned images of Mars' "quilted" surface. Similar polygon patterns form on icy ground in Earth's arctic regions.

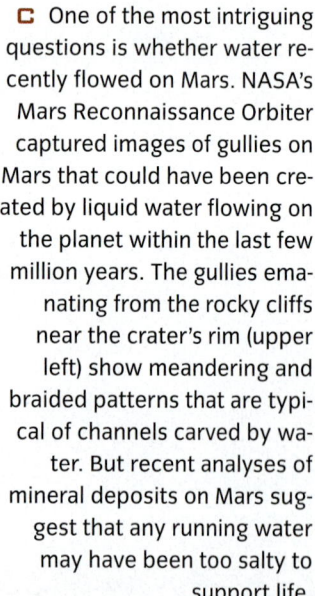

C One of the most intriguing questions is whether water recently flowed on Mars. NASA's Mars Reconnaissance Orbiter captured images of gullies on Mars that could have been created by liquid water flowing on the planet within the last few million years. The gullies emanating from the rocky cliffs near the crater's rim (upper left) show meandering and braided patterns that are typical of channels carved by water. But recent analyses of mineral deposits on Mars suggest that any running water may have been too salty to support life.

SUMMARY

1 Astronomy and the Scientific Revolution

1. The concept of a **heliocentric universe**, with Earth rotating on its own axis and other planets revolving around the Sun, provides a simple explanation of the retrograde motion of *planets*.

2. The seasons occur because Earth's axis of rotation is tilted.

3. Kepler developed the following three laws of planetary motion: (a) The planets revolve around the Sun in elliptical orbits, (b) in such a way that a line drawn from a planet to the Sun sweeps out equal areas in equal amounts of time. (c) For any planet, the square of the orbital period in years is proportional to the cube of the planet's average distance from the Sun. Newton realized that the planets are held in orbit by their gravitational attraction to the Sun.

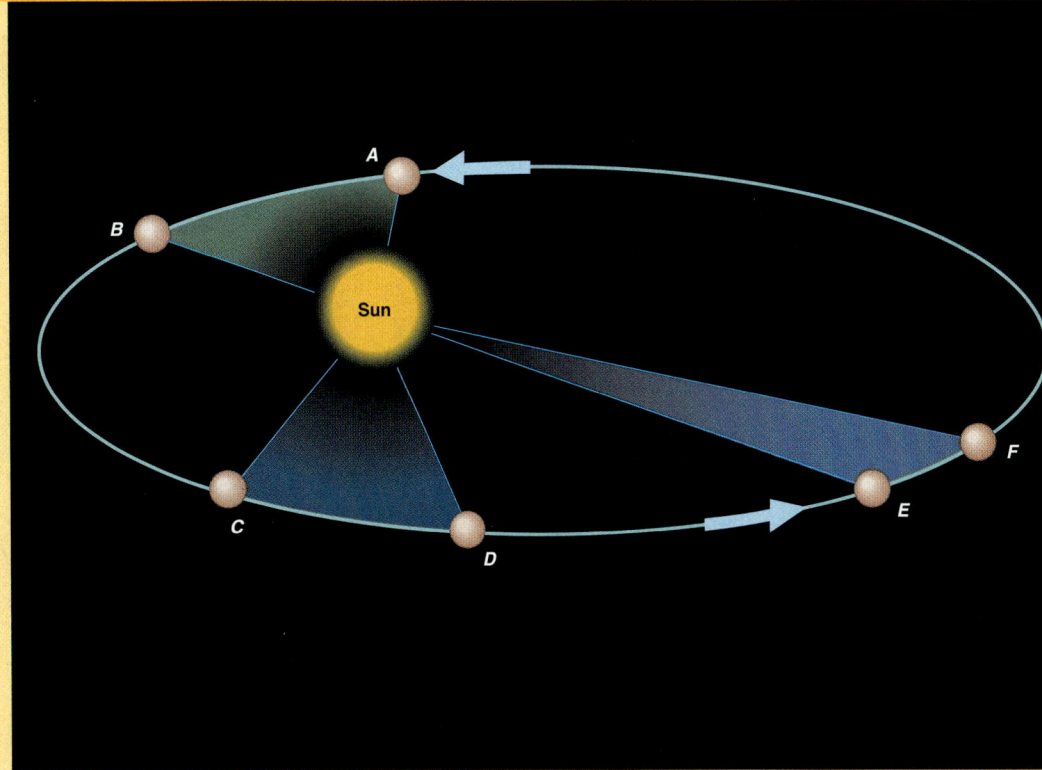

2 The Solar System

1. The *solar system* consists of the Sun, eight planets, many dwarf planets, numerous moons, and vast numbers of *asteroids*, millions of *comets*, and innumerable *meteoroids*, which can fall to Earth as **meteorites**.

2. The solar system was formed according to the **nebula hypothesis**, through the condensation of a solar nebula followed by *planetary accretion*, a process that was completed about 4.56 billion years ago.

3. Solar energy is generated in the Sun's core by the **proton–proton chain** of nuclear fusion reactions. Energy travels through the solar interior to the surface by *radiation* and *convection*. The Sun's outer layers are the *photosphere*, the *chromosphere*, and the corona.

4. The planets can be divided into two groups: the *terrestrial planets*, the four nearest the Sun, each a small, rocky mass with a high density; and the *Jovian planets*, the four outermost planets, each large and gassy.

5. The terrestrial planets have a metallic core, a mantle, and a crust.

6. The Jovian planets are shrouded by thick atmospheres that are rich in hydrogen and helium. Their cores are inferred to be rocky, like a terrestrial planet, surrounded by a thick layer of ice; above the ice is liquid hydrogen, which grades outward to the hydrogen-rich atmosphere.

3 Stars and Stellar Evolution

1. Stars are characterized on a **Hertzsprung-Russell diagram** by their surface temperature and luminosity. A star's fate is determined by its initial mass. Low-mass stars will evolve as *brown dwarfs*. Stars with a mass similar to our Sun will become *red giants* and then *white dwarfs*. Higher-mass stars will explode as **supernovae**, and their cores may become *neutron stars* or *black holes*.

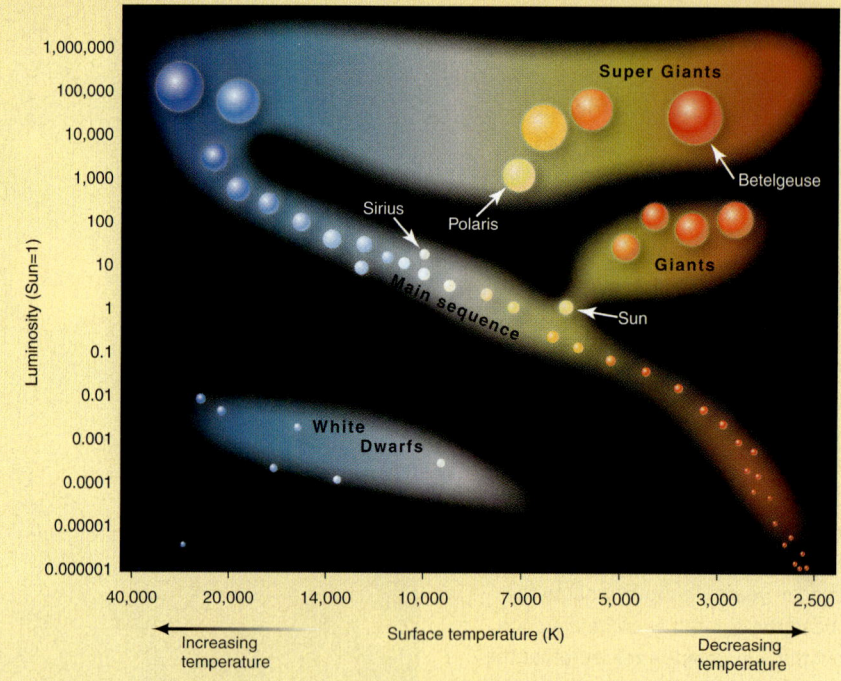

4 The Universe and How It Came to Be

1. The Milky Way is one of countless *galaxies*. These galaxies are receding from one another, and according to **Hubble's law**, the farther the galaxy is from Earth, the faster it is receding.

2. The **big bang theory** states that the universe began at a specific moment in time and has been expanding ever since. The recession of the galaxies, the discovery of the **cosmic microwave background radiation**, and the abundance of light elements in the universe are evidence for the big bang.

3. There are two main types of galaxies: *spirals* and *elliptical galaxies*. The Milky Way is a spiral galaxy.

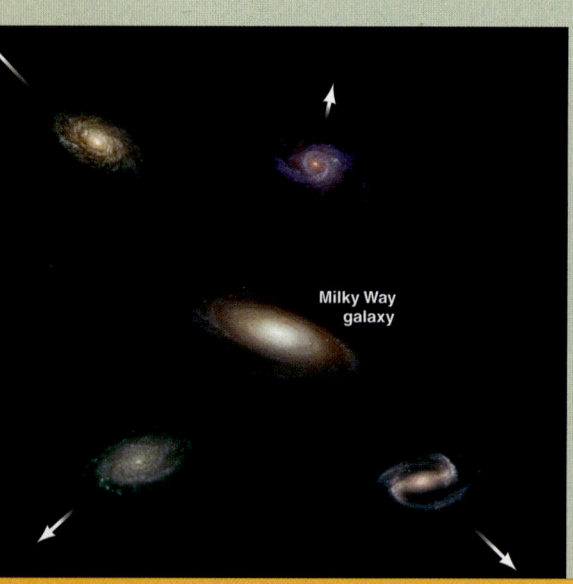

KEY TERMS

- geocentric p. 518
- heliocentric p. 518
- nebula hypothesis p. 525
- proton–proton chain p. 527
- meteorite p. 535
- exoplanet p. 537
- Hertzsprung-Russell (H-R) diagram p. 538
- supernova p. 539
- Hubble's law p. 542
- big bang theory p. 543
- cosmic microwave background (CMB) radiation p. 543

CRITICAL AND CREATIVE THINKING QUESTIONS

1. NASA scientists have focused more attention on Mars than any other planet in the solar system. One of the reasons for doing so is that Mars may once have had an environment on which life might have started. If you had a chance to seek evidence of life elsewhere on the planets or moons of the solar system, what would you choose as targets?

2. Earth scientists have a hypothesis that the last of the dinosaurs were driven to extinction at the end of the Cretaceous Period by the impact of a large meteorite. What might the consequences be if an equally large meteorite were to strike Earth today? Consider two possibilities: first, that the site of the impact be in the center of the Atlantic Ocean; and second, that the impact be in the fertile farming region of the midwestern United States.

3. Since the time of Aristotle, our view of our place in the universe has changed dramatically. Astronomical observations have played a major role in changing our view. How do you think our view of our place in the universe would change if scientists discovered evidence of intelligent life on a planet orbiting some other star?

4. The Hubble Space telescope has gathered exceptional evidence of stars and other things in our Milky Way galaxy, and in galaxies in the universe beyond. Imagine you are a member of Congress meeting with scientists who seek funds for an even more powerful eye-in-the sky. What scientific justifications would you ask of the scientists in order to decide whether to support or deny their request?

5. Discussions are now being held in both space science and political circles about a manned space mission to Mars. The mission would be very dangerous for astronauts and the expense would be great. How would you justify such a mission? If you don't agree with a manned mission, what would your objectives be for future unmanned missions?

What is happening in this picture?

- This Hubble image shows a possible exoplanet orbiting a star.

- What features would make an exoplanet hospitable for life?

- If it were possible to send an unmanned spaceship to a nearby star in order to inspect the planets in orbit around the star, what kind of measurements would you advise the space scientists to try to obtain as the spaceship flies past each planet?

SELF-TEST

1. Who first proposed a heliocentric model of the universe?
   a. Aristarchus
   b. Aristotle
   c. Copernicus
   d. Kepler

2. Which of the following was *not* discovered by Kepler?
   a. Planets' orbits around the Sun are elliptical due to the force of gravity.
   b. Planets change speed as they move along their orbits.
   c. Planets sweep out equal areas along their orbits, in equal times.
   d. Planets move such that the square of their period in years is proportional to the cube of the planet's average distance from the Sun.

3. What feature, shown here, did Galileo observe, supporting the heliocentric universe?
   a. phases of the Moon
   b. phases of Venus
   c. partial eclipse of the Moon
   d. transit of Venus

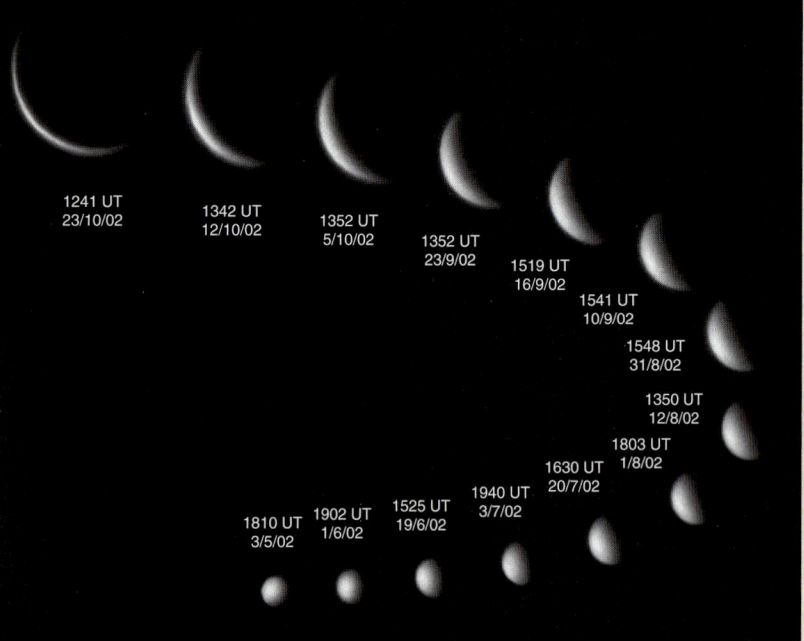

4. Label the following regions on this diagram of the Sun: (a) chromosphere, (b) corona, (c) core, (d) convection zone, (e) photosphere, (f) radiation zone.

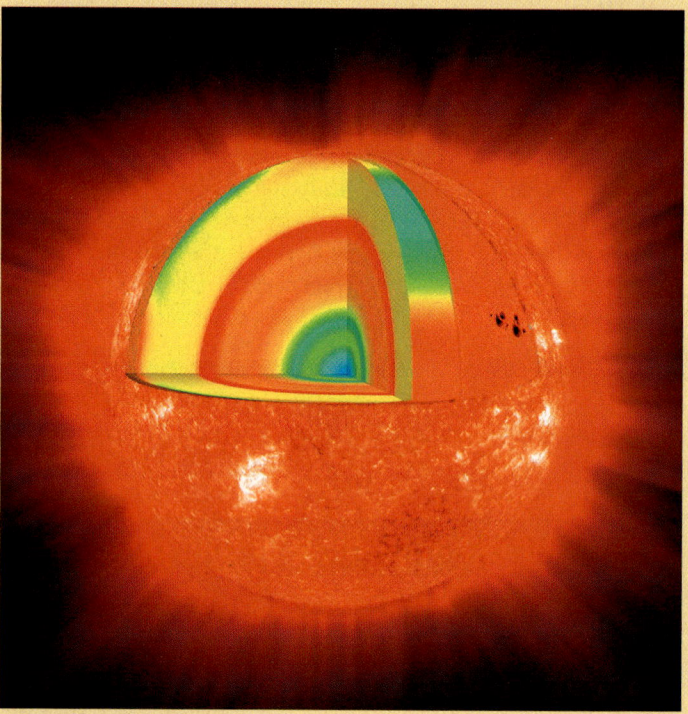

5. The Sun derives its energy from _____.
   a. hydrogen fission
   b. hydrogen fusion
   c. helium fission
   d. helium fusion

550 CHAPTER 17 Earth's Place in Space

6. Which of the following features is not shared by all terrestrial planets?
   a. an atmosphere
   b. a core
   c. a magnetic field
   d. a crust

7. The pressure inside Jupiter's core is so high that it is possible that _____.
   a. hydrogen gas could condense to liquid hydrogen
   b. hydrogen atoms could fuse to form helium atoms
   c. hydrogen becomes metallic
   d. (a) and (b)
   e. (a) and (c)
   f. (b) and (c)

8. The Jovian planets—Jupiter, Saturn, Uranus, and Neptune—all share which of the following features?
   a. molten cores
   b. metallic hydrogen sheaths surrounding their cores
   c. thick atmospheres
   d. all of the above

9. The asteroid belt lies between _____ and _____.
   a. Earth/Mars
   b. Mars/Jupiter
   c. Jupiter/Saturn
   d. Saturn/Uranus

10. Planetary _____ accounts for the existence of asteroids and meteoroids.
    a. condensation
    b. aggregation
    c. accretion
    d. clustering

11. Comets are thought to originate in _____.
    a. the Oort cloud
    b. the Kuiper belt
    c. the asteroid belt
    d. (a) and (b)
    e. (a) and (c)
    f. (b) and (c)

12. The Moon is thought to have been formed when a _____-sized object collided with Earth.
    a. Mercury
    b. Venus
    c. Mars
    d. Uranus

13. Why might the exoplanet Gliese 581c not be habitable?
    a. It probably does not have a rocky composition.
    b. Its atmosphere is probably poisonous.
    c. It is probably too cold.
    d. It is probably undergoing a runaway greenhouse effect.

14. What objects observed by Hubble enabled him to work out the distance to neighboring galaxies?
    a. quasars
    b. dwarf galaxies
    c. Cepheid variables
    d. supernovae

15. Which of the following does not serve as evidence for the big bang theory of the origin of the universe?
    a. the amount of light elements in the universe
    b. the direction of rotation of the Sun and the planets (excluding Venus)
    c. the recession of galaxies
    d. the cosmic microwave background

Appendix A
Periodic Table of the Elements

The periodic table lists the known **chemical elements**, the basic units of matter. The elements in the table are arranged left-to-right in rows in order of their **atomic number**, the number of protons in the nucleus. Each horizontal row, numbered from 1 to 7, is a **period**. All elements in a given period have the same number of electron shells as their period number. For example, each atom of hydrogen or helium has one electron shell, while each atom of potassium or calcium has four electron shells. The elements in each column, or **group**, share chemical properties. For example, the elements in column IA are very chemically reactive, whereas the elements in column VIIIA have full electron shells and thus are chemically inert.

Scientists now recognize up to 118 different elements; 92 occur naturally on Earth, and the rest (with the exception of element 117) have been produced synthetically using particle accelerators. Elements are designated by **chemical symbols**, which are the first one or two letters of the element's name in English, Latin, or another language.

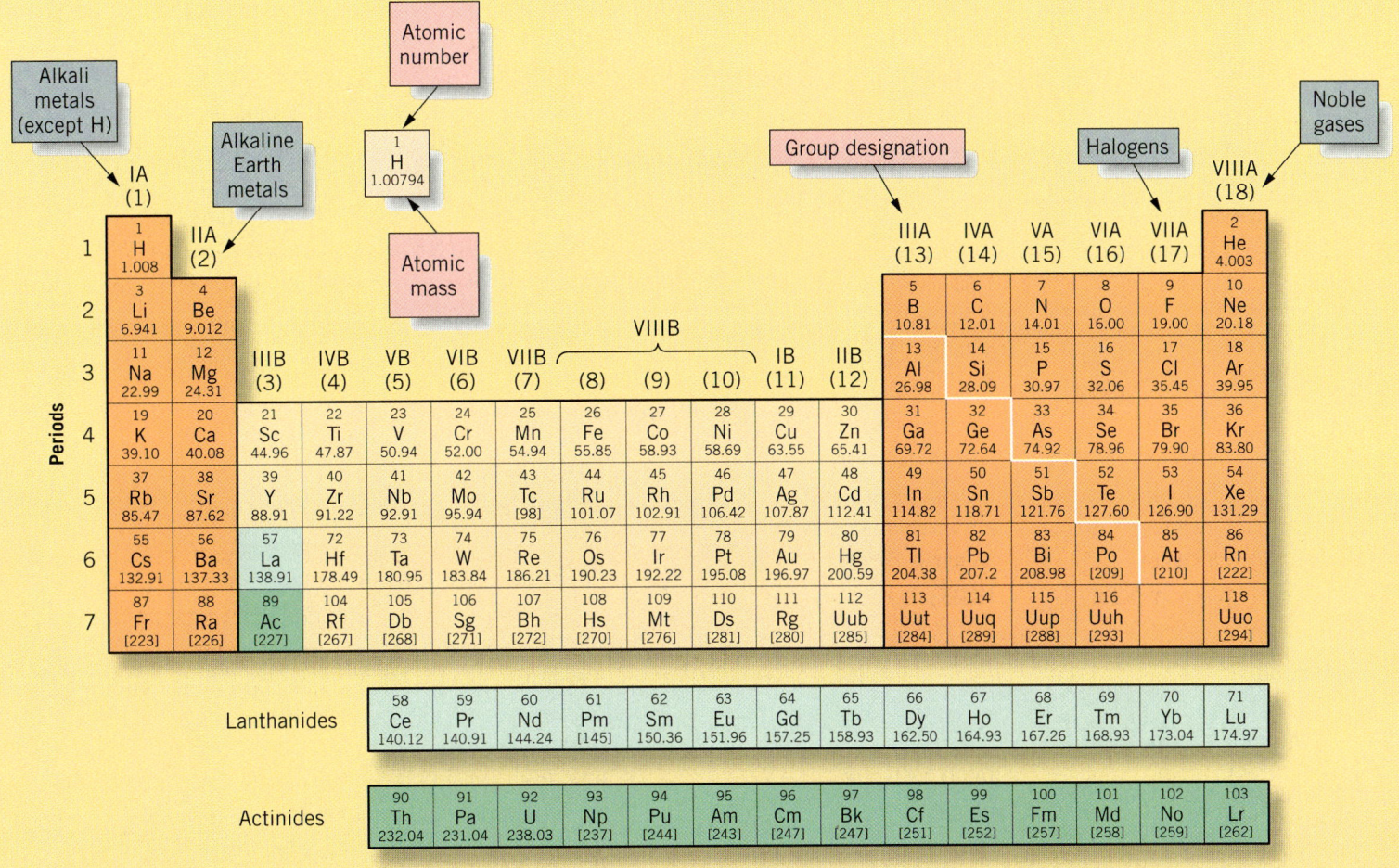

553

Appendix B
Units and Their Conversions

Commonly Used Units of Measure

Length

Metric Measure

1 kilometer (km)	= 1000 meters (m)
1 meter (m)	= 100 centimeters (cm)
1 centimeter (cm)	= 10 millimeters (mm)
1 millimeter (mm)	= 1000 micrometers (μm) (formerly called microns)
1 micrometer (μm)	= 0.001 millimeter (mm)
1 angstrom (Å)	= 10^{-8} centimeters (cm)

Nonmetric Measure

1 mile (mi)	= 5280 feet (ft) = 1760 yards (yd)
1 yard (yd)	= 3 feet (ft)
1 fathom (fath)	= 6 feet (ft)

Conversions

1 kilometer (km)	= 0.6214 mile (mi)
1 meter (m)	= 1.094 yards (yd)
	= 3.281 feet (ft)
1 centimeter (cm)	= 0.3937 inch (in)
1 millimeter (mm)	= 0.0394 inch (in)
1 mile (mi)	= 1.609 kilometers (km)
1 yard (yd)	= 0.9144 meter (m)
1 foot (ft)	= 0.3048 meter (m)
1 inch (in)	= 2.54 centimeters (cm)
1 inch (in)	= 25.4 millimeters (mm)
1 fathom (fath)	= 1.8288 meters (m)

Area

Metric Measure

1 square kilometer (km^2)	= 1,000,000 square meters (m^2)
	= 100 hectares (ha)
1 square meter (m^2)	= 10,000 square centimeters (cm^2)
1 hectare (ha)	= 10,000 square meters (m^2)

Nonmetric Measure

1 square mile (mi^2)	= 640 acres (ac)
1 acre (ac)	= 4840 square yards (yd^2)
1 square foot (ft^2)	= 144 square inches (in^2)

Conversions

1 square kilometer (km^2)	= 0.386 square mile (mi^2)
1 hectare (ha)	= 2.471 acres (ac)
1 square meter (m^2)	= 1.196 square yards (yd^2)
	= 10.764 square feet (ft^2)
1 square centimeter (cm^2)	= 0.155 square inch (in^2)
1 square mile (mi^2)	= 2.59 square kilometers (km^2)
1 acre (ac)	= 0.4047 hectare (ha)
1 square yard (yd^2)	= 0.836 square meter (m^2)
1 square foot (ft^2)	= 0.0929 square meter (m^2)
1 square inch (in^2)	= 6.4516 square centimeter (cm^2)

Volume

Metric Measure

1 cubic meter (m^3)	= 1,000,000 cubic centimeters (cm^3)
1 liter (l)	= 1000 milliliters (ml)
	= 0.001 cubic meter (m^3)
1 centiliter (cl)	= 10 milliliters (ml)
1 milliliter (ml)	= 1 cubic centimeter (cm^3)

Nonmetric Measure

1 cubic yard (yd^3)	= 27 cubic feet (ft^3)
1 cubic foot (ft^3)	= 1728 cubic inches (in^3)
1 barrel (oil) (bbl)	= 42 gallons (U.S.) (gal)

Conversions

1 cubic kilometer (km^3)	= 0.24 cubic miles (mi^3)
1 cubic meter (m^3)	= 264.2 gallons (U.S.) (gal)
	= 35.314 cubic feet (ft^3)
1 liter (l)	= 1.057 quarts (U.S.) (qt)
	= 33.815 ounces (U.S. fluid) (fl. oz.)
1 cubic centimeter (cm^3)	= 0.0610 cubic inch (in^3)
1 cubic mile (mi^3)	= 4.168 cubic kilometers (km^3)
1 acre-foot (ac-ft)	= 1233.46 cubic meters (m^3)
1 cubic yard (yd^3)	= 0.7646 cubic meter (m^3)
1 cubic foot (ft^3)	= 0.0283 cubic meter (m^3)
1 cubic inch (in^3)	= 16.39 cubic centimeters (cm^3)
1 gallon (gal)	= 3.784 liters (l)

Mass

Metric Measure

1000 kilograms (kg)	= 1 metric ton (also called a *tonne*) (m.t)
1 kilogram (kg)	= 1000 grams (g)

Nonmetric Measure

1 short ton (sh.t)	= 2000 pounds (lb)
1 long ton (l.t)	= 2240 pounds (lb)
1 pound (avoirdupois) (lb)	= 16 ounces (avoirdupois) (oz) = 7000 grains (gr)
1 ounce (avoirdupois) (oz)	= 437.5 grains (gr)
1 pound (Troy) (Tr. lb)	= 12 ounces (Troy) (Tr. oz)
1 ounce (Troy) (Tr. oz)	= 20 pennyweight (dwt)

Conversions

1 metric ton (m.t)	= 2205 pounds (avoirdupois) (lb)
1 kilogram (kg)	= 2.205 pounds (avoirdupois) (lb)
1 gram (g)	= 0.03527 ounce (avoirdupois) (oz) = 0.03215 ounce (Troy) (Tr. oz) = 15,432 grains (gr)
1 pound (lb)	= 0.4536 kilogram (kg)
1 ounce (avoirdupois) (oz)	= 28.35 grams (g)
1 ounce (avoirdupois) (oz)	= 1.097 ounces (Troy) (Tr. oz)

Pressure

Metric Measure

1 pascal (Pa)	= 1 newton/square meter (N/m^2)
1 kilogram-force square centimeter (kg/cm^2 or kgf/cm^2)	= 1 technical atmosphere (at) = 98,067 Pa = 0.98067 bar
1 bar	= 10^5 pascals (Pa) = 1.02 kilogram-force/square centimeter (kgf/cm^2 or at)

Nonmetric Measure

1 atmosphere (atm)	= 1.01325 bar = 14.696 lb/in^2 (psi)
1 pound per square inch (lb/in^2 or psi)	= 68.046 × 10^{-3} atmospheres (atm)

Conversions

1 kilogram-force/ square centimeter (kgf/cm^2 or at)	= 0.96784 atmosphere (atm) = 14.2233 pounds/square inch (lb/in^2 or psi) = 0.98067 bar = 98,067 Pa
1 bar	= 0.98692 atmosphere (atm) = 10^5 pascals (Pa) = 1.02 kilogram-force/square centimeter (kgf/cm^2)
1 Pa = 10^{-5} bar	= 10.197 × 10^{-6} kgf/cm^2 (or at) = 9.8692 × 10^{-6} atm = 145.04 × 10^{-6} lb/in^2 (or psi)
1 atm	= 101,325 Pa = 1.01325 bar

Temperature

Metric Measure

0 degrees Celcius (°C)	= freezing point of water at sea level
100 degrees Celcius (°C)	= boiling point of water at sea level
0 degrees Kelvin (K)	= −273.15°C = absolute zero
1°C	= 1 K (temperature increments)
273.15 K	= 0.0° C

Nonmetric Measure

Fahrenheit (°F)	= (K · 9/5) − 459.67
Fahrenheit (°F)	= (°C · 9/5) + 32

Conversions

degrees Kelvin (K)	= °C + 273.1
degrees Celcius (°C)	= K − 273.15
degrees Fahrenheit (°F)	= (°C · 9/5) + 32
degrees Celcius (°C)	= (°F − 32) · 5/9

Appendix C
Answers to Self-Tests

CHAPTER 1
1. c; **2.** d; **3.** See Figure 1.5; **4.** d; **5.** b; **6.** b; **7.** d; **8.** See Figure 1.9; **9.** c; **10.** b; **11.** See Figure 1.10; **12.** c; **13.** d; **14.** e; **15.** e

CHAPTER 2
1. d; **2.** See Figure 2.1; **3.** b; **4.** c; **5.** e; **6.** d; **7.** a; **8.** b; **9.** a; **10.** c; **11.** e; **12.** b; **13.** a; **14.** See Table 2.2; **15.** d

CHAPTER 3
1. See Figure 3.2; **2.** c; **3.** A = phaneritic-texture plutonic rock B = porphyritic-texture volcanic rock; **4.** b; **5.** d; **6.** a; **7.** b; **8.** b; **9.** a; **10.** e; **11.** d; **12.** d; **13.** c; **14.** c; **15.** c

CHAPTER 4
1. b ; **2.** c; **3.** d; **4.** e; **5.** c; **6.** a; **7.** See Figure 4.11; **8.** d; **9.** d; **10.** b; **11.** d; **12.** c; **13.** b; **14.** See Figure 4.17; **15.** c

CHAPTER 5
1. See Figure 5.1; **2.** a; **3.** d; **4.** c; **5.** d; **6.** b; **7.** d; **8.** d; **9.** b; **10.** c; **11.** d; **12.** c; **13.** b; **14.** b; **15.** See Figure 5.15.

CHAPTER 6
1. b; **2.** c; **3.** d; **4.** d; **5.** See Figure 6.7; **6.** b; **7.** d; **8.** b; **9.** See Figure 6.13; **10.** a; **11.** b; **12.** c; **13.** c; **14.** b; **15.** c

CHAPTER 7
1. a; **2.** c; **3.** d; **4.** c
5.

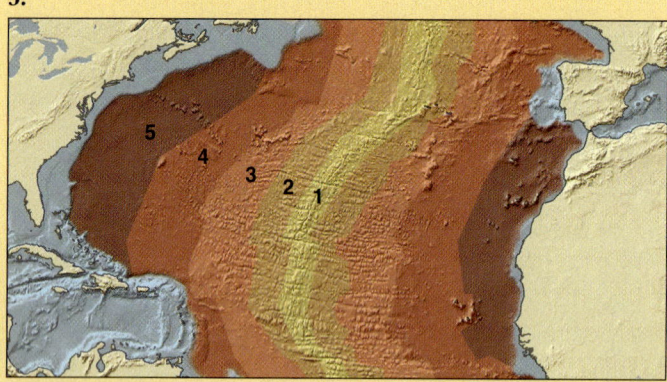

6. a
7.

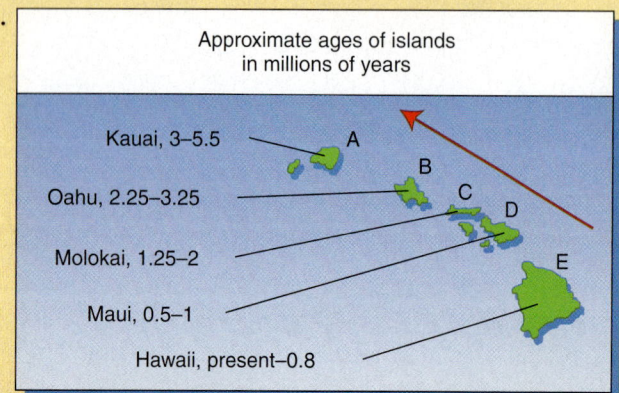

8. b; **9.** c; **10.** a; **11.** d; **12.** c; **13.** See Figure 7.12; **14.** See Figure 7.12; **15.** c

CHAPTER 8
1. c; **2.** b; **3.** d; **4.** b; **5.** a; **6.** a; **7.** See Figure 8.12; **8.** d; **9.** d; **10.** e; **11.** a; **12.** d; **13.** a; **14.** b; **15.** b

CHAPTER 9
1. a; **2.** b; **3.** b; **4.** b; **5.** c; **6.** a; **7.** d; **8.** b; **9.** b; **10.** c; **11.** c; **12.** c; **13.** See Figure 9.20; **14.** b; **15.** c

CHAPTER 10
1. b; **2.** c; **3.** c; **4.** c; **5.** c; **6.** See Figure 10.3; **7.** d; **8.** b; **9.** b; **10.** b; **11.** See Figure 10.9; **12.** b; **13.** a; **14.** b; **15.** d

CHAPTER 11
1. a; **2.** See Figure 11.2B; **3.** c; **4.** c; **5.** See Figure 11.8; **6.** e; **7.** c; **8.** b; **9.** e; **10.** e; **11.** b; **12.** d; **13.** d; **14.** a; **15.** d

CHAPTER 12
1. c; **2.** a; **3.** d ; **4.** c; **5.** a; **6.** c; **7.** d; **8.** b; **9.** b; **10.** d; **11.** a; **12.** b; **13.** b; **14.** a; **15.** b

CHAPTER 13
1. e; **2.** See Figure 13.3; **3.** a; **4.** a; **5.** See Figure in *What an Earth Scientist Sees*, page 401; **6.** d; **7.** b; **8.** d; **9.** See Figure 13.11; **10.** c; **11.** b; **12.** b; **13.** a; **14.** c; **15.** a

CHAPTER 14
1. See Figure 14.1B; **2.** a; **3.** c; **4.** See Figure 14.2; **5.** c; **6.** b; **7.** b; **8.** b; **9.** c; **10.** a; **11.** c; **12.** a; **13.** b; **14.** a; **15.** a

CHAPTER 15
1. c; **2.** c; **3.** a; **4.** c; **5.** b; **6.** See Figure 15.10; **7.** d; **8.** See Figure 15.12; **9.** a; **10.** b; **11.** c; **12.** a; **13.** c; **14.** b; **15.** a

CHAPTER 16
1. a; **2.** c; **3.** d; **4.** c; **5.** e; **6.** c; **7.** b; **8.** d; **9.** See Figure 16.7; **10.** b; **11.** c; **12.** c; **13.** c; **14.** e; **15.** d

CHAPTER 17
1. a; **2.** a; **3.** b; **4.** See Figure 17.11B; **5.** b; **6.** a; **7.** c; **8.** c; **9.** b; **10.** c; **11.** d; **12.** c; **13.** d; **14.** c; **15.** b

GLOSSARY

abrasion Wind erosion in which airborne particles chip small fragments off rocks that protrude above the surface.

adiabatic principle A principle of science that states that a gas cools as it expands and warms as it is compressed, provided that no heat flows into or out of the gas during the process.

air mass An extensive body of air in which temperature and moisture characteristics are fairly uniform over a large area.

alluvium Unconsolidated sediment deposited by a stream.

angiosperm A flowering, or seed-enclosed, plant.

anthropogenic Produced by human activities.

anticyclone A center of high atmospheric pressure.

aquiclude A layer of impermeable rock.

aquifer A body of rock or regolith that is water-saturated, porous, and permeable.

asthenosphere A layer of weak, ductile rock in the mantle that is close to melting but not actually molten.

atmospheric pressure Pressure exerted by the atmosphere at Earth's surface because of the force of gravity acting on the overlying column of air.

atom The smallest individual particle that retains the distinctive chemical properties of an element.

barometer An instrument that measures atmospheric pressure.

barrier island A long, narrow, sandy island lying offshore and parallel to a lowland coast.

batholith A large, irregularly shaped pluton that cuts across the layering of the rock into which it intrudes.

beach drift The movement of particles along a beach as they are driven up and down the beach slope by wave action.

beach Wave-washed sediment along a coast.

bed load Sediment that is moved along the bottom of a stream.

bedding The layered arrangement of strata in a body of sediment or sedimentary rock.

big bang theory The theory that the universe began to expand after an explosion of concentrated matter and energy.

biogenic sediment Sediment that is composed primarily of plant and animal remains, or precipitates as a result of biologic processes.

body wave A seismic wave that travels through Earth's interior.

bond The force that holds the atoms together in a chemical compound.

Bowen's reaction series The order in which minerals crystallize from, and subsequently react with, a cooling magma.

brittle deformation A permanent change in shape or volume, in which a material breaks or cracks.

cave and **cavern** Underground open space; a cavern is a system of connected caves.

cell The basic structural and functional unit of life; a complex grouping of chemical compounds enclosed in a porous membrane.

channel The clearly defined natural passageway through which a stream flows.

chemical sediment Sediment formed by the precipitation of minerals dissolved in lake, river, or seawater.

chemical weathering The decomposition of rocks and minerals by chemical and biochemical reactions.

clastic sediment Sediment formed from fragmented rock and mineral debris produced by weathering and erosion.

clay A family of hydrous aluminosilicate minerals. The term is also used for tiny mineral particles of any kind that have physical properties like those of the clay minerals.

cleavage The tendency of a mineral to break in preferred directions along bright, reflective plane surfaces.

climate The annual cycle of prevailing weather conditions at a given place based on statistics recorded over a long period.

coal A combustible rock formed from the lithification of plant-rich sediment.

compound A combination of atoms of one or more elements in a specific ratio.

compression A stress that acts in a direction perpendicular and *toward* a surface.

compressional wave A seismic body wave consisting of alternate pulses of compression and expansion in the direction of travel; P wave or primary wave.

condensation The process by which water changes from vapor into a liquid.

conduction The process by which heat moves through a solid body without deforming it.

conglomerate A clastic sedimentary rock with large fragments in a finer-grained matrix.

continental crust The older, thicker, and less dense part of Earth's crust; the bulk of Earth's land masses.

continental drift The slow, lateral movement of continents across Earth's surface.

convection A form of heat transfer in which hot material circulates from hotter to colder regions, loses its heat, and then repeats the cycle.

convection loop A circuit of moving fluid, such as air or water, created by unequal heating of the fluid.

convectional precipitation Precipitation that is induced when warm, moist air is heated at the ground surface, rises, cools, and condenses to form water droplets, raindrops, and, eventually, rainfall.

convergent margin A boundary along which two plates move toward one another.

core Earth's innermost compositional layer, where the magnetic field is generated and much geothermal energy resides.

Coriolis effect The effect of Earth's rotation, which acts like a force to deflect a moving object on the surface to the right in the Northern Hemisphere and to the left in the Southern Hemisphere.

correlation A method of equating the ages of strata that come from two or more different places.

cosmic microwave background (CMB) radiation Radiation left over from the big bang, detected in the microwave portion of the spectrum with a temperature of only 2.7 K.

creep The imperceptibly slow downslope granular flow of regolith.

crust The outermost compositional layer of the solid Earth; part of the lithosphere.

cryosphere The perennially frozen part of the hydrosphere.

crystal structure An arrangement of atoms or molecules into a regular geometric lattice. Materials that possess a crystal structure are said to be crystalline.

crystallization The process whereby mineral grains form and grow in a cooling magma (or lava).

cyclone A center of low atmospheric pressure.

deflation Wind erosion in which loose particles of sand and dust are removed by the wind, leaving coarser particles behind.

delta Sediment deposit built by a stream entering a body of standing water.

density The mass of material per unit volume.

deposition (1) The laying down of sediment.

deposition (2) The process by which water changes from a vapor into a solid.

desert An arid land that receives less than 250 mm of rainfall or snow equivalent per year, and is sparsely vegetated unless it is irrigated.

desertification Invasion of desert conditions into nondesert areas.

discharge (1) The amount of water passing by a point on the channel's bank during a unit of time.

discharge (2) The process by which subsurface water leaves the saturated zone and becomes surface water.

dissolution The separation of a material into ions in solution by a solvent, such as water or acid.

divergent margin A boundary along which two plates move apart from one another.

divide A topographic high that separates adjacent drainage basins.

DNA Deoxyribonucleic acid; a double-chain biopolymer that contains all the genetic information needed for organisms to grow and reproduce.

domain The broadest taxonomic category of living organisms; biologists today recognize three domains: bacteria, *Archaea*, and eukaryotes.

drainage basin The total area from which water flows into a stream.

ductile deformation A permanent but gradual change in the shape or volume of a material, caused by flowing or bending.

dune A hill or ridge of sand deposited by the wind.

Earth science The scientific study of all aspects of Earth.

Earth system science The study of Earth as a closed system composed of interacting open systems and how the open systems may be changed as a result of human activities.

earthquake A sudden motion in Earth caused by the abrupt release of slowly accumulated energy.

El Niño An episodic cessation of the normal upwelling of cold water off the coast of Peru.

elastic deformation A temporary change in the shape or volume from which a material rebounds after the deforming stress is removed.

elastic rebound theory The theory that continuing stress along a fault results in a buildup of elastic energy in the rocks, which is abruptly released when an earthquake occurs.

element The most fundamental substance into which matter can be separated by chemical means.

epicenter The point on Earth's surface that is directly above an earthquake's focus.

erosion The wearing away of bedrock and transport of loosened particles by a fluid, such as water.

eukaryotes Organisms composed of eukaryotic cells—that is, cells that have a well-defined nucleus and organelles.

evaporation The process by which water changes from a liquid to a vapor.

evaporite A rock formed by evaporation of lake water or seawater, followed by lithification of the resulting salt deposit.

evolution The theory that life on Earth has developed gradually, from one or a few simple organisms to more complex organisms.

exoplanet Short for *extrasolar planet*, a planet outside our solar system.

fault A fracture in the lithosphere along which movement has occurred.

feedback mechanisms Reactions that enhance (positive) or retard (negative) change in an open system.

flood An event in which a water body overflows its banks.

floodplain The relatively flat valley floor adjacent to a stream channel, which is inundated when the stream overflows its banks.

flow Any mass-wasting process that involves a flowing motion of regolith containing water and/or air within its pores.

focus The location where rupture commences and an earthquake's energy is first released.

foliation A planar arrangement of textural features in a metamorphic rock, which give the rock a layered or finely banded appearance.

fossil Remains of an organism from a past age, embedded and preserved in Earth's crust.

fractional crystallization Separation of crystals from liquids during crystallization.

fractional melt A mixture of molten and solid rock.

fractionation Separation of melted materials from the remaining solid matter during the course of melting.

front The surface of contact between two dissimilar air masses.

geocentric A model of the universe in which a stationary Earth is at the center and all other celestial bodies revolve around it.

geologic column The succession of strata, fitted together in relative chronological order.

geostrophic wind Wind at high levels above Earth's surface blowing parallel with a system of straight, parallel isobars.

geothermal gradient The rate at which temperature increases with depth below Earth's surface.

glaciation The covering of large land areas by glaciers or ice sheets.

glacier A semipermanent or perennially frozen body of ice, consisting largely of recrystallized snow, that moves under the pull of gravity.

gneiss A coarse-grained, high-grade metamorphic rock.

gradient The steepness of a stream channel.

greenhouse effect The accumulation of heat in the lower atmosphere through the absorption of longwave radiation from Earth's surface.

groundwater Subsurface water contained in pore spaces in regolith and bedrock.

gymnosperm A naked-seed plant.

habit The distinctive shape of a particular mineral.

half-life The time needed for half the parent atoms of a radioactive substance to decay into daughter atoms.

hardness A mineral's resistance to scratching.

heliocentric A model of the universe in which a stationary Sun is at the center and the planets revolve around it.

Hertzsprung-Russell (H-R) diagram A plot of stars according to their luminosity and surface temperature.

Hubble's law The observed relationship that the velocity of recession of a galaxy is proportional to its distance.

humidity The amount of water vapor in the air.

humus Partially decayed organic matter in soil.

hydrologic cycle A model that describes the movement of water through the reservoirs of the Earth system; the *water cycle*.

hypothesis A plausible, but yet to be proved, explanation for how something happens.

ice sheet The largest type of glacier on Earth, a continent-sized mass of ice that covers all or nearly all the land within its margins.

igneous rocks Rocks that form by cooling and solidification of molten rock.

infiltration The process by which water works its way into the ground through small openings in the soil.

insolation The flow of incoming solar radiation intercepted by an exposed surface, assuming a uniformly spherical Earth with no atmosphere.

intertropical convergence zone (ITCZ) A zone of convergence of air masses along the equatorial trough.

isobars Lines on a map drawn through all points having the same atmospheric pressure.

isotopes Atoms with the same atomic number and different mass numbers.

jet streams High-speed air flows in narrow bands within the upper-air westerlies and along certain other global latitude zones at high levels.

joint A fracture in a rock, along which no appreciable movement has occurred.

kingdom The second-broadest taxonomic category. There are six recognized kingdoms, including animals and plants.

lapse rate The rate at which air temperature decreases with increasing altitude.

latent heat Heat that is absorbed and stored in a gas or liquid during the processes of evaporation or condensation, melting or freezing, sublimation or deposition.

lava Molten rock that reaches Earth's surface.

limestone A sedimentary rock that consists primarily of the mineral calcite.

lithification The group of processes by which sediment is transformed into sedimentary rock.

lithosphere Earth's outermost rocky layer, comprising the crust and the uppermost part of the mantle.

littoral drift Transport of sediment parallel to the shoreline by the combined action of beach drift and longshore current transport.

load The suspended and dissolved sediment carried by a stream.

longshore current A current within the surf zone that flows parallel to the coast.

low-latitude rainforest Dense evergreen forest of low-latitude equatorial and tropical zones with abundant rainfall.

luster The quality and intensity of light that reflects from a mineral.

magma Molten rock, with any suspended mineral grains and dissolved gases, that forms when melting occurs in the crust or mantle.

magmatic differentiation The formation of many different kinds of igneous rock from a single magma.

magnetic reversal The period of time in which Earth's magnetic polarity reverses itself.

mantle The middle compositional layer of Earth, between the crust and the core.

marble The product of metamorphism formed by the recrystallization of limestone.

mass extinction A catastrophic episode in which a large fraction of living species become extinct within a geologically short time.

mass wasting The downslope movement of regolith and/or bedrock masses due to the pull of gravity.

mechanical weathering The breakdown of rock into solid fragments by physical processes that do not change the rock's chemical composition.

metamorphic rocks Rocks that have been altered by exposure to high temperature, high pressure, or both.

metamorphism The mineralogical, textural, chemical, and structural changes that occur in rocks as a result of exposure to elevated temperatures and pressures.

meteorite A fragment of extraterrestrial material that falls to Earth.

Milankovitch cycles Climate cycles that occur over tens to hundreds of thousands of years because of changes in Earth's orbit and tilt.

mineral A naturally formed, solid, inorganic substance with a characteristic crystal structure and a specific chemical composition.

molecule The smallest chemical unit that has all the properties of a particular compound.

moment magnitude A measure of earthquake strength based on the rupture size, rock properties, and amount of displacement on the fault surface.

moraine A ridge or pile of debris that has been, or is being, transported by a glacier.

natural resources Useful materials obtained from the lithosphere, atmosphere, hydrosphere, or biosphere.

natural selection The process by which individuals that are well adapted to their environment have a survival advantage, and pass on their favorable characteristics to their offspring.

nebula hypothesis A model that explains the formation of the solar system from a large cloud of gas and dust floating in space 4.56 billion years ago.

nonrenewable resource A resource that cannot be replenished or regenerated on the scale of a human lifetime.

normal fault A fault in which the block above the fault surface moves down relative to the block below.

numerical age The age when a rock layer or natural feature was formed, in years before the present.

ocean basins Regions of Earth's crust that are covered by seawater.

ocean conveyor belt *See* **thermohaline circulation**.

ocean current A persistent, dominantly horizontal flow of water in the ocean.

oceanic crust The thinner, denser, and younger part of Earth's crust, underlying the ocean basins.

ore deposit A localized concentration in the crust from which one or more minerals can be profitably extracted.

orographic precipitation Precipitation that is induced when moist air is forced over a mountain barrier.

ozone A form of oxygen with a molecule consisting of three atoms of oxygen (O_3).

P wave The first, or primary, wave to be detected by a seismograph.

paleomagnetism The study of rock magnetism in order to determine the intensity and direction of Earth's magnetic field in the past.

paleontology The study of fossils and the record of ancient life on Earth; the use of fossils for the determination of relative ages.

paleoseismology The study of prehistoric earthquakes.

peat A biogenic sediment formed from the accumulation and compaction of plant remains.

percolation The process by which groundwater seeps downward and flows under the influence of gravity.

periglacial Conditions that are near glacial.

permafrost Ground that is perennially below the freezing point of water.

permeability A measure of how easily a solid allows fluids to pass through it.

photosynthesis A chemical reaction whereby plants use light energy to induce carbon dioxide to react with water, producing carbohydrates and oxygen.

plate tectonics The movement and interactions of large fragments of Earth's lithosphere, called *plates*.

pluton Any body of intrusive igneous rock, regardless of size or shape.

plutonic rock An igneous rock formed underground from magma.

porosity The percentage of the total volume of a body of rock or regolith that consists of open space.

precipitation The process by which water that has condensed in the atmosphere falls back to the surface as rain, snow, or hail.

pressure A particular kind of stress in which forces acting on a body are the same in all directions.

pressure gradient A change of atmospheric pressure, measured along a line at right angles to the isobars.

prokaryotes Single-celled organisms with no distinct nucleus—that is, no membrane separates their DNA from the rest of the cell.

proton–proton chain A chain of nuclear reactions in which hydrogen nuclei (or protons) fuse together to form helium nuclei, releasing energy.

pyroclastic flow A stream of hot volcanic fragments (tephra) that are buoyed by heat and volcanic gases and flow very rapidly.

quartzite The product of metamorphism formed by recrystallization of sandstone.

radioactivity A process in which an element spontaneously transforms itself into another isotope of the same element, or into a different element.

radiometric dating The use of naturally occurring radioactive isotopes to determine the numerical ages of minerals, rocks, and fossils.

recharge Replenishment of groundwater.

reef A hard structure on a shallow ocean floor, usually but not always built by coral.

reflection The bouncing back of a wave from an interface between two different materials.

refraction The bending of a wave as it passes from one material into another material, through which it travels at a different speed.

regolith A loose layer of broken rock and mineral fragments that covers most of Earth's surface.

relative age The age of a rock layer, fossil, or other natural feature relative to another feature.

renewable resource A resource that can be replenished or regenerated on the scale of a human lifetime.

reverse fault A fault in which the block on top of the fault moves up and over the block on the bottom.

Richter magnitude scale A scale of earthquake intensity based on the heights, or amplitudes, of the seismic waves recorded on a seismograph.

rock A naturally formed, coherent aggregate of minerals and possibly other nonmineral matter.

rock cycle The set of crustal processes that form new rock, modify it, transport it, and break it down.

Rossby waves Horizontal undulations in the flow path of the upper-air westerlies; also known as *upper-air waves*.

S wave The second kind of body wave to be detected by a seismograph.

salinity A measure of the salt content of a solution.

saltation Sediment transport in which particles move forward in a series of short jumps along arc-shaped paths.

sand A sediment made of relatively coarse mineral grains.

sandstone A medium-grained clastic sedimentary rock in which the clasts are typically, but not necessarily, dominated by quartz grains.

schist A high-grade metamorphic rock with pronounced schistosity, in which individual mineral grains are large enough to be visible.

scientific method The way a scientist approaches a problem; steps include observing, formulating a hypothesis, testing, and evaluating results.

seafloor spreading The processes by which the seafloor splits and moves apart along a midocean ridge and new oceanic crust forms along the ridge.

sedimentary rocks Rocks that form from sediment under conditions of low pressure and low temperature near the surface.

seismic discontinuity A boundary inside Earth where the velocities of seismic waves change abruptly.

seismic wave An elastic shock wave that travels outward in all directions from an earthquake's source.

seismogram The record made by a seismograph.

seismograph An instrument that detects, measures, and records vibrations of Earth's surface.

seismology The scientific study of earthquakes and seismic waves.

shale A very fine-grained fissile or laminated sedimentary rock, consisting primarily of clay-sized particles.

shear A stress that acts in a direction *parallel* to a surface.

shear wave A seismic body wave in which rock is subjected to side-to-side or up-and-down forces, perpendicular to the direction of travel; S wave or secondary wave.

shield volcano A broad volcano with gently sloping sides, built of successive flows of low-viscosity lava, generally of basaltic composition.

slate A very fine-grained, low-grade, metamorphic rock with slaty cleavage; the product of metamorphism of shale.

slope failure The falling, slumping, or sliding of relatively coherent masses of rock.

soil horizon One of a succession of zones or layers within a soil profile, each with distinct physical, chemical, and biologic characteristics.

soil profile The sequence of soil horizons from the surface down to the underlying bedrock.

soil The uppermost layer of regolith, which can support rooted plants.

species A population of genetically and/or morphologically similar individuals that can interbreed and produce fertile offspring.

spring Where the water table intersects the land surface, allowing groundwater to flow out.

strain A change in the shape or volume of a rock in response to stress.

stratigraphy The science of rock layers and the processes by which strata are formed.

stratosphere The layer of atmosphere above the troposphere; here temperature increases slowly with altitude.

stratovolcano A volcano composed of solidified lava flows interlayered with pyroclastic material, generally of andesitic or rhyolitic composition. Such volcanoes have steep sides that curve upward.

streak A thin layer of powdered mineral made by rubbing a specimen on an unglazed fragment of porcelain.

stream A body of water that flows downslope along a clearly defined natural pathway.

stress The force acting on a surface, per unit area, which may be greater in certain directions than others.

subduction zone A boundary along which one plate of lithosphere descends into the mantle beneath another plate.

supernova A stellar explosion marking the death of a massive star.

surf The "broken" turbulent water found between a line of breakers and the shore.

surface creep Sediment transport in which the wind causes particles to roll along the ground.

surface runoff Precipitation that drains over the land or in stream channels.

surface wave A seismic wave that travels along Earth's surface.

suspended load Sediment that is carried in suspension by a flowing stream of water or wind.

suspension Sediment transport in which the wind carries very fine particles over long distances and periods of time.

system A portion of the universe that can be separated from the rest of the universe for the purpose of observing changes that happen in it.

tectonic cycle Movements and interactions of the lithosphere by which rocks are cycled from the mantle to the crust and back; includes earthquakes, volcanism, and plate motion, driven by convection in the mantle.

tension A stress that acts in a direction perpendicular to and *away* from a surface.

theory A hypothesis that has been tested and is strongly supported by experimentation, observation, and scientific evidence.

thermohaline circulation The rising and sinking of water driven by contrasts in water density created by differences in temperature and salinity; this circulation is also known as the *ocean conveyor belt*.

thrust fault A reverse fault that cuts Earth's surface at a shallow angle.

thunderstorms Intense local storms associated with a tall, dense cumulonimbus cloud containing very strong updrafts.

tide A regular, daily cycle of rising and falling sea level that results from the gravitational attraction between the Moon, the Sun, and Earth.

till A heterogeneous mixture of crushed rock, sand, pebbles, cobbles, and boulders deposited by a glacier.

trace fossil Fossilized evidence of an organism's life processes, such as tracks, footprints, and burrows.

trade winds Surface winds that blow from about 30° north and south latitude toward the intertropical convergence zone.

transform fault An approximately vertical fracture in the lithosphere along which two plates slide past each other.

transpiration The process by which water taken up by plants passes directly into the atmosphere.

tropical cyclone An intense traveling cyclone occurring in tropical and subtropical latitudes, accompanied by high winds and heavy rainfall.

troposphere The lowest layer of the atmosphere, in which temperature falls steadily with increasing altitude.

tsunami A train of sea waves traveling over the ocean surface, triggered by an earthquake or other disturbance of the seafloor.

turbidite A graded layer of sediment that is deposited by a turbidity current.

turbidity current A gravity-driven current consisting of a dilute mixture of sediment and water with a density greater than that of the surrounding water.

unconformity A substantial gap in a stratigraphic sequence that marks the absence of part of the rock record.

uniformitarianism The concept that the processes governing the Earth system have operated in a similar manner throughout Earth's history and that past events can be explained by phenomena and forces observable today.

viscosity The degree to which a substance resists flow; a less viscous liquid is runny and flows rapidly, whereas a more viscous liquid is thick and flows slowly.

volcanic rock An igneous rock formed from lava.

volcano A vent through which magma, rock debris, volcanic ash, and gases erupt from Earth's crust to its surface.

water table The top surface of the saturated zone.

wave cyclone Traveling cyclone of the midlatitudes involving interaction of cold and warm air masses along sharply defined fronts.

wave-cut cliff A coastal cliff cut by wave action at the base of a rocky coast.

weathering The chemical and physical breakdown of rock exposed to air, moisture, and living organisms.

Younger Dryas A cold period that occurred between about 11,000 and 10,000 years ago, during the generally mild epoch.

PHOTO CREDITS

Chapter 1
Page 2: Yann Arthus-Bertrand/Photo Researchers, Inc.; (inset) Courtesy NASA; NG Maps; page 3: (top) John S. Shelton; (bottom) James L. Amos/NG Image Collection; page 5: Tom & Susan Bean, Inc./ PO; (left) S.C. Porter; (center) John S. Shelton; (right) Stephen Porter; page 7: (inset) age fototstock/SuperStock, Inc., Courtesy NASA; page 9: Todd Gipstein/NG Image Collection; page 10: Annie Griffiths Belt/NG Image Collection; page 16: NG Maps; (top left) Marvin Mattelson/NG Image Collection; (center) Dorling Kindersley/Getty Images; (bottom left) Publiphoto/Photo Researchers, Inc.; page 17: (bottom right) Publiphoto/Photo Researchers, Inc.; (top) Publiphoto/Photo Researchers, Inc.; page 18: Frans Lanting/NG Image Collection; page 20: James L. Amos/NG Image Collection; page 21: (left) bildagentur-online.com/th-foto//Alamy; (right) © Nick Brooks; (inset) © Nick Brooks; page 23: (top) Michael Nichols/NG Image Collection; (bottom) James P. Blair/NG Image Collection; page 24: (top) Alberto Garcia/Corbis; (top inset) NG Maps; (bottom) David Rydevik; (bottom inset) NG Maps; page 26: NG Maps; 28: Todd Gipstein/NG Image Collection; page 30: (top) David Rydevik; (bottom) Michael Nichols/NG Image Collection; page 31: Thomas J. Abercrombie/NG Image Collection; page 33: Frans Lanting/NG Image Collection.

Chapter 2
Page 34: Reuters/Landov; (inset) NG Maps; page 35: (top) E.R. Degginger/Photo Researchers, Inc.; (top center) The Natural History Museum; (bottom center) © Breck P. Kent; (bottom) James L. Amos/NG Image Collection; page 37: (top) E.R. Degginger/Photo Researchers, Inc.; (center) Mark A. Schneider/Photo Researchers, Inc.; (bottom) JAMES P. BLAIR/NG Image Collection; page 39: (top left) C.D. Winters/Photo Researchers, Inc.; (top right) Corbis Digital Stock; (top center left) E.R. Degginger/Photo Researchers, Inc.; (top center right) PhotoDisc/Getty Images; (bottom center right) With permission of the Royal Ontario Museum cROM; (bottom center left) David Doubilet/NG Image Collection; (bottom left) Mark A. Schneider/Photo Researchers, Inc.; (bottom right) Corbis Digital Stock; page 40: (top left) MICHAEL S. QUINTON/NG Image Collection; (bottom left) TODD GIPSTEIN/NG Image Collection; (top right) ED GEORGE/NG Image Collection; (bottom right) NG Image Collection; page 41: (top left) JAMES P. BLAIR/NG Image Collection; (bottom left) Digital Vision; (top right) C.D. Winters/Photo Researchers, Inc.; (bottom right) Alice Millikan/iStockphoto; page 42: (top) Courtesy Michael Hochella; (bottom) The Natural History Museum; page 43: (top) Brian J. Skinner; (center) Brian J. Skinner; (bottom) Brian J. Skinner; (right) William Sacco; page 44: (top) William Sacco; (bottom) Brian J. Skinner; page 45: (top) © Breck Kent//Animals Animals; (bottom) William Sacco; page 46: William Sacco; (right) The Natural History Museum; page 47: (center) Aldo Tutino/Art Resource; (bottom left) cAP/Wide World Photos; (bottom right) Xinhua/Landov LLC; page 48: William Sacco; page 51: (top) Brian J. Skinner; (second from top) © Boltin Picture Library; (third from top) © Breck P. Kent; (center) © Breck P. Kent; (third from bottom) © Breck P. Kent; (second from bottom) © Breck P. Kent; (bottom) © Breck P. Kent; page 53: James L. Amos/NG Image Collection; page 55: Kevin Telmer; page 56: (top left) Javier Trueba/MSF/Photo Researchers, Inc.; (top right) Javier Trueba/MSF/Photo Researchers, Inc.; (bottom) Javier Trueba/MSF/Photo Researchers, Inc.; page 57: (top) With permission of the Royal Ontario Museum cROM; (center) William Sacco; page 58: (top) Kevin Telmer; (bottom) © Breck P. Kent; page 59: NG Image Collection; page 60: The Natural History Museum.

Chapter 3
Page 62: Richard Nowitz/NG Image Collection; (inset) Gerry Ellis/Minden Pictures/NG Image Collection; NG Maps; page 63: NG Image Collection; (top center) © J.D. Griggs//USGS; (center) Nicole Duplaix//USGS Image Collection; (bottom center) © Breck P. Kent; (bottom) © Ken Lucas/Visuals Unlimited; page 64: (top) NG Image Collection; (center) Brian Skinner; (bottom) NG Image Collection; page 65: (bottom left) Brian Skinner; (top) NG Image Collection; (center) Brian Skinner; page 67: (left) Kenneth Garrett/NG Image Collection; (center) Brian J. Skinner; (right) © Tony Waltham; page 68: (right) William Sacco; (left) Brian J. Skinner; page 69: Marc Moritsch/NG Image Collection; page 70: (top) Brian J. Skinner; (center) Brian J. Skinner; (bottom) Brian J. Skinner; page 71: (top left) Carsten Peter/NG Image Collection; (center) © J.D. Griggs//USGS; (top right) Schofield/Getty Images, Inc; page 73: (left) Gerals & Buff Corsi/Visuals Unlimited; (center left) Scientifica/Visuals Unlimited; (right) Ken Lucas/Visuals Unlimited; (center right) Emory Kristof/NG Image Collection; page 75: (left) Jim Richardson/NG Image Collection; (right) Courtesy Dr. Kenneth L. Finger and Chevron Corporation; page 76: Brian J. Skinner; page 77: Nicole Duplaix/NG Image Collection; page 78: (top left) S.C. Porter; (top right) © Stephen Trimble//DRK Photo; (bottom left) Medford Taylor/NG Image Collection; (bottom right) S.C. Porter; page 79: (top left) Minden Pictures, Inc.; (top right) © Tom Bean; (bottom) Courtesy The Field Museum. Photo by Mark Widhalm; page 81: (top) SIPA/NewsCom; (bottom) Lealisa Westerhoff/AFP/Getty Images; page 82: James L. Amos/NG Image Collection; page 83: Courtesy Jay Ague, Dept. of Geology and Geophysics at Yale University; page 85: (top left) William Sacco; (top center) William Sacco; (bottom center) William Sacco; (bottom) William Sacco; page 86: (right) © William Sacco; (left) © William Sacco; page 87: (top left) Brian J. Skinner; (top center) Brian J. Skinner; (top right) Brian J. Skinner; (bottom left) Brian J. Skinner; (bottom right) Brian J. Skinner; page 88: (left) © Breck P. Kent; (left inset) Runk/ Schoenberger/Grant Heilman Photography; (right) © A.J. Copley/Visuals Unlimited; (right inset) © Craig Johnson; page 89: J.A. Wilkinson//Valan Photos; page 90: (left) © Ken Lucas/Visuals Unlimited; page 91: (top left) Stacy Gold/NG Image Collection; (bottom left) George F. Mobley/NG Image Collection; (top right) Stephen Sharnoff/NG Image Collection; page 92: (top) NG Image Collection; (bottom) Marc Moritsch/NG Image Collection; page 93: © Stephen Trimble//DRK Photo; page 94: (top) © William Sacco; (bottom) J.A. Wilkinson//Valan Photos; page 95: Chris Rainier/NG Image Collection; page 96: (left) Brian J. Skinner; (right) cTony Waltham; (top right) © Lee Boltin/Boltin Picture Library; (center right) © Fred Hirschmann; (bottom right) Richard J. Stewart; page 97: (top) © Breck P. Kent; (top inset) © Runk/ Schoenberger/Grant Heilman Photography; (bottom) © A.J. Copley/Visuals Unlimited; (bottom inset) © Craig Johnson.

Chapter 4
Page 98: Sandy Felsenthal/NG Image Collection; (bottom inset) Associated Press; (top inset) Coinery/Alamy; page 99: (top) Maria Stenzel/NG Image Collection; (top center) Photo Researchers, Inc.; (bottom center) S.C. Porter; (bottom) © William E. Ferguson; page 100: © William E. Ferguson; page 101: Peter Essick/NG Image Collection; page 103: (top right) George F. Mobley/NG Image Collection; (bottom right) Maria Stenzel/NG Image Collection; (top left) Phil Schermeister/NG Image Collection; (bottom right) © Kenneth W. Fink//Ardea London; page 104: (left) Courtesy Brian J. Skinner; (right) Courtesy Brian J. Skinner; page 105: (top left) Brian J. Skinner; (bottom) Thomas J. Abercrombie/NG Image Collection; page 106: (top left) © Gordon Wiltsie/AlpenImages Ltd.; (top right) Stephanie Maze/NG Image Collection; (bottom) Medford Taylor/NG Image Collection; page 107: (top) Courtesy Amy Larson/Smith College; (center) Raymond Gehman/NG Image Collection; (bottom) © Frans Lanting//Minden Pictures, Inc.; page 110: (left) Photo Researchers, Inc.; (right) Photo Researchers, Inc.; page 112: (right) USDA/ Soil Conservation Service; (left) USDA/ Soil Conservation Service; page 116: (top) Raymond Gehman/NG Image Collection; (bottom) Jim Richardson/NG Image Collection; NG Maps; page 118: Michael Nichols/NG Image Collection; page 119: USGS; page 120: (top left) Brian J. Skinner; (top right) S.C. Porter; (bottom) S.C. Porter; page 121: (left) Bill Hatcher/NG Image Collection; (top) Josef Muench; (bottom) Dr. Marli Miller/Visuals Unlimited; page 123: (top left) George Plafker, U.S Geological Survey; (top right) W.E. Garrett R./NG Image Collection; page 124: (bottom left) © William E. Ferguson; (bottom right) Brian J. Skinner; page 125: (top) Paul Johnson/Index Stock; (center left) Walker Howell/NG

Image Collection; (center right) Stephen Sharnoff/NG Image Collection; page 126: (bottom) © Frans Lanting//Minden Pictures, Inc.; page 127: (top) Raymond Gehman/NG Image Collection; (bottom) Michael Nichols/NG Image Collection; page 128: Brian J. Skinner; Associated Press.

Chapter 5

Page 132: James L. Stanfield/NG Image Collection; (inset) Courtesy Brian Skinner; page 133: (top) Emory Kristof/NG Image Collection; (top center) cMarli Bryant Miller; (bottom center) Tyrone Turner/NG Image Collection; (bottom) Walter Meayers Edwards/NG Image Collection; page 136: (top left) Bill Curtsinger/NG Image Collection; (bottom left) Emory Kristof/NG Image Collection; (top right) © Miroslav Krob/AgeFotostock; (bottom right) Bates Littlehales/NG Image Collection; page 137: Annie Griffiths Belt/NG Image Collection; page 139: (top left) George F. Mobley/NG Image Collection; (right) James P. Blair/NG Image Collection; (bottom left) Nicolas Reynard/NG Image Collection; page 141: (bottom left) cMarli Bryant Miller; (right) Visible Earth/NASA; page 142: New York Times Graphics; page 143: (bottom) NASA; page 144: Wes C. Skiles/NG Image Collection; page 145: (left) © Earth Satellite Corporation; (right) © Earth Satellite Corporation; page 147: (inset) Tyrone Turner/NG Image Collection; 147: Tyrone Turner/NG Image Collection; page 150: (left) Peter Essick/NG Image Collection; (bottom left) NGS Maps; page 151: (main) NGS Maps; (bottom) NGS Maps; Ed Kashi/cCorbis; page 152: NG Maps; (top) Phil Schermeister/NG Image Collection; (bottom left) © 2005, Greg Reis; (bottom right) Craig Aurness/CORBIS; NG Maps; page 157: (bottom left) Walter Meayers Edwards/NG Image Collection; page 159: (left) Peter Morgan/Reuters/Corbis; (right) Associated Press; page 161: (left) Wes C. Skiles/NG Image Collection; (right) © Bruno Barbey/Magnum Photos, Inc.; page 162: (top left) Michael Nichols/NG Image Collection; (top right) Michael Nichols/NG Image Collection; (bottom left) Michael Nichols/NG Image Collection; (bottom right) Michael Nichols/NG Image Collection; NG Maps; page 163: Nicolas Reynard/NG Image Collection; page 165: Jim Tuten/Black Star.

Chapter 6

Page 168: Carsten Peter/NG Image Collection; inset John Burcham/NG Image Collection; NG Maps; page 171: GERRY ELLIS/MINDEN PICTURES/NG Image Collection; GORDON WILTSIE/NG Image Collection; page 172: (top left) Carsten Peter/NG Image Collection; (top right) Dr. Cynthia M. Beall & Dr. Melvyn C. Goldstein/NG Image Collection; (bottom left) Marc Moritsch/NG Image Collection; (bottom right) Annie Griffiths Belt/NG Image Collection; page 173: cAP/Wide World Photos; cAP/Wide World Photos; page 174: (top right) S.C. Porter; (bottom) Brian J. Skinner; page 175: (left) © Jim Richardson//Woodfin Camp & Associates; (right) S.C. Porter; page 176: (bottom) © Tom Bean//DRK Photo; page 177: (top left) Gerry Ellis/Minden Pictures; (top right) George Steinmetz/NG Image Collection; (bottom left) George Steinmetz/NG Image Collection; (bottom center) © John S. Shelton; (bottom right) Marc Moritsch/NG Image Collection; page 178: (top left) NASA/JPL/University of Arizona; (top right) NASA/JPL/University of Arizona; 179: Courtesy Stephen C. Porter; page 180: Steve McCurry/NG Image Collection; page 181: NOAA George E. Marsh Album; page 182: Courtesy NASA; page 184: (top) Melissa Farlow/NG Image Collection; (bottom) Frans Lanting/NG Image Collection; page 185: (center) John Lythgoe//Planet Earth Pictures; (bottom right) Peter Essick/NG Image Collection; (top right) Marli Miller/Visuals Unlimited; page 186: Courtesy NASA; page 188: Courtesy Stephen C. Porter; page 190: (center right) Chris Johns/NG Image Collection; (bottom left) Chris Johns/NG Image Collection; page 192: (top) Mark Burnett/Photo Researchers, Inc.; (center) William Thompson/NG Image Collection; (bottom) James P. Blair/NG Image Collection; page 193: (top left) Raymond Gehman/NG Image Collection; (top right) S.C. Porter; (bottom left) Sam Abell/NG Image Collection; (bottom right) CARY WOLINSKY/NG Image Collection; page 194: Maria Stenzel/NG Image Collection; page 195: (top left) Barry Tessman/NG Image Collection; (right) © Marli Bryant Miller; (bottom left) Gordon Wiltsie/NG Image Collection; NG Maps; page 197: Steve McCutcheon; page 199: (bottom right) CARY WOLINSKY/NG Image Collection.

Chapter 7

Page 200: Galen Rowell/Mountain Light Photography, Inc.; NG Maps; (left) US Navy; (right) US Navy; (top) Emory Kristof/NG Image Collection; page 205: Courtesy Willem van der Westhuizen; page 206: National Library of Wales; page 212: NGS Maps; page 215: (top) Emory Kristof/NG Image Collection; (center left) James L. Stanfield/NG Image Collection; (bottom left) Courtesy NASA; (right) NG Image Collection; page 216: NGS Maps; (bottom) NASA; page 224: DIANE COOK AND LEN JENSHEL/NG Image Collection; (top) Chris Johns/NG Image Collection; (bottom) Hawaii Center for Volcanology; page 227: Emory Kristof/NG Image Collection; page 229: Georg Gerster/Photo Researchers.

Chapter 8

Page 232: Bay Ismoyo/Getty; NG Maps; (top left) DigitalGlobe; (bottom left) DigitalGlobe; page 233: (top) © Lysaght/Liaison Agency, Inc./Getty Images; (center bottom) © Dane Penland/Courtesy Smithsonian Institution; page 235: (left) © John S. Shelton; (right) Winfield Parks/NG Image Collection; page 236: Peter Essick/NG Image Collection; page 239: (top left) © George Plafker, U.S. Geological Survey; (top right) Reza/NG Image Collection; (center right) © Lysaght/Liaison Agency, Inc./Getty Images; (bottom left) Courtesy DigitalGlobe; (bottom right) Courtesy DigitalGlobe; page 241: GRONDIN EMMANUEL/Maxppp/Landov LLC; page 243: cAP/Wide World Photos; page 244: (main) Eric Hanson/NG Image Collection; page 245: (top right) Peter Essick/NG Image Collection; (center right) Courtesy NOAA/NGDC; (bottom right) © Oshihara//Sipa Press; page 246: Jim Mendenhall/NG Image Collection; page 251: (top) Associated Press; (center top) Karen Kasmauski/NG Image Collection; (center) Karen Kasmauski/NG Image Collection; (bottom) Waldemar Lindgren/NG Image Collection; page 254: Sarah Leen/NG Image Collection; page 257: (top left) Courtesy Dan Schulze; (bottom left) © Dane Penland/Courtesy Smithsonian Institution; page 258: (right) Bryan & Cherry Alexander/Photo Researchers, Inc.; page 264: Emory Kristof/NG Image Collection; NG Maps; page 267: © Kevin Schafer//Tom Stack & Associates.

Chapter 9

Page 270: Photri/The Image Works; (bottom inset) Chris Newhall//USGS; (top inset) Chris Newhall//USGS; NG Maps; page 271: (top) Carsten Peter/NG Image Collection; (top center) Krafft Explorer/Photo Researchers, Inc.; (bottom center) Brian J. Skinner; (bottom) Raymond Gehman/NG Image Collection; page 272: (left) Paul Chesley/NG Image Collection; (center) Harry Glicken/USGS; (right) Harry Glicken//USGS; page 274: (top left) Frans Lanting/NG Image Collection; (center left) © John S. Shelton; (bottom left) Carsten Peter/NG Image Collection; (top right) © Reuters/Corbis-Bettmann; (bottom right) © Photodisc/SUPERSTOCK; page 275: (bottom) © Steve Vidler/eStock Photo; page 276: (top left) Courtesy Brian Skinner; (left) Photodisc/SUPERSTOCK; (right) Courtesy Brian Skinner; page 277: (left) S.C. Porter; (right) Courtesy Brian Skinner; Paul Chesley/NG Image Collection; page 278: (bottom) © G. Brad Lewis/Liaison Agency, Inc./Getty Images; (top) Roger Rossmeyer/NG Image Collection; page 279: Peter Turnley/Black Star; NG Maps; page 281: Grant Dixon/Lonely Planet Images/Getty; page 282: John Stanmeyer/NG Image Collection; (inset) NG Maps; page 284: NRSC Ltd./Science Photo Library/Photo Researchers, Inc.; page (left) © Science VU-ASIS/Visuals Unlimited; (right) Krafft Explorer/Photo Researchers, Inc.; page 289: (top left) © J.D. Griggs//USGS; (top right) © J.D. Griggs//USGS; page 291: (left) Brian J. Skinner; (right) Brian J. Skinner; page 295: (top) Gary Ladd Photography; (center) Tom Bean/DRK Photo; (bottom) Raymond Gehman/NG Image Collection; page 296: (bottom right) Lyn Topinka/Courtesy USGS; (bottom right) Robert Madden/NG Image Collection; (center right) Jim Richardson/NG Image Collection; (bottom left) P. Frenzen/USDA Forest Service, 1991; NG Maps; (top left) Photodisc/SUPERSTOCK; page 298: (top left) Brian J. Skinner;

(top right) Brian J. Skinner; (bottom) Gary Ladd Photography; page 300: (bottom) © Steve Vidler/eStock Photo; (top) Frans Lanting/NG Image Collection.

Chapter 10

Page 302: © Walter Imber; (inset) Jonathan Blair/NG Image Collection; NG Maps; page 303: (top) NG Image Collection; (center) ROBERT GIUSTI/NG Image Collection; (bottom) JAMES L. AMOS/NG Image Collection; page 305: (top) Annie Griffiths Belt/NG Image Collection; (center) Sam Abell/NG Image Collection; (bottom) CARY WOLINSKY/NG Image Collection; page 306: (top) Walter Meayers Edwards/NG Image Collection; (bottom) © Jeff Gnass; NG Maps; page 307: (left) Landform Slides; (right) Courtesy Lee Gerhard; page 309: Dr. K.Roy Gill; page 310: (top left) NG Image Collection; (top right) NG Image Collection; (bottom) NG Image Collection; page 311: (left) NG Image Collection; (right) Mark Gibson/Index Stock; page 313: Jonathan Blair/NG Image Collection; page 315: (top) MARK HALLETT/NG Image Collection; (center) ROBERT GIUSTI/NG Image Collection; (bottom) KAZUHIKO SANO/NG Image Collection; page 316: (top) KAM MAK/NG Image Collection; (center) KAM MAK/NG Image Collection; (bottom) KAM MAK/NG Image Collection; page 318: (center) American Institute of Physics; (center left) Mary Evans Picture Library/Photo Researchers, Inc.; (bottom right) Mary Evans/Photo Researchers, Inc.; (top) Astrid & Hanns-Frieder Michler/Photo Researchers, Inc.; (bottom) Patrick McFeeley/NG Image Collection; (far right) James P. Blair/NG Image Collection; page 324: (left) Enrico Ferorelli; (right) KENNETH GARRETT/NG Image Collection; page 326: © Sam Bowring; page 327: JAMES L. AMOS/NG Image Collection; page 328: (top) Ralph Lee Hopkins/NG Image Collection; (bottom) Brian J. Skinner; NG Maps; page 329: (top) CARY WOLINSKY/NG Image Collection; (bottom) KAM MAK/NG Image Collection; page 330: JAMES L. AMOS/NG Image Collection; page 331: NG Image Collection; page 332: (center) ROBERT GIUSTI/NG Image Collection.

Chapter 11

Page 334: Photo by psihoyos.com; page 335: (top) O. Louis Mazzatenta/NG Image Collection; (center bottom) Photo by psihoyos.com; (bottom) Chris Johns/NG Image Collection; page 337: (left) © Royalty-Free/CORBIS; page 341: (top left) O. Louis Mazzatenta/NG Image Collection; (center right) Kirk Moldoff/NG Image Collection; (bottom left) © WHOI, D. Foster/Visuals Unlimited; (bottom center) Photo courtesy of Colleen M. Cavanaugh, Department of Organismic and Evolutionary Biology, Harvard University; (bottom right) Woods Hole Oceanographic Institution; page 342: (left) O. Louis Mazzatenta/NG Image Collection; (right) Francois Gohier/Photo Researchers, Inc.; page 343: (left) © Dr. Jeremy Burgess/ SPL/Photo Researchers, Inc.; (right) © CNRI/ SPL/Photo Researchers, Inc.; page 344: (left) O. Louis Mazzatenta/NG Image Collection; (right) O. Louis Mazzatenta/NG Image Collection; page 345: (left) Robert Clark/NG Image Collection; (right) Mary Evans Picture Library/Photo Researchers, Inc.; page 346: (top) NG Maps; (center left) Ralph Lee Hopkins/NG Image Collection; (top right) © J. Dunning/VIREO; (top left) © Gerald & Buff Corsi/Visuals Unlimited; (center right) © Fritz Polking/Visuals Unlimited; (center right) © Joe & Mary Ann McDonald/Visuals Unlimited; (bottom left) Tierbild Okapia/Photo Researchers, Inc.; (bottom right) Eric Hosking/Photo Researchers, Inc.; page 348: (top left) Paul Zahl/NG Image Collection; (top right) Stephen St. John/NG Image Collection; (bottom left) Photo by psihoyos.com; (bottom right) Louie Psihoyos/NG Image Collection; page 349: (left) Courtesy Brian J. Skinner; (right) Robert Clark/NG Image Collection; page 350: Raymond Gehman/NG Image Collection; page 351: (top left) © Edward R. Degginger//Bruce Coleman, Inc.; (top right) Chris Johns/NG Image Collection; (bottom right) © Breck P. Kent/Animals Animals/Earth Scenes; (bottom right) © Theodore Clutter/Photo Researchers, Inc.; page 352: Courtesy Peabody Museum, Yale University; page 353: (left) Hans Fricke/NG Image Collection; (right) Tom McHugh/Photo Researchers, Inc.; page 354: (left) © Francois Gohier/Photo Researchers, Inc.; (right) Carnegie Museum of Natural History; page 356: (left) © Michael Rothman; (center) Lealisa Westerhoff/Getty Images; (right) © John Reader/Science Photo Library/Photo Researchers, Inc.; page 358: (top) © Chris Butler; page 359: (top) Jonathan Blair/NG Image Collection; (bottom) Courtesy NASA; page 360: Courtesy Dr. Richard Ernst; page 361: (top) Michael Melford/NG Image Collection; page 363: NGS Maps; page 364: O. Louis Mazzatenta/NG Image Collection; page 365: Paul Zahl/NG Image Collection; page 366: Lealisa Westerhoff/Getty Images; (right) © CNRI/SPL/Photo Researchers, Inc.; page 367: (top) Jonathan Blair/NG Image Collection; (bottom) Jonathan Blair/Corbis Images.

Chapter 12

Page 370: Pacific Stock/SuperStock; page 372: © Don Dixon; page 375: (top left) Copyright © 1995, David T. Sandwell. Used by permission.; (top right) Copyright © 1995, David T. Sandwell. Used by permission.; page 376: Ken MacDonald/Photo Researchers, Inc.; page 377: (top) Courtesy NASA; (center) Courtesy NASA; (bottom) Courtesy NASA; page 379: Stephen Porter; page 380: (left) Paul Nicklen/NG Image Collection; (right) Tim Laman/NG Image Collection; (bottom) Ingo Arndt/Minden Pictures; page 386: (top left) NG Maps; (center) NG Maps; (bottom) NG Maps; page 387: (top) NG Maps; (bottom right) NG Maps; (bottom left) NG Maps; page 389: (top) Courtesy NOAA; (center) Paul Nicklen/NG Image Collection; (bottom left) David Doubilet/NG Image Collection; (bottom right) Emory Kristof/NG Image Collection; NG Maps; page 390: (top) Copyright © 1995, David T. Sandwell. Used by permission.; page 391: Courtesy Otis B. Brown, Robert Evans, and M. Carle, University of Miami, Rosenstiel School of Marine and Atmospheric Science, Florida, and NOAA/Satellite Data Services Division.

Chapter 13

Page 394: Corbis Images; (inset) Kristoffer Tripplaar/NewsCom; page 395: (top center) Raymond Gehman/NG Image Collection; (bottom center) Tim Laman/NG Image Collection; (bottom) Vie De Lucia/NYT Pictures; page 396: Manley, W.F., 2002, Postglacial Flooding of the Bering Land Bridge: A Geospatial Animation: INSTAAR, University of Colorado, v1, http://instaar.colorado.edu/QGISL/bering_land_bridge; page 397: Manley, W.F., 2002, Postglacial Flooding of the Bering Land Bridge: A Geospatial Animation: INSTAAR, University of Colorado, v1, http://instaar.colorado.edu/QGISL/bering_land_bridge; page 398: (left) © Stephen Rose/Liaison Agency, Inc./Getty Images; (right) Stephen Crowley/New York Times Pictures; page 399: (left) Richard Nowitz/NG Image Collection; This is a MODIS 8-day composite image acquired from 23 October 2008 to 30 October 2008. The image was downloaded from the USGS GLOVIS web-site. This false color image was prepared by Larry Bonneau at the Center for Earth Observation at Yale University.; page 400: (top) Arnulf Husmo/Stone/Getty Images; (bottom) Dr. Susanne Lehner/DLR; page 401: (right) Patrick McFeeley/NG Image Collection; page 402: (top) Raymond Gehman/NG Image Collection; (bottom) © Nicholas DeVore III; page 404: Raymond Gehman/NG Image Collection; page 405: © G. R. Roberts; page 407: (top) David Alan Harvey/NG Image Collection; (center) Cotton Coulson/NG Image Collection; (bottom) Raymond Gehman/NG Image Collection; page 409: Tim Laman/NG Image Collection; page 411: (top) Digital Globe/HO/AFP/Getty Images; (bottom left) Courtesy DigitalGlobe; (bottom right) Courtesy DigitalGlobe; page 412: (left) Richard Reid/NG Image Collection; page 413: (top) Lander, James F. and Patricia A. Lockridge/U.S. Department of Commerce; (bottom) Damian Gadal/iStockphoto; page 414: James P. Blair/NG Image Collection; page 415: (top) James L. Stanfield/NG Image Collection; (bottom) Vie De Lucia/NYT Pictures; page 416: (left) Danita Delimont/Alamy; (right) Danita Delimont/Alamy; page 417: Danita Delimont/Alamy; page 418: (left) Thomas K. Gibson/Florida Keys National Marine Sanctuary; (top right) Phillip Dustan, College of Charleston; (bottom right) © Steven Frink; NG Maps; page 419: (top) Richard Nowitz/NG Image Collection; (bottom left) © Nicholas DeVore III; (bottom right) Raymond Gehman/NG Image Collection; page 420: Digital Globe/HO/AFP/Getty

Images; page 421: Bob Jordan/cAP/Wide World Photos; page 423: Stephen Crowley/New York Times Pictures.

Chapter 14
Page 424: Courtesy NASA; page 425: (top) Todd Gipstein/NG Image Collection; (center) John Dunn/Arctic Light/NG Image Collection; (bottom center) John Eastcott and Yva Momatiuk/NG Image Collection; (bottom) Dick Blume/Syracuse Newspapers//cAP/Wide World Photos; page 426: Todd Gipstein/NG Image Collection; page 427: Peter Hendrie/The Image Bank/Getty Images; page 430: (left) Courtesy NASA; (center) Courtesy NASA; (right) NASA Media Services; page 435: (left) John Dunn/Arctic Light/NG Image Collection; (right) Jeremy Woodhoue/Masterfile; page 440: Adalberto Rias Szalay/Sexto Sol/Photodisc/Getty Images; page 441: (top left) John Eastcott and Yva Momatiuk/NG Image Collection; (top right) Carsten Peter/NG Image Collection; (bottom left) John Eastcott and Yva Momatiuk/NG Image Collection; (bottom right) Todd Gipstein/NG Image Collection; page 442: (left) James P. Blair/NG Image Collection; page 443: Dick Blume/Syracuse Newspapers//cAP/Wide World Photos; page 447: Gandee Vasan/Getty Images; page 448: Warren Faidley/Weatherstock; page 450: (top) Joel Knain/NASA images; (bottom left) NASA image by Jeff Schmaltz, MODIS Rapid Response Team, Goddard Space Flight Center.; (bottom right) Courtesy Terri and Mike Lawson of California, USA/NOAA; NG Maps; page 452: (bottom left) Dick Blume/Syracuse Newspapers//cAP/Wide World Photos; (top right) John Eastcott and Yva Momatiuk/NG Image Collection; page 453: University of Wisconsin Space Science and Engineering Center; page 455: Todd Gipstein/NG Image Collection.

Chapter 15
Page 456: Jaques Descloitres/MODIS Land Rapid Response Team/NASA/Visible earth; NG Maps; page 457: (top) Nick Caloyianis/NG Image Collection; page 458: (bottom) Nick Caloyianis/NG Image Collection; page 461: © Gene Blevins/Los Angeles Daily News/cCorbis; page 470: Courtesy NASA; page 476: Carsten Peter/NG Image Collection; page 478: (bottom left) Joel Sartore/NG Image Collection; page 480: (top) NASA Earth Observatory; page 482: (left) Ping Amranand/SuperStock, Inc.; (right) AP/Wide World Photos; page 485: Steve Pace/Envision.

Chapter 16
Page 488: Science Faction/Getty Images; PHOTOPQR/LA DEPECHE DU MIDI/JEAN LOUIS PRADELS/NewsCom; page 490: NGS Maps; page 491: NGS Maps; page 492: (top right) Courtesy National Academies Press; (center right) Taylor S. Kennedy/NG Image Collection; (bottom right) Courtesy Rob Dunbar, Stanford University; (left) Peter Essick/NG Image Collection; page 493: NASA Images; page 494: Jeffrey Park; page 495: age fotostock/SuperStock, Inc.; page 498: (bottom left) John Dunn/NG Image Collection; (bottom center) Jim Richardson/NG Image Collection; (bottom right) Tim Laman/NG Image Collection; page 500: (top) William Albert Allard/NG Image Collection; (bottom) © Will & Deni McIntyre/Photo Researchers, Inc.; NG Maps; page 501: (top left) Kari Niemelainen/Alamy Images; page 502: (top) Peter Essick/NG Image Collection; (bottom) Sisse Brimberg/NG Image Collection; page 503: Gerd Ludwig/NG Image Collection; page 504: (top) Hinrich Baesemann/Landov LLC; (bottom) Maria Stenzel/NG Image Collection; page 505: (left) NASA graph by Robert Simmon, based on data provided by the NOAA; (right) NASA graph by Robert Simmon, based on data provided by the NOAA; page 506: (bottom) Ralph Lee Hopkins/NG Image Collection; (top right) Peter Essick/NG Image Collection; (bottom right) cAP/Wide World Photos; page 508: (top right) Peter Essick/NG Image Collection; (top left) Courtesy NASA; (bottom) Stockbyte/SuperStock, Inc.; page 509: (bottom right) cAP/Wide World Photos; (top right) Bertrand Gardel/Hemis/cCorbis; (bottom center) PhotoAlto/SuperStock, Inc.; (top center) James L. Stanfield/NG Image Collection; page 510: (top) Steven Kazlowski/Getty Images; (bottom) Steven Kazlowski/Peter Arnold, Inc.; NG Maps; page 511: (top) Peter Essick/NG Image Collection; (bottom) Jim Richardson/NG Image Collection; page 512: cAP/Wide World Photos; page 513: (bottom) S.C. Porter; page 514: (left) Nasa Images; (right) Jeffrey Park; (bottom) Peter Essick/NG Image Collection.

Chapter 17
Page 516: NASA Images; page 517: (top) © Mary Evans Picture Library; (top center) NASA Images; (bottom) NASA/WMAP Science Team; page 518: © Mary Evans Picture Library; page 520: (top) Bruce Dale/NG Image Collection; page 522: TBGS Observatory; page 523: NASA Images; page 525: NASA Media Services; page 526: C.R. O Dell/Rice University/NASA Media Services; page 528: (top left) Stock Image/SuperStock; (center) NASA/Visible earth; (bottom) NASA/Visible earth; page 529: (top) CONTACT gregm@sierra-remote.com; (center) NASA/Visible earth; (bottom) NASA/Visible earth; page 530: NASA/Visible earth; page 531: NASA/Visible earth; page 532: (left) NASA images; (right) NASA images; 533: (top) NASA images; (bottom) NASA images; page 534: (top left) NASA/JPL; (top right) NASA images; page 535: (left) Breck Kent; (right) Courtesy of the Canada Centre for Remote Sensing Department of Natural Resources Canada; NG Maps; page 537: Tohoku/Getty Images; page 541: (top) Richard Powell; (bottom) ESA AND NASA/NG Image Collection; page 544: (top) cAP/Wide World Photos; (center) NASA/WMAP Science Team; (bottom) NASA / WMAP Science Team; page 545: (left) NASA Images; (center) NASA/JPL-Caltech; (right) NOAO/AURA/Photo Researchers, Inc.; page 546: (top right) NASA Images; (left) NASA/JPL-Caltech/University of Arizon; (right) NASA Images; page 549: NASA Images; page 550: (left) TBGS Observatory; (right) NASA Media Services.

TEXT, TABLE, AND LINE ART CREDITS

Chapter 1
Figure 1.7: From Flint, Robert F. and Brian J. Skinner, *Physical Geology*, 2nd ed. Copyright 1977 John Wiley & Sons, Inc. Reprinted with permission of John Wiley & Sons, Inc.

Chapter 2
What an Earth Scientist Sees, top: Adapted from Mineral Information Institute, "How Many Minerals and Metals Does It Take to Make a Light Bulb?" Retrieved July 5, 2006, from *www.mii.org/lightbulb.html*. Reprinted by permission of The Mineral Information Institute, *www.mii.org*.

Chapter 5
Figure 5.7: From The New York Times Graphics, "Coastal Defenses Are Disappearing," August 30, 2005. Used by permission of The New York Times Agency. NOTE: Any future revisions, editions thereof in print and any other format require clearance by The New York Times Agency.

Chapter 6
Figure 6.6: Adapted from Pye, K. and L. Tsoar, *Aeolian Sand and Sand Dunes*, Figure 7.1, Chapman & Hall, 1990. Reprinted with kind permission of Springer Science and Business Media; Figure 6.7: Adapted from Hack, John T., *The Geographical Review*, Vol. 31, Figure 19, page 260, by permission of the American Geographical Society.

Chapter 7
Amazing Places: Adapted from Rubin, Ken, "Loihi Volcano—The Bathymetric Map of Loihi Seamount." Hawaii Center for Volcanology web site, 1997. Retrieved October 6, 2006, from *www.soest.hawaii.edu/GG/HCV/loihi.html*. Reprinted by permission of the Hawaii Center for Volcanology, Ken Rubin.

Chapter 8
Case Study: Satake, Kenji, "2004 Sumatra Earthquake," Figure S7, December 2004. Retrieved October 3, 2006, from *http://staff.aist.go.jp/kenji.satake/animation.gif*. Reprinted by permission of K. Satake, Geological Survey of Japan, AIST.

Chapter 9
Figure 9.8: Adapted from Sigurdsson, Haraldur, "Volcanic Pollution and Climate: The 1783 Laki Eruption," *Eos*, Vol. 63, pages 601–602, 1982. Copyright 1982, American Geophysical Union. Modified by permission of The American Geophysical Union.

Chapter 12
Figure 12.1: We have made every effort to secure proper permission for reproducing this image. However, it has not been possible to determine the current copyright owner. If you have information indicating who the copyright owner may be, please contact us at rflahive@wiley.com; Figure 12.11: From Anderson, Bruce and Alan Strahler, *Visualizing Weather and Climate*. Copyright 2009 John Wiley & Sons, Inc. Reprinted with permission of John Wiley & Sons, Inc.; Figure 12.12: National Weather Service.

Chapter 13
Figure 13.11: Drawn by E. Raisz.

Chapter 14
Figure 14.22: From Skaggs, R.H., *Proc. Assoc. American Geographers*, Vol. 6, Figure 2. Used by permission.

Chapter 15
Figure 15.16: Data from U.S. Dept. of Commerce; Figure 15.21: From Anderson, Bruce and Alan Strahler, *Visualizing Weather and Climate*. Copyright 2009 John Wiley & Sons, Inc. Reprinted with permission of John Wiley & Sons, Inc.; Figure 15.23: From Anderson, Bruce and Alan Strahler, *Visualizing Weather and Climate*. Copyright 2009 John Wiley & Sons, Inc. Reprinted with permission of John Wiley & Sons, Inc.

Chapter 16
What an Earth Scientists Sees (B): After Calder, Nigel, *The Weather Machine*, BBC Publications, London 1974. Used by permission of Mr. Nigel Calder.

Chapter 17
Figure 17.22: Adapted from Trefil, James and Robert M. Hazen, *The Sciences: An Integrated Approach*, 5th ed. Copyright 2007 John Wiley & Sons, Inc. Reprinted with permission of John Wiley & Sons, Inc.; Figure 17.22: Adapted from Trefil, James and Robert M. Hazen, *The Sciences: An Integrated Approach*, 5th ed. Copyright 2007 John Wiley & Sons, Inc. Reprinted with permission of John Wiley & Sons, Inc.

Line drawings in the following figures have been adapted from Murck, Barbara, Brian Skinner, and Dana Mackenzie, *Visualizing Geology*. Copyright 2008 John Wiley & Sons, Inc. Reprinted with permission of John Wiley & Sons, Inc.

Chapter 1: 1.4; *What an Earth Scientist Sees*; 1.5; 1.6; 1.8; 1.9; Table 1.1; 1.10; 1.15. **Chapter 2:** 2.1; 2.2; 2.3; 2.5; 2.10; 2.13; 2.14; Table 2.2; 2.15. **Chapter 3:** 3.6; 3.7; 3.13; 3.16; 3.17; *Amazing Places*. **Chapter 4:** *What an Earth Scientist Sees*; 4.4; 4.7; 4.10; 4.11; 4.13; 4.17; *Amazing Places*. **Chapter 5:** 5.1; 5.2; 5.3; 5.8; 5.11; 5.13; 5.15; 5.16; 5.17; 5.18; 5.19; 5.20; 5.21. **Chapter 6:** 6.1; 6.4; 6.5; 6.10; 6.13; 6.15; 6.16; 6.17; 6.19; *What an Earth Scientist Sees*. **Chapter 7:** 7.1; 7.2; 7.3; 7.4; 7.5; 7.6; 7.7; 7.9; *What an Earth Scientist Sees*; 7.11; 7.12; 7.13; 7.15. **Chapter 8:** 8.1; 8.3; 8.4; 8.6; 8.9; 8.10; 8.11; 8.12; 8.13; 8.15; *What an Earth Scientist Sees*; 8.16; 8.17; 8.18; *Amazing Places*. **Chapter 9:** 9.3; *Case Study*; 9.9; 9.13; 9.14; 9.15; 9.17; 9.20; 9.21. **Chapter 10:** 10.3; 10.7; 10.9; 10.11; 10.12; 10.13; 10.14; 10.15; 10.16; *Case Study*. **Chapter 11:** 11.1; 11.2; 11.3; 11.11; 11.19; 11.20; 11.22; *Amazing Places*. **Chapter 12:** 12.9; 12.10; 12.14; 12.15. **Chapter 13:** 13.3; 13.5; *What an Earth Scientist Sees*. **Chapter 14:** 14.1; 14.2; 14.3. **Chapter 15:** 15.13. **Chapter 16:** 16.1; 16.18. **Chapter 17:** 17.9; 17.17.

Line drawings in the following figures have been adapted from Strahler, Alan and Zeeya Merali, *Visualizing Physical Geography*. Copyright 2008 John Wiley & Sons, Inc. Reprinted with permission of John Wiley & Sons, Inc.

Chapter 3: 3.2. **Chapter 4:** 4.8. **Chapter 5:** 5.6; 5.14; 5.23. **Chapter 12:** *What an Earth Scientist Sees*; 12.13. **Chapter 13:** 13.7. **Chapter 14:** *What an Earth Scientist Sees*; 14.6; 14.7; 14.8; 14.10; 14.11; 14.12; 14.13; 14.15; 14.17; 14.19; 14.20; 14.21. **Chapter 15:** 15.2; 15.4; 15.5; 15.7; 15.8; 15.9; 15.10; 15.11; 15.12; 15.14; 15.15; 15.17; 15.18; 15.19; 15.25. **Chapter 16:** 16.7. **Chapter 17:** 17.4.

Line drawings in the following figures have been adapted from Skinner, Brian, Stephen Porter, and Daniel Botkin, *The Blue Planet: An Introduction to Earth System Science*. Copyright 1999 John Wiley & Sons, Inc. Reprinted with permission of John Wiley & Sons, Inc.

Chapter 1: 1.12. **Chapter 4:** 4.12. **Chapter 12:** 12.2; 12.8. **Chapter 14:** 14.5. **Chapter 15:** 15.3. **Chapter 17:** 17.2; 17.5; 17.6; 17.8.

Line drawings in the following figures have been adapted from Skinner, Brian, Stephen Porter, and Jeffrey Park, *Dynamic Earth: An Introduction to Physical Geology*. Copyright 2004 John Wiley & Sons, Inc. Reprinted with permission of John Wiley & Sons, Inc.

Chapter 7: 7.14; 7.16. **Chapter 8:** 8.12. **Chapter 9:** 9.5; 9.19. **Chapter 12:** 12.5; 12.6. **Chapter 16:** 16.4; 16.5.

INDEX

A

A horizon, of soil, 111–112
aa, lava flow, 289
ablation, in glacier formation, 188
abrasion
 definition of, 173
 erosion by, 102, 173–174
absorption
 of radiation in atmosphere, 434–437
 of seismic waves, 253–254
abyssal plain, of ocean floor, 204
accessory minerals, 50–52
accretion, in solar system formation, 524, 530, 535–536
accumulation, in glacier formation, 188
acid rain, and weathering, 102
active continental margins, 374
active galaxies, 545
adiabatic principle
 and cloud formation, 438–439, 446–447
 definition of, 438
 and orographic precipitation, 444
 and weather systems, 464
aerobic metabolism, 340, 343–344
aerosols
 in climate change, 495
 as condensation nuclei, 429, 440
 definition of, 426
 from volcanoes, 281
Africa
 future climate change, 509
 Mandara Lakes (Libya), 168–169
 Namib Desert (Namibia), 171
 rift valleys of, 211, 214, 216, 229
 Victoria Falls, 132–133
aftershocks, in earthquakes, 238–239
age, geologic
 of Earth, 325–327. *See also* numerical age; relative age
 of universe, 544
agglomerates, 276
 definition of, 71
aggregates, rock as, 64
agriculture
 hail damage, 449
 role in desertification, 180–182
 volcanism and, 281
 water use, 137, 149–151
air. *See also* air masses; atmosphere
 definition of, 426
 pressure and winds, 458–462
 role in soil, 109–110
 unstable, 447
air masses
 definition of, 470
 and fronts, 472–473, 498–499
 movement of, 445–449, 458–462, 470–473, 498–499
 types of, 471, 498–499
Alabama Mobile, 450
Alaska
 Barrow, 510
 Columbia Glacier, 185
 Glacier Bay National Park, 79
 Great Alaska Earthquake (1964), 123
 Hubbard Glacier, 190
 Matanuska Glacier, 120
 Yakutat Bay earthquake (1899), 235
albedo, in global energy system, 435
albite, crystallization of, 291–292
alfisols, soil order, 113–115
Allende Meteorite, 326–327
alluvial fans, 140–141
 deep-sea fans, 379
 in deserts, 178–179
alluvium
 definition of, 140
 landforms of, 140–142
Alps
 Findelen Glacier, 120
 metamorphic rock in, 65
 Similaun Glacier, 488–489
aluminum, in planet formation, 526
Alvarez, Walter and Luis, mass extinction theory, 357
amino acids, as building blocks, 340–341
ammonite fossils, 313
amphibians, in evolution, 353, 354
amphiboles
 crystallization of, 291–292
 structure of, 51
anaerobic metabolism, 340, 343
Andes Mountains (Argentina), rockslide in, 121
andesite, characteristics of, 70
andesitic magma, 290–292, 293–294
andisols, soil order, 113–115
angiosperms, definition of, 354
angular unconformities, in strata, 308–309
animals
 evolution of, 352–355
 populations and climate change, 507
anorthite, crystallization of, 291–292
Antarctic Bottom Water, 385
Antarctica
 Beacon Valley, 194
 ice sheets of, 186, 504
 ozone hole and, 430
 Queen Maude Land, 171
 Victoria Land, 103
anthropogenic effects on resources, 19–25
 climate change, 505–509
 definition of, 505
 desertification, 179–182
 human footprint, 362–363
 shoreline changes, 413–417
 soil management, 116, 168, 181–182
 on surface water, 148–151, 168
anticyclones
 characteristics of, 463–464
 definition of, 464
 and wave cyclones, 474–475
aphanitic texture, of rocks, 67–68
aquicludes, 156–157
 definition of, 156
aquifers, 156–157
 definition of, 156
 depletion of, 158–159
Archaeopteryx, 354–355
Archean Eon, 313
 life-forms of, 342–343
arches, landform, 406
arctic-front zone, 503
Arctic Ocean, characteristics of, 373–375
Arctic region, climate change effects in, 506
arêtes, landform, 191, 192
Argentina, Andes Mountains, 121
argon, in atmosphere, 426
arid regions, deserts in, 173
aridisols, soil order, 113–115
Aristarchus, on heliocentric universe, 518–519
Aristotle, on geocentric universe, 518
Arizona
 Painted Desert, 79
 Petrified Forest National Park, 348
arroyos, flooding in, 179
artesian wells, 156–157
arthropods, in evolution, 352
Artists Point (Colorado), ripple marks, 78
asbestos, habit of, 44
ash. *See* volcanic ash
Asia, future climate change, 509
Assateague Island (Virginia), beach erosion of, 414
asteroids, 534–535
asthenosphere. *See also* plate tectonics
 characteristics of, 14–15
 definition of, 15
 seismic discontinuities and, 255–256
 of terrestrial planets, 531
astronomical unit, 522
astronomy
 big bang theory and CMB, 540–545
 factors in climate change, 494–495
 geocentric vs. heliocentric universe, 518–522
 solar system, 524–537
 stars and stellar evolution, 538–540
Atlantic Ocean
 characteristics of, 373–375
 North Atlantic Oscillation, 387
atmosphere. *See also* air masses; atmospheric pressure; global atmospheric circulation; precipitation
 changes over time, 336–338
 cloud formation, 438–441, 446–447
 functions of, 424–425
 gases of, 426–427
 interactions with other systems, 183, 334–335, 384–388, 402–404, 456–457, 506–507
 moisture in, 431–434
 oceanic circulation and, 384–388
 solar radiation in, 434–437, 527
 thermal structure of, 427–431
atmospheric pressure. *See also* global atmospheric circulation; pressure gradients
 and adiabatic principle, 438–439
 definition of, 458
 in metamorphism, 83–85
atomic number, and chemical structure, 36–37
atoms
 and chemical structure, 36–39
 definition of, 36
 in radioactive decay, 319–321
auroras, 527, 529
Ausable Chasm (New York), mud cracks in shale, 78
Australia
 Bungle Bungle Range sandstone, 77
 Ediacara Hills fossils, 344
 future climate change, 509
 Gosses Bluff, 2–3

569

Australia *(cont.)*
   Hamersley Range banded iron formation, 76
   Uluru/Ayers Rock, 62–63
Australopithecus, 356–357
avalanches, debris, 122, 281, 296
Awash River (Ethiopia), sedimentary facies of, 81
axial rift, 374–375

B

B horizon, of soil, 111–112
backwash, of wave cycle, 403
Baja California (Mexico), coastal desert, 172
Banda Aceh (Indonesia), Sumatran tsunami and, 410–412
banded iron formations, 76, 338
bar, unit of atmospheric pressure, 458
barchans, dunes, 177–178
barometer, definition of, 459
barrier islands
   characteristics of, 406–408
   definition of, 406
barrier reefs, 409, 418
Barrow (Alaska), global warming effects, 510
basal sliding, of glaciers, 189–190
basalt
   characteristics of, 70
   flood basalts, 273, 274, 281, 360
   in magma, 290–292
   in oceanic crust, 18, 278
   texture of, 67
   uses of, 89
basalt plateaus, 273, 274, 281, 360
basaltic magma, 290–292
basins (ocean), characteristics of, 372–375
batholiths
   characteristics of, 293–294
   definition of, 293
bauxite, ore mineral, 124
bay barriers, landform, 406
Bay of Fundy (Canada), tidal range, 399
bays, currents and, 403, 404
beach drift, 402–403
   definition of, 402
   and erosion protection, 414–416
beaches
   characteristics of, 406–408
   definition of, 406
   erosion protection, 414–416
   pocket, 405
   wave-cut, 405
   wave erosion of, 402–404
Beacon Valley (Antarctica), patterned ground, 194
bed load
   definition of, 117
   and erosion, 117–118
bedding
   clues from, 77–81
   definition of, 77
benches (coastal), 405–406
benthic zones, 380–381
Bering land bridge, 396–397
beryl
   structure of, 51
   uses of, 90
big bang theory, definition of, 543

biogenic sediment
   definition of, 77
   as resource, 76, 90
   and sedimentary rock, 75–77
biologic resources
   definition of, 20
   use of, 20–21
biopolymers, as building blocks, 340–341
biosphere
   as Earth subsystem, 10–12
   interactions with other systems, 334–335, 506–507
biotic zones, 380–381
biotite, crystallization of, 291–292
birds, in evolution, 354–355
black holes, 540
Black Rapids Glacier (Alaska), 168–169
Black Sea Coast (Russia), beach erosion of, 416–417
black smoker vents, 376
body waves. *See also* P waves; S waves
   definition of, 247
   measurement of, 247–249
bonds (chemical), 38–39
   definition of, 38
Bora Bora, island, 9
Bowen, N.L., Bowen's reaction series, 291–292
Bowen's reaction series
   and crystallization, 291–292
   definition of, 291
Brahe, Tycho. *See also* Kepler, Johannes
   on heliocentric model, 521
braided channels, 138–141
Brazil, Sugarloaf Mountain, 106
breaking waves, 401–403
breakwaters, erosion protection, 414–415, 416–417
breezes, 462
brightness, of stars, 538–539
brittle deformation, definition of, 261
brown dwarfs (stars), 539
Bryce Canyon (Utah), sedimentary rock of, 64–65
Bungle Bungle Range (Australia), sandstone formations of, 77
Burgess Shale (Canada), fossils of, 361

C

C horizon, of soil, 111–112
calcareous ooze, 380–381
calcite
   in karst formation, 160–161
   in limestone, 77, 89
   in weathering, 102, 104
calcium carbonate. *See* calcite
calcium sulfate. *See* gypsum
calderas, of volcanoes, 276
California
   Death Valley, 141, 195
   Joshua Tree National Park, 101
   La Conchita mudflow, 129
   Mojave Desert, 501–502
   Mono Lake, 152
   Monterey Bay, 389
   orographic precipitation in, 445
   San Francisco earthquake (1906), 235–237
   Santa Ana winds and fires, 460–461
   Yosemite National Park, 103

Cambrian Period, Cambrian Explosion, 314, 316, 349–353
Cameroon, volcanic lake deaths, 278–279
Canada
   Acasta gneiss formation, 326
   Bay of Fundy tidal range, 399
   Burgess Shale, 361
   Northwest Territories diamonds, 34–35
   the Yukon, 120
Cape Cod (Massachusetts), as spit, 406
carbon
   in planet formation, 525
   in sediment, 338–339
carbon-14, decay of, 319, 320–321
carbon dioxide
   in atmosphere, 426–427
   and climate change, 505–507
   radiation absorption by, 434–437
carbonaceous chondrites, 326–327
Carboniferous Period, evolution in, 354
catastrophism, 304
catchments. *See* drainage basins
caverns
   definition of, 160
   formation of, 160–161
caves
   definition of, 160
   formation of, 160–161
cells, definition of, 340
cementation, in lithification, 73–74
Cenozoic era, 313–316
   climate trends since, 490–493
   evolution of life in, 355–357
centers, low- and high-pressure, 464
central rift valley, 374–375
Cepheid variables, variable stars, 541
Ceres, dwarf planet, 534
CFCs (chlorofluorocarbons), and ozone layer, 430–431
changes of state, and energy transfers, 431–432
channels
   definition of, 138
   modifications to, 149
   types of, 138–141
chemical differentiation, in planet formation, 13
chemical sediment
   definition of, 75
   as resource, 90
   and sedimentary rock, 75–77
chemical weathering, 102–105. *See also* dissolution
   definition of, 101
   resources formed by, 124
chemosynthesis, by extremophiles, 340–341
chert, biogenic rock, 77
Chicxulub crater (Mexico), meteorite impact, 357–359
China
   Great Sichuan Earthquake (2008), 243
   Guilin karst, 161
Chinook winds, 444, 461
chlorofluorocarbons (CFCs), and ozone layer, 430–431
chordates, in evolution, 353
chromosphere, of Sun, 527–529
cinder cones, of volcanic eruptions, 273, 276

circulation. *See* global atmospheric circulation; oceans, circulation in; thermohaline circulation
cirque glaciers, 184
cirques, landform, 191, 192
cirrus clouds, 440–441
clastic sediments
  definition of, 72
  lithification of, 72–74
clay
  as aquiclude, 156
  characteristics of, 109
  definition of, 109
  in shale formation, 72, 73
cleavage, definition of, 45
climate. *See also* climate change
  definition of, 491
  feedback mechanisms and, 11–12
  global types, 498–504
  ocean circulation regulation, 384–388
  and soil formation, 112, 114–115
  volcanoes and, 281
  and weathering, 105, 107–108
climate change
  anthropogenic effects, 505–509
  causes of, 495
  extreme regions as indicators of, 168
  temperature record, 492–493
  trends since Cenozoic Era, 490–493
  the Younger Dryas event, 495–497
closed systems
  characteristics of, 8
  Earth as, 8–12, 22
clouds
  along fronts, 472–473
  families and types, 440–441
  formation of, 438–441, 446–447
  and greenhouse effect, 436–437
  hole punch, 450
  and insolation, 434–435
  and precipitation, 442–444
coal
  definition of, 77
  formation of, 77
coastal-continental location, influence on climate, 498–499
coastal deserts, 172–173
coastal regions
  erosion protection in, 413–417
  hazards in, 410–412
  population concentration, 410
coastlines. *See* coastal regions; shorelines
cold fronts
  characteristics of, 472–473
  and wave cyclones, 474–475
collision zones, of tectonic plates, 214
color, in minerals, 46–48
Colorado, Artists Point, 78
Colorado River (United States), and delta, 10
Columbia Glacier (Alaska), 185
coma, of comets, 536
comets, characteristics of, 536–537
compaction, in lithification, 73–74
compounds (chemical), 37–39
  definition of, 37
compression, 260–263
  definition of, 260

compressional waves. *See also* P waves
  definition of, 247
  measurement of, 247–249
condensation
  and atmospheric moisture, 431–434
  condensation nuclei, 429, 440
  definition of, 135
  in hydrologic cycle, 134–135
condensation nuclei, 429, 440
conduction, definition of, 219
cone of depression, 158
confined aquifers, 156
confining stress, 260
conglomerate, 72, 73
  definition of, 72
construction, rock in, 89
contamination, of groundwater, 158–159
continental air masses
  continental antarctic (cAA), 470
  continental arctic (cA), 470, 503
continental crust
  chemical composition of, 49, 50
  definition of, 18
  in plate tectonics, 214–218, 222
continental drift. *See also* plate tectonics
  definition of, 202
  early evidence for, 203–207
  paleomagnetism evidence for, 207–210
  Wegener's hypothesis, 202–207
continental interior deserts, 172–173
continental polar (cP) air masses, 471, 476, 501, 503
continental rise, 204
continental shelf, and continental drift, 203–205
continental slope, and continental drift, 203–205
continental tropical (cT) air masses, 470
continents. *See also* continental drift; *specific continent*
  structure of edge, 202–204
  supercontinents, 202–207, 223, 339
continuous reaction series, of crystallization, 291–292
convection
  and cloud formation, 446–447
  definition of, 219
  and plate tectonics, 219–221
  in Sun, 527
convection cells
  in global circulation, 466–467
  of thunderstorms, 448
convection loops
  definition of, 460
  and global winds, 465–466
  and pressure gradients, 460–461
convection zone, of Sun, 527
convectional precipitation, 445
  and cloud formation, 446–447
  definition of, 446
convergence, 445
  and low pressure systems, 464
convergent margins/boundaries, 214–218
  definition of, 214
  earthquake characteristics of, 262
  volcanoes and, 212–213, 277–278
Copernicus, Nicolaus, astronomical theories of, 202, 519–521
coprolites, 348

coral, and temperature records, 492–493
coral reefs
  characteristics of, 409
  dangers to, 418
  dust storms and, 7
  Florida Keys Reef, 418
core, 13–15
  composition of, 259
  definition of, 14
  of Mercury, 530
  Oldham's liquid core theory, 253–254
Coriolis, Gaspard-Gustave de, and Coriolis effect, 463
Coriolis effect
  on air motion, 463–464, 466–467, 468, 469
  definition of, 463
  and ocean currents, 383–384
corona, of Sun, 527–529
correlation
  definition of, 312
  and geologic column, 312–316
  by magnetic reversals, 325
corundum, properties of, 46–47
cosmic microwave background (CMB) radiation, 543–544
  definition of, 543
counterradiation, and greenhouse effect, 436–437
covalent bonding, definition of, 39
Crab Nebula, 525
Crater Lake (Oregon), 276
craters, of volcanoes, 272, 276
creep, definition of, 122
crest, of flood, 146
Cretaceous Period
  evolution in, 354, 355
  mass extinction in, 357
crevasses, 189
crust. *See also* continental crust; oceanic crust
  chemical composition of, 49, 50–51
  definition of, 15
  in plate tectonics, 214–218, 222
  seismic discontinuities and, 255–256
  types of, 18
cryosphere
  components of, 183–186
  definition of, 183
crystal structure
  definition of, 41
  of minerals, 41–42, 43–46
crystal zonation, 291–292
crystallization
  definition of, 290
  fractional, 290–292
crystals
  crystal zonation, 291–292
  faces and habits, 43–44, 56
cumuliform clouds, 440–441
cumulus clouds, 440–441, 446–447
  along fronts, 472–473
Curie, Marie and Pierre, work on radioactivity, 318, 320
Curie point, and magnetism, 258–259
currents
  longshore, 402–403
  ocean, 382–388
  rip, 401

Index 571

currents (cont.)
  tidal, 404
  turbidity, 378–379
cut bank, 140
cyanobacteria, 342
cycles. See hydrologic cycle; Milankovitch cycles; rock cycle; tectonic cycle
cyclones
  characteristics of, 463–464
  definition of, 464
  tropical cyclones, 477–479, 480–481
  wave cyclones, 474–475, 480–481, 503
cyclonic precipitation, 445

D

Dana, James Dwight, explorer, 224
dark matter. See galaxies
Darwin, Charles
  on evolution and natural selection, 345–347
  on numerical age, 317, 318
Darwin's finches, 346
dating
  of human ancestors, 324
  of rocks and fossils, 320–325
daughter atoms, 319–321
Death Valley (California)
  alluvial fan in, 141
  extreme conditions of, 172, 195
debris avalanches, 122, 281, 296
Deccan Traps, mass extinction and, 281, 360
decompression melting, of rock, 287
deep-sea fans, 379
deep zone, of ocean, 382
deflation
  definition of, 174
  erosion by, 174–175
deformation, of rock, 260–261
delta coasts, characteristics of, 408
deltas, 140–142
  definition of, 408
density
  definition of, 48
  of Earth layers, 259
  of seawater, 382
deoxyribonucleic acid. See DNA
deposition. See also shorelines; streams and streamflow
  definition of, 72, 135
  in hydrologic cycle, 134–135
  of sand in shorelines, 402–404
desertification, 179–182
  definition of, 180
deserts, 168–182
  climate and, 170–171, 501–502
  definition of, 171
  desertification, 179–182
  dunes and other landforms, 175–179
  types of, 171–173
Devonian Period, evolution in, 352, 353
dew-point temperatures, 432, 434
diagenesis. See also metamorphism
  and sedimentary rock, 74
diamonds, 34–35
  formation of, 257
  volcanic activity and, 284
differential stress, 260
  in metamorphism, 84

differentiation, in planet formation, 530
dikes, plutons, 293, 294–295
dinosaurs
  in evolution, 355, 424
  mass extinction of, 357–359
diorite
  in batholiths, 293
  characteristics of, 70
discharge (stream), definition of, 138
discharge (subsurface water), 154–155
  definition of, 155
disconformities, in strata, 308
discontinuous reaction series, of crystallization, 292
dissolution
  definition of, 102
  and karst formation, 160–161
  and weathering, 102–105
distributaries, and deltas, 408
divergence, and high pressure systems, 464
divergent margins/boundaries, 214–218
  definition of, 214
  volcanoes and, 212–213, 278
divide, definition of, 143
DNA
  definition of, 340
  in early life, 340–341
dolostone
  biogenic rock, 77
  in karst formation, 160–161
domain, definition of, 342
Doppler effect, in astronomical studies, 542
drainage basins
  definition of, 142
  of streams, 142–143
drift, continental. See continental drift
drilling, in Earth study, 256–257
drought, and desertification, 179–182
dry adiabatic lapse rate, 439
  and orographic precipitation, 444
Dryas pollen, and climate change, 495
ductile deformation, definition of, 261
dunes, 175–178
  definition of, 175
Dust Bowl period, Great Plains of U.S.A., 181–182
dust storms
  ecological impacts of, 7, 118–119, 181
  formation of, 173
dwarf planets, 516, 534
dwarf stars, 538–539

E

E horizon, of soil, 111–112
early life on Earth. See life on Earth
early warning systems
  tornadoes and, 476
  tsunamis and, 243, 410–412
  volcanoes and, 270, 280
Earth
  age of, 325–327
  composition of, 13–15
  density of components, 259
  Oldham's liquid core theory, 253–254
  physical dimensions of, 259
  rotation of, 494–495, 520
  and solar system, 12–13, 326–327, 530–531
  study of interior, 252–259

  as terrestrial planet, 530–531, 536
  unique characteristics of, 15–18, 531
Earth resources. See also nonrenewable resources; renewable resources
  definition of, 20
  use of, 20–21
Earth science
  definition of, 4
  Earth system in, 7–12
  reasons for study, 22–25
  scientific method in, 4–6
  vs. Earth system science, 12
Earth system science, 7–12
  definition of, 12
  plate tectonics example, 222
  reasons for study, 22–25
  systems approach, 7–12
  vs. Earth science, 12
Earth systems, 7, 8–12, 22. See also atmosphere; biosphere; hydrosphere; lithosphere
  interactions in, 183, 334–335, 384–388, 402–404, 456–457, 506–507
earthquakes. See also seismology
  definition of, 234
  and design safety, 243–245
  elastic rebound theory and, 235–238
  Great Alaska Earthquake (1964), 123
  Great Sichuan Earthquake (2008), 243
  hazards from, 123, 238–241, 243–245
  locating, 248–250
  measurement of, 250–251
  plate margins and, 217–218
  and plate motion, 234–238
  prediction and forecasting, 240–242
  San Francisco earthquake (1906), 235–237
  Sumatra-Andaman (2004), 232–233, 237, 240–241
  Yakutat Bay earthquake (1899), 235
Easter Island, decline of Rapa Nui, 20, 22
ebb currents, 404
eccentricity, and climate change, 494–495
echo sounders, 373
Ediacara fauna, early multi-cellular life, 344
Ediacara Hills (Australia), fossils of, 344
Egypt, Nile River delta, 141, 408
Ehrlich, Paul, on population growth, 19
Ekman drift, 386
El Niño
  definition of, 384
  and extreme weather, 456–457
El Niño southern oscillation (ENSO), 384
elastic deformation, definition of, 261
elastic rebound theory, 235–238
  definition of, 235
electrical power, modern dependence on, 22–23
electrons, and chemical structure, 37–39
elements
  and chemical structure, 36–39
  definition of, 36
  of Earth's crust, 49, 50
  in solar system formation, 525–526
  in stellar evolution, 539
ellipses, law of, 521
elliptical galaxies, 545
energy exchange
  and changes of state, 431–432

ocean circulation and, 384–388
in Sun, 527
energy resources. *See also* fossil fuels
  modern dependence on, 22–23
entisols, soil order, 113–115
environmental changes
  clues in sediments, 77–81
  over Earth's history, 336–339
environmental temperature lapse rate, 428
  in unstable air, 447
eons, geologic, 312–316
epicenter, 218
  definition of, 248
  location of, 248–249
epochs, geologic, 312–316
equal areas, law of, 521
equatorial easterlies, upper-level winds, 466, 468, 499–500
equatorial zone, of climate, 499–501
equinoxes, and insolation, 465
eras, geologic, 312–316
Eris, dwarf planet, 516–517, 534
erosion. *See also* glaciers; mass wasting; shorelines; streams and streamflow
  definition of, 66, 100
  by ice, 119, 120
  resources formed by, 124
  and rock cycle, 66
  shoreline protection, 413–417
  of soil, 113, 116
  by storms, 398, 404, 410–412
  vs. weathering/mass wasting, 117
  by water, 117–118
  by waves, 402–404
  by wind, 173–175
erratics, in glacial till, 193
eskers, 191, 193
estuaries
  currents and, 403, 404
  definition of, 408
Ethiopia
  Awash River, 81
  Haddar region hominids, 324
eukaryotes
  definition of, 343
  of Proterozoic Eon, 343–344
Europe
  climate change history, 495–497
  future climate change, 508
eutrophication, in lakes, 144
evaporation
  and atmospheric moisture, 431–434
  definition of, 135
  in hydrologic cycle, 134–135
evaporites
  definition of, 75
  in karst formation, 160–161
evolution. *See also* Phanerozoic Eon, evolution of life in
  of arthropods, 352
  Cambrian Explosion, 349–352
  definition of, 345
  development of theory, 345–347
  of fishes and amphibians, 353
  fossil evidence for, 347–348
  of human family, 356–357
  of plants, 350–352, 355

reptiles, birds, and mammals, 354–355
  sea-to-land transition, 350
exfoliation, 102, 103
exoplanets, definition of, 537
extinction, mass extinctions, 357–360
extreme climatic regions. *See* deserts; glaciers; ice sheets
extremophiles, 340–341
extrusive igneous rocks. *See* volcanic rock
eye, of cyclone, 477

F
falls, slope failure, 121
farmland, erosion of, 116
faults
  definition of, 211
  and earthquake movement, 234–238
  stresses and, 262–263
  types of, 214–215, 218, 262–263
  vs. joints, 102
feedback mechanisms
  definition of, 11
  global warming example, 507
feldspar
  crystallization of, 291–292
  as gem, 47
  structure of, 51
  weathering of, 62–63, 104, 124
felsic rock, 69–70
Ferrel cells, convection cells, 466–467
Findelen Glacier (Alps), erosion by, 120
fire, as earthquake hazard, 238–239, 278
fishes, in evolution, 353
fissure eruptions, 273, 274, 280
fjord glaciers, 185
flash floods, 179
flood, definition of, 145
flood basalts, 273, 274, 281, 360
flood currents, 404
flooding, 145–149
  and barrier islands, 406
  definition of, 145
  flash floods, 179
  from hurricanes, 478
  prediction and prevention, 148–149
  resulting from volcanoes, 278, 281
floodplains, 140–141
  definition of, 140
  New Orleans built on, 482
Florida
  Florida Keys Reef, 418
  sinkholes in, 161, 165
flow (mass wasting), 121–122. *See also* slides; *specific type of flow*
  definition of, 121
  mudflows, 129
focus, 218
  definition of, 247
foggara, groundwater resources, 20–21
foliation
  definition of, 86
  in metamorphism, 86–88
follies, erosion and, 412
forecasting
  of earthquakes, 240–242
  Geostationary Operational Environmental Satellite (GOES), 453

weather models, 506–507
foreshocks, earthquake, 241
fossil fuels
  and climate change, 505, 507
  coal formation, 77
fossil tracks, in rock, 78, 79
fossils
  Burgess Shale (Canada), 361
  and climate change evidence, 493
  as continental drift evidence, 206–207
  and correlation, 310–312
  dating of, 320–325
  definition of, 347
  Ediacara Hills (Australia), 344
  formation of, 347–348
  fossil tracks, 78, 79
  Jura Mountains (Switzerland), 302–303
Foucault, Léon, Foucault's pendulum, 520
fractional crystallization, 290–292
  definition of, 290
fractional melt. *See also* fractional crystallization
  definition of, 287
  process of, 288
fractionation
  definition of, 287
  process of, 288
fresh water. *See also* groundwater; surface water
  in hydrologic cycle, 135–137
  in ice sheets, 186
fringing reefs, 409
fronts
  definition of, 472
  types of, 472–473
  and wave cyclones, 474–475, 503
frost wedging, 102, 103
fumaroles, 277

G
gabbro
  characteristics of, 70
  in sills, 294–295
galaxies
  characteristics of, 540–545
  types of, 545
galena, mineral, 42, 48
Galilei, Galileo, work of, 522
Galveston (Texas), hurricane of 1900, 394–395, 478
Garamantian culture, decline of, 20–22
gases. *See also* greenhouse effect
  of atmosphere, 426–427
  and volcanic activity, 272–275, 278–279, 337, 339
gems
  properties of, 46–47
  volcanoes and, 284
geocentric universe model
  definition of, 518
  history of, 518–522
geologic column, 312–316
  definition of, 312
Georgia, Providence Canyon, 116
geosphere, as Earth subsystem, 10–12
Geostationary Operational Environmental Satellite (GOES), forecasting tool, 453
geostrophic wind, definition of, 468

geothermal gradient, 385
   definition of, 285
geysers, 277
gibbsite, formation of, 124
glaciation, definition of, 493
Glacier Bay National Park (Alaska), sandbars in, 79
glacier ice, 187
   melting of, 506–507
Glacier National Park (Montana), 184
glaciers
   Black Rapids Glacier, 168–169
   and climate change evidence, 493
   definition of, 119, 183
   erosion and deposition by, 119, 120, 190–194, 206
   evidence of continental drift, 205, 206
   formation of, 186–187
   movement of, 188–190
   types of, 184–185
glassy texture, of rocks, 67
Gliese 581c, exoplanet, 537
global atmospheric circulation. *See also* weather systems
   cyclones and anticyclones, 463–464
   global wind patterns, 465–467
   pressure and air movement, 458–462, 470–473
   upper-level winds, 467–469
global climates, 498–504
   air mass movements and, 498–499
   high-latitude climates (Group III), 503–504
   low-latitude climates (Group I), 499–501
   midlatitude climates (Group II), 501–503
global energy system, 434–437
Global Positioning System (GPS), used in research, 210
global warming
   anthropogenic effects, 505–509
   and coral reef decline, 418
   feedback mechanisms in, 11–12
   importance of data on, 490–491
   and rising sea levels, 397–398
   and tropical cyclones, 479
gneiss
   Acasta formation, 326
   characteristics of, 65
   definition of, 86
   formation of, 82, 83, 85–87
   uses of, 89
GOES (Geostationary Operational Environmental Satellite), forecasting tool, 453
gold, as resource, 90, 124
Gosses Bluff (Australia), 2–3
Gould, Stephen Jay, on Darwin's work, 347
gradient (pressure). *See* pressure gradients
gradient (stream), 138–142
   definition of, 138
gradualism, in evolution, 347
Grand Canyon, as sedimentary record, 306, 328
granite
   in batholiths, 293
   characteristics of, 64, 70
   in continental crust, 18
   uses of, 89–90
granular flows, of regolith, 122
granules, on Sun, 527, 528

gravity
   and atmospheric pressure, 458
   and planetary motion, 521
Great Alaska Earthquake (1964), 123
Great Barrier Reef (Australia), 409
Great Glen Fault (Scotland), Loch Ness on, 264
Great Plains
   Dust Bowl period, 181–182
   hail damage in, 449
Great Salt Lake (Utah), 75
Great Sichuan Earthquake (2008), 243
greenhouse effect
   and climate change, 168, 495
   and counterradiation, 436–437
   definition of, 436
   troposphere and, 429
greenhouse gases. *See* greenhouse effect
Greenland
   climate of, 504
   gneiss formation, 82
groins, erosion protection, 414–416
grooves, glacial, 191–192, 206
groundwater, 153–161
   cave and cavern formation, 160–162
   definition of, 153
   depletion and contamination of, 158–159
   movement of, 154–156
   storage of, 156–157
   water table, 153–154
groundwater mining, 158
Gubbio (Italy), marl deposits of, 494
Gulf of Aden, and plate movement, 216
Gulf Stream, 385
gymnosperms, definition of, 352
gypsum, in weathering, 102, 104

H

habit, definition of, 44
Hadean Eon, 313
Hadley cells, convection cells, 466–467
hail, formation of, 444, 449
half-life
   in decay, 320–321
   definition of, 320
halite, in weathering, 102, 104
Halley, Edmund, on numerical age, 317, 318
Hamersley Range (Australia), banded iron formations, 76
hardness (mineral), 44–45
   definition of, 44
Hawaii
   Pacific Tsunami Warning Center, 412
   tectonic activity of, 224–225
   volcanic eruptions, 18, 272–274, 278
Hawaiian volcanic eruptions, 272–274
hazards
   coastal, 410–412
   from earthquakes, 123, 238–241, 243–245
   from hurricanes, 477–479
   need for study, 22–25
   from volcanoes, 278–284
headlands, 405
heat. *See also* latent heat
   in atmospheric energy exchange, 431–434
   in metamorphism, 83–85
heliocentric universe model
   definition of, 518

   history of, 518–522
helium
   in atmosphere, 426
   in big bang theory, 543
   in solar system formation, 525–526
   in Sun's reactions, 527, 539
helium burning, in stars, 539
Hertzsprung-Russell (H-R) diagram, definition of, 538
heterosphere, 429
high-latitude climates (Group III), 503–504
Hillary, Edmund, scaling of Mt. Everest, 200
Himalaya, tectonic origin of, 200–201, 214–215
histosols, soil order, 113–115
hole punch clouds, 450
hominids, evolution of, 356–357
Homo genus, 357
homosphere, 429
Hubbard Glacier (Alaska), 190
Hubble, Edwin, galaxy studies, 540–542
Hubble Space Telescope, 541, 545, 546
human effects on resources. *See* anthropogenic effects on resources
human footprint, 362–363
humans, emergence of, 356–357
humidity
   definition of, 432
   relative and specific, 432–434
humus
   definition of, 109
   role in soil, 109–110, 112–113
Hurricane Andrew (1992), storm surge of, 478
Hurricane Dennis (1999), damage from, 421
Hurricane Isabel (2003), shoreline damage from, 398
Hurricane Katrina (2005), flooding from, 147, 482
hurricanes. *See also specific hurricane*; tropical cyclones
   characteristics and hazards, 477–479
Hutton, James, and uniformitarianism, 304, 318
hydroelectric power, early history, 132
hydrogen
   in atmosphere, 426
   in big bang theory, 543
   in metallic form, 532
   in solar system formation, 525–526
   in Sun's reactions, 527
hydrographs, 146
hydrologic cycle, 134–137
   definition of, 134
   in Earth system, 66, 135–137
hydrologic stations, 146
hydrosphere
   changes over time, 336–338
   as Earth subsystem, 10–12, 135–137
   hydrologic cycle, 134–137
   interactions with other systems, 183, 334–335, 384–388, 402–404, 456–457, 506–507
hypothesis
   definition of, 4
   in scientific method, 4–6
   vs. theory, 211

I

ice. *See also* glaciers; ice sheets
   erosion by, 119, 120, 190–194, 206

formation of, 442–443
  as fossil preservative, 348
  ice caps, 184–185
  ice-core analysis, 492–493, 513
ice caps, 184–185
ice-core analysis, for temperature records, 492–493, 513
ice sheets, 183, 186
  in climate change, 496–497
  climate in region, 504
  definition of, 183
ice shelves, 186
ice storms, 443
ice wedges, landform, 194
Iceland
  Laki eruption (1783), 278, 281
  rift valley, 215
igneous rocks, 67–71. *See also* magma; volcanoes
  chemical composition, 69–71
  cooling and formation, 67–69
  definition of, 67
  in landscape, 64–65
  plutonic formations, 293–295
  radiometric dating of, 322, 324
  as resources, 89–90
  water movement through, 154–155
impact debris, from meteorite, 359
inceptisols, soil order, 113–115
Indian Ocean
  characteristics of, 373–375
  Sumatra-Andaman earthquake (2004), 232–233, 237, 240–241
Indian Ocean tsunami. *See* Sumatran tsunami (2004)
Indonesia. *See also* Sumatran tsunami (2004)
  Banda Aceh, 411
  Krakatau eruption (1883), 278, 280–281
  Mt. Merapi volcano, 282–283
  Sulawesi gold mining, 55
  Tambora eruption (1815), 281
Indus River (Pakistan), weathering and, 106
industrial uses
  rock in construction, 89
  of water, 149–151
inertia, seismographs and, 246
infiltration, definition of, 135
inselbergs, definition of, 125
insolation
  definition of, 434
  in global energy system, 434–435
  and seasons, 465
interbasin transfer, of water, 149
interglacial periods, 492
interior of Earth. *See also* core; mantle
  mantle/crust discontinuities, 255–256
  study of, 252–259
International Astronomical Union (IAU), on Pluto, 516, 534
intertropical convergence zone (ITCZ)
  climate of, 499–500
  definition of, 466
intrusive igneous rocks. *See* plutonic rock
ion exchange, in weathering, 104–105
ionic bonding, definition of, 39
ions, and chemical structure, 36–39
iron
  banded iron formations, 76

as core component, 259
in planet formation, 525–526
in star cycle, 539
in weathering, 105, 124
islands, future climate change, 509
isobars, definition of, 460
isolated systems, characteristics of, 8
isostasy, 211
isotherms, 286
isotopes
  definition of, 37
  in numerical age determination, 319–322
Italy, Gubbio marl deposits, 494

J
Japan, Mt. Fuji, 275
Java Trench, and Sumatran tsunami, 410–412
jet streams, 469–470
  definition of, 469
joints
  definition of, 101
  weathering and, 101–102
Joly, John, on numerical age, 317, 318, 376
Joshua Tree National Park (California), rock joints in, 101
Jovian planets, 13
  characteristics of, 524, 532–534
Jupiter
  characteristics of, 532
  motion of, 519
Jura Mountains (Switzerland), fossil limestone of, 302–303
Jurassic Period, evolution in, 354–355

K
kaolinite, formation of, 104
karst topography, 160–161
Kaskawulsh Glacier (Canada), 120, 193
Kepler, Johannes, planetary motion laws of, 521–522
Kettle Moraine State Forest (Wisconsin), 193
kettles, glacial landform, 191
kimberlite pipes, 257
kingdoms, definition of, 350
Kola Peninsula (Russia), deepest drilled hole, 256
Krakatau eruption (1883), 278, 280–281
krypton, in atmosphere, 426
Kuiper belt, 537

L
La Niña phenomenon, 384
laccoliths, 294
lagoons, landform, 407, 408
lahars, 270, 280–281
lakes
  characteristics of, 144
  oxbow, 139, 140, 141
Laki eruption (1783), 278, 281
laminar flow, in erosion, 117, 120
land breezes, 462
land bridge, Bering, 396–397
landforms
  coastal, 405–409
  desert, 175–179
  ice, 191–192, 194
  stream, 140–142
  volcanic, 276–277

landscape changes, from volcanic activity, 281
landslides, 121, 123
  coastal, 412
  as earthquake hazard, 238–239
  as hurricane hazard, 478
lapse rates
  adiabatic lapse rates, 439, 444
  definition of, 428
  in unstable air, 447
latent heat
  in cloud formation, 439
  definition of, 432
  of seawater, 385–388
  in thunderstorms, 447, 474
laterites, characteristics of, 124
Latin America, future climate change, 508
latitude, influence on climate, 498–499
Lauterbrunnen Valley (Alps), 65, 192
lava. *See also* magma
  composition and properties of, 287–289
  definition of, 272
  in seafloor spreading, 209–210
  in volcanic rock formation, 67–68
Law of Original Horizontality, 305, 307
  definition of, 6
laws
  of crystal angles, 43–44
  of planetary motion, 521–522
  of radioactive decay, 320
  scientific, 6
  of sediments, 305, 307
Le Verrier, Urbain, planetary orbit study, 516
lead-206, decay of, 321
Lechuguilla Cave (New Mexico), crystal formations, 162
levees, 140–141, 147
Libya, Mandara Lakes, 168–169
life on Earth
  early conditions, 339–340
  evolution and fossil record, 345–348
  first life-forms, 342–344
  mass extinctions and, 357–360
  in Phanerozoic Eon, 349–363
  properties of, 340–341
  sea-to-land transition, 350–353
life zone
  changes in, 336–339
  characteristics of, 10–11
  definition of, 10
lifting condensation level, 446, 447
Lightfoot, John, on date of Creation, 304
lightning, 447, 448–449, 474
limestone
  in building, 89
  definition of, 77
  in karst formation, 160–161
  metamorphism of, 88
limonite, formation of, 105
liquefaction (ground), as earthquake hazard, 238–239, 245
liquid core theory, 253–254
lithification, 72–74. *See also* metamorphism
  of chemical and biogenic sediments, 75–77
  definition of, 72
lithium, in big bang theory, 543
lithosphere. *See also* plate tectonics
  characteristics of, 14–15

Index   575

lithosphere *(cont.)*
  definition of, 15
  as Earth subsystem, 10–12
  interactions with other systems, 183, 334–335, 402–404
  seismic discontinuities and, 255–256
  of terrestrial planets, 531
littoral drift, 402–403
  definition of, 402
Lituya Bay (Alaska), tsunami, 412–413
load (stream), definition of, 138
loam, 109
local winds, 460–462
logarithmic scales, 250–251
long-term forecasting, of earthquakes, 240–242
longitudinal dunes, 177
longshore currents, 402–403
  definition of, 402
longshore drift, 403
longwave radiation, counterradiation, 436–437
Louisiana
  New Orleans, 147, 482
Love waves, 248
low-latitude climates (Group I), 499–501
low-latitude rainforests, definition of, 500
luminosity, of stars, 538–539
luster, 42–43
  definition of, 43

M

mafic rock, 69–70
magma
  composition and properties of, 287–289
  cooling and crystallization of, 290–292
  definition of, 67, 272
  in seafloor spreading, 209–210
  in volcanic eruptions, 272–277
magma chamber
  and later landforms, 276–277, 279
  in volcano structure, 280
magmatic differentiation, definition of, 290
magnetic fields
  of Earth, 207–208, 258–259, 322–323, 325
  of neutron stars, 539–540
magnetic polarity dating, 322–323, 325
magnetic poles, "wandering," 207–208
magnetic reversals, 322–323, 325
  definition of, 325
magnitude, of earthquakes, 250–251
main-sequence stars, 538
Maine, granite formations in, 64
mammals, in evolution, 354–355
Mandara Lakes (Libya), oasis, 168–169
manganese
  commercial uses of, 124
  in weathering, 105, 124
mantle, 13–15
  convection in, 219–221
  definition of, 14
  seismic discontinuities of, 255–256
mantle plumes, 221
mantling, 291–292. *See also* crystal zonation
marble
  in building, 89
  characteristics of, 88
  definition of, 88
Mariana Trench, tectonic origin of, 200–201

marine terraces, 406
maritime air masses
  maritime equatorial (mE), 470
  maritime polar (mP), 470, 501, 503
  maritime tropical (mT), 470, 474, 476, 501
marl deposits, of Gubbio, 494
Mars
  dunes on, 178
  motion of, 519
  NASA studies of, 546
  as terrestrial planet, 13–15, 18, 530–531
Maryland, Ocean City erosion, 414
mass extinctions, 357–360
  definition of, 357
  flood basalt eruptions and, 281, 360
mass wasting. *See also* erosion
  characteristics of, 119–122
  definition of, 119
  tectonics and, 122–123
Massachusetts, Cape Cod, 406
Matanuska Glacier (Alaska), glacial till, 120
Mauna Kea (Hawaii), volcanic eruption, 274
Mauna Loa (Hawaii), volcanic eruption, 18, 274
meandering channels, 138–141
mechanical weathering, 100–103
  definition of, 100
Mediterranean climate, 502
megathrust earthquakes, 234
melting (rock), 285–289
  absent in metamorphism, 82–84
meltwater lakes, in Younger Dryas event, 496–497
mercury, as environmental hazard, 55
Mercury (planet)
  motion of, 519
  as terrestrial planet, 530–531
mesopause, 428, 429
mesosphere (atmosphere), 428, 429
mesosphere (mantle)
  characteristics of, 14–15
  seismic discontinuities of, 255–256
Mesozoic era, 313–316
  evolution of life in, 354–355
metabolism
  advances in early life-forms, 343–344
  as requirement of life, 340–341
metallic bonding, definition of, 39
metamorphic rocks, 82–88
  definition of, 82
  factors in metamorphism, 82–84
  in landscape, 65
  as resources, 89–90
  stress and foliation, 84–88
  water movement through, 154–155
metamorphism
  definition of, 82
  factors in, 83–84
  mechanisms of, 82
  stress and effects, 84–88
meteorites
  ages of, 326–327
  characteristics of, 259
  definition of, 535
  and mass extinction, 357–359
meteoroids
  collision impacts, 535–536
  definition of, 12

methane
  in atmosphere, 426
  increases and global warming, 507
  radiation absorption by, 436–437
Mexico
  Baja California, 172
  Chicxulub crater, 357–359
  Naica Mine, 56
mica
  crystallization of, 291–292
  in metamorphic rock, 85, 87
  structure of, 45, 51
Mid-Atlantic Ridge
  and seafloor spreading, 208
  topography of, 374–375
midlatitude climates (Group II), 501–503
midocean ridges
  definition of, 204
  in seafloor spreading, 208–210
  topography of, 374–375
  volcanic flow, 278
Milankovitch cycles
  and climate change, 494
  definition of, 495
Milky Way, characteristics of, 540–541, 545
Miller, Stanley, amino acid generation, 340–341
millibar, unit of atmospheric pressure, 458
mineral resources, 21–22, 124
  environmental concerns, 53–55
  rock sources, 90
mineralization, in fossils, 348
mineraloids
  definition of, 41
  as gems, 47
minerals. *See also* mineral resources; rocks; *specific mineral*
  chemical composition/structure, 40–42
  definition of, 36
  families of, 49–52
  inclusion criteria, 36, 40–41
  physical properties of, 42–48
  in rock melting, 285–288
mining
  Naica Mine crystals, 56
  problems with, 53–55
Mississippi River (U.S.A.)
  coastline changes of, 142–143
  delta of, 408
  flooding of, 145, 147
Modified Mercalli Intensity Scale, for earthquakes, 250
Moho (Mohorovičíc) discontinuity, 255
Mohs scale, of mineral hardness, 44–45
moisture. *See also* air masses
  atmospheric, 431–434
Mojave Desert (California), 501–502
molds, in fossil formation, 348
molecules, definition of, 38
mollisols, soil order, 113–115
moment magnitude, definition of, 250
monadnocks, definition of, 125
Mono Lake (California), renewal of, 152
monsoon winds, 500
Montana, Glacier National Park, 184
Monterey Bay (California), 389
Moon
  crust dust, 326

formation of, 535–536
and tides, 398–399
moraines
   definition of, 191
   in glacial landscapes, 191, 193
Mt. Pinatubo
   eruption damage from, 24, 270–271, 281
   warning system success, 280
mountain breezes, 462
mountains, orographic precipitation and, 444–446
Mt. Fuji (Japan), volcano, 275
Mt. Mazama (Oregon), stratovolcano, 276
Mt. Merapi (Indonesia), 282–283
Mt. Monadnock (New Hampshire), 125
Mt. Pelée eruption (1902), 278
Mt. St. Helens eruption (1980), 272, 274, 276, 281, 296
Mt. Vesuvius eruption (79 A.D.), 278
mud cracks, in rock, 78
mud flats, and salt marsh formation, 404
mudflows
   mass wasting, 129
   mudslides, 270, 281
mudslides, 270, 281
mudstone, formation of, 73
mummification, 348
mutations, in evolution, 347

N

Naica Mine (Mexico), crystal formations, 56
Namib Desert (Namibia), 171
NASA, studies of Mars, 546
National Weather Service (NWS), tornado warning system, 476
natural hazards. *See* hazards
natural resources
   definition of, 20
   human historical use of, 20–22, 23
natural selection
   characteristics of, 345–347
   definition of, 345
Navajo Sandstone (U.S.A.), 91
nebula hypothesis, 524–526. *See also* planetary accretion hypothesis
   definition of, 525
nebulae, 524–526
   Hubble's studies of, 540–541
negative feedback mechanisms, in Earth systems, 11–12
neon, in atmosphere, 426
Nepal, Himalaya of, 200–201
Neptune, characteristics of, 532, 534
neutron stars, 539–540
neutrons, and chemical structure, 36–39
New Hampshire
   Mt. Monadnock, 125
   Old Man of the Mountain, 98–99
New Mexico, Carlsbad Caverns National Park, 162
New Orleans (Louisiana), Hurricane Katrina flooding, 147, 482
New York, Ausable Chasm, 78
New Zealand, future climate change, 509
Newton, Isaac, on gravity, 523
newton (N), unit of force, 458
nickel, as core component, 259

Nile River (Egypt), delta of, 141, 408
nitrogen, in atmosphere, 426
nitrogen-14, decay of, 320–321
nitrous oxide, in atmosphere, 426
nonconformities, in strata, 308
nonrenewable resources
   definition of, 21
   human historical use of, 20–22, 23
   ore deposits as, 53–54
nor'easters, wave cyclones, 480
Norgay, Tenzing, scaling of Mt. Everest, 200
normal faults, definition of, 262
North America
   climate change history, 495–497
   future climate change, 508
North Atlantic Deep Water, 385
North Atlantic Oscillation, 387
North Carolina
   Ocracoke Island, 78
   Outer Banks, 406–407, 421
   Pilot Mountain, 107
North Pole, and magnetic pole, 207, 325
Northern Hemisphere, distribution of oceans, 373
notochord, 353
nuclear fusion
   in big bang theory, 543
   of Sun, 526–527
nuclei, and chemical structure, 36–39
nucleotides, 341
numerical age, 319–325
   definition of, 309
   early methods of determination, 317–318
   of Earth, 325–327
   magnetic polarity dating and, 322–323, 325
   radioactivity and, 319–322

O

O horizon, of soil, 110–112
oases, 168–169
observation methods, of Earth's interior, 258–259
obsidian, texture of, 67
occluded fronts
   characteristics of, 472–473
   and wave cyclones, 474–475
ocean basins
   characteristics of, 372–375
   definition of, 373
Ocean City (Maryland), beach erosion of, 414
ocean conveyor belt, 384–388, 496–497
ocean currents, 382–388
   characteristics of, 382–384
   and climate regulation, 385–388
   definition of, 382
   and hurricane activity, 478
oceanic crust. *See also* continental crust; crust; oceans
   definition of, 18
   in plate tectonics, 214–218, 222
   in seafloor spreading, 208–210
   volcanic flow and, 278
oceanic subtropical high-pressure belt, climate of, 499
oceans
   basins, 372–375
   circulation in, 382–388, 496–497
   composition and movements, 376–381

density layers of, 382
global temperature trends and, 490–491
in hydrologic cycle, 135–137
as open system, 317
Ocracoke Island (North Carolina), 78
Old Man of the Mountain (New Hampshire), rock formation, 98–99
Oldham, Richard Dixon, on liquid core, 253
olivine
   in Allende Meteorite, 327
   crystallization of, 291–292
   as gem, 47
   polymorph form, 256
   structure of, 51
On the Origin of Species, 345–347
   publication of, 317, 318, 345
open systems
   characteristics of, 8–9
   in Earth system, 9–12
   hydrologic cycle, 135–137
   ocean as, 317
orbital harmony, law of, 522
orbits, planetary, 518–523
ore deposits
   concerns with, 53–55
   definition of, 53
ore minerals, 90
Oregon, Crater Lake, 276
organelles, of eukaryotes, 343
organic molecules, generation of, 340–341
organisms, and soil formation, 112
Orion Nebula, proplyds in, 526
orographic precipitation, 444–446, 503
   definition of, 445
Ötzi the Iceman, mummified body, 488
Outer Banks (North Carolina)
   characteristics of, 406–407
   hurricane damage, 421
overland flow, 138
oxbow lakes, 139, 140, 141
oxidation, in weathering, 105
oxide minerals, characteristics of, 50
oxisols, soil order, 113–115
oxygen. *See also* ozone
   in atmosphere, 426–427
   as component of crust, 49, 50–52
   and early life, 337–339, 349–350, 424
   in planet formation, 525
ozone
   in atmosphere, 426, 429–431
   definition of, 429
   and early life, 338, 349–350
ozone layer, structure and function, 429–431

P

P waves
   behavior of, 252–255
   definition of, 247
   measurement of, 247–249
Pacific Decadal Oscillation (PDO), 386
Pacific Ocean
   characteristics of, 373–375
   El Niño/La Niña events, 384, 387, 456–457
   Pacific Decadal Oscillation (PDO), 386
Pacific Tsunami Warning Center, 412
pahoehoe, lava flow, 289
Painted Desert (Arizona), fossilized sandstone, 79

Pakistan, Indus River, 106
Paleogene Period, evolution in, 355
paleomagnetism
　definition of, 207
　magnetic polarity dating, 258–259, 322–323, 325
paleontology. *See also* evolution; fossils; Phanerozoic Eon
　correlation, 310–312
　definition of, 311
paleoseismology
　definition of, 241
　in earthquake forecasting, 241–242
paleosols, soil horizons, 493
Paleozoic era, 313–316
　proliferation of life in, 349–353
Palisades, gabbro sill formation, 294
Pangaea
　breakup of, 339
　and continental drift theory, 202–207
parabolic dunes, 177
parent atoms, 319–321
Pascal, Blaise, on air pressure and altitude, 459
pascal (Pa), unit of pressure, 458
passive continental margins, 374
patterned ground, landform, 194
pauses, between atmospheric layers, 428
PDO (Pacific Decadal Oscillation), 386
peat, definition of, 77
pegmatite, definition of, 69
pelagic zones, 380–381
Penzias, Arno, and CMB radiation, 544
percolation
　definition of, 155
　of groundwater, 155
periglacial conditions, 191, 194
　definition of, 191
permafrost
　characteristics of, 504
　definition of, 194
　landforms of, 194
permeability
　definition of, 155
　and groundwater movement, 154–157
Permian Period, mass extinction in, 354, 357
Petrified Forest National Park (Arizona), 348
petroleum, as natural resource, 21–22
phaneritic texture, of rocks, 69
Phanerozoic Eon. *See also* Cenozoic Era; Mesozoic Era; Paleozoic Era
　diversity of species in, 345
　eras of, 349
　evolution of life in, 349–363
　fossil record of, 313
photic zone, 380
photosphere, of Sun, 527–528
photosynthesis
　and atmospheric oxygen levels, 337–339
　definition of, 337
phyla, in Cambrian Period, 350
phyllite, in metamorphism, 85–87
Piccard, Jacques, Mariana Trench research, 200
piedmont glaciers, 185
Pilot Mountain (North Carolina), rock formation, 107
placer deposits, 124
plagioclase feldspar, crystallization of, 291–292

planetary accretion hypothesis, 530, 535–536. *See also* nebula hypothesis
planets
　accretion hypothesis, 530, 535–536
　dwarf, 516, 534
　Jovian, 13, 524, 532–534
　orbits of, 519–521
　terrestrial, 13–18, 524, 530–531
planktic zone, 380
plants
　and climate change, 507
　in evolution, 350–352, 355
plate margins/boundaries, 212–213
　earthquakes and, 217–218, 234–238
　types, 214–217
plate tectonics. *See also* continental drift; earthquakes
　definition of, 15, 211
　earthquakes and, 217–218, 234–238
　major world plates, 211–213
　mantle convection mechanism, 219–221
　margin types and movement, 214–217
　role in Earth geography, 15–18
　supercontinents and, 223
　tectonic cycle, 222
　and terrestrial planets, 531
　theory of, 211–218
Plinian volcanic eruptions, 273, 274
plumes
　mantle, 221
　water, 377
Pluto, reclassification, 516–517, 534
plutonic rock
　characteristics of, 69–71
　definition of, 67
　formations, 293–295
plutons, definition of, 293
pocket beaches, 405
point bar, 140
polar cells, convection cells, 466–467
polar deserts, 171, 173
polar easterlies, winds, 467
polar-front jet stream, 469
polar-front zone, climate of, 501–503
polar glaciers, 183
polar low, upper-air pattern, 466, 468
polarity, of magnetic field, 207, 258–259, 322–323, 325
pollution, water, 149
polymerization, of silicates, 51, 52
polyps, in coral reef formation, 409
population growth, concerns of, 19, 149
pore fluid, in rocks, 83, 87
porosity
　definition of, 154
　and groundwater movement, 154–157
porphyritic texture, of rocks, 68
positive feedback mechanisms, in Earth systems, 11–12
potassium feldspar, dissolution of, 104
power
　dependence on electrical, 22–23
　hydroelectric, 132
Precambrian time, 314, 316
precession, and climate change, 494–495
precipitation, 442–449. *See also specific climate type*
　convectional, 446–447

　definition of, 135
　in desert definition, 171
　orographic, 444–446, 503
　in thunderstorms, 447–449, 474
　types of, 442–444
　and water vapor content, 429, 432
precursor phenomena, of earthquakes, 240–241
prediction
　of earthquakes, 240–242
　of flooding, 148–149
　of volcanoes, 284
pressure
　and adiabatic principle, 438–439
　definition of, 260
　in metamorphism, 83–85
　and rock melting, 285–288
　vs. stress, 84
pressure gradients
　definition of, 460
　and upper-level winds, 467–469
primary atmosphere, 337
primary effects, of volcanoes, 278–279
primary hazards, in earthquakes, 238–239
principle of cross-cutting relationships, 307
principle of faunal and floral succession, 311–312
principle of lateral continuity, 307
principle of stratigraphic superposition, 306–307
principles, scientific, 6
prokaryotes
　in Archean Eon, 342–343
　definition of, 342
prominences, solar, 528
Proterozoic Eon, 313
　life-forms of, 343–344
proton-proton chain, definition of, 527
protons, and chemical structure, 36–39
Providence Canyon (Georgia), soil management problems, 116
pulsars, neutron stars, 539–540
pumice, 68, 299
punctuated equilibrium, in evolution, 347
pyroclastic flow, 273, 274
　definition of, 273
　hazards from, 278
pyroclasts, definition of, 71
pyrolusite, formation of, 105
pyroxene, crystallization of, 291–292

Q

quartz
　as gem, 47
　in metamorphic rock, 85–88
　structure of, 51
　weathering of, 62–63, 124
quartzite, definition of, 88
quasars, 545
Queen Maude Land (Antarctica), polar desert, 171

R

radiation. *See also* ozone layer; ultraviolet radiation
　in global energy system, 434–437
radiation zone, of Sun, 527
radioactive decay, 319–322
radioactivity
　definition of, 318

578　Index

and numerical age, 319–322, 324
radiometric dating, 320–322, 324
    definition of, 320
radiosonde, weather balloon, 458
rain. *See also* precipitation; rainfall
    formation of, 442–443
    in weathering, 102–105
rainfall
    deserts, 171, 179
    measurement of, 444
    in severe storms, 478–479
rainshadow deserts, 172–173
rainshadows, and orographic precipitation, 444–446
rainwater, and weathering, 102–105
Rayleigh waves, 248
recharge (groundwater), 154–155
    definition of, 155
recrystallization, in lithification, 73–74, 83–88
recurrence interval, flooding and, 148
red clay, in sediments, 381
red giants (stars), 538–539
Red Sea, and plate movement, 216
reefs, 409, 418
    definition of, 409
reflection
    definition of, 254
    of seismic waves, 253–254
refraction
    definition of, 254
    of seismic waves, 253–254
regolith. *See also* soil
    definition of, 15, 100
    and erosion, 117–119
    and mass wasting, 121–122
    and weathering, 100–108
Reid, Harry Fielding, elastic rebound theory, 235
relative age. *See also* geologic column
    definition of, 304
    stratigraphic principles of, 304–309
relative humidity, 433
remote sensing, 252. *See also specific technique*
renewable resources
    definition of, 21
    human historical use of, 20–22, 23
replication, as requirement of life, 340–341
reproduction, sexual, 349–350
reptiles, in evolution, 354–355
reservoirs, in hydrologic cycle, 135–137
residual soils, 111–112
resources, human historical use of, 20–22, 23
resurgent domes, of volcanoes, 276, 296
retrograde motion, of planets, 519
reverse faults, definition of, 262
rhyolite, characteristics of, 70
rhyolitic magma, 290–292
ribonucleic acid. *See* RNA
Richter, Charles, on earthquake prediction, 240
Richter magnitude scale, 250–251
    definition of, 250
rift valleys
    East African valleys, 211, 214, 216, 229
    oceanic, 374–375
rifting, 214–218
rings, of Jovian planets, 532–533
ripple marks, in rock, 78
RNA, in early life, 340–341

rock cycle, definition of, 66
rock families, 64–65
rock-forming minerals, definition of, 50
rocks. *See also* igneous rocks; metamorphic rocks; sedimentary rocks; soil
    characteristics of, 64–65
    as continental drift evidence, 203–206
    definition of, 36, 64
    in Earth composition, 13–18
    elastic rebound theory, 235–238
    igneous, 64–65, 67–71, 89–90, 154–155, 293–295, 299, 322
    melting of, 285–289
    metamorphic, 65, 82–88, 89–90, 154–155
    oldest found on Earth, 325–327, 372
    radiometric dating of, 322, 324
    as resources, 89–91
    rock cycle, 66
    rock-forming minerals, 50
    sedimentary, 64–65, 72–81, 89–90, 322–325
    stresses and deformations, 260–263
rocky coasts
    characteristics of, 405–406
    erosion hazards, 412
Rodinia, supercontinent, 223
rogue waves, 400
root wedging, 102, 103
Rossby waves, definition of, 469
rotation of Earth, 494–495, 520
run-up, tidal, 399
runoff. *See* surface runoff
Russia, Black Sea Coast, 416–417

S

S waves
    behavior of, 252–255
    definition of, 247
    measurement of, 247–249
safety, earthquake hazards and, 243–245
Saffir-Simpson scale, of hurricane intensity, 478–479
Sahel, desertification of, 179–180
salinity
    definition of, 376
    and ocean density, 382, 384–385
salt, in weathering, 102
salt marshes, formation of, 404
saltation
    definition of, 117, 173
    in erosion, 117–118, 173, 176
San Andreas Fault (California), 235, 236
    instability of, 215
San Francisco earthquake (1906), 235–237
sand
    as aquifer, 156
    characteristics of, 109
    definition of, 109
    in shoreline erosion, 402–403
sandstone
    as aquifer, 156
    characteristics of, 65
    definition of, 72
    examples of, 77, 79, 91
    formation of, 73, 91
    metamorphism of, 88
Santa Ana winds, 460–461

saturation, 432–434
    and cloud formation, 439
Saturn
    characteristics of, 532–534
    motion of, 519
savannah, 501
schist
    definition of, 86
    formation of, 85–87
schistosity, 86–87
    definition of, 86
scientific method
    definition of, 4
    seafloor spreading example, 210
    steps and application of, 4–6
scientific revolution, Copernicus' role in, 520–521
Scotland
    Great Glen Fault, 264
    Siccar Point, 309
sea breezes, 462
sea ice, 186
sea landforms, 406
sea level, changes in, 396–399, 507
seafloor hydrothermal vents, life in, 341
seafloor spreading
    definition of, 209
    hypothesis of, 208–210
seas and gulfs, major, 373
seasons
    Earth-Sun relations, 520
    insolation and, 465
seawater. *See also* ocean currents; oceans
    composition, 376–381
    depth and volume, 373, 397–398
    salinity and density, 382
secondary atmosphere, 337
secondary effects, of volcanoes, 278, 279–280
secondary hazards, in earthquakes, 238–241
sediment cores, for seafloor research, 380
sedimentary rocks, 72–81
    chemical and biogenic sediments, 75–77
    dating of, 322–325
    definition of, 72
    environmental clues in, 77–81
    in landscape, 64–65
    lithification of, 72–74
    marl deposits, 494
    as resources, 89–90
    sedimentary facies, 80–81
    water movement through, 154–155
sediments
    biogenic, 77
    chemical, 75–76
    clastic sediment, 72–74
    climate change evidence in, 490–494
    landforms of, 140–142
    Law of Original Horizontality, 6
    seafloor, 380–381
    and shoreline erosion, 401–404
    in turbidity currents, 378–379
    water movement through, 154–155
seismic creep, 235
seismic discontinuities
    core/mantle, 253–254
    crust/mantle, 255–256
    definition of, 253
    wave behavior and, 253–254

seismic gaps, 242
seismic methods, study of Earth's interior, 252–254
seismic tomography, 254
seismic waves
 behavior in Earth's interior, 252–256
 definition of, 235
 measurement of, 247–249
 mechanism of, 235–238
seismogram
 definition of, 246
 measurement with, 246–251
seismograph
 definition of, 246
 measurement with, 246–251
seismology, 246–251
 definition of, 234
 locating earthquakes, 248–249
 magnitude scales, 250–251
 seismic waves, 247–248
 seismograph/seismogram design, 246
 in study of Earth's interior, 252–256
semiarid regions, deserts in, 171, 173, 179–180
sexual reproduction, as evolutionary advantage, 349–350
shadow zones, and seismic waves, 253–254
shale
 as aquiclude, 156
 Burgess Shale, 361
 definition of, 72
 formation of, 73
 metamorphism of, 85–87
shear, 260–263
 definition of, 260
shear waves. See also S waves
 behavior of, 252–255
 definition of, 247
 measurement of, 247–249
sheet flow, 138
sheet jointing, 102, 103
shelf break, 203–204
shield volcanoes
 characteristics of, 273, 274, 278
 definition of, 273
shorelines
 erosion protection, 413–417
 types and landforms, 405–409
 wave action and erosion, 401–404, 412
short-term prediction, of earthquakes, 240–241
Siberian Traps, mass extinction and, 281, 360
Siccar Point (Scotland), strata of, 309
silica. See also silicon
 igneous rock content, 69–71
 in magma and lava, 273, 274, 287–289
 silicate minerals, 50–52
silicate minerals, characteristics of, 50–52
siliceous ooze, 380–381
silicon. See also silica
 as component of crust, 49, 50–52
 in magma and lava, 287
 in planet formation, 525–526
sills, plutons, 293, 294–295
silt, characteristics of, 109
siltstone, formation of, 72–73
Silurian Period, evolution in, 350–351, 352
Similaun Glacier (Alps), 488–489
sinkholes, formation of, 160–161, 165

slate
 definition of, 86
 formation of, 85–87
slides
 during hurricanes, 478
 landslides, 121, 123, 238–239, 412, 478
 mudslides, 270, 281
 rockslides, 106, 121
 slope failure, 121
slip faces, 176
slope failures, definition of, 121
slumps, slope failure, 121
slurry flows, of regolith, 122
Smith, William, and stratigraphy, 311–312
snow
 albedo of, 435
 formation of, 442–443
 in glacier formation, 187
snowfall, measurement of, 444
sodium chloride. See halite
soil. See also erosion
 characteristics of, 109–110
 definition of, 15, 100
 as Earth resource, 20–22
 formation factors, 111–115
 management of, 116, 181–182
 profiles and horizons, 110–113
 volcanic enrichment of, 281
 weathering and, 100
soil horizons, 110–113
 definition of, 110
soil orders, 113
soil profiles, 110–113
 definition of, 110
soil texture, 109
solar flares, from Sun, 527
solar system, 524–536
 asteroids, meteorites, and Moon, 326–327, 535–537
 components of, 12–13
 origin of, 524–526
 other systems, 537
 planets, 530–534
 reclassification of planets, 516–517
 Sun in, 524–525, 526–529
solar wind, from Sun, 527, 529
solifluction, slurry flow, 122
source regions, for air masses, 470–471, 498–499
Southern Hemisphere, distribution of oceans, 373
space, Earth in, 12–13
spatter cones, of volcanic eruptions, 273, 276
species
 definition of, 345
 and evolution, 345–347
specific humidity, 433–434
spheres, Earth subsystems, 10–11
spicules, on Sun, 527
spiral galaxies, 541, 545
spits, 403, 406–407
spodosols, soil order, 113–115
spreading centers. See also plate margins/boundaries
 plate margins, 214
springs
 definition of, 156
 origin of, 157

thermal, 277, 284
stalactites, 160–162
stalagmites, 160–162
star dunes, 177
stars and stellar evolution, 538–540
stationary fronts, characteristics of, 472–473
Steno, Nicolaus
 on fossil ages, 310
 law of crystal angles, 43–44
stocks, characteristics of, 293–294
stomata, of plants, 350
storm surges, 147, 478–479
storms. See also cyclones
 dust, 7, 118–119, 173, 181
 erosion and, 398, 404, 410–412
 ice, 443
 increases from global warming, 507–509
 thunderstorms, 447–449, 474
 tornadoes, 476
straight channels, 138–140
strain, definition of, 260
strata. See also sedimentary rocks
 correlation and, 310–312
 definition of, 307
 and stratigraphy, 304–309
stratiform clouds, 440–441
stratigraphic time scale. See geologic column
stratigraphy, 304–309
 definition of, 304
 fossils and correlation, 310–312
 and geologic column, 312–316
 by magnetic reversals, 325
 unconformities in, 308–309
stratopause, 428, 429
stratosphere
 characteristics of, 428–429
 definition of, 429
stratovolcanoes. See also volcanoes
 definition of, 273
 formation of, 275
stratus clouds, 440–441
streak
 definition of, 46
 in minerals, 46–48
streams and streamflow, 138–144. See also Flooding
 channel modifications, 149
 channels and flow, 138–140
 definition of, 138
 deposits and landforms, 140–142
 drainage basins, 142–143
 lakes and, 144
stress
 definition of, 260
 faults and, 262–263
 in metamorphism, 84–88
 types of, 260–261
 vs. pressure, 84
striations, glacial, 191–192
strike-slip faults, 262. See also transform faults
stromatolites, and photosynthesis, 342
Strombolian volcanic eruptions, 273, 274
subduction, 214–218. See also earthquakes
 conduction and, 219–222
 subduction trenches, 374
 subduction zones, 214–218
subduction trenches, 374

subduction zones, 214–218
  definition of, 214
sublimation
  by chinook winds, 444
  in hole punch clouds, 450
subsidence (land), 148, 158, 398
subtropical deserts, 171–172
subtropical zone, of climate, 499–501
Sugarloaf Mountain (Brazil), rock formation, 106
Sumatra-Andaman earthquake (2004), 232–233, 237, 240–241
Sumatran tsunami (2004), 24, 237, 240–241
  damage from, 410–412
Sun
  and global winds, 465
  insolation, 434–435, 465
  and solar system, 12–13, 524–525
  structure and characteristics, 526–529
  and tides, 398–399
  ultraviolet radiation dangers, 429–431
sunspots, 528
supercontinents
  Pangaea, 202–207, 339
  Rodinia, 223
  theory of cycle, 223
supercooling, of water, 442, 450
supernovae
  definition of, 539
  and planet creation, 525
surf
  definition of, 401
  and erosion, 401–404
surface creep, definition of, 173
surface layer, of ocean, 382
surface runoff. See also streams and streamflow
  definition of, 135
surface water, 145–152. See also groundwater
  flooding, 145–149
  use and access, 149–151
surface waves
  definition of, 247
  measurement of, 247–249
surges, glacial, 190
suspended load
  definition of, 117
  and erosion, 117–118
suspension
  definition of, 173
  and dust storm formation, 173
sustainable farming, 116
swash, of wave cycle, 403
Switzerland
  Findelen Glacier, 120
  Jura Mountains, 302–303
  Lauterbrunnen Valley (Alps), 65, 192
systems. See also Earth systems
  characteristics of, 7–8
  definition of, 7

T

talus slope, 103
Tambora eruption (1815), 281
tar, as fossil preservative, 348
tectonic cycle
  definition of, 222
  in Earth system, 66

tectonic forces. See also plate tectonics
  and mass wasting, 122–123
  and stratigraphy, 306–307, 308–309
  and weathering, 105–106
temperate glaciers, 183
temperature
  atmospheric profiles, 427–429
  and climate change, 490–491, 505–509
  Equator-poles gradient, 467
  humidity and, 432–434
  influences on, 498
  in metamorphism, 83–85
  and ocean circulation, 384–388
  records, 492–493
  and rock melting, 285–289
tension, 260–263
  definition of, 260
tephra, 275–276. See also pyroclastic flow
  definition of, 71
terraces (coastal), 405–406
terrestrial planets. See also Earth
  composition of, 13–15
  life on, 531
  origin and characteristics of, 530–531
tertiary effects, of volcanoes, 281
Texas, Galveston hurricane(1900), 394–395, 478
textures (rocks), 65
  of igneous rocks, 67–69
  of metamorphic rocks, 85–87
theory
  definition of, 6
  evolution example, 347
  in scientific method, 6
  vs. hypothesis, 211
thermal springs, 277, 284
thermocline, of ocean, 382
thermohaline circulation
  definition of, 384
  in Younger Dryas event, 496–497
thermosphere, 428, 429
Thompson , William (Lord Kelvin), on numerical age, 317–318
thrust faults, definition of, 262
thunder, definition of, 474
thunderstorms, 447–449
  characteristics of, 448–449, 474
  definition of, 447
tidal bore, 399
tidal inlets, 406, 407
tidal range, 399
tides, 398–399
  definition of, 398
till
  definition of, 191
  in glacial landscapes, 191, 193
tilt (axis of rotation), and climate change, 494–495
time, and soil formation, 113
tipping point, in climate change, 507
tombolos, landform, 406
topography
  karst, 160–161
  of ocean basins, 374–375
  and soil formation, 113
  and weathering, 105–106
tornadoes, 476
Tornbough, Clyde, discovery of Pluto, 516

tourmaline, as gem, 47
trace fossils, definition of, 348
trade winds
  climate of region, 499–501
  definition of, 467
transform faults, 214–215, 218
  definition of, 214, 263
  Loch Ness on, 264
transitional zone, of ocean, 382
transpiration
  definition of, 135
  in hydrologic cycle, 134–135
transverse dunes, 177–178
tree-ring analysis, for temperature records, 492–493
tremors, tectonic, 237–238
trenches
  Mariana Trench, 200–201
  worldwide, 212–213, 214–216
triangulation, of earthquake epicenter, 248–249
Triassic Period, evolution in, 354
tributary streams, 142–143
tropical cyclones
  characteristics of, 477–479, 481
  definition of, 477
  vs. wave cyclones, 480–481
tropical zone, of climate, 499–501
Tropics of Cancer and Capricorn, 499
tropopause, 428, 429
troposphere
  characteristics of, 428–429
  definition of, 428
  and wind patterns, 465–469
tsunamis
  characteristics of, 410–413
  definition of, 410
  as earthquake hazard, 238–241
  Krakatau (1883), 278, 280–281
  Lituya Bay (1958), 412–413
  Sumatra-Andaman (2004), 24, 232–233, 237, 240–241, 410–412
tuff, 276
  definition of, 71
tundra
  climate of, 504
  periglacial conditions of, 191, 194
turbidites, 378–379
  definition of, 379
turbidity currents, 378–379
  definition of, 378
turbulent flow, in erosion, 117
Twin Towers (New York City), tremors from September 11 collapse, 238
typhoons. See tropical cyclones

U

ultisols, soil order, 113–115
ultraviolet radiation, and atmosphere, 426, 429–431
Uluru/Ayers Rock (Australia), 62–63
unconfined aquifers, 156
unconformities, 308–309
  definition of, 309
uniform stress, 260
uniformitarianism, definition of, 304

universe
    big bang theory and evidence, 543–545
    dark matter and energy, 544–545
    discovery of galaxies, 540–542
upper-air westerlies, upper-level winds, 466, 468–469
upper-level winds, 467–469
upwelling, of ocean water, 384–385, 409
uranium-238, decay of, 321
Uranus, characteristics of, 532, 534, 536
urbanization, effects on water, 148
Ussher, James, on date of Creation, 304
Utah
    Bingham Canyon Copper Mine, 53
    Bonneville Salt Flats/Great Salt Lake, 75
    Bryce Canyon sedimentary rock, 64–65
    Zion National Park sandstone, 91

V

vadose zone, 153, 160
valley breezes, 462
valley glaciers, 184, 192
Van der Waals bonding, definition of, 39
vegetation, and weathering, 105, 107
veins, in rock, 83, 90
ventifacts, 173
Venus
    characteristics of, 13–15, 18, 530–531, 536
    motion of, 519, 536
vertisols, soil order, 113–115
Victoria Falls (Africa), 132–133
Victoria Land (Antarctica), talus slope, 103
Virginia, Assateague Island, 414
viscosity
    definition of, 273
    in rock melting, 289
    and volcanic eruptions, 273–275
volcanic ash, 273–275
    damage from, 270–271
    definition of, 71
volcanic necks, 294–295
volcanic pipes, 294–295
volcanic rock
    characteristics of, 67–71, 299
    definition of, 67
    in sea floor, 208–210
volcanic vents, 294
volcaniclastic sediments, 73
volcanoes. *See also* magma; volcanic ash; volcanic rock
    climatic effects of, 281
    definition of, 272
    effects on atmosphere, 337
    eruption types, 272–276
    fissure eruptions, 273, 274, 280
    Hawaiian islands, 18, 224–225
    hazards from, 278–284
    and hot spots, 221
    Krakatau eruption (1883), 278, 280–281
    Laki eruption (1783), 278, 281
    landforms of, 276–277
    locations of, 212–213, 277–278
    mass extinctions and, 360

Mount Pinatubo eruption (1991), 24, 270–271, 280, 281
Mt. Pelée eruption (1902), 278
Mt. St. Helens eruption (1980), 272, 274, 276, 281, 296
Mt. Vesuvius eruption (79 A.D.), 278
plutonism and, 293–295
prediction of, 284
primary effects of, 278–279
secondary effects of, 278, 279–281
on terrestrial planets, 531
tertiary effects, 281
von Laue, Max, and crystal structure, 44
Vulcanian volcanic eruptions, 273, 274

W

Waggoner, Paul, on population growth, 19
Wallace, Alfred Russell, on natural selection, 347
Walsh, Donald, Mariana Trench research, 200
warm fronts
    characteristics of, 472–473
    and wave cyclones, 474–475
Washington state, Mt. St. Helens eruption (1980), 272, 274, 276, 281, 296
water. *See also* groundwater; surface water
    in atmosphere. *See* water vapor
    erosion by, 117–118
    historical use as resource, 20–22
    hydrologic cycle, 66, 134–137
    major bodies of, 373
    origin of, 372
    properties of, 431–432
    as requirement for life, 531
    and rock melting, 285–287
    role in soil, 109–110
    use and access, 149, 159
    volcanic gas origin, 339
    in weathering, 102–105, 108
water cycle. *See* hydrologic cycle
water pollution, from human projects, 149
water rights, 159
water table, 153–154
    definition of, 153
water vapor
    atmospheric component, 426, 429
    as humidity, 432–434
    properties of, 431–432
    radiation absorption by, 434–437
watershed. *See* drainage basins
wave-cut beaches, 405–406
wave-cut cliffs, 405–406
    definition of, 405
wave cyclones
    characteristics of, 474–475, 503
    definition of, 474
    vs. tropical cyclones, 480–481
waves. *See also* erosion; shorelines
    characteristics of, 400–404
    seismic. *See* seismic waves
weather. *See also* climate; weather systems
    forecasting models, 506–507
    global warming effects, 398
    and pressure systems, 464
    troposphere and, 428–429, 465–469

weather systems, 470–481
    air masses and fronts, 470–473
    thunderstorms, 447–449, 474
    tornadoes, 476
    tropical cyclones, 477–479, 481
    wave cyclones, 474–475, 503
weathering, 100–108. *See also* erosion; soil
    chemical weathering, 102–105, 124
    definition of, 15, 66, 100
    factors affecting, 105–108
    mechanical weathering, 100–103
    resources formed by, 124
    and rock cycle, 66
    vs. erosion, 117
Wegener, Alfred, on continental drift, 202–207
wells, depletion problems, 158–159
wet adiabatic lapse rate, 439
    and orographic precipitation, 444
wetlands, in lake life cycle, 144
white dwarfs (stars), 538–539
Wilkinson Map Anisotropy Probe (WMAP), and CMB radiation, 544
Wilson, J. Tuzo, and plate tectonics, 223
Wilson, Robert W., and CMB radiation, 544
Wilson cycle, 223
winds
    characteristics of, 460–461
    definition of, 460
    erosion by, 118, 173–175, 402–404
    global patterns, 465–467
    local, 444, 460–462
    monsoon, 500
    and oceanic circulation, 383–384
    in severe storms, 448–449, 474, 476, 478–479
    of stratosphere, 429
    upper-level, 467–469
    and wave action, 402–404
Wisconsin, Kettle Moraine State Forest, 193
Wolcott, Charles, discovery of Burgess Shale, 361
World Ocean, 373
    map of currents, 383
    seafloor sediments, 380–381
World Trade Center (New York City), tremors from September 11 collapse, 238

X

xenoliths
    as Earth study tool, 256–257
    relative age and, 307

Y

Yakutat Bay (Alaska), earthquake (1899), 235
Yellowstone National Park (U.S.A.), glacial erosion in, 193
Yosemite National Park (California), rock formation in, 103
Younger Dryas event
    definition of, 496
    rapid climate change of, 495–497
Yukon Territory (Canada), Kaskawulsh Glacier, 120, 193

Z

Zion National Park (Utah), Navajo sandstone of, 91